Lectures on Analytic Differential Equations

Lectures on Analytic Differential Equations

Yulij Ilyashenko
Sergei Yakovenko

Graduate Studies
in Mathematics
Volume 86

American Mathematical Society
Providence, Rhode Island

2000 *Mathematics Subject Classification.* Primary 34A26, 34C10;
Secondary 14Q20, 32S65, 13E05.

For additional information and updates on this book, visit
www.ams.org/bookpages/gsm-86

Library of Congress Cataloging-in-Publication Data

Il'iashenko, IU. S.
Lectures on analytic differential equations / Yulij Ilyashenko, Sergei Yakovenko.
p. cm. — (Graduate studies in mathematics, ISSN 1065-7339 ; v. 86)
Includes bibliographical references and index.
ISBN 978-0-8218-3667-5 (alk. paper)
1. Differential equations, Nonlinear. 2. Riemann-Hilbert problems. 3. Analytics spaces. 4. Dynamics. I. Yakovenko, S. II. Title.

QA372.I445 2007
515'.355–dc22 2007060785

10 9 8 7 6 5 4 3 2 1 13 12 11 10 09 08

To Helen and Anna, for their infinite patience
and unrelenting support during these long years...

Contents

Preface

The branch of mathematics which deals with ordinary differential equations can be roughly divided into two large parts, *qualitative theory of differential equations* and the *dynamical systems theory*. The former mostly deals with systems of differential equations on the plane, the latter concerns multidimensional systems (diffeomorphisms on two-dimensional manifolds and flows in dimension greater than two and up to infinity). The former can be considered as a relatively orderly world, while the latter is the realm of chaos.

A key problem, in some sense a paradigm influencing the development of dynamical systems theory from its origins, is the problem of turbulence: how a deterministic nature of a dynamical system can be compatible with its apparently chaotic behavior. This problem was studied by the precursors and founding fathers of the dynamical systems theory: L. Landau, H. Hopf, A. Kolmogorov, V. Arnold, S. Smale, D. Ruelle and F. Takens. Currently this is one of the principal challenges on the crossroad between mathematics, physics and computer science. Dynamical systems theory heavily uses methods and tools from topology, differential geometry, probability, functional analysis and other branches of mathematics.

The qualitative theory of differential equations is mostly associated with autonomous systems on the plane and closely related to analytic theory of ordinary differential equations. The principal theme is investigation of local and global topological properties of phase portraits on the plane. One of the main problems of the whole area is Hilbert's sixteenth problem, the question on the number and position of limit cycles of a polynomial vector field on the plane. In a very broad sense this can be assessed as the question: to which extent properties of polynomials defining a differential equation are inherited by its absolutely transcendental (and sometimes very weird) solutions.

Another major part of analytic theory of differential equations is the linear theory. Here the key problem is Hilbert's twenty-first problem, also known as the Riemann–Hilbert problem, which has a long dramatic history and was solved "only yesterday". Discussion of this problem constitutes an important part of this book.

The qualitative theory of differential equations was essentially created in the works by H. Poincaré who discovered that differential equations belong not only to the realm of analysis, but also to geometry. Deriving geometric properties of solutions directly from the equations defining them, was his principal idea. These ideas were further developed in each of the two branches separately, but their present appearance looks very different.

Differential equations brought into existence such areas of mathematics as topology and Lie groups theory. In turn, the analytic theory of differential equations is not a closed area, but rather provides a source of applications and motivation for other disciplines. In this book we stress using complex analysis, algebraic geometry and topology of vector bundles, with some other interesting links briefly outlined at the appropriate places.

On the frontier between differential equations and the singularity theory, lies the notion of a normal form, one of the central concepts of this book. The first chapter contains the basics of formal and analytic normal form theory. The tools developed in this chapter are systematically used throughout the book. The study of phase portraits of composite singular points requires elaboration of the blowing-up technique, another classical tool known for over a century. The famous Bendixson desingularization theorem is proved in our textbook by transparent methods.

A new approach to local problems of analysis, based on the notion of algebraic and analytic solvability, was suggested by V. Arnold and R. Thom around forty years ago. In Chapter II we treat from this point of view the local theory of singular points of planar vector fields. It is proved that the stability problem and the problem of topological classification of planar vector fields are algebraically solvable in all cases except for the center/focus dichotomy. This dichotomy is algebraically unsolvable, as is proved in the same chapter. Besides these topics, the chapter contains local analysis of singular points of holomorphic foliations: the proof of the C. Camacho–P. Sad theorem on existence of analytic separatrices through singular points, integrability via the local holonomy group as discovered by J.-F. Mattei and R. Moussu, and demonstration of the Bautin theorem on small limit cycles of quadratic vector fields.

The third chapter deals with the linear theory. Somewhat paradoxically, application of normal forms of nonlinear systems to investigation of linear systems considerably simplifies exposition of many classical results. The chapter contains a succinct derivation of some positive and negative results on solvability of the Riemann–Hilbert problem.

Chapter IV deals with a new direction in the theory of normal forms, the functional moduli of analytic classification of resonant singularities. The main working tool used in this study is an almost complex structure and quasiconformal maps. The latter already played a revolutionizing role in the nearby theory of holomorphic dynamics. The main basic facts from these theories are briefly summarized in this chapter. The chapter ends with the proof of the "easy version" of the finiteness theorem for limit cycles of analytic vector fields, with an additional assumption that all singular points are hyperbolic saddles. The proof illustrates the power of using local normal forms in the solution of problems of a global nature.

Chapter V is concerned with the global theory of polynomial differential equations on the real and complex plane, bridging between algebraic, "almost algebraic" and essentially transcendental questions.

The chapter begins with the solution of the Poincaré problem on the maximal degree which can have an algebraic solution of a polynomial differential equation (a relatively recent spectacular result due to D. Cerveau, A. Lins Neto and M. Carnicer). The second section focuses on the interaction between the theory of Riemann surfaces and global theory of differential equations. We describe the topology of stratification of the complex projective plane by level curves of a generic bivariate polynomial, including derivation of the Picard–Lefschetz formulas for the Gauss–Manin connexion. This is the main working tool for deriving certain inequalities for the number of zeros of complete Abelian integrals, a question very closely related to Hilbert's sixteenth problem. Finally, we discuss generic properties of complex foliations that are very often drastically different from their real counterparts. For instance, finiteness of limit cycles on the real plane is in sharp contrast with a typically infinite number of the complex limit cycles, and the structural stability of real phase portraits counters rigidity of a generic complex foliation.

Some basic facts from complex analysis in several variables frequently used in the book, are recalled in the Appendix.

Almost all sections are ended by the problem lists. Together with easy problems, sometimes called exercises, the lists contain difficult ones, lying on the frontier of the current research.

The book was not intended to serve as a comprehensive treatise on the whole analytic theory of ordinary differential equations. The selection of topics was based on the personal taste of the authors and restricted by the size of the book. We do not even mention such classical areas as the theory of Riccati and Painlevé equations, the Malmquist theorem, integral representations and transformations. We omit completely the differential Galois theory, resurgent functions introduced by Ecalle and the fewnomial theory invented by A. Khovansky. Nevertheless, the subjects covered in the book constitute in our opinion a connected whole revolving around few key problems that play an organizing role in the development of the entire area.

Exposition of each topic begins with basic definitions and reaches the present-day level of research on many occasions. Traditionally, the proofs of many results of analytic theory of differential equations are very technically involved. Whenever available, we tried to preface formulas by motivations and avoid as much as possible all cumbersome and nonrevealing computations.

The book is primarily aimed at graduate students and professionals looking for a quick and gentle initiation into this subject. Yet experts in the area will find here several results whose complete exposition was never published before in books. On the other hand, undergraduate students should be able to read at least some parts of the book and get introduced into the beautiful area that occupies a central position in modern mathematics.

* * *

The idea to write this book, especially the chapter on linear systems, was to a large extent inspired by the recent dramatic achievements by our dear friend and colleague **Andrei Bolibruch**, who solved one of the most challenging problems of analytic theory of ordinary differential equations, the Riemann-Hilbert problem. Andrei read several first drafts of this chapter and his comments and remarks were extremely helpful.

On November 11, 2003, at the age 53, after a long and difficult struggle, Andrei Andreevich Bolibruch succumbed to the grave disease. This book is a posthumous tribute to his mathematical talents, artistic vision and impeccable taste with which he always chose problems and solved them.

* * *

When the work on this book (which took a much longer time than initially expected) was essentially over, another similar treatise appeared. In 2006 Henryk Żołądek published the fundamental monograph [**Żoł06**] titled very tellingly "The Monodromy Group". The scope of both books is surprisingly similar, though the symmetric difference is also very large. Yet most of the subjects which simultaneously occur in the two books are treated in rather different ways. This gives a reader a rare opportunity to choose the exposition that is closer to his/her heart: the mathematics can be the same but our ways of speaking about it differ.

* * *

Acknowledgements. Many people helped us in different ways to improve the manuscript. Our colleagues F. Cano, D. Cerveau, C. Christopher, A. Glutsyuk, L. Gavrilov, J. Llibre, C. Li, F. Loray, V. Kostov, V. Katsnelson, Y. Yomdin explained us delicate points of mathematical constructions and gave useful advices concerning the exposition.

We are grateful to all those who read preliminary versions of separate sections and spotted endless errors and typos, among them T. Golenishcheva-Kutuzova, Yu. Kudryashov, A. Klimenko, D. Ryzhov and M. Prokhorova. Needless to say, the responsibility for all remaining errors is entirely ours.

The AMS editorial staff was extremely patient and helpful in bringing the manuscript to its final form, including computer graphics. Our profound gratitude goes to Luann Cole, Lori Nero and especially to Sergei Gelfand for wise application of moderate physical pressure to ensure the delivery of the book.

Last but not least, we are immensely grateful to Dmitry Novikov who assisted us on all stages of the preparation of the manuscript. Without long discussions with him the book would certainly look very different.

During the preparation of the book Yulij Ilyashenko was supported by the grants NSF no. 0100404 and no. 0400495. Sergei Yakovenko is incumbent of The Gershon Kekst Professorial Chair. His research was supported by the Israeli Science Foundation grant no. 18-00/1 and the Minerva Foundation.

Chapter I

Normal forms and desingularization

1. Analytic differential equations in the complex domain

For an open domain $U \subseteq \mathbb{C}^n$ we denote by $\mathcal{O}(U)$ the complex linear space of functions holomorphic in U (see Appendix). The space of vector-valued holomorphic functions is denoted by

$$\mathcal{O}^m(U) = \underbrace{\mathcal{O}(U) \times \cdots \times \mathcal{O}(U)}_{m \text{ times}} = \mathcal{O}(U) \otimes_{\mathbb{C}} \mathbb{C}^m.$$

1A. Differential equations, solutions, initial value problems. Let $U \subseteq \mathbb{C} \times \mathbb{C}^n$ be an open domain and $F = (F_1, \dots, F_n)\colon U \to \mathbb{C}^n$ a holomorphic vector function. An *analytic ordinary differential equation* defined by F on U is the vector equation (or the *system* of n scalar equations)

$$\frac{dx}{dt} = F(t,x), \qquad (t,x) \in U \subseteq \mathbb{C} \times \mathbb{C}^n, \quad F \in \mathcal{O}^n(U). \tag{1.1}$$

The *solution* of this equation is a parameterized holomorphic curve, the holomorphic map $\varphi = (\varphi_1, \dots, \varphi_n)\colon V \to \mathbb{C}^n$, defined in an open subset $V \subseteq \mathbb{C}$, whose graph $\{(t, \varphi(t))\colon t \in V\}$ belongs to U and whose complex "velocity vector" $\frac{d\varphi}{dt} = \left(\frac{d\varphi_1}{dt}, \dots, \frac{d\varphi_n}{dt}\right) \in \mathbb{C}^n$ at each point t coincides with the vector $F(t, \varphi(t)) \in \mathbb{C}^n$.

The graph of φ in U is called the *integral curve*. From the real point of view it is a 2-dimensional smooth surface in $\mathbb{R}^{2n+2}$. Note that from the beginning we consider only holomorphic solutions which may be, however, defined on domains of different size.

The equation is *autonomous*, if F is independent of t. In this case the image $\varphi(V) \subseteq \mathbb{C}^n$ is called the *phase curve*. Any differential equation (1.1) can be "made" autonomous by adding a fictitious variable $z \in \mathbb{C}$ governed by the equation $\frac{dz}{dt} = 1$.

If $(t_0, x_0) = (t_0, x_{0,1}, \ldots, x_{0,n}) \in U$ is a specified point, then the *initial value problem*, sometimes also called the *Cauchy problem*, is to find an integral curve of the differential equation (1.1) passing through the point (t_0, x_0), i.e., a solution satisfying the condition

$$\varphi\colon V \to \mathbb{C}^n, \qquad \varphi(t_0) = x_0 \in \mathbb{C}^n. \tag{1.2}$$

In what follows we will often denote by a dot the derivative with respect to the complex variable t, $\dot{x}(t) = \frac{dx}{dt}(t)$.

The first fundamental result is the local existence and uniqueness theorem.

Theorem 1.1. *For any holomorphic differential equation* (1.1) *and every point* $(t_0, x_0) \in U$ *there exists a sufficiently small polydisk* $D_\varepsilon = \{|t - t_0| < \varepsilon,\ |x_j - x_{0,j}| < \varepsilon,\ j = 1, \ldots, n\} \subseteq U$, *such that the solution of the initial value problem* (1.2) *exists and is unique in this polydisk.*

This solution depends holomorphically on the initial value $x_0 \in \mathbb{C}^n$ *and on any additional parameters, provided that the vector function* F *depends holomorphically on these parameters.*

From the real point of view, Theorem 1.1 asserts existence of $2n$ functions of *two* independent real variables whose graph is a *surface* in $\mathbb{C}^{n+1} \cong \mathbb{R}^{2n+2}$, with the tangent plane spanned by two real vectors $\operatorname{Re} F, \operatorname{Im} F$. To derive this theorem from the standard results on existence, uniqueness and differentiability of solutions of *smooth* ordinary differential equations in the *real domain*, one should use rather deep results on *integrability of distributions*; see Remark 2.10 below. Rather unexpectedly, the direct proof is *simpler* than in the real case in the part concerning dependence on initial conditions. This proof is given in the next subsection. The main idea of this proof, as well as many other proofs below, is the *contracting map principle*.

1B. Contracting map principle. Consider the linear space $\mathcal{A}(D_\rho)$ of functions holomorphic in the polydisk D_ρ and continuous on its closure,

$$\mathcal{A}(D_\rho) = \{f\colon D_\rho \to \mathbb{C} \text{ holomorphic in } D_\rho \text{ and continuous on } \overline{D_\rho}\}. \tag{1.3}$$

This space is naturally equipped with the supremum-norm,

$$\|f\|_\rho = \max_{z \in D_\rho} |f(z)|, \qquad z = (z_1, \ldots, z_n) \in \mathbb{C}^n, \tag{1.4}$$

and thus naturally a subspace of the *complete normed* (i.e., Banach) space $C(\overline{D_\rho})$ of continuous complex-valued functions. Though holomorphic functions may have very complicated boundary behavior and thus $\mathcal{A}(U) \subsetneq \mathcal{O}(U)$, they are continuous and therefore for any smaller domain U' relatively compact in U (i.e., when $\overline{U'} \Subset U$), there is an obvious inclusion $\mathcal{A}(U') \supset \mathcal{O}(U)$.

Theorem 1.2. *The space $\mathcal{A}(D_\rho)$ and its vector counterparts $\mathcal{A}^m(D_\rho) = \bigoplus_{m \text{ times}} \mathcal{A}(D_\rho)$ are complete (Banach) spaces.*

Proof. Any fundamental sequence in $\mathcal{A}(D_\rho)$ is by definition fundamental in the Banach space $C(\overline{D_\rho})$ and has a uniform limit in the latter space. By the Weierstrass compactness principle [**Sha92**], this limit is again holomorphic in D_ρ, i.e., belongs to $\mathcal{A}(D_\rho)$. □

A map F of a metric space $\mathcal{M}$ into itself is called *contracting*, if for some positive real number $\lambda < 1$ and all $u, v \in \mathcal{M}$ the inequality $\operatorname{dist}(F(u), F(v)) \leqslant \lambda \operatorname{dist}(u, v)$ holds. A point $w \in \mathcal{M}$ is *fixed* (by F), if $F(w) = w$.

Theorem 1.3 (Contracting map principle). *Any contracting map $F \colon M \to M$ of a complete metric space M has a unique fixed point in $\mathcal{M}$.*

This fixed point is the limit of any sequence of iterations $u_{k+1} = F(u_k)$, $k = 0, 1, 2, \ldots$ beginning with an arbitrary initial point $u_0 \in M$.

Proof. For any initial point $u_0 \in M$, the sequence u_k, $k = 1, 2, \ldots$ is fundamental, since $\operatorname{dist}(u_k, u_{k+1}) \leqslant \lambda^k \operatorname{dist}(u_0, u_1)$ and by the triangle inequality $\operatorname{dist}(u_k, u_l) \leqslant \operatorname{dist}(u_0, u_1)\lambda^k/(1-\lambda)$ for any $k < l$. By completeness assumption, the sequence u_k converges to a limit $w \in M$. Since F is continuous, passing to the limit in the identity $u_{k+1} = F(u_k)$ yields $w = F(w)$. If w_1, w_2 are two fixed points, then $\operatorname{dist}(w_1, w_2) \leqslant \lambda \operatorname{dist}(F(w_1), F(w_2)) = \lambda \operatorname{dist}(w_1, w_2)$ which is possible only if $\operatorname{dist}(w_1, w_2) = 0$, i.e., when $w_1 = w_2$. □

1C. Picard operators and their contractivity. The exposition below is based on [**Arn78**, §31] with minor modifications.

Consider the equation (1.1) defined in a domain U. Denote by $D_\varepsilon = \{|z - x_0| < \varepsilon, |t - t_0| < \varepsilon\} \subset \mathbb{C}^{n+1}$ a polydisk centered at the point $(t_0, x_0) \in U$ and small enough to belong to U.

Definition 1.4. The *Picard operator* $\mathbf{P}$ associated with the differential equation (1.1) and the initial value $(t_0, z_0) \in U$, is the operator $f \mapsto \mathbf{P}f$ defined by the integral formula

$$(\mathbf{P}f)(s, z) = z + \int_{t_0}^{s} F(t, f(t, z))\, dt \tag{1.5}$$

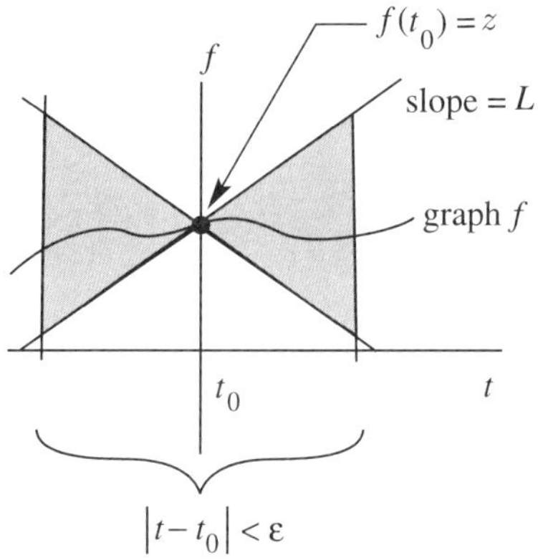

Figure I.1. Domain of definition of Picard iterations (in the intersection with the hyperplane $z = \text{const}$)

for all vector functions $f(t, z)$ the expression in the right hand side makes sense.

We will now construct a complete metric space invariant by $\mathbf{P}$, on which this operator is contracting. Denote by L_0 and L_1 the bounds for the magnitude of F and its Lipschitz constant in U: for any $(t, x), (t, x') \in U$,

$$|F(t,z)| \leqslant L_0, \qquad |F(t,z) - F(t,z')| \leqslant L_1\,|z - z'|. \tag{1.6}$$

Denote by $\mathcal{M}$ the subspace of the space $\mathcal{A}^n(D_\varepsilon)$ which consists of the functions satisfying the additional inequality

$$|f(t,z) - z| \leqslant L_0\,|t - t_0|. \tag{1.7}$$

This space is complete in the metric induced by the norm $\|\cdot\|_\varepsilon$ inherited from the ambient space $\mathcal{A}^n(D_\varepsilon)$ (Exercise 1.3).

Lemma 1.5. *If the polydisk D_ε is sufficiently small, the Picard operator* $\mathbf{P}$ *given by the integral* (1.5), *is well defined and contracting on* $\mathcal{M}$.

More precisely, for sufficiently small ε its contraction factor λ does not exceed εL_1, where L_1 is the Lipschitz constant for F in U.

Proof. Explicit majorizing of the integral shows that

$$|\mathbf{P}f(s,z) - z| \leqslant L_0 \int_{t_0}^{s} |dt| \leqslant L_0\,|s - t_0| \leqslant L_0\varepsilon,$$

so if ε is chosen sufficiently small, the operator $\mathbf{P}$ is well defined on $\mathcal{M}$ and maps this space into itself. For any two vector functions f, f' defined on such a small polydisk D_ε, we have by virtue of the same estimate

$$\|\mathbf{P}f - \mathbf{P}f'\| = \sup_{|s-t_0|<\varepsilon} \int_{t_0}^{s} L_1\,|f(t,z) - f'(t,z)|\,|dt| \leqslant \varepsilon L_1\,\|f - f'\|.$$

If $\varepsilon L_1 < 1$, the operator $\mathbf{P}$ is contracting. □

Proof of Theorem 1.1. Assume ε is so small that the $\varepsilon L_1 < 1$ so that by Lemma 1.5, the Picard operator $\mathbf{P}$ is contracting. By Theorem 1.2 the fixed point of this operator (which exists by Theorem 1.3 and Lemma 1.5) is a *holomorphic* vector function $f\colon D_\varepsilon \to \mathbb{C}^n$ that satisfies the integral equation

$$f(s,z) = z + \int_{t_0}^{s} F(t, f(t,z))\,dt, \qquad |s-t_0| < \varepsilon, \quad |z-x_0| < \varepsilon. \tag{1.8}$$

For each fixed z, the function $\varphi_z(t) = f(t,z)$ clearly satisfies both the initial condition (1.2) with $x_0 = z$ and the differential equation (1.1). By construction, it depends holomorphically on the initial condition z.

To prove holomorphic dependence on additional parameters, one can treat them as fictitious dependent variables. Assume that the vector function $F = F(t,x,y)$ depends holomorphically on additional parameters $y \in \mathbb{C}^m$, and consider the initial value problem (recall that the dot means the derivative $\frac{d}{dt}$)

$$\begin{cases} \dot x = F(t,x,y), & x(t_0) = x_0, \\ \dot y = 0, & y(t_0) = y_0. \end{cases} \tag{1.9}$$

The solution of this initial value problem is a function $f(t,x,y,x_0,y_0)$ holomorphically depending on all variables. □

Remark 1.6. For a differential equation with the right hand side $F(t,x)$ the *shifted solution* $x'(t) = x(t-y)$, $y \in \mathbb{C}^1$, satisfies the shifted equation $\dot x' = F(t-y, x')$ which analytically depends on the parameter y. By Theorem 1.1, this shows that solutions of the initial value problem depend holomorphically also on the t-component of the initial point $(t_0, x_0) \in U$.

1D. Principal example: exponential formula for linear systems. The proof of the existence theorem is *constructive*: the solution of a differential equation is obtained as the uniform limit of its *Picard approximations*, iterations of the Picard operator.

In the simplest case of a differential equation with *constant* (i.e., independent of t,x,y) right hand side $F = \text{const} \in \mathbb{C}^n$ the Picard approximations stabilize immediately: if $f_0(t,v) = v$, then $f_1(t,v) = f_2(t,v) = \cdots = v + (t-t_0)F$.

A *linear system with constant coefficients* is the system of equations

$$\dot x = Ax, \qquad x \in \mathbb{C}^n, \quad A \in \operatorname{Mat}(n, \mathbb{C}) \tag{1.10}$$

where $A = \|a_{ij}\|$ is a constant (independent of t and x) $(n \times n)$-matrix with complex entries. Reasoning by induction, one can see that the Picard approximations for the solution of (1.10) which start with the constant initial term $f_0(t, v) = v$, have the form

$$f_k(t, v) = \left(E + tA + \tfrac{t^2}{2!}A^2 + \cdots + \tfrac{t^k}{k!}A^k\right) v. \tag{1.11}$$

Indeed,

$$\begin{aligned}\mathbf{P} f_k(t, v) &= v + \int_0^t A \cdot \left(E + sA + \cdots + \tfrac{s^k}{k!}A^k\right) v\, ds \\ &= Ev + \left(tA + \cdots + \tfrac{t^{k+1}}{(k+1)!}A^{k+1}\right) v = f_{k+1}(t, v).\end{aligned}$$

These formulas motivate the following fundamental object.

Definition 1.7 (matrix exponential). For an arbitrary constant matrix $A \in \operatorname{Mat}(n, \mathbb{C})$ its *exponential* $\exp A$ is the sum of the infinite (matrix) series

$$\exp A = E + A + \frac{1}{2!}A^2 + \cdots + \frac{1}{k!}A^k + \cdots. \tag{1.12}$$

Since $|A^k| \leqslant |A|^k$ and since the factorial series $\sum_{k \geqslant 0} r^k/k!$ converges absolutely for all values $r \in \mathbb{R}$, the matrix series (1.12) converges absolutely on the complex linear space $\operatorname{Mat}(n, \mathbb{C}) \cong \mathbb{C}^{n^2}$ for any finite n.

Note that for any two *commuting* matrices A, B their exponents satisfy the *group identity*

$$\exp(A + B) = \exp A \cdot \exp B = \exp B \cdot \exp A. \tag{1.13}$$

This can be proved by substituting A, B instead of two scalars a, b into the formal identity obtained by expansion of the law $e^a e^b = e^{a+b}$.

The explicit formula (1.11) for Picard approximations for the linear system (1.10) immediately proves the following theorem.

Theorem 1.8. *The solution of the linear system $\dot{x} = Ax$, $A \in \operatorname{Mat}(n, \mathbb{C})$, with the initial value $x(0) = v$ is given by the matrix exponential,*

$$x(t) = (\exp tA)\, v, \qquad t \in \mathbb{C}, \quad v \in \mathbb{C}^n. \quad \square \tag{1.14}$$

Remark 1.9. Computation of the matrix exponential can be reduced to computation of a matrix polynomial of degree $\leqslant n - 1$ and exponentials of eigenvalues of A. Indeed, assume that A has a Jordan normal form $A = \Lambda + N$, where $\Lambda = \operatorname{diag}\{\lambda_1, \ldots, \lambda_n\}$ is the diagonal part and N the upper-triangular (nilpotent) part *commuting* with Λ. Then $\exp \Lambda$ is a diagonal matrix with the exponentials of the eigenvalues of A on the diagonal, $N^n = 0$

by nilpotency, and therefore

$$\exp[t(\Lambda+N)] = \exp t\Lambda \cdot \exp tN$$
$$= \begin{pmatrix} \exp t\lambda_1 & & \\ & \ddots & \\ & & \exp t\lambda_n \end{pmatrix} \cdot \left(E + tN + \frac{t^2}{2!}N^2 + \cdots + \frac{t^{n-1}}{(n-1)!}N^{n-1} \right). \tag{1.15}$$

This provides a practical way of solving linear systems with constant coefficients: components of any solution in any basis are linear combinations of *quasipolynomials* $t^k \exp t\lambda_j$, $0 \leqslant k \leqslant n-1$ with complex coefficients.

Remark 1.10 (Liouville–Ostrogradskii formula). By direct inspection of the formula (1.15) we conclude that

$$\forall A \in \operatorname{Mat}(n, \mathbb{C}) \qquad \det \exp A = \exp \operatorname{tr} A. \tag{1.16}$$

Indeed, $\det \exp A = \det \exp \Lambda \cdot \det \exp N = \prod_{i=1}^n \exp \lambda_i \cdot 1 = \exp \operatorname{tr} \Lambda = \exp \operatorname{tr} A$, since the matrix polynomial $\exp N$ is upper triangular with units on the diagonal.

1E. Flow box theorem. Let $f(t, x_0)$ be the holomorphic vector function solving the initial value problem (1.2) for the differential equation (1.1).

Definition 1.11. The *flow map* for a differential equation (1.1) is the vector function of $n+2$ complex variables (t_0, t_1, v) defined when $(t_0, x) \in U$ and $|t_0 - t_1|$ is sufficiently small, by the formula

$$(t_0, t_1, v) \mapsto \Phi_{t_0}^{t_1}(v) = f(t_1, v), \tag{1.17}$$

where $f(t, v)$ is the fixed point of the Picard operator $\mathbf{P}$ as in (1.8) associated with the initial point t_0.

In other words, $\Phi_{t_0}^{t_1}(v)$ is the value $\varphi(t)$ which takes the solution of the initial value problem with the initial condition $\varphi(t_0) = v$, at the point t_1 sufficiently close to t_0.

Example 1.12. For a linear system (1.10) with constant coefficients, the flow map is linear:

$$\Phi_{t_0}^{t_1}(v) = [\exp(t_1 - t_0)A]\, v.$$

This map is defined for *all* values of t_0, t_1, v.

By Theorem 1.1, Φ is a holomorphic map. Since the solution of the initial value problem is unique, it obviously must satisfy the functional equation

$$\Phi_{t_1}^{t_2}(\Phi_{t_0}^{t_1}(x)) = \Phi_{t_0}^{t_2}(x) \tag{1.18}$$

for all t_1, t_2 sufficiently close to t_0 and all x sufficiently close to x_0. Since for any x the vector function $t \mapsto \varphi_x(t) = \Phi^t_{t_0}(x)$ is a solution of (1.1), we have

$$\frac{\partial}{\partial t}\bigg|_{t=t_0,\, x=x_0} \Phi^t_{t_0}(x) = -\frac{\partial}{\partial t_0}\bigg|_{t=t_0,\, x=x_0} \Phi^t_{t_0}(x) = F(t_0, x_0).$$

From the integral equation (1.8) it follows that

$$\Phi^t_{t_0}(x_0) = x_0 + (t - t_0)F(t_0, x_0) + o(|t - t_0|), \tag{1.19}$$

and therefore the Jacobian matrix of Φ with respect to the x-variable is

$$\left(\frac{\partial \Phi^t_{t_0}(x)}{\partial x}\right)_{t=t_0,\, x=x_0} = E. \tag{1.20}$$

Differential equations can be transformed to each other by various transformations. The most important is the (bi)holomorphic equivalence, or holomorphic conjugacy.

Definition 1.13. Two differential equations, (1.1) and another such equation

$$\dot{x}' = F'(t', x'), \qquad (t', x') \in U', \tag{1.21}$$

are *conjugated* by the biholomorphism $H \colon U \to U'$ (the *conjugacy*), if H sends any integral trajectory of (1.1) into an integral trajectory of (1.21).

Two systems are *holomorphically equivalent* in their respective domains, if there exists a biholomorphic conjugacy between them.

Clearly, biholomorphically conjugate systems are indistinguishable in everything that concerns properties invariant by biholomorphisms. Finding a simple system biholomorphically equivalent to a given one, is therefore of paramount importance.

Theorem 1.14 (Flow box theorem). *Any holomorphic differential equation (1.1) in a sufficiently small neighborhood of any point is biholomorphically conjugated by a suitable biholomorphic conjugacy $H \colon (t, x) \mapsto (t, h(t, x))$ preserving the independent variable t, to the trivial equation*

$$\dot{x}' = 0. \tag{1.22}$$

Proof of the theorem. Consider the map $H' \colon \mathbb{C}^{n+1} \to \mathbb{C}^{n+1}$ which is defined near the point (t_0, x_0) using the flow map (1.17) for the equation (1.1),

$$H' \colon (t, x') \mapsto (t, \Phi^t_{t_0}(x')), \qquad (t, x') \in (\mathbb{C}^{n+1}, (t_0, x_0)).$$

By construction, it takes the lines $x' =$ const parallel to the t-axis, into integral trajectories of the equation (1.1), while preserving the value of t.

The Jacobian matrix $\partial H'(t, x')/\partial(t, x')$ of the map H' at the point (t_0, x_0) has by (1.20) the form $\left(\begin{smallmatrix} 1 & \\ * & E \end{smallmatrix}\right)$ and is therefore invertible.

Thus H' restricted on a sufficiently small neighborhood of the point (t_0, x_0), is a biholomorphic conjugacy between the trivial system (1.21), whose solutions are exactly the lines $x' = \text{const}$, and the given system (1.1). The inverse map also preserves t and conjugates (1.1) with (1.21). $\square$

1F. Vector fields and their equivalence. The above constructions after small modification become more transparent in the *autonomous* case, when the vector function $x \mapsto F(x)$ which is now independent of t, can be considered as a *holomorphic vector field* on its domain $U \subseteq \mathbb{C}^n$. The space of vector fields holomorphic in a domain $U \subseteq \mathbb{C}^n$ will be denoted by $\mathcal{D}(U)$, while the notation $\mathcal{D}(\mathbb{C}^n, x_0)$ is reserved for the space of germs of holomorphic vector fields at a specific point $x_0 \in \mathbb{C}^n$, usually the origin, $x_0 = 0$.

In the autonomous case, translation of the independent variable preserves solutions of the equation

$$\dot{x} = F(x), \qquad F\colon U \to \mathbb{C}^n, \tag{1.23}$$

so the flow map $\Phi_{t_0}^{t_1}$ actually depends only on the difference $t = t_1 - t_0$ and hence will be denoted simply by $\Phi^t(\cdot) = \Phi_0^t(\cdot)$. The functional identity (1.18) takes the form

$$\Phi^t(\Phi^s(x)) = \Phi^{t+s}(x), \qquad t, s \in (\mathbb{C}, 0),\ x \in (\mathbb{C}^n, x_0), \tag{1.24}$$

which means that the maps $\{\Phi^t\}$ form a one-parametric *pseudogroup* of biholomorphisms. ("Pseudo" means that the composition in (1.24) is not always defined. The pseudogroup is a true group, $\Phi^t \circ \Phi^s = \Phi^{t+s}$, if the maps Φ^t are globally defined for all $t \in \mathbb{C}$. For more details on pseudogroups see §6**D**).

For autonomous equations it is natural to consider biholomorphisms that are *time-independent*.

Definition 1.15. Two holomorphic vector fields, $F \in \mathcal{D}(U)$ and $F' \in \mathcal{D}(U')$ defined in two domains $U, U' \subseteq \mathbb{C}^n$, are *biholomorphically equivalent* if there exists a biholomorphic map $H\colon U \to U'$ conjugating their respective flows,

$$H \circ \Phi^t = {\Phi'}^t \circ H \tag{1.25}$$

whenever both sides are defined. The biholomorphism H is said to be a *conjugacy* between F and F'.

A conjugacy H maps *phase* curves of the first field into phase curves of the second field; in a similar way the suspension

$$\operatorname{id} \times H\colon (\mathbb{C}, 0) \times U \to (\mathbb{C}, 0) \times U', \qquad (t, x) \mapsto (t, H(x)),$$

maps *integral* curves of the two fields into each other. Differentiating the identity (1.25) in t at $t = 0$, we conclude that the differential $dH(x)$ of a holomorphic conjugacy sends the vector $v = F(x)$ of the first field, attached to a point $x \in U$, to the vector $v' = F'(x')$ of the second field *at the appropriate point* $x' = H(x)$. In the coordinates this property takes the form of the identity

$$H_*(x) \cdot F(x) = F'(H(x)), \qquad H_*(x) = \left(\tfrac{\partial H}{\partial x}\right) = \left(\tfrac{\partial h_i}{\partial x_j}\right), \tag{1.26}$$

in which the Jacobian matrix $H_*(x) = \left(\frac{\partial H}{\partial x}\right)$ is involved. The formula (1.26) is sometimes used as the alternative *definition* of the holomorphic equivalence. The third (algebraic, in some sense most natural) way to introduce this equivalence is explained in the next section.

1G. Vector fields as derivations. It is sometimes convenient to define vector fields in a way independent of the coordinates. Each vector field $F = (F_1, \dots, F_n)$ in a domain $U \subset \mathbb{C}^n$ defines a *derivation* $\mathbf{F} \in \operatorname{Der} \mathcal{O}(U)$ of the $\mathbb{C}$-algebra $\mathcal{O}(U)$ of functions holomorphic in U, by the formula

$$\mathbf{F}f(x) = \sum_{j=1}^{n} F_j(x)\,\frac{\partial f}{\partial x_j}. \tag{1.27}$$

We often identify the holomorphic vector field F with the components F_i with the corresponding differential operator $\mathbf{F} = \sum F_j \frac{\partial}{\partial x_j}$.

Derivations can be defined in purely algebraic terms as $\mathbb{C}$-linear maps of the algebra $\mathcal{O}(U)$ satisfying the Leibnitz identity,

$$\mathbf{F}(fg) = f(\mathbf{F}g) + (\mathbf{F}f)g.$$

Indeed, any $\mathbb{C}$-linear operator with this property in any coordinate system $(x_1, \dots, x_n)$ defines n functions $F_j = \mathbf{F}x_j$ and (obviously) sends all constants to zero. Any analytic function f can be written $f(x) = f(a) + \sum_1^n h_j(x)\,(x_j - a_j)$ with $h_j(a) = \frac{\partial f}{\partial x_j}(a)$. Applying the Leibnitz rule, we conclude that $(\mathbf{F}f)(a) = \sum_j F_j h_j(a) + 0 \cdot \mathbf{F}h_j = \sum_j F_j \frac{\partial f}{\partial x_j}(a)$, as claimed.

A similar algebraic description can be given for holomorphic maps. With any holomorphic map $H\colon U \to U'$ between two domains $U, U' \subseteq \mathbb{C}^n$ one can associate the *pullback operator* $\mathbf{H}\colon \mathcal{O}(U') \to \mathcal{O}(U)$, acting on $f' \in \mathcal{O}(U')$ by composition, $(\mathbf{H}f')(x) = f'(H(x))$. This operator is a *homomorphism* of commutative $\mathbb{C}$-algebras, a $\mathbb{C}$-linear map respecting multiplication (i.e., $\mathbf{H}(f'g') = \mathbf{H}f' \cdot \mathbf{H}g'$ for any $f', g' \in \mathcal{O}(U')$). Conversely, any continuous homomorphism $\mathbf{H}$ between the two algebras is induced by a holomorphic map $H = (h_1, \dots, h_n)$ with $h_i = \mathbf{H}x_i$, where $x_i \in \mathcal{O}(U')$ are the coordinate functions (restricted on U'). The map H is a biholomorphism if and only if $\mathbf{H}$ is an isomorphism of $\mathbb{C}$-algebras.

In this language the action of biholomorphisms on vector fields can be described as a simple *conjugacy of operators*: two derivations $\mathbf{F}$ and $\mathbf{F}'$ of the algebras $\mathcal{O}(U)$ and $\mathcal{O}(U')$ respectively, are said to be conjugated by the biholomorphism $H: U \to U'$, if

$$\mathbf{F} \circ \mathbf{H} = \mathbf{H} \circ \mathbf{F}' \tag{1.28}$$

as two $\mathbb{C}$-linear operators from $\mathcal{O}(U')$ to $\mathcal{O}(U)$.

Another advantage of this invariant description is the possibility of defining the *commutator* of two vector fields naturally, as the commutator of the respective differential operators. One can immediately verify that $[\mathbf{F}, \mathbf{F}'] = \mathbf{F}\mathbf{F}' - \mathbf{F}'\mathbf{F}$ satisfies the Leibnitz identity as soon as $\mathbf{F}, \mathbf{F}'$ do, and hence corresponds to a vector field. In coordinates the commutator takes the form

$$[F, F'] = \left(\frac{\partial F'}{\partial x}\right) F - \left(\frac{\partial F}{\partial x}\right) F'. \tag{1.29}$$

Example 1.16. For any two $\mathbf{F} = Ax$, $\mathbf{F}' = A'x$ *linear* vector fields, their commutator $[\mathbf{F}, \mathbf{F}']$ is again a linear vector field with the linearization matrix $A'A - AA'$. It coincides (modulo the sign) with the usual matrix commutator $[A, A']$.

1H. Rectification of vector fields. A straightforward counterpart of the Flow box Theorem 1.14 for holomorphic vector fields holds only if the field is nonvanishing.

Definition 1.17. A point x is a *singular point* (*singularity*) of a holomorphic vector field F, if $F(x_0) = 0$. Otherwise the point is *nonsingular.*

Theorem 1.18 (Rectification theorem). *A holomorphic vector field F is holomorphically equivalent to the constant vector field $F'(x') = (1, 0, \dots, 0)$ in a sufficiently small neighborhood of any nonsingular point.*

Proof. The flow Φ' of the constant vector field F' can be immediately computed: $(\Phi')^t(x') = x' + t \cdot (1, 0, \dots, 0)$. Consider any affine hyperplane $\Pi \subset U$ passing through x_0 and transversal to $F(x_0)$ and the hyperplane $\Pi' = \{x_1' = 0\}$. Let $t = x_1' \colon \mathbb{C}^n \to \mathbb{C}$ be the function equal to the first coordinate in $\mathbb{C}^n$, so that $(\Phi')^{-t}(x') \in \Pi'$. Let $h' \colon \Pi' \to \Pi$ be *any* biholomorphism (e.g., linear invertible map). Then the map

$$H' = \Phi^t \circ h \circ (\Phi')^{-t}, \qquad t = t(x'),$$

is a holomorphic map that sends any (parameterized) trajectory of F', passing through a point $x' \in \Pi'$, to the parameterized trajectory of F passing through $x = h(x')$. Being composition of holomorphic maps, H' is also holomorphic, and coincides with h' when restricted on Π'. It remains to notice that the differential $dH'(x_0)$ maps the vector $(1, 0, \dots, 0)$ transversal to Π', to the vector $F(x_0)$ transversal to Π. This observation proves that H' is invertible in some sufficiently small neighborhood U of x_0, and the inverse map H conjugates F in U with F' in $H(U)$. □

1I. One-parametric groups of holomorphisms. The Rectification theorem from §1 can be formulated in the language of germs as follows: *Two germs of holomorphic vector fields at nonsingular points are always holomorphically equivalent to each other.* In particular, any germ of a holomorphic vector field at a nonsingular point is holomorphically equivalent to the germ of a nonzero constant vector field.

Because of this "triviality" of local description of nonsingular vector fields, we will mostly be interested in germs of vector fields at the *singular points.* The first result is existence of germs of the flow maps Φ^t at the singular point, for all values of $t \in \mathbb{C}$.

Denote by $\operatorname{Diff}(\mathbb{C}^n, 0)$ the group of germs of holomorphic self-maps $H\colon (\mathbb{C}^n, 0) \to (\mathbb{C}^n, 0)$ equipped with the operation of composition (which is always defined).

Proposition 1.19. *If $F \in \mathcal{D}(\mathbb{C}^n, 0)$ is the germ of a holomorphic vector field which is singular (i.e., $F(0) = 0$), then the germs of the flow maps $\Phi^t(\cdot)$ are defined for all $t \in \mathbb{C}$ and form a one-parametric subgroup of the group* $\operatorname{Diff}(\mathbb{C}^n, 0)$ *of germs of biholomorphic self-maps: $\Phi^t \circ \Phi^s = \Phi^{t+s}$ for any $t, s \in \mathbb{C}$.*

Proof. The existence of the flow maps Φ^t for all sufficiently small $t \in (\mathbb{C}, 0)$, the possibility of their composition, and validity of the group identity for such small t all follow from Theorem 1.1 and the fact that $\Phi^t(x_0) = x_0$.

For an arbitrary large value of $t \in \mathbb{C}$ we may define Φ^t as the composition of germs of the flow maps Φ^{t_i}, $i = 1, \dots, N$, taken in any order, where the complex numbers t_i are sufficiently small to satisfy conditions of Theorem 1.1 but added together give t. From the local group identity it follows that the definition does not depend on the particular choice of t_i and preserves the group property. □

Remark 1.20. Every germ of a self-map $H \in \operatorname{Diff}(\mathbb{C}^n, 0)$ uniquely defines an *auto*morphism $\mathbf{H} \in \operatorname{Aut} \mathcal{O}(\mathbb{C}^n, 0)$ of the commutative algebra of holomorphic germs acting by substitution, $\mathbf{H}f = f \circ H$.

Proposition 1.19 translates into the algebraic language as follows: for any derivation $\mathbf{F} \in \operatorname{Der} \mathcal{O}(\mathbb{C}^n, 0)$ of the algebra of holomorphic germs there exist a one-parametric subgroup $\{\mathbf{H}^t\colon t \in \mathbb{C}\} \subset \operatorname{Aut} \mathcal{O}(\mathbb{C}^n, 0)$ of automorphisms of this algebra, such that $\frac{d}{dt}\big|_{t=0} \mathbf{H}^t = \mathbf{F}$.

For the reasons to be explained below in §3**C**, the subgroup of automorphisms $\mathbf{H}^t$ is often referred to as the *exponent*, $\mathbf{H}^t = \exp(t\mathbf{F})$, of the derivation **F**. Respectively, the flow (germs of self-maps) will be sometimes denoted by the exponent, $\Phi^t = \exp(tF)$, of the corresponding vector field F.

Exercises and Problems for §1.

Exercise 1.1. Let $a \in U$ be a nonsingular point of a holomorphic vector field $F \in \mathcal{D}(U)$. A trajectory of the vector field is the projection of the graph of the solution into the domain of the field along the time axis.

Prove that the trajectory passing through a is either the line $x = a$, or can be represented as the graph of a function $y = \varphi_a(x)$ having an *algebraic* ramification point of some finite order ν. Express ν in terms of orders of the components of the field F at a.

Exercise 1.2. Let $P\colon (\mathbb{C}^n, 0) \to (\mathbb{C}^{n-1}, 0)$ be a holomorphic epimorphism (i.e., map of rank $n-1$) constant along trajectories of an analytic vector field $F \in \mathcal{D}(\mathbb{C}^n, 0)$. Construct explicitly the rectifying chart for F.

Exercise 1.3. Prove that the space $\mathcal{M}$ of functions satisfying the inequality (1.7), is indeed complete.

Exercise 1.4. Two linear vector fields in $\mathbb{C}^n$ are holomorphically equivalent in some domains containing the origin. Prove that these fields are *linear* equivalent, i.e., that there exists a linear map $H \in \mathrm{GL}(n, \mathbb{C})$ conjugating them.

Exercise 1.5. Prove that if two germs of vector fields at a singular point are analytically equivalent, then the eigenvalues of these fields at the singular point coincide.

Exercise 1.6. Prove that the vector field $F(z) = z^2 \frac{\partial}{\partial z}$ is holomorphic on the Riemann sphere $\mathbb{P}^1 = \mathbb{C} \cup \{\infty\}$. Compute the flow of this field.

Problem 1.7. Give a complete analytic classification of the holomorphic flows on the Riemann sphere $\mathbb{P}^1$ (i.e., construct a list, finite or infinite, of flows such that every holomorphic flow in analytically equivalent to one of the flows from the list, while any two different flows in the list are *not* holomorphically equivalent.

Exercise 1.8. Prove that the constant holomorphic vector fields $\frac{\partial}{\partial z}$ on the two tori $\mathbb{T}_1 = \mathbb{C}/(\mathbb{Z} + i\mathbb{Z})$ and $\mathbb{T}_2 = \mathbb{C}/(\mathbb{Z} + 2i\mathbb{Z})$, are not holomorphically equivalent.

2. Holomorphic foliations and their singularities

By the Existence/Uniqueness Theorem 1.1, any open connected domain $U \subseteq \mathbb{C}^n$ with a holomorphic vector field F defined on it, can be represented as the disjoint union of connected phase curves passing through all points of U. The Rectification Theorem 1.18 provides a local model for the geometric object called *foliated space* of simply *foliation*. A systematic treatment of foliations can be found, for instance, in [**Tam92, CC03**].

2A. Principal definitions. Speaking informally, a foliation is a partition of the phase space into a continuum of connected sets called *leaves*, which locally look as the family of parallel affine subspaces.

Definition 2.1. The *standard holomorphic foliation* of dimension n (respectively, of codimension m) of a polydisk $B = \{(x, y) \in \mathbb{C}^n \times \mathbb{C}^m : |x| < 1, |y| < 1\}$ is the representation of B as the disjoint union of n-disks, called (standard) *plaques*,

$$B = \bigsqcup_{|y|<1} L_y, \qquad L_y = \{|x| < 1\} \times \{y\} \subseteq B. \tag{2.1}$$

Definition 2.2. A holomorphic foliation $\mathcal{F}$ of a domain $U \subset \mathbb{C}^{n+m}$ (or, more generally, a complex analytic manifold U of dimension $n + m$) is the partition $U = \bigsqcup_\alpha L_\alpha$ of the latter into the disjoint union *connected* subsets L_α, called *leaves*, which locally is biholomorphically equivalent to the standard holomorphic foliation by plaques.

The latter phrase means that each point $a \in U$ admits an open neighborhood $B' \ni a$ and a biholomorphism $H : B' \to B$ of B' onto the standard polydisk B, which sends the *local leaves*, the connected components of the intersections $L_\alpha \cap B'$, to the plaques of the standard foliation,

$$\forall \alpha\ \exists Y = Y(\alpha) : \qquad H(L_\alpha \cap B') = \bigsqcup_{y \in Y(\alpha)} L_y. \tag{2.2}$$

Sometimes the local leaves will also be referred to as the *plaques* of the foliation near a point a: the plaques constitute biholomorphic images of n-disks, parameterized by a small m-disk. Note that different plaques may belong to the same leaf of the global foliation.

Remark 2.3. The definition of foliation admits several flavors. In the weakest settings the standard foliations are families of parallel balls slicing the real cylinder in $\mathbb{R}^{n+m}$ (the formulas remain the same as in (2.1)), while the local equivalencies H are simply homeomorphisms or smooth maps of low or high differentiability (up to C^∞ or even real analytic). In particular, we will call the *topological foliation* a partition of the space U into disjoint subsets L_α which is locally *homeomorphic* to the standard foliation (in the sense (2.2) with H being a homeomorphism).

Moreover, one can require *different* regularity of H along the leaves and in the transversal direction. We will not deal with such exotic cases until §28.

Remark 2.4 (important). The space of plaques of a foliation is naturally parameterized by points of a polydisk. Yet the index set $Y(\alpha)$ in (2.2) can be rather complicated (e.g., dense), since the global behavior of leaves outside the ball B' can be rather complicated. Yet in all of our applications all sets $Y(\alpha)$ will be at most countable.

The global space of leaves may have a very complicated structure even topologically (non-Hausdorff), therefore for indexing the leaves we use "abstract" sets without any additional structure.

Definition 2.5. Two holomorphic foliations $\mathcal{F}$ and $\mathcal{F}'$ defined on the respective holomorphic manifolds U, U', are called *holomorphically equivalent* or *topologically equivalent*, if there exists a biholomorphism $H: U \to U'$ (respectively, a homeomorphism) which maps (necessarily biholomorphically or homeomorphically, depending on the context) the leaves of $\mathcal{F}$ to those of $\mathcal{F}'$: $H(L_\alpha) = L'_{\alpha'}$ for some indices α, α'.

Note that this definition is *global*.

Everywhere below U stands for a holomorphic manifold or an open domain in $\mathbb{C}^n$. The following result is an obvious reformulation of the Rectification theorem in the language of foliations.

Proposition 2.6. *For any holomorphic vector field $F \in \mathcal{D}(U)$ without singularities in U, the partition of U into maximal integral curves of F forms a holomorphic foliation $\mathcal{F}_F$ of (complex) dimension 1 and codimension $n-1$.* □

We say that the foliation $\mathcal{F}_F$ is generated by the vector field F. Speaking about foliations rather than about vector fields means that the parametrization of solutions by the (complex) time is to be ignored.

Proposition 2.7. *Two holomorphically equivalent vector fields $F \in \mathcal{D}(U)$ and $F' \in \mathcal{D}(U')$ generate two holomorphically equivalent one-dimensional foliations.*

Conversely, if the foliations $\mathcal{F}, \mathcal{F}'$ generated by two nonsingular vector fields, are holomorphically equivalent by a biholomorphism $H: U \to U'$, then there exists a nonvanishing holomorphic function $\rho \in \mathcal{O}(U)$ such that

$$\rho(x) \cdot H_*(x) \cdot F(x) = F'(H(x)), \qquad \rho(x) \neq 0 \quad \text{in } U; \tag{2.3}$$

cf. with (1.26) *and Definition 1.15.*

Proof. The first assertion is obvious immediately. To prove the second, it is sufficient to show that two vector fields generating *the same* holomorphic one-dimensional foliation, differ by a nonvanishing holomorphic scalar factor ρ. This is obvious for the standard foliation: the first component must be nonzero while all other components are identically zero. □

2B. Foliations and integrable distributions. For a given holomorphic foliation $\mathcal{F}$ of dimension n and codimension m, the tangent spaces to leaves at different points are n-dimensional complex spaces in an obvious sense analytically depending on the point.

Such a geometric object is called *distribution.* To define formally subspaces analytically depending on parameters, one can choose between the language of holomorphic vector fields and that of holomorphic differential forms.

Definition 2.8. A (holomorphic nonsingular) n-dimensional distribution in a domain $U \subseteq \mathbb{C}^{n+m}$ is either

- a tuple of n holomorphic vector fields $F_1, \dots, F_n \in \mathcal{D}(U)$, linearly independent at every point of U, or
- tuple of m holomorphic 1-forms $\omega_1, \dots, \omega_m \in \Lambda^1(U)$, linearly independent at every point of U so that $\omega_1 \wedge \cdots \wedge \omega_m \in \Lambda^k(U)$ is nonvanishing.

Two tuples of the same type $\{F_j\}$ and $F'_j\}$ (resp., $\{\omega_i\}$ and $\{\omega'_i\}$ define the same distribution, if $F'_j = \sum_k c_{jk}(x)F_k$, resp., $\omega'_i = \sum_k c'_{ik}(x)\omega_k)$ for some holomorphic functions $c_{jk}(x), c'_{ik}(x)$. The forms and the fields defining the same distribution must be dual to each other, $\omega_i \cdot F_j = 0$ for all i, j.

A one-dimensional distribution is usually called a *line field.*

Clearly, any holomorphic foliation defines the corresponding tangent distribution of the same dimension. The converse in general is not true unless $n = 1$.

A holomorphic n-dimensional distribution is called *integrable* in U, if it is tangent to leaves of a nonsingular holomorphic foliation in U.

Theorem 2.9 (Frobenius integrability criteria). *A distribution defined by a tuple of holomorphic vector fields is integrable, if and only if the commutator of any two vector fields belongs to the same distribution, i.e., if*

$$[F_i, F_j] = \sum_{k=1}^{n} c_{ijk} F_k, \qquad c_{ijk} \in \mathcal{O}(U). \tag{2.4}$$

A distribution defined by a tuple of holomorphic 1-forms is integrable, if and only if the ideal spanned by these forms in the exterior algebra $\Lambda^\bullet(U)$ over $\mathcal{O}(U)$, is closed by the exterior derivative, i.e., if

$$d\omega_i = \sum_{k=1}^{m} c'_{ik}\, \omega_k \wedge \eta_k, \qquad \eta_k \in \Lambda^1(U),\ c_{ik} \in \mathcal{O}(U). \tag{2.5}$$

We will not prove this theorem. Its proof can be derived from the local existence theorem for holomorphic vector fields in the same way as it is done, *mutatis mutandis*, in the C^∞-smooth case in [**War83**].

Remark 2.10. The Frobenius integrability condition trivially holds for $n = 1$. On the other hand, from the real point of view the holomorphic vector field F corresponds to a 2-dimensional distribution generated by *two* vector fields $F_1 = F$ and $F_2 = iF$, $i = \sqrt{-1}$, in $\mathbb{R}^{2n} \cong \mathbb{C}^n$. The Frobenius integrability condition for this distribution reduces, as one can easily verify, to the Cauchy–Riemann identities between the real and imaginary parts of the components of the holomorphic vector field F.

Remark 2.11. In the (complex) 2-dimensional case where $U \subseteq \mathbb{C}^2$ that will be our principal object of studies later, the only nontrivial possibility is a one-dimensional distribution that is automatically integrable. It can be defined either by one vector field $F \in \mathcal{D}(U)$ or by one Pfaffian form $\omega \in \Lambda^1(U)$. For many reasons the Pfaffian presentation is more convenient.

2C. Holonomy. The notion of holonomy intends to be a replacement of the flow of the vector fields in the case where the natural parametrization of the solutions is absent or ignored.

Definition 2.12. A (parameterized) *cross-section* to a leaf L of a foliation $\mathcal{F}$ of codimension m on U at a point $a \in U$ is a holomorphic map $\tau\colon (\mathbb{C}^m, 0) \to (U, a)$ transversal to L. Very often we identify the cross-section with the image of the map τ.

If $\mathcal{F}$ is a *standard* foliation and τ, τ' any two cross-sections (at different, in general) points a, a' of the leaf, say $L_0 = \{y = 0\}$, then any other leaf L_α sufficiently close to L_0 intersects each cross-section exactly once. This defines in a unique way the holomorphic *correspondence map* $\Delta_{\tau,\tau'}\colon (\tau, a) \to (\tau', a')$: points with the same y-components are mapped into each other. In the charts on τ, τ' defined by the parameterizations, the correspondence map becomes the germ of a holomorphic map from $\operatorname{Diff}(\mathbb{C}^m, 0)$.

The correspondence maps obviously satisfy the identity

$$\Delta_{\tau,\tau''} = \Delta_{\tau',\tau''} \circ \Delta_{\tau,\tau'} \tag{2.6}$$

for any three cross-sections τ, τ', τ'' to the same leaf of the standard foliation.

Taking a biholomorphic image of this construction, we arrive at the following conclusion. For any two cross-sections τ, τ' to two sufficiently close points on the same leaf, there exists a uniquely defined correspondence map $\Delta_{\tau,\tau'}$ between the cross-sections that satisfies the identity (2.6) for any third cross-section which is also sufficiently close.

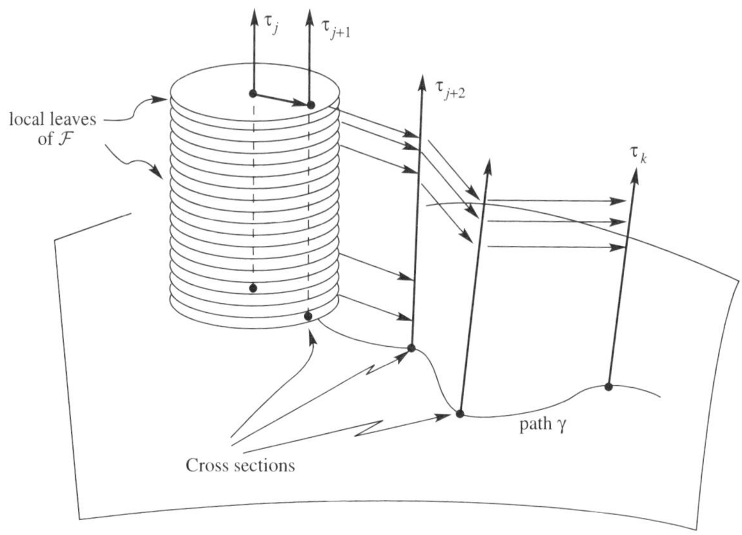

Figure I.2. Construction of the holonomy map for a foliation over a given path γ connecting two points on the leaf. The cross-sections τ_j are chosen close enough

Globalization of this construction associates the correspondence map not with just a pair of cross-sections to the same leaf, but rather with a *path* connecting the base points of these cross-sections. Let L be a leaf of a holomorphic foliation $\mathcal{F}$, τ, τ' two cross-sections cutting L at the points $a, a' \in L$, and $\gamma\colon [0,1] \to L$ an (oriented) path connecting $a = \gamma(0)$ with $a' = \gamma(1)$.

Since the segment $[0,1]$ and its image are compact, one can cover them by finitely many open sets U_j in such a way that in each set the foliation is locally trivial (biholomorphically equivalent to the standard foliation). One can insert between the cross-sections τ, τ' sufficiently many intermediate cross-sections τ_j, $j = 1, \dots, k$, $\tau_0 = \tau$, $\tau_k = \tau'$, at some intermediate points of the curve γ such that every two consecutive cross-sections τ_j, τ_{j+1} belong to the same domain U_j (for this purpose one has to choose $\tau_j \subset U_{j-1} \cap U_j$. Let $\Delta_{\tau_j, \tau_{j+1}}$ be the corresponding local correspondence maps as defined earlier. The composition

$$\Delta_\gamma = \Delta_{\tau_{k-1}, \tau_k} \circ \cdots \circ \Delta_{\tau_0, \tau_1} : (\tau, a) \to (\tau', a') \tag{2.7}$$

is a holomorphic map (more precisely, a germ) from $\mathrm{Diff}(\mathbb{C}^m, 0)$, also called the *correspondence map along the path* γ.

The identity (2.6) means that the correspondence map Δ_γ in fact does not depend on the choice of the intermediate cross-sections τ_j. Moreover, Δ_γ depends on the *homotopy class* of the path γ (with fixed endpoints) rather than on the path itself. Indeed, for another sufficiently close path

γ' connecting *the same endpoints*, we can choose cross-sections $\tau'_1, \dots, \tau'_{k-1}$ sufficiently close to the respective cross-sections τ_j for all $j = 1, \dots, k-1$ (the two extreme cross-sections coincide). Then one can use the identities (2.6) to show that the composition $\Delta_{\gamma'} = \Delta_{\tau'_{k-1},\tau'_k} \circ \dots \circ \Delta'_{\tau'_0,\tau_1} : (\tau, a) \to (\tau', a')$ coincides with Δ_γ, since $\tau'_0 = \tau_0$ and $\tau'_k = \tau_k$.

Remark 2.13. The construction of holonomy maps corresponds to what in the classical parlance was called "continuation of solutions of differential equations over a path": a specific solution (corresponding to the leaf) was explicitly or implicitly singled out together with a certain path on it, and all nearby solutions were "continued over the path" on the selected solution.

Choosing another pair of cross-sections at the same endpoints (or another parametrization of the same cross-sections) results in composition of Δ_γ with two biholomorphisms from left and right, so using suitable charts, one can always bring any particular correspondence map Δ_γ to be the identity map. The situation changes completely if there is more than one homotopically distinct path connecting the same endpoints, or, what is the same, when one considers *closed paths.*

Let $a \in L$ be a point on the leaf L of a holomorphic foliation, $\tau\colon (\mathbb{C}^m, 0) \to (U, a)$ a cross-section at a, and $\gamma \in \pi_1(L, a)$ a closed loop considered modulo the homotopic equivalence.

Definition 2.14. The *holonomy self-map* $\Delta_\gamma\colon (\tau, a) \to (\tau, a)$ is the holomorphic holonomy correspondence map associated with a closed path $\gamma \in \pi_1(L, a)$.

The *holonomy group* of the foliation $\mathcal{F}$ along the leaf $L \in \mathcal{F}$ is the image of the fundamental group $\pi_1(L, a)$ in the group of germs of holomorphic self-maps $\operatorname{Diff}(\tau, a)$.

The holonomy group is defined as a subgroup in $\operatorname{Diff}(\mathbb{C}^m, 0)$ modulo a simultaneous conjugacy of all holonomy maps, independently of the choice of the cross-section τ or even the base point $a \in L$. It is an obvious invariant of a foliation which carries almost all information on behavior of leaves of the foliation, adjacent to L.

Proposition 2.15. *Assume that two holomorphic foliations $\mathcal{F}$, $\mathcal{F}'$ are topologically or holomorphically conjugate by a homeomorphism (resp., biholomorphism) H. If $L \in \mathcal{L}$ is a leaf mapped by H into a leaf $L' \in \mathcal{F}'$, then for any choice of the points $a \in L$, $a' \in L'$ and the corresponding cross-sections τ, τ' the corresponding holonomy groups $G \subset \operatorname{Diff}(\tau, a)$ and $G' \subset \operatorname{Diff}(\tau', a')$ are topologically (resp., holomorphically) conjugate: there exists the germ of a map $h\colon (\tau, a) \mapsto (\tau', a')$, holomorphic or continuous respectively, such* *that h conjugates each element of g with some element $g' \in G'$ and respects the group law.*

Proof. Let τ be a cross-section to L at a and $\tau' = H(\tau)$ (with the induced chart), then the assertion is a tautology: the restriction $h = H|_\tau$ realizes the required conjugacy between G and G'. Any other choice of a' and τ' results in replacing G' by a holomorphically conjugate group. $\square$

However, the inverse statement is in general wrong (see Exercise 2.10).

Definition 2.16. Let $\mathcal{F}$ be a holomorphic foliation on a complex manifold U, and $B \subseteq U$ an arbitrary subset. The *saturation* of B by leaves of $\mathcal{F}$ is the union of all leaves that intersect B:

$$\operatorname{Sat}(B, \mathcal{F}) = \bigcup_{L \in \mathcal{F},\ L \cap B \neq \varnothing} L.$$

In general, saturations of even simple sets can be rather complicated. Yet some basic things can be guaranteed. The following can be considered as a generalization of the theorem on continuous dependance of solutions of differential equations on initial conditions.

Lemma 2.17. *Saturation of an open set is open. In particular, saturation of a neighborhood of any point on each leaf contains an open neighborhood of the leaf.* $\square$

From this observation we can derive a corollary that will be used later. Let $G \subset \operatorname{Diff}(\tau, a)$ be a finitely generated subgroup. A germ of an analytic function $u \in \mathcal{O}(\tau, a)$ is called G-invariant, if $u \circ g = u$ for all germs of self-maps $g \in G$.

Lemma 2.18. *Any germ of a holomorphic function $u \in \mathcal{O}(\tau, a)$ which is invariant by the holonomy group $G \subseteq \operatorname{Diff}(\tau, a)$, uniquely extends as a holomorphic function defined in some open neighborhood V of the leaf L and constant along all leaves of the foliation $\mathcal{F}$ in V.*

Proof. Let $a' \in L$ be any point on L, connected by a path $\gamma\colon [0, 1] \to L$ with the base point a. The holonomy map $\Delta_{a,a'}$ allows us to translate (analytically continue) the germ u, considered as a function from $\mathcal{O}(U, a)$ constant along the local plaques of $\mathcal{F}$, to the germ $u' \in \mathcal{O}(U, a')$, also constant along the local plaques. This extension depends on the choice of the path γ, yet for a different choice of this path γ' the result will differ by the continuation of the germ $u \circ g$, where g is the holonomy map associated with the loop $\gamma' \circ \gamma^{-1} \in \pi_1(L, a)$. Yet since u by assumption is G-invariant, the result will be the same and thus correctly defined for an arbitrary point $a' \in L$. $\square$

Remark 2.19. Most holonomy groups do not admit nonconstant invariant functions. Exceptions correspond to *integrable foliations*; see §11.

2D. Singular foliations. The holonomy group may be nontrivial only for a leaf of the foliation which has a nontrivial fundamental group. Such leaves, in general difficult to find for arbitrary holomorphic foliations, can be easily found for *foliations with singularities*, or *singular foliations*. Starting from this moment, we consider only one-dimensional foliations unless explicitly stated otherwise.

A holomorphic vector field $F \in \mathcal{D}(U)$ defines a nonsingular holomorphic foliation on the complement to its singular locus $\Sigma = \Sigma_F = \{x \in U : F(x) = 0\}$ by Proposition 2.6. This singular locus can be an arbitrary analytic subset of U. However, very often the foliation can be extended from U on a bigger open subset eventually containing a part of Σ.

If $U \subset U'$ are two domains and $\mathcal{F}'$ a foliation on the larger domain, then $\mathcal{F}'$ can be restricted on U: by definition, this means the foliation whose leaves are *connected components* of the intersections $L'_\alpha \cap U$ for all leaves $L'_\alpha \in \mathcal{F}'$.

Theorem 2.20. *Let U be a connected open domain in $\mathbb{C}^n$ and $0 \not\equiv F \in \mathcal{D}(U)$ a holomorphic vector field with the singular locus $\Sigma \subset U$.*

Then there exists an analytic subset $\Sigma' \subseteq \Sigma$ of complex codimension $\geqslant 2$ in U and the foliation $\mathcal{F}'$ of $U \smallsetminus \Sigma'$ whose restriction on $U \smallsetminus \Sigma$ coincides with the foliation generated by the initial vector field F.

Proof. The assertion needs the proof only when Σ is an analytic hypersurface (a complex analytic set of codimension 1).

Consider an arbitrary *smooth point* $a \in \Sigma$ of the singular locus Σ: nonsmooth points already form an analytic subset $\Sigma_1 \subset \Sigma$ of codimension $\geqslant 2$ in U. Locally near this point Σ can be described by one equation $\{f = 0\}$ with f holomorphic and $df(a) \neq 0$. Let $\nu \geqslant 1$ be the maximal power such that all components $F_1, \dots, F_n$ of the vector field F are divisible by f^ν. By construction, the vector field $f^{-\nu} F$ extends analytically on Σ near a and its singular locus is a *proper* analytic subset $\Sigma_2 \subset \Sigma$ (locally near a). Since the germ of Σ at a is smooth hence irreducible, such a subset necessarily has codimension $\geqslant 2$ respective to the ambient space.

The union $\Sigma' = \Sigma_1 \cup \Sigma_2$ has codimension $\geqslant 2$ and in $U \smallsetminus \Sigma'$ the field locally represented as $f^{-\nu} F$ is nonsingular and thus defines a holomorphic foliation $\mathcal{F}'$ extending $\mathcal{F}$ on the neighborhood of all points of Σ. □

Remark 2.21. If U is two-dimensional, the holomorphic vector field F can be replaced by the distribution defined by an appropriate holomorphic 1-form $\omega \in \Lambda^1(U)$ with the singular locus Σ which consists of isolated points only (the singular locus of a holomorphic 1-form is the common zero of its coefficients).

Theorem 2.20 means that when speaking about holomorphic foliations with singularities, generated by holomorphic vector fields, one can always assume that the singular locus has codimension $\geqslant 2$; in particular, singularities of holomorphic foliations on the plane (and more generally, on holomorphic surfaces) are *isolated points*. The inverse statement is also true, as was observed in [**Ily72b**].

Theorem 2.22 ([**Ily72b**]). *Assume that $\Sigma \subset U \subseteq \mathbb{C}^n$ is an analytic subset of codimension $\geqslant 2$ and $\mathcal{F}$ a holomorphic nonsingular 1-dimensional foliation of $U \smallsetminus \Sigma$ which does not extend on any part of Σ.*

Then near each point $a \in \Sigma$ the foliation $\mathcal{F}$ is generated by a holomorphic vector field F with the singular locus Σ.

Proof. One can always assume that the local coordinates near a are chosen so that the line field tangent to leaves of $\mathcal{F}$, is not everywhere parallel to the coordinate x_1-plane. Then this line field is spanned by the *meromorphic* vector field $G = (1, G_2, \dots, G_n)$, where $G_j \in \mathcal{M}(U \smallsetminus \Sigma)$ are *meromorphic* functions in $U \smallsetminus \Sigma$. By E. Levi's theorem, any meromorphic function can be meromorphically extended on analytic subsets of codimension 1 [**GH78**, Chapter III, §2]. Therefore we may assume that G_j are in fact meromorphic in U. Decreasing if necessary the size of U, each G_j can be represented as the ratio $G_j = P_j/Q_j$, where $P_j, Q_j \in \mathcal{O}(U)$ are holomorphic in U and the representation is irreducible.

Let $\Sigma_j = \{P_j = Q_j = 0\}$, $j = 2, \dots, n$: by irreducibility, Σ_j is of codimension $\geqslant 2$, so $\bigcup_{j \geqslant 2} \Sigma_j$ is also of codimension $\geqslant 2$. Multiplying the field G by the product of denominators $Q_2 \cdots Q_n$, we obtain a holomorphic vector field tangent to the same foliation; cancelling a nontrivial common factor for the components of this field as in Theorem 2.20, we arrive at yet another holomorphic field F, also tangent to $\mathcal{F}$, such that the singular locus $\Sigma' = \operatorname{Sing}(F)$ of this field has codimension $\geqslant 2$.

It remains to show that the singular locus Σ' *coincides* with Σ locally in U. In one direction it is obvious: if Σ' is *smaller* than Σ, this means that $\mathcal{F}$ is analytically extended as a nonsingular holomorphic foliation to some parts of Σ, contrary to the assumption that Σ is the minimal singular locus. Assume that Σ' is *larger* than Σ, i.e., there exists a nonsingular point $b \in U \smallsetminus \Sigma$ of $\mathcal{F}$, at which F vanishes. Since the foliation $\mathcal{F}$ is biholomorphically equivalent to the standard foliation near b, in the suitable chart F is parallel to the first coordinate axis, so that singular points of F are zeros of its first component. On the other hand, by construction Σ' is of codimension $\geqslant 2$ and hence cannot be the zero locus of any holomorphic function. The contradiction proves that $\Sigma' \cap U$ cannot be larger than $\Sigma \cap U$. □

Example 2.23. The vector field $\frac{\partial}{\partial x} + e^{1/x} \frac{\partial}{\partial y}$ is analytic outside the line $\Sigma = \{x = 0\}$ of codimension 1 on the plane and defines a holomorphic foliation in $\mathbb{C}^2 \setminus \Sigma$. This foliation cannot be defined by a vector field holomorphically extendable on Σ, which shows that the condition on the codimension in Theorem 2.22 cannot be relaxed.

Together Theorems 2.20 and 2.22 motivate the following concise definition. Since we will never consider in this book holomorphic foliations of dimension other than 1, this is explicitly included in the definition.

Definition 2.24. A *singular holomorphic foliation* in a domain U (or a complex analytic manifold) is a holomorphic foliation $\mathcal{F}$ with complex one-dimensional leaves in the complement $U \setminus \Sigma$ to an analytic subset Σ of codimension $\geqslant 2$, called the *singular locus* of $\mathcal{F}$.

Usually we will assume that the singular locus Σ is maximal, i.e., the foliation cannot be analytically extended on any set larger than $U \setminus \Sigma$.

The second part of Proposition 2.7 motivates the following important definition.

Definition 2.25. Two holomorphic vector fields $F \in \mathcal{D}(U)$, $F' \in \mathcal{D}(U')$ with singular loci Σ, Σ' of codimension $\geqslant 2$ are *holomorphically orbitally equivalent* if the singular foliations $\mathcal{F}, \mathcal{F}'$ they generate, are holomorphically equivalent, i.e., there exists a biholomorphism $H \colon U \to U'$ which maps Σ into Σ' and is a biholomorphism of foliations outside these loci.

Proposition 2.7 remains valid also for *singular* holomorphic foliations: if two such foliations are holomorphically equivalent, then the corresponding vector fields are orbitally equivalent, i.e., related by the identity (2.3) with the holomorphic function ρ nonvanishing in U.

Indeed, from Proposition 2.7 it follows that for the holomorphically orbitally equivalent fields there exists a holomorphic function ρ satisfying (2.3) and nonvanishing outside $\Sigma = \operatorname{Sing}(F)$. Since Σ has codimension $\geqslant 2$, ρ must be nonvanishing *everywhere* on U.

Changing only one adjective in Definition 2.25 (requiring that H be merely a homeomorphism), we obtain the definition of *topologically* orbitally equivalent vector fields. This weaker equivalence cannot be translated into a formula similar to (2.3), since homeomorphisms in general do not act on the vector fields.

2E. Complex separatrices. Foliations with isolated singularities may have *multiply-connected* leaves, i.e., leaves with a nontrivial holonomy group.

Recall that a (singular) analytic curve $S \subset U$ is a complex analytic set of complex dimension 1 at its smooth points. Intrinsic structure of *irreducible* components of analytic curves is relatively easy. This result can be found, e.g., in [**Chi89**, §6].

Theorem 2.26. *The germ of an irreducible analytic curve* $S \subset (\mathbb{C}^n, 0)$ *admits a holomorphic* injective *map*

$$\gamma\colon (\mathbb{C}^1, 0) \to (\mathbb{C}^n, 0), \qquad t \mapsto \gamma(t) \in S. \tag{2.8}$$

The map γ is called *local uniformization*, or *local parametrization* of analytic curves. It is obviously nonconstant, and without loss of generality one may assume that the derivative $\frac{d}{dt}\gamma(t)$ is nonvanishing outside the origin $t = 0$. The local parametrization is defined uniquely modulo a biholomorphism: for any other injective parametrization γ' there exists $h \in \operatorname{Diff}(\mathbb{C}^1, 0)$ such that $\gamma' = \gamma \circ h$ (cf. with Exercise 2.1).

Let $\mathcal{F}$ be a singular holomorphic foliation on an open domain U with the singular locus Σ.

Definition 2.27. A *complex separatrix* of a singular holomorphic foliation $\mathcal{F}$ at a singular point $a \in \operatorname{Sing}(\mathcal{F})$ is a local leaf $L \subset (U, a) \smallsetminus \Sigma$ whose closure $L \cup \{a\}$ is the germ of an analytic curve.

Since the leaves are by definition connected, the closure is irreducible (as a germ) at any it's point, hence (by the above uniformization arguments) the complex separatrix is topologically a punctured disk near the singularity. The fundamental group of the separatrix is nontrivial (infinite cyclic), thus the holomorphic map generating the *local holonomy group* is an invariant of the singular foliation. Note that the leaves are naturally oriented by their complex structure, so the loop generating the local fundamental group is uniquely defined modulo free homotopy.

In other words, every singular point that admits a complex separatrix, produces at least one holomorphic germ of a self-map that is an analytic invariant of the foliation. Later, in §14 we will show that *every* planar foliation (on a complex 2-dimensional manifold) has at least one separatrix through each singularity. Besides, by blow-up (desingularization) and Poincaré compactification, two related operations discussed in detail in §8 and §25**A** respectively, one can often *create* multiply-connected leaves of singularities *extending* a given singular foliation.

The rest of this section consists of a few examples important for future applications.

Example 2.28. Consider first the singular foliation spanned by a *diagonal* linear system

$$\dot{x} = Ax, \qquad A = \operatorname{diag}\{\lambda_1, \dots, \lambda_n\}, \quad \lambda_j \neq 0. \tag{2.9}$$

This foliation has an isolated singularity (of codimension n) at the origin, and all coordinate axes are complex separatrices.

Consider the first coordinate axis $S_1 = \{x_2 = \cdots = x_n = 0\}$ and the separatrix $L_1 = S_1 \smallsetminus \{0\}$. The loop $\gamma = \{|x_1| = 1\}$ parameterized counterclockwise is the canonical generator of L_1. Choose the affine hyperplane $\tau = \{x_1 = 1\} \subset \mathbb{C}^n$ as the cross-section to S_1 at the point $(1, 0, \dots, 0) \in S_1$. A solution of the system (the parameterized leaf of the foliation) passing through the point $(1, b_2, \dots, b_n) \in \tau$ is as follows:

$$\mathbb{C}^1 \ni t \mapsto x(t) = (\exp \lambda_1 t, b_2 \exp \lambda_2 t, \dots, b_n \exp \lambda_n t) \in \mathbb{C}^n.$$

The image of the straight line segment $[0, 2\pi i/\lambda_1] \subset \mathbb{C}$ on the t-plane coincides with the loop γ when $b = 0$ (i.e., on the separatrix S_1) and is uniformly close to this loop on all leaves near S_1. The endpoints $x(2\pi i/\lambda_1)$ all belong to τ and hence the holonomy map $M_1 \colon \mathbb{C}^{n-1} \to \mathbb{C}^{n-1}$ is linear diagonal,

$$b \mapsto M_1 b, \qquad M_1 = \operatorname{diag}\{2\pi i \lambda_j/\lambda_1\}_{j=2}^{n}. \tag{2.10}$$

The other holonomy maps M_k for the canonical loops on the separatrices S_k parallel to the kth axis, are obtained by obvious relabelling of the indices.

Particular cases of this result are of special importance.

Example 2.29. Consider an *integrable* planar foliation given by the Pfaffian equation $\omega = 0$ with an *exact* form $\omega = du$, $u \in \mathcal{O}(\mathbb{C}^2, 0)$. If u has a Morse critical point, then in suitable analytic coordinates (x, y) the germ u takes the form $u = xy$, hence the foliation is given by the *linear* form $x\,dy + y\,dx = 0$ corresponding to the vector field $\dot{y} = y$, $\dot{x} = -x$. The holonomy operators corresponding to the two coordinate axes, are both *identical*.

Integrable foliations with more degenerate singularities will be treated in detail in §11.

Example 2.30. Let $n = 2$. Consider the vector field $F = (x+y)\frac{\partial}{\partial x} + y\frac{\partial}{\partial y}$ corresponding to a linear vector field with a nontrivial Jordan matrix. The corresponding singular foliation has only one complex separatrix, the punctured axis $S = \{y = 0\}$.

Consider the standard cross-section $\tau = \{x = 1\}$. Solutions of the differential equation with the initial condition (x_0, y_0) can be written explicitly,

$$x(t) = (x_0 + t y_0) \exp t, \quad y = y_0 \exp t.$$

Let $t(y_0)$ be another moment of complex time when the solution close to the separatrix again crosses τ after continuing along a path close to the standard loop on the separatrix; by definition, this means that we consider the initial point with $x_0 = 1$ and $x(t(y_0)) \equiv 1$, i.e., $1 + t(y_0) y_0 = 1/\exp t(y_0)$.

If the holonomy map is linear, then $y(t(y_0)) = \lambda y_0$ identically in y_0, i.e., $\exp t(y_0) = \lambda$ is a constant. Substituting this into the previous identity, we obtain $1 + t(y_0)y_0 = 1/\lambda$. This is impossible in the limit $y_0 \to 0$ unless $\lambda = 1$. On the other hand, $\lambda = 1$ is also impossible since $t(y_0) \not\equiv 0$.

Thus the holonomy map cannot be linear. The principal term of this map in a more general setting is computed in Proposition 27.14.

This example shows that a linear foliation may have nonlinear (and even nonlinearizable) holonomy.

2F. Suspension of a self-map. The construction of holonomy associates with any loop γ on a leaf $L \in \mathcal{F}$ of a holomorphic foliation $\mathcal{F}$ the holomorphic self-map Δ_γ. Very often the inverse problem appears: given an invertible holomorphic self-map f, construct a foliation for which this self-map would be the holonomy, associated with a loop on a leaf.

We will show that in absence of additional constraints on the phase space M and the leaf L, this problem is always trivially solvable. The construction is well known in the real analysis as *suspension* of a map to a flow.

Theorem 2.31. *Any biholomorphic germ $f \in \operatorname{Diff}(\mathbb{C}^n, 0)$ can be realized as the holonomy map along a loop on the leaf of a holomorphic foliation on an $(n+1)$-dimensional holomorphic manifold M^{n+1}.*

Construction of the foliation. For simplicity we discuss only the case $n = 1$: the general case requires only minimal modifications.

Consider the segment $[0, 1] \subset \mathbb{C}$ and let U be its ε-neighborhood, $\varepsilon < \frac{1}{2}$. In the Cartesian product $\widetilde{M} = U \times (\mathbb{C}, 0)$ with the coordinates (z, w) consider the trivial foliation $\mathcal{F}_0$ by "horizontal lines" $\{w = \text{const}\}$.

Any self-map from $f \in \operatorname{Diff}(\mathbb{C}^1, 0)$ can be considered as a map $f\colon (\tau_0, 0) \to (\tau_1, 0)$, $w \mapsto f(w)$, between the cross-sections $\tau_0 = \{z = 0\}$ and $\tau_1 = \{z = 1\}$. The latter can be extended as a holomorphic invertible map $\mathbf{f}\colon (z, w) \mapsto (z+1, f(w))$ between the open sets $M_0 = \{|z| < \varepsilon\} \times (\mathbb{C}, 0) \subset \widetilde{M}$ and $M_1 = \{|z - 1| < \varepsilon\} \times (\mathbb{C}, 0) \subset \widetilde{M}$. By construction, this map preserves the restriction of the foliation $\mathcal{F}_0$ on the open sets M_i.

The quotient space $M = \widetilde{M}/\mathbf{f}$ is defined as the topological space obtained from $\widetilde{M}$ by identification of all points a and $\mathbf{f}(a)$. This space inherits the structure of an (abstract) holomorphic manifold (the charts are inherited from those on M). Moreover, since $\mathbf{f}$ preserves the foliation, the quotient manifold M carries a well defined foliation $\mathcal{F}$. Two different cross-sections $\tau_0, \tau_1 \subset \widetilde{M}$ after identification become a single cross-section τ to the leaf L of the foliation $\mathcal{F}$ obtained from the zero leaf $\{w = 0\} \in \mathcal{F}_0$, and the segment $[0, 1]$ on this leaf becomes a closed loop on L. The holonomy of the foliation

$\mathcal{F}$, associated with the loop $\gamma \subset L$, by construction coincides with the map f which is transformed into the self-map. □

The construction can be modified by a number of ways, while keeping the principal idea the same. If $\widetilde{M}$ is a manifold with a foliation $\mathcal{F}_0$ on it, and $\mathbf{f}\colon M_0 \to M_1$ is a biholomorphic map between open subsets of $\widetilde{M}$, which is an automorphism of the foliation $\mathcal{F}_0$, then the quotient space $M = \widetilde{M}/\mathbf{f}$ is a new manifold with a different topology, which carries a holomorphic foliation on it.

Exercises and Problems for §2.

Exercise 2.1. Let $S \subset (\mathbb{C}^n, 0)$ be the germ of an irreducible analytic curve and γ an injective analytic parametrization. Prove that any other holomorphic map $\gamma'\colon (\mathbb{C}^1, 0) \to (\mathbb{C}^n, 0)$ with the range in S differs from γ by a holomorphic map $h\colon (\mathbb{C}^1, 0) \to (\mathbb{C}^1, 0)$ so that $\gamma' = \gamma \circ h$.

Problems 2.2–2.7 together constitute a proof of the Frobenius Theorem 2.9.

Problem 2.2. Prove that vector fields generating an integrable distribution, are *in involution*, i.e., always satisfying condition (2.4).

Prove that Pfaffian forms generating an integrable distribution, are in involution, i.e., satisfy the conditions (2.5).

Problem 2.3. Prove that two holomorphic vector fields $F, F' \in \mathcal{D}(M)$ on a holomorphic manifold M, have identically zero commutator, $[F, F'] \equiv 0$, if and only if their flows $\exp(tF), \exp(t'F') \in \operatorname{Diff}(M)$ commute for all complex values of $t, t' \in \mathbb{C}$.

Formulate and prove an analog of this result for *incomplete* vector fields (i.e., when the flows are not globally defined for all values of t, t', as in the case where $U \subseteq \mathbb{C}^2$ is a noninvariant planar domain).

Problem 2.4. Prove that any tuple of everywhere linearly independent commuting vector fields generates an integrable distribution tangent to leaves of a holomorphic foliation.

Problem 2.5. Let $F_1, \dots, F_k$ be holomorphic everywhere linearly independent vector fields in involution (i.e., satisfying condition (2.4)).

Construct another tuple of holomorphic vector fields $F_1', \dots, F_k'$ spanning the same distribution, such that the fields $[F_i', F_j'] \equiv 0$ for all $1 \leqslant i, j \leqslant k$.

Prove that vector fields in involution generate an integrable distribution.

Problem 2.6. Prove that for any differential 1-form ω and two vector fields F, G on a manifold M,

$$d\omega(F, G) = F\,\omega(G) - G\,\omega(F) - \omega([F, G]) \tag{2.11}$$

(the right hand side contains the evaluation of ω on the fields F, G and $[F, G]$ and their derivatives along G and F).

Problem 2.7. Prove that a tuple of everywhere linearly independent 1-forms satisfying (2.5), defines an integrable distribution.

Exercise 2.8. Prove that a nonvanishing Pfaffian form ω in $\mathbb{C}^3$ defines an integrable distribution, if and only if $\omega \wedge d\omega = 0$.

Problem 2.9. Prove that each holonomy operator g corresponding to any separatrix of an integrable foliation $du = 0$ with an analytic potential $u \in \mathcal{O}(x, y)$, is periodic: some iterated power of g is identity.

Exercise 2.10. Construct two foliations having leaves with holomorphically conjugated holonomy groups, which are themselves not holomorphically conjugate in neighborhoods of the leaves.

Exercise 2.11. Is it always possible to rectify *simultaneously* two nonsingular vector fields? Two *commuting* nonsingular vector fields? Give a simple sufficient condition guaranteeing such simultaneous rectification.

Exercise 2.12. Consider the foliation $\{\omega = 0\}$ on $\mathbb{C}^2 = \{(z, t)\}$ defined by a *meromorphic* Pfaffian 1-form

$$\omega = \frac{dz}{z} - \sum_{j=0}^{n} \frac{\lambda_j \, dt}{t - a_j}, \qquad \lambda_j \in \mathbb{C}, \ \sum_{0}^{n} \lambda_j = 0,$$

and its extension on $\mathbb{C} \times \mathbb{P}^1$.

Prove that the projective line $L = \{0\} \times \mathbb{P}^1$ is a separatrix of this foliation carrying singular points $(0, a_j)$, $j = 0, \dots, n$. Compute the holonomy group of the leaf $L \smallsetminus$ (singular points).

Exercise 2.13. The same question about the foliation on $\mathbb{C}^m \times \mathbb{P}^1$ defined by the vector Pfaffian form

$$dz - \Omega z = 0, \qquad \Omega = \sum_{0}^{n} \frac{A_j \, dt}{t - a_j},$$

where $A_j \in \operatorname{Mat}(m, \mathbb{C})$ are *commuting* matrix residues of the meromorphic matrix 1-form Ω.

Problem 2.14. Consider the Riccati equation

$$\frac{dz}{dt} = a(t)\, z^2 + b(t)\, z + c(t), \qquad a, b, c \in \mathcal{M}(\mathbb{P}) \cong \mathbb{C}(t), \tag{2.12}$$

with meromorphic coefficients a, b, c having poles only on the finite point set $\Sigma \subseteq \mathbb{P}$. Is it true that solutions of this equation can be continued along any path on the t-plane, avoiding the singular locus Σ?

Prove that equation (2.12) defines a singular holomorphic foliation $\mathcal{F}$ on the compactified phase space $\mathbb{P}^1 \times \mathbb{P}^1$, which is transversal to any "vertical" projective line $\{t = a\}$, $a \notin \Sigma$. Show that each leaf of $\mathcal{F}$ can be continued over any path in the t-sphere, avoiding the singular locus. Prove that the induced transformation between any two cross-sections $\{t = a\} \times \mathbb{P}^1$ and $\{t = b\} \times \mathbb{P}^1$, $a, b \notin U$, is a well-defined Möbius transformation (fractional linear map $z \mapsto \frac{\alpha z + \beta}{\gamma z + \delta}$ with $\alpha\delta - \beta\gamma \neq 0$). Does $\mathcal{F}$ always possess a separatrix?

Exercise 2.15. How many separatrices a *homogeneous* vector field of degree r on $\mathbb{C}^2$ may have? How many separatrices a *generic* homogeneous vector field has?

3. Formal flows and embedding theorem

The assumption on convergence of Taylor series for the right hand sides of differential equations and their respective solutions is a very serious restriction: if it holds, then one can use various geometric tools as described in §2. However, considerable information can be gained without the convergence assumption, on the level of *formal power (Taylor) series.* For natural reasons, the corresponding results have more algebraic flavor.

In this section we introduce the class of formal vector fields and formal morphisms (self-maps), and prove that the flow of any such formal field can be correctly defined as a formal automorphism. The correspondence "field $\mapsto$ flow" can be inverted for maps with unipotent linearization: as was shown by F. Takens in 1974, any such formal map can be embedded in a unique formal flow [**Tak01**]. In §4 we establish classification of formal vector fields by the natural action of formal changes of variables.

3A. Formal vector fields and formal self-maps. For convenience, we will always assume that all Taylor series are centered at the origin, and use the standard multi-index notation: for $\alpha = (\alpha_1, \dots, \alpha_n) \in \mathbb{Z}_+^n$ we denote $|\alpha| = \alpha_1 + \cdots + \alpha_n$ and $\alpha! = \alpha_1! \cdots \alpha_n!$.

Definition 3.1. A *formal* (Taylor) series at the origin in $\mathbb{C}^n$ is the expression

$$f = \sum_{\alpha} c_\alpha x^\alpha, \qquad \alpha \in \mathbb{Z}_+^n, \quad c_\alpha \in \mathbb{C}. \tag{3.1}$$

The minimal degree $|\alpha|$ corresponding to a nonzero coefficient c_α, is called the *order* of f.

The set of all formal series is denoted by $\mathbb{C}[[x]] = \mathbb{C}[[x_1, \dots, x_n]]$. It is a commutative *infinite-dimensional* algebra over $\mathbb{C}$ which contains as a proper subset the algebra of germs of holomorphic functions, isomorphic to the algebra $\mathbb{C}\{x_1, \dots, x_n\}$ of *converging* series.

Definition 3.2. The *canonical basis* of $\mathbb{C}[[x]]$ is the collection of all monomials x^α, $\alpha \in \mathbb{Z}_+^n$, ordered in the following way: (i) all monomials of lower degree $|\alpha|$ precede all monomials of higher degree, and (ii) all monomials of the same degree are ordered lexicographically. This order will be denoted `deglex`-order.

Since the series may diverge, evaluation of $f(x_0)$ at any point $x_0 \in \mathbb{C}^n$ other than $x_0 = 0$, makes no sense. However, without risk of confusion we will denote the free term of a series $f \in \mathbb{C}[[x]]$ by $f(0)$ and the coefficient c_α by $\frac{1}{\alpha!}\frac{\partial^\alpha f}{\partial x^\alpha}(0)$. Under these agreements the Taylor formula becomes a *definition* of the Taylor series $f = \sum_{\alpha \geqslant 0} \frac{1}{\alpha!}\frac{\partial^\alpha f}{\partial x^\alpha}(0)\, x^\alpha$. Sometimes we write

$f(x)$ as an indication of the formal variables $x = (x_1, \dots, x_n)$ in which the series f depends.

All formal partial derivatives $\partial^\alpha f/\partial x^\alpha$ of a formal series f are well defined in the class $\mathbb{C}[[x]]$ as termwise derivatives.

The subset of $\mathbb{C}[[x]]$ which consists of formal series without the free term, is (as one can easily verify) a maximal ideal of the commutative ring $\mathbb{C}[[x]]$; it will be denoted by

$$\mathfrak{m} = \{f \in \mathbb{C}[[x]] \colon f(0) = 0\} = \Big\{ \sum_{|\alpha| \geqslant 1} c_\alpha x^\alpha \Big\}.$$

The maximal ideal is *unique* (again a simple exercise). In other words, the ring $\mathbb{C}[[x]]$ is a *local ring*.

For any finite $k \in \mathbb{N}$ the space of kth order jets can be described as the quotient

$$J^k(\mathbb{C}^n, 0) = \mathbb{C}[[x_1, \dots, x_n]]/\mathfrak{m}^{k+1}.$$

As a quotient ring, the affine finite-dimensional $\mathbb{C}$-space $J^k(\mathbb{C}^n, 0)$ inherits the structure of a commutative $\mathbb{C}$-algebra.

Definition 3.3. The *truncation* of formal series to a finite order k is the canonical projection map $j^k \colon \mathbb{C}[[x]] \to J^k(\mathbb{C}^n, 0)$, $f \mapsto f \bmod \mathfrak{m}^{k+1}$.

The name comes from the natural identification of $J^k(\mathbb{C}^n, 0)$ with polynomials of degree $\leqslant k$ in the variables $x_1, \dots, x_n$. If $l > k$ is a higher order, then $\mathfrak{m}^{l+1} \subset \mathfrak{m}^{k+1}$ so that the truncation operator j^k naturally "descends" as the projection $J^l(\mathbb{C}^n, 0) \to J^k(\mathbb{C}^n, 0)$ which will also be denoted by j^k.

In other words, a formal Taylor series $f \in \mathbb{C}[[x]]$ uniquely defines the k-jet $j^k f$ of any finite order k so that $\mathbb{C}[[x_1, \dots, x_n]]$ is in a sense the limit of the jet spaces $J^k(\mathbb{C}^n, 0)$ as $k \to \infty$. We will sometimes refer to formal series as *infinite* jets and write $\mathbb{C}[[x_1, \dots, x_n]] = J^\infty(\mathbb{C}^n, 0)$.

The canonical monomial basis in $\mathbb{C}[[x]]$ projects into canonically `deglex`-ordered monomial bases in all jet spaces $J^k(\mathbb{C}^n, 0)$. Convergence in $\mathbb{C}[[x]]$ is defined via finite truncations.

Definition 3.4. A sequence $\{f_j\}_{j=1}^\infty \subset \mathbb{C}[[x]]$ is said to be convergent, if and only if all its truncations $j^k f_j$ converge in the respective finite-dimensional k-jet space $J^k(\mathbb{C}^n, 0)$ for any finite $k \geqslant 0$.

Remark 3.5 (important). All formal algebraic constructions described above can be implemented over the field $\mathbb{R}$ rather than $\mathbb{C}$ as the ground field. Moreover, for future purposes we will need the algebra $\mathfrak{A}[[x]]$ of formal power series in the indeterminates $x = (x_1, \dots, x_n)$ with the coefficients belonging to more general $\mathbb{C}$- or $\mathbb{R}$-algebras $\mathfrak{A}$. The principal examples are the algebras $\mathfrak{A} = \mathbb{C}[\lambda_1, \dots, \lambda_m]$ of polynomials in auxiliary indeterminates or the algebra $\mathcal{A} = \mathcal{O}(U)$ of holomorphic functions of additional variables $\lambda_1, \dots, \lambda_m$.

After introducing the algebra of "formal functions" we can define formal vector fields and formal maps via their algebraic (functorial) properties as in §1**G**.

With any *vector formal series* $F = (F_1, \dots, F_n)$ (n-tuple of elements from $\mathbb{C}[[x]]$) one can associate a derivation $\mathbf{F} = \sum_1^n F_j \partial/\partial x_j \in \operatorname{Der} \mathbb{C}[[x]]$ of the algebra $\mathbb{C}[[x]]$, a $\mathbb{C}$-linear application satisfying the Leibnitz rule (cf. with (1.27)),

$$\mathbf{F}\colon \mathbb{C}[[x]] \to \mathbb{C}[[x]], \qquad \mathbf{F}(gh) = g\,(\mathbf{F}h) + h\,(\mathbf{F}g).$$

Conversely, any derivation $\mathbf{F} \in \operatorname{Der} \mathbb{C}[[x]]$ is of the form $\mathbf{F} = \sum_1^n F_j \partial/\partial x_j$ with the components $F_j = \mathbf{F}x_j$. By *formal vector fields*, we mean both realizations, $F \in \mathbb{C}[[x]]^n$ or $\mathbf{F} \in \operatorname{Der} \mathbb{C}[[x]]$. The field F is said to have *singularity* (at the origin), if all these series are without free terms, $F_j(0) = 0$, $j = 1, \dots, n$.

The collection of formal vector fields will be denoted $\mathcal{D}[[\mathbb{C}^n, 0]]$. It is a $\mathbb{C}$-linear (infinite dimensional) space which possesses additional algebraic structures of the *module* over the ring $\mathbb{C}[[x]]$. The *commutator* (Lie bracket) of formal fields is defined in the usual way as $[\mathbf{F}, \mathbf{G}] = \mathbf{FG} - \mathbf{GF}$.

In a parallel way, a vector formal series $H = (h_1, \dots, h_n) \in \mathbb{C}[[x]]^n$ can be identified with an *automorphism* $\mathbf{H} \in \operatorname{Aut} \mathbb{C}[[x]]$ of the algebra $\mathbb{C}[[x]]$ if $H(0) = 0$, i.e., $h_j \in \mathfrak{m}$. Under this assumption, for any formal series $f = \sum_\alpha c_\alpha x^\alpha \in \mathbb{C}[[x]]$ one can correctly define the *substitution*

$$\mathbf{H}f(x) = f(H(x)) = \sum_{\alpha \geqslant 0} c_\alpha h^\alpha = \sum_{\alpha \geqslant 0} c_\alpha h_1^{\alpha_1}(x) \cdots h_n^{\alpha_n}(x). \tag{3.2}$$

Indeed, any k-truncation of $f(H(x))$ is completely determined by the k-truncations of f and H. We will say that $\mathbf{H}$ is *tangent to identity*, if $j^1\mathbf{H} = \mathrm{id}$.

The operator $\mathbf{H}$ defined by (3.2), is an *automorphism* of the algebra $\mathbb{C}[[x]]$, a $\mathbb{C}$-linear map respecting the multiplication,

$$\mathbf{H}\colon \mathbb{C}[[x]] \to \mathbb{C}[[x]], \qquad \mathbf{H}(fg) = \mathbf{H}f \cdot \mathbf{H}f.$$

Conversely, any homomorphism preserving convergence in $\mathbb{C}[[x]]$ is of the form $f \mapsto f \circ H$ for an appropriate vector series $H \in \mathbb{C}[[x]]^n$ with the *components* $h_j = \mathbf{H}x_j \in \mathbb{C}[[x]]$. By a *formal map* we mean either H or $\mathbf{H}$, depending on the context. If $\mathbf{H}$ is an homomorphism, then it must map the maximal ideal $\mathfrak{m} \subset \mathbb{C}[[x]]$ into itself and hence $h_j(0) = 0$, $j = 1, \dots, n$, which can be abbreviated to $H(0) = 0$.

If $\mathbf{H}$ is invertible (an isomorphism of the algebra $\mathbb{C}[[x]]$), we say it is a *formal isomorphism* of $\mathbb{C}^n$ at the origin. The collection of such isomorphisms forms a *group* denoted by $\operatorname{Diff}[[\mathbb{C}^n,0]]$ with the operation of composition. The latter can be defined either via substitution of the series, or as the composition of the operators acting on $\mathbb{C}[[x]]$.

Since the maximal ideal $\mathfrak{m}$ is preserved by any formal map $\mathbf{H} \in \operatorname{Diff}[[\mathbb{C}^n,0]]$ and any *singular* formal vector field $\mathbf{F} \in \mathcal{D}[[\mathbb{C}^n,0]]$, $F(0)=0$,

$$\mathbf{H}(\mathfrak{m}) = \mathfrak{m}, \qquad \mathbf{F}(\mathfrak{m}) \subseteq \mathfrak{m},$$

truncation of the series at the level of k-jets commutes with the action of $\mathbf{H}$ and $\mathbf{F}$, therefore defining correctly the isomorphism $j^k\mathbf{H}\colon J^k(\mathbb{C}^n,0) \to J^k(\mathbb{C}^n,0)$ and derivation $j^k\mathbf{F}\colon J^k(\mathbb{C}^n,0) \to J^k(\mathbb{C}^n,0)$ respectively, which can be identified with the k-jets of the formal map H and the formal vector field F. We wish to stress that $j^k\mathbf{F}$ is defined as an automorphism of the finite-dimensional jet space only if $F(0)=0$.

3B. Inverse function theorem. For future purposes we will need the formal inverse function theorem.

Theorem 3.6. *Let H be a formal map with the linearization matrix $A = \left(\frac{\partial H}{\partial x}\right)(0)$ which is nondegenerate. Then H is invertible in* $\operatorname{Diff}[[\mathbb{C}^n,0]]$.

If $A = E$ is the identity matrix and $H = (h_1,\dots,h_n)$, $h_i(x) = x_i + v_i(x) \bmod \mathfrak{m}^{k+1}$, where v_i are homogeneous polynomials of degree $k \geqslant 2$, then the formal inverse map $H^{-1} = (h'_1,\dots,h'_n)$ has the components $h'_i(x) = x_i - v_i(x) \bmod \mathfrak{m}^{k+1}$.

Clearly, the first assertion of the theorem follows from the second assertion applied to the formal map $A^{-1}H$.

Recall that a finite-dimensional linear operator $A\colon \mathbb{C}^n \to \mathbb{C}^n$ is *unipotent*, if $A-E$ is nilpotent, $(A-E)^n = 0$.

Lemma 3.7. *If $H \in \operatorname{Diff}[[\mathbb{C}^n,0]]$ is a formal map with the identical linearization matrix $\left(\frac{\partial H}{\partial x}\right)$, then its truncation $j^k\mathbf{H}$ considered as an automorphism of the finite-dimensional jet algebras $J^k(\mathbb{C}^n,0)$, is a unipotent map for any finite order k.*

Proof. For any monomial x^α from the canonical basis, $\mathbf{H}x^\alpha = x^\alpha+$(higher order terms)$= x^\alpha+$(linear combination of monomials of higher `deglex`-order). □

Proof of Theorem 3.6. Consider the homomorphism $\mathbf{H} \in \operatorname{Aut}\mathbb{C}[[x]]$ and denote $\mathbf{N} = \mathbf{H} - \mathbf{E}$ the formal "finite difference" operator ($\mathbf{E} = \mathrm{id}$ denotes the identical operator), $\mathbf{N}f = f \circ H - f$ (in the sense of the substitution of formal series). By Lemma 3.7, all finite truncations $j^k\mathbf{N}$ are nilpotent.

Define the operator H^{-1} as the series

$$\mathbf{H}^{-1} = \mathbf{E} - \mathbf{N} + \mathbf{N}^2 - \mathbf{N}^3 \pm \cdots . \tag{3.3}$$

This series converges (in fact, stabilizes) after truncation to any finite order because of the above nilpotency, hence by definition converges to an operator on $\mathbb{C}[[x]]$ satisfying the identities $\mathbf{H}\circ\mathbf{H}^{-1}=\mathbf{H}^{-1}\circ\mathbf{H}=\mathbf{E}$. It is an homomorphism of algebra(s), since for any $a,b\in\mathbb{C}[[x]]$ and their images $a'=\mathbf{H}a$, $b'=\mathbf{H}b$ which also can be chosen arbitrarily, we have $\mathbf{H}(ab)=a'b'$ and therefore

$$\mathbf{H}^{-1}(a'b')=\mathbf{H}^{-1}\mathbf{H}(ab)=ab=(\mathbf{H}^{-1}a')(\mathbf{H}^{-1}b').$$

Direct computation of the components of the inverse map yields

$$h_i'=\mathbf{H}^{-1}x_i=x_i-\mathbf{N}x_i+\cdots=x_i-(h_i(x)-x_i)+\cdots=x_i-v_i(x)+\cdots$$

as asserted by the theorem. □

The above formal construction is the algebraization of the recursive computation of the Taylor coefficients of the formal inverse map $H^{-1}(x)$. Note that stabilization of truncations of the series (3.3) means that computation of the terms of any finite degree k of the components h_i' of the inverse map is achieved in a finite (depending on k) number of steps.

3C. Integration and formal flow of formal vector fields. Consider an (autonomous) formal ordinary differential equation

$$\dot{x}=F(x),\qquad F=(F_1,\dots,F_n)\in\mathcal{D}[[\mathbb{C}^n,0]]\cong\mathbb{C}[[x]]^n \tag{3.4}$$

with a *formal* right hand side part F. Since evaluation of a formal series at any point other than the origin makes no sense, the "standard" definition of solutions can at best be applied to constructing a solution with the initial condition $x(0)=0$. Yet in the most interesting case where $F(0)=0$, this solution is trivial, $x(t)\equiv 0$.

The alternative, suggested by Remark 1.20, is to define a *one-parametric subgroup of formal self-maps* $\{H^t\colon t\in\mathbb{C}\}\subset\operatorname{Diff}[[\mathbb{C}^n,0]]$ satisfying the condition

$$H^t\circ H^s=H^{t+s}\qquad\forall t,s\in\mathbb{C},\qquad H^0=E. \tag{3.5}$$

Together with the group $\{H^t\}$ of self-maps we always consider the corresponding one-parameter group of automorphisms $\{\mathbf{H}^t\}\subset\operatorname{Aut}\mathbb{C}[[x]]$.

This subgroup is said to be *holomorphic*, if all finite truncations j^kH^t depend holomorphically on t. For a holomorphic subgroup the derivative

$$\mathbf{F}=\left.\frac{d\mathbf{H}^t}{dt}\right|_{t=0}=\lim_{t\to0}t^{-1}(\mathbf{H}^t-\mathbf{E})\colon\mathbb{C}[[x]]\to\mathbb{C}[[x]] \tag{3.6}$$

is a formal vector field,

$$\begin{aligned}\mathbf{F}(fg) &= \frac{d}{dt}\bigg|_{t=0} \mathbf{H}^t(fg) = \frac{d}{dt}\bigg|_{t=0} [(\mathbf{H}^t f)(\mathbf{H}^t g)] \\ &= \left[\frac{d}{dt}\bigg|_{t=0} (\mathbf{H}^t f)\right](\mathbf{H}^0 g) + (\mathbf{H}^0 f)\left[\frac{d}{dt}\bigg|_{t=0} (\mathbf{H}^t g)\right] \\ &= g\,\mathbf{F}f + f\,\mathbf{F}g.\end{aligned}$$

Definition 3.8. A holomorphic one-parametric subgroup of formal self-maps $\{H^t\} \subseteq \operatorname{Diff}[[\mathbb{C}^n, 0]]$ is a *formal flow* of the formal vector field F corresponding to the derivation $\mathbf{F} \in \operatorname{Der}\mathbb{C}[[x]]$, if the corresponding group of automorphisms $\{\mathbf{H}^t\}$ satisfies the identity

$$\mathbf{F} = \frac{d\mathbf{H}^t}{dt}\bigg|_{t=0} \in \operatorname{Der}\mathbb{C}[[x]]. \tag{3.7}$$

The formal field F is called the *generator* of the subgroup $\{H^t\}$.

The above observation means that *any analytic one-parametric subgroup* of formal maps is always a formal flow of some formal field F (3.7). The following theorem is a formal analog of Proposition 1.19 showing that, conversely, *any* formal vector field F generates an holomorphic one-parametric subgroup of formal self-maps $\{H^t\} \subset \operatorname{Diff}[[\mathbb{C}, 0]]$.

Denote by $\mathbf{F}^m$ the iterated composition $\mathbf{F} \circ \cdots \circ \mathbf{F} \colon \mathbb{C}[[x]] \to \mathbb{C}[[x]]$ (m times) and consider the exponential series

$$\mathbf{H}^t = \exp t\mathbf{F} = \mathbf{E} + t\mathbf{F} + \frac{t^2}{2!}\mathbf{F}^2 + \cdots + \frac{t^m}{m!}\mathbf{F}^m + \cdots. \tag{3.8}$$

Theorem 3.9. *Any singular formal vector field F admits a formal flow $\{H^t\}$. This flow is defined by the series* (3.8) *which converges for all values of $t \in \mathbb{C}$ and depends analytically on t.*

Proof. We have to show that this series converges and its sum is an isomorphism of the algebra $\mathbb{C}[[x]]$ for any $t \in \mathbb{C}$. Then the identity (3.7) will follow automatically by the termwise differentiation of the series (3.8).

Convergence of the series (3.8) can be seen from the following argument. Let k be any finite order. Truncating the series (3.8), i.e., substituting $j^k\mathbf{F}$ instead of $\mathbf{F}$, we obtain a matrix formal power series. This series is always convergent: for an arbitrary choice of the norm $|\cdot|$ on the finite-dimensional space $J^k(\mathbb{C}^n, 0)$ the norm of the operator $j^k\mathbf{F}$ is finite, $|j^k\mathbf{F}| = r < +\infty$, and hence the series (3.8) is majorized by the convergent scalar series $1 + |t|r + |t|^2r^2/2! + \cdots = \exp|t|r < +\infty$ for any finite $t \in \mathbb{C}$; cf. with Definition 1.7. Denote its sum by $\exp j^k\mathbf{F} \colon J^k(\mathbb{C}^n, 0) \to J^k(\mathbb{C}^n, 0)$.

Truncations $\exp j^k\mathbf{F}$ for different orders k agree in common terms: if $l > k$, then $j^k(\exp t\, j^l\mathbf{F}) = \exp t\, j^k\mathbf{F}$. This allows us to define the sum of the series $\exp t\mathbf{F}$ as a linear operator $\mathbf{H}^t\colon \mathbb{C}[[x]] \to \mathbb{C}[[x]]$ via its finite truncations of all orders.

The group property $\mathbf{H}^{t+s} = \mathbf{H}^t \circ \mathbf{H}^s$ equivalent to the group property (3.5), follows from the formal identity $\exp(t+s) = \exp t \cdot \exp s$, since $t\mathbf{F}$ and $s\mathbf{F}$ obviously commute. It remains to show that $\mathbf{H}^t$ is an algebra *homomorphism*, i.e., $\mathbf{H}^t(fg) = \mathbf{H}^t f\, \mathbf{H}^t g$ for any two series $f, g \in \mathbb{C}[[x]]$.

By the iterated Leibnitz rule, for any $f, g \in \mathbb{C}[[x]]$,

$$\mathbf{F}^k(fg) = \sum_{p+q=k} \tfrac{(p+q)!}{p!q!}\, \mathbf{F}^p f \cdot \mathbf{F}^q g.$$

Substituting this identity into the exponential series, we have

$$\begin{aligned}\mathbf{H}^t(fg) &= \sum_k \tfrac{t^k}{k!}\, \mathbf{F}^k(fg) = \sum_k \sum_{p+q=k} \tfrac{t^{p+q}}{p!q!}\, \mathbf{F}^p f \cdot \mathbf{F}^q g \\ &= \Bigl(\sum_p \tfrac{t^p}{p!}\, \mathbf{F}^p f\Bigr) \cdot \Bigl(\sum_q \tfrac{t^q}{q!}\, \mathbf{F}^q g\Bigr) = \mathbf{H}^t f \cdot \mathbf{H}^t g. \qquad \square\end{aligned}$$

Motivated by the series (3.8), we will often use the exponential notation: if F is a formal or analytic vector field with a singular point at the origin, we will denote by $\exp tF$ the time t flow (formal or analytic) of this field.

3D. Embedding in the flow and matrix logarithms.

Definition 3.10. A holomorphic germ $H \in \operatorname{Diff}(\mathbb{C}^n, 0)$ or a formal self-map $H \in \operatorname{Diff}[[\mathbb{C}^n, 0]]$ is said to be *embeddable*, if there exists a holomorphic germ of a vector field F (resp., a formal vector field $F \in \mathcal{D}[[\mathbb{C}^n, 0]]$) such that H is a time one (resp., formal time one) flow map of F, i.e., $H = \exp F$.

For a linear system $\dot x = Ax$ with constant coefficients, the flow consists of *linear* maps $x \mapsto (\exp tA)x$; see (1.12). Therefore for a *linear* map $x \mapsto Mx$, $M \in \operatorname{GL}(n, \mathbb{C})$, it is natural to consider the embedding problem in the class of linear vector fields $F(x) = Ax$. Then the problem reduces to finding a *matrix logarithm*, a matrix solution of the equation

$$\exp A = M, \qquad A, M \in \operatorname{Mat}(n, \mathbb{C}). \tag{3.9}$$

Clearly, the necessary condition for solvability of this equation is *nondegeneracy* of M. It also turns out to be sufficient for matrices over the field $\mathbb{C}$.

Lemma 3.11. *For any nondegenerate matrix $M \in \operatorname{Mat}(n, \mathbb{C})$, $\det M \neq 0$, there exists the matrix logarithm $A = \ln M$, a complex matrix satisfying the equation* (3.9)

Proof. We give two constructions of matrix logarithms for nondegenerate matrices.

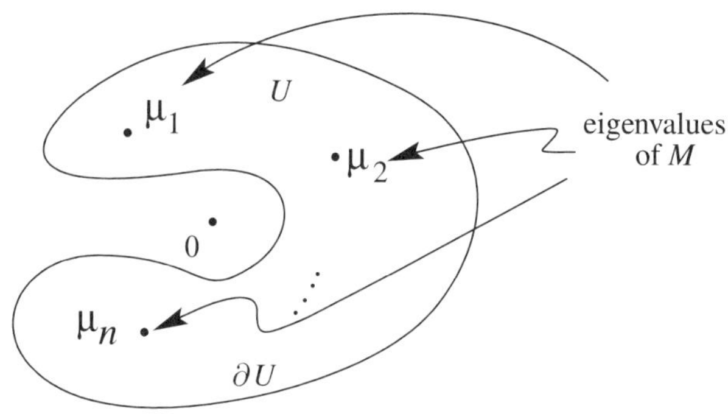

Figure I.3. Construction of the integral representation of the matrix logarithm for a nondegenerate matrix with the given spectrum

1. First, for a scalar matrix $M = \lambda E$, $0 \neq \lambda \in \mathbb{C}$, the logarithm can be defined as $\ln M = (\ln \lambda)\, E$, for any choice of $\ln \lambda$. A matrix having a single nonzero eigenvalue of high multiplicity has the form $M = \lambda(E + N)$, where N is a nilpotent (upper-triangular) matrix. Its logarithm can be defined using the formal series for the scalar logarithm as follows:

$$\ln M = \ln(\lambda E) + \ln(E + N) = (\ln \lambda)\, E + N - \tfrac{1}{2} N^2 + \tfrac{1}{3} N^3 - \cdots \qquad (3.10)$$

(the sum is finite). This formula gives a well-defined answer by virtue of the formal identity $\exp(x - \frac{x^2}{2} + \frac{x^3}{3} \pm \dots) = 1 + x$, since the matrices E and N commute.

An arbitrary matrix M can be reduced to a block diagonal form with each block having a single eigenvalue. The block diagonal matrix formed by logarithms of individual blocks solves the problem of computing the matrix logarithm in the general case.

2. The second proof uses the integral representation for analytic matrix functions. For any function $f(x)$ complex analytic in a domain $U \subset \mathbb{C}$ bounded by a simple curve ∂U and any matrix M with all eigenvalues in U, the value $f(M)$ can be defined by the contour integral

$$f(M) = \frac{1}{2\pi i} \oint_{\partial U} f(\lambda)(\lambda E - M)^{-1}\, d\lambda \qquad (3.11)$$

[**Gan59**, Ch. V, §4]. In application to $f(x) = \ln x$ we have to choose a simple loop ∂U containing all eigenvalues of M inside U but the origin $\lambda = 0$ outside (cf. with Fig. I.3). Then in the domain U one can unambiguously select a branch of complex logarithm $\ln \lambda$ which can be substituted into the integral representation.

To prove that the integral representation gives the same answer as before, it is sufficient to verify it only for the diagonal matrices, when the inverse can be computed explicitly. The advantage of this formula is the possibility of bounding the norm $|\ln M|$ defined by the above integral, in terms of $|M|$ and $|M^{-1}|$. □

Remark 3.12. The matrix logarithm is *by no means unique.* In the first construction one has the freedom to choose branches of logarithms of eigenvalues arbitrarily and independently for different Jordan blocks. In the second construction besides choosing the branch of the logarithm, there exists a freedom to choose the domain U (i.e., the loop ∂U encircling all the eigenvalues of M but not the origin).

Remark 3.13. There is a natural obstruction for extracting the matrix logarithm in the class of *real* matrices. If $\exp A = M$ for some real matrix A, then M can be connected with the identity E by a path of nondegenerate matrices $\exp tA$, in particular, M should be orientation-preserving. If M is nondegenerate but orientation-reverting, it has no real matrix logarithm.

However, there are more subtle obstructions. Consider the real matrix $M = \begin{pmatrix} -1 & 1 \\ & -1 \end{pmatrix}$ with determinant 1. If $M = \exp A$, then by (1.16) $\exp \operatorname{tr} A = 1$ so that for a real matrix necessarily $\operatorname{tr} A = 0$. The two eigenvalues cannot be simultaneously zero, since then the exponent will have the eigenvalues both equal to 1. Therefore the eigenvalues must be different, in which case the matrix A and hence its exponent M must be diagonalizable. The contradiction shows impossibility of solving the equation $\exp A = M$ in the class of real matrices.

3E. Logarithms and derivations. Inspired by the construction of the matrix exponential, one can attempt to prove that for any formal map $H \in \operatorname{Diff}[[\mathbb{C}^n, 0]]$ there exists a formal vector field F whose formal time one flow coincides with H, as follows.

Consider an arbitrary finite order k and the k-jet $\mathbf{H}_k = j^k\mathbf{H}$ considered as an isomorphism of the finite-dimensional $\mathbb{C}$-algebra $\mathfrak{F}^k = J^k(\mathbb{C}^n, 0)$. By Lemma 3.11, there exists a linear map $\mathbf{F}_k \colon \mathfrak{F}^k \to \mathfrak{F}^k$ such that $\exp \mathbf{F}_k = \mathbf{H}_k$.

Assume that for some reasons

(i) jets of the logarithms $\mathbf{F}_k$ of different orders agree after truncation, i.e., $j^k\mathbf{F}_l = \mathbf{F}_k$ for $l > k$, and
(ii) each $\mathbf{F}_k$ is a *derivation* of the commutative algebra $\mathfrak{F}^k$, thus a k-jet of a vector field.

Then together these jets would define a derivation $\mathbf{F}$ of the algebra $\mathfrak{F} = \mathbb{C}[[x]]$.

The first objective can be achieved if $\mathbf{F}_k$ are truncations of some polynomial or infinite series. There is only one such candidate, the *logarithmic series* $\ln \mathbf{H} \colon \mathbb{C}[[x]] \to \mathbb{C}[[x]]$, obtained from the formal series for

$\ln(1+x) = x - \frac{1}{2}x^2 + \frac{1}{3}x^3 \mp \cdots$ by substitution,

$$\ln \mathbf{H} = (\mathbf{H}-\mathbf{E}) - \frac{1}{2}(\mathbf{H}-\mathbf{E})^2 + \frac{1}{3}(\mathbf{H}-\mathbf{E})^3 \mp \cdots \tag{3.12}$$

(cf. with (3.10)). Until the end of this section we use the notation $\ln \mathbf{H}$ only in the sense of the series (3.12).

The series for $\ln \mathbf{H}$ does not converge everywhere even in the finite-dimensional case: the domain of convergence contains the ball $|\mathbf{H}-\mathbf{E}| < 1$ and all unipotent finite-dimensional matrices, but most certainly *not* the matrix $-\mathbf{E}$. Besides that difficulty, it is absolutely not clear why the formal logarithm of an isomorphism of the commutative algebra $\mathbb{C}[[x]]$, even if it converges, must be a derivation: no simple arguments similar to the one used in the proof of Theorem 3.9, exist (sometimes this circumstance is overlooked).

Let $\mathfrak{F}$ be a commutative $\mathbb{C}$-algebra of finite dimension n over $\mathbb{C}$ and $\mathbf{H}$ an automorphism of $\mathfrak{F}$.

Theorem 3.14. *The series* (3.12) *converges for all unipotent automorphisms* $\mathbf{H}$ *of a finite dimensional algebra* $\mathfrak{F}$ *and its sum* $\mathbf{F} = \ln \mathbf{H}$ *in this case is a derivation of this algebra.*

Proof using the Lie group tools. Consider the matrix Lie group $\mathfrak{T} \subset \mathrm{GL}(n,\mathbb{C})$ of upper-triangular matrices with units on the principal diagonal and the corresponding Lie algebra $\mathfrak{t} \subset \mathrm{Mat}(n,\mathbb{C})$ of *strictly* upper-triangular matrices.

The exponential series (3.8) and the matrix logarithm (3.12) restricted on $\mathfrak{t}$ and $\mathfrak{T}$ respectively, are *polynomial* maps defined everywhere. They are mutually inverse: for any $\mathbf{F} \in \mathfrak{t}$ and $\mathbf{H} \in \mathfrak{T}$ we have $\ln \exp \mathbf{F} = \mathbf{F}$ and $\exp \ln \mathbf{H} = \mathbf{H}$. This follows from the identities $\ln e^z = z$, $e^{\ln w} = w$ expanded in the series. In particular, exp is surjective.

For any Lie subalgebra $\mathfrak{g} \subseteq \mathfrak{t}$ and the corresponding Lie subgroup $\mathfrak{G} \subseteq \mathfrak{T}$ the exponential map $\exp\colon \mathfrak{g} \to \mathfrak{G}$ is the restriction of (3.8) on $\mathfrak{g}$.

By [**Var84**, Theorem 3.6.2], the exponential map remains surjective also on $\mathfrak{G}$, i.e., $\exp \mathfrak{g} = \mathfrak{G}$. We claim that in this case the logarithm maps $\mathfrak{G}$ *into* $\mathfrak{g}$. Indeed, if $\mathbf{H} \in \mathfrak{G}$ and $\mathbf{H} = \exp \mathbf{F}$ for some $\mathbf{F} \in \mathfrak{g}$, then $\ln \mathbf{H} = \ln \exp \mathbf{F} = \mathbf{F} \in \mathfrak{g}$.

The assertion of the theorem arises if we take $\mathfrak{G} = \mathfrak{T} \cap \mathrm{Aut}(\mathfrak{F})$ to be the Lie subgroup of *triangular automorphisms* of $\mathfrak{F} \cong \mathbb{C}^n$ and $\mathfrak{g} = \mathfrak{t} \cap \mathrm{Der}(\mathfrak{F})$ of *triangular derivations* of the commutative algebra $\mathfrak{F}$. □

Remark 3.15. Surjectivity of the exponential map for a subgroup of the triangular group $\mathfrak{T}$ is a delicate fact that follows from the nilpotency of the Lie algebra $\mathfrak{t}$. Indeed, by the Campbell–Hausdorff formula, $\exp \mathbf{F} \cdot \exp \mathbf{G} =$

$\exp \mathbf{K}$, where $\mathbf{K} = \mathbf{K}(\mathbf{F}, \mathbf{G})$ is a series which in the nilpotent case is a polynomial map $\mathfrak{t} \times \mathfrak{t} \to \mathfrak{t}$ defined everywhere. Thus the image $\exp \mathfrak{g}$ is a *Lie subgroup* in $\mathfrak{G} \subseteq \mathfrak{T}$ for *any* subalgebra $\mathfrak{g}$, containing a small neighborhood of the unit $\mathbf{E}$. It is well known that any such neighborhood generates (by the group operation) the whole connected component of the unit, so that $\exp \mathfrak{g}$ coincides with this component. If $\mathfrak{G}$ is simply connected, then $\exp \mathfrak{g} = \mathfrak{G}$ as asserted.

Without nilpotency the answer may be different: as follows from Remark 3.13, for two Lie algebras $\mathfrak{gl}(n, \mathbb{R}) \subset \mathfrak{gl}(n, \mathbb{C})$ and the respective Lie groups $\mathrm{GL}(n, \mathbb{R}) \subset \mathrm{GL}(n, \mathbb{C})$, the exponent is surjective on the ambient (bigger) group but *not* on the subgroup.

Remark 3.16. Using similar arguments, one can prove that for an arbitrary automorphism $\mathbf{H} \in \mathrm{Aut}(\mathfrak{F})$ *sufficiently close to the unit* $\mathbf{E}$, the logarithm $\ln \mathbf{H}$ given by the series (3.12) is a derivation, $\ln \mathbf{H} \in \mathrm{Der}(\mathfrak{F})$. This follows from the fact that ln and exp are mutually inverse on sufficiently small neighborhoods of $\mathbf{E}$ and 0 respectively. However, the size of this neighborhood depends on $\mathfrak{F}$.

3F. Embedding in the formal flow. Based on Theorem 3.14, one can prove the following general result obtained by F. Takens in 1974; see [**Tak01**].

Theorem 3.17. *Let $H \in \mathrm{Diff}[[\mathbb{C}^n, 0]]$ be a formal map whose linearization matrix $A = \frac{\partial H}{\partial x}(0)$ is unipotent, $(A - E)^n = 0$.*

Then there exists a formal vector field $F \in \mathcal{D}[[\mathbb{C}^n, 0]]$ whose linearization is a nilpotent matrix N, such that H is the formal time 1 map of F.

Proof. As usual, we identify the formal map with an automorphism $\mathbf{H}$ of the algebra $\mathfrak{F} = \mathbb{C}[[x_1, \dots, x_n]]$ so that its finite k-jets $j^k\mathbf{H}$ become automorphisms of the finite dimensional algebras $\mathfrak{F}^k = J^k(\mathbb{C}^n, 0)$. Without loss of generality we may assume that the matrix A is upper-triangular so that the isomorphism $\mathbf{H}$ and all its truncations $j^k\mathbf{H}$ in the canonical `deglex`-ordered basis becomes upper-triangular with units on the diagonal: the jets $j^k\mathbf{H}$ are finite-dimensional upper-triangular (unipotent) automorphisms of the algebras $\mathfrak{F}^k$.

Consider the infinite series (3.12) together with its finite-dimensional truncations obtained by applying the operation j^k to all terms. Each such truncation is a logarithmic series for $\ln j^k\mathbf{H}$ which converges (actually, stabilizes after finitely many steps) and its sum is a derivation $j^k\mathbf{F}$ of $\mathfrak{F}^k$ by Theorem 3.14. Clearly, different truncations agree on the lower order terms, thus $\ln \mathbf{H}$ converges in the sense of Definition 3.4 to a derivation $\mathbf{F}$ of $\mathfrak{F}$. This derivation corresponds to the formal vector field F as required. □

Exercises and Problems for §3.

Problem 3.1. Let $F \in \mathcal{D}[[\mathbb{C}^n,0]]$ be a formal vector field corresponding to the derivation $\mathbf{F} \in \operatorname{Der}\mathbb{C}[[x]]$, and $\{H^t\} \subset \operatorname{Diff}[[\mathbb{C}^n,0]]$ its formal flow corresponding to the one-parametric group of automorphisms $\{\mathbf{H}^t\} \subset \operatorname{Aut}\mathbb{C}[[x]]$ related by the identity (3.7).

Prove that in this case $\frac{d}{dt}H^t = F \circ H^t$ for any t on the level of vector formal series.

Exercise 3.2. Consider the derivation $\mathbf{F} = \frac{\partial}{\partial x}$ on the algebra $\mathbb{C}[x]$ of univariate polynomials. Prove that the exponential series $\exp t\mathbf{F}$ is well defined for all values of $t \in \mathbb{C}$ as an automorphism of $\mathbb{C}[x]$, but is not defined if the algebra $\mathbb{C}[x]$ is replaced by the algebras $\mathbb{C}[[x]]$ or $\mathcal{O}(\mathbb{D})$, where $\mathbb{D} = \{|x| < 1\}$ is the unit disk.

Problem 3.3. Prove that the integral representation (3.11) coincides with the standard definition of a matrix function $f(M)$ in the case where f is a (scalar) polynomial.

Exercise 3.4. Find *all* complex logarithms of the matrix $M = \begin{pmatrix} -1 & 1 \\ & -1 \end{pmatrix}$ (i.e., solutions of the equation $\exp A = M$).

4. Formal normal forms

In the same way as holomorphic maps act on holomorphic vector fields by conjugacy (1.26), formal maps act on formal vector fields. In this section we investigate the *formal normal forms*, to which a formal vector field can be brought by a suitable formal isomorphism.

Definition 4.1. Two formal vector fields F, F' are *formally equivalent*, if there exists an invertible formal self-map H such that the identity (1.26) holds on the level of formal series.

The fields are formally equivalent if and only if the corresponding derivations $\mathbf{F}, \mathbf{F}'$ of the algebra $\mathbb{C}[[x]]$ are conjugated by a suitable isomorphism $\mathbf{H} \in \operatorname{Diff}[[\mathbb{C}^n,0]]$ of the formal algebra: $\mathbf{H} \circ \mathbf{F}' = \mathbf{F} \circ \mathbf{H}$.

Obviously, two holomorphically equivalent (holomorphic) germs of vector fields are formally equivalent. The converse is in general not true, as the formal self-maps may be divergent.

4A. Formal classification theorem. Formal classification of formal vector fields strongly depends on its principal part, in particular, on properties of the linearization matrix $A = \left(\frac{\partial F}{\partial x}\right)(0)$ when the latter is nonzero (cases with $A = 0$ are hopelessly complicated if the dimension is greater than one).

We start with the most important example and introduce the definition of a resonance as a certain arithmetic (i.e., involving integer coefficients) relation between complex numbers.

Definition 4.2. An ordered tuple of complex numbers $\lambda = (\lambda_1, \dots, \lambda_n) \in \mathbb{C}^n$ is called *resonant*, or, more precisely, *additive resonance* if there exist nonnegative integers $\alpha = (\alpha_1, \dots, \alpha_n) \in \mathbb{Z}_+^n$ such that $|\alpha| \geqslant 2$ and the *resonance* identity occurs,

$$\lambda_j = \langle \alpha, \lambda \rangle, \qquad |\alpha| \geqslant 2. \tag{4.1}$$

Here $\langle \alpha, \lambda \rangle = \alpha_1\lambda_1 + \cdots + \alpha_n\lambda_n$. The natural number $|\alpha|$ is the *order* of the resonance.

A square matrix is resonant, if the collection of its eigenvalues (with repetitions if they are multiple) is resonant. A formal vector field $F = (F_1, \dots, F_n)$ at the origin is resonant if its linearization matrix $A = \left(\frac{\partial F}{\partial x}\right)(0)$ is resonant.

Though resonant tuples $(\lambda_1, \dots, \lambda_n)$ can be dense in some parts of $\mathbb{C}^n$ (see §5**A**), their measure is zero.

Theorem 4.3 (Poincaré linearization theorem). *A nonresonant formal vector field $F(x) = Ax + \cdots$ is formally equivalent to its linearization $F'(x) = Ax$.*

The proof of this theorem is given in the sections §4**B**–§4**C**. In fact, it is the simplest particular case of a more general statement valid for resonant formal vector fields that appears in §4**D**.

4B. Induction step: homological equation. The proof of Theorem 4.3 goes by induction. Assume that the formal vector field F is already partially normalized, and contains no terms of order less than some $m \geqslant 2$:

$$F(x) = Ax + V_m(x) + V_{m+1}(x) + \cdots,$$

where $V_m, V_{m+1}, \dots$ are arbitrary homogeneous vector fields of degrees $m, m+1$, *etc.*

We show that in the assumptions of the Poincaré theorem, the term V_m can be removed from the expansion of F, i.e., that F is formally equivalent to the formal field $F'(x) = Ax + V'_{m+1} + \cdots$. Moreover, the corresponding conjugacy can be in fact chosen as a polynomial of the form $H(x) = x + P_m(x)$, where P_m is a homogeneous vector polynomial of degree m. The Jacobian matrix of this self-map is $E + \left(\frac{\partial P_m}{\partial x}\right)$.

The conjugacy H with these properties must satisfy the equation (1.26) on the formal level. Keeping only terms of order $\leqslant m$ from this equation and using dots to denote the rest, we obtain

$$\left(E + \frac{\partial P_m}{\partial x}\right)(Ax + V_m + \cdots) = A(x + P_m(x)) + V'_m(x + P_m(x)) + \cdots.$$

The homogeneous terms of order 1 on both sides coincide. The next non-trivial terms appear in the order m. Collecting them, we see that in order to meet the condition $V'_m = 0$, the vector of homogeneous terms P_m must satisfy the commutator identity

$$[\mathbf{A}, P_m] = -V_m, \qquad \mathbf{A}(x) = Ax, \tag{4.2}$$

where $\mathbf{A} = Ax$ is the linear vector field, the principal part of F, and the homogeneous vector polynomials P_m and V_m are considered as vector fields on $\mathbb{C}^n$. The left hand side of (4.2) is the commutator, $[\mathbf{A}, P](x) = \left(\frac{\partial P}{\partial x}\right) Ax - AP(x)$.

Conversely, if the condition (4.2) is satisfied by P_m, the polynomial map $H(x) = x + P_m(x)$ conjugates $F = \mathbf{A} + V_m + \cdots$ with the (formal) vector field $F'(x) = \mathbf{A} + \cdots$ having no terms of degree m.

Definition 4.4. The identity (4.2), considered as an equation on the unknown homogeneous vector field P_m, is called the *homological equation.*

4C. Solvability of the homological equation. Solvability of the homological equation depends on invertibility of the operator ad_A of commutation with the linear vector field $\mathbf{A}$.

Let $\mathcal{D}_m$ be the linear space of all homogeneous vector fields of degree m (we will be interested only in the case $m \geqslant 2$). This linear space has the *standard monomial basis* consisting of the fields

$$F_{k\alpha} = x^\alpha \frac{\partial}{\partial x_k}, \qquad k = 1, \dots, n, \ |\alpha| = m. \tag{4.3}$$

We shall order elements of this basis lexicographically so that x_i precedes x_j if $i < j$, but $\frac{\partial}{\partial x_j}$ precedes $\frac{\partial}{\partial x_i}$. To that end, we assign to each formal variable $x_1, \dots, x_n$ pairwise different positive weights $w_1 > \cdots > w_n$ that are *rationally independent.* This assignment extends on all monomials and monomial vector fields if the symbol $\frac{\partial}{\partial x_j}$ is assigned the weight $-w_j$. Now the monomial vector fields can be arranged in the decreasing order of their weights: the independence condition guarantees that the only different vector monomials having the same weight can be $x^\alpha \cdot x_j \frac{\partial}{\partial x_j}$ and $x^\alpha \cdot x_k \frac{\partial}{\partial x_k}$ with the same α and $j \neq k$. The order between these monomials is not essential for future exposition.

The operator

$$\operatorname{ad}_A \colon P \mapsto [\mathbf{A}, P], \qquad (\operatorname{ad}_A P)(x) = \left(\frac{\partial P}{\partial x}\right) \cdot Ax - AP(x), \tag{4.4}$$

preserves the space $\mathcal{D}_m$ for any $m \in \mathbb{N}$.

Lemma 4.5. *If A is nonresonant, then the operator* ad_A *is invertible. More precisely, if the coordinates $x_1, \dots, x_n$ are chosen such that A has the upper-triangular Jordan form, then* ad_A *is lower-triangular in the respective standard monomial basis ordered in the decreasing weight order.*

Proof. The assertion of the lemma is completely transparent when A is a diagonal matrix $\Lambda = \operatorname{diag}\{\lambda_1, \dots, \lambda_n\}$. In this case each $F_{k\alpha} \in \mathcal{D}_m$ is an eigenvector for $\operatorname{ad}_\Lambda$ with the eigenvalue $\langle \lambda, \alpha \rangle - \lambda_k$. Indeed, by the Euler identity,

$$F_{k\alpha} = x^\alpha \begin{pmatrix} 0 \\ \vdots \\ 1 \\ \vdots \\ 0 \end{pmatrix}, \qquad \left(\frac{\partial F_{k\alpha}}{\partial x} \right) = x^\alpha \begin{pmatrix} & 0 & \\ & \vdots & \\ \frac{\alpha_1}{x_1} & \cdots & \frac{\alpha_n}{x_n} \\ & \vdots & \\ & 0 & \end{pmatrix},$$

so that in the diagonal case $\Lambda F_{k\alpha} = \lambda_k F_{k\alpha}$, and $\left(\frac{\partial F_{k\alpha}}{\partial x} \right) \Lambda x = \langle \lambda, \alpha \rangle F_{k\alpha}$. Being diagonal with nonzero eigenvalues, $\operatorname{ad}_\Lambda$ is invertible.

To prove the lemma in the general case where A is in the upper-triangular Jordan form, we consider the weight introduced above.

The operator $\operatorname{ad}_\Lambda$ with the diagonal matrix Λ preserves the weights, since all vector monomials are eigenvectors for it.

On the other hand, the monomial vector field $\mathbf{J}_j = x_j \frac{\partial}{\partial x_{j+1}}$ with the upper-diagonal constant matrix J_j acts by increasing weight. Indeed,

$$\left[x^\alpha \frac{\partial}{\partial x_k}, x_j \frac{\partial}{\partial x_{j+1}} \right] = x^\alpha \left[\frac{\partial}{\partial x_k}, x_j \frac{\partial}{\partial x_{j+1}} \right] + \alpha_{j+1} x^\alpha \frac{x_j}{x_{j+1}} \cdot \frac{\partial}{\partial x_k}.$$

The second term, if present, has higher weight than $F_{k\alpha} = x^\alpha \frac{\partial}{\partial x_k}$, since $w_j > w_{j+1}$. The first term is nonzero only if $j - k$, and in this case reduces to $x^\alpha \frac{\partial}{\partial x_{k+1}}$, which also has higher weight than $F_{k\alpha}$.

It remains to notice that an arbitrary matrix A in the upper-triangular Jordan normal form is the sum of the diagonal part Λ and a linear combination of matrices $J_1, \dots, J_{n-1}$. The operator ad_A linearly depends on A, so ad_A is equal to $\operatorname{ad}_\Lambda$ modulo a linear combination of the weight-increasing operators ad_{J_j}. Therefore, if the monomial fields $F_{k\alpha}$ are ordered in the decreasing order of their weights, as in the standard basis, then the operator ad_A is lower-triangular with the diagonal part $\operatorname{ad}_\Lambda$. □

Proof of Theorem 4.3. Now we can prove the Poincaré linearization theorem. By Lemma 4.5, the operator ad_A is invertible and therefore the homological equation (4.2) is always solvable no matter what the term

$V = V_m$ is. Repeating this process inductively, we can construct an infinite sequence of polynomial maps $H_1, H_2, \dots, H_m, \dots$ and the formal fields $F_1 = F$, $F_2, \dots, F_m, \dots$ such that $F_m = Ax +$ (terms of order m and more), and the transformation $H_m = \mathrm{id} +$(terms of order m and more) conjugates F_m with F_{m+1}. Thus the composition $H^{(m)} = H_m \circ \cdots \circ H_1$ conjugates the initial field F_1 with the field F_{m+1} without nonlinear terms up to order m.

It remains to notice that by construction of H_{m+1} the composition $H^{(m+1)} = H_{m+1} \circ H^{(m)}$ has the same terms of order $\leqslant m$ as $H^{(m)}$ itself. Thus the limit

$$H = H^{(\infty)} = \lim_{m\to\infty} H^{(m)}$$

(the infinite composition) exists in the class of formal morphisms. By construction, H_*F cannot contain any nonlinear terms and hence is linear, as required. □

Remark 4.6. The formal map linearizing a nonresonant formal vector field and tangent to the identity, is unique. Indeed, otherwise there would exist a *nontrivial* formal map $\mathrm{id} + h$ which conjugates the linear field with itself,

$$\left(\frac{\partial h}{\partial x}\right) Ax = Ah(x), \qquad \text{i.e.,} \qquad \mathrm{ad}_A h = 0.$$

But in the nonresonant case the commutator ad_A is injective, hence $h = 0$.

Thus the only formal maps conjugating a linear field with itself, are linear maps $x \mapsto Bx$, with the matrix B commuting with A, $[A, B] = 0$.

4D. Resonant normal forms: Poincaré–Dulac paradigm. The inductive construction linearizing nonresonant vector fields, can be used to *simplify* the resonant ones.

In this *resonant* case the operator $\mathrm{ad}_A = [\mathbf{A}, \cdot]$ of commutation with the linear part may be no longer surjective and in general the condition $V_m' = 0$, meaning absence of terms of order m after the transformation, cannot be achieved.

In the presence of resonances one can choose in each linear space $\mathcal{D}_m$ a complementary (transversal) subspace $\mathcal{N}_m$ to the image of the operator ad_A, so that

$$\mathcal{D}_m = \mathcal{N}_m + \mathrm{ad}_A(\mathcal{D}_m) \tag{4.5}$$

(the sum should not necessarily be direct).

Theorem 4.7 (Poincaré–Dulac paradigm). *If the subspaces $\mathcal{N}_m \subset \mathcal{D}_m$ are transversal to the image of the commutator ad_A as in (4.5), then any formal vector field $F(x) = Ax + \cdots$ with the linearization matrix A is formally conjugated to some formal vector field whose all nonlinear terms of degree m belong to the subspace $\mathcal{N}_m$.*

Proof. If the transformation $H_m(x) = x + P_m$ conjugates the field $F(x) = Ax + \cdots + V_m(x) + \cdots$ with another field $F'(x) = Ax + \cdots + V'_m(x) + \cdots$ *with the same* $(m-1)$*-jet* on the level of terms of order m, then instead of the homological equation (4.2) in the case $V'_m \neq 0$, the correction term P_m must satisfy the equation

$$\operatorname{ad}_A P_m = V'_m - V_m. \tag{4.6}$$

If $\mathcal{N}_m$ satisfies (4.5), then (4.6) can always be solved with respect to P_m for any V_m provided that V'_m is suitably chosen from the subspace $\mathcal{N}_m$.

The transform H_m does not affect the lower order terms and hence the process can be iterated for larger values of m exactly as in the nonresonant case. As a result, one can prove that any formal vector field F is formally equivalent to a formal field containing only terms belonging to the "complementary" parts $\mathcal{N}_m$ for all $m = 2, 3, \ldots$.

The rest of the proof of Theorem 4.7 is the same as that of the Poincaré–Dulac theorem. □

The choice of the transversal subspaces $\mathcal{N}_m$ depends on ad_A, hence on the matrix A itself.

Example 4.8. Assume that the matrix $A = \Lambda = \operatorname{diag}\{\lambda_1, \ldots, \lambda_n\}$ is diagonal. In this case the operator $\operatorname{ad}_\Lambda$ was already shown to be diagonal in the vector monomial basis, eventually with some zeros among the eigenvalues. For diagonal operators on finite-dimensional space the kernel and the image are complementary subspaces, so one may choose $\mathcal{N}_m = \ker \operatorname{ad}_L \subset \mathcal{D}_m$. The kernel of the diagonal operator $\operatorname{ad}_\Lambda$ can be immediately described.

Definition 4.9. A *resonant vector monomial* corresponding to the resonance $\lambda_k - \langle \lambda, \alpha \rangle = 0$, is the monomial vector field $F_{k\alpha} = x^\alpha \frac{\partial}{\partial x_k}$; see (4.3).

The kernel $\ker \operatorname{ad}_\Lambda$ consists of linear combinations of resonant monomials. From the discussion above it follows immediately that a formal vector field with diagonal linear part Λx is formally equivalent to the vector field with the same linear part and only resonant monomials among the nonlinear terms.

Actually, the assumption on diagonalizability is redundant. The following statement is one of the most popular formal classification results.

Theorem 4.10 (Poincaré–Dulac theorem)**.** *A formal vector field is formally equivalent to a vector field with the linear part in the Jordan normal form and only resonant monomials in the nonlinear part.*

Proof. Assume that the coordinates are already chosen so that the linearization matrix A is Jordan upper-triangular.

Choose the subspace $\mathcal{N}_m$ as the linear span of all resonant monomials, $\mathcal{N}_m = \bigoplus_{\langle\lambda,\alpha\rangle-\lambda_k=0} \mathbb{C}\cdot F_{k\alpha}$.

By Lemma 4.5, the operator $L_m = \operatorname{ad}_A|_{\mathcal{D}_m}$ is upper triangular with the expressions $\langle\lambda,\alpha\rangle - \lambda_k = 0$ on the diagonal. By the choice of $\mathcal{N}_m$, whenever zero occurs on the diagonal of L, the corresponding basis vector was included in $\mathcal{N}_m$. This obviously means (4.5). The rest is the Poincaré–Dulac paradigm. □

4E. Belitskii theorem. The choice of the "resonant normal form" (i.e., of the family of subspaces $\mathcal{N}_m$) in Theorem 4.10, is excessive in the sense that the *dimension* of these spaces (the number of parameters in the normal form) is not minimal. For example, if A is a nonzero nilpotent Jordan matrix, then *all* monomials are resonant in the sense of Definition 4.9, whereas the image of ad_A is clearly nontrivial. We describe now one possible *minimal* choice, introduced by G. Belitskii [**Bel79**, Ch. II, §7].

Consider the standard Hermitian structure on the space $\mathbb{C}^n$, so that the basis vectors $e_j = \frac{\partial}{\partial x_j}$ form an orthonormal basis.

For any natural $m \geqslant 1$ denote by $\mathcal{H}_m$ the complex linear space of all homogeneous polynomials of degree m. We introduce the *standard Hermitian structure* in $\mathcal{H}_m$ in such a way that the normalized monomials $\varphi_\alpha = \frac{1}{\sqrt{\alpha!}}\,x^\alpha$ form an orthonormal basis,

$$(\varphi_\alpha, \varphi_\beta) = \delta_{\alpha\beta}, \qquad \varphi_\alpha = \frac{1}{\sqrt{\alpha!}}\,x^\alpha, \quad \alpha,\beta\in\mathbb{Z}_+^n,\ |\alpha| = |\beta| = m. \tag{4.7}$$

Here, as usual, $\alpha! = \alpha_1!\cdots\alpha_n!$ for $\alpha = (\alpha_1,\dots,\alpha_n)$, $0! = 1$ and $\delta_{\alpha\beta}$ is the standard Kronecker symbol.

Then the linear space $\mathcal{D}_m$ of homogeneous vector fields of degree m can be naturally identified with the tensor product $\mathcal{D}_m = \mathcal{H}_m \otimes_{\mathbb{C}} \mathbb{C}^n$ and inherits the standard Hermitian structure for which the monomials $\varphi_\alpha\otimes e_k = \frac{1}{\sqrt{\alpha!}}\,F_{\alpha k}$ form an orthonormal basis.

Given a matrix $A \in \operatorname{Mat}(n,\mathbb{C})$, denote by A^* the *adjoint matrix* obtained from A by transposition and complex conjugacy: $a^*_{ij} = \bar a_{ji}$. If $\mathbf{A}(x) = Ax$ is the corresponding linear vector field on $\mathbb{C}^n$ and, respectively, $\mathbf{A}^*(x) = A^*x$, then both $\mathbf{A}, \mathbf{A}^*$ act as linear differential operators, $\mathbf{A} = \sum a_{ij}x_i\frac{\partial}{\partial x_j}$ and $\mathbf{A}^* = \sum \bar a_{ji}x_i\frac{\partial}{\partial x_j}$, on $\mathcal{H}_m$. Furthermore, the commutation operators $\operatorname{ad}_A = [\mathbf{A},\cdot]$ and $\operatorname{ad}_{A^*} = [\mathbf{A}^*,\cdot]$ are linear operators on $\mathcal{D}_m$.

The following statement claims that the operators in each pair are mutually adjoint (dual to each other) with respect to the standard Hermitian structures on the respective spaces.

Lemma 4.11.

1. The derivation $\mathbf{A}^*\colon \mathcal{H}_m \to \mathcal{H}_m$ *is adjoint to the derivation* $\mathbf{A}$ *(with respect to the standard Hermitian structure) and vice versa.*

2. The commutator $\operatorname{ad}_{A^*} = [\mathbf{A}^*, \cdot]\colon \mathcal{D}_m \to \mathcal{D}_m$ *is adjoint to the commutator* $\operatorname{ad}_A = [\mathbf{A}, \cdot]$ *(with respect to the standard Hermitian structure) and vice versa.*

Proof. 1. The identity $(\mathbf{A}f, g) = (f, \mathbf{A}^*g)$ for any pair of polynomials $f, g \in \mathcal{H}_m$ "linearly" depends on the matrix A: if it holds for two matrices $A, A' \in \operatorname{Mat}(n, \mathbb{C})$, then it also holds for their combination $\lambda A + \lambda' A'$ with any two complex numbers $\lambda, \lambda' \in \mathbb{C}$.

Thus it is sufficient to verify the assertion for the monomial derivations $\mathbf{A} = x_i \frac{\partial}{\partial x_j}$ and $\mathbf{A}^* = x_j \frac{\partial}{\partial x_i}$.

If $i = j$, then $\mathbf{A} = \mathbf{A}^* = x_i \frac{\partial}{\partial x_i}$ is diagonal in the orthonormal basis $\{\varphi_\alpha\}$ with the real eigenvalues $\lambda_\alpha = \alpha_i = \alpha_j \in \mathbf{Z}_+$, and hence is self-adjoint.

Otherwise both $\mathbf{A}$ and $\mathbf{A}^*$ can be represented as permutations of the basic vectors composed with the diagonal operators. If β is the multi-index obtained from α by the operation

$$\beta_k = \begin{cases} \alpha_k, & k \neq i, j, \\ \alpha_i + 1, & k = i, \\ \alpha_j - 1, & k = j, \end{cases} \qquad \alpha_k = \begin{cases} \beta_k, & k \neq i, j, \\ \beta_i - 1, & k = i, \\ \beta_j + 1, & k = j, \end{cases}$$

then $\beta!/\alpha! = (\alpha_i + 1)/\alpha_j = \beta_i/\alpha_j$ and

$$\mathbf{A}\varphi_\alpha = \frac{\alpha_j}{\sqrt{\alpha!}}\, x^\beta = \alpha_j \frac{\sqrt{\beta!}}{\sqrt{\alpha!}}\, \varphi_\beta = \alpha_j \frac{\sqrt{\beta_i}}{\sqrt{\alpha_j}}\, \varphi_\beta = \sqrt{\alpha_j \beta_i}\, \varphi_\beta.$$

Reciprocally, $\mathbf{A}^*\varphi_\beta = \beta_i\, x^\alpha/\sqrt{\beta!} = \cdots = \sqrt{\beta_i \alpha_j}\, \varphi_\alpha$. But since the vectors φ_α form an orthonormal basis,

$$(\mathbf{A}\varphi_\alpha, \varphi_\beta) = (\varphi_\alpha, \mathbf{A}^*\varphi_\beta) = \sqrt{\beta_i \alpha_j} \in \mathbb{R}$$

and all other matrix entries in the basis $\{\varphi_\alpha\}$ are zeros. Therefore the derivations $\mathbf{A}$ and $\mathbf{A}^*$ are mutually adjoint on $\mathcal{H}_m$.

2. Using the structure of the tensor product $\mathcal{D}_m = \mathcal{H}_m \otimes \mathbb{C}^n$, one can represent the commutators as follows:

$$\operatorname{ad}_A = \mathbf{A} \otimes E - \operatorname{id} \otimes A.$$

Indeed, for any element φv, where $\varphi \in \mathcal{H}_m$ is a polynomial and $v \in \mathbb{C}^n$ a vector considered as a constant vector field on $\mathbb{C}^n$, by the Leibnitz rule

$$[\mathbf{A}, \varphi v] = (\mathbf{A}\varphi)v + \varphi[\mathbf{A}, v] = (\mathbf{A}\varphi)v - \varphi\, Av.$$

Obviously, because of the choice of the Hermitian structure on $\mathcal{H}_m \otimes \mathbb{C}^n$, the operator $\mathrm{id} \otimes A$ is adjoint to $\mathrm{id} \otimes A^*$ whereas the adjoint to $\mathbf{A} \otimes E$ is the tensor product of the adjoint to $\mathbf{A}$ by the identity. By the first statement of the lemma, the former is equal to $\mathbf{A}^*$, so that the adjoint to $[\mathbf{A}, \cdot]$ is $\mathbf{A}^* \otimes E - \mathrm{id} \otimes A^*$ which coincides with $[\mathbf{A}^*, \cdot] = \mathrm{ad}_{A^*}$. □

Theorem 4.12 (G. Belitskii [**Bel79**]; see also [**Dum93, Van89**]). *A formal vector field $F(x) = Ax + V_2(x) + \cdots$ with the linearization matrix A is formally equivalent to a vector field $F'(x) = Ax + V_2'(x) + \cdots$ whose nonlinear part commutes with the linear vector field $\mathbf{A}^*(x) = A^*x$:*

$$[F' - \mathbf{A}, \mathbf{A}^*] = 0. \tag{4.8}$$

If the vector field F is real (i.e., has only real Taylor coefficients, in particular, A is real), then both the formal normal form and the conjugating transformation can be chosen real.

Proof. The proof is based on the following well-known observation: if L is a linear endomorphism of a complex Hermitian or real Euclidean space H into itself, then the image of L and the kernel of its Hermitian (resp., Euclidean) adjoint L^* are orthogonal complements to each other:

$$(\mathrm{img}\, L)^\perp = \ker L^*.$$

It follows then that $\ker L^*$ is complementary to $\mathrm{img}\, L$ in H.

Indeed, $\xi \in (\mathrm{img}\, L)^\perp$ if and only if $(\xi, Lv) = 0$ for all $v \in H$, which means that any vector v is orthogonal to $L^*\xi$. This is possible if and only if $L^*\xi = 0$.

Applying this observation to the operator $L_m = \mathrm{ad}_A$ restricted on any space $\mathcal{D}_m$ and using Lemma 4.11, we see that the subspaces $\mathcal{N}_m = \ker \mathrm{ad}_{A^*} |_{\mathcal{D}_m}$ are orthogonal (hence complementary) to the image of L_m and therefore satisfy the assumption (4.5) of Theorem 4.7. Therefore all *nonlinear* terms $V_2, V_3, \ldots$ can be chosen to commute with $\mathbf{A}^*(x) = A^*x$, which is in turn possible if and only if their formal sum, equal to $F - \mathbf{A}$, commutes with $\mathbf{A}^*$.

In the real case one has to replace the Hermitian spaces $\mathcal{H}_m$, $\mathbb{C}^n$ and $\mathcal{D}_m = \mathcal{H}_m \otimes_{\mathbb{C}} \mathbb{C}^n$ by their real (Euclidean) counterparts ${}^{\mathbb{R}}H_m$, $\mathbb{R}^n$ and ${}^{\mathbb{R}}\mathcal{D}_m = {}^{\mathbb{R}}\mathcal{H}_m \otimes_{\mathbb{R}} \mathbb{R}^n$. Then for any real matrix A the image of the commutator ad_A and the kernel of ad_{A^*}, where A^* is a transposed matrix, are orthogonal and hence complementary. Then the homological equation $\mathrm{ad}_A P_m = V_m' - V_m$ can be solved with respect to $P_m \in {}^{\mathbb{R}}\mathcal{D}_m$ and $V_m' \in \ker \mathrm{ad}_{A^*} \cap {}^{\mathbb{R}}\mathcal{D}_m$ when $V_m \in {}^{\mathbb{R}}\mathcal{D}_m$. The Poincaré–Dulac paradigm does the rest of the proof. □

This general statement immediately implies a number of corollaries.

Example 4.13. If A is a diagonal matrix with the spectrum $\{\lambda_1, \dots, \lambda_n\}$, then A^* is also diagonal with the conjugate eigenvalues $\{\bar\lambda_1, \dots, \bar\lambda_n\}$. As was already noted, restriction of ad_{A^*} on $\mathcal{D}_m$ is diagonal with the eigenvalues $\langle\bar\lambda, \alpha\rangle - \bar\lambda_k = \overline{\langle\lambda, \alpha\rangle - \lambda_k}$. Its kernel consists of the same resonant monomials as defined previously, so in this case Theorem 4.12 yields the usual Poincaré–Dulac form.

Sometimes, diagonalization of the linear part is nonconvenient (especially for real vector fields). In such a case Theorem 4.12 may yield a simple real normal form.

Example 4.14. If $I = \left(\begin{smallmatrix} 0 & 1 \\ -1 & 0 \end{smallmatrix}\right) = -I^*$ is the matrix of rotation on the *real* plane $\mathbb{R}^2$ with the coordinates (x, y), then $\ker \operatorname{ad}_{I^*} = \ker \operatorname{ad}_I$ and the *entire* formal normal form, including the linear part, commutes with the rotation vector field $\mathbf{I} = x\frac{\partial}{\partial y} - y\frac{\partial}{\partial x}$. Any such rotationally symmetric real vector field must necessarily be of the form

$$f(x^2+y^2)\left(x\tfrac{\partial}{\partial x} + y\tfrac{\partial}{\partial y}\right) + g(x^2+y^2)\left(x\tfrac{\partial}{\partial y} - y\tfrac{\partial}{\partial x}\right), \tag{4.9}$$

where $f(r), g(r) \in \mathbb{R}[[r]]$ are two real formal series in one variable. Indeed, A commutes with itself and the radial (Euler) vector field $\mathbf{E} = x\frac{\partial}{\partial x} + y\frac{\partial}{\partial y}$ which form a basis at all nonsingular points; a linear combination $f\mathbf{E} + g\mathbf{I}$ with f, g scalar coefficients, commutes with $\mathbf{I}$ if and only if $\mathbf{I}f = \mathbf{I}g = 0$, that is, if f and g are constants on all circles $x^2 + y^2 = r^2$.

The linear part is of the prescribed form if $f(0) = 0$, $g(0) = 1$. Since g is formally invertible, the normal form (4.9) is formally orbitally equivalent to the formal vector field

$$\begin{aligned} F' &= \mathbf{I} + f(x^2+y^2)\mathbf{E}, \qquad & f &\in \mathbb{R}[[u]], \quad f(0) = 0, \\ \mathbf{I} &= x\tfrac{\partial}{\partial y} - y\tfrac{\partial}{\partial x}, \qquad & \mathbf{E} &= x\frac{\partial}{\partial x} + y\frac{\partial}{\partial y}, \end{aligned} \tag{4.10}$$

with a formal series $f(u)$ in the resonant monomial $u = x^2 + y^2$.

Note that the "standard" demonstration of this result via preliminary diagonalization of A requires that all subsequent Poincaré–Dulac transformations be preserving the complex conjugacy, which is an additional independent condition.

The same observation explains why the normal form is so often explicitly integrable.

Corollary 4.15. *Assume that the matrix $A \neq 0$ is* normal, *i.e., it commutes with the adjoint matrix A^*. Then the vector field can be formally transformed to a field which commutes with the (nontrivial) linear vector field $\mathbf{A}^*$.* □

Indeed, in this case from (4.8) and $[\mathbf{A}, \mathbf{A}^*] = 0$ it follows that $[F, \mathbf{A}^*] = 0$. This observation allows us to decrease the dimension of the system; cf. with §4**J**.

Remark 4.16. We wish to stress that *there is no distinguished Hermitian structure on* $\mathbb{C}^n$. One can choose this structure arbitrarily and only then the standard Hermitian structure appears on $\mathcal{H}_m$ and $\mathcal{D}_m$. Thus the assumption of this corollary is not restrictive, in particular, it always holds whenever A is diagonalizable.

4F. Parametric case. The Poincaré–Dulac method of normalization of any finite jet or the entire Taylor series, involves only the *polynomial (ring)* operations (additions, subtractions and multiplications) with the Taylor coefficients of the original field, *except for inversion of the operator* ad_A. This allows us to construct formal normal forms depending on parameters.

Definition 4.17. A formal series $f \in \mathbb{C}[[x]]$ is said to depend *polynomially* on finitely many parameters $\lambda = (\lambda_1, \dots, \lambda_m) \in \mathbb{C}^n$, if each coefficient depends polynomially on λ,

$$f = \sum_\alpha c_\alpha x^\alpha, \qquad c_\alpha \in \mathbb{C}[\lambda].$$

No assumption on the degrees $\deg c_\alpha$ is made.

The formal series $f = \sum c_\alpha x^\alpha \in \mathbb{C}[[x]]$ depends (strongly) *analytically* on the parameters *in a domain* $\lambda \in U$, if each coefficient c_α of this series depends on the parameters analytically in the *common domain* $U \subseteq \mathbb{C}^m$, $c_\alpha \in \mathcal{O}(U)$.

The formal series $f = \sum c_\alpha x^\alpha$ *weakly analytically* depends on the parameters $\lambda \in (\mathbb{C}^m, 0)$, if each coefficient c_α is a germ of analytic function, $c_\alpha \in \mathcal{O}(\mathbb{C}^m, 0)$. In contrast to the previously defined analytic dependence, intersection of all domains where representatives of the germs c_α are defined, can reduce to the single point $\lambda = 0$.

We will use the common name *semiformal series* to denote elements from the algebras $\mathfrak{A}[[x]]$ in the above three cases when $\mathfrak{A} = \mathbb{C}[\lambda]$, $\mathfrak{A} = \mathcal{O}(U)$ and $\mathfrak{A} = \mathcal{O}(\mathbb{C}^n, 0)$ respectively.

Theorem 4.18 (Formal normal form with parameters).

1. *If the vector field (holomorphic or formal)* $F = F(\cdot, \lambda) = \mathbf{A}(\lambda) + F_2(\lambda) + \cdots$ *depends weakly analytically on parameters* $\lambda \in (\mathbb{C}^m, 0)$, *then by a formal transformation one can bring the field to the formal normal form* F' *satisfying the condition*

$$[F' - \mathbf{A}, \mathbf{A}^*(0)] = 0, \tag{4.11}$$

where $\mathbf{A}(0)$ *is the linear vector field corresponding to* $\lambda = 0$, *and* $\mathbf{A}^*(0)$ *its adjoint linear field. Both the formal normal form* F' *and the transformation* H *reducing* F *to* F' *can be chosen weakly analytically depending on the parameters* $\lambda \in (\mathbb{C}^m, 0)$ *in the sense of Definition 4.17. If* F *was real, then also* F' *and* H *can be chosen real.*

2. *If the linear part* $\mathbf{A}(\lambda) \equiv \mathbf{A}(0) \equiv \mathbf{A}$ *is constant (does not depend on* λ*) and the field itself depends polynomially or strongly analytically on the parameters* $\lambda \in U$, *then both the normal form* (4.11) *and the corresponding normalizing transformation can be chosen polynomially (resp., strongly analytically) depending on the parameters in the same domain.*

Proof. We start with a very general observation, basically, a geometrical reformulation of the Implicit Function theorem.

If $L\colon X \to Y$ is a linear map between vector spaces, which is *transversal* to a subspace $Z \subseteq Y$, then for any analytic or polynomial map $y\colon \lambda \mapsto y(\lambda)$, $\lambda \in U$ or $\lambda \in \mathbb{C}^n$, one can find two maps $x\colon \lambda \mapsto x(\lambda) \in X$ and $z\colon \lambda \mapsto z(\lambda) \in Z$, such that $Lx(\lambda) + z(\lambda) = y(\lambda)$. If in addition L also depends on λ and is transversal to Z for $\lambda = 0$, then the solutions still can be found, but only locally for the parameter values $\lambda \in (\mathbb{C}^m, 0)$ sufficiently close to the origin. In this case analyticity of $x(\lambda), z(\lambda)$ in the larger domain U or polynomiality in general may fail.

This observation can be applied to the homological operator $L = \operatorname{ad}_A$ acting in the space $X = \mathcal{D}_m$, and the subspace $Y = \mathcal{N}_m$ of homogeneous vector fields commuting with $\mathbf{A}^*(0)$. Holomorphic (polynomial) solvability of the homological equation on each step guarantees the possibility of transforming the field to the normal form with the required properties. $\square$

Remark 4.19 (Warning). The difference between constant and nonconstant linearization matrices is rather essential in what concerns the size of the common domain of analyticity of all Taylor coefficients of the normal form and/or conjugating transformation.

Suppose that all coefficients of the analytic family $F(\lambda)$ of formal vector fields are defined and holomorphic in some *common* domain U (e.g., the field is analytic in $D \times U$, where D is a small polydisk).

If the linearization matrix of $F(\lambda)$ does not depend on the parameters, then by the second assertion of Theorem 4.18, one may remove from the expansion of F all terms that are nonresonant (i.e., the terms that do not commute with the linear field $\mathbf{A}^*$ which is independent of the parameters). All coefficients of all series (the normal form and the conjugacy) will be holomorphic in the maximal natural domain U.

All the way around, if the linearized field $\mathbf{A}(\lambda)$ depends on parameters, then by a formal transformation one can eliminate all terms that are resonant with respect to $\mathbf{A}(0)$. The coefficients of the normal form and the transformation will still be analytically dependent on λ, but their domains should be expected to shrink as the degree of the corresponding terms grow.

Indeed, assume that the linear field $\mathbf{A}(0)$ is nonresonant. Then the formal normal form guaranteed by the first assertion of Theorem 4.18 is *linear*, $F' = \mathbf{A}(\lambda)$. Yet clearly for some values of the parameter λ which are arbitrarily close to $\lambda = 0$, the spectrum of the matrix $A(\lambda)$ can become resonant, hence it will be impossible to eliminate completely all terms of the corresponding order. The apparent contradiction is easily explained: the domain of analyticity of the coefficient of a high order cannot be so large as to include values of the parameter corresponding to resonances of that order. Note that if $A(0)$ is nonresonant, then the possible order of resonances occurring for $A(\lambda)$ necessarily grows to infinity as $\lambda \to 0$.

4G. Formal classification of self-maps. Besides formal vector fields, formal isomorphisms act also on themselves by conjugacy: if

$$G(x) = Mx + V_2(x) + \cdots \in \operatorname{Diff}[[\mathbb{C}^n, 0]], \qquad \det M \neq 0, \tag{4.12}$$

is a formal self-map, then another formal self-map $H \in \operatorname{Diff}[[\mathbb{C}^n, 0]]$ transforms G to $G' = H \circ G \circ H^{-1}$. In the same way as before, one may ask if all nonlinear terms $V_2, V_3, \ldots$ can be removed from the expansion by applying a suitable formal conjugacy.

The strategy is the same as described in §4**B**. The polynomial transformation $H(x) = x + P_m(x)$ with a vector homogeneous nonlinearity P_m of degree m conjugates $G(x)$ as in (4.12) with a self-map $G'(x) = G(x) + R_m(x) + \cdots$, in which R_m is a homogeneous vector polynomial of order m, implicitly defined by the identity

$$G(x) + P_m(G(x)) = G(x + P_m(x)) + R_m(x + P_m(x)) + \cdots . \tag{4.13}$$

After collection of terms of order m this yields the equation

$$P(Mx) - MP(x) = R(x), \qquad P = P_m,\ R = R_m, \tag{4.14}$$

which we can attempt to solve with respect to P. This is the multiplicative analog of the homological equation (4.2). The operator

$$S_M \colon \mathcal{D}_m \to \mathcal{D}_m, \qquad P(x) \mapsto MP(x) - P(Mx), \tag{4.15}$$

can be studied by methods similar to the operator ad_A. If M is a diagonal matrix with the diagonal entries $\mu_1, \ldots, \mu_n$, then all monomials $F_{k\alpha}$ of the standard basis in $\mathcal{D}_m$ are eigenvectors for S_M with the eigenvalues $\mu_j - \mu^\alpha = \mu_j - \mu_1^{\alpha_1} \cdots \mu_n^{\alpha_n}$. If all these expressions are nonzero, the operators S_M will always be invertible and hence the formal self-map G will be formally linearizable. If some of the expressions $\mu_j - \mu^\alpha$ are zeros, then one can transform G to a nonlinear normal form. All these results can be obtained in exactly the same way as for the formal vector fields.

Definition 4.20. A *multiplicative* resonance between nonzero complex numbers $\mu = (\mu_1, \ldots, \mu_n) \in (\mathbb{C}^*)^n$ is an identity of the form

$$\mu_j - \mu^\alpha = 0, \qquad |\alpha| \geqslant 2,\ j = 1, \ldots, n. \tag{4.16}$$

A nondegenerate matrix $M \in \operatorname{GL}(n, \mathbb{C})$ and a formal self-map $G(x) = Mx + \cdots \in \operatorname{Diff}[[\mathbb{C}^n, 0]]$ are nonresonant if there are no multiplicative resonances between the eigenvalues of M. A *multiplicative resonant monomial* corresponding to the resonance (4.16), is the vector whose jth component is x^α and all others are zeros.

Theorem 4.21 (Poincaré–Dulac theorem for self-maps). *Any invertible formal self-map is formally equivalent to a formal self-map whose linear part is in the Jordan normal form, and the nonlinear part contains only resonant monomials with complex coefficients. In particular, a nonresonant formal self-map is formally conjugated to the linear map $G'(x) = Mx$.* □

Rather obviously, Theorem 4.21 can be further elaborated and an analog of Belitskii Theorem 4.12 established. However, we will not deal with these matters and concentrate from now on on vector fields and automorphisms in low dimension (2 for fields, 1 for self-maps) which will be the principal tool in the rest of the book.

* * *

4H. Cuspidal points. One important case where Theorem 4.12 is considerably stronger than the Poincaré–Dulac Theorem 4.10 is that of vector fields with *nilpotent* linear parts, which are "maximally nondiagonalizable". In this case *all* monomials will be resonant and Theorem 4.10 is void. We will only consider the planar case where the linear part is the vector field $J = y\frac{\partial}{\partial x} \in \operatorname{Mat}(2,\mathbb{R})$ (the linearization matrix is a nilpotent Jordan cell of size 2). From Theorem 4.12 we can immediately derive the following corollary.

Theorem 4.22. *A vector field on the plane with the linear part $J = y\frac{\partial}{\partial x}$ is formally equivalent to the vector field*

$$J + b(x)E + a(x)\frac{\partial}{\partial y}, \qquad a, b \in \mathbb{C}[[x]],\ E = x\frac{\partial}{\partial x} + y\frac{\partial}{\partial y}, \tag{4.17}$$

with the formal series $a, b \in \mathbb{C}[[x]]$ in one variable x starting with terms of order 2 and 1 respectively.

Proof. We need only to describe the kernel of the operator ad_{J^*}, where $J^* = x\frac{\partial}{\partial y}$ is the "adjoint" vector field. The kernel of the operator $\operatorname{ad}_{J^*} = [x\frac{\partial}{\partial y}, \cdot\,]$ restricted on $\mathcal{D}_m$ can be immediately computed. Indeed,

$$[x\frac{\partial}{\partial y}, u\frac{\partial}{\partial x} + v\frac{\partial}{\partial y}] = xu_y\frac{\partial}{\partial x} + (xv_y - u)\frac{\partial}{\partial y},$$

and the commutator vanishes only if both u and hence v_y depend only on x. Since both u, v must be homogeneous of degree m, we conclude that

$$\ker \operatorname{ad}_{J^*}\big|_{\mathcal{D}_m} = \beta(x^m\frac{\partial}{\partial x} + x^{m-1}y\frac{\partial}{\partial y}) + \alpha x^m\frac{\partial}{\partial y} = \beta x^m\,(x\frac{\partial}{\partial x} + y\frac{\partial}{\partial y}) + \alpha x^m\frac{\partial}{\partial y}$$

for some constants $\alpha = \alpha_m$ and $\beta = \beta_m$ which will be the coefficients of the respective series a, b. □

Yet the complementary subspaces $\mathcal{N}_m$ may be chosen in a different way, not necessary as prescribed by Theorem 4.12. This may be more convenient for some applications.

Theorem 4.23. *The planar formal vector field with the linear part $J = y\frac{\partial}{\partial x}$, is formally equivalent to the vector field*

$$J + [yb(x) + a(x)]\frac{\partial}{\partial y}, \tag{4.18}$$

where $a(x)$ and $b(x)$ are two formal series of orders 2 and 1 respectively.

Proof. We reduce this assertion directly to the general Poincaré–Dulac paradigm. The image of ad_J in $\mathcal{D}_m$ can be complemented by the 2-dimensional space $\mathcal{N}'_m$ of vector fields $(\alpha x^m + \beta x^{m-1}y)\frac{\partial}{\partial x}$, as noted in [**Arn83**, §35 D]. Indeed, the condition $[y\frac{\partial}{\partial x}, f\frac{\partial}{\partial x} + g\frac{\partial}{\partial y}] = u\frac{\partial}{\partial x} + v\frac{\partial}{\partial y}$ takes the form of the system of linear partial differential equations

$$yf_x - g = u, \qquad yg_x = v.$$

While it can be not solvable for some u, v, the system of equations

$$yf_x - g = u, \quad yg_x + \alpha x^m + \beta x^{m-1}y = v \tag{4.19}$$

can be always resolved for any pair of homogeneous polynomials $u, v \in \mathbb{C}[x, y]$ of degree m and the constants α, β. To see this, apply $y\frac{\partial}{\partial x}$ to the first equation:

$$y^2 f_{xx} = yu_x + v - \alpha x^m - \beta x^{m-1}y.$$

The equation $y^2 f_{xx} = w$ is uniquely solvable for any monomial w divisible by y^2. On the other hand, the constants α, β can be found to guarantee that the terms proportional to x^m and $x^{m-1}y$ in the right hand side of this equation vanish. This choice automatically guarantees solvability of the second equation in (4.19) as well. The constants found in this way, appear as coefficients of the respective series a, b. □

4I. Vector fields with zero linear parts. If the formal vector field F starts with kth order terms, $F(x) = V_k(x) + V_{k+1}(x) + \cdots$, $k \geqslant 2$, then application of the formal transformation $H(x) = x + P_2(x)$ conjugates F with the vector field $F'(x) = V_k + V'_{k+1} + \cdots$ with the same (nonlinear) principal part V_k, if

$$V_k(x) + V_{k+1}(x) + \left(\frac{\partial P_2}{\partial x}\right) V_k(x) + \cdots = V_k(x + P_2(x)) + V'_{k+1}(x + P_2(x)) + \cdots$$

which after collecting the homogeneous terms of order $k+1$ yields

$$[V_k, P_2] = V_{k+1} - V'_{k+1}.$$

If this equation is resolved for a suitably chosen V'_{k+1} (e.g., equal to zero if that is possible), one can pass to terms of order $k+2$ by applying a transform of the form $H(x) = x + P_3(x)$ which does not affect the terms of order V_k and V_{k+1} and so on. As a result, one has to resolve in each order the homological equation

$$\operatorname{ad}_{V_k} P_m = V_{m+k-1} - V'_{m+k-1} \tag{4.20}$$

with respect to the homogeneous vector field P_m of degree m. As before, complete elimination of all nonprincipal terms of orders $k+1$ and more, is possible if the operator ad_{V_k} is surjective, otherwise it will be necessary to introduce the "normal subspaces" $\mathcal{N}_{m+k-1} \subset \mathcal{D}_{m+k-1}$ complementary to the image $\operatorname{ad}_{V_k}(\mathcal{D}_m) \subseteq \mathcal{D}_{m+k-1}$ and choose the components V'_{m+k-1} of the formal normal form from these subspaces.

In contrast to the case $k = 1$ discussed earlier, the operator ad_{V_k} *increases the degrees*, i.e., acts between *different* spaces, the dimension of the target space in general being higher than that of the source space. Thus the number of parameters in the normal form will be infinite. A notable exception is the one-dimensional case $\dim x = 1$.

Theorem 4.24. *A nonzero formal vector field from* $\mathcal{D}[[\mathbb{C}, 0]]$ *is formally equivalent to one of the vector fields of the form*

$$\bigl(x^{k+1} + ax^{2k+1}\bigr)\frac{\partial}{\partial x}, \qquad k \in \mathbb{N},\ a \in \mathbb{C}. \tag{4.21}$$

Proof. Any nonzero formal vector field on $\mathbb{C}^1$ starts with the term $a_{k+1}x^{k+1}\frac{\partial}{\partial x}$, $a_{k+1} \neq 0$. One can make a_{k+1} equal to 1 by a linear transformation $x \mapsto cx$, if the ground field is $\mathbb{C}$.

In this case all spaces $\mathcal{D}_m$ are one-dimensional, and the commutator with the principal term $x^{k+1}\frac{\partial}{\partial x}$ can be immediately computed:

$$\bigl[x^{k+1}\tfrac{\partial}{\partial x}, x^m\tfrac{\partial}{\partial x}\bigr] = (k - m + 1)x^{k+m}\tfrac{\partial}{\partial x}. \tag{4.22}$$

This operator is surjective for all $m \neq k+1$. Thus only the term $x^{2k+1}\frac{\partial}{\partial x}$ cannot be eliminated. □

Note that over the field of reals $\mathbb{R}$ the normal form is different: if k is *even*, then by the *real* homothety one can make the principal coefficient only ± 1,

$$\bigl(\pm x^{k+1} + ax^{2k+1}\bigr)\frac{\partial}{\partial x}, \qquad k \in \mathbb{N},\ a \in \mathbb{R}.$$

For odd k the fields with different signs are equivalent (transformed into each other by the symmetry $x \mapsto -x$).

Remark 4.25. In fact, the above arguments show that *any two formal vector fields on the line having a zero of multiplicity $k+1$ at the origin and common $(2k+1)$-jet, are formally equivalent.*

It is sometimes more convenient instead of the polynomial normal form (4.21) to use the *rational* formal normal form

$$\frac{x^{k+1}}{1-ax^k}\cdot\frac{\partial}{\partial x},\qquad k\in\mathbb{N},\ a\in\mathbb{C}. \tag{4.23}$$

This (rational) field is *analytically* equivalent to the field (4.21) with the same a. On the other hand, two vector fields in the normal form (4.23) with different values of a cannot be equivalent, as will be shown in §6$\mathbf{B}_2$.

Theorem 4.26. *Any self-map $x \mapsto x + x^{k+1} + \dots$, $k \in \mathbb{N}$, tangent to identity, is formally equivalent to:*

(1) *the time one map of the polynomial vector field* (4.21),

(2) *the time one map of the rational vector field* (4.23),

(3) *the polynomial map $x \mapsto x + x^{k+1} + ax^{2k+1}$,*

with the same complex parameter $a \in \mathbb{C}$ which is the formal invariant of the classification together with the order $k+1$.

Proof. One can prove this result in exactly the same way as Theorem 4.24, namely, modifying the Poincaré–Dulac paradigm for the equation (4.13) and using the computation from Proposition 6.11 below.

Yet one can circumvent this parallel construction by reference to the formal embedding Theorem 3.17. Indeed, any formal self-maps from $\mathrm{Diff}[[\mathbb{C},0]]$ tangent to the identity with some order $k+1$ can be represented as a time one formal flow of a formal vector field from $\mathcal{D}[[\mathbb{C},0]]$. This field in turn can be brought to one of the two formal normal forms or to the formal (nonpolynomial!) field generating the polynomial normal form. □

4J. Formal normal forms of elementary singular points on the real plane. In this section we summarize the (orbital) formal normal forms for all planar (i.e., for $n = 2$) *real* vector fields with nonzero linear parts. Recall that two formal vector fields $F, F' \in \mathcal{D}[[\mathbb{R}^2,0]]$ are called *orbitally formally equivalent*, if there exist an invertible real formal series $\varphi \in \mathbb{R}[[x,y]]$, $\varphi(0,0) \neq 0$, such that F is formally equivalent to $\varphi \cdot F'$, and the corresponding formal self-map has all real coefficients, i.e., belongs to the group $\mathrm{Diff}[[\mathbb{R}^2,0]]$. We use everywhere the term *singularity* to denote jets or germs of analytic vector fields or formal vector fields at the origin, depending on the context.

Definition 4.27. A singularity of the planar vector field is *elementary*, if at least one of the eigenvalues $\lambda_{1,2}$ of its linearization matrix is nonzero.

The only nonelementary singularity that has nonzero linearization matrix with both zero eigenvalues, is called *cuspidal*, or *nilpotent* singularity.

Real elementary points can be of several types that exhibit essentially different properties.

Definition 4.28. An elementary singularity is a *resonant node*, if the ratio of its eigenvalues is a natural or inverse natural number. The singularity is a *resonant saddle*, if both eigenvalues are real and their ratio is negative rational. A singularity is *elliptic*, if $\lambda_{1,2} = \pm i\omega$, $\omega > 0$. Finally, the singularity is a *saddle-node*, if exactly one eigenvalue is zero.

Proposition 4.29 (Formal normal forms of planar singularities). *By a real orbital formal transformation from the group* $\operatorname{Diff}[[\mathbb{R}^1, 0]] \times \operatorname{Diff}[[\mathbb{R}^2, 0]]$ *any real formal vector field* $\mathcal{D}[[\mathbb{R}^2, 0]]$ *appearing in Table I.1, can be brought to the normal form from the right column of this table.*

Proof. Most of these results are particular cases of the general results proved earlier for the ground field $\mathbb{C}$, modulo the following obvious remark. If the linear part of the vector field can be brought into its Jordan normal form by a *real* linear transformation, then all results of the formal classification remain valid if the ground field is replaced by $\mathbb{R}$. The only nontrivial case where a real matrix cannot be normalized over $\mathbb{R}$ is that of the *elliptic* singular points whose linear part is linear rotation $\omega x\frac{\partial}{\partial y} - \omega y\frac{\partial}{\partial x}$, with the eigenvalues $\pm i\omega$. From the complex point of view this is a resonant saddle, yet diagonalization of this matrix requires enlarging the ground field. The alternative treatment of the elliptic case is explained in Example 4.14.

The assertion concerning saddle-nodes is a combination of the Poincaré–Dulac theorem and Theorem 4.24. While the condition $\lambda_2 = 0$ is not a resonance, it implies infinitely many resonances $\lambda_j = \lambda_j + m$ for any $m \in \mathbb{N}$. By the Poincaré–Dulac theorem, the field is formally equivalent to the field $xf(y)\frac{\partial}{\partial x} + yg(y)\frac{\partial}{\partial y}$ with $f(0) \neq 0$ and $g(0) = 0$ (otherwise the singular point cannot be elementary degenerate). Dividing by the invertible series $f(y)$ one can assume that $f \equiv 1$ and the variables (formally) separate. It remains to make the formal change of the variable y which puts the one-dimensional vector field $g(y)\frac{\partial}{\partial y}$ into the normal form (4.21).

The saddle case is analyzed similarly: the identity $\langle \lambda, m\rangle = 0$ itself is not a resonance, but its integer multiple can be added to the right hand side of each of the identities $\lambda_1 = \lambda_1$ or $\lambda_2 = \lambda_2$, thus producing infinitely many resonances. Without loss of generality we assume that $\lambda_1 = -p$, $\lambda_2 = q$, $p, q \in \mathbb{N}$. Clearly, there are no other resonances and the Poincaré–Dulac normal form looks like $-pxf(u)\frac{\partial}{\partial x} + qy\,g(u)\frac{\partial}{\partial y}$, $f(0) = g(0) = 1$, where $u = x^p y^q$ is the resonant monomial. Passing to an orbitally equivalent system, one can assume that $f \equiv 1$.

Type	**Conditions**	**Formal normal form**
Nonresonant	$[\lambda_1 : \lambda_2] \notin \mathbb{Q}$ or $\lambda_1 = \lambda_2 \neq 0$	Linear
Resonant node	$[\lambda_1 : \lambda_2] = [r : 1]$, $r \in \mathbb{N}$, $r \geqslant 2$	$\dot{x} = rx + ay^r$, $\dot{y} = y$ $a \in \mathbb{C}$ formal invariant.
Resonant saddle (orbital)	$[\lambda_1 : \lambda_2] = -[p : q]$, $p, q \in \mathbb{N}$, not formally orbitally linearizable	$\dot{x} = -px$, $\dot{y} = \quad qy(1 \pm u^r + au^{2r})$, $u = x^q y^p$, $r \in \mathbb{N}$, $a \in \mathbb{R}$ formal orbital invariants
Elliptic points (orbital)	$\lambda_{1,2} = \pm i\omega$, not formally orbitally linearizable	$\dot{x} = \quad y \pm x(u^r + au^{2r})$, $\dot{y} = -x \pm y(u^r + au^{2r})$, $u = x^2 + y^2$, $a \in \mathbb{R}$ formal orbital invariant
Saddle-node (orbital classification)	$\lambda_1 \neq 0$, $\lambda_2 = 0$, formally isolated singularity	$\dot{x} = x$, $\dot{y} = \pm y^{r+1} + ay^{2r+1}$, $r \in \mathbb{N}$, $a \in \mathbb{R}$ formal orbital invariants
Cuspidal (nilpotent) point (nonelementary)	Nonvanishing linearization matrix with two zero eigenvalues	$\dot{x} = y$, $\dot{y} = a(x) + yb(x)$, $a, b \in \mathbb{R}[[x]]$ two formal series, $\operatorname{ord} a \geqslant 2$, $\operatorname{ord} b \geqslant 1$.
One-dimensional degenerate vector field	$\lambda = 0$, formally isolated singularity	$\dot{x} = \pm x^{r+1} + ax^{2r+1}$, or $\dot{x} = \pm \dfrac{x^{r+1}}{1 - ax^r}$, $r \in \mathbb{N}$, $a \in \mathbb{C}$ formal invariants

Table I.1. Formal normal forms for real vector fields. All rows of the table, except the last one, refer to planar formal vector fields and give orbital formal normal forms.

The field in the Poincaré–Dulac normal form admits the projection $\mathbb{R}^2 \to \mathbb{R}^1$, $(x, y) \mapsto u = x^p y^q \in \mathbb{R}^1$. The projected system has the form

$$\dot{u} = uF(u), \qquad F(u) = g(u) - 1, \tag{4.24}$$

called the *quotient equation*. By a suitable formal transformation $u \mapsto u' = u(1 + h(u))$, $h(0) = 0$, the quotient vector field can be brought to the form (4.21), corresponding to $g(u) = 1 + u^{k-1} + au^{2k-1}$. It remains to observe that any formal transformation of the variable $u \mapsto u[1 + h(u)]$, $h(0) = 0$, can be "covered" by the transformation $(x, y) \mapsto (x', y'(x, y))$, where

$$x' = x, \qquad y' = y[1 + h(x^p y^q)]^{1/q} \in \mathbb{R}[[x, y]],$$

re-expanding the invertible series in square brackets into the binomial series. This transformation brings the initial field into the required formal normal form.

The same construction almost literally applies to the elliptic case: the infinite formal normal form (4.10) admits projection onto the u-axis with $u = x^2 + y^2$, and the quotient equation takes the form $\dot{u} = 2uf(u)$. We leave it as an exercise to prove that any formal line transformation $u \mapsto u[1 + h(u)]$, $h(0) = 0$, can also be "covered" by a suitable *real* plane formal transformation. $\square$

Remark 4.30. If necessary, the polynomial normal forms from Table I.1 can be replaced by rational normal forms involving the rational normal form for one-dimensional quotient vector fields.

Note also that all normal forms of *elementary* singularities from this table are integrable: the quotient equation can be explicitly integrated in quadratures (especially easily if it has the rational normal form (4.23)). After this integration the variables x and y always separate. This integrability will be repeatedly used in the rest of the book to produce explicit computations with normal forms.

The cuspidal normal form is the famous Liénard system, corresponding to one of the simplest nonlinear and nonintegrable vector fields for which questions on the number of *limit cycles* is highly nontrivial. The Liénard system is sometimes written under the form

$$\dot{x} = y - f(x), \qquad \dot{y} = -g(x),$$

or as a second order scalar differential equation.

Remark 4.31. The dynamic (full, nonorbital) formal normal form contains more parameters than indicated in Table I.1. For instance, for the saddle-node the formal normal form is

$$\begin{cases} \dot{x} = x(\lambda_1 + b_1 y + \cdots + b_k y^k), \\ \dot{y} = y^{k+1}k + ay^{2k+1}, \qquad \lambda_1, b_1, \dots, b_k, a \in \mathbb{C}. \end{cases} \tag{4.25}$$

To prove this formula, we reduce the vector field to the form $xf(y)\frac{\partial}{\partial x} + g(y)\frac{\partial}{\partial y}$ as above and then by a suitable change of the variable y only put g into the standard form $g(y) = y^{k+1} + ay^{2k+1}$. The function $f(x)$ can be further simplified by transformations of the form $(x, y) \mapsto (h(y)x, y)$, $h(0) \neq 0$, preserving the second component: one immediately verifies that such a transformation results in replacing the series $f = f(y) \in \mathbb{C}[[y]]$ by another series

$$f' = f + \frac{g}{y} \cdot \frac{dh}{dy} = f + (y^{k+1} + ay^{2k+1}) \frac{d}{dy} \ln h.$$

Since g begins with terms of order $k+1$, the difference between f and f' is necessarily k-flat (the logarithmic derivative $\frac{d}{dy} \ln h$ in the above formula is a well defined formal series from $\mathbb{C}[[y]]$ since $h(0)$ is nonvanishing). On the other hand, if the difference $f - f'$ is divisible by y^{k+1}, the quotient can be represented as the logarithmic derivative of a suitable series $h \in \mathbb{C}[[y]]$. Thus all terms of order $k+1$ and above can be eliminated from f by the formal transformation.

A similar result can be formulated for resonant saddles and elliptic singularities.

Exercises and Problems for §4.

A complex tuple $\lambda = (\lambda_1, \dots, \lambda_n) \in \mathbb{C}^n$ is called single-resonant, if all resonances between the components of this tuple follow from a single integer identity

$$\langle \alpha, \lambda \rangle = 0, \qquad \alpha \in \mathbb{Z}_+^n, \ \alpha \neq 0. \tag{4.26}$$

Problem 4.1. Describe the formal normal form of a vector field with a single-resonant spectrum of the linearization matrix. Show that this normal form is integrable in quadratures.

Problem 4.2. Describe all linear maps that preserve the formal normal form in Problem 4.1.

Problem 4.3. Describe the real formal normal forms for vector fields in $\mathbb{R}^3$ with the spectrum $0, \pm i\omega$.

Problem 4.4. The same question for fields in $\mathbb{R}^4$ with the spectrum $\pm i\omega_1$, $\pm i\omega_2$, if the ratio ω_1/ω_2 is irrational.

Problem 4.5. Describe symmetries of the formal normal forms in the Problems 4.3 and 4.4.

Exercise 4.6. Prove that if F is a resonant formal vector field, then $\exp tF$ is a multiplicative resonant formal self-map for any $t \neq 0$. Is the inverse true?

Problem 4.7. Construct a formal normal form for vector fields in $\mathbb{C}^3$ with the nilpotent Jordan linear part $J = y\frac{\partial}{\partial x} + z\frac{\partial}{\partial y}$.

Answer: $J + a(x,u)E + b(x,u)F + c(x,u)F'$, where $E = x\frac{\partial}{\partial x} + y\frac{\partial}{\partial y} + z\frac{\partial}{\partial z}$ is the Euler field in three dimensions, $F = x\frac{\partial}{\partial y} + y\frac{\partial}{\partial z}$, $F' = \frac{\partial}{\partial z}$, and $u = u(x,y,z) = 2xz - y^2$.

Exercise 4.8. Find a formal normal form for a saddle-nodal self-map with the spectrum $(1, \mu)$, $|\mu| \neq 1$, in two dimensions.

Problem 4.9. Give a complete proof of the Poincaré–Dulac theorem for self-maps (Theorem 4.21).

Problem 4.10. Prove that the formal normal form of any vector field in the Poincaré domain is integrable in quadratures.

5. Holomorphic normal forms

5A. Poincaré and Siegel domains. To linearize a given (say, nonresonant) vector field, on each step of the Poincaré–Dulac process one has to compute the inverse of the operator $\mathrm{ad}_A = [\mathbf{A}, \cdot\,]$ on the spaces of homogeneous vector fields. To that end, one has to divide the Taylor coefficients by the *denominators*, expressions of the form $\lambda_j - \langle \alpha, \lambda \rangle \in \mathbb{C}$ with $\alpha \in \mathbb{Z}_+^n$, $|\alpha| \geqslant 2$, that may a priori be *small* even in the nonresonant case where ad_A is invertible. These denominators associated with the spectrum λ of the linearization matrix A, behave differently as $|\alpha|$ grows to infinity, in the following two different cases.

Definition 5.1. The *Poincaré domain* $\mathfrak{P} \subset \mathbb{C}^n$ is the collection of all tuples $\lambda = (\lambda_1, \dots, \lambda_n)$ such that the convex hull of the point set $\{\lambda_1, \dots, \lambda_n\} \subset \mathbb{C}$ does not contain the origin inside or on the boundary.

The Siegel domain $\mathfrak{S}$ is the complement to the Poincaré domain in $\mathbb{C}^n$.

The *strict Siegel domain* is the set of tuples for which the convex hull contains the origin strictly inside.

Sometimes we say about *tuples*, *spectra* or even germs of vector fields at singular points as being of Poincaré (resp., Siegel) type.

Proposition 5.2. *If λ is of Poincaré type, then only finitely many denominators $\lambda_j - \langle \alpha, \lambda \rangle$, $\alpha \in \mathbb{Z}_+^n$, $|\alpha| \geqslant 2$, may actually vanish.*

Moreover, nonzero *denominators are bounded away from the origin: the latter is an isolated point of the set of all denominators $\{\lambda_j - \langle \alpha, \lambda \rangle \,|\, j = 1, \dots, n,\ |\alpha| \geqslant 2\}$.*

On the contrary, if λ is of Siegel type, then either there are infinitely many vanishing denominators, or the origin $0 \in \mathbb{C}$ is their accumulation point (these two possibilities are not mutually exclusive).

Proof. If the convex hull of $\{\lambda_1, \dots, \lambda_n\} \subset \mathbb{C}$ does not contain the origin, by the convex separation theorem there exists a real linear functional $\ell\colon \mathbb{C}^2 \to \mathbb{R}$ such that $\ell(\lambda_j) \leqslant -r < 0$ for all λ_j, and hence $\ell(\langle \alpha, \lambda \rangle) \leqslant -r|\alpha|$. But then for any denominator we have

$$\ell(\lambda_j - \langle \alpha, \lambda \rangle) \geqslant \ell(\lambda_j) + |\alpha| r \to +\infty \qquad \text{as } |\alpha| \to \infty.$$

Since ℓ is bounded on any small neighborhood of the origin $0 \in \mathbb{C}$, the first two assertions are proved.

To prove the last assertion, notice that in the Siegel case there are either two or three numbers, whose linear combination with positive (real) coefficients is zero, depending on whether the origin lies on the boundary or in the interior of the convex hull. We give the proof in the second case, more difficult and more generic (the proof for the first case is simpler).

If the origin lies inside a triangle formed by the eigenvalues, then modulo their re-enumeration and a (nonconformal) affine transformation of the complex plane $\mathbb{R}^2 \cong \mathbb{C}$, we may assume without loss of generality that $\lambda_1 = 1$, $\lambda_2 = +i$ and $-\lambda_3 \in \mathbb{R}^2_+ = \mathbb{R}_+ + i\mathbb{R}_+$. In this case all "fractional parts" $-\mathbb{N}\lambda_3 \bmod \mathbb{Z} + i\mathbb{Z}$ of natural multiples of $-\lambda_3$ either form a finite subset of the 2-torus $\mathbb{R}^2/\mathbb{Z}^2$ (in which case all points of this set correspond to infinitely many vanishing denominators), or are uniformly distributed along some 1-torus, or dense. In both latter cases the point $(0,0) \in \mathbb{R}^2/\mathbb{Z}^2$ is the accumulation point of the "fractional parts" which are affine images of the denominators. □

Corollary 5.3. *If the spectrum of the linearization matrix A of a formal vector field belongs to the Poincaré domain, then the resonant formal normal form for this field established in Theorem 4.10, is polynomial.* □

Remark 5.4. Resonant tuples $\lambda \in \mathbb{C}^n$ are dense in the Siegel domain $\mathfrak{S}$ and not dense in the Poincaré domain $\mathfrak{P}$. The proof of this fact can be found in [**Arn83**].

5B. Holomorphic classification in the Poincaré domain. In the Poincaré domain, the normalizing series reducing vector fields or holomorphic maps to their Poincaré–Dulac normal forms, always converge.

Theorem 5.5 (Poincaré normalization theorem). *A holomorphic vector field with the linear part of Poincaré type is holomorphically equivalent to its polynomial Poincaré–Dulac formal normal form.*

In particular, if the field is nonresonant, then it can be linearized by a holomorphic transformation.

We prove this theorem first for vector fields with a diagonal nonresonant linear part $A = \operatorname{diag}\{\lambda_1, \dots, \lambda_n\}$. The resonant case will be addressed later in §5**C**. The classical proof by Poincaré was achieved by the so-called *majorant method.* In the modern language, it takes a more convenient form of the contracting map principle in an appropriate functional space, the *majorant space.*

Definition 5.6. The *majorant operator* is the nonlinear operator acting on formal series by replacing all Taylor coefficients by their absolute values,

$$\mathbf{M}\colon \sum_{\alpha\in\mathbb{Z}_n^+} c_\alpha\, z^\alpha \mapsto \sum_{\alpha\in\mathbb{Z}_n^+} |c_\alpha|\, z^\alpha.$$

The action of the majorant operator naturally extends on all formal objects (vector formal series, formal vector fields, formal transformations, *etc.*).

Definition 5.7. The *majorant ρ-norm* is the functional on the space of formal power series $\mathbb{C}[[z_1, \dots, z_n]]$, defined as

$$\llbracket f \rrbracket_\rho = \sup_{|z|<\rho} |\mathbf{M}\, f(z)| = |\mathbf{M} f(\rho, \dots, \rho)| \leqslant +\infty. \tag{5.1}$$

For a formal vector function $F = (F_1, \dots, F_n)$ the majorant norm is

$$\llbracket F \rrbracket_\rho = \llbracket F_1 \rrbracket_\rho + \dots + \llbracket F_n \rrbracket_\rho. \tag{5.2}$$

The majorant space $\mathcal{B}_\rho$ is the subspace of formal (vector) functions from $\mathbb{C}[[x]]$ having finite majorant ρ-norm.

Proposition 5.8. *The space $\mathcal{B}_\rho$ with the majorant norm $\llbracket \cdot \rrbracket_\rho$ is complete.*

Proof. If $\rho = 1$, this is obvious: $\mathcal{B}_1$ is the space of infinite absolutely converging sequences $\{c_\alpha\}$, and hence is isomorphic to the standard Lebesgue space ℓ^1 which is complete. The general case of an arbitrary ρ follows from the fact that the correspondence $f(\rho x) \leftrightarrow f(x)$ is an isomorphism between $\mathcal{B}_\rho$ and $\mathcal{B}_1$. □

Remark 5.9. The space $\mathcal{B}_\rho$ is closely related but not coinciding with the space $\mathcal{A}_\rho = \mathcal{A}(D_\rho)$ of functions, holomorphic in the polydisk $D_\rho = \{|z| < \rho\}$, continuous on its closure and equipped with the usual sup-norm $\|f\|_\rho = \max_{|z|<\rho} |f(z)|$.

Clearly, $\mathcal{B}_\rho \subset \mathcal{A}_\rho$, since a series from $\mathcal{B}_\rho$ is absolutely convergent on the closed polydisk $\overline{D_\rho}$. Conversely, if f is holomorphic in D_ρ and continuous on the boundary, then by the Cauchy estimates, the Taylor coefficients c_α of f satisfy the inequality

$$|c_\alpha| \leqslant \|f\|_\rho \cdot \rho^{-|\alpha|}, \qquad \alpha \in \mathbb{Z}^n_+.$$

Though the series $\llbracket f \rrbracket_\rho = \sum |c_\alpha| \rho^{|\alpha|}$ may still diverge, any other norm $\llbracket f \rrbracket_{\rho'}$ with $\rho' < \rho$, will already be finite:

$$\llbracket f \rrbracket_{\rho'} \leqslant \|f\|_\rho \cdot \sum_{\alpha \in \mathbb{Z}^n_+} \delta^{|\alpha|} < C \, \|f\|_\rho, \qquad C = C(\delta, n), \quad \delta = \rho'/\rho < 1.$$

To construct a counterexample showing that indeed $\mathcal{A}_\rho \supsetneqq \mathcal{B}_\rho$, consider a convergent but not absolutely convergent Fourier series $\sum_{k \in \mathbb{Z}_+} c_k e^{ikt}$ in one real variable t and let $f(z) = \sum c_k z^k$. Such a series converges at all points of the boundary $|z| = 1$ and represents a function from $\mathcal{A}(D_1)$, but by construction its 1-norm is infinite. Details can be found in [**Edw79**, §10.6]

The important properties of the majorant spaces and norms concern operations on functions. We will use the notation $f \ll g$ for two vector series from $\mathbb{C}^n[[x]]$ with positive coefficients, if each coefficient of f is no greater than the corresponding coefficient of g. In a similar way the notation $x \ll y$ will be used to denote the componentwise set of inequalities between two vectors $x, y \in \mathbb{R}^n$. If $f \in \mathbb{R}^n[[x]]$ is a (vector) series with *nonnegative* coefficients, then it is monotonous: $f(x) \ll f(y)$ if $x \ll y$.

Lemma 5.10. *1. For any two series $f, g \in \mathbb{C}[[x]]$ and any ρ,*

$$\llbracket fg \rrbracket_\rho \leqslant \llbracket f \rrbracket_\rho \cdot \llbracket g \rrbracket_\rho, \tag{5.3}$$

provided that all norms are finite.

2. If $G \ll G'$, are two formal series from $\mathbb{R}^n[[x]]$ and F is a series with nonnegative coefficients, then $F \circ G \ll F \circ G'$.

3. If $F, G \in \mathbb{C}^n[[z_1, \dots, z_n]]$ are two formal vector series, $F(0) = G(0) = 0$, then for their composition we have

$$\llbracket F \circ G \rrbracket_\rho \leqslant \llbracket F \rrbracket_\sigma, \qquad \sigma = \llbracket G \rrbracket_\rho. \tag{5.4}$$

Proof. The first two statements are obvious: all Taylor coefficients of the product or composition are obtained from the coefficients of entering terms by operations of addition and multiplication only. In particular, $\mathbf{M}(fg) \ll \mathbf{M}\, f \cdot \mathbf{M}\, g$. Evaluating both parts at $\boldsymbol{\rho} = (\rho, \dots, \rho)$ proves the first statement.

Since all binomial coefficients are nonnegative (in fact, natural numbers), we have $\mathbf{M}(F \circ G) \ll (\mathbf{M}\, F) \circ (\mathbf{M}\, G)$. Evaluating at $\boldsymbol{\rho} = (\rho, \dots, \rho)$ yields $\mathbf{M}\, G(\boldsymbol{\rho}) = y \ll \boldsymbol{\sigma} = (\sigma, \dots, \sigma)$, where $\sigma = \llbracket G \rrbracket_\rho$. By monotonicity, $\|F \circ G\|_\rho = \big((\mathbf{M}\, F) \circ (\mathbf{M}\, G)\big)(\boldsymbol{\rho}) \ll \mathbf{M}\, F(y) \ll \mathbf{M}\, F(\boldsymbol{\sigma}) = \llbracket F \rrbracket_\sigma$. The last statement is proved. □

Lemma 5.11. *If $\Lambda \in \operatorname{Mat}(n, \mathbb{C})$ is a nonresonant diagonal matrix of Poincaré type, then the operator $\operatorname{ad}_\Lambda$ has a bounded inverse in the space of vector fields equipped with the majorant norm.*

Proof. The formal inverse operator $\operatorname{ad}_\Lambda^{-1}$ is diagonal,

$$\operatorname{ad}_\Lambda^{-1} \colon \sum_{k,\alpha} c_{k\alpha} x^\alpha \frac{\partial}{\partial x_k} \longmapsto \sum_{k,\alpha} \frac{c_{k\alpha}}{\lambda_k - \langle \alpha, \lambda \rangle}\, x^\alpha \frac{\partial}{\partial x_k}.$$

In the Poincaré domain the absolute values of all denominators are bounded from below by a positive constant $\varepsilon > 0$, therefore *any* majorant ρ-norm is increased by no more than ε^{-1}:

$$\llbracket \operatorname{ad}_\Lambda^{-1} \rrbracket_\rho \leqslant \Big(\inf_{j,\alpha} |\lambda_j - \langle \alpha, \lambda \rangle| \Big)^{-1} < +\infty.$$

This proves that $\operatorname{ad}_\Lambda$ has the bounded inverse. □

Remark 5.12. A diagonal operator of the form $\sum_\alpha c_\alpha z^\alpha \mapsto \sum \mu_\alpha c_\alpha x^\alpha$ with bounded entries, $\sup_\alpha |\mu_\alpha| < +\infty$, which is always defined and bounded in the majorant norm, may be *not defined* or defined but unbounded on the holomorphic space $\mathcal{A}(D_\rho)$; see Remark 5.9. The "real" counterexample is even simpler: the operator which multiplies odd coefficients by -1, sends the series $1 - x^2 + x^4 - \cdots$, converging and bounded on $[-1, 1]$, into an unbounded function.

Let $F = (F_1, \dots, F_n) \in \mathcal{D}(\mathbb{C}^n, 0)$ be a holomorphic vector function defined in some polydisk near the origin. The *operator of argument shift* is the operator

$$S_F \colon h(x) \mapsto F(x + h(x)), \tag{5.5}$$

acting on holomorphic vector fields $h \in \mathcal{D}(\mathbb{C}^n, 0)$ without the free term, $h(0) = 0$. We want to show that S_F is in some sense strongly contracting. The formal statement looks as follows.

Consider the one-parameter family of majorant Banach spaces $\mathcal{B}_\rho$ as in Definition 5.7 indexed by the real parameter $\rho \in (\mathbb{R}_+, 0)$. We consider $\mathcal{B}_{\rho'}$ as a subspace in $\mathcal{B}_\rho$ for all $0 < \rho < \rho'$ (the natural embedding $\mathrm{id}_{\rho',\rho} \colon \mathcal{B}_{\rho'} \to \mathcal{B}_\rho$ is continuous).

Let S be an operator defined on all of these spaces for all sufficiently small values of ρ, considered as a family of operators $S_\rho \colon \mathcal{B}_\rho \to \mathcal{B}_\rho$ which commute with the "restriction operators" $\mathrm{id}_{\rho',\rho}$ for any $\rho < \rho'$, but we will omit the subscript in the notation of $S_\rho = S$.

Definition 5.13. The operator $S \cong \{S_\rho\}$ is *strongly contracting*, if

(1) $[\![S(0)]\!]_\rho = O(\rho^2)$ and
(2) S is Lipschitz on the ball $B_\rho = \{[\![h]\!]_\rho \leqslant \rho\} \subset \mathcal{B}_\rho$ of the majorant ρ-norm (with the same ρ), with the Lipschitz constant no greater than $O(\rho)$ as $\rho \to 0$.

Note that any strongly contracting operator takes the balls B_ρ strictly into themselves, since the center of the ball is shifted by $O(\rho^2)$ and the diameter of the image $S(B_\rho)$ does not exceed $2\rho\, O(\rho) = O(\rho^2)$.

The involved definition of strong contraction intends to make the formulation of the following claim easy.

Lemma 5.14. *Assume that the germ $F \colon (\mathbb{C}^n, 0) \to (\mathbb{C}^n, 0)$ is holomorphic and its linearization is zero, $\left(\frac{\partial F}{\partial x}\right)(0) = 0$.*

Then the operator of argument shift (5.5) *is strongly contracting.*

Proof. First note that S_F takes $h = 0$ into $F(x)$; the latter function has ρ-norm $O(\rho^2)$ for all sufficiently small ρ, since F begins with quadratic terms.

Next we compute the Lipschitz constant for $S = S_F$ restricted on the ball $B_\rho \subseteq \mathcal{B}_\rho$. If $h, h' \in \mathbb{C}^n[[x_1, \dots, x_n]]$ are two vector fields, then the difference

$$g = Sh - Sh' = F \circ (\mathrm{id} + h) - F \circ (\mathrm{id} + h')$$

can be represented as the integral

$$g(x) = \int_0^1 \left(\frac{\partial F}{\partial x}\right) \bigl(x + \tau h(x) + (1-\tau)h'(x)\bigr) \cdot (h(x) - h'(x))\, d\tau.$$

By Lemma 5.10, since $\tau \in [0,1]$, we have

$$\llbracket g \rrbracket_\rho \leqslant \llbracket \tfrac{\partial F}{\partial x} \rrbracket_\sigma \cdot \llbracket h - h' \rrbracket_\rho, \qquad \sigma = \llbracket x + \tau h(x) + (1-\tau)h'(x) \rrbracket_\rho.$$

The norm σ is no greater than $\llbracket x \rrbracket_\rho + \max(\llbracket h \rrbracket_\rho, \llbracket h' \rrbracket_\rho) = (n+1)\rho$ if both h, h' are from the ρ-ball B_ρ. On the other hand, if F is a holomorphic vector function without free and linear terms, its Jacobian matrix is holomorphic without free terms and hence its σ-norm is no greater than $C\sigma$ for all sufficiently small $\sigma > 0$. Collecting everything together, we see that S_F is Lipschitz on the ρ-ball B_ρ, with the Lipschitz constant (contraction rate) not exceeding $(n+1)C\rho$, so S_F is strongly contracting. □

Proof of Theorem 5.5 (nonresonant case). Now we can prove that a holomorphic vector field with diagonal nonresonant linearization matrix Λ of Poincaré type is holomorphically linearizable in a sufficiently small neighborhood of the origin. The proof serves as a paradigm for a more technically involved proof required for the resonant case.

A holomorphic transformation $H = \operatorname{id} + h$ conjugates the linear vector field Λx (the normal form) with the initial nonlinear field $\Lambda x + F(x)$, if and only if

$$\Lambda h(x) - \left(\tfrac{\partial h}{\partial x}\right) \Lambda x = F\bigl(x + h(x)\bigr). \tag{5.6}$$

Using the operators introduced earlier, this can be rewritten as the identity

$$\operatorname{ad}_\Lambda h = S_F h, \qquad S_F\, h = F \circ (\operatorname{id} + h), \quad \operatorname{ad}_\Lambda = [\Lambda, \cdot\,]. \tag{5.7}$$

We will show in an instant that the operator $\operatorname{ad}_\Lambda^{-1} \circ S_F$ restricted on the space $\mathcal{B}_\rho$ has a fixed point h, if $\rho > 0$ is sufficiently small,

$$h = \bigl(\operatorname{ad}_\Lambda^{-1} \circ S_F\bigr)(h), \qquad h \in \mathcal{B}_\rho. \tag{5.8}$$

Applying to both parts the operator $\operatorname{ad}_\Lambda$, we conclude that h solves (5.7) and therefore $\operatorname{id} + h$ conjugates the linear field Λx with the nonlinear field $\Lambda x + F(x)$ in the polydisk $\{|x| < \rho\}$.

Consider this operator $\operatorname{ad}_\Lambda^{-1} \circ S_F$ in the space $\mathcal{B}_\rho$ with sufficiently small ρ. The operator $\operatorname{ad}_\Lambda^{-1}$ is bounded by Lemma 5.11; its norm is the reciprocal to the smallest small divisor and is independent of ρ. On the other hand, the argument shift operator S_F is strongly contracting with the contraction rate (Lipschitz constant) going to zero with ρ as $O(\rho)$. Thus the composition will be contracting on the ρ-ball B_ρ in the ρ-majorant norm with the contraction rate $O(1) \cdot O(\rho) = O(\rho) \to 0$. By the contracting map principle, there exists a unique fixed point of the operator equation (5.8) in the space $\mathcal{B}_\rho$

which is therefore a holomorphic vector function. The corresponding map $H = (\mathrm{id} + h)^{-1}$ linearizes the holomorphic vector field. $\square$

5C. Resonant case: polynomial normal form. Modification of the previous construction allows us to prove that a resonant holomorphic vector field in the Poincaré domain can be brought into a *polynomial* normal form.

Consider a holomorphic vector field $F(x) = Ax + V(x)$ with the linearization matrix A having eigenvalues in the Poincaré domain, and nonlinear part V of order $\geqslant 2$ (i.e., 1-flat) at the origin. Without loss of generality (passing, if necessary, to an orbitally equivalent field cF, $0 \neq c \in \mathbb{C}$), one may assume that the eigenvalues of A satisfy the condition

$$1 < \operatorname{Re} \lambda_j < r \qquad \forall j = 1, \dots, n \tag{5.9}$$

with some natural $r \in \mathbb{N}$.

Theorem 5.15 (A. M. Lyapunov, H. Dulac). *If the eigenvalues of the linearization matrix A of a holomorphic vector field $F(x) = Ax + V(x)$ satisfy the condition* (5.9) *with some integer $r \in \mathbb{N}$, then the holomorphic vector field $F(x)$ is locally holomorphically equivalent to any holomorphic vector field with the same r-jet.*

Proof. A holomorphic conjugacy $H = \mathrm{id} + h$ between the fields F and $F + g$ satisfies the functional equation $\left(\frac{\partial H}{\partial x}\right) F = (F+g) \circ H$ which can be expanded to

$$\left(\frac{\partial h}{\partial x}\right) Ax - Ah = (V \circ (\mathrm{id} + h) - V) + g \circ (\mathrm{id} + h) - \left(\frac{\partial h}{\partial x}\right) V. \tag{5.10}$$

Consider the three operators,

$$T_V : h \mapsto V \circ (\mathrm{id} + h) - V, \quad S_g : h \mapsto g \circ (\mathrm{id} + h), \qquad \Psi : h \mapsto \left(\frac{\partial h}{\partial x}\right) V.$$

Using these three operators, the differential equation (5.10) can be written in the form

$$\operatorname{ad}_A h = Th + Sh + \Psi h, \tag{5.11}$$

where $T = T_V$, $S = S_g$ and, as before in (5.7), ad_A is the commutator with the *linear* field $\mathbf{A}(x) = Ax$. The key difference with the previous case is two-fold: first, because of the resonances, the operator ad_A is *not invertible* anymore, and second, since the field F is nonlinear, the additional operator Ψ occurs in the right hand side. Note that this operator is a derivation of h, thus is unbounded in *any* majorant norm $[\![\cdot]\!]_\rho$.

Let $\mathcal{B}_{m,\rho} = \{f : j^m f = 0\} \cap \mathcal{B}_\rho$ be a subspace of m-flat series in the Banach space $\mathcal{B}_\rho$, equipped with the same majorant norm $[\![\cdot]\!]_\rho$. Since V is 1-flat, all three operators T, S, Ψ map the subspace $\mathcal{B}_{m,\rho}$ into itself for any $m > 1$.

Moreover, by Lemma 5.14, the argument shift operator S is strongly contracting, regardless of the choice of m. The "finite difference" operator T_V differs from the argument shift, S_V by the constant operator $V = T(0)$ which does not affect the Lipschitz constant. Since $\llbracket V \rrbracket_\rho = O(\rho^2)$, the operator T is also strongly contracting.

The operator ad_A preserves the order of all monomial terms and hence also maps $\mathcal{B}_{m,\rho}$ into itself for all m, ρ, and is *invertible* on these spaces if m is sufficiently large. Indeed, if $|\alpha| > r+1$, then by (5.9) $\operatorname{Re}(\langle \alpha, \lambda\rangle - \lambda_j) > 0$, and all denominators in the formula

$$\operatorname{ad}_A^{-1}|_{\mathcal{B}_{m,\rho}} : \sum_{|\alpha|\geqslant m} c_{k\alpha}\, x^\alpha \frac{\partial}{\partial x_j} \longmapsto \sum_{|\alpha|\geqslant m} \frac{c_{k\alpha}}{\langle \alpha, \lambda\rangle - \lambda_j}\, x^\alpha \frac{\partial}{\partial x_j} \tag{5.12}$$

are nonzero if $m \geqslant r+1$, and the restriction of ad_A^{-1} on $\mathcal{B}_{m,\rho}$ is bounded. Moreover,

$$\llbracket \operatorname{ad}_A^{-1} h \rrbracket_\rho \leqslant O(1/m)\, \llbracket h \rrbracket_\rho \tag{5.13}$$

uniformly over all $h \in \mathcal{B}_{m,\rho}$ of order $m \geqslant r+1$.

Thus the two compositions, $\operatorname{ad}_A^{-1} \circ S$ and $\operatorname{ad}_A^{-1} \circ T$, are strongly contracting. To prove the theorem, it remains to prove that the *linear* operator $\operatorname{ad}_A^{-1} \circ \Psi \colon \mathcal{B}_{m,\rho} \to \mathcal{B}_{m,\rho}$ is strongly contracting when m is larger than $r+1$.

Consider the $\llbracket\cdot\rrbracket_\rho$-normalized vectors $h_{k\beta} = \rho^{-|\beta|} x^\beta \frac{\partial}{\partial x_k}$ for all $k = 1, \dots, m$ and all $|\beta| \geqslant m$ spanning the entire space $\mathcal{B}_{m,\rho}$. We prove that

$$\llbracket \operatorname{ad}_A^{-1} \Psi h_{k\beta} \rrbracket_\rho = O(\rho) \qquad \text{as } \rho \to 0 \tag{5.14}$$

uniformly over $|\beta| \geqslant m$ and all k. Since $\operatorname{ad}_A^{-1} \circ \Psi$ is linear, this would imply that $\operatorname{ad}_A^{-1} \circ \Psi$ is strongly contracting.

The direct computation yields

$$\Psi h_{k\beta} = \sum_{i=1}^n \rho^{-|\beta|} \frac{\beta_i}{x_i}\, x^\beta\, V_i \frac{\partial}{\partial x_k}.$$

Since V is 1-flat, $\llbracket V_i \rrbracket_\rho = O(\rho^2)$; substituting this into the definition of the majorant norm, we obtain

$$\llbracket \Psi h_{k\beta} \rrbracket_\rho \leqslant \sum_i \beta_i\, \rho^{-1} O(\rho^2) = \beta_i\, O(\rho),$$

where $O(\rho)$ is uniform over all β. Since the order of the products $\frac{x^\beta}{x_i} V_i$ is at least $|\beta|+1$, by (5.13) we have

$$\llbracket \operatorname{ad}_A^{-1} \Psi h_{k\beta} \rrbracket_\rho \leqslant \frac{\beta_i}{|\beta|}\, O(\rho) = O(\rho)$$

uniformly over all β with $|\beta| \geqslant m \geqslant r+1$. Thus the last remaining composition $\operatorname{ad}_A^{-1} \circ \Psi$ is also strongly contracting, which implies existence of a solution for the fixed point equation

$$h = \operatorname{ad}^{-1} \circ (T + S + \Psi)h$$

equivalent to (5.11), in a sufficiently small polydisk $\{|x| < \rho\}$. □

Now one can easily complete the proof of the holomorphic normalization theorem in the Poincaré domain in the resonant case.

Proof of Theorem 5.5 (resonant case). By the Poincaré–Dulac normalization process, one can eliminate all nonresonant terms up to any finite order m by a polynomial transformation. By Theorem 5.15, m-flat holomorphic terms can be eliminated by a holomorphic transformation if m is large enough (depending on the spectrum of the linearization matrix). □

Remark 5.16. In the Poincaré domain one can prove an even stronger claim: if a holomorphic vector field depends analytically on finitely many additional parameters $\lambda \in (\mathbb{C}^m, 0)$ and belongs to the Poincaré domain for $\lambda = 0$, then by a holomorphic change of variables holomorphically depending on parameters, the field can be brought to a polynomial normal form involving only resonant terms. In such a form this assertion is formulated in [**Bru71**]. The proof can be achieved by minor adjustment of the arguments used in the demonstration of Theorem 5.15.

5D. Holomorphic normal forms for self-maps. In the same way as the formal theory for vector fields $\mathcal{D}[[\mathbb{C}^n, 0]]$ and maps $\operatorname{Diff}[[\mathbb{C}^n, 0]]$ are largely parallel (see §4**G**), the analytic theory of vector fields and biholomorphisms are also parallel.

The additive resonance conditions $\lambda_j - \langle \alpha, \lambda \rangle \neq 0$ correspond to the multiplicative resonance conditions $\mu_j^{-1}\mu^\alpha \neq 1$. The additive Poincaré condition (Definition 5.1) requires that (eventually after a rotation) all eigenvalues λ_j of the vector field lie to one side of the imaginary axis. Its multiplicative counterpart requires that all eigenvalues μ_j of the map must be to one side of the unit circle. Such maps are automatically contracting or expanding, and admit at most finitely many multiplicative resonance relations between the eigenvalues.

The result parallel to the Poincaré Theorem 5.5 takes the following form. Let $M \in \operatorname{GL}(n, \mathbb{C})$ be a matrix in the upper triangular Jordan normal form with the eigenvalues $\mu_1, \dots, \mu_n \in \mathbb{C}^*$. The Poincaré–Dulac normal form is a map

$$f\colon \mathbb{C}^n \to \mathbb{C}^n, \qquad x \mapsto f(x) = Mx + \sum_{\substack{\alpha \in \mathbb{Z}_+,\ |\alpha| \geqslant 2 \\ \mu_j = \mu^\alpha}} x^\alpha \mathbf{e}_j, \tag{5.15}$$

where $\mathbf{e}_j \in \mathbb{C}^n$ is the jth basis vector. If M is in the *multiplicative Poincaré domain*, i.e., if the eigenvalues are all of modulus less than one or all of modulus greater than one, then the normal form (5.15) is polynomial (contains finitely many terms).

The general result for holomorphic self-maps in the Poincaré domain has the following form.

Theorem 5.17. *A holomorphic invertible map $f \in \operatorname{Diff}(\mathbb{C}^n, 0)$ with the spectrum $\mu_1, \dots, \mu_n$ inside the unit disk, $0 < |\mu_j| < 1$, $j = 1, \dots, n$, is analytically equivalent to its polynomial Poincaré–Dulac formal normal form* (5.15). □

In the important particular case of one-dimensional maps, the multiplicative Poincaré condition holds automatically if the map is *hyperbolic*, i.e., if its multiplicator μ has modulus different from one. This automatically guarantees that resonances are impossible, and hence the Poincaré–Dulac normal form (5.15) is linear. The corresponding result was proved by E. Schröder (1870) and A. Kœnigs (1884).

Theorem 5.18. *A holomorphic germ $f\colon (\mathbb{C}, 0) \to (\mathbb{C}, 0)$, $f(x) = \mu x + O(x^2)$, is analytically linearizable if $|\mu| \neq 1$.*

If $f = f_t$ depends analytically on additional parameter $t \in U \subseteq \mathbb{C}^p$, the linearizing chart can also be chosen analytically depending on this parameter as soon as the respective multiplier μ_t remains off the unit circle.

Because of its importance, we will give an independent proof of this result by the path method in §5**F** below. Yet another (shortest known) proof is outlined in Problem 5.6.

5E. Linearization in the Siegel domain: Siegel, Brjuno and Yoccoz theorems (micro-survey). In the Siegel domain the denominators $\lambda_j - \langle \alpha, \lambda \rangle$ are not separated from zero, hence even in the nonresonant case the operator $\operatorname{ad}_A = [\mathbf{A}, \cdot\,]$ of commutation with the linear part of the field has unbounded inverse ad_a^{-1}. Yet since the operator S_F is *strongly contracting*, the equation (5.7) can be solved with respect to h by Newton-type iterations, provided that the small denominators $|\lambda_j - \langle \alpha, \lambda \rangle\,|$ do not approach zero too fast as $|\alpha| \to \infty$.

The corresponding technique is known under the general name of *KAM theory* (after A. Kolmogorov, V. Arnold and J. Moser). The issue is very classical; accurate formulations and proofs can be found in many excellent sources, e.g., [**CG93, Arn83**]. We formulate only the basic results.

Definition 5.19. A tuple of complex numbers $\lambda \in \mathbb{C}^n$ from the Siegel domain $\mathfrak{S}$ is called *Diophantine*, if the small denominators decay no faster than polynomially with $|\alpha|$, i.e.,

$$\exists C, N < +\infty \text{ such that } \quad \forall \alpha \in \mathbb{Z}_+^n, \quad |\lambda_j - \langle \alpha, \lambda \rangle|^{-1} \leqslant C\,|\alpha|^N. \tag{5.16}$$

Otherwise the tuple (vector, collection) is called *Liouvillean.*

Liouvillean vectors are scarce: the set of points $\lambda \in \mathbb{C}^n$ satisfying violating the condition 5.16 with a given N, has Lebesgue measure zero in $\mathfrak{S} \subset \mathbb{C}^n$ if $N > (n-2)/2$; see [**Arn83**].

Theorem 5.20 (Siegel theorem). *If the linearization matrix Λ of a holomorphic vector field is nonresonant of Siegel type and has Diophantine spectrum, then the field is holomorphically linearizable.*

Thus the majority (in the sense of Lebesgue measure) of germs of holomorphic vector fields are analytically linearizable. Yet one may further relax sufficient conditions for convergence of linearizing series in the Siegel domain.

Definition 5.21. A nonresonant collection $\lambda \in \mathbb{C}^n$ is said to satisfy the *Brjuno condition*, if the small denominators decrease *sub-exponentially*,

$$|\lambda_j - \langle \alpha, \lambda \rangle|^{-1} \leqslant Ce^{|\alpha|^{1-\varepsilon}}, \qquad \text{as } |\alpha| \to \infty, \tag{5.17}$$

for some finite C and positive $\varepsilon > 0$.

Theorem 5.22 (Brjuno theorem). *A holomorphic vector field with nonresonant linearization matrix of Siegel type satisfying the Brjuno condition, is holomorphically linearizable.*

On the other hand, if the denominators $|\lambda_j - \langle \alpha, \lambda \rangle|$ accumulate to zero too fast, e.g., *super-exponentially*, then the corresponding germs are in general nonlinearizable (cf. with Remark 5.33 below).

Analogs of the Siegel and Brjuno theorems hold for holomorphic germs. The most important case is that of one-dimensional *conformal germs* from the group $\operatorname{Diff}(\mathbb{C}^1, 0)$. Such germs belong to the Siegel domain if and only if their multiplicator μ belongs to the unit circle, $\mu = \exp 2\pi i l$, with some $l \in \mathbb{R}$; they are nonresonant if l is an irrational number. The Diophantine and Brjuno conditions translate for this case as assumptions that this irrational number $l \in \mathbb{R} \smallsetminus \mathbb{Q}$ does not admit abnormally accurate rational approximations.

For instance, if the complex number $\mu = \exp 2\pi i l$, $l \in \mathbb{R}$, satisfies the *multiplicative Brjuno condition*

$$|\mu^k - 1|^{-1} < Ce^{k^{1-\varepsilon}}, \qquad C < +\infty,\ \varepsilon > 0, \tag{5.18}$$

then any holomorphic map $(\mathbb{C}, 0) \to (\mathbb{C}, 0)$, $z \mapsto \mu z + z^2 + \cdots$, is holomorphically linearizable. The sufficient arithmetic condition (5.18) turns out to also be *necessary* in the following sense.

Theorem 5.23 (J.-C. Yoccoz [**Yoc88, Yoc95**]). *If the complex number* $\mu = \exp 2\pi i l$, $l \in \mathbb{R}$, *violates the multiplicative Brjuno condition* (5.18), *then there exists a nonlinearizable holomorphic germ* $(\mathbb{C}, 0) \to (\mathbb{C}, 0)$, $z \mapsto \mu z + f(z)$, $f(z) = z^2 + \cdots$.

In fact, in the assumptions of this theorem for almost all complex numbers $w \in \mathbb{C}$ the germ $f_w(z) = \mu z + wf(z)$ is analytically nonlinearizable; cf. with Theorem 5.29 below and [**PM01**].

Remark 5.24. The condition on the rate of convergence of small denominators can be reformulated in terms of the growth rate of coefficients of decomposition of the irrational number $l \in \mathbb{R} \setminus \mathbb{Q}$ into the continuous fraction. This is a more standard way of formulating the Brjuno condition in the recent literature.

If a resonance occurs in the Siegel case, then the situation turns out to be even more complicated: a resonant conformal germ $f \in \operatorname{Diff}(\mathbb{C}, 0)$ with multiplicator $\mu \in \exp 2\pi i \mathbb{Q}$ is almost never analytically equivalent to its polynomial Poincaré–Dulac formal normal form described in Theorem 4.26. This result and its numerous developments are explained in detail in §21.

A two-dimensional analytic orbital classification of Siegel resonant vector fields (saddle-nodes and resonant saddles from Table I.1) is at least as difficult as the analytic classification of resonant germs from $\operatorname{Diff}(\mathbb{C}, 0)$. Indeed, in §7 we will show that the corresponding foliations have leaves with nontrivial (infinite cyclic) fundamental group, whose holonomy is generated by Siegel resonant germs from $\operatorname{Diff}(\mathbb{C}, 0)$. The details can be found in Chapter IV; see §22.

Somewhat unexpectedly, the cuspidal points behave better than their less degenerate brethren. In [**SZ02**] H. Żołądek and E. Stróżyna proved that one can always reduce a holomorphic planar vector field near a cuspidal singular point to a *holomorphic* normal form (4.17) (i.e., with converging series $a(x), b(x) \in \mathcal{O}(\mathbb{C}, 0)$) by a biholomorphic transformation. The direct and difficult proof from [**SZ02**] was recently replaced by beautiful geometric arguments by F. Loray [**Lor06**]. This proof, based on *nonlocal uniformization technique*, is split into a series of problems in §23 (Problems 23.6–23.13).

5F. Path method. In this section we outline another very powerful *analytic* method of reducing holomorphic vector fields and self-maps to their normal forms. This method is called *path method* (méthode de chemin, homotopy method) since it consists of connecting the initial object (field, self-map) with its normal form by a path (usually a line segment) and then looking for a flow of a nonautonomous vector field that would conjugate with each other all objects in this parametric family.

We *illustrate* the path method by proving two relatively simple results, analytic reducibility of one-dimensional holomorphic vector fields (cf. with Theorem 4.24) and one-dimensional hyperbolic self-maps to their normal form. Both results, however, can be proved by shorter arguments; see Problems 5.5 and 5.6 below.

Theorem 5.25. *Any analytic vector field $F(x) = x^{k+1}(1+\cdots)\frac{\partial}{\partial x} \in \mathcal{D}(\mathbb{C},0)$ is analytically conjugate to its polynomial formal normal form $F_0(x) = (x^{k+1} + ax^{2k+1})\frac{\partial}{\partial x}$.*

Proof. Without loss of generality we may assume from the beginning, that the jet of F of any specified order is already reduced to the normal form. Thus we can assume that the field $F = F_1$ is given as $F_0(x) + R(x)\frac{\partial}{\partial x}$, where R is as flat at the origin, as necessary. It will be sufficient to require that the function $R(x)$ has zero of multiplicity $2k+2$ at the origin, $R(x) = x^{2k+2}S(x)$, $S \in \mathcal{O}(\mathbb{C},0)$. We want to show that for all values of an auxiliary complex parameter z from some domain $U \subseteq \mathbb{C}$ containing the segment $[0,1]$, the vector fields $F_z(x) = F_0(x) + zR(x)\frac{\partial}{\partial x}$ are holomorphically equivalent to each other. This, in particular, would imply that F_0 and F_1 are holomorphically equivalent, which would immediately imply the assertion of the theorem.

Consider the planar domain $(\mathbb{C},0)\times U$ and the vector field on it,

$$\mathbf{F} = \mathbf{F}_0 + zR(x)\cdot\frac{\partial}{\partial x} + 0\cdot\frac{\partial}{\partial z}, \qquad \mathbf{F}_0 = (x^{k+1} + ax^{2k+1})\frac{\partial}{\partial x}, \tag{5.19}$$

which is the suspension of the above parametric family of vector fields on the line.

Consider another planar vector field $\mathbf{H} \in \operatorname{Diff}\big((\mathbb{C}^1,0)\times U\big)$, $U \subseteq \mathbb{C}$, which has the form

$$\mathbf{H} = h(x,z)\frac{\partial}{\partial x} + 1\cdot\frac{\partial}{\partial z}, \qquad h(0,z)\equiv 0. \tag{5.20}$$

Lemma 5.26 (Path method paradigm). *If there exists a holomorphic vector field $\mathbf{H} \in \operatorname{Diff}\big((\mathbb{C}^1,0)\times U\big)$ of the form* (5.20) *which* commutes *with* $\mathbf{F}$,

$$[\mathbf{F},\mathbf{H}] = 0, \tag{5.21}$$

then all germs of vector fields $F_z \in \mathcal{D}(\mathbb{C}^1,0)$ are holomorphically equivalent to each other for all values of $z \in U$.

Proof. If the vector fields $\mathbf{F}$ and $\mathbf{H}$ commute, then the flow of the vector field $\mathbf{H}$ commutes with the flow of $\mathbf{F}$ and hence the flow maps of $\mathbf{H}$ are symmetries of the field $\mathbf{F}$.

Because of the special structure of $\mathbf{H}$, its flow sends the lines $\{z = \text{const}\}$ into each other, each time fixing the origin $\{x=0\}$. Thus the flow $\exp\mathbf{H}$ maps $\{z=0\}$ into $\{z=1\}$, is defined in some neighborhood of the origin and conjugating $\mathbf{F}|_{z=0} = F_0$ with $\mathbf{F}|_{z=1} = F_1$. □

Now we can complete the proof of Theorem 5.25, showing that in the assumptions of the theorem, such a vector field $\mathbf{H}$ indeed exists. The *homological equation* (5.21) is equivalent to a partial differential equation on the function H,

$$f\cdot\frac{\partial h}{\partial x} - h\cdot\frac{\partial f}{\partial x} = -R, \qquad f(x,z) = x^{k+1} + ax^{2k+1} + zR(x). \tag{5.22}$$

Yet in fact this equation can be considered as a linear first order ordinary (with respect to the x-variable) nonhomogeneous differential equation analytically depending on the parameter $z \in U$. The solution of the corresponding *homogeneous* equation is immediate,

$h_0(x,z) = f(x,z)$, using the ansatz $h(x,z) = s(x,z)h_0(x,z)$ we obtain the equation (recall that $R = x^{2k+2}S$),

$$f^2 \cdot \frac{\partial s}{\partial z} = -R(x), \quad \text{i.e.} \quad \frac{\partial s}{\partial x} = -\frac{S(x)}{\left(1 + ax^k + x^{k+1}S(x)\right)^2}. \tag{5.23}$$

Integration of the right hand side with the initial condition $s(0,z) = 1$ yields a solution $s = s(x,z)$ holomorphic at $x = 0$ for all $z \in U$. The vector field $\mathbf{H} = s(x,z)\mathbf{F} + 1 \cdot \frac{\partial}{\partial z}$ satisfies all conditions imposed by Lemma 5.26 and allows us to construct a holomorphic conjugacy between F_0 and F_1. □

Obviously, the polynomial normal form (4.21) can be replaced by the rational normal form (4.23).

Remark 5.27. Besides holomorphic differential vector fields, one may consider *meromorphic differential 1-forms* on the complex line (or, more precisely, their germs at the origin): the set of all such forms is naturally denoted by $\Lambda^1(\mathbb{C},0) \otimes \mathfrak{M}(\mathbb{C},0)$.

The group $\operatorname{Diff}(\mathbb{C},0)$ acts on such forms, so one can establish normal forms. Yet instead of developing parallel theory, one can use duality: a 1-form $\omega \in \Lambda^1(\mathbb{C},0)$ and a vector field $F \in \mathcal{D}(\mathbb{C},0)$ are called *dual*, if $\omega(F) \equiv 1$. Holomorphic transform of a dual pair is again a dual pair.

Meromorphic (i.e., with a pole at the origin) 1-forms have two obvious invariants that cannot be changed by holomorphic transformations: the *order of the pole* and the *residue* at this point.

The form dual to the rational vector field (4.23) is $\frac{dx}{x^{k+1}} - a\frac{dx}{x}$, and the formal invariant $a \in \mathbb{C}$ is the residue of this form (modulo the sign). This observation explains the role of the formal invariant.

As yet another application of the path method, we give an independent proof of the Schröder–Kœnigs Theorem 5.18.

Consider the analytic self-map $f \in \operatorname{Diff}(\mathbb{C},0)$, $f(x) = \mu x + r(x)$, with the multiplicator $\mu \in \mathbb{C}^*$, $|\mu| < 1$ and analytic nonlinearity $r(x) = O(x^2)$.

As before, we embed f into an analytic one-parameter deformation $f_z(x) = \mu x + zr(x)$ with a complex parameter $z \in U \subseteq \mathbb{C}$, $[0,1] \subseteq U$, and suspend it to the planar self-map $\mathbf{f} \in \operatorname{Diff}(\mathbb{C}^2,0)$,

$$\mathbf{f}\colon (x,z) \mapsto (\mu x + zr(x), z), \qquad (x,z) \in (\mathbb{C}^1,0) \times U. \tag{5.24}$$

The following lemma is a reformulation of the main paradigm of the path method (Lemma 5.26) for the current context.

Lemma 5.28. *If a vector field* $\mathbf{H}$ *as in* (5.20) *is preserved by the self-map* $\mathbf{f}$, *i.e.,*

$$\mathbf{f}_* \cdot \mathbf{H} = \mathbf{H} \circ \mathbf{f}, \qquad \mathbf{f}_* = \frac{\partial \mathbf{f}(x,z)}{\partial(x,z)}, \tag{5.25}$$

then all self-maps f_z *for all* $z \in U$, *are analytically equivalent, in particular,* $f_1 = f$ *is analytically equivalent to the linear map* f_0. *The conjugacy is achieved by the flow of the field* $\mathbf{H}$ *restricted on the lines* $\{z = \text{const}\}$. □

The proof of Lemma 5.28 almost literally reproduces that of Lemma 5.26 and is skipped. In order to prove Theorem 5.18, we need only to show that the homological equation (5.25) is solvable.

Alternative proof of Theorem 5.18. The identity (5.25) reduces to a single scalar linear nonhomogeneous functional equation

$$\frac{\partial f_z(x)}{\partial x} \cdot h(x,z) - h(f_z(x),z) = r(x). \tag{5.26}$$

This equation can be solved in two steps, solving first the corresponding homogeneous equation $(\frac{\partial f}{\partial x})u - u \circ f = 0$, and then looking for a solution of (5.26) in the form $h = su$, similar to the way the equation (5.22) was solved.

The homogeneous equation can be rewritten as a fixed point statement,

$$h = \left(\frac{\partial f}{\partial x}\right)^{-1} \cdot (h \circ f), \qquad f = f_z \in \operatorname{Diff}(\mathbb{C}, 0). \tag{5.27}$$

It has a trivial (zero) solution, yet we can restrict the operator occurring in the right hand side, on the subspace of functions tangent to identity, $h(x) = x + O(x^2)$.

Without loss of generality we may assume (passing to a sufficiently small neighborhood of the origin which is rescaled to the unit disk) that all maps f_z satisfy the inequalities

$$\begin{gathered} \left|\frac{\partial f}{\partial z}\right| \geqslant \mu_-, \qquad |f(x)| < \mu_+|x|, \qquad \forall x \in D_1 = \{|x| \leqslant 1\}, \\ 0 < \mu_- < |\mu| < \mu_+ < 1. \end{gathered} \tag{5.28}$$

Here $\mu_\pm$ are two positive constants which can be assumed to be arbitrarily close to $|\mu| < 1$.

First we show that the operator $\Phi\colon h \mapsto (\frac{\partial f}{\partial x})^{-1} \cdot (h \circ f)$ restricted on the subspace

$$\mathcal{M} = \{u \in \mathcal{A}(D_1) \colon u(0) = 0,\ \tfrac{du}{dx} = 1\}$$

of holomorphic functions tangent to the identity at the origin, is contracting in the sense of the usual supremum-norm $\|u\| = \max_{x \in D_1} |u(x)|$. Clearly, $\Phi(\mathcal{M}) \subseteq \mathcal{M}$.

Indeed, since Φ is linear, it is sufficient to show that $\|\Phi q\| < \lambda \|q\|$ for any $q \in \mathcal{A}(D_1)$ having a second order zero at the origin and some λ strictly between 0 and 1. Note that for any such function $q(x)$, we have the inequality $|q(x)| \leqslant \|q\| \cdot |x|^2$: it is sufficient to apply the maximum modulus principle to the holomorphic ratio $q(x)/x^2$. Then from (5.28) it immediately follows that

$$\|\Phi q\| \leqslant \max_{|x| \leqslant 1} \frac{1}{\mu_-} \|q\| \cdot |f(x)|^2 \leqslant \frac{\mu_+^2}{\mu_-} \cdot \max_{|x| \leqslant 1} \|q\|\, |x|^2 \leqslant \frac{\mu_+^2}{\mu_-} \cdot \|q\|.$$

Since the ratio μ_+^2/μ_- can be made arbitrarily close to $|\mu| < 1$, the operator Φ restricted on $\mathcal{M}$ is contracting and hence has a holomorphic fixed point u analytically depending on z and any additional parameters (if present).

Now a solution of the nonhomogeneous equation can be found using the ansatz $h = su$. Substituting this ansatz into the equation (5.26), we obtain the *Abel-type equation*

$$s - s \circ f = -R(x), \qquad R = R_z(x) = \frac{r(x)}{(\frac{\partial f}{\partial z}) \cdot u(x,z)}, \qquad f = f_z(x). \tag{5.29}$$

The function $R_z(x)$ is holomorphic and vanishes at the origin $x = 0$ for all values of x, since f_z has a simple zero and $r(x)$ has a double zero at the origin.

The formal solution of the equation (5.29) is given by the series

$$s = -\sum_{k=0}^{\infty} R \circ f^{\circ k}, \qquad s = s(\cdot, z),\ f = f_z,\ R = R_z, \tag{5.30}$$

which is well defined because f is contracting. Moreover, since R vanishes at the origin, we have $|R_z(x)| < C|x|$ for some $C < \infty$ and all $x \in D_1$. Combining this with the uniform bounds $|f^{\circ k}(x)| \leqslant \mu_+^k |x|$ implied by (5.28), we conclude that the series (5.30) converges uniformly on D_1 and hence its sum is a holomorphic function vanishing at $x = 0$. The holomorphic vector field $\mathbf{H} = s(x,z)u(x,z)\frac{\partial}{\partial x} + 1 \cdot \frac{\partial}{\partial z}$ solves the equation (5.25).

The alternative proof of Theorem 5.18 is complete. □

* * *

5G. Divergence dichotomy. As follows from the Poincaré, Siegel and Brjuno theorems, for most linear parts the linearizing series converges, and in the remaining cases the linearizing series *may diverge.* On the other hand, no matter how "bad" the linearization and its eigenvalues are, there are always nonlinear systems that can be linearized (e.g., linear systems in nonlinear coordinates). It turns out that in some precise sense for a given linear part, the convergence/divergence pattern is common for *most* nonlinearities.

Consider a *parametric* nonlinear system

$$\dot{x} = Ax + z\,f(x), \qquad x \in \mathbb{C}^n, \quad z \in \mathbb{C}, \tag{5.31}$$

holomorphic in some neighborhood of the origin with the *nonresonant* linearization matrix A and the nonlinear part linearly depending on the auxiliary complex parameter $z \in \mathbb{C}$. For such systems for each value of the parameter $z \in \mathbb{C}$ there is a *unique* (by Remark 4.6) formal series $H_z(x) = x + h_z(x) \in \operatorname{Diff}[[x,z]]$ linearizing (5.31). This series may converge for some values of z while diverging for the rest. It turns out that there is a strict alternative: either the linearizing series *converges for all values of z without exception*, or on the contrary *the series H_z diverges for all z outside a rather small exceptional set $K \Subset \mathbb{C}$.*

The exceptional sets are small in the sense that their (electrostatic) *capacity* is zero. The notion of capacity is formally introduced below in §**5H**, where some of its basic properties are collected. We mention here only that zero capacity implies zero Lebesgue measure for any compact set.

Theorem 5.29 (Divergence dichotomy, Yu. Ilyashenko [**Ily79a**], R. Perez Marco [**PM01**]). *For any nonresonant linear family* (5.31) *one has the following alternative*:

(1) *Either the linearizing series $H_z \in \operatorname{Diff}[[\mathbb{C}^n,0]]$ converges for all values of $z \in \mathbb{C}$ in a symmetric polydisk $\{|x| < r\}$ of a positive radius $r = r(z) > 0$ decreasing as $O(|z|^{-1})$ as $z \to \infty$, or*

(2) *The linearizing series H_z diverges for all values of z except for a set $K_f \Subset \mathbb{C}$ of capacity zero.*

The proof is based on the following property of polynomials, which can be considered as a quantitative uniqueness theorem for polynomials. If K is a set of positive capacity and $p \in \mathbb{C}[z]$ a polynomial vanishing on K, then by definition p vanishes identically. One can expect that if p is small on K, then it is also uniformly small on *any* other compact subset, in particular, on all compact subsets of $\mathbb{C}$.

Theorem 5.30 (Bernstein inequality). *If $K \Subset \mathbb{C}$ is a set of positive capacity, then for any polynomial $p \in \mathbb{C}[z]$ of degree $r \geqslant 0$,*

$$|p(z)| \leqslant \|p\|_K \exp(rG_K(z)), \tag{5.32}$$

where $\|p\|_K = \max_{z\in K} |p(z)|$ is the supremum-norm of p on K, and $G_K(z)$ is the nonnegative Green function of the complement $\mathbb{C} \smallsetminus K$ with the source at infinity; see (5.36).

We postpone the proof of this theorem until §5**H** and proceed with deriving Theorem 5.29 from the Bernstein inequality.

Lemma 5.31. *Formal Taylor coefficients of the formal series linearizing the field* (5.31) *are polynomial in z.*

More precisely, every monomial x^α, $|\alpha| \geqslant 2$, enters into the vector series h_z with the coefficient which is a polynomial of degree $\leqslant |\alpha| - 1$ in z.

Proof. The equation determining $h = h_z$ is of the form

$$\left(\frac{\partial h_z}{\partial x}\right)(Ax + z\,f(x)) = Ah_z(x). \tag{5.33}$$

Collecting the terms of degree m in x, we obtain for the corresponding mth homogeneous (vector) components $h_z^{(m)}$, $f^{(l)}$, the recurrent identities

$$\left(\frac{\partial h_z^{(m)}}{\partial x}\right) Ax - Ah_z^{(m)} = -z \sum_{k+l=m,\, l\geqslant 2} \left(\frac{\partial h_z^{(k+1)}}{\partial x}\right) f^{(l)}.$$

From these identities it obviously follows by induction that each $h_z^{(m)}$ is a polynomial of degree $m-1$ in z for all $m \geqslant 1$ (recall that f does not depend on z). □

Proof of Theorem 5.29. Assume that the formal series $H_z(x) = x+h_z(x)$ linearizing the field $F_z(x) = Ax + z\,f(x)$ converges for values of z belonging to some set $K^* \subset \mathbb{C}$ of positive capacity.

Consider the subsets $K_{c\rho} \Subset \mathbb{C}$, $\rho > 0$, $c < +\infty$, defined by the condition

$$z \in K_{c\rho} \iff |h_z^{(m)}(0)| \leqslant c\rho^{-m} \quad \forall m \in \mathbb{N}.$$

By this definition, $K^* = \bigcup_{c,\rho} K_{c\rho}$, since a Taylor series converges if and only if satisfies some Cauchy-type estimate. Each of the sets $K_{c\rho}$ obviously is a compact subset of $\mathbb{C}$, being an intersection of semialgebraic compact sets.

The compacts $K_{c\rho}$ are naturally nested: $K_{c'\rho'} \subseteq K_{c\rho}$ if $\rho' > \rho$ and $c' < c$. Passing to a countable sub-collection, one concludes that the set K of positive capacity is a countable union of compacts $K_{c\rho}$. By Proposition 5.35 (see below), one of these compacts must also be of positive capacity. Denote this compact by $K = K_{c\rho}$; by its definition,

$$|h_z^{(m)}| \leqslant c\rho^{-m}, \qquad \forall z \in K, \ \forall m \in \mathbb{N}.$$

Since the capacity of K is positive, Theorem 5.30 applies. By this theorem and Lemma 5.31, the polynomial coefficients of the series h_z for *any* $z \in \mathbb{C}$ satisfy the inequalities

$$|h_z^{(m)}| \leqslant c\rho^{-m} \exp[(m-1)G_K(z)] \leqslant c(\rho/\exp G_K(z))^{-m}, \quad \forall z \in \mathbb{C}, \ \forall m \in \mathbb{N}.$$

This means that the series h_z converges for any $z \in \mathbb{C}$ in the symmetric polydisk $\{|x| < \rho/\exp G_K(z)\}$. Together with the asymptotic growth rate $G_K(z) \sim \ln|z| + O(1)$ as $z \to \infty$ (see (5.36)) this proves the lower bound on the convergence radius of H_z. □

The dichotomy established in Theorem 5.29 may be instrumental in constructing "nonconstructive" examples of diverging linearization series. Consider again the nonresonant case where the homological equation $\operatorname{ad}_A g = f$ is always formally solvable.

Theorem 5.32 ([**Ily79a**]). *Assume that the formal solution $g \in \mathcal{D}[[\mathbb{C}^n, 0]]$ of the homological equation $\operatorname{ad}_A g = f$ is divergent.*

Then the series linearizing the vector field $F_z(x) = Ax + z\,f(x)$, diverges for most values of the parameter z, eventually except for a zero capacity set.

Proof. Assume the contrary, that the linearizing series H_z converges for a positive capacity set. By Theorem 5.29, it converges then for all values of z, in particular, h_z is holomorphic in some small polydisk $\{|x| < \rho', \ |z| < \rho''\}$.

Differentiating (5.33) in z, we see that the derivative $g(x) = \frac{\partial h_z(x)}{\partial z}\big|_{z=0}$ is a converging solution of the equation $(\frac{\partial g}{\partial x})Ax - Ag = f$, contrary to the assumption of the theorem. □

Remark 5.33. The divergence assumption appearing in Theorem 5.32 can be easily achieved. Assume that A is a diagonal matrix with the spectrum $\{\lambda_j\}_1^n$ such that the differences $|\lambda_j - \langle\lambda, \alpha\rangle|$ decrease faster than *any* geometric progression $\rho^{|\alpha|}$ for any nonzero ρ. Assume also that the Taylor coefficients of f are bounded *from below* by *some* geometric progression. Then the series $\operatorname{ad}_A^{-1} f$ diverges.

It remains to observe that a set of positive measure is necessarily of positive capacity (Proposition 5.35), hence divergence guaranteed in the assumptions of Theorem 5.32, occurs for almost all z in the measure-theoretic sense, as stated in [**Ily79a**].

5H. Capacity and Bernstein inequality. The brief exposition below is based on [**PM01**] and the encyclopedic treatise [**Tsu59**].

Recall that the function $\ln|z-a|^{-1} = -\ln|z-a|$ is the electrostatic potential on the z-plane $\mathbb{C} \cong \mathbb{R}^2$, created by a unit charge at the point $a \in \mathbb{C}$ and harmonic outside a. If μ is a nonnegative measure (charge distribution) on the compact $K \Subset \mathbb{C}$, then its potential is the function represented by the integral $u_\mu(z) = \int_K \ln|z-a|^{-1}\,d\mu(a)$ and the energy of this measure is

$$E_\mu(K) = \iint_{K\times K} \ln|z-w|^{-1}\,d\mu(z)\,d\mu(w).$$

This energy can be either infinite for all measures, or $E_\mu(K) < +\infty$ for some nonnegative measures. In the latter case one can show that among all nonnegative measures normalized by the condition $\mu(K) = 1$, the (finite) minimal energy $E^*(K) = \inf_{\mu(K)=1} E_\mu(K)$ is achieved by a unique *equilibrium distribution* μ_K. The corresponding potential $u_K(z)$ is called the *conductor potential* of K.

Definition 5.34. The (harmonic, electrostatic) *capacity* of the compact K is either zero (when $E_\mu = +\infty$ for any charge distribution on K) or $\exp(-E^*(K)) > 0$ otherwise;

$$\varkappa(K) = \begin{cases} 0, & \text{if } \forall\mu\ E_\mu(K) = +\infty, \\ \sup\limits_{\mu(K)=1,\ \mu\geqslant 0} \exp(-E_\mu(K)), & \text{otherwise.} \end{cases} \tag{5.34}$$

Proposition 5.35. *Capacity of compact sets possesses the following properties*:

(1) *Countable union of zero capacity sets also has capacity zero.*

(2) $\varkappa(K) \geqslant \sqrt{\operatorname{mes}(K)/\pi e}$, *where* $\operatorname{mes}(K)$ *is the Lebesgue measure of* K, *in particular, if* K *is a set of positive measure, then* $\varkappa(K) > 0$.

(3) *If* K *is a Jordan curve of positive length, then* $\varkappa(K) > 0$.

Proof. All these assertions appear in [**Tsu59**] as Theorems III.8, III.10 and III.11 respectively. □

Proposition 5.36. *For compact sets of positive capacity, the conductor potential is harmonic outside* K, *and*

$$\begin{aligned} u_K \leqslant \varkappa^{-1}(K), \qquad & u_K|_K = \varkappa^{-1}(K) \quad \textit{a.e.}, \\ u_K(z) = -\ln|z| + O(|z|^{-1}) \quad & \textit{as } z\to\infty. \end{aligned} \tag{5.35}$$

Proof. [**Tsu59**, Theorem III.12] □

As a corollary, we conclude that for sets of the positive capacity there exists the Green function

$$G_K(z) = \varkappa^{-1}(K) - u_K(z) = \ln|z| + \varkappa^{-1}(K) + o(1) \quad \text{as } z\to\infty, \tag{5.36}$$

nonnegative on $\mathbb{C}\smallsetminus K$, vanishing on K and asymptotic to the fundamental solution of the Laplace equation with the source at infinity.

Proof of Theorem 5.30 (Bernstein inequality). Since the assertion is invariant by multiplication by scalars, it is sufficient to prove for monic polynomials only.

Let $p(z) = z^r + \cdots$ be a monic polynomial of degree r. Consider the function

$$g(z) = \ln|p(z)| - \ln\|p\|_K - rG_K(z), \qquad z \in \mathbb{C} \smallsetminus K.$$

We claim that this function is nonpositive, $g \leqslant 0$ outside K. Indeed, g is negative near infinity since $g(z) = -\ln\|p\|_K - r\varkappa^{-1}(K) + o(1)$ as $z \to \infty$ by (5.36). On K we have the obvious inequality $\ln|p(z)| \leqslant \ln\|p\|_K$, and the Green function G_k has zero limit on K by (5.35). By construction, the function g is harmonic in $\mathbb{C} \smallsetminus K$ outside the isolated zeros of p where it tends to $-\infty$. By the maximum principle, the function g is nonpositive everywhere, $\ln|p(z)| \leqslant \ln\|p\|_K + rG_K(z)$ for all $z \in \mathbb{C} \smallsetminus K$. After passing to exponents this nonpositivity proves the theorem. □

Example 5.37. Assume that $K = [-1, 1]$ is the unit segment. Its complement is conformally mapped into the exterior of the unit disk $D = \{|w| < 1\}$ by the function $z = \frac{1}{2}(w + w^{-1})$, $w = z + \sqrt{z^2 - 1}$. The Green function G_D of the exterior is $\ln|w|$. Thus we obtain the explicit expression for G_K,

$$G_K = \ln\left|z + \sqrt{z^2 - 1}\right|,$$

which implies the classical form of the Bernstein inequality,

$$|p(z)| \leqslant \left|z + \sqrt{z^2 - 1}\right|^{\deg p} \max_{-1 \leqslant z \leqslant +1} |p(z)|. \tag{5.37}$$

Exercises and Problems for §5.

Problem 5.1. Prove that if h is a solution for the homogeneous homological equation (5.27) with a hyperbolic map f, then $H = h(x)\frac{\partial}{\partial x} \in \mathcal{D}(\mathbb{C}, 0)$ is a vector field that only by a constant factor differs from the generator of the self-map f: $f = \exp cH$, for some $c \in \mathbb{C}$.

Problem 5.2. Supply a detailed proof of the Poincaré theorem for self-maps (Theorem 5.17).

Exercise 5.3. Let $l \in \mathbb{R}$ be an irrational number whose rational approximations have only sub-exponential accuracy,

$$|l - \tfrac{p}{q}| > Ce^{-q^{1-\varepsilon}} \qquad \text{for some } C, \varepsilon > 0, \tag{5.38}$$

and $\mu = \exp 2\pi i l$. Prove that for any holomorphic right hand side f the homological equation

$$h \circ \mu - \mu h = f, \qquad f \in \mathcal{O}(\mathbb{C}, 0), \tag{5.39}$$

has an analytic (convergent) solution $h \in \mathcal{O}(\mathbb{C}, 0)$.

Exercise 5.4. Let $l \in \mathbb{R}$ be an irrational number which admits infinitely many exponentially accurate rational approximations p/q such that $|l - \frac{p}{q}| < e^{-q}$. Prove that for some right hand sides f the homological equation (5.39) has only divergent solutions (cf. with Remark 5.33).

Problem 5.5. Let $\mathbf{F} = F(x)x\frac{\partial}{\partial x} \in \mathcal{D}(\mathbb{C}^1, 0)$ be the germ of a holomorphic vector field at a singular point of multiplicity $k+1 \geqslant 2$ at the origin, $F(x) = x^{k+1}(1+o(1))$, and $\mathbf{F}' = \mathbf{F} + \mathbf{o}(x^{2k+1}) \in \mathcal{D}(\mathbb{C}^1, 0)$ is another such germ with the same $2k+1$-jet.

(i) Prove that these two germs are analytically equivalent if and only if two meromorphic 1-forms ω and ω', dual to $\mathbf{F}$ and $\mathbf{F}'$ respectively, are holomorphically equivalent (cf. with Remark 5.27).

(ii) Show that in the assumptions of the problem, the orders of the poles and the Laurent parts of the 1-forms ω and ω' coincide so that the difference $\omega - \omega'$ is holomorphic.

(iii) Passing to the primitives and denoting by $a_k, \dots, a_1, a_0$ the common Laurent coefficients of the forms ω, ω', prove that the equation

$$\frac{a_k}{y^k} + \cdots + \frac{a_1}{y} + a_0 \ln y + O(y) = \frac{a_k}{x^k} + \cdots + \frac{a_1}{x} + a_0 \ln x + O(z)$$

with holomorphic terms $O(y)$ and $O(x)$, admits a holomorphic solution $y = y(x)$ tangent to identity (substitute $y = ux$ and apply the implicit function theorem to the function $u(x)$ with $u(0) = 1$).

Problem 5.6 (Yet another proof of Schröder–Kœnigs theorem; cf. with [**CG93**]). Let $f \in \operatorname{Diff}(\mathbb{C}, 0)$ be a contracting hyperbolic holomorphic self-map, $f(z) = \lambda z + \cdots$, $|\lambda| < 1$, and $g(z) = \lambda z$ its linearization (the normal form).

Prove that the sequence of iterations $h_n = g^{-\circ n} \circ f^{\circ n}$ is defined and converges in some small disk around the origin. The limit $h = \lim h_n$ conjugates f and g.

Problem 5.7. Prove Theorem 5.5 along the same lines (M. Villarini).

6. Finitely generated groups of conformal germs

Thus far we have studied classification and certain dynamic properties of *single* germs of vector fields and biholomorphisms. However, in §2**C** we introduced an important invariant of foliation, the holonomy group of a leaf $L \in \mathcal{F}$ with nontrivial fundamental group $\pi_1(L, a)$, $a \in L$. By construction, the holonomy is a representation of $\pi_1(L, a)$ by conformal germs $\operatorname{Diff}(\tau, a)$, where τ is a cross-section to L at a, and the holonomy group G is identified with the image of that representation. Usually if the fundamental group of a leaf of a holomorphic foliation is finitely generated, then so is the group G. We will consider only the case of holomorphic foliations on complex 2-dimensional surfaces, thus dealing only with finitely generated subgroups of the group $\operatorname{Diff}(\mathbb{C}, 0)$ of conformal germs.

In this section we study classification problems for *finitely generated groups* of conformal germs and their dynamic properties, focusing on the properties which will be later used in §11 and §28. In much more detail the theory is treated in the recent monograph [**Lor99**].

6A. Equivalence of finitely generated groups of conformal germs. The following definition is inspired by Proposition 2.15.

Definition 6.1. Two finitely generated subgroups $G, G' \subseteq \operatorname{Diff}(\mathbb{C}, 0)$ are called analytically (topologically, formally) equivalent if one can choose two systems of generators $G = \langle f_1, \dots, f_n \rangle$ and $G' = \langle f'_1, \dots, f'_n \rangle$ which are simultaneously conjugated by the germ of a holomorphic map (homeomorphism, formal series) h so that $h \circ f_j = f'_j \circ h$ for all $j = 1, \dots, n$.

Remark 6.2. If the generators of two groups are simultaneously conjugated as below, then the groups are isomorphic in the group theoretic sense. Indeed, any relation between generators of one group is automatically true in the second groups and vice versa, since the identical germ $\mathrm{id} \in \operatorname{Diff}(\mathbb{C}, 0)$ can be conjugated only to itself. Thus both groups are isomorphic to the quotients of the free group on n generators by the isomorphic sets of relations.

Example 6.3. Two conformal germs f and g from $\operatorname{Diff}(\mathbb{C}, 0)$ are analytically, topologically or formally equivalent if and only if the cyclic (commutative) subgroups $\{f^{\circ\mathbb{Z}}\}$ and $\{g^{\circ\mathbb{Z}}\}$ of $\operatorname{Diff}(\mathbb{C}, 0)$ generated by these germs are equivalent in the corresponding sense. In particular, they must be both finite or both infinite.

It turns out that some very important information on the analytic structure of the group is encoded in its algebraic properties.

Example 6.4. A generic single conformal germ can be linearized. However, simultaneous linearization (analytic, formal or topological) of two or more germs is possible *only if the group generated by these germs is commutative.* Indeed, the subgroup generated by any finite number of linear germs $f_j \colon z \mapsto \mu_j z$ in $\operatorname{Diff}(\mathbb{C}, 0)$ is commutative.

The "derivative map"

$$T\colon \operatorname{Diff}(\mathbb{C}, 0) \to \mathbb{C}^*, \qquad Tg = \tfrac{dg}{dz}(0) \in \mathbb{C}^*, \tag{6.1}$$

associating with any germ g its multiplicator at the fixed point at the origin, is a group homomorphism: by the chain rule of differentiation, $T(g \circ f) = Tg \cdot Tf = Tf \cdot Tg$ with the kernel equal to the normal subgroup of germs tangent to the origin, denoted by $\operatorname{Diff}_1(\mathbb{C}, 0)$:

$$\operatorname{Ker} T = \operatorname{Diff}_1(\mathbb{C}, 0) = \{g \in \operatorname{Diff}(\mathbb{C}, 0) \colon g(z) = z + O(z^2)\}. \tag{6.2}$$

Definition 6.5. Elements of the subgroup $\operatorname{Diff}_1(\mathbb{C}, 0)$ tangent to identity, are called *parabolic germs.*

The *parabolic subgroup* $\operatorname{Diff}_1(\mathbb{C},0)$ is *filtered* by the order of contact with the identity:

$$\begin{aligned} \operatorname{Diff}_1(\mathbb{C},0) \smallsetminus \{\mathrm{id}\} &= \mathscr{A}_1 \sqcup \mathscr{A}_2 \sqcup \mathscr{A}_3 \sqcup \cdots, \\ \mathscr{A}_p &= \{g \in \operatorname{Diff}_1(\mathbb{C},0) : g(z) = z \cdot (1 + az^p + \cdots),\ a \neq 0\}. \end{aligned} \tag{6.3}$$

The natural index p in the above formulas will be referred to as the *level* of a conformal germ $g \in \mathscr{A}_p$: this parameter is slightly more convenient to use than the order of tangency between the germ and identity, equal to $p+1$. One can easily verify that the level is invariant (does not change by conjugacy $g \mapsto h \circ g \circ h^{-1}$, $h \in \operatorname{Diff}(\mathbb{C},0)$).

Example 6.6. If the group G has no nontrivial parabolic germs, i.e., $G \cap \operatorname{Diff}_1(\mathbb{C},0) = \{\mathrm{id}\}$, then T is injective and hence G is necessarily commutative as a group isomorphic to a subgroup of the commutative group $\mathbb{C}^*$. Moreover, if G is analytically or formally linearizable, then each element g can be conjugated only with the linear germ $x \mapsto \nu_g x$, $\nu_g = Tg \in \mathbb{C}^*$, since the multiplicators of g and $h \circ g \circ h^{-1}$ necessarily coincide. Yet we wish to stress that being *algebraically* isomorphic to a subgroup of $\mathbb{C}^*$ (e.g., an infinite cyclic subgroup) is not sufficient for linearizability of the group, even on the formal level.

A simple *sufficient* condition for simultaneous linearizability (and hence commutativity) of a finitely generated group is its finiteness.

Theorem 6.7 (Bochner linearization theorem)**.** *Any finite subgroup $G \subseteq \operatorname{Diff}(\mathbb{C},0)$ can be linearized: there exists a biholomorphism $h \in \operatorname{Diff}(\mathbb{C},0)$ such that all germs $h \circ g \circ h^{-1}$ are linear,*

$$\forall g \in G \qquad h \circ g \circ h^{-1}(x) = \nu_g x, \qquad \nu_g = Tg \in \mathbb{C}^*. \tag{6.4}$$

Proof. Define the germ of the analytic *function* $h \in \mathcal{O}(\mathbb{C},0)$ by the formula

$$h = \sum_{g \in G} (Tg)^{-1} \cdot g$$

in any chart on $(\mathbb{C},0)$ (note that the addition makes sense only in $\mathcal{O}(\mathbb{C},0)$, but not in $\operatorname{Diff}(\mathbb{C},0)$). The germ h has the linear part $Th = \sum_g 1 = |G| \neq 0$ and is therefore invertible.

By the chain rule T, for any germ $f \in G$ we have

$$\begin{aligned} h \circ f = \sum_{g \in G} (Tg)^{-1} \cdot (g \circ f) = Tf \cdot \sum_{g \in G} \bigl(T(g \circ f)\bigr)^{-1} \cdot (g \circ f) \\ = Tf \cdot \sum_{g' \in G} (Tg')^{-1} \cdot g' = Tf \cdot h, \end{aligned}$$

which means that h conjugates f with the multiplication by $\nu_f = Tf$. $\square$

This linearization theorem implies a simple but useful corollary. Recall that for nonhyperbolic germs with multiplicators on the unit circle the problem of convergence of linearizing transformations is in general very difficult for the nonresonant case; see §5**E**. The resonant case turns out to be unexpectedly simple.

Theorem 6.8. *A* resonant *conformal germ* $f\colon z \mapsto \mu z + \cdots \in \operatorname{Diff}(\mathbb{C}, 0)$ *with* $\mu \in \exp 2\pi i\mathbb{Q}$*, is formally linearizable if and only if it is analytically linearizable.*

Proof. Only one direction of the equivalence is nontrivial. Assume that h is a formal germ linearizing the germ f. Since the multiplicator μ is a root of unity, $(h \circ f \circ h^{-1})^{\circ n} = h \circ f^{\circ n} \circ h^{-1} = \mathrm{id}$ for some finite order n. This means that the formal series h conjugates the holomorphic germ $f^{\circ n}$ with the identity. Yet the only *holomorphic map formally equivalent to identity* is the identity itself, hence $f^{\circ n} = \mathrm{id}$ and thus f is periodic (generates a finite group). By Theorem 6.7, f is analytically linearizable. □

One can replace finiteness of the group in the Linearization Theorem 6.7 by the assumption that *all elements of this group have finite order.*

Theorem 6.9. *A finitely generated subgroup of germs* $G \subset \operatorname{Diff}(\mathbb{C}, 0)$ *whose elements all have finite order, is analytically linearizable and finite, hence commutative and cyclic.*

Proof of Theorem 6.9. If the group is noncommutative, then it contains an element $\mathrm{id} \neq f \in \operatorname{Diff}_1(\mathbb{C}, 0)$ (cf. with Example 6.6). Such an element always has an infinite order in contradiction with our assumptions: if $f(z) = z + cz^{p+1} + \cdots$, $c \neq 0$, then $f^n(z) = z + nc\, z^{p+1} + \cdots \neq \mathrm{id}$. Thus G must be commutative.

A commutative group generated by finitely many elements of finite orders, is itself finite. By Theorem 6.7, the group G is analytically conjugate to a finite multiplicative subgroup of $\mathbb{C}^*$. All such subgroups are cyclic and generated by appropriate primitive roots of unity. □

6B. First steps of formal classification. In this subsection we study *formal* classification of finitely generated groups of conformal germs.

$6\mathbf{B}_1$. *Solvable and metabelian groups.* Recall that the commutator $[G, G]$ of an (abstract) group G is the group generated by all commutators of pairs of elements $[f, g] = f \circ g \circ f^{-1} \circ g^{-1}$; it is a subgroup in G. Moreover, since $T[f, g] = 1$, the commutator $[G, G]$ is a subgroup in $\operatorname{Diff}_1(\mathbb{C}, 0)$.

A group is *solvable*, if the decreasing chain of iterated commutators stabilizes on the trivial group:

$$\begin{gathered} G^0 \supseteq G^1 \supseteq G^2 \supseteq \cdots \supseteq G^{\ell-1} \supsetneq G^\ell = \{\mathrm{id}\}, \\ G^0 = G, \qquad G^{k+1} = [G^k, G^k],\ k = 0, 1, 2, \dots. \end{gathered} \tag{6.5}$$

If G is commutative (abelian), then $\ell = 1$. Solvable groups with $\ell = 2$ are called *metabelian*: their first commutators are commutative.

While for arbitrary groups the index ℓ may take any finite value, for subgroups of $\operatorname{Diff}(\mathbb{C}, 0)$ the only possibilities are $\ell = 0, 1$ (abelian and metabelian respectively) or $\ell = \infty$ (for nonsolvable groups). In other words, we have the following alternative.

Theorem 6.10 (Tits alternative for groups of conformal germs). *A finitely generated subgroup $G \subset \operatorname{Diff}(\mathbb{C}, 0)$ is either metabelian (commutative or noncommutative), or nonsolvable.*

To prove this result, we start with a simple computation (in part explaining, why the level is more convenient to deal with than the order of tangency with the identity).

Proposition 6.11. *For two germs of different levels $p \neq q$ their commutator has the level $p + q$. More specifically, if $f(z) = z + az^{p+1} + \cdots$, $g(z) = z + bz^{q+1} + \cdots$ with $p, q > 0$, then*

$$[f, g](z) = z + ab(p - q)z^{p+q+1} + \cdots. \tag{6.6}$$

Proof. The identity (6.6) is an assertion on the leading term of the germ of the *function* $\left(f \circ g \circ f^{-1} \circ g^{-1}\right)(z) - z \in \mathcal{O}(\mathbb{C}, 0)$ in any holomoprphic chart z. This leading term is not changed if we change the local coordinate from z to $t = f^{-1} \circ g^{-1}(z)$. In the new chart $z = (g \circ f)(t)$ and the leading term of the difference $(f \circ g)(t) - (g \circ f)(t)$ can be computed directly:

$$\begin{aligned} (f \circ g)(t) - (g \circ f)(t) &= t(1 + bt^q + \cdots)(1 + at^p(1 + bt^q + \cdots)^p + \cdots) \\ &\quad - t(1 + at^p + \cdots)(1 + bt^q(1 + at^p + \cdots)^q + \cdots) \\ &= ba\, pt^{p+q+1} - ab\, qt^{q+p+1} + \cdots. \end{aligned}$$

This proves (6.6). □

Remark 6.12. A similar (even easier) computation with $q = 0$ yields the following: if $g(z) = bz + \cdots$, $b \neq 1$, and f as above, then

$$[f, g](z) = z + a(b^p - 1)z^{p+1} + \cdots. \tag{6.7}$$

This computation immediately implies the following alternative for groups of conformal germs tangent to identity.

Lemma 6.13. *A finitely generated subgroup G of* $\operatorname{Diff}_1(\mathbb{C},0)$ *is either commutative or nonsolvable.*

Proof. If $G = G^0$ contains two germs of different *positive* levels $p \neq q$, $p,q > 0$ then it also contains the germ of level $p+q$ (again different from both p and q). Proceeding this way, we construct infinitely many germs of different levels, all belonging to the commutator $G_1 = [G,G]$. Thus G^1 also contains at least two germs of different levels which allows us to conclude that all iterated commutators $G^k = [G^{k-1}, G^{k-1}]$ are nontrivial.

If all germs in G are of the same level $p \geqslant 1$, then the group is in fact commutative. Indeed, in this case the commutator of any two germs $f,g \in G$, if nontrivial, must have the level *strictly greater* than p (again by (6.6)), which again leads to nonsolvability. Hence $[f,g]$ should be identity and the group G commutative. □

Theorem 6.10 is now one step away.

Proof of Theorem 6.10. For *any* group $G \subseteq \operatorname{Diff}(\mathbb{C},0)$ its commutator $G^1 = [G,G]$ belongs to $\operatorname{Diff}_1(\mathbb{C},0) = \ker T$ and therefore can be either trivial (and then G is commutative) or commutative (and then G is metabelian noncommutative) or nonsolvable (and then G is also nonsolvable), by Lemma 6.13. □

Remark 6.14. The same argument shows that if G is a subgroup of $\operatorname{Diff}(\mathbb{C},0)$ *disjoint* from $\operatorname{Diff}_1(\mathbb{C},0)$ (apart from the identical germ), then G is necessarily commutative, as $[G,G] \subseteq G \cap \operatorname{Diff}_1(\mathbb{C},0) = \{\mathrm{id}\}$; cf. with Example 6.6.

6**B**$_2$. *Centralizers and symmetries.* Solvable subgroups admit a rather accurate classification on the level of formal equivalence: unless formally linearizable, they are all formally equivalent to subgroups of (twisted) flows of certain nonhyperbolic vector fields. To establish this fact, we need a description of symmetries of parabolic germs.

A *centralizer* of an element g in a group G is the set $Z(g) \subseteq G$ of all elements $f \in G$ commuting with g: $Z(g) = \{f \in G \colon [f,g] = 0\}$. One can instantly verify that the centralizer is a subgroup of G, but in general this subgroup does not have to be commutative.

A parallel notion for the vector fields is a *symmetry*: a germ $g \in \operatorname{Diff}(\mathbb{C},0)$ is called a symmetry of a vector field $F \in \mathcal{D}(\mathbb{C},0)$ (interpreted as a derivation $\mathbf{F}$ of the algebra $\mathcal{O}(\mathbb{C},0)$), if $g^*\mathbf{F} = \mathbf{F}g^*$, in other words, if g transforms F into itself. We will (lacking a better term) call $g \in \operatorname{Diff}(\mathbb{C},0)$ an *orbital symmetry* of a vector field $F \in \mathcal{D}(\mathbb{C},0)$, if g conjugates F with its constant multiple λF, $\lambda \in \mathbb{C}^*$. The construction is identical in the formal context (i.e., for operators on the ring $\mathbb{C}[[z]]$).

If g is a symmetry of F, then g commutes with any flow map $f^t = \exp tF$. In general, mere commutativity of g and $f = \exp F$ is not sufficient for g to be a symmetry of F. Nevertheless, if f is parabolic, the inverse holds.

Recall (Theorem 3.17) that any parabolic germ $f \in \operatorname{Diff}_1(\mathbb{C}, 0)$ is *formally* embeddable: there exists a formal vector field $F \in \mathcal{D}[[\mathbb{C}, 0]]$ such that $f = \exp F$. Without loss of generality we may assume that F is brought to the formal normal form,

$$F = F_{p,a} = z^{p+1}(1 + az^p)\tfrac{\partial}{\partial z}, \qquad a \in \mathbb{C}, \quad p \in \mathbb{N}, \tag{6.8}$$

where p is equal to the level of f (Theorem 4.24).

Lemma 6.15. *If $g \in \operatorname{Diff}(\mathbb{C}, 0)$ is a symmetry of a parabolic germ or a formal series $f = \exp F \in \operatorname{Diff}_1(\mathbb{C}, 0)$, then g is also the symmetry of the field F.*

Proof. Let $\mathfrak{A}$ be the algebra of analytic germs $\mathcal{O}(\mathbb{C}, 0)$ or formal series $\mathbb{C}[[z]]$ respectively (depending on the context).

Consider the operators (automorphisms) $\mathbf{g}, \mathbf{f} \in \operatorname{Aut} \mathfrak{A}$, corresponding to the self-maps g and f, and denote by $\mathbf{F} \in \operatorname{Der} \mathfrak{A}$ the derivation corresponding to the field $F \in \mathcal{D}(\mathbb{C}, 0)$. If g is a symmetry of f, then $\mathbf{g}$ commutes with $\mathbf{f}$.

The derivation $\mathbf{F}$ can be restored from the isomorphism $\mathbf{f}$ by the formal logarithmic series (3.12),

$$\mathbf{F} = (\mathbf{f} - \mathrm{id}) - \frac{1}{2}(\mathbf{f} - \mathrm{id})^2 + \frac{1}{3}(\mathbf{f} - \mathrm{id})^3 \mp \cdots,$$

which stabilizes on the level of any finite order jets, since the difference $\mathbf{f} - \mathrm{id}$ is nilpotent; cf. with Theorem 3.14.

If $\mathbf{g}$ commutes with $\mathbf{f}$, then by the above identity $\mathbf{g}$ commutes also with $\mathbf{F}$, that is, the self-map g is a symmetry of the corresponding vector field F. □

Symmetries (generalized and orbital) of a nonhyperbolic vector field can be easily described. Without loss of generality we can consider only vector fields in the polynomial normal form (6.8).

Proposition 6.16. *A symmetry group of a vector field $F = F_{p,a}$ is the subgroup $G_{p,a} \subset \operatorname{Diff}(\mathbb{C}, 0)$ of the form*

$$G_{p,a} = \{b \cdot \exp tF_{p,a} \colon b \in \mathbb{C}^*, \ b^p = 1, \ t \in \mathbb{C}\} \cong \mathbb{Z}_p \times \mathbb{C}. \tag{6.9}$$

A nontrivial orbital symmetry g with $\lambda \neq 1$ may exist only if $a = 0$ (i.e., if the field is homogeneous), and then the orbital symmetry group is *the semi-direct product,*

$$G'_{p,0} = \{b \cdot \exp F_{p,0} \colon b \in \mathbb{C}^*,\ t \in \mathbb{C}\} \cong \mathbb{C}^* \rtimes \mathbb{C}. \tag{6.10}$$

Note that the groups $G_{p,a}$ and $G'_{p,0}$ indeed consist of symmetries (resp., orbital symmetries) of the field $F_{p,a}$ in the normal form (6.8). Thus the description given by Proposition 6.16, is exact.

Corollary 6.17. *The centralizer $Z(f)$ of a parabolic element $f \in \operatorname{Diff}_1(\mathbb{C},0)$ of level p in the group $\operatorname{Diff}(\mathbb{C},0)$ is formally equivalent to the group $G_{p,a} \cong \mathbb{Z}_p \times \mathbb{C}$ of germs of the form* (6.9).

Proof. This follows from Proposition 6.16 and Lemma 6.15. □

Corollary 6.18. *The centralizer of any parabolic element $f \in \operatorname{Diff}(\mathbb{C},0)$ is a commutative subgroup in* $\operatorname{Diff}(\mathbb{C},0)$. □

Remark 6.19. The orbital symmetry group $G'_{p,0}$ is solvable but nonabelian: the composition law for this group has the form

$$(b,t) \circ (b',t') = (bb', tb'^{-p} + t') \neq (b',t') \circ (b,t). \tag{6.11}$$

Yet the commutator $[G'_{p,0}, G'_{p,0}]$ consists of all flow maps and hence is commutative.

Proof of Proposition 6.16. Instead of the polynomial normal form (6.8), we will use the rational normal form

$$F'_{p,a} = \frac{z^{p+1}}{1 - az^p} \cdot \frac{\partial}{\partial z} \tag{6.12}$$

with the same $p \in \mathbb{N}$ and $a \in \mathbb{C}$: the fields $F_{p,a}$ and $F'_{p,a}$ are analytically equivalent; see Remark 4.25.

Let $g \in \operatorname{Diff}(\mathbb{C},0)$ be an analytic germ, given in some chart z by the germ of the function $w = g(z) \in \mathcal{O}(\mathbb{C},0)$. This germ will be an orbital symmetry of $F'_{p,a}$ if and only if the function $w(z)$ satisfies the ordinary differential equation

$$\frac{dw}{dz} \cdot \frac{z^{p+1}}{1 - az^p} = \lambda \cdot \frac{w^{p+1}}{1 - aw^p}. \tag{6.13}$$

This differential equation has separating variables and can be immediately integrated by reducing it to the Pfaffian form:

$$\frac{(1 - az^p)\,dz}{z^{p+1}} = \lambda \cdot \frac{(1 - aw^p)\,dw}{w^{p+1}}.$$

Note that the equality between two meromorphic 1-forms is possible only if their residues at the origin, equal to $-a$ and $-\lambda a$ respectively, coincide. Thus a nontrivial ($\lambda \neq 1$) orbital symmetry is possible only for a homogeneous vector field (with $a = 0$).

To find all genuine symmetries (with $\lambda = 1$), we integrate the above identity and obtain the equality

$$\frac{1}{pz^p} + a \ln z = \frac{1}{pw^p} + a \ln w - t, \tag{6.14}$$

where $t \in \mathbb{C}$ is a constant of integration. Replacing the germ g by another germ $g \circ (\exp tF)$, we can without loss of generality assume that the constant of integration is equal to zero, $t = 0$. Since the germ g is analytic, the solution w can be represented under the form $w(z) = z\,u(z)$, with an analytic nonvanishing function $u(\cdot)$. Substituting this ansatz into the above formula, we arrive at the identity

$$\frac{1}{pz^p}(1 - u(z)^{-p}) = a \ln u(z), \qquad u(0) \neq 0.$$

The right hand side is holomorphic at the origin, whereas the left hand side has a pole unless $u^p \equiv 1$, i.e., $u(z) \equiv b$ is a constant (root of unity). Then we necessarily have $a = 0$ and the map g must be linear (modulo a flow map, as mentioned above). □

6**B**$_3$. *Formal classification of solvable subgroups.* The formal classification of *cyclical* abelian groups coincides with that of their generators and was given in §4**I** (Theorem 4.26). The first nontrivial classification problem concerns *noncyclical* abelian groups.

Theorem 6.20. *A commutative group G which contains no nontrivial parabolic germs, is formally linearizable, i.e., formally equivalent to a subgroup of linear maps $\mathbb{C}^* \subset \operatorname{Diff}(\mathbb{C}, 0)$.*

Proof. If G contains a germ with a nonresonant multiplicator $\mu \notin \exp 2\pi\mathbb{Q}$, then such a germ is formally linearizable. By Remark 6.14, the group must be commutative, yet any germ commuting with the linear map $z \mapsto \mu z$ is itself linear, as it follows immediately from (6.7).

Thus the only remaining possibility is that $TG \subseteq \exp 2\pi i\mathbb{Q}$. But all such germs must be periodic, since their appropriate iteration powers must be parabolic. By Theorem 6.9, this group is analytically linearizable. □

We note that the multiplicative group $\mathbb{C}^*$ can be described in a way similar to (6.9) as the flow group of any *hyperbolic* germ of vector field, e.g., $F(z) = z$,

$$\mathbb{C}^* = \{g(z) = (\exp t) \cdot z \colon t \in \mathbb{C}\} \subset \operatorname{Diff}(\mathbb{C}, 0). \tag{6.15}$$

Theorem 6.21 (classification of abelian nonlinearizable groups). *If a finitely generated group G is commutative and contains a nontrivial parabolic element of some level p, then G is formally equivalent to a subgroup of the group $G_{p,a} \cong \mathbb{Z}_p \times \mathbb{C}$ as in (6.9) for some complex $a \in \mathbb{C}$.*

Proof of the theorem. Because of the commutativity of the group G, it must belong to the centralizer (in $\operatorname{Diff}(\mathbb{C},0)$) of its nontrivial parabolic element f which is described in Corollary 6.17. □

Theorem 6.22 (classification of noncommutative metabelian groups). *Any metabelian noncommutative group G is formally equivalent to a subgroup of the group $G'_{p,0}$ for some finite level p.*

Proof. 1. The parabolic subgroup $G_1 = G \cap \operatorname{Diff}_1(\mathbb{C},0)$ must be commutative by Lemma 6.13 and nontrivial by Remark 6.14. Therefore G_1 belongs to the centralizer (in $\operatorname{Diff}_1(\mathbb{C},0)$) of any its nontrivial element $f \in G_1$ and hence is formally equivalent to a subgroup of $\exp(\mathbb{C}F) = \{\exp tF : t \in \mathbb{C}\}$. Without loss of generality we assume from the very beginning that $G_1 \subseteq \exp(\mathbb{C}F)$, where F is a vector field in the formal normal form (6.8).

2. Since G is noncommutative, there exists another element $h \in G$ not commuting with f. Indeed, the centralizer of f in the bigger group $\operatorname{Diff}(\mathbb{C},0)$ is still *commutative* by Corollary 6.18. Since G is noncommutative, $G \setminus Z(f) \neq \varnothing$.

3. The subgroup $G_1 = G \cap \operatorname{Diff}_1(\mathbb{C},0)$ of parabolic elements of G is a normal subgroup, hence $h \circ G_1 \circ h^{-1} \subseteq G_1 \subseteq \exp(\mathbb{C}F)$. Thus we conclude that $f' = h \circ f \circ h^{-1} = \exp \lambda F$; by our choice of h, the constant λ is different from 1. In other words, h is a nontrivial orbital symmetry of the field F.

By the second assertion of Proposition 6.16, F must be homogeneous and h must belong to the subgroup $G'_{p,0}$ as in (6.10).

4. Any other element $h' \in G$ may either commute with f or not. In the first case by Corollary 6.17 we conclude that $h' \in G_{p,0} \subsetneqq G'_{p,0}$. In the second case $h' \in G'_{p,0}$ by the arguments of step 3 above. □

Remark 6.23. From the proof of Theorem 6.22 it immediately follows that a metabelian noncommutative group is *analytically* equivalent to a subgroup of $\mathbb{C} \cdot \exp(\mathbb{C}F_{p,0})$ for some p, if at least one parabolic germ from G is analytically embeddable.

6C. Integrable germs. Finitely generated groups may possess certain symmetry. Because of the intimate connections with the geometry of foliations, such groups are called *integrable.*

Definition 6.24. A *symmetry group* of the germ of an analytic function $u \in \mathcal{O}(\mathbb{C},0)$ is the subgroup $S_u = \{g \in \operatorname{Diff}(\mathbb{C},0) : u \circ g = u\}$ of holomorphisms preserving u.

Conversely, we say that an analytic germ u is the *first integral* of a group $G \subseteq \operatorname{Diff}(\mathbb{C},0)$, if $G \subseteq S_u$. The group G is said then to be *integrable.*

If G is cyclic and generated by a holomorphism g, then we say that u is a *first integral* of g. The germ g is *integrable* if it admits a nontrivial holomorphic first integral.

Proposition 6.25. *An holomorphism is periodic if and only if it is integrable.*

More precisely, $h \in \operatorname{Diff}(\mathbb{C}, 0)$ admits a first integral $u(z) = cz^m + \cdots$, $c \neq 0$, if and only if $h^k = \mathrm{id}$, where k divides m.

Proof. A periodic holomorphism h is linearizable by Theorem 6.7 and any linear map $x \mapsto \nu x$, $\nu^k = 1$, has the first integrals $u(z) = z^m$ for all m divisible by k (the case $m = 0$ is trivial and has to be excluded).

Conversely, if h is integrable and $u(z) = z^m + \cdots$ is the integral, then every level set $M_c = \{u(z) = c\} \subseteq (\mathbb{C}, 0)$ in a sufficiently small neighborhood of 0 consists of exactly m points that are permuted by h. By the Lagrange theorem, $h|_{M_c}$ is of period $k = k(c)$ that divides m. Let k be the minimal value such that the set of k-periodic points is infinite. Then the kth iterate of h is identity by the uniqueness theorem. □

From this proposition and Theorem 6.9 we immediately derive the following necessary condition of integrability.

Corollary 6.26. *An integrable group is finite cyclic (commutative).* □

Remark 6.27. Any germ of a holomorphic function $u(z) = cz^m + \cdots$ of finite order m admits a cyclic symmetry group of order m. The group is generated by the germ of a self-map f which is the linear rotation by the primitive root of unity of order m in the holomorphic chart $w = z \cdot (c + \cdots)^{1/m}$, in which the function itself becomes a monomial.

* * *

Thus far we concentrated on commutative (finite or infinite) and metabelian groups, which are relatively tame. As was already shown, they admit simple formal classification based on the formal type of a single nontrivial parabolic element from the group. Topological classification of solvable groups is also relatively simple and can be derived from Theorem 21.2 (see §21) which claims that the only topological invariant of a parabolic germ is its level. The analytic classification of solvable groups of germs can be reduced to that of the nontrivial parabolic element as above. The corresponding analytic theory is developed in §21 and involves *nonpolynomial normal forms*; see Chapter IV. In summary,

(1) dynamics of solvable groups is relatively simple, in particular,

(2) they have no limit cycles, and

(3) their analytic classification is much finer than the formal one, and the latter in turn is finer than topological classification.

For nonsolvable groups all of these properties fail. In the remaining part of this section we will show that a *generic* (nonsolvable) finitely generated group:

(1) has dense orbits, among which

(2) there exist countably many (properly defined) complex limit cycles. Moreover,

(3) generic groups are *rigid*: two such groups can be topologically equivalent if and only if they are analytically equivalent.

These phenomena will again manifest themselves for singular holomorphic foliations on $\mathbb{P}^2$: the subject will be treated in detail in §28.

The term *generic* in application to finitely generated pseudogroups will mean the following. We fix the number n (usually 2 or more) of generating germs and say that a certain property is *generic*, if it holds for all n-tuples of germs whose jets of some finite order r belong to a "massive" (say, open dense or full measure) subset of the total jet space $\bigoplus_{n \text{ times}} J^r(\mathbb{C}, 0)$.

Example 6.28. A generic group with $n \geqslant 2$ generators is noncommutative and, moreover, nonsolvable.

Indeed, both generators generically are hyperbolic (their multiplicators are off the unit circle). Since the above definition of genericity does not depend on the choice of the chart, without loss of generality, we may assume that one of the generators, f_1, is linear hyperbolic. The group will be noncommutative if the second generator f_2 in this chart is nonlinear (the second Taylor coefficient of f_2 is nonzero).

The commutator $h = [f_1, f_2]$ will be a parabolic element which is generically of level 1 (i.e., tangent to identity with a quadratic nonlinearity). Another parabolic element of level 1 is the commutator $[f_1, h]$. One can show that generically $[[f_1, h], h]$ will be nonzero and hence, by (6.6), have level 2 or more, which would imply nonsolvability.

Usually we will omit routine checks that a certain collection of requirements is fulfilled for a generic finite generated group: in more details various properties determined by finite or infinite order jets, will be discussed in §10, where the notion of *decidable* properties will be introduced.

6D. Dynamics generated by finitely generated groups of germs: pseudogroups. We need first to introduce a proper language for describing *dynamical properties* of finitely generated groups of conformal germs.

If a group G acts (in an abstract manner) on a space X, then the *orbit* of a point $x \in X$ is defined as the subset $G(x) = \{g \cdot x \colon g \in G\} \subseteq X$. However, if the elements of the group are not defined on the whole space X, then the definition of an orbit requires appropriate modification.

This caveat is especially important when G is the holonomy group of a holomorphic foliation. By the very definition of holonomy, if a point $a \in \tau$ on the cross-section belongs to the domain of the holonomy map Δ_γ, then the points a and $b = \Delta_\gamma(a)$ belong to the same leaf of the foliation. Thus orbits of the holonomy group understood as images of all well-defined holonomy maps, describe intersection of leaves of the foliation with a fixed cross-section.

We introduce a relaxed notion of a *pseudogroup* which differs from a group by the fact that the composition is not always defined. For our purposes it is sufficient to define pseudogroups of holomorphic maps whose domains are open subsets of $\mathbb{C}$ containing a common fixed point (the origin); the modification for the general case can be made following the same lines.

Definition 6.29. Let U be a neighborhood of the origin in $\mathbb{C}$ and $G \subseteq \operatorname{Diff}(\mathbb{C}, 0)$ an *arbitrary* subgroup of the group of germs. A *pseudogroup* $\varGamma$ associated with G is a collection of pairs (f_α, U_α), indexed by some index set $\alpha \in A$, such that $U_\alpha \subseteq U$ is an open set containing the origin, $f_\alpha \colon U_\alpha \to U$ is a holomorphic map defined (at least) in U_α and the group G consists of the germs at the origin of all maps f_α from the pseudogroup $\varGamma$.

Composition of two elements (f_α, U_α) and (f_β, U_β) is defined as the pair $(f_\alpha \circ f_\beta, U_{\alpha\beta})$ if and only if $U_{\alpha\beta} \subseteq U_\beta$ and $f_\beta(U_{\alpha\beta}) \subseteq U_\alpha$.

In other words, each conformal germ $\widehat{f} \in G$ (in particular, the neutral element $\widehat{\mathrm{id}} \in G$) is represented by many different maps f_α with different, in general, domains (of course, the maps coincide on the pairwise intersections of their domains).

The natural way to associate a pseudogroup $\varGamma$ with any *finitely generated* group $G = \langle \widehat{f}_1, \dots, \widehat{f}_r \rangle \subset \operatorname{Diff}(\mathbb{C}, 0)$ is as follows (we temporarily use the hats to distinguish between germs at the origin and holomorphic maps). Choose any collection of *representatives* $f_j^\pm \colon U_j \to \mathbb{C}$, $j = 0, \dots, r$, of the germs $\widehat{f}_1^{\pm 1}, \dots, \widehat{f}_r^{\pm 1}$ generating G. Then with an arbitrary word $w = (w_{j_n}^\pm w_{j_{n-1}}^\pm \cdots w_{j_2}^\pm w_{j_1}^\pm) \in \mathfrak{F}_r$ (an element in the free group on r symbols, written from right to left) we can associate the conformal map f_w as the composition $f_{j_n}^\pm \circ f_{j_{n-1}}^\pm \circ \cdots \circ f_{j_2}^\pm \circ f_{j_1}^\pm$ defined in the maximal domain U_w on which *all partial compositions*

$$f_{j_1} = f_{j_1}^\pm, \quad f_{j_2 j_1} = f_{j_2}^\pm \circ f_{j_1}^\pm, \quad \dots, \quad f_{j_n \dots j_1} = f_{j_n}^\pm \circ \cdots \circ f_{j_2}^\pm \circ f_{j_1}^\pm$$

are well defined. Associating this domain (obviously, open and containing the origin) with the map $f_{j_n \cdots j_1} = f_w$ representing the respective germ $\widehat{f}_w$, we obtain a pseudogroup. Choosing a different collection of the initial domains $U_1, \dots, U_r$ formally results in a different pseudogroup, though most properties would not be affected.

If there are nontrivial identities in the group G, then the same germ admits several representatives with eventually different domains. To distinguish between such elements, we will remember together with each element $(f_\alpha, U_\alpha) \in \Gamma$ of the pseudogroup Γ the corresponding word w_α in the free group $\mathfrak{F}_r$. The corresponding collection of triples

$$\{(f_w, U_w, w) \colon w \in \mathfrak{F}_r,\ f_w \in \mathcal{O}(U_w)\} = \Lambda_G$$

will be called the *pseudogroup associated with the finitely generated group of conformal germs*. A triple (element of the pseudogroup) is *nontrivial*, if the corresponding word w is nontrivial in $\mathfrak{F}_r$, even if $f_w = \operatorname{id}|_{U_w}$. Yet in most cases we will omit the third component to simplify the notation.

Remark 6.30. In order to avoid technical problems, we will always assume that if (f, U) belongs to a pseudogroup Γ, then all restrictions $(f|_V, V)$ for $V \subseteq U$, also belong to Γ.

For a pseudogroup Γ the notion of an orbit of a point can be introduced without any complications. A "periodic" orbit is naturally called a cycle.

Definition 6.31. The orbit of a point $x \in U$ by a pseudogroup Γ is the set $\Gamma(x)$ of all points $f_\alpha(x)$ for all elements $(f_\alpha, U_\alpha) \in \Gamma$ such that $x \in U_\alpha$.

Definition 6.32. The point $x \neq 0$ is called a *cycle*, if it is fixed by a nontrivial element (f_α, U_α) of the pseudogroup, i.e., $x \in U_\alpha$ and $f_\alpha(x) = x$ (thus for a cyclic group *all* points are cycles). The cycle is *limit* (in full, a complex limit cycle of a pseudogroup), if x is an *isolated* fixed point of f_α in U_α.

The notion of equivalence of groups of conformal germs translates naturally into equivalence of pseudogroups. Two pseudogroups Γ, Γ' are equivalent if there exists a conformal biholomorphism $h \colon (U, 0) \to (U', 0)$ such that Γ' consists of all pairs $\bigl(h \circ f_\alpha \circ h^{-1}, h(U_\alpha)\bigr)$ such that $(f_\alpha, U_\alpha) \in \Gamma$ (subject to the technical convention from Remark 6.30). Clearly, equivalent pseudogroups have identical dynamical properties.

6E. Periodic orbits and periodic germs. To illustrate the usefulness of the notion of a pseudogroup, we establish simple dynamic properties of periodic (and aperiodic) germs.

Periodicity of a germ $\widehat{g} \in \operatorname{Diff}(\mathbb{C}, 0)$ (meaning that $\widehat{g}^n = \mathrm{id}$) implies that all g-orbits are periodic (cycles) for any representative g of $\widehat{g}$. The inverse statement is less obvious.

Let $\widehat{g} \in \operatorname{Diff}(\mathbb{C}, 0)$ be a conformal germ that admits a representative g defined in an open set V containing the origin. For any set $U \subseteq V$ consider the restriction $g|_U$ and the "cyclical" pseudogroup Γ_U generated by the element (g, U). For an arbitrary point $x \in U$ denote by $\Gamma(x|U)$ the Γ_U-orbit of the point x: by definition,

$$\Gamma(x|U) = \{g^n(x) \colon n \in \mathbb{Z}, \text{ and for all } k \text{ between } 0 \text{ and } n,\ g^k(x) \in U\}.$$

The orbit may be finite, in which case it consists of the consecutive iterates

$$g^{-n}(x), g^{-n+1}(x), \dots, g^{-1}(x), x, g(x), \dots, g^{m-1}(x), g^m(x)$$

for some $n, m \geqslant 0$, or infinite in one or both directions. We consider only *maximal* orbits, i.e., assume that $g^{-n-1}(x)$ and $g^{m+1}(x)$ already do not belong to U if n (resp., m) is finite. Note that the infinite orbit may consist of finitely many *distinct* points (if and only if the orbit is periodic).

Consider the integer-valued function $\nu(x)$ defined as the length of the maximal orbit,

$$\nu(x) = \nu(x|U) = \max\{m + n \colon g^{-n}(x), \dots, x, \dots, g^m(x) \in U\}. \tag{6.16}$$

If the orbit is infinite, we set $\nu(x) = +\infty$. By construction, *the function ν is constant along orbits of g.*

The continuity of g implies semicontinuity of the function ν: if U is open and $\nu(x) < +\infty$, then for all $y \in U$ sufficiently close to x, $\nu(y) \geqslant \nu(x)$. Conversely, if U is a closed subset of V and $\nu(x)$ is finite, then for all $y \in U$ sufficiently close to x, $\nu(y) \leqslant \nu(x)$. In the latter case if x is a point of discontinuity for ν, then the orbit $\Gamma(x|U)$ intersects the boundary ∂U. All these properties are immediate (Exercise 6.2).

Lemma 6.33. *If the germ $g \in \operatorname{Diff}(\mathbb{C}, 0)$ is aperiodic, i.e., if the cyclic group $G = \{g^{\mathbb{Z}}\}$ is infinite, then for any small open domain $U \ni 0$ there are uncountably many infinite aperiodic orbits $\Gamma(x|U)$.*

Proof. Consider an arbitrary *closed* circular disk $D_\rho = \{|x| \leqslant \rho\}$ and its boundary circle $K_\rho = \partial D_\rho$, $\rho > 0$.

1. We prove that there are uncountably many points on D_ρ with infinite orbits in D_ρ. To that end, we will show that on each circle K_r, $r \leqslant \rho$, there is at least one point with an infinite orbit in $D_r \subseteq D_\rho$. Since the number of different circles which can intersect any given orbit is at most countable, this will prove that the number of infinite orbits in uncountable.

Assume that all points on the circumference K_r have finite orbits in D_r, i.e., the corresponding length function $\nu(\cdot) = \nu_{D_r}(\cdot)$ takes only finite values on K_r. Since K_r is compact, this means that ν is bounded from above on K_r, and all orbits intersecting the boundary, are finite and no longer than some finite number $N \in \mathbb{N}$.

On the other hand, since $g(0) = 0$, the orbit of the origin $x = 0$ is infinite and $\nu(0) = +\infty$. Because of the semicontinuity of ν on the connected disk D_r, the function ν must have a discontinuity point $y \in D_r \smallsetminus K_r$ with the value $\nu(y)$ strictly greater than N somewhere in the interior of D_r. Yet this means that the orbit $\Gamma(y|D_r)$ which is longer than N, intersects the boundary K_r. Since ν is constant along orbits, this contradicts the choice of N as the upper bound of ν on K_r.

2. To complete the proof of the lemma, note that the set of points with infinite orbits is the union of periodic points and the infinite aperiodic orbits. For each finite n, the n-periodic points inside D_r are roots of the equation $g^n(x) - x = 0$ and form a finite subset of D_r by the uniqueness theorem for the analytic germ g^n. The union of these finite sets is at most countable. Therefore the complement, the union of infinite aperiodic orbits in D_r, is uncountable. □

Thus we have the following alternative.

Theorem 6.34. *Any finitely generated group $G \subset \operatorname{Diff}(\mathbb{C}, 0)$ is either integrable, or any pseudogroup associated with G has uncountably many infinite aperiodic orbits.*

Proof. If G includes an aperiodic germ g, then this germ has uncountably many aperiodic orbits by Lemma 6.33. Conversely, if all elements of G are of finite order, then by Theorem 6.9 the group is finite and cyclic, hence linearizable. Its integrability follows from Proposition 6.25. □

6F. Closure of a pseudogroup and density of orbits. Once a group of conformal germs is replaced by the pseudogroup, one can define the notions of convergence, closure, *etc.*

Definition 6.35. A sequence of elements $\{(f_j, U_j)\}_{j=1}^{\infty}$ of a pseudogroup Γ *converges* to an element $(g_*, U_*) \in \Gamma$, if $U_* \subseteq U_j$ for all j (starting from some number) and the restrictions $f_j|_{U_*}$ converge uniformly to g_* there.

The *closure* of a pseudogroup Γ is the collection of all limits of converging sequences of elements from Γ.

Clearly, the closure is again a pseudogroup, denoted by $\overline{\Gamma}$. The following statement is proved by the standard approximation arguments.

Proposition 6.36. *Let Γ be a pseudogroup of conformal maps and $\overline{\Gamma}$ its closure. If the orbit $\overline{\Gamma}(x)$ of some point is dense in an open domain U, then the orbit $\Gamma(x)$ of the initial pseudogroup is also dense there.* □

This proposition is especially useful when the closure of a pseudogroup contains a sub-pseudogroup with dense orbits. This happens, as we will show, when the group of germs G contains a pair of hyperbolic germs with the multiplicators generating a dense lattice of points in $\mathbb{C}$.

By Schröder–Kœnigs Theorem 5.18, a hyperbolic germ is always linearizable: there exists a biholomorphism h conjugating g with the linear map $x \mapsto \mu x$. Replacing the pseudogroup Γ by an equivalent one, we may assume from the very beginning that Γ contains a linear hyperbolic map.

Consider again the multiplicator homomorphisms $T\colon G \to \mathbb{C}^*$ and $T\colon \Gamma \to \mathbb{C}^*$ mapping each germ $f_\alpha \in G$ (resp., element $(f_\alpha, U_\alpha) \in \Gamma$) into its multiplicator at the origin (defined independently of the choice of a chart). Denote the image of this application by Λ_G or Λ_Γ respectively: this is a multiplicative subgroup of $\mathbb{C}^*$.

Theorem 6.37. *If the pseudogroup $\Gamma = \{(f_\alpha, U_\alpha)\}$ contains a linear hyperbolic map $(\mu_0 x, U_0)$, $|\mu_0| \neq 1$, then the closure $\overline{\Gamma}$ contains also all linear maps $(\mu_\alpha x, \frac{1}{2}U_\alpha)$ for all $\mu_\alpha = df_\alpha(0)/dx \in \Lambda_\Gamma$.*

Proof. Since Γ is a pseudogroup associated with a *group* G which together with each germ contains its inverse, without loss of generality, we may assume that the multiplicator μ_0 denoted for brevity by μ, is *contracting* as a linear map, $|\mu| < 1$.

Let $(g, V) \in \Gamma$ be an arbitrary element. By Remark 6.30 we may assume that V is a circular disk, so that $\mu V \subset V$. Hence all the elements (g_n, V), $g_n = \mu^{-n} \circ g \circ \mu^n$, belong to Γ for all $n \geqslant 0$ (i.e., these compositions are all defined in V).

Expanding g into the Taylor series converging in V as $g(x) = \sum_{k=1}^{\infty} a_k x^k$, $a_1 = \lambda$, we conclude that the kth Taylor coefficient of g_n is $a_k \mu^{(k-1)n}$. As $n \to +\infty$, this tends to zero for all $k \geqslant 2$, which means that the elements g_n converge uniformly to the linear map $(\lambda x, V')$ for any $V' \Subset V$. □

In the future we will often require the following condition imposed either on groups of conformal germs, or on the associated pseudogroups of maps.

Definition 6.38. A finitely generated group $G \subset \operatorname{Diff}(\mathbb{C}^1, 0)$ (resp., a pseudogroup Γ) satisfies the *density condition* if the multiplicative subgroup Λ_G (resp., Λ_Γ) generated by multiplicators of all germs (resp., maps) is dense in the multiplicative group $\mathbb{C}^*$:

$$\overline{\Lambda_G} = \mathbb{C} \supset \mathbb{C}^*, \qquad \text{resp.,} \qquad \overline{\Lambda_\Gamma} = \mathbb{C} \supset \mathbb{C}^*. \tag{6.17}$$

Example 6.39 (Closed subgroups of $\mathbb{C}^*$). Consider the exponential map $\exp 2\pi i\colon \mathbb{C} \to \mathbb{C}^*$, $\lambda \mapsto \exp(2\pi i\lambda)$, which is a topological nonramified covering, and finitely generated (multiplicative) subgroup $\subseteq \mathbb{C}^*$. The preimage of

G by this map is a *lattice of points* L, a $\mathbb{Z}$-module in $\mathbb{C} \cong \mathbb{R}^2$, which always contains the unity (and hence all the integers $\mathbb{Z}$): this lattice is generated by 1 and logarithms $\lambda_j = \frac{1}{2\pi i} \ln \mu_j$, $j = 1, \dots, n$, of the generators $\mu_1, \dots, \mu_n$ of G. Obviously, G is dense in $\mathbb{C}^*$ if and only if L is dense in $\mathbb{C}$.

The *closed* lattices of points in $\mathbb{C}$ can be easily described: they can be only discrete (i.e., contain 0 as an isolated point) isomorphic to $\mathbb{Z}+\lambda_1\mathbb{Z}$, the union of parallel translates of a line $\mathbb{Z}+\lambda_1\mathbb{R}$ or $\mathbb{R}+\lambda_1\mathbb{Z}$, and the whole plane $\mathbb{C}$. The latter case is generic if the number of generators is three or more (it is sufficient to demand that the three generators are nonresonant, i.e., $n_0 + n_1\lambda_1 + n_2\lambda_2$ never represents zero for a nontrivial choice of coefficients n_0, n_1, n_2). This implies that generically a multiplicative subgroup generated by two generators $\mu_1, \mu_2 \in \mathbb{C}^*$ is dense.

Other types of closed sublattices of points in $\mathbb{C}$ produce the following closed subgroups of $\mathbb{C}^*$ (for the sake of completeness we include $\mathbb{C}^*$ as well):

(1) $\mathbb{C}^*$ (the entire group);

(2) $\mathbb{Z}_p \times \mathbb{R}_+^*$, $1 \leqslant p < \infty$ (finite number of spirals that eventually degenerate into rays);

(3) $2^{\mathbb{Z}} \times \mathbb{T}$ (infinite many circles with the radii forming a geometric progression);

(4) $\mathbb{Z}_p \times 2^{\mathbb{Z}}$ (finite number of complex geometric progressions).

In this list $2^{\mathbb{Z}}$ stands for the infinite cyclic multiplicative subgroup of $\mathbb{C}^*$, $\mathbb{T} = \{|\mu| = 1\} \cong \mathbb{R}/\mathbb{Z}$ is the unit circle (considered as the a multiplicative group) and $\mathbb{Z}_p \subset \mathbb{T}$ is the group of roots of unity of degree p.

Example 6.39 explains why the density condition (6.17) is *generic*: for any $n \geqslant 2$ the tuple of germs $(f_1, \dots, f_n)$, $f_j \in \operatorname{Diff}(\mathbb{C}, 0)$ such that the group generated by $f_j^{\pm 1}$ satisfies (6.17), form a dense subset in the space of all tuples $\bigl(\operatorname{Diff}(\mathbb{C}, 0)\bigr)^n$.

Corollary 6.40 (Density theorem for generic pseudogroups). *If a pseudogroup Γ satisfies the density condition* (6.17), *then for any $x \in U$ there exists a small neighborhood V such that the orbit $\Gamma(x)$ is dense in V.*

Proof of the corollary. If two multiplicators generate a dense subgroup of $\mathbb{C}^*$, then both of them should be hyperbolic (off the unit circle), hence the closure of Γ contains two linear maps $x \mapsto \mu_j x$, $j = 1, 2$, defined in two disks U_1, U_2.

By Proposition 6.36, to show density of orbits of the initial pseudogroup Γ it is sufficient to prove that the orbits of the pseudogroup generated by the two linear maps are dense.

Given any linear map $x \mapsto \mu x$, we can find a product $\prod \mu_{1,2}^{\pm 1}$ approximating μ with any degree of accuracy by (6.17). Using commutativity of $\mathbb{C}^*$, we can always rearrange the factors in this product so that all contracting terms come first and the expanding terms after. This would guarantee that the composition of the respective linear maps is well defined in the pseudogroup. □

6G. Abundance of limit cycles for generic pseudogroups. Under the density assumptions one can show that a *noncommutative* pseudogroup has infinitely many distinct complex limit cycles accumulating to the origin.

Theorem 6.41. *A noncommutative finitely generated pseudogroup of conformal maps meeting the density condition, possesses infinitely many limit cycles accumulating to the origin.*

Both assumptions of the theorem (density condition and noncommutativity) are obviously generic.

Proof. Consider the maps of the pseudogroup Γ in the canonical chart linearizing one (hence all) hyperbolic germs belonging to it. Being noncommutative, Γ contains a nonidentical map (f, U) with multiplicator 1. Rescaling the canonical chart, we may assume that U is large enough to contain the unit disk: $\mathbb{D} = \{|z| \leqslant 1\} \Subset U$, $f(0) = 0$, $f(z) - z \not\equiv 0$. The ratio $f(z)/z$ is a nonconstant holomorphic function that takes at least two distinct values, 1 (at the origin) and $\mu \neq 1$ at some other point a. Without loss of generality we may assume that $|a| < \frac{1}{3}$.

By Theorem 6.37 and the density condition (6.17), the closure of the pseudogroup contains the linear map $g(z) = \mu z$ defined on $\frac{1}{2}D$, i.e., there exists an element $h \in \Gamma$ of the pseudogroup approximating g with arbitrary high accuracy on $\frac{1}{2}\mathbb{D}$. The function $f(z) - g(z)$ has at least two isolated roots ($z = 0$ and $z = a$) in $\frac{1}{2}\mathbb{D}$; by the argument principle, $f(z) - h(z)$ has at least the same number of roots in this domain if the approximation is accurate enough. In other words, the map $f^{-1} \circ h$ has at least two isolated fixed points in $\frac{1}{2}\mathbb{D}$, one at the origin, the other elsewhere. The latter point is the limit cycle.

Clearly, this construction can be repeated with U replaced by an arbitrarily small neighborhood of the origin. This shows that limit cycles accumulate to the origin, as asserted by the theorem. □

Remark 6.42. Theoretically, all limit cycles constructed in the proof of Theorem 6.41, can belong to a single orbit of the pseudogroup.

6H. Rigidity of finitely generated groups of conformal germs. The term "rigidity" will repeatedly appear in this book in connection with various phenomena sharing a common feature that "weaker equivalence implies stronger equivalence". Rigidity appears in the context where we consider objects for which there is a hierarchy of equivalence relations of various strength (topological, differentiable, holomorphic). An object is rigid when any weaker equivalence between it and any other object means that the two objects are strongly equivalent.

One rigidity-type result was already observed in Remark 4.6, when weaker (formal) equivalence automatically implies analytic equivalence.

In general, however, rigidity deals with an interplay between topological and stronger (smooth, analytic, *etc.*) classifications. In this more restricted context rigidity means that there is no way to change the weak structure (topology) in a nontrivial way (i.e., without changing all other, finer structures of the object).

Example 6.43. The sphere is rigid in the class of Riemann surfaces: any other Riemann surface topologically equivalent to the sphere, is conformally equivalent to it.

All the way around, the tori are not rigid: a two-dimensional torus $\mathbb{T}^2$ has conformal invariants.

The simplest rigidity-type property can be observed for finitely generated groups of conformal germs.

Definition 6.44. A finitely generated group of germs $G \subset \operatorname{Diff}(\mathbb{C},0)$ is called *rigid*, or *topologically rigid*, if any germ of a homeomorphism h topologically conjugating G with another group $G' \subset \operatorname{Diff}(\mathbb{C},0)$ is necessarily conformal, $h \in \operatorname{Diff}(\mathbb{C},0)$.

Sufficient conditions for rigidity are the same as for Theorem 6.41 on abundance of limit cycles.

Theorem 6.45 (Rigidity theorem for groups of conformal germs). *A noncommutative finitely generated pseudogroup Γ of conformal maps meeting the density condition, is rigid.*

Moreover, if Γ_t' is a family of pseudogroups analytically depending on a complex parameter $t \in U \subset \mathbb{C}^p$ and topologically equivalent to the pseudogroup Γ with the above listed properties for all t, then there exists a holomorphic conjugacy $h_t\colon (\mathbb{C}^1,0) \to (\mathbb{C}^1,0)$ between Γ_t' and Γ which analytically depends on t.

We start the proof of Theorem 6.45 by analyzing topological conjugacies between dense subgroups of the commutative multiplicative group $\mathbb{C}^*$.

Proposition 6.46. *Let $G, G' \subseteq \mathbb{C}^*$ be two finitely generated dense subgroups topologically conjugated by a homeomorphism $h\colon (\mathbb{C},0) \to (\mathbb{C},0)$ preserving the orientation.*

Then $h(z) = cz\,|z|^\beta$ for some complex numbers $c \in \mathbb{C}^*$ and $\beta \in \mathbb{C}$.

Proof. The topological conjugacy between the groups means that there exists a multiplicative group isomorphism $A\colon G \to G' \subseteq \mathbb{C}^*$ and a homeomorphism $h\colon (\mathbb{C},0) \to (\mathbb{C},0)$ such that

$$h(\mu z) = A(\mu)h(z), \qquad \forall z \in (\mathbb{C},0), \quad \forall \mu \in G. \tag{6.18}$$

1. We claim first that $A(\mu) = \mu\,|\mu|^\beta$ for some complex value of $\beta \in \mathbb{C}$.

The automorphism A satisfies the multiplicativity condition

$$A(\mu\nu) = A(\mu)A(\nu), \qquad \forall \mu,\nu \in G, \text{ hence } A(1) = 1, \tag{6.19}$$

and the functional equation (6.18) implies immediately that both A and its inverse are *continuous* as complex-valued functions of μ. Thus without loss of generality we may assume that A (resp., A^{-1}) is defined on the closed subgroups $\overline{G}$, resp., $\overline{G'}$, i.e., $A\colon \mathbb{C}^* \to \mathbb{C}^*$ is a homeomorphism.

2. Intuitively, the functional equation (6.19) becomes additive after passing to logarithms. Yet since the logarithm is multivalued, one has to exercise some extra care. In the special case where $\overline{G} = \mathbb{C}^*$, one can choose the *continuous* branch of logarithm. More precisely, there exists a continuous complex function $\widehat{A}\colon \mathbb{C} \to \mathbb{C}$ which covers the automorphism $A\colon \mathbb{C}^* \to \mathbb{C}^*$ by the exponential map:

$$A(\exp 2\pi i w) = \exp 2\pi i \widehat{A}(w), \qquad w \in \mathbb{C}. \tag{6.20}$$

If normalized by the condition $\widehat{A}(0) = 0$, it becomes uniquely defined and the multiplicative identity (6.19) implies that $\widehat{A}$ is additive modulo $\mathbb{Z}$,

$$\widehat{A}(\lambda + \lambda') = \widehat{A}(\lambda) + \widehat{A}(\lambda') \bmod \mathbb{Z} \qquad \forall \lambda, \lambda' \in \mathbb{C}.$$

The integer number $N(\lambda,\lambda') = \widehat{A}(\lambda+\lambda') - \widehat{A}(\lambda) - \widehat{A}(\lambda')$ is a continuous, hence constant function of $(\lambda,\lambda') \in \mathbb{C}^2$: because of the normalizing condition $\widehat{A}(0) = 0$ we have $N(0,0) = 0$ and therefore the application $\widehat{A}$ is truly additive,

$$\widehat{A}(\lambda + \lambda') = \widehat{A}(\lambda) + \widehat{A}(\lambda'), \qquad \forall \lambda, \lambda' \in \mathbb{C}. \tag{6.21}$$

This additivity implies that $m \cdot \widehat{A}(\frac{1}{m}\lambda) = \widehat{A}(\lambda)$ for any natural m; since $\widehat{A}$ is one-to-one, we conclude that $\widehat{A}(\frac{1}{m}\lambda) = \frac{1}{m}\widehat{A}(\lambda)$ and therefore $\widehat{A}$ is $\mathbb{Q}$-linear map of $\mathbb{C}$ into itself. Finally, the continuity of $\widehat{A}$ means that $\widehat{A}$ is an $\mathbb{R}$-linear automorphism of $\mathbb{C} \cong \mathbb{R}^2$. Any such automorphism necessarily has the form

$$\widehat{A}(\lambda) = a\lambda + b\overline{\lambda}, \qquad \text{for some } a, b \in \mathbb{C}, \ |a| \neq |b|. \tag{6.22}$$

The orientation is preserved if $|a| > |b|$ and reverted otherwise.

The $\mathbb{R}$-linear map $\widehat{A}$ covers the multiplicative map A by the logarithmic covering $\lambda \mapsto \mu = \exp 2\pi i\lambda$. Therefore $\widehat{A}(1)$ must be an integer number $n \in \mathbb{Z}$. This means that

$$\widehat{A}(\lambda) = n\lambda + \tfrac{1}{2}\beta(\lambda - \overline{\lambda}), \tag{6.23}$$

for some $\beta \in \mathbb{C}$, and hence

$$A(\mu) = \exp 2\pi i \left\{ n \cdot \frac{\ln \mu}{2\pi i} + \frac{\beta}{2}\left[\frac{\ln \mu}{2\pi i} - \overline{\left(\frac{\ln \mu}{2\pi i}\right)}\right]\right\}$$
$$= \mu^n \exp \tfrac{\beta}{2}\left[\ln \mu + \ln \bar{\mu}\right] = \mu^n\,|\mu|^\beta.$$

It remains to notice that by (6.18) the map A must be an orientation-preserving homeomorphism, which leaves only one possibility, $n = 1$. This proves the equality $A(\mu) = \mu\,|\mu|^\beta$.

3. If the homeomorphism h solving the functional equation (6.18) is represented under the form $h(z) = z\,|z|^\beta f(z)$ with the same β as before and some complex-valued function f continuous on $(\mathbb{C}^*, 0)$, then from the functional equation (6.18) we obtain after cancellation of all terms the trivial "functional equation" on f,

$$f(\mu z) = f(z), \qquad \forall \mu \in G,\ z \in (\mathbb{C}, 0).$$

Since G is dense in $\mathbb{C}^*$, we conclude that $f(z) \equiv c \neq 0$ must be a constant. This completes the proof of the proposition. □

Remark 6.47. Passing to a different chart in the preimage or the image, one can always assume that $c = 1$.

Remark 6.48. If h is a homeomorphism *reverting* the orientation and topologically conjugating G with G' as in (6.18), then $A(\mu) = \bar{\mu}\,|\mu|^\beta$ and $h(z) = c\bar{z}\,|z|^\beta$. This corresponds to the case $n = -1$ in (6.23). To prove that, it is sufficient to replace $h(z)$ by $\bar{h}(z) = \overline{h(\bar{z})}$ which is an orientation-preserving homeomorphism conjugating the two groups $\bar{G}, \bar{G}' \subseteq \mathbb{C}^*$ obtained from G and G' by conjugation with the mirror symmetry $z \mapsto \bar{z}$.

Remark 6.49. From the proof of Proposition 6.46 it follows that two dense multiplicative subgroups $\langle \mu_1, \dots, \mu_n \rangle$ and $\langle \mu_1', \dots, \mu_n' \rangle$ are topologically conjugate if and only if there exists an $\mathbb{R}$-linear map $\widehat{A}\colon \mathbb{C} \to \mathbb{C}$ which takes 1 to 1 and establishes a one-to-one correspondence between logarithms of the generators modulo integers for *some* choice of the branch of logarithm:

$$\widehat{A}\lambda_j = \lambda_j' \bmod \mathbb{Z}, \qquad \lambda_j = \frac{\ln \mu_j}{2\pi i} \bmod \mathbb{Z}, \quad \lambda_j' = \frac{\ln \mu_j'}{2\pi i} \bmod \mathbb{Z}. \tag{6.24}$$

From this observation and the topological invariance of the holonomy at infinity we can already conclude that the topological classification of certain classes of foliations is nondiscrete.

Now we can prove the main result of this subsection, the Rigidity theorem for finitely generated groups of conformal germs.

Proof of Theorem 6.45. Let $G = \langle f_1, \dots, f_n \rangle$ and $G' = \langle f_1', \dots, f_n' \rangle$ be two topologically conjugated noncommutative groups of germs with G satisfying the density condition 6.17. Without loss of generality we may assume that f_1 is hyperbolic, and the corresponding multiplicator μ_1 has modulus less than 1.

1. Consider the germ $f_1' \in G'$ conjugated by h with f_1. This germ is also hyperbolic, moreover, we can easily see that $|\mu_1'| < 1$. Indeed, choose representatives of f_1, f_1' defined in small topological disks U and $U' = h(U)$ respectively, so small that $f_1(U) \Subset U$. Then $f_1'(U') \Subset U'$ and by the Schwarz lemma, $|\mu_1'| < 1$.

2. If a homeomorphism h conjugates G with G', then a representative of h conjugates (topologically) the corresponding pseudogroups Γ and Γ' and also the respective closures $\overline{\Gamma}$ and $\overline{\Gamma'}$. By Theorem 6.37, each closure contains a dense subgroup of the multiplicative group $\mathbb{C}^*$. Thus a representative of h topologically conjugates two dense subgroups as in Proposition 6.46.

3. Applying Proposition 6.46, we obtain an explicit description of the conjugating homeomorphism h: there exist holomorphic charts on U and U' (linearizing the hyperbolic germs f_1, f_1' respectively) in which $h(z) = z\,|z|^{\beta}$.

4. If the commutator $[G, G]$ is nontrivial, it must contain a parabolic germ $f(z) = z + az^{n+1} + \cdots \in \operatorname{Diff}_1(\mathbb{C}, 0)$ which is conjugated by a homeomorphism $h(z) = z\,|z|^{\beta}$ with another parabolic element $f'(z') = z' + a'z'^{n'+1} + \cdots \in [G', G']$. Clearly, $n = n'$ since this number is a topological invariant of germs (related to the number of petals; see §21). We will show that $\beta = 0$ so that $h(z) = z$.

To see this, we substitute the explicit form $h(z) = z\,|z|^{\beta}$ found in Proposition 6.46 into the conjugacy equation $h \circ f = f' \circ h$. After substitution, division by $z\,|z|^{\beta}$ and subtraction of 1 from each side we obtain the equation

$$z^n[a + \tfrac{\beta}{2}(a + \bar{a})] + \cdots = a'z^n\,|z|^{n\beta} + \cdots$$

where the dots stay for the terms decreasing as $|z| \to 0$ faster than the terms explicitly written on each side. Note that the principal term of left hand side after restriction on each circle $|z| = r$ is a trigonometric polynomial of degree n, whereas the principal part of the right hand side is a trigonometric polynomial of the same degree n only if $\beta = 0$. Thus $h(z) = z$ is a linear, hence holomorphic, map.

5. To prove that the conjugacy h analytically depends on additional parameters t, we proceed as follows. The condition that $h(z) = cz$ conjugates any generator $f_{j,t}$ with $f_{j,t}'$ translates into an infinite number of analytic

conditions on c and t. Thus the entire set $Q = \{(t,c)\colon f_{j,t}(cz) = cf'_{j,t}(z),\ j = 1,\dots,r\}$ is analytic near the point $(0,1)$. If $f_t(z) = z + a_t z^{n+1} + \cdots$ is conjugate with $f'_t(z) = z + a'_t z^{n+1} + \cdots$, and $a_0 a'_0 \neq 0$, then $Q \subseteq \{c^n = a'_t/a_t\}$ (equating the coefficients before z^{n+1}). The latter analytic set consists of n analytic branches $c = c_k(t)$, $k = 1,\dots,n$. Since these branches are locally irreducible and $Q \cap \{t = \text{const}\}$ is nonempty for all t, the set Q contains at least one such branch. This branch gives the holomorphic dependence of $h(z) = c(t)\,z$ on t.

The proof of Theorem 6.45 is complete. □

6I. Relaxing the genericity assumptions. Though the assumptions of noncommutativity and density required in Theorems 6.41 and 6.45 are generic, they fail for some important classes of finitely generated groups. For instance, the density condition fails for groups with resonant multiplicators; such groups constitute a dense subspace in $\mathrm{Diff}(\mathbb{C}^1, 0)$.

Yet the assumptions of the above theorems can be relaxed to an *open and dense* condition of *nonsolvability*; see Example 6.28. We formulate here without proofs several results in this direction.

Theorem 6.50 (A. Shcherbakov [**Shc84**], I. Nakai [**Nak94**]). *A nonsolvable finitely generated group $G \subset \mathrm{Diff}(\mathbb{C}^1, 0)$ is rigid.*

The following result suggests that certain rigidity-like properties may occur even in the infinite cyclic subgroups. Recall that a germ is called elliptic, if its multiplicator μ has modulus one: $\mu = \exp 2\pi i\varphi$, $\varphi \in \mathbb{R}/\mathbb{Z}$.

Theorem 6.51 (V. A. Naĭshul [**Naĭ82**]; see also [**GLCP96**]). *Suppose that two elliptic germs of conformal maps $f, f' \in \mathrm{Diff}(\mathbb{C}, 0)$ are topologically conjugate by an orientation-preserving homeomorphism. Then the multiplicators of f and f' coincide.*

This theorem is relatively easy in the case where the multiplicators are roots of unity or Diophantine irrationalities (cf. with §5**E**). The real difficulties occur in the Cremer case.

Nonsolvability turns out also sufficient for existence of infinitely many limit cycles.

Theorem 6.52 (A. Shcherbakov [**Shc86**]; see also [**BLL97**] and [**SRO98**]). *A nonsolvable finitely generated group $G \subset \mathrm{Diff}(\mathbb{C}^1, 0)$ possesses infinitely many complex limit cycles accumulating to the origin.*

The density of orbits obviously fails under the single nonsolvability assumption. For instance, if a group $G \subset \mathrm{Diff}(\mathbb{C}^1, 0)$ consists of real (i.e., preserving $\mathbb{R}$) germs with positive multiplicators, then any orbit starting in the upper (resp., lower) open half-plane, remains in the same half-plane forever and hence cannot be dense. Yet this is the only possible deviation from the density pattern.

Theorem 6.53 (I. Nakai [**Nak94**], a weaker result was proved in [**Shc82**]). *If G is a nonsolvable subgroup of $\mathrm{Diff}(\mathbb{C}^1, 0)$, then there exits the germ of a real analytic curve $K \subsetneq (\mathbb{C}^1, 0) \cong (\mathbb{R}^2, 0)$ invariant by G, such that orbits of any pseudogroup Γ associated with G are dense in the connected components (sectors) of $(\mathbb{C}^1, 0) \smallsetminus K$.*

Exercises and Problems for §6.

Exercise 6.1. Prove the rigidity assertions from Example 6.43.

Exercise 6.2. Prove the properties of the length function ν introduced in §6**E**.

Problem 6.3. Compute all holonomy maps of an integrable foliation $\{du = 0\}$, $u \in \mathcal{O}(\mathbb{C}^2, 0)$, if $u = \prod u_j^{p_j}$ is the primary decomposition of the holomorphic germ u with irreducible factors u_j and natural exponents $p_j \in \mathbb{N}$.

Problem 6.4. Prove that a formally integrable holomorphic self-map (or a finitely generated group G of holomorphic germs of self-maps from $\operatorname{Diff}(\mathbb{C}, 0)$) is also analytically integrable; cf. with Theorem 6.8.

Suggestion. Use the formal chart in which $\widehat{u}(z) = z^m$.

Problem 6.5. Prove that an (orbital) symmetry of a holomorphic vector field on $(\mathbb{C}, 0)$ is necessary holomorphic itself.

Problem 6.6. Construct a finitely generated subgroup $G \subset \operatorname{Diff}(\mathbb{C}, 0)$, whose orbits are dense in each of the two half-planes $\{\pm \operatorname{Im} z > 0\}$ separately, yet both half-planes are invariant by G.

Generalize this example and find a group whose orbits are dense in each of $2p$ invariant sectors in $(\mathbb{C}, 0)$ for any $p > 1$ (cf. with Theorem 6.53).

Problem 6.7 (formal rigidity of generic groups). Assume that two finitely generated subgroups $G, G' \subseteq \operatorname{Diff}(\mathbb{C}, 0)$ are *formally* equivalent and one of these groups contains a hyperbolic germ. Prove that in such case G and G' are holomorphically equivalent, moreover, any formal conjugacy between them is necessarily holomorphic (convergent).

7. Holomorphic invariant manifolds

In this short section we show that under rather weak conditions one can eliminate enough nonresonant terms to ensure existence of *holomorphic invariant (sub)manifolds.* Recall that a holomorphic submanifold $W \subset (\mathbb{C}^n, 0)$ is invariant for a holomorphic vector field F, if the vector $F(x)$ is tangent to W at any point $x \in W$. Traditionally the prefix 'sub' is omitted, though it plays an important role: in §14 we will discuss invariant analytic subvarieties that are *not* submanifolds because of their singularity.

7A. Invariant manifolds of hyperbolic singularities. Suppose that the spectrum $S \subset \mathbb{C}$ of linearization matrix A of a holomorphic vector field consists of two parts $S^{\pm} \subset \mathbb{C}$ separated by a real line (i.e., each part belongs to an open half-plane bounded by the line). In this case no eigenvalue from one part can be equal to a linear combination of eigenvalues from the other part with nonnegative coefficients,

$$\begin{gathered} \lambda_j^- - \sum \alpha_i \lambda_i^+ \neq 0, \qquad \lambda_i^+ - \sum \alpha_j \lambda_j^- \neq 0, \\ \lambda_i^+ \in S^+, \quad \lambda_j^- \in S^-, \quad \alpha_i, \alpha_j \in \mathbb{Z}_+, \end{gathered} \tag{7.1}$$

(we say that there are no *cross-resonances* between the two parts). Without loss of generality A can be assumed to be in the block diagonal form. By the Poincaré–Dulac theorem, there exists a formal transformation eliminating all nonresonant terms corresponding to the nonzero cross-combinations (7.1). The corresponding formal normal form has two invariant manifolds coinciding with the corresponding coordinate subspaces.

Moreover, all denominators (7.1) are obviously bounded from below. Therefore one can expect that the corresponding transformation converges and the invariant manifolds will exist in the analytic category. This is indeed the case, though the accurate proof is organized along different lines.

Theorem 7.1 (Hadamard–Perron theorem for holomorphic flows). *Assume that the linearization operator of a holomorphic vector field $Ax + F(x)$ has a transversal pair of invariant subspaces $L^{\pm}$ such that the spectra of A restricted on these subspaces are separated from each other.*

Then the vector field has two holomorphic invariant manifolds $W^{\pm}$ tangent to the subspaces $L^{\pm}$.

However, the proof of this result is indirect. We start by formulating and proving a counterpart of Theorem 7.1 for biholomorphisms.

Definition 7.2. A holomorphic self-map $H \in \operatorname{Diff}(\mathbb{C}^n, 0)$, $x \mapsto Mx + h(x)$, $h(0) = \frac{\partial h}{\partial x}(0) = 0$, is said to be *hyperbolic* if no eigenvalue of the linearization matrix $M \in \operatorname{GL}(n, \mathbb{C})$ has modulus 1.

For a matrix M without eigenvalues on the unit circle, we denote $L^{\pm} \subseteq \mathbb{C}^n$ two invariant subspaces such that the restriction $M|_{L^-}$ is contracting (in a suitable Hermitian metric) and $M|_{L^+}$ expanding (i.e., $M^{-1}|_{L^+}$ is contracting).

To define invariant manifolds for biholomorphisms we need to be careful and replace sets by their germs at the fixed points. Otherwise it would be necessary to give different definitions for expanding and contracting submanifolds.

Definition 7.3. A holomorphic submanifold W passing through a fixed point of a biholomorphism $H\colon (\mathbb{C}^n, 0) \to (\mathbb{C}^n, 0)$ is *invariant*, if *the germ* of $H(W)$ at the origin coincides with *the germ* of W.

Theorem 7.4 (Hadamard–Perron theorem for biholomorphisms). *A hyperbolic holomorphism in a sufficiently small neighborhood of the fixed point at the origin has two holomorphic invariant submanifolds W^+ and W^-.*

These manifolds pass through the origin, transversal to each other and are tangent to the corresponding invariant subspaces $L^{\pm}$ of the linearized map $x \mapsto Mx$.

The dimensions of the invariant manifolds are necessarily equal to the dimension of the corresponding subspaces. The manifold W^+ is called *unstable manifold*, whereas W^- is referred to as the *stable manifold*, because the restriction of H on these manifolds is unstable and stable respectively.

Proof. The linearization matrix M of a holomorphic biholomorphism $H\colon (\mathbb{C}^n, 0) \to (\mathbb{C}^n, 0)$ can be put into the block diagonal form. Choosing appropriate system of local holomorphic coordinates $(x, y) \in (\mathbb{C}^k, 0) \times (\mathbb{C}^l, 0)$, $k + l = n$, one can always assume that the map H has the form

$$H\colon \begin{pmatrix} x \\ y \end{pmatrix} \longmapsto \begin{pmatrix} Bx + g(x,y) \\ Cy + h(x,y) \end{pmatrix}, \qquad (x, y) \in (\mathbb{C}^k, 0) \times (\mathbb{C}^l, 0). \tag{7.2}$$

Here the square matrices B, C and the nonlinear terms g, h of order $\geqslant 2$ satisfy the conditions

$$\begin{aligned} |B| \leqslant \mu, \quad |C^{-1}| \leqslant \mu, &\qquad \mu < 1, \\ |f(x,y)| + |g(x,y)| < |x|^2 + |y|^2, &\qquad \text{for } |x| < 1,\ |y| < 1. \end{aligned} \tag{7.3}$$

with some *hyperbolicity parameter* $\mu < 1$.

It is sufficient to prove the existence of the *stable* manifold only; the unstable manifold for H is the stable manifold of the inverse map H^{-1} which is also hyperbolic.

The stable manifold W^+ tangent to $L^+ = \{(x, 0)\}$ is necessarily the graph of a holomorphic vector function $\varphi\colon \{|x| \leqslant \varepsilon\} \to \{|y| \leqslant \varepsilon\}$ defined in a sufficiently small polydisk, $\varphi(0) = 0$, $\frac{\partial\varphi}{\partial x}(0) = 0$. For this manifold to be invariant, the function φ must satisfy the functional equation

$$\varphi\bigl(Bx + g(x, \varphi(x))\bigr) = C\varphi(x) + h(x, \varphi(x)). \tag{7.4}$$

This equation can be transformed to the fixed point form as follows:

$$\varphi = \mathcal{H}\varphi, \qquad (H\varphi)(x) = C^{-1}\varphi\bigl(Bx + g(x, \varphi(x))\bigr) - h(x, \varphi(x)). \tag{7.5}$$

All assertions of Theorem 7.4 follow from the contracting map principle and the following Lemma 7.5. $\square$

The "linearization" (removal of all nonlinear terms of order 2 and higher) of the operator $\mathcal{H}$ at the "point" $\varphi = 0$ results in the operator

$$\varphi(x) \mapsto C^{-1}\varphi(Bx), \qquad |B|,\ |C^{-1}| \leqslant \mu < 1,$$

which is obviously contracting. Lemma 7.5 shows that nonlinear terms do not affect this property.

Denote by $\mathcal{A}_\varepsilon$ the Banach space of functions holomorphic in the open disk of radius $\varepsilon > 0$ and continuous on the closure.

Lemma 7.5. *Under the assumptions* (7.3), *the nonlinear operator* $\mathcal{H}$ *has the following properties*:

(1) *$\mathcal{H}$ is well defined for φ in the ball $\mathcal{B}_\varepsilon = \{\varphi\colon \sup_{|x|<\varepsilon} |\varphi(x)| < \varepsilon\}$ inside the space $\mathcal{A}_\varepsilon$, and takes this ball into itself,*

(2) *the subset $\mathcal{B}^1_\varepsilon$ of functions in $\mathcal{B}_\varepsilon$ with the Lipschitz constant $\leqslant 1$, is preserved by $\mathcal{H}$,*

(3) *the operator $\mathcal{H}$ is contracting on $\mathcal{B}^1_\varepsilon$,*

provided that the value $\varepsilon > 0$ is sufficiently small.

Proof. To prove the first assertion, note that $|Bx + g(x, \varphi(x))| < \mu|x| + |x|^2 + |\varphi|^2 < \mu\varepsilon + 2\varepsilon^2 < \varepsilon$ for $|x| < \varepsilon$, if ε is sufficiently small. Thus the composition occurring in the definition of $\mathcal{H}$ makes perfect sense and $\mathcal{H}\varphi$ is well defined. For the same reason, $|\varphi|$ never exceeds $\mu\varepsilon + 2\varepsilon^2 < \varepsilon$ which means that $\mathcal{B}_\varepsilon$ is taken by $\mathcal{H}$ into itself.

The Jacobian matrix $J(x) = \frac{\partial\varphi}{\partial x}$ is transformed into $J' = C^{-1}J(\cdots)(B + \frac{\partial g}{\partial x} + \frac{\partial g}{\partial y}J) + (\frac{\partial h}{\partial x} + \frac{\partial h}{\partial y}J)$. Since the terms g, h are of order $\geqslant 2$, their derivatives vanish at the origin and therefore the Jacobian is no greater (in the sense of the matrix norm) than $(\mu^2 + O(\varepsilon))|J|$. As $\mu < 1$, this proves the assertion about the Lipschitz constant.

To prove the last assertion that $\mathcal{H}$ is contractive, notice that the operator $\varphi(x) \mapsto h(x, \varphi(x))$ is strongly contracting:

$$|h(x, \varphi_1(x)) - h(x, \varphi_2(x))| \leqslant \left|\tfrac{\partial h}{\partial y}\right| |\varphi_1(x) - \varphi_2(x)| \leqslant O(\varepsilon)\|\varphi_1 - \varphi_2\|_\varepsilon. \tag{7.6}$$

Consider the operator $\varphi \mapsto \mathcal{G}\varphi = \varphi(Bx + g(x, \varphi))$ and the difference of the values it takes on two functions $\varphi_1, \varphi_2 \in \mathcal{B}^1_\varepsilon$: by the triangle inequality,

$$\begin{aligned} |\mathcal{G}\varphi_1(x) - \mathcal{G}\varphi_2(x)| &= |\varphi_1(Bx + g_1(x)) - \varphi_2(Bx + g_2(x))| \\ &\leqslant |\varphi_1(Bx + g_2(x)) - \varphi_2(Bx + g_2(x))| \\ &\quad + |\varphi_1(Bx + g_1(x)) - \varphi_1(Bx + g_2(x))|, \end{aligned}$$

where we denoted $g_i(x) = g(x, \varphi_i(x))$ for brevity. The first term does not exceed $\|\varphi_1 - \varphi_2\|_\varepsilon$. Since the vector function $\varphi_1 \in \mathcal{B}^1_\varepsilon$ has Lipschitz constant 1, the second term does not exceed $|g_1(x) - g_2(x)| = |g(x, \varphi_1(x)) - g(x, \varphi_2(x))|$. Similarly to (7.6), this part is no greater than $O(\varepsilon)\|\varphi_1 - \varphi_2\|_\varepsilon$. Finally, we conclude that $\mathcal{G}$ is Lipschitz on $\mathcal{B}^1_\varepsilon$: $\|\mathcal{G}\varphi_1 - \mathcal{G}\varphi_2\|_\varepsilon \leqslant (1 + O(\varepsilon))\|\varphi_1 - \varphi_2\|$.

Adding all terms together for $\mathcal{H} = C^{-1}\mathcal{G} - h(x, \cdot)$, we conclude that if $\varphi_{1,2} \in \mathcal{B}^1_\varepsilon$, then

$$\|\mathcal{H}\varphi_1 - \mathcal{H}\varphi_2\|_\varepsilon \leqslant (\mu + O(\varepsilon))\, \|\varphi_1 - \varphi_2\|_\varepsilon.$$

Since $\mu < 1$, the operator $\mathcal{H}$ is contracting on the closed subset $\mathcal{B}^1_\varepsilon$ of the complete metric space $\mathcal{B}_\varepsilon \subset \mathcal{A}_\varepsilon$. □

Remark 7.6. Characteristically for the proofs based on the contracting map principle, the germs of invariant manifolds are automatically unique.

Now we can derive Theorem 7.1 from Theorem 7.4.

Proof. Passing if necessary to an orbitally equivalent field, one may assume that the linearization $A = \operatorname{diag}\{A_+, A_-\}$ is block diagonal with the spectra of the blocks are *separated by the imaginary axis.*

Consider the flow maps $\Phi^t = \exp tF\colon (\mathbb{C}^n, 0) \to (\mathbb{C}^n, 0)$ for $t = 1/k$, $k = 1, 2, \ldots$. Each of them is a biholomorphism with the linear part $x \mapsto \exp tAx$ whose eigenvalues are the corresponding exponentials $\{\exp t\lambda_i : \lambda_i \in S\}$ separated by the unit circle $\{|\lambda| = 1\}$. In the assumptions of the theorem, each flow map Φ^t is hyperbolic for the specified values of $t \in 1/\mathbb{N}$. By Theorem 7.4, the map Φ^t has a pair of invariant manifolds $W_t^\pm$, tangent to the corresponding invariant subspaces $L^\pm$ common for all $t \in \mathbb{R}$.

Apriori, the invariant subspaces $W_t^\pm$ do not have to coincide. However, $\left(\Phi^{1/k}\right)^k = \Phi^1$, therefore manifolds invariant for $\Phi^{1/k}$, are invariant also for Φ^1. Since the invariant manifolds for the latter map are unique, we conclude that all the maps $\Phi^{1/k}$ leave the pair $W^\pm = W_1^\pm$ invariant.

In other words, an analytic trajectory $x(t)$ of the vector field which begins on, say, W^-, $x(0) \in W^-$, remains on W^- for $t = 1/k$. Since isolated zeros of analytic functions cannot have accumulation points, $x(t)$ is on W^- for all (sufficiently small) values of $t \in (\mathbb{C}, 0)$. Then W^- is invariant for the vector field $Ax + F(x)$. The proof for W^+ is similar. □

Remark 7.7. Intersection of invariant manifolds is again an invariant manifold. This observation allows us to construct small-dimensional invariant manifolds for holomorphic vector fields. For instance, if the linearization matrix A has a simple eigenvalue $\lambda_1 \neq 0$ such that $\lambda_1/\lambda_j \notin \mathbb{R}_+$ for all other eigenvalues λ_j, $j = 2, \ldots, n$, then the vector field has a *one-dimensional holomorphic invariant manifold* (curve) tangent to the corresponding eigenvector.

The Hadamard–Perron theorem for holomorphic flows, as formulated above, is the nearest analog of the Hadamard–Perron theorem for smooth flows in $\mathbb{R}^n$. There are known stronger results in this direction; see [**Bib79**].

7B. Hyperbolic invariant curves for saddle-nodes. Consider a holomorphic vector field on the plane $(\mathbb{C}^2, 0)$ with the saddle-node at the origin. Recall that by Definition 4.28, this means that exactly one of the eigenvalues is zero, while the other eigenvalue must be nonzero. The null space (line) of the linearization operator is called the *central* direction. The direction of eigenvector with the nonzero eigenvalue is referred to as *hyperbolic*.

The nonzero eigenvalue cannot be separated from the null one, thus the Hadamard–Perron theorem cannot be applied. However, the invariant manifold (smooth holomorphic curve) tangent to the eigenvector with nonzero eigenvalue, exists and is unique in this case as well. As before, we start with the case of biholomorphisms with one contracting eigenvalue $|\mu| < 1$ and the other eigenvalue equal to 1. For obvious reasons, such maps are called *saddle-node* biholomorphisms.

Any saddle-node biholomorphism $H\colon (\mathbb{C}^2, 0) \to (\mathbb{C}^2, 0)$ can be brought into the form

$$H\colon \begin{pmatrix} x \\ y \end{pmatrix} \longmapsto \begin{pmatrix} \mu x + g(x,y) \\ y + y^2 + h(x,y) \end{pmatrix}, \qquad \mu \in (0,1) \subset \mathbb{R}, \tag{7.7}$$

with g, h holomorphic nonlinear terms of order $\geqslant 3$, by a suitable holomorphic choice of coordinates x, y. Indeed, all other quadratic terms are nonresonant and can be removed (Exercise 4.8).

Theorem 7.8. *The biholomorphism* (7.7) *has a unique holomorphic invariant manifold* (*curve*) *tangent to the eigenvector* $(1, 0) \in \mathbb{C}^2$.

Proof. The manifold $W = \operatorname{graph}\varphi$ is invariant for the saddle-node self-map H of the form (7.7) if the function φ satisfies the functional equation

$$\varphi(\mu x + g(x, \varphi(x))) = \varphi(x) + \varphi^2(x) + h(x, \varphi(x)). \tag{7.8}$$

This equation can be represented under the fixed point form $\mathcal{H}\varphi = \varphi$ using the operator $\mathcal{H}$ defined as follows:

$$(\mathcal{H}\varphi)(x) = \varphi\big(\mu x + g(x, \varphi(x))\big) - \varphi^2(x) - h(x, \varphi(x)). \tag{7.9}$$

This operator is no longer contracting: its linearization at $\varphi = 0$ is the operator $\varphi(x) \mapsto \varphi(\mu x)$ which keeps all constants fixed. To restore the contractivity, we have to restrict this operator on the subspace of functions vanishing at the origin, with the norm $\|\varphi\|' = \sup_{x\neq 0} \frac{|\varphi(x)|}{|x|}$. Technically it is more convenient to substitute $\varphi(x) = x\psi(x)$ into the functional equation (7.8) and bring it back to the fixed point form. As a result, we obtain the equation

$$(\mu x + g(x, x\psi(x)) \cdot \psi\big(\mu x + g(x, x\psi(x))\big) = x\psi(x) + x^2\psi^2(x) + h(x, x\psi(x)),$$

which yields the nonlinear operator $\mathcal{H}'$,

$$(\mathcal{H}'\psi)(x) = \big(\mu + g'(x, \psi(x))\big) \cdot \psi\big(\mu x + g(x, x\psi(x))\big) - x\psi^2(x) - h'(x, \psi). \tag{7.10}$$

Here the holomorphic functions $g'(x,y) = g(x, xy)/x$, $h'(x,y) = h(x,xy)/x$ are of order $\geqslant 2$ at the origin.

The proof of Lemma 7.5 carries out almost literally for the operator $\mathcal{H}'$ as in (7.10), proving that it is contractible on the space of functions $\psi\colon \{|x| < \varepsilon\} \to \{|y| < \varepsilon\}$ with respect to the usual supremum-norm on sufficiently small discs. $\square$

Completely similar to derivation of Theorem 7.1 from Theorem 7.4 in the hyperbolic case, Theorem 7.8 implies the following result concerning holomorphic saddle-nodes.

Theorem 7.9. *A holomorphic vector field on the plane* $(\mathbb{C}^2, 0)$ *having a saddle-node singularity* (*one eigenvalue zero, another nonzero*) *at the origin, admits a unique holomorphic nonsingular invariant curve passing through the singular point and tangent to the hyperbolic direction.* □

This curve is called the *hyperbolic invariant manifold.*

It is important to conclude this section by the explicit example showing that the other invariant manifold, the *central manifold* tangent to the central direction, may not exist in the analytic category. Note, however, that the formal invariant manifold always exists and is unique: this follows from the formal orbital classification of saddle-nodes (Proposition 4.29).

Example 7.10 (L. Euler). The vector field

$$x^2\frac{\partial}{\partial x} + (y - x)\frac{\partial}{\partial y} \tag{7.11}$$

has vertical hyperbolic direction $\frac{\partial}{\partial y}$ and the central direction $\frac{\partial}{\partial x} + \frac{\partial}{\partial y}$. The central manifold, if it exists, must be represented as the graph of the function $y = \varphi(x)$, $\varphi(x) = x + \sum_{k\geqslant 2} c_k x^k$. However, this series diverges, as was noticed already by L. Euler. Indeed, the function φ must be the solution to the differential equation

$$\frac{d\varphi}{dx} = \frac{\varphi(x) - x}{x^2}$$

which implies the recurrent formulas for the coefficients,

$$k\, c_k = c_{k+1}, \qquad k = 1, 2, \dots, \qquad c_1 = 1.$$

The factorial series with $c_k = (k-1)!$ has zero radius of convergence, hence no analytic central manifold exists.

However, sufficiently large "pieces" of the central manifold for the saddle-node can be shown to exist; see §22**I**.

Exercises and Problems for §7.

Exercise 7.1. Prove that a nonresonant hyperbolic self-holomorphism is analytically linearizable on its holomorphic invariant manifolds W^+ and W^-.

Problem 7.2. Prove that if a hyperbolic self-map analytically depends on additional parameters (and remains hyperbolic for all values of these parameters), then the invariant manifolds $W^\pm$ also depend analytically on the parameters.

Problem 7.3. Formulate and prove a parallel statement for a saddle-node.

Exercise 7.4. Describe possible number and relative position of analytic separatrices of elementary planar singularities of holomorphic vector fields.

Problem 7.5. Assume that the first k eigenvalues $\lambda_1, \dots, \lambda_k$ from the spectrum of a holomorphic vector field $F \in \mathcal{D}(\mathbb{C}^n, 0)$, $k \leqslant n$, are real, and the others are not.

Prove that the field F has a holomorphic k-dimensional invariant manifold tangent to the coordinate plane generated by the first k basis vectors.

8. Desingularization in the plane

Reasonably complete analysis of singular points of holomorphic vector fields using holomorphic normal forms and transformations by biholomorphisms, is possible under the assumption that the linear part is not very degenerate. The degenerate cases have to be treated by transformations that can alter the linear part. Such transformations, necessarily not holomorphically invertible, are known by the name *desingularization*, *resolution of singularities*, *sigma-process* or *blow-up*. Very roughly, the idea is to consider a holomorphic map $\pi \colon M \to (\mathbb{C}^2, 0)$ of a holomorphic surface (2-dimensional manifold) M that squeezes (blows down) a complex 1-dimensional curve $D \subset M$ to the single point $0 \in \mathbb{C}^2$, while being one-to-one between $M \smallsetminus D$ and $(\mathbb{C}^2, 0) \smallsetminus \{0\}$. The second circumstance allows us to pull back local objects (functions, curves, foliations, 1-forms, *etc.*) from $(\mathbb{C}^2, 0)$ to M and then extend them on D. These pullbacks are called desingularizations, or blow-ups of the initial objects; sometimes M is itself called the blow-up of (the neighborhood of) the point $0 \in \mathbb{C}^2$.

In this section we develop some basic algebraic geometry necessary to deal with desingularizations and introduce the notion of multiplicity of an isolated singularity of a foliation.

Using desingularization one can ultimately simplify singularities of holomorphic foliations in dimension 2. The main result of this section, the fundamental Desingularization Theorem 8.14 asserts that by a suitable blow-up any singular holomorphic foliation in a neighborhood of a singular point can be resolved into a singular foliation, defined in a neighborhood of a union $D = \bigcup_i D_i$ of one or more transversally intersecting holomorphic curves D_i, which has only *elementary* singularities on D.

8A. Polar blow-up. We start with a transcendental but geometrically more transparent construction in the real domain.

Definition 8.1. The *polar blow-down* is the map P of the real cylinder $C = \mathbb{R} \times \mathbb{S}^1 \to \mathbb{R}^2$ onto the plane $\mathbb{R}^2$,

$$P \colon (r, \varphi) \mapsto (r \cos \varphi, r \sin \varphi). \tag{8.1}$$

This map is a real analytic diffeomorphism between the open half-cylinder $C_+ = \{r > 0\} \subset C$ and the punctured plane $\mathbb{R}^2 \smallsetminus \{0\}$. The

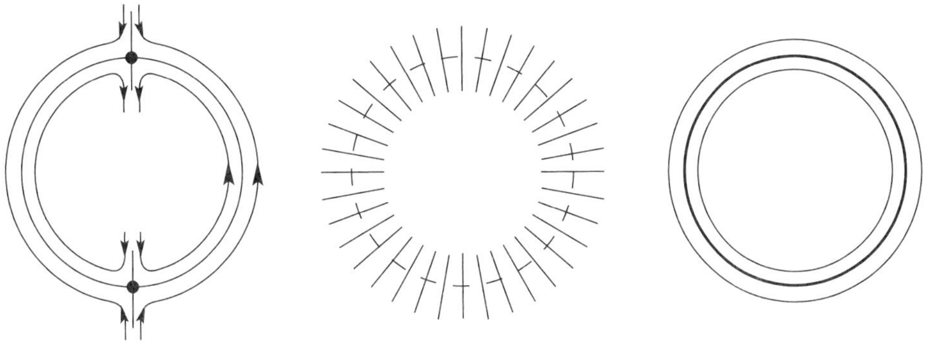

(i) (ii) (iii)

Figure I.4. Trigonometric blow-up of a nonsingular (i) and singular (ii), (iii) foliations

image of the narrow band $C = (\mathbb{R}, 0) \times \mathbb{S}^1$ (cylinder) is a double covering of the small neighborhood of the origin $\{|x| < \varepsilon\}$ except for the central equator $S = \{r = 0\} \subset C$, also called the *exceptional divisor*. The latter is squeezed into one point, the origin $0 \in \mathbb{R}^2$.

The map P pulls back functions and differential 1-forms from $(\mathbb{R}^2, 0)$ on C (in noninvariant terms, passing to the polar coordinates and ignoring the inequality $r > 0$). However, the pullback $P^*\omega \in \Lambda^1(C)$ of any 1-form $\omega \in \Lambda^1(\mathbb{R}^2, 0)$ always has a nonisolated singularity on S. In the real analytic case one can always divide $P^*\omega$ by a suitable natural power r^ν so that the 1-form $\widetilde{\omega} = r^{-\nu} P^*\omega \in \Lambda^1(C)$ still remains real analytic but has only isolated singularities on S.

Consider now the singular foliation $\mathcal{F}$ defined by the Pfaffian equation $\{\omega = 0\}$ on $(\mathbb{R}^2, 0) \smallsetminus \{0\}$. As P is one-to-one outside the origin, $P^{-1}(\mathcal{F})$ is a foliation of $C \smallsetminus S$ with the Pfaffian equation $\{P^*\omega = 0\}$. Since r is nonvanishing outside S, the foliation $\widetilde{\mathcal{F}}$ can be defined by the Pfaffian equation $\{\widetilde{\omega} = 0\}$ which has only isolated singularities on S and thus extends $P^{-1}(\mathcal{F})$ as a singular foliation on C.

Definition 8.2. The line field defined by the Pfaffian distribution $r^{-\nu} P^*\omega = 0$ with isolated singularities and the corresponding singular foliation $\widetilde{\mathcal{F}}$ on C are called the *trigonometric blow-up* of the distribution $\omega = 0$ and the corresponding foliation $\mathcal{F}$ respectively.

Intuitively, the singular point is "stretched" to the central circle so that the complicated behavior of leaves near the singularity can be studied in different "parts" separately. The first examples show that sometimes the singularity may even disappear.

Example 8.3.

(i) The form $dx = 0$ defining a nonsingular foliation, after trigonometric blow-up becomes $\cos\varphi\, dr - r\sin\varphi\, d\varphi$ and has two isolated singular points $(0,0)$ and $(0,\pi)$ on $\mathbb{R}\times\mathbb{S}^1$. Both these points are nondegenerate saddles. The exceptional circle without these points is the leaf of the blow-up foliation.

(ii) The form $\omega = y\,dx - x\,dy$ defines the "radial" singular foliation on $\mathbb{R}^2$. The pullback $P^*\omega = -r^2\,d\varphi$, has a nonisolated singularity on $r = 0$, but after division the form $\widetilde{\omega} = r^{-2}P^*\omega = d\varphi$ defines the nonsingular "parallel" foliation $\{\varphi = \text{const}\}$. All leaves of this foliation cross the exceptional circle S transversally.

(iii) The form $x\,dx + y\,dy = \frac{1}{2}d(x^2+y^2)$ which defines the foliation of $\mathbb{R}^2$ by the circles $x^2 + y^2 = \text{const}$, pulls back as the line field $r\,dr = 0$ which after division also becomes a nonsingular form dr on C. The exceptional circle is a leaf of the blow-up foliation carrying no singular points.

The map P can be complexified and the above examples generalized. However, the complexification will also be a two-fold covering, which is not natural geometrically. Besides, using the trigonometric functions $\sin\varphi, \cos\varphi$ makes the corresponding formulas nonalgebraic.

There is an algebraic version of the map P, called the *sigma-process, monoidal transformation*, or simply the *blow-up* without the adjective trigonometric.

8B. Algebraic blow-up (σ-process). It is not so easy to construct a holomorphic 2-dimensional manifold M and a holomorphic map $\sigma\colon M \to \mathbb{C}^2$ such that (i) the preimage of the origin is a compact irreducible holomorphic curve $S \subset M$ and (ii) the map σ is one-to-one between $M \smallsetminus S$ and $\mathbb{C}^2 \smallsetminus \{0\}$. These two requirements together imply rather specific properties of M and S; cf. with Remark 8.6 below.

One such construction goes as follows. Consider the *canonical map* from $\mathbb{C}^2 \smallsetminus \{0\}$ to the projective line $\mathbb{P}^1$ that associates each point $(x,y) \neq (0,0)$ different from the origin, with the line $\{(tx,ty)\colon t \in \mathbb{C}\}$ passing through this point. The graph of this map is a complex 2-dimensional surface in the complex 3-dimensional manifold (the Cartesian product) $\mathbb{C}^2 \times \mathbb{P}^1$. The graph is not closed; to construct the closure, one has to add the *exceptional curve* $\mathbb{E} = \{0\} \times \mathbb{P}^1 \subset \mathbb{C}^2 \times \mathbb{P}^1$. The result is a nonsingular surface which we will denote by $\mathbb{M}$: by construction it is embedded in the complex 3-dimensional space $\mathbb{C}^2 \times \mathbb{P}^1$ and carries the compact curve (complex projective line, Riemann sphere) $\mathbb{E} \cong \mathbb{P}^1 \cong \mathbb{S}^2$ on it. The Cartesian projection $\mathbb{C}^2 \times \mathbb{P}^1 \to \mathbb{C}^2$ on the first component, after restriction on $\mathbb{M}$ becomes a holomorphic map

$$\sigma\colon \mathbb{M} \to \mathbb{C}^2, \qquad \sigma(\mathbb{E}) = \{0\} \in \mathbb{C}^2,$$

which is by construction one-to-one between $\mathbb{M} \smallsetminus \mathbb{E}$ and $\mathbb{C}^2 \smallsetminus \{0\}$.

Definition 8.4. The map $\sigma\colon \mathbb{M} \to \mathbb{C}^2$ between two 2-dimensional complex manifolds is called the (standard) *monoidal* map. The analytic curve $\mathbb{E} \subset \mathbb{M}$ is referred to as the (standard) *exceptional divisor*. The inverse map $\sigma^{-1}\colon \mathbb{C}^2 \smallsetminus \{0\} \to \mathbb{M} \smallsetminus \mathbb{E}$ is called the (standard) blow-up map, or simply the blow-up. A less frequently used term for the map σ is *blow-down*.

To see why $\mathbb{M}$ is a nonsingular manifold (and justify the assertions on the closure and smoothness), we produce a convenient ("standard") atlas on $\mathbb{M}$. Let z, w be two affine charts on the Riemann sphere $\mathbb{P}^1$, which take the line passing through the point $(x, y) \neq (0, 0)$ into the numbers $z = y/x$ and $w = x/y$ respectively: by construction, $w = 1/z$. These charts induce two affine charts in the respective domains V_1, V_2 on the Cartesian product $\mathbb{C}^2 \times \mathbb{P}^1$. In these charts the graph of the canonical map is given by the equations

$$y - xz = 0, \qquad \text{resp.,} \qquad x - wy = 0, \qquad (x, y) \neq (0, 0).$$

The surfaces defined by these equations, clearly remain nonsingular after extension on the line $\{x = 0,\ y = 0\} \subseteq \mathbb{C}^3$. Moreover, the functions (x, z) in the chart V_1 and (y, w) in chart V_2 respectively, become two coordinate systems (charts) on $\mathbb{M}$, defined in the two domains $U_i = \mathbb{M} \cap V_i$, $i = 1, 2$. The transition map between these charts is a monomial transformation

$$y = zx, \quad w = 1/z, \qquad \text{and reciprocally,} \qquad x = wy, \quad z = 1/w. \tag{8.2}$$

Thus $\mathbb{M}$ is indeed a nonsingular 2-dimensional complex analytic manifold. It remains to observe that the map $\sigma\colon \mathbb{M} \to \mathbb{C}^2$ in these charts is polynomial, hence globally holomorphic: $\sigma|_{U_i} = \sigma_i$, $i = 1, 2$, where

$$\sigma_1\colon (x, z) \mapsto (x, xz), \qquad \text{resp.,} \qquad \sigma_2\colon (y, w) \mapsto (yw, y). \tag{8.3}$$

The exceptional divisor $\mathbb{E}$ in the respective charts is given by the equations

$$\mathbb{E} \cap U_1 = \{x = 0\}, \qquad \text{resp.,} \qquad \mathbb{E} \cap U_2 = \{y = 0\}.$$

Remark 8.5. The formulas (8.2) and (8.3) are *real* algebraic, thus defining at the same time the real counterpart of the above construction. The real projective line $\mathbb{R}P^1$ is diffeomorphic to the circle $\mathbb{S}^1$, so the surface ${}^{\mathbb{R}}\mathbb{M}$ is constructed as a submanifold of the cylinder $\mathbb{R}^2 \times \mathbb{S}^1$. This submanifold is homeomorphic to the Möbius band. Having this analogy in mind, we will often refer to $\mathbb{M}$ as the *complex Möbius band*.

Remark 8.6. Nontriviality of the construction becomes even more striking in the complex domain. Note that the exceptional divisor cannot be *globally* defined by a single equation $\{f = 0\}$ with a function f holomorphic on $\mathbb{M}$ near $\mathbb{E}$. Indeed, if such a function exists, it would uniquely define a function

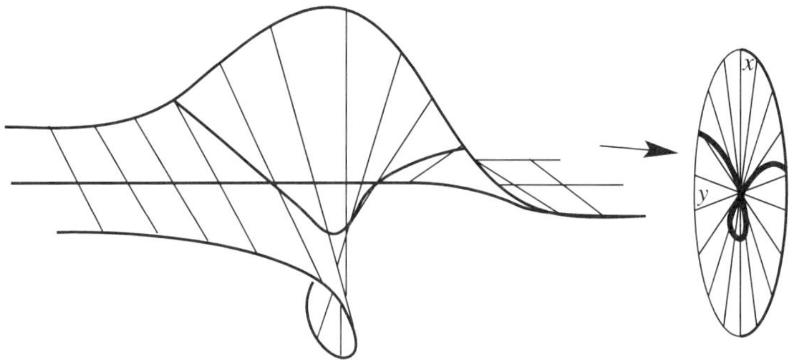

Figure I.5. Real Möbius band and its projection on $\mathbb{R}^2$ which is one-to-one outside the origin and blows down the circle $\mathbb{R}P^1 \cong \mathbb{S}^1$ into the origin

$f \circ \sigma^{-1}$ on $(\mathbb{C}^2, 0) \smallsetminus \{0\}$. This function is holomorphic and nonvanishing outside the origin and, since the point has codimension 2 in $\mathbb{C}^2$, f extends holomorphically at the origin. But the zero locus of a holomorphic function cannot have codimension 2—contradiction.

Similar arguments show that $\mathbb{E}$ is *exceptional* in the following sense: it sits rigidly inside $\mathbb{M}$ and cannot be deformed. Indeed, since $\mathbb{E}$ is compact, any deformation $\mathbb{E}'$ (a manifold uniformly close to $\mathbb{E}$) should necessarily also be compact, hence its image $\sigma(\mathbb{E}')$ should be a compact subset of $(\mathbb{C}^2, 0)$. This is impossible unless this image is a point, since σ is one-to-one outside the origin. The only remaining possibility is $\sigma(\mathbb{E}') = \{0\}$, i.e., $\mathbb{E}' = \mathbb{E}$.

Remark 8.7. These properties of the map $\sigma\colon (\mathbb{M}, S) \to (\mathbb{C}^2, 0)$ may seem to be caused by the artificial construction. However, one can prove that the construction of blow-up is natural and unique in the following sense. Consider *any* holomorphic map $\sigma'\colon (\mathbb{M}', \mathbb{E}') \to (\mathbb{C}^2, 0)$ defined in a neighborhood of a compact holomorphic curve $\mathbb{E}'$, which maps $\mathbb{E}'$ to a single point and is one-to-one on the complement $\mathbb{M}' \smallsetminus \mathbb{E}'$.

Assume that $\mathbb{E}'$ is irreducible. Then σ' is necessarily equivalent to the standard monoidal map σ: there exists a biholomorphic map $H\colon (\mathbb{M}, \mathbb{E}) \to (\mathbb{M}', \mathbb{E}')$ such that $\sigma = \sigma' \circ H$. (Without this assumption σ' can be equivalent to a *composition* of several monoidal maps.) In particular, the construction does not depend on the choice of the local coordinates (x, y) near the origin. The proof of these facts in the algebraic category can be found in [**Sha94**, Chapter IV, §3.4].

Using the local model provided by the standard monoidal transformation σ, we can construct a global map blowing up any finite set of points Σ on any two-dimensional complex manifold (surface) M.

Proposition 8.8. *Let M be a complex surface and $\Sigma \subset M$ a finite point set on it.*

Then there exists a holomorphic surface M' and a holomorphic map $\pi\colon M' \to M$ such that the preimage of any point from Σ is a Riemann sphere $\mathbb{E}_p = \pi^{-1}(p) \cong \mathbb{P}^1$ whereas π is one-to-one between $M' \smallsetminus \bigcup_{p\in\Sigma} \mathbb{E}_p$ and $M \smallsetminus \Sigma$.

Restriction of π on a small tubular neighborhood of each exceptional sphere $\mathbb{E}_p$ is equivalent to the standard monoidal map $\sigma\colon (\mathbb{M},\mathbb{E}) \to (\mathbb{C}^2, 0)$ restricted on a neighborhood of the exceptional divisor $\mathbb{E}$.

The surface M' and the map π are unique modulo a biholomorphic isomorphism and the right equivalence respectively. As follows from Remark 8.7, the requirement that $\mathbb{E}_p$ are biholomorphically equivalent to the Riemann sphere, can be relaxed to a mere irreducibility.

The inverse map $\pi^{-1}\colon M \smallsetminus \Sigma \to M'$ is called the *simple blow-up* of the locus (finite point set) Σ. The map π itself is sometimes called a simple blow-down.

Proof of Proposition 8.8. If $M = \mathbb{C}^2$ is the standard plane, then one might try to prove the possibility of *simultaneous blow-up of several points*, constructing a suitable polynomial map by interpolation.

Yet in the category of abstract holomorphic manifolds the construction of the map π from local monoidal transformations is trivial (tautological). Consider an atlas of charts $\{U_\alpha\}$ on M including special charts U_p identifying neighborhoods of each point $p \in \Sigma$ with a neighborhood $(\mathbb{C}^2, 0)$ of the origin. Without loss of generality we can assume that all other charts do not intersect the locus Σ. The manifold M can be then described as the quotient space of the disjoint union, $M = \bigsqcup_\alpha U_\alpha / \sim$ by the equivalence relationship $\sim$ (images of the same points in different charts are identified). The manifold M' in these terms can be described as follows. Replace each special chart U_p by the neighborhood $U'_p = (\mathbb{M},\mathbb{E})_p$, and consider again the disjoint union $\bigsqcup_\alpha U'_\alpha$ with $U'_\alpha = U_\alpha$ when the chart does not intersect Σ. The equivalence relationship $\sim$ lifts to an equivalence relationship $\sim'$ on the new disjoint union (all nonsingular points have unique preimages in U'_α). The quotient space $M' = \bigsqcup_\alpha U'_\alpha / \sim'$ by construction is a manifold. There are natural holomorphic maps $\pi\colon U'_\alpha \to U_\alpha$ which coincide with the monoidal map σ if the chart U_α was special, and identical otherwise. Clearly these maps agree with the equivalences $\sim, \sim'$ and hence define a holomorphic map $\pi\colon M' \to M$ with the required local properties. □

8C. Blow-up of analytic curves and singular foliations. As any holomorphic map, the standard monoidal map $\sigma\colon (\mathbb{M},\mathbb{E}) \to (\mathbb{C}^2, 0)$ carries holomorphic functions and forms (by pullback) and analytic subsets (by preimages) from $(\mathbb{C}^2, 0)$ to the surface $\mathbb{M}$. However, the results are quite degenerate on the exceptional divisor $\mathbb{E}$.

The alternative is to carry the objects from the *punctured* plane $\mathbb{C}^2 \smallsetminus \{0\}$ to the *complement* $\mathbb{M} \smallsetminus \mathbb{E}$ of the exceptional divisor, and then *extend* them in one way or another on $\mathbb{E}$. The result is called the *blow-up* (*desingularization*) of the initial object.

The accurate construction is slightly different for analytic curves and for (singular) holomorphic foliations.

8**C**$_1$. *Blow-up of analytic curves.* Recall that σ^{-1} is a well-defined holomorphic map of $\mathbb{C}^2 \smallsetminus \{0\}$ to $\mathbb{M} \smallsetminus \mathbb{E}$.

Definition 8.9. The *blow-up* of an analytic curve $\gamma \subseteq (\mathbb{C}^2, 0)$ is the closure (in $\mathbb{M}$) of the preimage of the *punctured* curve $\gamma \smallsetminus \{0\}$:

$$\widetilde{\gamma} = \overline{\sigma^{-1}(\gamma \smallsetminus \{0\})}. \tag{8.4}$$

We have to verify that the result $\widetilde{\gamma}$ is an analytic curve in $\mathbb{M}$. The proof is obtained by *explicitly computing* the blow-up.

Proposition 8.10. *The blow-up of any analytic curve is again an analytic curve in* $(\mathbb{M}, \mathbb{E})$ *intersecting the exceptional divisor* $\mathbb{E}$ *only at isolated points.*

Proof. The equation of the blow-up in $\mathbb{M}$ is obtained by pulling back the equation of γ and cancelling out all terms vanishing identically on $\mathbb{E}$. However, because of the special properties of $\mathbb{E}$ in $\mathbb{M}$ (see Remark 8.6), it can be done only locally.

Consider any holomorphic germ f defining γ and its pullback $f' = \sigma^* f = f \circ \sigma \in \mathcal{O}(\mathbb{M})$. For each point $a \in \mathbb{E}$ the germ of $f' \in \mathcal{O}(\mathbb{M}, a)$ in the local ring $\mathcal{O}(\mathbb{M}, a)$ vanishes identically on $\mathbb{E}$ and can be divided by the maximal power g^ν, $\nu \geqslant 1$, where $g \in \mathcal{O}(\mathbb{M}, a)$ is any irreducible germ locally defining $\mathbb{E} = \{g = 0\}$ near a. After division we obtain the germ $\widetilde{f} = g^{-\nu} f \in \mathcal{O}(\mathbb{M}, a)$ with the following properties:

(1) *outside* $\mathbb{E}$ the germs (at a) of the loci $\sigma^{-1}(\gamma) = \{f' = 0\}$ and $\widetilde{\gamma} = \{\widetilde{f} = 0\}$ coincide,

(2) $\widetilde{f}|_{\mathbb{E}} \not\equiv 0$, hence $\mathbb{E} \not\subseteq \widetilde{\gamma}$.

If the germ $\widetilde{f} \in \mathcal{O}(\mathbb{M}, a)$ is invertible, then the germs of both $\widetilde{\gamma}$ and γ at a are both empty. If $\widetilde{f}$ is noninvertible, then $\widetilde{\gamma} = \sigma^{-1}(\gamma \smallsetminus \{0\}) \cup \{a\}$, that is, the *analytic curve* $\widetilde{\gamma}$ is a one-point closure of the preimage of $\gamma \smallsetminus \{0\}$. □

The blow-up can be alternatively described as the smallest analytic curve $\widetilde{\gamma} \subset \mathbb{M}$ such that $\sigma(\widetilde{\gamma}) = \gamma$. Note that in general this curve can be nonconnected.

8**C**$_2$. *Blow-up of foliations.* Let $\mathcal{F}$ be a singular holomorphic foliation of $(\mathbb{C}^2, 0)$ defined by a holomorphic Pfaffian form $\omega \in \Lambda^1(\mathbb{C}^2, 0)$. By definition, this means that $\mathcal{F}$ is a nonsingular holomorphic foliation of the punctured neighborhood $(\mathbb{C}^2, 0) \smallsetminus \{0\}$. Its preimage $\sigma^{-1}(\mathcal{F})$ is a nonsingular foliation of $\mathbb{M} \smallsetminus \mathbb{E}$ generated by the 1-form $\sigma^* \omega$. But since $\operatorname{codim} \mathbb{E} = 1$, by Theorem 2.20 this preimage foliation can be extended as a singular holomorphic foliation $\sigma^* \mathcal{F}$ with isolated singular points on $\mathbb{E}$.

Definition 8.11. The *blow-up of a singular foliation* $\mathcal{F}$ of $(\mathbb{C}^2, 0)$ is the singular holomorphic foliation $\widetilde{\mathcal{F}} = \sigma^*\mathcal{F}$ of $\mathbb{M}$ extending the preimage foliation $\sigma^{-1}(\mathcal{F})$ of $\mathbb{M} \smallsetminus \mathbb{E}$.

One may have two apriori possibilities for the blow-up $\widetilde{\mathcal{F}}$: either the exceptional divisor $\mathbb{E}$ is a *separatrix* of $\widetilde{\mathcal{F}}$, or different points of $\mathbb{E}$ belong to different leaves of $\widetilde{\mathcal{F}}$. In the latter case leaves of $\widetilde{\mathcal{F}}$ cross $\mathbb{E}$ transversally at almost all points, with the exception of finitely many *tangency points* and isolated singularities of $\widetilde{\mathcal{F}}$.

Definition 8.12. A singular point of a holomorphic foliation $\mathcal{F}$ on $(\mathbb{C}^2, 0)$ is called *nondicritical*, if the exceptional divisor $\mathbb{E} = \sigma^{-1}(0)$ is a separatrix of the blow-up $\sigma^*\mathcal{F}$ by the simple monoidal transformation σ.

Otherwise the singular point is called *dicritical*.

It will be shown that the "generic" singularities of a given order are nondicritical, whereas dicritical singularities correspond to certain degeneracy of the principal homogeneous terms of the vector field defining the foliation.

Remark 8.13. The previous arguments can be carried out *verbatim* for any holomorphic nonconstant map $\pi\colon (M, D) \to (\mathbb{C}^2, 0)$ squeezing a holomorphic curve $D = \pi^{-1}(0)$ (eventually, singular or reducible) into the single point at the origin and one-to-one between $M \smallsetminus D$ and $(\mathbb{C}^2, 0) \smallsetminus \{0\}$. Any holomorphic foliation $\mathcal{F}$ on $(\mathbb{C}^2, 0)$ can be pulled back as a foliation $\pi^{-1}(\mathcal{F})$ on $M \smallsetminus D$ and then extended on D everywhere except for finitely many points. The resulting singular foliation on M will be denoted by $\pi^*\mathcal{F}$ and referred to as a *desingularization*, or *blow-up* of $\mathcal{F}$ by the map π.

8D. Desingularization theorem. It turns out that singular points of *any* holomorphic foliation can be completely simplified by iterated blow-ups. The following result was first discovered by Ivar Bendixson [**Ben01**] in 1901 and improved and generalized by S. Lefschetz [**Lef56, Lef68**], A. F. Andreev [**And62, And65a, And65b**] and A. Seidenberg [**Sei68**]. A. van den Essen simplified the proof considerably in [**vdE79**]; see also [**MM80**]. In [**Dum77**] F. Dumortier obtained a generalization of this theorem for smooth rather than analytic foliations and showed that tangencies can also be eliminated. Recently O. Kleban in [**Kle95**] computed the number of iterates of simple blow-ups required to *desingularize* completely an isolated singularity of a holomorphic foliation.

Recall (see Definition 4.27) that a singularity of the foliation $\mathcal{F}$ defined by the Pfaffian equation $\omega = 0$, $\omega = f\,dx + g\,dy$ with the coefficients $f, g \in \mathcal{O}(\mathbb{C}^2, 0)$ without common factors, is *elementary*, if the linearization matrix

$A = \partial F(0,0)/\partial(x,y)$ of the dual vector field $F = -g\frac{\partial}{\partial x} + f\frac{\partial}{\partial y}$ has at least one nonzero eigenvalue.

Theorem 8.14 (I. Bendixson, A. Andreev, A. Seidenberg, S. Lefschetz, F. Dumortier). *For any singularity of a holomorphic foliation $\mathcal{F}$ one can construct a holomorphic surface M with an analytic curve D on it and a holomorphic map $\pi\colon (M, D) \to (\mathbb{C}^2, 0)$, one-to-one between $M \smallsetminus D$ and $(\mathbb{C}^2, 0) \smallsetminus \{0\}$, such that the blow-up $\pi^*\mathcal{F}$ has only elementary singularities on D.*

More precisely, the map π resolving the singularity can be constructed as a composition of finitely many simple blow-downs.

The vanishing divisor $D = \pi^{-1}(0)$ is the union of finitely many projective lines intersecting transversally, $D = \bigcup D_j$, $D_j \cong \mathbb{P}^1$, $D_i \pitchfork D_j$.

In this section we give the *constructive* proof of this result, based on the idea of van den Essen [**vdE79, MM80**]. This idea is to introduce the *multiplicities* of isolated singularities of holomorphic foliations and monitor their decrease under blow-ups.

Detailed inspection of this algorithm yields the following estimate for the complexity of the desingularization map. It is formulated in terms of *multiplicity* of a singular point of holomorphic foliation, which will be introduced in §8**G**–§8**I**.

Theorem 8.15. *The number of simple blow-ups required to resolve an isolated singularity of multiplicity μ, does not exceed $2\mu + 1$.*

A stronger result was achieved by O. Kleban in [**Kle95**]. He proved that besides resolving all singularities into elementary, in at most $\mu + 2$ steps one can eliminate all *tangency points* between the foliation $\pi^*\mathcal{F}$ and the vanishing divisor D (Theorem 8.37).

8E. Blow-up in an affine chart: computations. Let $\omega = f\,dx + g\,dy \in \Lambda^1(\mathbb{C}^2, 0)$ be a holomorphic 1-form having an isolated singularity of order n. By definition, this means that the Taylor expansions of the coefficients f, g of this form begin with homogeneous polynomials f_n, g_n of degree n and at least one of these two homogeneous polynomials does not vanish identically:

$$\operatorname{ord}_0 \omega = \min\{\operatorname{ord}_0 f, \operatorname{ord}_0 g\}.$$

Consider the pullback $\sigma^*\omega$ on the complex Möbius band $\mathbb{M}$ in the coordinates (x, z) in the chart U_1. In this chart the exceptional divisor $\mathbb{E}$ is defined by the equation $\{x = 0\}$ and the map σ takes the form $\sigma_1\colon (x, z) \mapsto (x, xz)$

and pulls back the form ω to $\omega_1 = \sigma_1^*\omega$ as follows:

$$\begin{aligned} \omega_1 &= [f(x,xz) + zg(x,xz)]\,dx + xg(x,xz)\,dz \\ &= x^{-1}[(\sigma_1^* h)\,dx + (\sigma_1^* g')\,dz], \qquad (8.5) \\ h = xf + yg, &\qquad g' = x^2 g, \qquad h, g' \in \mathcal{O}(\mathbb{C}^2, 0). \end{aligned}$$

Both coefficients of the form ω_1 are divisible at least by x^n. However, the second coefficient is in fact even divisible by x^{n+1}. On the other hand, the first coefficient can "*accidentally*" also be divisible by x^{n+1}, if the homogeneous polynomial $h_{n+1} = xf_n + yg_n$ vanishes identically.

In order to extend the foliation $\widetilde{\mathcal{F}} = \sigma_1^{-1}(\mathcal{F})$ on the line $\mathbb{E} = \{x = 0\}$ in the chart U_1, we have to divide the coefficients of the form (8.5) by the *maximal possible* power of x so that the result will be not identically zero on $\mathbb{E}$. Thus we have two cases which correspond to dicritical and nondicritical singularities; cf. with Definition 8.12.

Proposition 8.16. *The singularity is nondicritical, if*

$$\operatorname{ord}_0(xf + yg) = 1 + \operatorname{ord}_0 \omega, \qquad (8.6)$$

and dicritical, if

$$\operatorname{ord}_0(xf + yg) > 1 + \operatorname{ord}_0 \omega. \qquad (8.7)$$

The *homogeneous* polynomial $h_{n+1} = xf_n + yg_n$ of degree $n+1$ will play an important role in computations pertinent to the dicritical case. It will be referred to as the *tangent form* for lack of a better name. The roots of h_{n+1} can be identified with the points of the projective line $\mathbb{P}$ *globally* isomorphic to the exceptional divisor $\mathbb{E}$.

Proof of the proposition. 1. In the first case (8.6) the blow-up of $\mathcal{F}$ in the chart U_1 is given by the Pfaffian equation with isolated singularities

$$\widetilde{\omega}_1 = 0, \qquad \widetilde{\omega}_1 = [h_{n+1}(1,z) + x(\cdots)]\,dx + x[g_n(1,z) + x(\cdots)]\,dz, \quad (8.8)$$

where f_n, g_n and $h_{n+1} = xf_n + yg_n$ are the homogeneous bivariate polynomials from $\mathbb{C}[x,y]$ as above and the dots denote some holomorphic functions of x and z.

The line $\mathbb{E} = \{x = 0\}$ is integral for the line field $\widetilde{\omega}_1 = 0$, so the exceptional divisor $\mathbb{E}$ in the nondicritical case is a *separatrix* of the blow-up foliation $\widetilde{\mathcal{F}}$. The singular locus $\Sigma = \operatorname{Sing}(\sigma^*\mathcal{F})$ consists of the isolated roots of the equation

$$\Sigma = \{x = 0,\ z = z_j\}, \qquad h_{n+1}(1, z_j) = 0. \qquad (8.9)$$

Their number (counted with multiplicities) is equal to $\deg_z h_{n+1}(1,z)$ which can be *less* than $n+1$ if the homogeneous polynomial $h_{n+1}(x,y)$ is divisible by x. In the latter case the point $z = \infty \in \mathbb{P}^1$ is singular and should be studied in the second affine chart U_2 on $\mathbb{M}$. Globally the singular locus $\Sigma \subset \mathbb{P}^1$ is defined by the tangent form h_{n+1} as the *projective* locus in the homogeneous coordinates $\{(x : y) \in \mathbb{P}^1 : h_{n+1}(x,y) = 0\}$. There is a simple sufficient condition guaranteeing that a point $a \in \Sigma$ is elementary (Proposition 8.18 below).

2. In the second case (8.7) the tangent form vanishes identically, $h_{n+1} \equiv 0$, and the Pfaffian form with isolated singularities which defines the blow-up foliation in the affine chart U_1, is

$$\widetilde{\omega}_1 = 0, \qquad \widetilde{\omega}_1 = [h_{n+2}(1,z) + x(\cdots)]\,dx + [g_n(1,z) + x(\cdots)]\,dz. \tag{8.10}$$

Outside the set $T = \{g_n(1,z) = 0\} \subset \mathbb{E}$ the form $\widetilde{\omega}_1$ is nonsingular and *transversal* to $\mathbb{E}$, which means that the leaves of the blow-up foliation cross $\mathbb{E}$ transversally outside T. Note that $g_n \not\equiv 0$; otherwise the condition $h_{n+1} \equiv 0$ would mean that $f_n \equiv 0$ in violation of the assumption that the order of ω is exactly equal to n.

The points of T may correspond to either *tangency points* if $h_{n+2}(1,z)$ does not vanish (and hence the point is nonsingular), or true singularities if both $g_n(1,z)$ and $h_{n+2}(1,z)$ vanish simultaneously there. □

Remark 8.17. If the singularity is nondicritical and the tangent form $h_{n+1}(1,z)$ has degree $n+1$ and only simple roots, the exceptional divisor $\mathbb{E}$ carries exactly $n+1$ singular points of $\widetilde{\mathcal{F}}$. The fundamental group of the complement $\mathbb{E} \smallsetminus \Sigma$ is generated by small loops around these singularities. Hence the holonomy group of the foliation $\widetilde{\mathcal{F}}$ along the leaf $\mathbb{E} \smallsetminus \Sigma$ is generated by $n+1$ germs $g_0, \dots, g_n \in \operatorname{Diff}(\mathbb{C}^1, 0)$ subject to a single relationship $g_0 \circ \cdots \circ g_n = \mathrm{id}$. This group is sometimes referred to as the *vanishing holonomy group* of the initial singular point of the foliation $\mathcal{F}$. Later, in §23**D**, we will discuss necessary and sufficient conditions for a group generated by $n+1$ conformal germs to be a vanishing holonomy group of a foliation satisfying the above assumptions.

Another computation will be required in the proof of the Desingularization theorem.

Proposition 8.18. *Each* simple (*nonmultiple*) *linear factor $ax + by$ of the tangent form $h_{n+1} = xf_n + yg_n$ corresponds to an elementary singularity $z = -a/b$ (resp., $w = -b/a$) of the blow-up foliation.*

Proof. In the assumptions of the proposition, the singularity is obviously nondicritical and without loss of generality we may assume that the factor is simply y, and $h_{n+1}(1,z) = zu(z)$ and $u(0) = 1$.

The vector spanning the same line field as the distribution (8.8), has the form

$$\dot{z} = z + ax + \mathfrak{m}^2, \qquad \dot{x} = -bx + \mathfrak{m}^2,$$

where a, b are some two complex numbers and $\mathfrak{m}^2$ denote functions of order $\geqslant 2$. The linearization matrix $\left(\begin{smallmatrix}1 & * \\ 0 & *\end{smallmatrix}\right)$ of this field has nonzero eigenvalue 1 for the eigenvector tangent to $\mathbb{E}$. □

8F. Divisors. To proceed with the demonstration of the desingularization theorem, we first introduce a convenient algebraic formalism for counting analytic subvarieties (points and analytic hypersurfaces) with certain integer *multiplicities* (positive or negative). While this formalism cannot be easily extended for subvarieties of intermediate dimensions, for the two extreme dimensions (zero and maximal, i.e., codimension 1) the theory is as complete as possible.

The integer multiplicity can be easily attached to analytic subvarieties of codimension one (hypersurfaces) using the fact that the ring of germs of analytic functions admits unique irreducible factorization. This construction leads to the notion of a *divisor*, introduced and discussed in this section. Multiplicity of zero-dimensional sets (isolated points) can be introduced in a different way via codimension of the respective ideals as explained in §8**G** as the *intersection multiplicity* of two analytic curves. Behavior of these multiplicities under blow-up is studied in §8**H**–§8**I**.

8**F**$_1$. *Definitions.* A *divisor* on a complex manifold M is a finite union of irreducible analytic hypersurfaces (analytic subsets of codimension 1) with assigned integer multiplicities (coefficients). By this definition, each divisor D is a formal sum $\sum_\gamma k_\gamma\,\gamma$ where the summation is formally over *all* irreducible subvarieties of codimension 1, but only finitely many integer coefficients $k_\gamma \in \mathbb{Z}$ can be in fact nonzero. Divisors form an Abelian group denoted by $\operatorname{Div}(M)$ with the operation denoted additively, $\left(\sum k_\gamma\,\gamma\right) + \left(\sum k'_\gamma\,\gamma\right) = \sum (k_\gamma + k'_\gamma)\,\gamma$. The divisor is called *effective* if all k_γ are nonnegative; any divisor can be formally represented as a formal *difference* of two effective divisors. The *support* of a divisor is the union of all subvarieties entering into D with nonzero coefficients,

$$|D| = \bigcup_{k_\gamma \neq 0} \gamma \cong \sum_{k_\gamma \neq 0} \gamma,$$

which can be alternatively thought of as either the point set or an effective divisor with all k_γ being just 0 or 1.

If M is one-dimensional, divisors are finite point sets with integer multiplicities attached to each point. We will be interested here in the two-dimensional case where M is a holomorphic surface and the divisors are unions of irreducible curves counted with multiplicities.

8**F**$_2$. *Divisors and meromorphic functions.* Each holomorphic function $f \in \mathcal{O}(M)$ defines an effective divisor D_f called the *divisor of zeros* of f as follows. The support $|D_f|$ is the zero locus $Z_f = \{f = 0\} \subseteq M$, and if the germ of f at a point $a \in M$ has the irreducible factorization $f = \prod f_j^{\nu_j}$ in the local ring $\mathcal{O}(M, a)$, then the component $D_j = D_{f_j}$ of D_f is assigned the multiplicity $\nu_j \geqslant 0$:

$$D_f = \sum_j \nu_j D_j, \qquad D_j = D_{f_j} = \{f_j = 0\}.$$

This definition allows us to assign the multiplicity ν_j to each irreducible component $D_j \subseteq D_f$ near the point a only, but the answer is obviously locally constant as a varies along D_j. Since D_j is connected, the result does not depend on a, moreover, one can always choose a being a smooth point on D_j.

For a meromorphic function $h = f/g$ the divisor D_h is defined as the formal difference,

$$D_{f/g} = D_f - D_g.$$

It obviously does not depend on the choice of the representation.

Conversely, any divisor can be associated with a meromorphic function, albeit only locally. Let $D = \sum k_\gamma \gamma$ be a divisor on M. Then M can be covered by a union of open domains $\{U_\alpha\}$ so that in each domain U_α each hypersurface $\gamma \subseteq |D|$ is represented by a holomorphic equation $\{f_{\alpha,\gamma} = 0\}$ with the differential $df_{\alpha,\gamma}$ nonvanishing outside a set of codimension 2 on γ. The divisor D locally in U_α is defined by the meromorphic function $f_\alpha = \prod_\gamma f_{\alpha,\gamma}^{k_\gamma} \in \mathcal{M}(U_\alpha)$. The collection $\{f_\alpha\}$ is called a *meromorphic* 1-*cochain* defining the divisor D.

Consider the pairwise intersections $U_{\alpha\beta} = U_\alpha \cap U_\beta$ and the ratios $g_{\alpha\beta} = f_\alpha / f_\beta$ in these intersections. These ratios are holomorphic and nonvanishing, since both f_α and f_β define the same divisor in the intersection $U_{\alpha\beta}$. The collection of holomorphic invertible functions $g_{\alpha\beta}$ is called the *holomorphic cochain* (more precisely, holomorphic 2-cochain) defining the divisor D. Addition of divisors corresponds to multiplication of the holomorphic cochains: if D, D' are two divisors defined by the holomorphic cochains $\{g_{\alpha\beta}\}$ and $\{g'_{\alpha\beta}\}$, then the sum $D + D'$ is defined by the cochain $\{g_{\alpha\beta} g'_{\alpha\beta}\}$.

Note that some divisors *may not be definable* by a single *global* equation on M, e.g., the exceptional divisor $\mathbb{E}$ on the complex Möbius band $\mathbb{M}$; see Remark 8.6.

With respect to holomorphic maps, divisors behave like analytic functions, i.e., they are *pulled back* by such maps. Let $\pi \colon M' \to M$ be a nonconstant holomorphic map between two connected manifolds of the same dimension and $D = \sum k_\gamma \gamma$ a divisor on M defined by the meromorphic cochain $\{f_\alpha\}$.

Definition 8.19. The *preimage* (pullback) $\pi^{-1}(D)$ of a divisor $D \in \operatorname{Div}(M)$ is the divisor on M' which in the open domains (charts) $U'_\alpha = \pi^{-1}(U)$ is defined by the meromorphic cochain $f'_\alpha = \pi^* f_\alpha \in \mathcal{M}(U'_\alpha)$.

Since π^* is a ring homomorphism, taking preimages commutes with addition/subtraction of divisors: for any two divisors D, D' on M,

$$\pi^{-1}(D \pm D') = \pi^{-1}(D) \pm \pi^{-1}(D').$$

In other words, $\pi^{-1}\colon \operatorname{Div}(M) \to \operatorname{Div}(M')$ is a homomorphism of Abelian groups.

Example 8.20. Preimage of the sum of n different straight lines $\sum_1^n \ell_j$ associated with the function $f(x,y) = \prod l_j \in \mathcal{O}(\mathbb{C}^2, 0)$ (the product of n different linear factors) by the monoidal map $\sigma\colon \mathbb{M} \to \mathbb{C}^2$ is the divisor $n\mathbb{E} + \sum_1^n \widetilde{\ell}_j$, where $\mathbb{E}$ is the exceptional divisor and $\widetilde{\ell}_j$ the blow-ups of the lines ℓ_j.

8G. Intersection multiplicity and intersection index. In this section we define the multiplicity of intersection of two divisors (curves) at an isolated point and the global intersection index between divisors. More details can be found in [**vdE79, MM80, Chi89**]. The theorem on equivalence of different definitions appears in [**AGV85**, §5], and the intersection theory in the algebraic context is explained in [**Sha94**, Chapter IV].

We start with the particular case of effective divisors and define first the local multiplicity of their intersection at a common point, say, the origin in $\mathbb{C}^2$. Let $f, g \in \mathcal{O}(\mathbb{C}^2, 0)$ be two holomorphic germs and D_f, D_g the respective *effective* divisors in $(\mathbb{C}^2, 0)$. We say that the intersection D_f and D_g is *isolated* at the origin, if $|D_f| \cap |D_g| \cap (\mathbb{C}^2, 0) = \{0\}$ (in the sense of germs of analytic sets). The intersection is isolated if and only if no irreducible component enters both divisors with positive coefficient, i.e., f, g have no common irreducible factors in the ring of germs $\mathcal{O}(\mathbb{C}^2, 0)$. In this case we can give several equivalent definitions of the intersection multiplicity $\mu = D_f \overset{0}{\cdot} D_g$ between D_f and D_g at the origin $a = 0$.

8$\mathbf{G}_1$. *Algebraic construction.* Consider the ideal $I_{f,g} = \langle f, g\rangle \subset \mathcal{O}(\mathbb{C}^2, 0)$ generated by these germs in the local ring of germs, and the quotient *local algebra* $Q_{f,g} = \mathcal{O}(\mathbb{C}^2, 0)/I_{f,g}$ as a linear space over $\mathbb{C}$. The *algebraic multiplicity of intersection* is defined as the dimension of the local algebra (codimension of the ideal),

$$\begin{aligned} D_f \overset{0}{\cdot} D_g &= \dim_{\mathbb{C}} Q_{f,g} = \operatorname{codim}_{\mathcal{O}(\mathbb{C}^2,0)} I_{f,g}, \\ I_{f,g} &= \langle f, g\rangle \subset \mathcal{O}(\mathbb{C}^2, 0), \qquad Q_{f,g} = \mathcal{O}(\mathbb{C}^2, 0)/I_{f,g}. \end{aligned} \tag{8.11}$$

By definition, the equality $\dim Q_{f,g} = \mu < +\infty$ means that there exist the germs $e_1, \dots, e_\mu$ which are a basis of the local algebra so that any other germ $u \in \mathcal{O}(\mathbb{C}^2, 0)$ admits the representation

$$u = \sum_1^{\mu} c_i\, e_i + af + bg, \qquad c_1, \dots, c_\mu \in \mathbb{C}, \quad a, b \in \mathcal{O}(\mathbb{C}^2, 0), \tag{8.12}$$

and the constant coefficients c_i are defined uniquely. By this definition, the multiplicity of intersection depends only on the ideal $\langle f, g \rangle$.

8**G**$_2$. *Geometric construction.* The pair of analytic functions (f, g) considered as coordinate functions, defines a holomorphic map $P = P_{f,g} \colon (\mathbb{C}^2, 0) \to (\mathbb{C}^2, 0)$. If the intersection D_f and D_g is isolated, the preimage of the $P^{-1}(0,0) = (0,0)$ is a single point. Maps with such properties have an integer topological invariant, the degree. Consider a small 3-dimensional real sphere $\mathbb{S}^3_\rho = \{|x|^2 + |y|^2 = \rho\} \subset \mathbb{C}^2 \cong \mathbb{R}^4$ and the "normalization" of P, the map $\widehat{P} = \widehat{P}_{f,g} \colon (x, y) \mapsto P(x,y)/|P(x,y)|$. The normalized map $\widehat{P}$ is not analytic, only differentiable, and its range is the unit sphere $\mathbb{S}^3_1 = \{|z|^2 + |w|^2 = 1\}$. Restricting $\widehat{P}$ on a sufficiently small sphere $\mathbb{S}^3_\rho$, we obtain thus a map between two spheres has an invariant, the *topological degree*, which can be described as the number of preimages (counted with the sign determined by the orientation) of a generic point in the target sphere. This degree is the same for all sufficiently small choices of $\rho > 0$.

The *geometric multiplicity* of intersection between D_f and D_g at the origin is defined as the topological degree of the map $\widehat{P}$,

$$\begin{aligned} D_f \overset{0}{\cdot} D_g &= \operatorname{top\,deg}_0 \widehat{P}_{f,g}, \qquad \widehat{P}_{f,g} \colon \mathbb{S}^3_\rho \to \mathbb{S}^3_1, \\ \widehat{P}_{f,g} &\colon (x, y) \mapsto \frac{\big(f(x,y), g(x,y)\big)}{|f(x,y)|^2 + |g(x,y)|^2}. \end{aligned} \tag{8.13}$$

8**G**$_3$. *Deformational construction.* Let the positive number $\rho > 0$ be so small that the system of equations $\{f = 0,\ g = 0\}$ has a unique solution $\{x = y = 0\}$ in the ball $B_\rho = \{|x|^2 + |y|^2 < \rho\}$ (as before, $f, g \in \mathcal{A}(B_\rho)$ are holomorphic representatives of the initial germs). Then for almost all sufficiently small (relative to ρ) complex values $a, b \in \mathbb{C}$, $|a|, |b| < \varepsilon$, the holomorphic level curves $\{f = a\}$ and $\{g = b\}$ are smooth inside B_ρ and intersect transversally. This follows from the Sard lemma: it is sufficient to require a be a regular value for f and b a regular value of g restricted on the nonsingular curve $\{f = a\}$. The transversality implies that the intersection $\{f = a\} \cap \{g = b\} \cap B_\rho$ consists of isolated points. The *deformational multiplicity of intersection* between D_f and D_g at the origin is the number of these points:

$$D_a \overset{0}{\cdot} D_g = \#\{f = a\} \cap \{g = b\} \cap B_\rho \quad \text{for generic } (a, b) \in (\mathbb{C}^2, 0). \tag{8.14}$$

Apriori it is not clear why this definition makes sense and the above number is the same for *all* generic combinations (a, b).

8**G**$_4$. *Definition of multiplicity.* One of the central results of the singularity theory claims that the three definitions of multiplicity lead to the same answer.

Theorem 8.21. *For a pair of germs $f, g \in \mathcal{O}(\mathbb{C}^2, 0)$ without common factors in the irreducible decomposition, all three definitions* (8.11), (8.13) *and* (8.14) *lead to the same* finite *number $\mu = \mu_{f,g} \in \mathbb{Z}_+$.* □

The proof of this theorem can be found in [**AGV85**, §5].

Definition 8.22. The common value established in Theorem 8.21 is called the multiplicity of intersection between the divisors D_f and D_g at the origin.

Remark 8.23. The ideas behind the proof of Theorem 8.21 are rather natural and can be explained as follows.

Coincidence between the geometric and deformational definitions is actually the theorem about the sum of indices of singular points of a vector field $P_{f-a,g-b} \in \mathcal{D}(B_\rho)$ with the coordinates $(f - a, g - b)$ in the ball B_ρ that is equal to the degree of this vector field on the boundary of the ball. Important is the fact that each transversal intersection in the complex domain corresponds to a singular point of index $+1$ (unlike the real case where the index can be of positive and negative sign). The degree of the vector field $P_{f-a,g-b}$ on the boundary is an integer-valued function of a, b that is continuous, hence it must be a constant equal to the limit, the degree of $P_{f-0,g-0}$ which is the geometric multiplicity (8.13). This argument can be made into a rigorous proof that the geometric and deformational multiplicities coincide.

If in the definition of the algebraic multiplicity we replace the germs f, g by the holomorphic functions $f - a$ and $g - b$ considered as elements from the ring $\mathcal{A}(B_\rho)$ for some positive ρ, then the quotient algebra $\mathcal{A}(B_\rho)/\langle f - a, g - b\rangle$ is isomorphic to the algebra of functions on μ distinct points, where μ is the deformational multiplicity given by (8.14). It requires some effort to prove that the dimension of the quotient algebra remains the same, first in the limit as $(a, b) \to 0 \in \mathbb{C}^2$, and then in the limit $\rho \to 0^+$. The latter is exactly the algebraic multiplicity.

A convenient tool for computation of the intersection multiplicity is the following lemma. Assume that the divisor D_f is irreducible (i.e., the germ f is irreducible in the local ring $\mathcal{O}(\mathbb{C}^2, 0)$. In this case D_f can be locally parameterized by an injective nonconstant map $\tau\colon (\mathbb{C}^1, 0) \to (\mathbb{C}^2, 0)$ such that $0 \equiv f \circ \tau \in \mathcal{O}(\mathbb{C}^1, 0)$ (see Theorem 2.26).

Lemma 8.24. *The intersection of an irreducible local divisor D_f with another effective local divisor D_g is isolated if and only if the germ $g \circ \tau$ is not identically zero, and the multiplicity $D_f \cdot D_g$ of this intersection is equal to the order* $\operatorname{ord}_0(f \circ \tau)$.

Proof. Consider a regular value b of the function g on the curve $\gamma = \{f = 0\}$ and the corresponding intersection locus $Z_{0b} = \{f = 0,\ g = b\}$ inside B_ρ. We prove first that the intersection multiplicity $\mu = D_f \cdot D_g$ is equal to the number $\# Z_{0b}$ of the points in this locus.

To that intermediate end, consider the coefficient $h \in \mathcal{A}(B_\rho)$ of the 2-form $df \wedge dg = h\, dx \wedge dy$. This coefficient cannot vanish identically on γ:

by irreducibility of f, the differential $df|_\gamma$ vanishes only at the origin, hence $h \equiv 0$ would mean that dg is proportional to df at all points of γ, therefore $dg|_\gamma \equiv 0$ and $g|_\gamma$ is a constant. As $g(0) = 0$, this constant is necessarily equal to zero, in contradiction with our assumptions that Z_{00} consists of a single point at the origin. Thus $h|_\gamma \not\equiv 0$, and one can assume without loss of generality that ρ is so small that $h|_\gamma$ is nonvanishing outside the origin.

Nonvanishing of h at all points $Z_{0b} \subseteq \gamma$ for $b \neq 0$ means that the restriction of f on the curve $\{g = b\}$ has simple roots at exactly these points. Any small perturbation $f - a$ will have exactly the same number $\#Z_{ab} = \#Z_{0b}$ of complex roots on $\{g = b\}$ which is by deformational definition of multiplicity equal to μ.

The points from Z_{0b} are τ-parameterized by the small roots of the holomorphic function of one variable $(g - b) \circ \tau = g \circ \tau - b$ which is a small perturbation of the function $g \circ \tau$. It remains to observe that a small perturbation of a germ of order μ in $\mathcal{O}(\mathbb{C}^1, 0)$ is a function that has exactly μ roots in a sufficiently small neighborhood of the origin. □

Another application of Theorem 8.21 is the following additivity of the intersection multiplicity.

Proposition 8.25. *For any three* effective *divisors D, D', D'' on $(\mathbb{C}^2, 0)$, such that $D \cap (|D'| \cup |D''|)$ is a single point 0, the intersection multiplicities satisfy the equality*

$$D \overset{\circ}{\cdot} (D' + D'') = D \overset{\circ}{\cdot} D' + D \overset{\circ}{\cdot} D''. \tag{8.15}$$

Proof. Let D', D'' and D be the divisors of the germs f, g and h respectively, which are identified with their representatives holomorphic in a sufficiently small ball B_ρ. Then the divisor $D' + D''$ is that of the product fg.

By the deformational construction, for a generic combination of the values $(a', a'', b) \in (\mathbb{C}^3, 0)$, the intersections $Z'_{a'b} = \{f = a',\ h = b\}$ and $Z''_{a''b} = \{g = a'',\ h = b\}$ are transversal and consist of $\mu' = D \overset{\circ}{\cdot} D'$ and $\mu'' = D \overset{\circ}{\cdot} D''$ points respectively. Excluding only finitely many values of b, one may assume without loss of generality that $Z'_{a'b}$ and $Z''_{a''b}$ are disjoint: this happens if the level curve $\{h = b\}$ avoids the common points of $\{f = a'\}$ and $\{g = a''\}$. In these assumptions, the number of transversal intersections between the curve $\{h = b\}$ and the reducible curve $\{(f - a')(g - a'') = 0\}$ is exactly equal to $\mu' + \mu''$.

The function $(f - a')(g - a'')$ is not a perturbation of the form $fg - a$ that appears in the deformational construction. Yet because of the continuity, the degree of the vector fields $P_{fg-a,h-b}$, $P_{(f-a')(g-a''),h-b}$ and $P_{fg,h}$ on the boundary of the ball B_ρ are the same if a, a', a'' and b are all sufficiently small compared to ρ. Thus by the geometric definition of the multiplicity, we conclude that $D \overset{0}{\cdot} (D' + D'') = \mu' + \mu''$. $\square$

8**G**$_5$. *Intersection form between arbitrary global divisors.* Using Proposition 8.25, one can extend the formulas for the multiplicity of intersections for arbitrary (not necessarily effective) divisors, by the standard construction.

For a pair of local divisors, an effective divisor D' and an arbitrary divisor D represented as a difference of two effective divisors $D = D_1 - D_2$, we define the multiplicity of intersection (always at the origin) as

$$D' \cdot D = D' \cdot D_1 - D' \cdot D_2. \tag{8.16}$$

If $D = D_3 - D_4$ is another representation, then by definition $D_1 + D_4 = D_2 + D_3$, so that by Proposition 8.25, $D' \cdot D_1 + D' \cdot D_4 = D' \cdot D_2 + D' \cdot D_3$ and hence $D_3 \cdot D' - D_4 \cdot D'$ coincides with $D' \cdot D_1 - D' \cdot D_2$, which means that the definition is self-consistent. Multiplicity of intersection of two noneffective divisors is defined by iterating this construction twice, and the additivity law (8.15) holds automatically for any three divisors.

Consider now the general case of divisors on an arbitrary *complex analytic* surface M. Two divisors D, D' on M are said to have *isolated intersection*, if $|D| \cap |D'|$ is a finite point set.

Definition 8.26. The *intersection index* between two divisors D, D' with isolated intersection is the sum of all intersection multiplicities:

$$D \cdot D' = \sum_{a \in M} D \overset{a}{\cdot} D', \qquad \text{if } |D| \cap |D'| \text{ is a finite set.} \tag{8.17}$$

The summation in (8.17) is formally extended over all points in M, but only points from $|D| \cap |D'|$ may contribute nonzero terms.

The intersection index is a bilinear (over $\mathbb{Z}$) symmetric form $\operatorname{Div}(M) \times \operatorname{Div}(M) \to \mathbb{Z}$, also called *intersection index*, defined on pairs of divisors with isolated intersection,

$$\begin{gathered} D, D' \longmapsto D \cdot D', \qquad \text{when } |D| \cap |D'| \text{ is finite set,} \\ D \cdot (D' \pm D'') = D \cdot D' \pm D \cdot D'', \qquad (D, D') = (D', D). \end{gathered} \tag{8.18}$$

Defined in this way, the intersection index generalizes the notion of the number of intersection points counted with multiplicities. Its functoriality (behavior by holomorphic maps) is studied in the next subsection.

8H. Blow-up and intersection index. The intersection index is well defined and invariant by *biholomorphisms*: if $\pi \colon M' \to M$ is a biholomorphism, then

$$\begin{gathered} \pi^{-1}(D) \cdot \pi^{-1}(D') = D \cdot D', \\ D, D' \in \operatorname{Div}(M), \quad \pi^{-1}(D), \pi^{-1}(D') \in \operatorname{Div}(M') \end{gathered} \tag{8.19}$$

for any two divisors D, D' on M with an isolated intersection. However, if σ is a *blow-up* then the preimage of the point $\{0\}$ is the exceptional divisor which therefore belongs to the preimage of *any* divisor. Hence $\sigma^{-1}(D)$ and $\sigma^{-1}(D')$ necessarily have nonisolated intersection even if $|D| \cap |D'| = \{0\}$: this intersection always contains the exceptional divisor $\mathbb{E}$ with a positive multiplicity if D, D' were effective; see Example 8.20.

One can attempt to *extend* the intersection form on pairs of divisors $R, R' \in \operatorname{Div}(C)$ which have no *nonexceptional* common components, i.e., when

$$|R| \cap |R'| \subseteq S, \tag{8.20}$$

so that the identity (8.19) would hold also when π is a blow-up. We shall see that only one such extension is possible.

Remark 8.27. Theorem 8.21 can be interpreted as the *local continuity* of the intersection index. For instance, consider an effective divisor D defined by a family of local equations $\{f_\alpha = 0\}$ in suitable charts U_α. If another family $\{f'_\alpha \in \mathcal{O}(U_\alpha)\}$ is a sufficiently small perturbation of $\{f_\alpha \in \mathcal{O}(U_\alpha)\}$ also has nonvanishing holomorphic ratios $f'_\alpha / f'_\beta \in \mathcal{O}(U_\alpha \cap U_\beta)$, it defines a small perturbation D' of the divisor D as explained in §8$\mathbf{F}_2$. By the deformational construction, the intersection index of D and D' with any other divisor D'' is the same, $D \cdot D'' = D' \cdot D''$ (while multiplicities of particular intersection points may of course change).

Thus in principle one might wish to define the *self-intersection index* for any divisor D by perturbing it slightly to become a divisor D_ε and let by definition $D \cdot D = \lim_{D_\varepsilon \to D} D \cdot D_\varepsilon$. For instance, if D is defined by a *global* equation $D = D_f$ for some $f\colon M \to \mathbb{C}$, then one can choose $D_\varepsilon = D_{f-\varepsilon}$: since different level curves are disjoint, $D \cdot D_e = 0$ for all $\varepsilon \neq 0$, and hence we have the identity $D \cdot D = 0$. On the other hand, if $M = \mathbb{P}^2$ is the projective plane and D is a line on it, then $D \cdot D = 1$.

Yet the self-intersection index of the exceptional divisor $\mathbb{E}$ cannot be obtained in this way because of the rigidity of $\mathbb{E}$ inside the Möbius band $\mathbb{M}$ (Remark 8.6). Moreover, we will see that in order to preserve (8.19), one has to assign the self-intersection index $\mathbb{E} \cdot \mathbb{E}$ the *negative* value -1 (note that the intersection index between any two *different* divisors is always nonnegative).

Example 8.28. Consider two divisors defined by two lines $\ell_{1,2}$ transversally crossing at the origin in $(\mathbb{C}^2, 0)$. Their preimages by the standard monoidal map $\sigma\colon \mathbb{M} \to (\mathbb{C}^2, 0)$ consist of the blow-ups $\widetilde{\ell}_{1,2}$ and the exceptional divisor:

$$\sigma^{-1}(\ell_j) = \mathbb{E} + \widetilde{\ell}_j, \qquad j = 1, 2;$$

cf. with Example 8.20. Note that both blow-ups $\widetilde{\ell}_{1,2}$ are smooth, intersect $\mathbb{E}$ transversally, hence $\widetilde{\ell}_j \cdot \mathbb{E} = 1$, and are *disjoint*, so $\widetilde{\ell}_1 \cdot \widetilde{\ell}_2 = 0$. If the preimages were to have the same intersection index $\sigma^{-1}(\ell_1)\cdot\sigma^{-1}(\ell_2) = \ell_1\cdot\ell_2 = 1$, then we would have the identity

$$1 = \ell_1\cdot\ell_2 = \mathbb{E}\cdot\mathbb{E} + \mathbb{E}\cdot(\widetilde{\ell}_1 + \widetilde{\ell}_2) + \widetilde{\ell}_1\cdot\widetilde{\ell}_2 = \mathbb{E}\cdot\mathbb{E} + 1 + 1 + 0,$$

which leaves only one possibility, $\mathbb{E}\cdot\mathbb{E} = -1$.

Theorem 8.29. *The intersection form between divisors on* $\mathbb{M}$ *can be uniquely extended for pairs of divisors satisfying* (8.20) *as a symmetric bilinear form with the following properties:*

$$\mathbb{E}\cdot\mathbb{E} = -1, \tag{8.21}$$

$$\sigma^{-1}(D)\cdot\mathbb{E} = 0, \qquad \forall D \in \operatorname{Div}(\mathbb{C}^2, 0), \tag{8.22}$$

$$\sigma^{-1}(D)\cdot\sigma^{-1}(D') = D\cdot D', \qquad \forall D, D' \in \operatorname{Div}(\mathbb{C}^2, 0) \tag{8.23}$$

(*the last condition holds only for pairs of divisors* $D, D' \in \operatorname{Div}(\mathbb{C}^2, 0)$ *having isolated intersection*).

Proof. We need to prove that the rule (8.21) if adopted as an axiom and combined with bilinearity, would imply the identities (8.22) and (8.23) for arbitrary divisors $D, D' \in \operatorname{Div}(\mathbb{C}^2, 0)$. Because of the bilinearity and symmetry, it is sufficient to complete the proof when the divisor $D = D_f$ is a curve defined by a holomorphic germ $f \in \mathcal{O}(\mathbb{C}^2, 0)$.

Denote by $n = \operatorname{ord}_0 f$ the order of the holomorphic germ $f = f_n + f_{n+1} + \cdots$. Without loss of generality we may assume that the principal homogeneous part f_n is *not* divisible by x, so that $f_n(x,y) = cy^n + \cdots$, $c \neq 0$ (otherwise an affine change of coordinates should first be made). In the chart U_1 we have

$$\begin{aligned}\sigma_1^* f(x,z) &= x^n f_n(1,z) + x^{n+1}(1,z) + \cdots = x^n[f_n(1,z) + x f_{n+1} + \cdots] \\ &= x^n \widetilde{f}(x,z), \qquad \widetilde{f}(0,z) = f_n(1,z) \neq 0,\end{aligned}$$

so that by definition of the preimage of divisors,

$$\sigma^{-1}(D_f) = n\mathbb{E} + \widetilde{D}_f, \qquad \widetilde{D}_f = D_{\widetilde{f}}, \quad n = \operatorname{ord}_0 f. \tag{8.24}$$

As a curve, $|\widetilde{D}_f|$ is the blow-up of the curve $|D_f|$, since the function $\widetilde{f}$ does not vanish identically on $\mathbb{E}$. Occurrence of the term $n\mathbb{E}$ stresses the difference between preimage of the divisor and blow-up of its support curve.

The intersection between $\widetilde{D}_f$ and $\mathbb{E}$ is isolated and consists of the roots of the polynomial $f_n(1,z)$ of degree exactly n. If $a = (0, a')$ is such a point, then the multiplicity of intersection $\widetilde{D}_f \overset{a}{\cdot} \mathbb{E}$ at this point is equal to the multiplicity of the root of $f_n(1,z)$ at $z = a' \in \mathbb{C}$, since $\widetilde{f}(x,z) = f_n(1,z) \bmod \langle x\rangle$ and the quotient rings $\mathcal{O}(\mathbb{C}^2, a)/\langle x, \widetilde{f}\rangle$ and $\mathcal{O}(\mathbb{C}^1, a')/\langle f_n(1,\cdot)\rangle$ are naturally isomorphic. Adding the contributions of all points together, we obtain

$$\widetilde{D}_f\cdot\mathbb{E} = \deg_z f_n(1,z) = \operatorname{ord} f = n. \tag{8.25}$$

Using the axiom (8.21), we obtain from (8.24) by linearity

$$\sigma^{-1}(D_f) \cdot \mathbb{E} = (-1) \cdot n + \widetilde{D}_f \cdot \mathbb{E} = -n + n = 0.$$

The proof of (8.22) is complete.

To prove (8.23) we assume that the analytic curve $D = D_f$ is irreducible and parameterized by an injective holomorphic map $\tau\colon (\mathbb{C}^1, 0) \to (\mathbb{C}^2, 0)$, $t \mapsto \big(x(t), y(t)\big)$.

By Lemma 8.24, the intersection multiplicity $D_f \mathbin{\mathring{:}} D_g$ is equal to the multiplicity (order) $\operatorname{ord}_0 g \circ \tau$ of the root $t = 0$ of the composition $g \circ \tau$.

If $D_f = \gamma$ is an irreducible curve parameterized by τ, then the map $\widetilde{\tau}\colon t \mapsto \sigma^{-1} \circ \tau$, $t \neq 0$, parameterizes the points of $\sigma^{-1}(\gamma) \smallsetminus \mathbb{E}$. It obviously extends holomorphically at the origin and becomes a map $\widetilde{\tau}\colon (\mathbb{C}^1, 0) \to C$ parameterizing the blow-up curve $\widetilde{D}_f = \widetilde{\gamma}$.

If $D' = D_g$ is an arbitrary divisor (reducible or not), then using Lemma 8.24 twice we obtain

$$\begin{aligned} D_g \cdot D_f &= \operatorname{ord}_0 g \circ \tau = \operatorname{ord}_0 g \circ \sigma \circ \sigma^{-1} \circ \tau = \operatorname{ord}_0 (\sigma^* g) \circ \widetilde{\tau} \\ &= D_{\sigma^* g} \cdot \widetilde{D}_f = \sigma^{-1}(D_g) \cdot \widetilde{D}_f. \end{aligned}$$

Combining this with (8.24) and (8.22), we obtain

$$\begin{aligned} \sigma^{-1}(D_g) \cdot \sigma^{-1}(D_f) &= \sigma^{-1}(D_g) \cdot (n\mathbb{E} + \widetilde{D}_f) \\ &= n\,\sigma^{-1}(D_g) \cdot \mathbb{E} + \sigma^{-1}(D_g) \cdot \widetilde{D}_f \\ &= 0 + D_g \cdot D_f = D_g \cdot D_f. \end{aligned}$$

The proof of (8.23) is complete when D is irreducible. As was already mentioned, the proof in the general case follows from bilinearity of the intersection index. □

As a corollary to Theorem 8.29, we obtain a simple formula for the intersection index between *blow-ups* of two analytic curves.

Corollary 8.30. *For any pair of two holomorphic curves $\gamma, \gamma' \subseteq (\mathbb{C}^2, 0)$ of orders m and m', and their blow-ups $\widetilde{\gamma}, \widetilde{\gamma}' \subset (\mathbb{M}, \mathbb{E})$, the intersection indices are related by the formula*

$$\gamma \cdot \gamma' = \widetilde{\gamma} \cdot \widetilde{\gamma}' + mm'. \tag{8.26}$$

Proof. By (8.24), on the level of divisors

$$\sigma^{-1}(\gamma) = m\mathbb{E} + \widetilde{\gamma}, \qquad \sigma^{-1}(\gamma') = m'\mathbb{E} + \widetilde{\gamma}'.$$

Using bilinearity, we conclude that

$$\widetilde{\gamma} \cdot \widetilde{\gamma}' = (\sigma^{-1}(\gamma) - m\mathbb{E}) \cdot (\sigma^{-1}(\gamma') - m'\mathbb{E}) = \gamma \cdot \gamma' - 0m - 0m' + (-1)\,mm'$$

by virtue of the three rules (8.21), (8.22) and (8.23). □

Example 8.31. If γ, γ' are two *smooth* (of order 1) analytic curves, then their intersection multiplicity decreases by 1 after blow-up. Since in the smooth case the intersection multiplicity is equal to the order of tangency between γ and γ' minus 1, *the order of tangency between smooth curves is also decreased by one by blow-up.*

8I. Blow-up and multiplicity of singular foliations. Consider a singular holomorphic foliation $\mathcal{F}$ defined by the Pfaffian equation $\{\omega = 0\}$, $\omega \in \Lambda^1(\mathbb{C}^2, 0)$ or a holomorphic vector field $F \in \mathcal{D}(\mathbb{C}^2, 0)$ near an isolated point at the origin. Denote by n the *order* of the form ω at the origin: by definition, it means that

$$\omega = f\,dx + g\,dy = (f_n + f_{n+1} + \cdots)\,dx + (g_n + g_{n+1} + \cdots)\,dy \tag{8.27}$$

and the homogeneous polynomials f_n, g_n of lowest degree n do not vanish identically: $f_n\,dx + g_n\,dy \neq 0$. The assumption that the singularity is isolated means that the intersection of the coordinate divisors D_f and D_g is isolated.

Definition 8.32. The *multiplicity* $\mu_0(\omega)$ of the singular point of the form (8.27) at the origin is the intersection multiplicity $D_f \,\mathring{\cdot}\, D_g$ between the respective divisors.

The multiplicity $\mu_a(\mathcal{F})$ of a singular foliation $\mathcal{F}$ at a point a is the multiplicity of any holomorphic form ω tangent to $\mathcal{F}$ and having an isolated singular point at a.

Consider a small perturbation F_ε of the vector field, say, $(f - \varepsilon_1)\frac{\partial}{\partial x} + (g - \varepsilon_2)\frac{\partial}{\partial y}$. If the vector field F_ε has only nondegenerate singularities and $\varepsilon \in (\mathbb{C}^2, 0)$ is sufficiently small, then the number of these singular points is exactly equal to the multiplicity by Theorem 8.21. By this definition, multiplicities of *nonsingular* points are equal to zero.

The definition of multiplicity does not depend on the choice of local coordinates used for writing the coefficients of the form. This follows from the deformational interpretation of the multiplicity. An alternative argument is as follows: changing the coordinates results in replacing the coefficients (f, g) of the form by another tuple of functions (f', g') belonging to the same ideal $\langle f, g\rangle$. If the change of coordinates is invertible, the two ideals are equal and so are the local algebras.

Our immediate goal is to compare the total multiplicity of all singularities of a foliation $\mathcal{F}$ and its blow-up $\widetilde{\mathcal{F}} = \pi^*\mathcal{F}$ for a simple blow-up π. Clearly, it is sufficient to consider the case where $\mathcal{F}$ has an isolated singularity on $(\mathbb{C}^2, 0)$ and the blow-up is the standard monoidal transformation $\sigma\colon (\mathbb{M}, \mathbb{E}) \to (\mathbb{C}^2, 0)$. The answer is different in the dicritical and nondicritical cases.

Consider the singular foliation $\mathcal{F}$ determined by 1-form $\omega = f\,dx + g\,dy$ of order n as in (8.27) and denote $\widetilde{\mathcal{F}}$ its blow-up as defined in Definition 8.11.

Theorem 8.33. *Let $\mathcal{F}$ be a singular foliation on $(\mathbb{C}^2, 0)$ and $\widetilde{\mathcal{F}}$ its blow-up. Then in all cases except for the dicritical singularity of order* 1,

$$\sum_{a \in S} \mu_a(\widetilde{\mathcal{F}}) = \mu_0(\mathcal{F}) - k(k-2) + n. \tag{8.28}$$

Here $n = \operatorname{ord}_0 \omega$, $m = \operatorname{ord}_0(xf + yg) \geqslant n+1$ (with the equality occurring in the nondicritical case) and

$$k = \min(n+2, m) = \begin{cases} n+1, & \text{in the nondicritical case,} \\ n+2, & \text{in the dicritical case.} \end{cases} \tag{8.29}$$

In the nondicritical case the formula (8.28) implies

$$\sum_{a} \mu_a(\widetilde{\mathcal{F}}) = \mu_0(\mathcal{F}) - (n^2 - n - 1). \tag{8.30}$$

In the dicritical case of order $n > 1$ the formula (8.28) yields

$$\sum_{a} \mu_a(\widetilde{\mathcal{F}}) = \mu_0(\mathcal{F}) - (n^2 + n). \tag{8.31}$$

In the dicritical case of order $n = 1$ we have $\mu_0(\mathcal{F}) = 1$ whereas the blow-up foliation $\widetilde{\mathcal{F}}$ is nonsingular, therefore

$$\sum_{a} \mu_a(\widetilde{\mathcal{F}}) = 0 = 1 - 1 = \mu_0(\mathcal{F}) - n^2. \tag{8.32}$$

Corollary 8.34. *If $n > 1$, then the total number of singularities of $\widetilde{\mathcal{F}}$ counted with their multiplicities, hence the multiplicity of every particular singularity, is strictly smaller than the multiplicity of the initial singularity,*

$$\sum_{a \in S} \mu_a(\widetilde{\mathcal{F}}) < \mu_0(\mathcal{F}). \quad \square \tag{8.33}$$

Proof of Theorem 8.33. We start with a convenient choice of the affine chart to work in. Making an affine transformation if necessary, we will then be able to assume without loss of generality that this chart is the standard affine chart U_1 with the coordinates (x, z).

First, we can assume that the only point *not* covered by the affine chart, is nonsingular for the blow-up foliation $\widetilde{\mathcal{F}}$. In the nondicritical case this is equivalent to assuming that the principal homogeneous part $h_{n+1} = xf_n + yg_n$ is not divisible by x.

Moreover, we can always assume in addition that the intersection of the divisors D_g and D_h is isolated: since $h = xf + yg$, this happens if and only if g is *not divisible* by x. To ensure that, we will assume that g_n is *not* divisible by x. Unlike the previous assumptions which can always be achieved by a suitable affine transformation, this last assumption can be achieved in all cases *except* for the dicritical case of order $n = 1$. In the latter case we always have $g_1(x, y) = x$ since the linear part of the corresponding vector field is a scalar matrix which remains scalar in any affine coordinates.

In the affine chart $U_1 \cong \mathbb{C}^2$ with the coordinates (x, z) the pullback of the form ω by the monoidal map $\sigma\colon (x, z) \mapsto (x, xz)$ was computed in (8.5). Technically it is more convenient to pull back the form $x\omega \in \Lambda^1(\mathbb{C}^2, 0)$: the fact that it has a nonisolated singularity does not matter, as the pullback will be in any case divided by a suitable power of x when extended on the exceptional divisor. The advantage is that the coefficients of the 1-*form* $\sigma_1^*(x\omega) = (\sigma_1^* h)\, dx + \sigma_1^*(x^2 g)\, dz$ are pullbacks of two *holomorphic germs* h and $g' = x^2 g$.

To extend the form $\sigma_1^*(x\omega)$ on the exceptional divisor $\mathbb{E} = \{x = 0\}$, one has to divide the coefficients $\sigma_1^* h$ and $\sigma_1^* g'$ by the maximal positive power x^k of the function x which is the local (relative to the chart U_1) equation of the exceptional divisor. Depending on whether the initial singularity is dicritical or not, we have two possibilities for this maximal order k, given by (8.29). The intersection multiplicity between $x^{-k}\sigma_1^* h$ and $x^{-k}\sigma_1^* g'$ at any point on the line $x = 0$ will then be the multiplicity of the corresponding singularity of the blow-up foliation.

On the language of the divisors the total multiplicity of all singular points of $\widetilde{\mathcal{F}}$ on the exceptional divisor $\mathbb{E}$ reduces to computation of the intersection index between the divisors $\sigma^{-1}(D_h) - k\mathbb{E}$ and $\sigma^{-1}(D_{x^2 g}) - k\mathbb{E} = \sigma^{-1}(D_g) - (k-2)\mathbb{E}$ in the open domain $U_1 \subset \mathbb{M}$. However, by our assumption that the point not covered by U_1 is nonsingular, we may extend the summation over all singular points on $\mathbb{E}$ using bilinearity and the rules established in Theorem 8.29:

$$\begin{aligned}\sum_a \mu_a(\widetilde{\mathcal{F}}) &= (\sigma^{-1}(D_h) - k\mathbb{E}) \cdot (\sigma^{-1}(D_{x^2 g}) - k\mathbb{E}) \\ &= (\sigma^{-1}(D_h) - k\mathbb{E}) \cdot \big(\sigma^{-1}(D_g) - (k-2)\mathbb{E}\big) \\ &= \sigma^{-1}(D_h) \cdot \sigma^{-1}(D_g) + k(k-2)\, \mathbb{E} \cdot \mathbb{E} \\ &= D_h \cdot D_g - k(k-2).\end{aligned} \tag{8.34}$$

It remains to compute the intersection index between two divisors $D_h, D_g \subset (\mathbb{C}^2, 0)$ at the origin, where $h = xf + yg$, and express it via $D_f \cdot D_g$. Using the fact that the intersection multiplicity depends only on the ideal generated by these germs, we obtain

$$D_h \overset{0}{\cdot} D_g = D_{xf+yg} \overset{0}{\cdot} D_g = D_{xf} \overset{0}{\cdot} D_g = D_x \overset{0}{\cdot} D_g + D_f \overset{0}{\cdot} D_g.$$

The multiplicity of intersection $D_x \overset{0}{\cdot} D_g$ is equal to the order of the function $\operatorname{ord}_0 g(0, y)$. If g_n is not divisible by x, this order is equal to n, so that ultimately

$$D_h \overset{0}{\cdot} D_g = \mu_0(\mathcal{F}) + n, \qquad n = \operatorname{ord}_0 \mathcal{F}.$$

Putting everything together, we obtain the formula (8.28). □

8J. Desingularization of cuspidal points. Multiplicity of isolated singularities of order $n > 1$ goes down after blow-up (dicritical or not). To prove the desingularization theorems, we need to show that the only nonelementary points of order 1, the cuspidal points, can be desingularized in finitely many steps. Note that since the order of a cuspidal point is 1, the total multiplicity of all singularities which appear after blow-up (nondicritical) goes up by 1 by (8.30). We will show that for cuspidal points the multiplicity decreases after *two* consecutive blow-ups if it was three or higher, whereas a cusp of multiplicity 2 after three blow-ups gets desingularized into elementary points.

Without loss of generality we may assume that the lower order terms of the form ω are brought to the normal form

$$\begin{aligned} \omega = y\,dy + [f(x) + yg(x)]\,dx, \qquad & f, g \in \mathbb{C}[[x]], \\ \operatorname{ord}_0 f = \mu \geqslant 2, \qquad & \operatorname{ord}_0 g > 0. \end{aligned} \tag{8.35}$$

(cf. with (4.18)). In fact, we need only terms of order 2 for the analysis below. The number $\mu \geqslant 2$ is the multiplicity of the singular point (8.35).

The quadratic tangent form $xf_1 + yg_1$ for (8.35) is equal to y^2. It is nonzero (hence the singularity is nondicritical) and the only singular point after blow-up is the point $z = 0$ in the chart U_1, where the blow-up of ω takes the form

$$xz\,dz + (ax + bx^2 + cxz + z^2)\,dx + \mathfrak{m}^3 \otimes \Lambda^1, \tag{8.36}$$

where a, b are the leading coefficients of $f(x) = ax^2 + bx^3 + \cdots$ ($a \neq 0$ if and only if $\mu = 2$) and c the leading coefficient of $g(x) = cx + \cdots$. Here and below the notation $\mathfrak{m}^k$ is used to denote a collection of terms of order $\geqslant k$ and the tensor product stands for the 1-form with third order coefficients.

Further arguments are different for *simple cusp* with $\mu = 2$ and *higher cusps* with $\mu > 2$.

8J$_1$. *Simple cusp.* We show that after three consecutive blow-ups the simple cusp of multiplicity $\mu = 2$ can be blown up into three nondegenerate singularities.

If $\mu = 2$, then without loss of generality one may assume that $a = 1$. The order of the singularity (8.36) which appears after the first blow-up, is again 1 so it is a simple cusp, its multiplicity (by (8.30) with $n = 1$) is $3 = 2 + 1$ and the tangent form is $x^2 \not\equiv 0$. After the *second* blow-up (substitution

$x = uz$ and division by z) the cuspidal singular point (8.36) is transformed into the foliation defined by the form

$$uz\,dz + (u+z)(u\,dz + z\,du) + \mathfrak{m}^3 \otimes \Lambda^1, \tag{8.37}$$

which has a unique singularity at $u = 0$. The order of this singularity is now 2 and multiplicity is equal to $4 = 3 + 1$ by (8.30) (again with $n = 1$).

The tangent form for (8.37), $uz^2 + 2uz(u+z) = uz(2u+3z)$, is the product of three different (simple) linear factors which means that after the *third* blow-up the foliation will have three singular points of total multiplicity $3 = 4 - 1$ (again by (8.30) yet this time with $n = 2$). This leaves only one combination of multiplicities $1, 1$ and 1 respectively, meaning that all three points are nondegenerate (hence elementary). One can show by direct computation that all three points are resonant saddles.

8**J**$_2$. *Higher cusp.* In this case already after the first blow-up the form (8.36) has order 2, multiplicity $\mu+1$ by (8.30) and the tangent form $xz^2 + x(bx^2 + cxz + z^2) = x(bx^2 + cxz + 2z^2)$ which is divisible by x but not a power of x. In other words, after the *second* blow-up there will appear at least *two* distinct points (three if $c^2 \neq 8b$) of *total* multiplicity μ by (8.30). This means that each of these two points has multiplicity at most $\mu - 1$ after *two* consecutive blow-ups.

Proof of Desingularization Theorems 8.14 and 8.15. We construct a sequence of blow-ups that would resolve completely an isolated singularity. The algorithm is very simple: starting from the initial singularity of a foliation $\mathcal{F} = \mathcal{F}_0$ at the origin $0 \in M_0 \cong (\mathbb{C}^2, 0)$, we construct a simultaneous simple blow-up $\pi_k \colon M_k \to M_{k-1}$, $k = 1, 2, \dots$, of all *nonelementary* singular points $\Sigma_{k-1} \subset M_{k-1}$ of the foliation $\mathcal{F}_{k-1}$ obtained on the previously constructed surface M_{k-1}.

The assertion on the vanishing divisor D (preimage of the origin) can be easily verified inductively. If $\gamma \subset M$ is a nonsingular curve biholomorphically equivalent to $\mathbb{P}^1$ and $a \in \gamma$ a center of blow-up $\pi \colon M' \to M$, then by Example 8.31 the blow-up $\pi^*\gamma$ will again be a nonsingular curve $\widetilde{\gamma}$ biholomorphically equivalent to γ and therefore again equivalent to $\mathbb{P}^1$ (note that the topology of embedding of $\widetilde{\gamma}$ in M' may change). If γ, γ' intersect transversally, then their blow-ups will be disjoint and both transversal to the exceptional divisor $\pi^{-1}(a) \subset M'$ created by π. Thus the assertion on the vanishing divisor reproduces itself inductively and holds at any moment. The proof of Theorem 8.14 is complete.

To prove Theorem 8.15, it remains to estimate the number of simple blow-ups before the algorithm terminates, i.e., before all singularities become elementary. Note that all singularities appearing in the process, can be organized in a tree graph with branches connecting each singularity with its descendants appearing by the simple blow-up. Take the longest branch in this tree, $0 = a_0$, $a_1 \in \Sigma_1$, $a_2 \in \Sigma_2$, *etc.* We claim that, with the possible exception of the last three steps, the multiplicity of singularities a_i decreases at least by one every step or, at worst, every two steps. Denoting by μ_i the respective multiplicities, we already know that:

(1) if a_i is of order > 1, then $\mu_{i+1} < \mu_i$ by Corollary 8.34;

(2) if a_i is of order 1 and is neither elementary nor simple cusp, then $\mu_{i+2} < \mu_i$ by §8$\mathbf{J}_2$;

(3) if a_i is a simple cusp, then the branch terminates after three more steps by §8$\mathbf{J}_1$.

These inequalities constrain the maximal length of the branch by $2(\mu-1)+3 = 2\mu+1$. The proof of Theorems 8.14 and 8.15 is complete. □

8K. Concluding remarks: elimination of resonant nodes and dicritical tangencies. Elementary singular points can also be to some extent simplified by blow-up. For instance, a nondegenerate singularity with the eigenvalues λ_1, λ_2, defined by the Pfaffian equation

$$x\,dy + \lambda y\,dx + \cdots = 0, \qquad \lambda = -\lambda_1/\lambda_2 \neq -1,$$

is "split" by the blow-up into two singularities which are both nondegenerate when $\lambda \neq -1$. The corresponding negative ratios of eigenvalues will be $\lambda+1$ and $(\lambda^{-1}+1)^{-1}$.

The case $\lambda = -1$ corresponds either to the dicritical node $x\,dy + y\,dx + \cdots = 0$ or to the Jordan node $(x+y)\,dy + y\,dx + \cdots = 0$. The former singularity *disappears* after blow-up, while the latter produces an elementary singular point whose hyperbolic eigenspace is *transversal* to the exceptional divisor (the corresponding tangent form is y^2).

Combining these observations, one can make additional blow-ups on top of the desingularization achieved in Theorem 8.14 and *eliminate all resonant nodes with natural ratios of eigenvalues.* Indeed, such points correspond to negative natural values $\lambda = -n$ which can be increased by 1 in $n-1$ steps until the parameter λ reaches the threshold value $\lambda = -1$ (all other singularities appearing in the process will be resonant saddles with $\lambda = n/(n-1)$). On the next step the singularity either disappears or becomes a saddle-node.

In another development, one can refine the assertion of the Desingularization Theorem 8.15 to *eliminate tangency points* between the foliation $\pi^*\mathcal{F}$ and the vanishing divisor D. We briefly outline here the required adjustments.

The tangency order between two smooth curves $\{f = 0\}$ and $\{g = 0\}$ is by definition the multiplicity of intersection $D_f \mathbin{\mathring{\cdot}} D_g$ minus 1: if two curves intersect transversally, the tangency order is 0, for a true tangency it is always positive.

The *tangency order* between a foliation $\mathcal{F}$ defined by the Pfaffian equation $\omega = 0$ and a *smooth* analytic curve $\gamma = \{f = 0\}$ at a point a is defined only when γ is *not* a leaf or separatrix of $\mathcal{F}$.

If a is nonsingular for $\mathcal{F}$, then the tangency order $\tau_a(\mathcal{F}, \gamma)$ is by definition the tangency order between γ and the leaf of $\mathcal{F}$ passing through a. If γ is defined by the equation $\{f = 0\}$ locally near a, then one can easily verify that

$$\tau_a(\mathcal{F}, \gamma) = D_{\omega\wedge df} \overset{a}{\cdot} D_f, \tag{8.38}$$

where $D_{\omega\wedge df}$ is the divisor of zeros of the 2-form $\omega \wedge df = \rho(x, y)\, dx \wedge dy$ identified with its coefficient ρ, $D_{\omega\wedge df} = D_\rho$.

Indeed, if the tangency order is k, then after choosing suitable local coordinates one can assume that $\omega = dy$ (recall that a is nonsingular) and $\gamma = \{f = 0\}$, $f(x, y) = y - b(x)$, $\operatorname{ord}_0 b = k + 1$. The expression in the right hand side of (8.38) will be then equal to the order of $\sigma(x, y) = db(x)/dx$ restricted on the smooth curve γ parameterized by x, i.e., to $k = \operatorname{ord}_0 b - 1$.

In the case where a is a singular point, one can use (8.38) as a *definition* of the tangency order. The important property of the tangency order thus defined, is the following one.

Proposition 8.35. *If a is a hyperbolic singular point of $\mathcal{F}$ which is not a resonant node, and L is a* separatrix *of the foliation $\mathcal{F}$ passing through it, then the order of tangency between L and any other smooth curve γ is by* 1 *greater than the order of tangency between $\mathcal{F}$ and γ,*

$$\gamma \overset{0}{\cdot} L = \tau(\mathcal{F}, \gamma) + 1.$$

Proof. We can assume that the local coordinates are chosen so that the separatrix L is a coordinate axis, $L = \{y = 0\}$. Then $\omega = \lambda y(1 + \mathfrak{m})\, dx + (x + \mathfrak{m}^2)\, dy$, where λ is the negative ratio of eigenvalues.

A curve γ tangent to $\{y = 0\}$ with order $k \geqslant 0$, is defined by the equation $y - b(x) = 0$, $\operatorname{ord}_0 b = k + 1$. Direct computation of (8.38) yields

$$\tau_0(\mathcal{F}, \gamma) = \operatorname{ord}_{x=0}[\lambda b(x)(1 + \mathfrak{m}) - b'(x)(x + \mathfrak{m}^2)] = k + 1$$

if $\lambda \neq k + 1$, i.e., if the singular point is not a resonant node with the ratio of eigenvalues $-1 : (k + 1)$. $\square$

Using the tangency order, one can combine the equalities (8.31) and (8.32) into a single identity valid for both $n > 1$ and $n = 1$. Assume that the origin is a *dicritical* singularity of a holomorphic foliation $\mathcal{F}$. Denote by Σ the singular locus of its blow-up $\widetilde{\mathcal{F}}$ and by T the collection of the tangency points between $\widetilde{\mathcal{F}}$ and the exceptional divisor.

Proposition 8.36. *If the singularity is dicritical of any order $n \geqslant 1$, then*

$$\sum_{a\in\Sigma} \mu_a(\widetilde{\mathcal{F}}) + \sum_{b\in T} \tau_b(\widetilde{\mathcal{F}}, S) = \mu_0(\mathcal{F}) - n^2. \tag{8.39}$$

Proof. When $n > 1$, the equality (8.39) follows from (8.31) and the observation that the order of tangency between $\widetilde{\mathcal{F}}$ given by the Pfaffian equation $x^{-n}[(\cdots)\,dx + g(x, xz)\,dz]$ and $\mathbb{E} = \{x = 0\}$ at any point is equal to the order of the root of the function $x^{-n}g(x, xz) = g_n(1, z) + \cdots$ restricted on $\mathbb{E}$. The total multiplicity of all roots of $g_n(1, z)$ is equal to n, which proves (8.39) for $n > 1$. For $n = 1$ this formula is proved by direct inspection: there are neither singular nor tangency points after blow-up, whereas the initial multiplicity $\mu_0(\mathcal{F})$ is equal to 1. □

Behavior of tangency points after blow-up can be easily controlled: by (8.26), the intersection multiplicity between two *smooth* analytic curves decreases by 1 after blow-up. Using this fact, one can achieve by elementary inductive arguments the following improvement of the Desingularization Theorem 8.14.

Theorem 8.37 (**[Kle95]**). *In the formulation of the Desingularization Theorem 8.14 one can always guarantee that the dicritical components of the vanishing divisor $D = \pi^{-1}(0)$ carry no tangency points with the foliation $\pi^*\mathcal{F}$ (in particular, no singularities of the latter).*

The number of simple blow-ups necessary to desingularize the singular point of multiplicity μ in this strong sense does not exceed $\mu + 2$.

Exercises and Problems for §8.

Exercise 8.1. Compute blow-ups of:

(1) a smooth analytic curve passing through 0,

(2) several lines through 0 crossing each other by nonzero angles,

(3) the cusp $y^2 - x^3 = 0$.

Exercise 8.2. What happens after blow-up of a *nonsingular* point of a vector field?

Exercise 8.3. What happens after blow-up of a *homogeneous* vector field?

Problem 8.4. Give direct algebraic proof of Proposition 8.25 based on constructing the basis for the local algebra $Q_{fg,h}$ from the bases of the local algebras $Q_{f,h}$ and $Q_{g,h}$ respectively.

Exercise 8.5. Compute the ratios of eigenvalues for all three nondegenerate singular points obtained by complete desingularization of the simple cuspidal point described in §8$\mathbf{J}_1$.

Problem 8.6. Prove that any holomorphic vector field $F = (F_1, F_2)$ with an *isolated* singular point at the origin $0 \in \mathbb{C}^2$ satisfies the *Lojasiewicz condition*: there exist finite positive C and M such that $|F(x)| > C|x|^M$ for all $x \in (\mathbb{C}^2, 0) \smallsetminus \{0\}$.

Problem 8.7. Prove that consecutive desingularization of a *rational node*, a singularity with the ratio of eigenvalues $\lambda = p/q \in \mathbb{Q}$, $p, q \neq 1$, necessarily involves a dicritical blow-up on some step. How many standard simple blow-ups are required to obtain a singular point whose subsequent blow-up is dicritical?

Problem 8.8. Suppose that the complete desingularization of an isolated singularity of multiplicity μ does not involve neither cusps nor dicritical blow-ups. Give an upper bound for the number of blow-ups in the desingularization, better than in Theorem 8.15.

Problem 8.9. Suppose that a nice blowing up of an isolated singular point of a planar analytic vector field is completely nondicritical and has at most one noncorner singular point. Prove that all the characteristic numbers (ratios of the eigenvalues) of the singular points of the nice blowing up are rational.

Chapter II

Singular points of planar analytic vector fields

In this chapter we apply the analytic tools developed earlier in Chapter I, to the study of singular points of planar, mostly real analytic vector fields.

9. Planar vector fields with characteristic trajectories

A real analytic vector field F on the plane or, more generally, a real analytic 2-dimensional manifold U (*surface*) defines a *real analytic* foliation $\mathcal{F}_F$ by real analytic curves on the complement to the zero locus $\Sigma_F = \{F = 0\}$. Note that the leaves of foliations generated by vector fields are naturally *oriented* by the field F. By definition, we say that $\mathcal{F}$ is a *singular* real analytic foliation, if in a neighborhood of any point it is locally defined by a real analytic vector field with complex analytic singularities, as described in §2**D**. Sometimes, when a real analytic foliations is originally defined by a vector field, the foliation is called the *phase portrait* of the field.

Topological ("qualitative") description of singularities of planar vector fields was essentially achieved in the middle of the twentieth century and its results exposed in several monographs. In this and the following section we will describe the results, placing special emphasis on the effectiveness and "algebraicity" of the algorithms which allow us to decide the topological types of the singular points.

The "real topological theory", while intuitively rather obvious, still would require developing of an appropriate technique that would lead us too far away from the main theme of the book. Thus in many cases we had to compress accurate elementary topological reasoning into "sketches" of the proofs. The interested reader is advised to consult the encyclopedic treatises [**ALGM73, Har82**] and the classical but recently reprinted book [**NS60**] and transform these sketches into accurate proofs. The survey [**AI85**], especially Chapter III, may be instrumental in finding accurate references.

9A. First steps of topological classification: Poincaré types and saddle-nodes. Two vector fields F and F' defined on two surfaces U and U' respectively, are *topologically (orbitally) equivalent* if there exists an orientation-preserving homeomorphism $H\colon U \to U'$ mapping Σ_F to $\Sigma_{F'}$ and the leaves of $\mathcal{F}$ to the leaves of $\mathcal{F}'$ *respecting the orientations.* Two germs of vector fields are topologically equivalent, if they admit topologically equivalent representatives.

One of the principal problems of the local theory of analytic differential equations on the plane is topological classification of germs of isolated singularities of planar analytic vector fields. The initial steps of this classification were implemented by H. Poincaré who gave topological classification of nondegenerate *linear* planar vector fields (a degenerate singularity cannot be linear and isolated simultaneously). Poincaré introduced the topological types listed in Table II.1 and proved that any linear vector field is topologically equivalent to one of the first three types listed in the table.

	Type	**Eigenvalues**	**Normal form**
1	Saddle	$\lambda_1\lambda_2 < 0$	$x\frac{\partial}{\partial x} - y\frac{\partial}{\partial y}$
2	Nodes	$\lambda_1\lambda_2 > 0,\ \operatorname{Re}\lambda_{1,2} \neq 0$	$\pm(x\frac{\partial}{\partial x} + y\frac{\partial}{\partial y})$
3	Center	$\lambda_{1,2} = \pm i\omega$	$-y\frac{\partial}{\partial x} + x\frac{\partial}{\partial y}$
4	Saddle-nodes	$\lambda_1 = 0 \neq \lambda_2$	$x^2\frac{\partial}{\partial x} \pm y\frac{\partial}{\partial y}$

Table II.1. Topological types of planar elementary singularities of real analytic vector fields: $\lambda_{1,2}$ are the eigenvalues of the linearization. The nodes and saddle-nodes with different signs, are not equivalent.

For *nonlinear nondegenerate* analytic singularities no new types arise: except for one case (center), any analytic (and even smooth) germ of vector field is topologically equivalent to its linear part. This follows from the Grobman–Hartman topological linearization theorem for hyperbolic singularities [**Gro62, Har82**]. A vector field whose linearization is a center, may

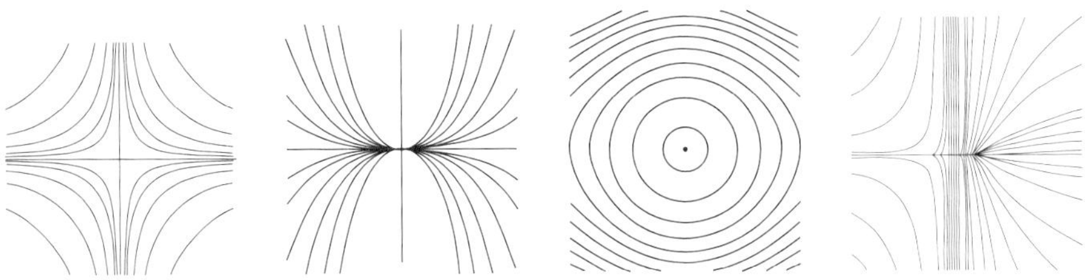

Figure II.1. Poincaré types of phase portraits: saddle, node, center and saddle-node

be center or *focus* (see Definition 9.10); we shall explore this issue in depth in §10**C** below.

Degenerate *elementary* singularities exhibit only one new topological type, the *saddle-node* (see Table II.1). We summarize these results as follows.

Theorem 9.1 (see [**ALGM73**]). *Any elementary singularity of a planar real analytic vector field is topologically equivalent to one of the six types listed in Table II.1.*

The linear part uniquely determines (via the conditions described in the second column of the table) the topological type for all nondegenerate singularities except for the case of purely imaginary eigenvalues, which may correspond to a center or a focus (topological node).

A degenerate elementary singularity can be topologically equivalent to a saddle-node if the multiplicity is even, or saddle or node if the multiplicity is odd. □

9B. Sectorial decomposition of nonelementary singularities. Any isolated singularity can be resolved into elementary ones by Theorem 8.14. Blowing down the corresponding two-dimensional surfaces with foliations on them, one can obtain description of phase portraits of degenerate singularities in terms of *sectors* which were introduced by I. Bendixson (1901); see also [**Har82**, Ch. VII, §8] and [**Per01**, §2.11].

Definition 9.2. A "standard sector" (which may be parabolic, hyperbolic or elliptic) is the germ of a standard oriented foliation $\mathcal{F}_p$, $\mathcal{F}_h$ or $\mathcal{F}_e$ defined on the quadrant $\{x \geqslant 0,\ y \geqslant 0\} \smallsetminus \{0\} \subseteq (\mathbb{R}^2, 0)$ by the vector fields

(i) $F_p = x\frac{\partial}{\partial x} + y\frac{\partial}{\partial y}$ (parabolic),

(ii) $F_h = x\frac{\partial}{\partial x} - y\frac{\partial}{\partial y}$ (hyperbolic),

(iii) $F_e = z^4\frac{\partial}{\partial z}$, $z = x + iy$ (elliptic)

respectively; see Fig. II.2.

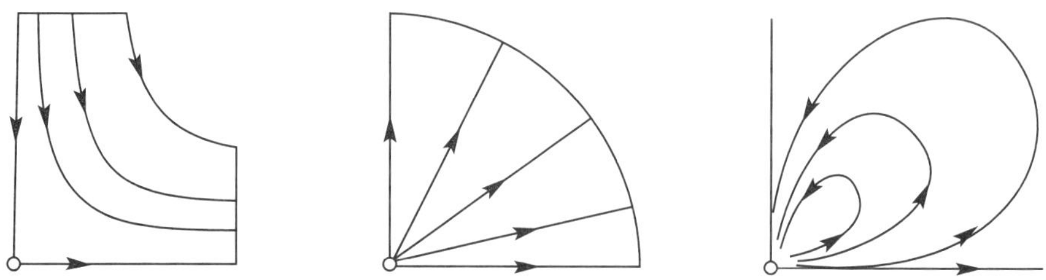

Figure II.2. Hyperbolic, parabolic and elliptic sectors

Since the boundary rays of the sector are leaves of the standard foliations, standard sectors of different types in the quadrant can be inscribed in sectors of arbitrary positive opening and attached to each other in a cyclical order to produce an oriented foliation on $(\mathbb{R}^2, 0) \smallsetminus \{0\}$ (this construction can be implemented in any smoothness, though in general not in the real analytic category). If a nonelementary singularity of a planar vector field is topologically equivalent to a foliation obtained by such a surgery, so that the boundaries of the sectors are characteristic trajectories, we say that the singularity admits *sectorial decomposition*.

Remark 9.3 (Warning). Boundaries between sectors of a sectorial decomposition are not uniquely defined; see Exercise 9.9.

9C. Monodromic singularities, characteristic orbits, limit cycles. Not all singularities, though, admit sectorial decomposition (e.g., centers). To give sufficient conditions for existence of sectorial decomposition, we need the following definitions which take the simplest form after a blow-up.

Let $\mathcal{F}$ be a foliation generated by a real analytic vector field on $(\mathbb{R}^2, 0)$, and $\mathcal{F}'$ its blow-up, the foliation on the (real standard) Möbius band $M = {}^{\mathbb{R}}\mathbb{M}$, as described in Remark 8.5.

Definition 9.4. A leaf L of the foliation $\mathcal{F}$, represented as a parameterized curve $\gamma\colon (-T, +\infty) \to (\mathbb{R}^2, 0)$, $t \mapsto \gamma(t)$, is called a *characteristic orbit* (or *characteristic curve*), if $\lim_{t\to+\infty} \gamma(t) = 0$ and the preimage $\widetilde{\gamma}(t) = \sigma^{-1}(\gamma(t))$ has a well-defined limit $a = \lim_{t\to+\infty} \widetilde{\gamma}(t)$ on the exceptional divisor $C \subset M$.

In other words, the characteristic orbit is a semi-infinite trajectory which tends in the corresponding limit (as the times tends to plus or minus infinity) to the singular point along a certain direction with the slope $a \in C \cong \mathbb{R}P^1$.

Remark 9.5 (Warning). Blow-up σ of a real analytic oriented foliation $\mathcal{F}$ is another real analytic oriented foliation $\mathcal{F}'$, yet orientation of the preimage leaf $L' = \sigma^{-1}(L) \in \mathcal{F}'$ may differ from the one induced from its source $L \in \mathcal{F}$ (consider the standard node given by the Euler field and its nonsingular dicritical blow-up).

Assume that the foliation $\mathcal{F}$ is nondicritical and $a \in M$ a nonsingular point of $\mathcal{F}'$ on the central circle $C = \sigma^{-1}(0) \subset M$. Consider a cross-section $\tau\colon (\mathbb{R}^1, 0) \to (M, a)$ to $\mathcal{F}'$ at a. The standard holonomy map associated with such a choice, may be not defined if C carries singular points. Yet for certain types of singularities one can still define the monodromy map without going into the complex domain.

Definition 9.6. The foliation $\mathcal{F}$ is called *monodromic*, if all nonexceptional leaves $L \not\subseteq C$ crossing τ sufficiently close to a, cross again this section at least once in the future and in the past.

Example 9.7. Among foliations generated by linear planar vector fields, only centers and foci are monodromic.

For monodromic foliations one may define the germ of the *monodromy map* $\Delta = \Delta_\tau\colon (\tau, a) \to (\tau, a)$ as the first return map associated with the cross-section τ. Yet (mainly for historical reasons) the monodromy is defined as the *second return map*.

Indeed, because of the topology of the real Möbius band, any trajectory (leaf) close to the central circle C, cuts the cross-section $\tau\colon (\mathbb{R}, 0) \to (C, a)$ from two sides so that the signs of the local coordinate, corresponding to consecutive intersections, alternate. After two turns around the band, any leaf of the monodromic singularity again crosses the section τ from the same side; see Fig. II.3.

Definition 9.8. The monodromy map of a monodromic singularity is the return map for the positive semi-section $\tau_+\colon (\mathbb{R}_+, 0) \to (C, a)$ to the central circle at a nonsingular point of the latter.

This map coincides with the *square* (second iteration) of the holonomy of the real exceptional divisor $C \subset M$ if this divisor carries no real singular points of the foliation.

Remark 9.9. The property of being monodromic is *not* invariant by topological equivalence: a focus is monodromic, while a node which is topologically equivalent to the focus, is not.

It is a simple exercise to show that an isolated singularity of a planar real analytic foliation is a *center*, i.e., topologically equivalent to the corresponding field from Table II.1, if and only if all nonsingular leaves are closed (compact, homeomorphic to the circle).

Definition 9.10. A singularity is *focus*, if it is a monodromic topological node.

A monodromic singularity is a focus, if and only if some neighborhood of it is free from periodic orbits (Problem 9.3).

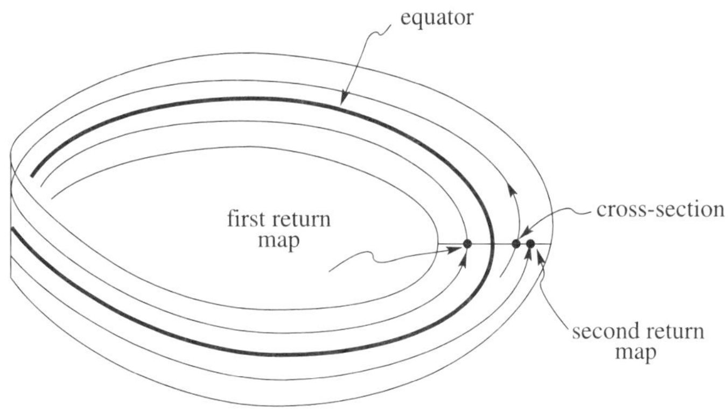

Figure II.3. First and second return maps for foliations on the real Möbius band

Finally, we define one of the most elusive objects in the analytic theory of analytic differential equations.

Definition 9.11. A *limit cycle* of a planar vector field is an isolated periodic trajectory (isolated compact leaf of the corresponding foliation).

A periodic orbit which has an annular neighborhood filled by periodic trajectories, is called an *identical cycle.*

In other words, a periodic trajectory of a vector field is a limit cycle, if it has an annular neighborhood free from other periodic trajectories. Identical cycles are closely related to centers: indeed, each periodic orbit sufficiently close to the center, is an identical cycle.

For cycles of real analytic vector fields, there is no third possibility.

Theorem 9.12. *Every periodic orbit of a real analytic vector field is either a limit cycle, or an identical cycle.*

Proof. Any cycle is a multiply-connected leaf of the corresponding foliation, hence on any cross-section one can define the holonomy map $\Delta \in \operatorname{Diff}(\mathbb{R}^1, 0)$. This map is real analytic by construction, and any other periodic orbit corresponds to a fixed point of Δ. Yet a real analytic one-dimensional self-map which differs from identity, may have only isolated fixed points by the uniqueness theorem. □

9D. Principal alternative and topological classification of singularities with a characteristic orbit. Clearly, monodromic singularities cannot have characteristic orbits, as the latter would constitute a barrier for spiralling along the central circle C. The inverse assertion is also true so that we have the following *principal alternative.*

Theorem 9.13 (Principal alternative). *An isolated singularity of a planar real analytic foliation is either monodromic, or has a characteristic trajectory.*

The foliation is monodromic if and only if after a complete desingularization it exhibits only topological saddles (degenerate or hyperbolic) at the corner points of the exceptional divisor, and all blow-ups leading to this desingularization are nondicritical.

Singularities with characteristic orbits can be described in combinatorial terms.

Theorem 9.14 ([**ALGM73**]). *A singular planar real analytic foliation with a characteristic orbit admits a sectorial decomposition into finitely many standard sectors separated by characteristic orbits.*

Sketch of the proof of Theorem 9.13. Consider the complete desingularization of the singular real analytic foliation $\mathcal{F}$. This is a singular foliation $\mathcal{F}'$ on the real analytic surface M with the exceptional divisors $D \subset M$ on it, such that D is the union of transversally intersecting real analytic circles $D_1, \dots, D_m$, and all singular points of $\mathcal{F}'$ are elementary and belong to D.

If one of the blow-ups leading to $\mathcal{F}'$ was dicritical, then there are infinitely many smooth real analytic leaves of $\mathcal{F}'$ crossing D; after blowing down they become characteristic orbits of $\mathcal{F}$. Thus we can concentrate only on the case where D is a union of separatrices of all singular points of $\mathcal{F}'$.

If $\mathcal{F}'$ has a *noncorner singular point*, that is, a point $a \in D$ that belongs to only one component D_i of the divisor D, then it must admit at least one more characteristic orbit not belonging to D, since all elementary singularities from Table II.1 have at least *two pairs* of such characteristic orbits (except for the center, yet the center cannot occur in a nondicritical blow-up). In a similar way, a node or saddle-node, even occurring at a corner (transversal intersection of two components D_i and D_j), implies existence of a characteristic orbit of $\mathcal{F}'$ outside D. Clearly, the blow-down of this leaf is a characteristic orbit of $\mathcal{F}$.

Thus the only case where $\mathcal{F}$ has no apparent characteristic orbits, is the case where after complete desingularization the foliation $\mathcal{F}'$ has only saddle singularities at the corner points. We show that in this case $\mathcal{F}$ is monodromic.

Indeed, consider a hyperbolic sector of the corner saddle and any pair of the cross-sections τ, τ' to the sides of this sector at the points a, a'. We will assume that the "positive" semi-intervals τ_+, τ'_+ are inside the sector.

Then the leaves of the standard hyperbolic foliation, corresponding to hyperbolas $\{xy = \text{const} > 0\}$ in the positive quadrant, establish one-to-one smooth correspondence between the positive semi-sections, which continuously extends at the vertex by associating a with a'.

In a similar way, two cross-sections τ, τ' to the same smooth component D_i admit real analytic correspondence between them provided that one of the two arcs of D_i connecting the base points, is free from other singularities of $\mathcal{F}'$. Note that the only case where *both* such arcs are nonsingular and hence the construction ambiguous, corresponds to a foliation

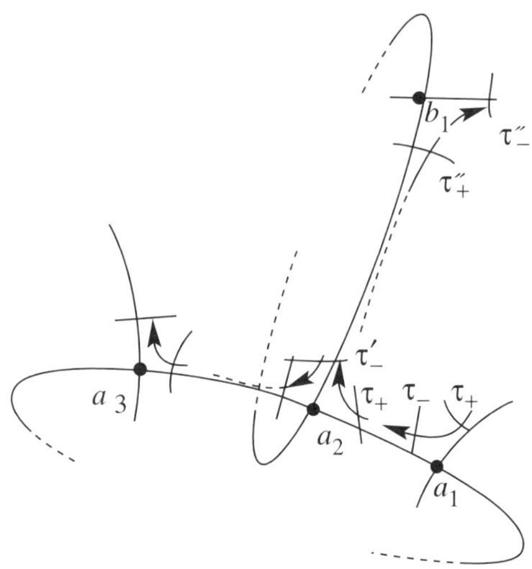

Figure II.4. Monodromic singularity: the monodromy is formed by consecutive traversing of hyperbolic sectors of corner singular points

which becomes nonsingular after the first blow-up; all subsequent blow-ups create at least one corner singularity on each divisor.

Thus we see that, by constructing a complete collection of $2m$ cross-sections (involving $4m$ semi-sections) and starting from an arbitrary semi-section $\tau = \tau_1$, one can uniquely determine the sequence of semi-sections $\tau_2, \tau_3, \dots$ in such a way that the correspondence maps $\Delta_i : \tau_i \to \tau_{i+1}$ are well defined. Since the total number of the cross-sections is finite, all of them will be ultimately traversed. The resulting composition $\Delta = \Delta_{4m} \circ \dots \circ \Delta_1$ is a well-defined map which coincides with the monodromy map associated with the initial semi-section; see Fig. II.4. □

Sketch of the proof of Theorem 9.14. For elementary singularities the assertion of the theorem is obvious. The general case is proved by induction in the number of blow-ups required for complete desingularization.

Consider the real monoidal map $\sigma : (M, C) \to (\mathbb{R}^2, 0)$. The blow-down of a sector of each type with a vertex at a singular point $a \in C$ is again the sector of the same type with the vertex at the origin, provided that both boundary curves of the sector are off the exceptional divisor.

New sectors can be formed by two characteristic orbits γ, γ' landing at different (adjacent on C) singular or tangency points a, a' on C. In this case the new sector is formed by squeezing two sectors between γ and C (resp., γ' and C). The list of possibilities for the nondicritical case may apriori consist of 6 possibilities (pp), (pe), ..., corresponding to different types of sectors formed at the corresponding intersections $\gamma \cap C$ and $\gamma' \cap C$. Yet for obvious topological reasons only the combinations (pp), (ph) and (hh) should be considered, since elliptic sectors cannot be adjacent to exceptional divisor C. After blow-down these pairs of sectors form elliptic, parabolic and hyperbolic sectors at the origin.

In a similar way one can construct sectorial decomposition of tangency points in the dicritical blow-down. In the trivial case where C carries only one singular point (at which the characteristic trajectory lands) and no points of contact, after blowing down we obtain topologically nonsingular foliation which can be thought as having two hyperbolic sectors. □

Sectorial decomposition allows us to associate with any real analytic foliation a finite word in the three-letter alphabet {p,h,e}, defined modulo a cyclic permutation. This word will be provisionally referred to as the *sectorial description.* The following result is rather obvious (see Problem 9.13), but gives a reasonably accurate description of foliations with a characteristic orbit.

Theorem 9.15. *Two real analytic foliations with the same sectorial descriptions are topologically equivalent (in the real domain).* □

Note, however, that not all "words" may be obtained from real analytic foliations, besides, some words correspond to topologically equivalent foliations, thus the "sectorial description" is *not* a classification (see Problem 9.12).

9E. Three questions. The topological results from §9**D** naturally suggest the following questions.

Question 1. How effective is the principal alternative? Can one determine, whether the singularity is monodromic or characteristic, by a finite order jet of a real analytic singularity? What is the order of such a jet?

Theorem 9.13 reduces the answer to this question to investigation of the complete desingularization. Since the desingularization procedure is effective, one can expect an explicit affirmative answer to the first question.

Assuming the singularity has a characteristic orbit, the next natural question is to determine its topological type.

Question 2. How constructive is the sectorial decomposition? In particular, is it determined by a finite order jet? Of what order?

For similar reasons, Theorems 9.14 and 9.15 raise hopes that the answer to the second question is also affirmative.

Finally, we have the last remaining case of monodromic singularities. Such singularities can be centers, foci or more complicated singularities. The following question reflects our belief that real analytic vector fields behave nicely.

Question 3. Is it true that a monodromic real analytic singular foliation with an isolated singularity is either center or focus (i.e., topologically equivalent to a node)? Is the topological type determined by a finite order jet?

These questions will be collectively referred to as *decidability of local classification problems* for germs of analytic vector fields. We will develop a suitable language and address them in this and the following sections.

Figure II.5. "Ill" behavior of C^∞-smooth vector fields: (a) nonmonodromic singularity without characteristic orbit, (b) infinitely many sectors, (c) infinitely many alternating periodic and aperiodic orbits

9F. Three nightmares. Definitions of characteristic orbit and monodromic singularity are usually given without resorting to blow-up in a form that applies to only C^1-smooth vector fields.

The principal alternative (Theorem 9.13 without reference to the complete desingularization) also holds under much less stringent regularity assumptions, while our proof uses heavily the existence of this desingularization. A simple direct proof, valid for only C^2-smooth vector fields and requiring only one blow-up, can be found in [**NS60**, §3], [**Har82**, Ch. VIII]; see Problem 9.14.

Yet the decidability questions raised in §9**E** turn out to have negative answers if the regularity is relaxed and real analytic vector fields are replaced by C^∞-smooth vector fields. First, the principal alternative may fail for such fields.

Example 9.16 (Nonmonodromic singularity without characteristic orbits)**.** Consider a function of one real variable, defined on the interval $(-1,1)$, which tends to $+\infty$ at both endpoints. Shifting the graph of this function in the vertical direction, one can construct a foliation without singular points on the infinite strip $[-1,1]\times\mathbb{R}$ tangent to the two border lines of the strip which are themselves the leaves. Rolling this strip (say, by the exponential map of the plane $\mathbb{R}^2\cong\mathbb{C}^1$), a foliation on the annulus $\{1\leqslant|z|\leqslant 2\}$ can be constructed. Finally, assembling together countably many homothetic copies of such annulus, we obtain a foliation shown in Fig. II.5(b).

This foliation is neither monodromic (it simply admits no cross-section to the exceptional divisor) nor does it have characteristic orbits.

Such an example can be constructed in the class of foliations generated by C^∞-smooth vector fields flat at the origin (i.e., the Taylor series is identically zero), bur cannot occur for real analytic foliations. Indeed, in this case the foliation is tangent to a line passing through the origin at infinitely many points accumulating to the origin, yet the line itself is not invariant.

The second nightmare shows that sectorial decomposition fails for C^∞-smooth vector fields.

Example 9.17 (Infinitely many sectors). The singular point schematically pictured on Fig. II.5(b), has infinitely many alternating hyperbolic and parabolic sectors.

Finally, for monodromic singularities the center-focus alternative may fail because of coexistence of infinitely many periodic and aperiodic trajectories. If trajectories of both types accumulate to the origin, then this singularity is neither center nor focus.

Example 9.18. Let $Z \subseteq (\mathbb{R}_+, 0)$ be any (relatively) closed subset. There exists a C^∞-smooth function φ flat at the origin and nonnegative, $\varphi \geqslant 0$, whose zero locus coincides with Z. Starting from this function, one can construct a C^∞-smooth monodromic vector field whose monodromy map differs from identity by φ, $\Delta(x) = x + \varphi(x)$. If both Z and $(\mathbb{R}_+, 0) \setminus Z$ accumulate to the origin, the corresponding singularity is neither center nor focus; see Fig. II.5(c).

If the foliation is real analytic, the monodromy map is necessarily real analytic at all *interior* points of the semi-interval $(\mathbb{R}_+, 0)$, which means that the set Z in this case may consist only of isolated points eventually accumulating to the origin. For some types of vector fields this accumulation is impossible for relatively simple reasons; see §10 below. Yet it is very difficult to prove that such accumulation is impossible for arbitrary analytic vector fields (the so-called *Nonaccumulation theorem*; see [**Ily91, Eca92, Ily02**] and §24**D**).

9G. Algebraicity of the decision. The proof of Theorem 9.13 is constructive: to decide, whether the singularity is monodromic or possesses a characteristic orbit, one has to construct complete desingularization and verify position (corner or noncorner) and topological types of all elementary singularities. These operations involve only algebraic manipulations with finitely many Taylor coefficients (arithmetic operations, sign testing and solution of algebraic equations). In this and the next section we formalize the respective notion of *algebraic decidability* and show that the principal alternative is indeed algebraically decidable, answering thus the first question from §9**E**.

We start with describing "decidable" subsets in affine finite-dimensional spaces. Without going into deep discussion on the general nature of computability, we postulate the class of semialgebraic sets as the only reasonable class of subsets of $\mathbb{R}^n$ or $\mathbb{C}^n$, which can be finitely presented. For any such set, one can imagine an "algorithm" involving only algebraic computations and sign tests, that in a finite number of steps allows us to decide, whether a given input (point) belongs to the set or not.

Definition 9.19. A subset of $\mathbb{R}^n$ is called *real semialgebraic* if it can be defined by finitely many polynomial equalities and inequalities of the form $p(x) = 0$, $p(x) < 0$ or $p(x) \leqslant 0$, where $p \in \mathbb{R}[x_1, \dots, x_n]$.

Semialgebraic sets form a Boolean algebra (their finite unions and intersections are obviously semialgebraic). What is more important, the class of semialgebraic sets is closed by taking complements and affine projections (and, more generally, polynomial maps).

Theorem 9.20 (A. Tarski–A. Seidenberg; see [**vdD88**]). *Affine projection of a semialgebraic set is again semialgebraic.*

Semialgebraic spaces are *decidable*: any such set can be defined by a finite formula involving polynomial equalities and inequalities over $\mathbb{R}[x_1, \dots, x_n, y_1, \dots, y_m]$ with "auxiliary" variables $y_1, \dots, y_m$, the logical operations "and", "or", "not", and the *quantifiers* $\forall y_i, \exists y_j$ which tie down the auxiliary variables. The Tarski–Seidenberg theorem asserts that all quantifiers can be effectively eliminated, meaning that the decision process is fully constructive.

Consider a subset M in the space, say, of germs of real or complex analytic vector fields at the origin on the plane $\mathcal{D} = \mathcal{D}(\mathbb{C}^2, 0)$ (other analytic objects, e.g., germs of functions, self-maps, can be treated exactly the same). Note that for any finite order n the space $J^n = J^n\mathcal{D}(\mathbb{C}^2, 0)$ of n-jets of such vector fields is a finite-dimensional complex affine space. In our constructions the set M will be defined by some *properties* of vector fields (e.g., topological type, multiplicity, order, existence of analytic separatrix, *etc.*), so we will often speak of *properties* of vector fields.

First we formalize the assertion that some property M is determined by a finite order jet.

Definition 9.21. A jet $g \in J^n$ of order n is said to be *sufficient for the set M* (resp., for the corresponding property), if all germs having this jet, either belong to M or to its complement $\mathcal{D} \smallsetminus M$:

$$(j^n)^{-1}(g) \subseteq M \text{ or } (j^n)^{-1}(g) \subseteq \mathcal{D} \smallsetminus M, \quad (j^n)^{-1}(g) = \{F : j^n F = g\}.$$

Definition 9.22. The set M is said to be *decidable at the level of n-jets*, if there exists a subset $M^{(n)} \subseteq J^n\mathcal{D}(\mathbb{C}^2, 0)$ such that $F \in M$ if and only if $j^n F \in M^{(n)}$.

The property is *algebraically decidable* (on the level of n-jets), if the set $M^{(n)}$ is semialgebraic in the affine space $J^n\mathcal{D}(\mathbb{C}^2, 0)$.

In other words, the set (or the respective property) is algebraically decidable at the level of n-jets, if all such jets are sufficient. This is a relatively rare opportunity: in most cases when M is described by its topological or analytic properties, there always are some jets that are insufficient to guarantee whether or not their representatives belong to M; see §10.

9H. Decidability of multiplicity. We illustrate the notion of decidability by showing that the multiplicity of an isolated singularity is a "decidable" function of analytic germs.

Theorem 9.23. *For any finite μ the set M_μ of holomorphic vector fields having multiplicity $\leqslant \mu$ at the origin, is algebraically decidable at the level of n-jets with $n = \mu$.*

Proof. First we show that if F is a germ of multiplicity $\leqslant \mu$, then its μ-jet is sufficient in the sense that any germ F' with the same μ-jet also has the same multiplicity. To prove that, we use the definition of the multiplicity as the dimension of the quotient local algebra, $\mu = \dim_{\mathbb{C}} \mathcal{O}_0 / \langle F_1, F_2 \rangle$, where $F_{1,2}$ are the coordinate functions of the germ F of the vector field.

Indeed, by [**AGV85**, Lemma 1, §5.5], any power $x^a y^b$ of order $a+b \geqslant \mu+1$ belongs to the ideal of any finite codimension μ. Thus any analytic germ of the form $F_i' = F_i + o((|x|+|y|)^\mu)$, $i = 1,2$, belongs to the ideal $\langle F_1, F_2 \rangle$ and hence $\langle F_1', F_2' \rangle = \langle F_1, F_2 \rangle$. Clearly, the arguments are symmetric and all germs with the same μ-jet generate the same ideals and hence have the same multiplicity.

Thus we can define the set $M_\mu^{(\mu)}$ as the set of polynomial vector fields of degree μ, having a singularity of multiplicity $\leqslant \mu$ at the origin. Regardless of the local coordinates, if the Taylor polynomial (truncation) of F belongs to $M_\mu^{(\mu)}$, then the corresponding μ-jet is sufficient for M_μ.

It remains to prove that $M_\mu^{(\mu)}$ is semialgebraic in the space of μ-jets $J^\mu \mathcal{D}(\mathbb{C}^2, 0)$. Consider the affine space $\mathcal{D}_\mu \cong \mathbb{C}^N$, $N = N(\mu)$, of polynomial vector fields of degree μ. The polynomial formula with quantifiers

$$\forall \varepsilon > 0 \ \exists y \in \mathbb{C}^2, \exists\, x_1, \dots, x_{\mu+1} \in \mathbb{C}^2, \qquad \prod_{i<j} |x_i - x_j| \neq 0 \ \&\ \{|x_i|, |y| < \varepsilon\} \ \&\ F(x_i) = y,$$

after substitution $F(x) = \sum_{|\alpha| \leqslant \mu} a_\alpha x^\alpha$, $a_\alpha \in \mathbb{C}^2$ defines a subset in $\mathcal{D}_\mu \cong \{a_\alpha\}_{|\alpha| \leqslant \mu}$ whose elements are polynomial vector fields having the singularity of multiplicity $\geqslant \mu+1$ (or nonisolated) at the origin, i.e., the complement to $M_\mu^{(\mu)}$. By the Tarski–Seidenberg Theorem 9.20, the set $M_\mu^{(\mu)}$ is semialgebraic. □

Remark 9.24. If a certain set (property) M is algebraically decidable at the level of n-jets, then for trivial reasons it is algebraically decidable at the level of any higher order jets.

9I. Algebraic decidability of the principal alternative. We prove now that the principal alternative is algebraically decidable *after restriction on the subspace of analytic germs of any given finite multiplicity.* These results develop the ideas put forward in the seminal paper [**Dum77**].

Theorem 9.25. *For each multiplicity $\mu \in \mathbb{N}$ there exists a finite order $n = n(\mu) \in \mathbb{N}$ and two disjoint semialgebraic subsets $C^{(n)}, M^{(n)} \subseteq J^n(\mathcal{D}(\mathbb{R}^2, 0))$ in the space of n-jets of planar vector fields, such that a field F of multiplicity $\leqslant \mu$ at the origin has a characteristic orbit (resp., is monodromic) if and only if its jet $j^n F$ belongs to $C^{(n)}$ (resp., $M^{(n)}$).*

Sketch of the proof of Theorem 9.25. By Theorem 9.23 and Remark 9.24, in all sufficiently high order jet spaces $J^n = J^n\mathcal{D}(\mathbb{R}^2, 0)$ there exists semialgebraic subsets sufficient for the corresponding singularities to have multiplicity $\leqslant \mu$. By Theorem 8.15, any such singularity can be completely resolved into elementary singularities in no more than $2\mu+1$ *steps* (consecutive simple blow-ups of finite sets).

As follows from Theorem 9.13, to decide between characteristic and monodromic cases, it is sufficient to identify ("recognize") the location and topological types of these elementary singularities which appear after complete desingularization.

Nondegenerate singularities (saddles and nodes) can be recognized looking at their 1-jets; the criteria (inequalities for the discriminants of characteristic polynomials of degree 2) are obviously semialgebraic in the elements of the linearization matrices.

Degenerate isolated elementary singularities of finite multiplicity μ can be saddles, nodes or saddle-nodes. To decide between these two types, one has to know the jet of order μ, as will be independently shown in §10**B**. The test condition is polynomial.

Finally, the decision on whether a given nonelementary singularity has a dicritical blow-up or not, depends on the terms of lower order (and is obviously expressed by an algebraic condition involving these terms). Since the order of a singularity cannot exceed its multiplicity (as follows from [**AGV85**, Lemma 1, §5.5] already cited in the proof of Theorem 9.23), we arrive at the following intermediate conclusion: *existence of a characteristic orbit can be expressed as a semialgebraic condition on the jets of order $\leqslant \mu + 1$ at all singularities that appear in the process of complete desingularization.*

Inspection of the process shows that the multiplicities and hence orders of all intermediate singularities which appear in the process, do not exceed $\mu + 1$. Thus all information sufficient to determine uniquely the desingularization process and the topological types of elementary singularities that appear after this construction terminates, is contained in a sufficiently high order jet of the initial singularity. The order $n = n(\mu)$ of this jet should be so large as to determine uniquely $(\mu + 1)$-jets at all intermediate singularities on each of at most $2\mu + 1$ steps; cf. with Theorem 8.15.

Consider an isolated singularity of order ν (hence of multiplicity $\geqslant \nu$) and its blow-up. The corresponding transformation of the Pfaffian equation involves a change of variables from (x, y) to (x, z), $z = y/x$, and division by an appropriate power of x, more precisely, by $x^{\nu-1}$ in the nondicritical case and by x^{ν} in the dicritical case respectively. This construction implies that jets of order $k + \nu$ (respectively, $k + \nu + 1$) at the initial point determine uniquely the jet of order k at any singularity that appears on the exceptional divisor after blow-up. Clearly, the formulas describing the transformation on the level of jets, are (real) algebraic.

Iterating these arguments, one obtains an upper bound for the order $n(\mu)$ of the initial jet that encodes all $(\mu+1)$-jets on all $2\mu+1$ steps of the desingularization process. In other words, all representatives of n-jets of vector fields of multiplicity μ have the same desingularization schemes and the same jets of order $\mu+1$ at all elementary singular points of multiplicity $\leqslant \mu+1$ that appear after complete desingularization.

Based on this information and a algebraic algorithm of detecting topological types of elementary singularities which will be discussed in more details in §10, one can apply Theorem 9.13 to obtain explicitly the semialgebraic conditions necessary and sufficient for existence of a characteristic orbit. □

9J. Topologically sufficient jets. Theorems 9.14 and 9.15 together imply that the topological type of a real analytic singular foliation $\mathcal{F}$ with a characteristic orbit is uniquely determined by its complete desingularization. Recall that the latter is a map $\pi\colon (M,D)\to(\mathbb{R}^2,0)$ between some 2-dimensional surface with an exceptional divisor D with normal crossings, and a small neighborhood of the origin on the plane. The singular real analytic foliation $\mathcal{F}'=\pi^*\mathcal{F}$ has only elementary singular points, all of them on D. In addition we will also assume in this subsection that the dicritical components of the exceptional divisor D have no interior tangency points (in particular, carry no singularities). This can always be achieved by additional blow-up of such tangencies; see Theorem 8.37. We show in this subsection, how this claim can be translated into the language of sufficient jets.

Definition 9.26. An m-jet of a planar vector field is called *topologically sufficient*, if any two real analytic vector fields extending this jet, are topologically equivalent to each other.

Theorem 9.27 (O. Kleban [**Kle95**]). *For an isolated singularity of a planar vector field of multiplicity μ, its $2\mu+2$-jet is topologically sufficient.*

Sketch of the proof. The same arguments as were used in the proof of Theorem 9.25, show that the map $\pi\colon (M,D)\to(\mathbb{R}^2,0)$ implementing the complete desingularization (in the above mentioned strong sense) is completely defined by a jet of some finite (depending on μ) order of the initial vector field. A higher order jet determines uniquely the topological types of all elementary and corner singularities of the strong desingularization of the initial foliation, so that any other foliation $\mathcal{G}$ with the same jet is desingularized by the same map π and the preimage $\mathcal{G}'=\pi^*\mathcal{G}$ has topologically equivalent singularities at all corresponding points. Moreover, the homeomorphisms conjugating the respective singularities, can be chosen *identical on the vanishing divisor*.

It remains to notice that (a) two foliations $\mathcal{F}',\mathcal{G}'$ on M with topologically equivalent elementary singularities at the same points, are topologically equivalent globally on M, and (b) topologically equivalent singularities have the same sectorial decomposition. Detailed proofs of these results can be found in [**Dum77, Kle95**]. Reference to Theorem 9.15 completes the proof. □

9K. Conclusion. The results established in this section prove that for any finite value of μ, the affine space $J^n=J^n\mathcal{D}(\mathbb{R}^2,0)$ of n-jets of planar vector fields for $n \geqslant 2\mu + 2$ admits decomposition as the disjoint union of three semialgebraic subsets

$$J^n = C^{(n)} \sqcup M^{(n)} \sqcup Z^{(n)}, \qquad C^{(n)} = \bigcup_{\alpha=1}^{N} C_\alpha^{(n)}.$$

Here $C^{(n)}$ is the subset of n-jets sufficient to guarantee existence of the characteristic orbit; different components $C_\alpha^{(n)}$ correspond to topologically different germs of vector fields, $M^{(n)}$ consists of n-jets sufficient to guarantee that all their representatives are monodromic, and $Z^{(n)}$ is the collection of jets whose representatives have multiplicity $\geqslant \mu + 1$ or are nonisolated. The codimension of the "nonsufficient" set $Z^{(n)}$, where the topological type cannot yet be specified, tends to infinity together with n. The polynomial equalities and inequalities defining the components, depend only on the components of the $2\mu+2$-jets and hence stabilize as n grows to infinity with the fixed μ.

Exercises and Problems for §9.

Problem 9.1. Prove that all *linear* real vector fields are topologically equivalent (at the origin) to the three Poincaré types.

Exercise 9.2. Find *minimal* (involving minimal number of sectors) sectorial decomposition of all standard normal forms from Table II.1.

Problem 9.3. Prove that an isolated monodromic singularity which has an open neighborhood free from closed leaves, is homeomorphic to a node, i.e., is a focus.

Exercise 9.4. Construct a cycle of C^∞ smooth planar vector field, which is neither limit nor identical.

Exercise 9.5. Assume that after a single blow-up σ of a foliation $\mathcal{F}$ the foliation $\sigma^*\mathcal{F}$ on the real Möbius band carries two nodal elementary singularities. Describe the topological type of $\mathcal{F}$ in terms of sectors.

Exercise 9.6. Describe all topologically nonequivalent phase portraits of *generic* vector fields of order 2 on the plane.

Exercise 9.7. Give an example of degenerate monodromic singularity.

Exercise 9.8. Show that a simple cusp on the real plane admits a characteristic orbit.

Exercise 9.9. Show that in the sectorial description of any singularity every letter 'e' occurs between two letters 'p'.

Exercise 9.10. Find two sectorial descriptions of different length, which correspond to topologically equivalent foliations.

Exercise 9.11. Show that the sectorial description of a real analytic foliation cannot consist of exactly three hyperbolic sectors.

Problem 9.12. Under what restrictions does a sectorial description (considered as a word in the three-letter alphabet) correspond to a $\mathbb{C}^\infty$-smooth foliation on $(\mathbb{R}^2, 0)$?

Problem 9.13. Give detailed proofs of Theorems 9.14 and 9.15.

Problem 9.14. Prove the principal alternative for isolated singularities of C^2-smooth planar vector fields directly (cf. with [**NS60**]).

Problem 9.15. Construct explicitly C^∞-smooth vector fields whose phase portraits exhibit pathologies as in Examples 9.16 and 9.18.

Exercise 9.16. Prove that the foliation $\mathcal{F}$ of the real plane $(\mathbb{R}^2, 0)$ by the level curves given by the complex equation $\operatorname{Im} z^{3/2} = \text{const}$, $z \in (\mathbb{C}^1, 0) \cong (\mathbb{R}^2, 0)$, cannot be complexified, i.e., there cannot exist a foliation ${}^{\mathbb{C}}\mathcal{F}$ on $(\mathbb{C}^2, 0)$, whose leaves intersect the real plane $\mathbb{R}^2 \subseteq \mathbb{C}^2$ by leaves of the foliation $\mathcal{F}$.

The following problem shows that the analog of the Tarski–Seidenberg theorem fails if semialgebraic sets are replaced by *semianalytic sets*, subsets of affine space, defined locally near each point of this space by equalities and inequalities involving *real analytic* functions.

Problem 9.17. Consider the one-dimensional semianalytic subset of $\mathbb{R}^3$ (curve), defined by the analytic equations $\{xz = -1,\ y(y - e^z) = 0,\ z > 0\}$. Prove that its projection on the (x, y)-plane parallel to the z-axis is not semianalytic.

10. Algebraic decidability of local problems and center-focus alternative

The previous section gives a partial affirmative answer to the decidability Questions 1 and 2 from §9**E**. For every given finite value of the multiplicity μ, existence of a characteristic trajectory and topological classification of foliations having such a trajectory are algebraically decidable in jets of a certain finite (depending on μ) order. Rather obviously, with the natural parameter μ growing to infinity, the number of different topological types also grows to infinity and hence one cannot get rid of this parameter in the formulations, at least for Question 2.

On the other hand, the role of multiplicity when discussing decidability of the center-focus alternative (Question 3 from the same section) seems to be marginal. Already the center-focus problem for vector fields whose linear part is rotation, is nontrivial, as we shall see below. Besides, simplest examples show that for any n the set of centers is not decidable on the level of n-jets. Indeed, by adding arbitrarily high order terms one can destroy center making it into stable or unstable focus (Exercise 10.1).

Thus we arrive at the problem of defining decidable (semialgebraic) subsets in the *infinite-dimensional* jet space $J^\infty\mathcal{D}(\mathbb{R}^2, 0) \cong \mathcal{D}[[\mathbb{R}^2, 0]]$.

The general notion of *algebraic decidability* was introduced by V. Arnold in [**Arn70a, Arn70b**]; see also [**Arn83**, §37]. Arnold proved algebraic decidability of several natural problems of local analysis, yet noticed that for sufficiently advanced problems this algebraic decidability may ultimately fail. For instance, the Lyapunov stability problem for singularities in dimension $n \geqslant 3$ and topological classification of holomorphic singular foliations in $(\mathbb{C}^2, 0)$ are algebraically undecidable. In §10**G** we show that the stability problem is not algebraically decidable already for planar analytic vector fields (i.e., for $n = 2$).

We discuss decidability of the topological classification for elementary real analytic planar singularities. We show that topological classification of degenerate elementary singularities (saddle/node/saddle-node trichotomy) and center-focus alternative for elliptic vector fields are algebraically decidable in the strongest sense of this notion (introduced later). On the other hand, we prove that for general monodromic singularities the center-focus alternative is *not* algebraically decidable.

10A. Decidability in the jet spaces: the language. Two holomorphic germs of analytic functions $f, f' \in \mathcal{O}(\mathbb{C}^n, 0)$ are said to be n-equivalent at the origin, if their difference vanishes with order $n+1$ at this point. The n-jet (at the origin) is the equivalence class with respect to this n-equivalence. The space of all jets has the natural structure of linear $n+1$-dimensional space over $\mathbb{C}$; in any local coordinates $x_1, \dots, x_n$ n-jets of functions can be identified with (Taylor) polynomials of degree $\leqslant n$.

This construction can be modified for various other classes of objects (vector fields, differential forms, complex or real, and even in the infinitely-smooth nonanalytic case).

The space of germs of real analytic vector fields $\mathcal{D}(\mathbb{R}^2, 0)$ (or, what is the same in the planar case, the space of germs of real analytic 1-forms $\Lambda^1(\mathbb{R}^2, 0)$) is infinite-dimensional and thus the decidability of subsets of this space cannot be defined in terms of semialgebraic sets. Yet this infinite-dimensional space is naturally endowed with infinitely many projections j^k associating with each germ its k-jet at the singular point. The jets of any finite order form a finite-dimensional space with the natural affine structure. Thus one can define decidable sets of germs in terms of decidability of their jet projections.

Consider a subset M in a space of all analytic germs $\mathcal{G}$, for example, in the space of germs of 1-forms $\mathcal{G} = \Lambda^1(\mathbb{R}^2, 0) = \Lambda^1$. By $J^k(\mathcal{G})$ we will denote the finite-dimensional space of k-jets of germs from $\mathcal{G}$.

Definition 10.1. A set $M \subset \mathcal{G}$ is *algebraically decidable to codimension* $r \in \mathbb{N}$, if for some jet order k there exist two disjoint semialgebraic subspaces $S_k^{\pm} \subseteq J^k(\mathcal{G})$ such that:

(1) any germ whose k-jet belongs to S_k^+, necessarily belongs to M;

(2) any germ whose k-jet belongs to S_k^-, necessarily belongs to the complement $\mathcal{G} \smallsetminus M$;

(3) the complement $N_k = J^k(\mathcal{G}) \smallsetminus (S_k^+ \cup S_k^-)$, automatically semialgebraic, has codimension $\geqslant r$ in $J^k(\mathcal{G})$.

Jets from the subsets $S_k^{\pm}$ are referred to as *sufficient* k-jets, while the complementary set N_k consists of *neutral* jets.

Algebraic decidability of a certain property $M \subset \mathcal{G}$ means that the corresponding set can be approximated from two sides by "cylindrical" semialgebraic subspaces in subspaces of k-jets,

$$\mathcal{S}_k^+ \subseteq M \subseteq \mathcal{G} \smallsetminus \mathcal{S}_k^-, \qquad \mathcal{S}_k^{\pm} = (j^k)^{-1}(S_k^{\pm}),$$

so that the "accuracy" of this approximation, $\mathcal{N}_k = \mathcal{G} \smallsetminus (\mathcal{S}_k^+ \cup \mathcal{S}_k^-)$, has a well-defined codimension that is at least r. It is the codimension r rather than the order k of the jets that plays the central role in this definition.

Definition 10.2. A subset $M \subset \mathcal{G}$ of the space of germs is algebraically decidable to *infinite codimension* (or simply *decidable*), if it is algebraically decidable to any finite codimension r.

According to this definition, for a decidable property (set) M there exists an *infinite sequence* of two-sided semialgebraic cylindrical approximations for M,

$$\begin{aligned} \mathcal{S}_0^+ \subseteq \mathcal{S}_1^+ \subseteq \cdots \subseteq \mathcal{S}_{k+1}^+ \subseteq \cdots \subseteq M \subset \\ \cdots \subseteq (\mathcal{G} \smallsetminus \mathcal{S}_{k+1}^-) \subseteq (\mathcal{G} \smallsetminus \mathcal{S}_k^-) \subseteq \cdots \subseteq (\mathcal{G} \smallsetminus \mathcal{S}_0^-), \end{aligned}$$

such that the codimension of the decreasing differences $\mathcal{N}_k = \mathcal{G} \smallsetminus (\mathcal{S}_k^+ \cup \mathcal{S}_k^-)$ grows to infinity:

$$\mathcal{G} \supseteq \mathcal{N}_1 \supseteq \cdots \supseteq \mathcal{N}_k \supseteq \mathcal{N}_{k+1} \supseteq \cdots, \qquad \operatorname{codim}_{\mathcal{G}} \mathcal{N}_k \to +\infty.$$

In particular, this codimension condition holds if stabilization occurs and $\mathcal{N}_k = \varnothing$ for some k. As before, the sets $\mathcal{S}_k^{\pm}, \mathcal{N}_k$ are cylindrical, that is, preimages of respective semialgebraic subsets $S_k^{\pm}$ and N_k in $J^k(\mathcal{G})$.

The intersection $\mathcal{N}_\infty = \bigcap_{k \geqslant 0} \mathcal{N}_k$, which may be empty even if all $\mathcal{N}_k$ are nonzero, may also be nontrivial, since the space of germs $\mathcal{G}$ is infinite-dimensional.

Definition 10.3. The subset $M \subseteq \mathcal{G}$ is *ultimately* (algebraically) decidable, if the intersection $\mathcal{N}_\infty$ entirely belongs either to M or to its complement.

Speaking in terms of algorithms, a set of germs ("property") $M \subseteq \mathcal{G}$ is decidable (i.e., algebraically decidable to infinite codimension), if there exists an algorithm that allows for any given germ $g \in \mathcal{G}$ to verify whether it belongs to M or not. This algorithm must be algebraic, meaning that it tests conditions expressed by polynomial equalities and inequalities on Taylor coefficients. On each step either the decision is made, whether $g \in M$ or $g \notin M$, or the computations should be continued involving higher order Taylor coefficients. The algorithm should terminate for almost all germs except for an eventual set of infinite codimension. The set is *ultimately* decidable, if all germs on which the algorithm never stops, belong to M or its complement simultaneously.

Remark 10.4. The definition of decidability admits possible variations. Clearly, the constructions remain the same for any other types of germs (vector fields, functions, self-maps *etc.*) and various types of properties.

In particular, instead of just two sets, M and its complement $\mathcal{G} \smallsetminus M$, one can consider a partition of the total space of germs into finitely many sets (types) $M_1, \dots, M_m$, $m \geqslant 2$, pairwise disjoint. The decision problem in this context is to determine the type of a given germ $g \in \mathcal{G}$. The "classification scheme" into the types $M_1, \dots, M_m$ is algebraically decidable, if for any $t = 1, \dots, m$ and any $k = 1, 2, \dots$, there can be constructed pairwise disjoint semialgebraic subsets $S_k^t \in J^k(\mathcal{G})$ of "sufficient jets", i.e., $\mathcal{S}_k^t = (j^k)^{-1}(S_k^t) \subseteq M_t$, which exhaust J^k in the sense that the complement $N_k = J^k(\mathcal{G}) \smallsetminus \bigcup_t S_k^t$ of neutral ("undecided") jets has codimension growing to infinity together with k. The decidability is *ultimate*, if the intersection $\mathcal{N}_\infty = \bigcap_{k \geqslant 0} (j^k)^{-1}(N_k)$ belongs to *only one* of the sets $M_1, \dots, M_m$ (classification types).

The classification problems are seldom decidable in the whole set of germs $\mathcal{G}$; however, some parts of the respective subsets (and sometimes large parts) can be.

Let $\mathcal{B} \subset \mathcal{G}$ be a subset in the space of germs, defined by semialgebraic conditions on some finite order jet. This means that for some finite l there is a semialgebraic subset $B_l \subset J^k(\mathcal{G})$ such that $\mathcal{B} = (j^l)^{-1}(B_l)$.

Definition 10.5. A subset M is decidable (resp., ultimately decidable) *relative* to a semialgebraic set $\mathcal{B}$, if the corresponding sufficient sets $S_k^{\pm}$ are semialgebraic in the intersection with $B_k = \{j^k g : j^l g \in B_l\}$ for all $k \geqslant l$.

When speaking about classification problems or alternatives, discussing relative decidability means that from the outset the problem is *restricted* only on a subclass of germs already defined by some semialgebraic conditions on their l-jets. In this setting the *relative* (ultimate or not) decidability means that the property is determined by algebraic conditions imposed on the higher order jets. In such cases we will say about (un)decidability of an alternative *for the specific class.* For example, the center-focus alternative is undecidable in general, but decidable (and even ultimately decidable) for germs with nondegenerate linear part; see §10**C**.

10B. Topological classification of degenerate elementary singularities on the plane. An isolated degenerate elementary singular point of a real analytic vector field on the real plane $(\mathbb{R}^2, 0)$ may be of three topological types: saddle-node, topological node or topological saddle, represented by the three standard models as described in §9**A**. We show that this classification is algebraically decidable to infinite codimension and even ultimately decidable. This classification problem constitutes perhaps the simplest nontrivial example of algebraic decidability.

To fit the formal settings, we consider the subspace $\mathcal{B}_{\text{elem}} = \mathcal{B}$ of germs of holomorphic 1-forms having one zero and one nonzero eigenvalue of the linearization: on the level of 1-jets this subspace is determined by the semialgebraic conditions $\det A = 0$, $\operatorname{tr} A \neq 0$ on the linearization matrix A of the corresponding vector field. Without loss of generality we may assume that A is already reduced to the diagonal form, so that

$$\mathcal{B} = \{\omega \colon j^1\omega = y\,dx\} \subset \Lambda^1(\mathbb{R}^2, 0) = \mathcal{G}.$$

The three subsets of $\mathcal{B}$, corresponding to different topological types, will be denoted M_S (saddles), M_N (nodes), M_{SN} (saddle-nodes). However, for the sake of completeness one has to introduce the fourth class $M_I \subseteq \mathcal{B}$ of germs having a nonisolated singularity (such germs become nonsingular after division by a noninvertible function $y + \cdots$). By Theorem 9.1,

$$\mathcal{B} = M_S \sqcup M_N \sqcup M_{SN} \sqcup M_I. \tag{10.1}$$

Theorem 10.6. *The problem of topological classifications of degenerate elementary singular points of analytic vector fields on the real plane is ultimately algebraically decidable.*

Formally the assertion of the theorem means that the partition (10.1) is ultimately decidable in the sense explained in Remark 10.4. The proof occupies the rest of §10**B**.

The proof is organized as follows: for every order k, we construct explicitly partition of jets of order k into jets sufficient for saddles, nodes and saddle-nodes and the neutral jets, and show that this partition is semialgebraic. Then conditions on the codimension will be verified. Finally, we verify that the germs with neutral jets of all orders, constitute the class M_I of the partition.

Denote by $N_k \subseteq J^k = J^k(\Lambda^1)$ the collection of k-jets of 1-forms $y\,dx + \cdots \in \mathcal{B}$, which are orbitally linearizable (equivalent to the linear jet $y\,dx$): in suitable coordinates, any germ ω with $j^k\omega \in N_k$, takes the form

$$\omega = f(x,y)(y\,dx + \omega'), \qquad \operatorname{ord}_0 \omega' \geqslant k+1, \quad f(0,0) \neq 0. \tag{10.2}$$

Denote by $S_k = \mathcal{B} \smallsetminus N_k$ the complement. We first claim that all jets from this complement are topologically sufficient.

Lemma 10.7. *The jets from the set S_k are topologically sufficient. More precisely, germs with the k-jet in $S_k = \mathcal{B} \smallsetminus N_k$ have one of the three "isolated" topological types,*

$$(j^k)^{-1}(S_k) \subseteq M_S \sqcup M_N \sqcup M_{SN}.$$

Sketch of the proof. This lemma is a refinement of the last assertion of Theorem 9.1. We briefly indicate the arguments which after proper elaboration yield an accurate proof.

If $j^k\omega \notin N_k$, then the 1-form ω by a formal orbital transformation can be brought to the polynomial form

$$\omega = (\pm x^m + a\,x^{2m-1})\,dy + y\,dx, \qquad 2 \leqslant m \leqslant k.$$

We claim that ω is a saddle-node, saddle or node depending on the parity of m and the sign of the leading coefficient. Since these data are uniquely determined by the k-jet, this proves sufficiency of the latter.

By the *center manifold theorem* [**Kel67**], there exists an invariant curve C tangent to the axis $y = 0$ (in general, this center manifold is only finitely smooth, but in the planar case one can prove its C^∞-smoothness; see [**Ily85**]). The curve C has a flat tangency with the axis $y = 0$ at the origin.

Consider the real analytic planar vector field $F = (\pm x^m + \cdots)\frac{\partial}{\partial x} - y\frac{\partial}{\partial y}$ generating the distribution $\{\omega = 0\}$. Its restriction on the center manifold C is a smooth vector field whose topological type is determined by the order m and the sign of the principal coefficient. This restriction is topologically equivalent to the field $\pm x^m \frac{\partial}{\partial x}$ by an orientation-preserving homeomorphism of the x-axis.

By the *Pugh–Shub–Shoshitaishvili reduction principle* [**PS70b, Šoš72, Šoš75**] (see also [**Tak71**]), the vector field F is topologically orbitally equivalent to hyperbolic suspension of its restriction on the center manifold. In our case this means that the germ F is topologically orbitally equivalent to the vector field

$$F' = -y\frac{\partial}{\partial y} \pm x^m \frac{\partial}{\partial x}.$$

Topological classification of these fields is obvious. □

Remark 10.8. The description of the jet sets S_k and N_k can be reformulated in terms of multiplicity. The k-jet of a germ $\omega \in \mathcal{B}$ is topologically sufficient, if and only if its multiplicity $\mu_0(\omega)$ is no greater than k.

Proof of Theorem 10.6. First we show that for any k the components of the set S_k corresponding to jets topologically sufficient for saddles, nodes and saddle-nodes, are semialgebraic. To that end we will prove that the complement N_k is semialgebraic. Semialgebraicity of N_k follows from its definition as a normal form for a suitable polynomial action.

Indeed, consider the orbital action of all k-jets of self-maps $H \in \operatorname{Diff}[[\mathbb{R}^2, 0]]$ tangent to the identity, on the linear 1-form $y\,dx$, that is, all 1-forms $\omega' = f(X,Y) \cdot Y\,dX$, with $X, Y, f \in \mathbb{R}[X,Y]$, $f(0,0) = 1$, $X = x + \cdots$, $Y = y + \cdots$, after truncation at the level of k-jets. By definition, the orbit of this action coincides with N_k. Without loss of generality we may assume that $\deg X, Y, f \leqslant k$ (higher order terms will in any case be truncated). On the other hand, the coefficients of ω' are polynomial in the coefficients of the polynomials X, Y, f that can be arbitrary. Thus the set N_k is the polynomial image of a finite-dimensional affine space. By the Tarski–Seidenberg theorem, this image is semialgebraic in $J^k(\Lambda^1)$. Clearly, N_k is also closed and the codimension of this set grows to infinity as $k \to \infty$.

The sufficiency sets S_k are therefore algebraic as complements of the semialgebraic sets N_k. Each sufficiency set consists of three different parts (sufficiency components), $S_k = S_{k,N} \sqcup S_{k,S} \sqcup S_{k,SN}$. In principle one can prove semialgebraicity of each component separately, using the same method. Yet in our case this step can be replaced by general arguments.

The different sufficiency components belong to different connected components of the set S_k, since it is impossible to deform continuously a saddle to a node or a saddle-node. But it is known [**vdD88**] that *a connected component of a semialgebraic set is itself a semialgebraic set.* Thus partition of topological sufficiency components the level of jets of any order is semialgebraic and the codimension of the neutral jets grows to infinity. The algebraic decidability of the topological classification is proved.

To show the *ultimate* decidability, we use Remark 10.8. By this remark, the germs with neutral k-jets must have multiplicity at least k. Thus the real analytic germs whose jets of any order are insufficient, have infinite multiplicity, i.e., a exhibit a nonisolated singularity at the origin. By definition, such germs form a separate class M_I. □

10C. Generalized elliptic points and alternative. Ultimate decidability of degenerate elementary singular points is in a sense a model problem serving to illustrate the concepts and use of some important tools. On the contrary, the problem of distinction between center and focus traditionally, since the times of Poincaré, is one of the most challenging in the qualitative theory of ordinary differential equations on the plane. We discuss this problem (in terms of algebraic decidability) for *generalized elliptic* singularities for which the principal homogeneous terms guarantee absence of characteristic trajectories, so that generalized elliptic singularities are always monodromic. For these singularities the center-focus is easily proved to be a valid alternative (i.e., accumulation of periodic orbits implies center). In this section we show that the alternative for generalized elliptic singularities is *ultimately algebraically decidable if the principal homogeneous part is* *fixed.* Yet if the principal part is considered as a variable parameter, the boundary between stable and unstable foci is nonalgebraic, as will be shown in §10**G**. This undecidability was first conjectured by A. Brjuno and proved in [**Ily72a**]. We give a simple proof here.

Everywhere in this section we use the Pfaffian forms. Consider the real singular foliation $\omega = 0$ defined by the real analytic Pfaffian form whose expansion into homogeneous components begins with terms of order n,

$$\begin{aligned} \omega = \omega_n + \omega_{n+1} + \cdots, \quad & \omega_k = p_k(x,y)\,dx + q_k(x,y)\,dy, \quad n \geqslant 1, \\ p_k, q_k \in \mathbb{R}[x,y], \quad & \deg p_k = \deg q_k = k, \quad k = n, n+1, \dots. \end{aligned} \tag{10.3}$$

Definition 10.9. The singular point is called *generalized elliptic*, if the *real* homogeneous polynomial $h_{n+1} = yp_n + xq_n \in \mathbb{R}[x,y]$ is nonvanishing except at the origin,

$$h_{n+1}(x,y) \equiv xp_n(x,y) + yq_n(x,y) \neq 0 \qquad \text{for } (x,y) \in \mathbb{R}^2 \smallsetminus (0,0). \tag{10.4}$$

Consider the complexification of a singularity (10.3) and its subsequent blow-up. By definition, this is a singular holomorphic foliation $\mathcal{F}'$ defined in a small complex neighborhood of the exceptional divisor $\mathbb{E} = \mathbb{P}^1$ in the complex 2-dimensional surface $\mathbb{M}$ (complex Möbius band). This surface is covered by the two charts, (x,z), $z = y/x$, and (y,w), $w = x/y$ respectively. In the chart (x,z) the foliation $\mathcal{F}'$ is defined by the Pfaffian form

$$\begin{aligned} \omega' = & \big(h_{n+1}(1,z) + xh_{n+2}(1,z) + x^2h_{n+3}(1,z) + \cdots\big)\,dx \\ & + x\left(q_n(1,z) + xq_{n+1}(1,z) + x^2q_{n+2}(1,z) + \cdots\right)\,dz. \end{aligned} \tag{10.5}$$

Here $h_{k+1} = xp_k + yq_k$ are homogeneous polynomials of degree $k+1$ in two variables; see §8**E**, in particular, (8.8).

The singular points of $\mathcal{F}'$ on the exceptional divisor are roots of the polynomial $p_n(1,z) + zq_n(1,z) = x^{-(n+1)}h_{n+1}(x,xz)$. For a generalized elliptic singularity this polynomial is not identically zero, hence the blow-up is always nondicritical in the sense of Definition 8.12. Then Definition 10.9 guarantees that there are no singular points of $\mathcal{F}'$ on the real line $\mathbb{R} \subset \mathbb{E}$ in the chart (x,z). For similar reasons the point $z = \infty$ (mapped as $w = 0$ in the second chart) is also nonsingular.

Thus we obtain an invariant description of generalized elliptic singularities.

Corollary 10.10 (invariant definition of generalized elliptic singularities). *A real analytic singularity is generalized elliptic if and only if it is nondicritical and after the blow-up all its singularities on the exceptional divisor are off the real projective line (equator) $\mathbb{R}P^1 \subset \mathbb{E} \subset \mathbb{M}$.* □

Elliptic singularity whose linearization matrix is $\left(\begin{smallmatrix} 0 & 1 \\ -1 & 0 \end{smallmatrix}\right)$, after blow-up has two singular points at $z = \pm i$.

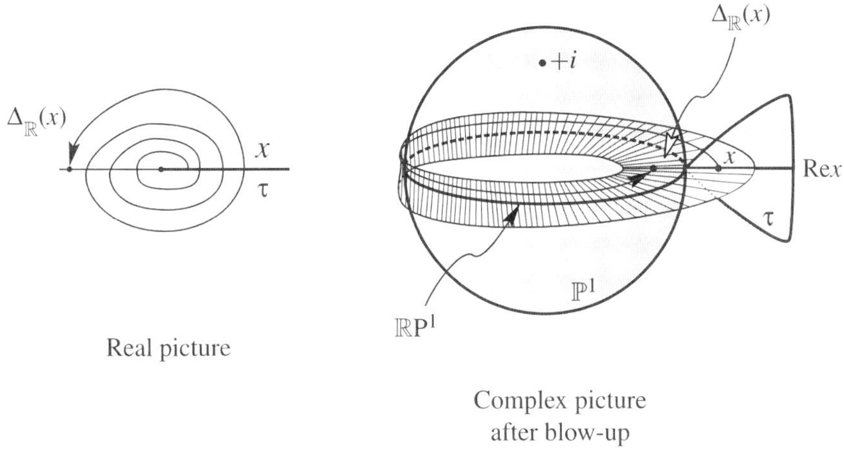

Figure II.6. Real equator and its complexification

The real equator $\mathbb{R}P^1 \cong \mathbb{S}^1$ is a closed loop on the Riemann sphere $\mathbb{S}^2 \cong \mathbb{E}$, which is "visible" as the real line $\mathbb{R}$ in the affine chart $\mathbb{C} \subset \mathbb{E}$. Thus the holonomy map $\Delta_{\mathbb{R}}$ along this loop is well defined for the foliation $\mathcal{F}'$, e.g., for the cross-section $\tau = \{z = 0\}$ with the coordinate x as a local chart on it. As the form ω was *real* analytic, the blow-up is a well-defined real singular foliation on the Möbius band which is the neighborhood of its central circle. The holonomy map $\Delta_{\mathbb{R}}$ is therefore real analytic.

Note, however, that this loop *does not* belong entirely to any of the two canonical charts: to compute the holonomy, one has to "continue" across infinity $z = \infty$, that is, pass to the other chart.

Still this difficulty can be easily avoided after complexification: if the singularity is generalized elliptic, the holonomy can be computed in the chart (x, z) as the result of analytic continuation along the semi-circular loop $[-R, R] \cup \{|z| = R, \operatorname{Im} z > 0\}$ homotopic to the real equator on the sphere.

The holonomy operator $\Delta_{\mathbb{R}}$ is visible on the real plane $(\mathbb{R}^2, 0)$ before the blow-up: the cross-section τ blows down as the x-axis on the (x, y)-plane. By construction, $(\Delta_{\mathbb{R}}(x), 0)$ is the first point of intersection with the x-axis of a solution starting at $(x, 0)$, after continuation counterclockwise. The monodromy map (as it is defined in §9**C**) is the *square* $\Delta_{\mathbb{R}} \circ \Delta_{\mathbb{R}}$ of the holonomy map.

Definition 10.11. The holonomy map $\Delta_{\mathbb{R}}$ (as well as its complexification) will be called the *semi-monodromy* of a generalized elliptic singular point.

This description of the semi-monodromy via holonomy immediately allows us to prove analyticity of it and the classical monodromy by referring to the standard results from §2**C**.

Theorem 10.12. *The semi-monodromy of a generalized elliptic singular point is real analytic on* $(\mathbb{R}, 0)$, *in particular, at the origin.*

If the Pfaffian form or the vector field depends analytically on additional parameters, the respective semi-monodromy depends analytically on these parameters as far as the singularity remains generalized elliptic. □

As a consequence, repeating verbatim the arguments proving Theorem 9.12, we immediately obtain the center-focus alternative for generalized elliptic point.

Corollary 10.13. *If infinitely many periodic orbits accumulate to a generalized elliptic singularity, then this singularity is a center, i.e., nonsingular trajectories are periodic.* □

10D. Computation of the holonomy map. Corollary 10.13 means that decision between center and focus is the true alternative for generalized elliptic points (no third possibility exists). It is equivalent to deciding whether $\Delta_{\mathbb{R}}$ is periodic with period 2. The latter alternative is ultimately algebraically decidable *in terms of the coefficients of the map* $\Delta_{\mathbb{R}}$; see Problem 10.4. Thus decidability of the center-focus alternative is reduced to *algebraic computability* of the Taylor coefficients of $\Delta_{\mathbb{R}}$ via the Taylor coefficients of the form ω.

We compute explicitly the coefficients of the semi-monodromy map. This computation is pretty much standard (see [**Bau54**]), yet in most sources it is carried out in the polar coordinates, corresponding to the trigonometric blow-up, which obscures their algebraic nature.

The Pfaffian equation $\omega = 0$ with the form ω which after blow-up takes the form (10.5), can be rewritten using the convergent expansion

$$dx = x\theta_1 + x^2\theta_2 + x^3\theta_3 + \cdots . \tag{10.6}$$

Here θ_i are *rational* (meromorphic)[1] 1-forms on the exceptional divisor,

$$\theta_i = R_i(z)\,dz \in \Lambda^1(\mathbb{E}) \otimes \mathfrak{M}(\mathbb{E}), \qquad i = 1, 2, \dots,$$

which are holomorphic (nonsingular) outside the polar locus

$$\Sigma = \{z \in \mathbb{C} \colon h_{n+1}(1, z) = 0\} \subset \mathbb{E}.$$

[1]The tensor multiplication $\otimes\mathfrak{M}$ by the algebra of meromorphic functions transforms spaces of holomorphic objects to their meromorphic counterparts, for instance, $\mathcal{D}(\mathbb{C}^k, 0) \otimes \mathfrak{M}(\mathbb{C}^k, 0)$ is the space of germs of meromorphic vector fields at the origin of the complex k-space, while $\Lambda^k(T) \otimes \mathfrak{M}(T)$ would denote the space of meromorphic k-forms on a manifold T, *etc.*

The expansion (10.6) can be obtained by division of both parts of (10.5) by the holomorphic function $\sum_{j\geqslant 0} x^j\, h_{n+1+j}(1,z)$ nonvanishing on the line $\{x=0\}\smallsetminus\Sigma$. In particular,

$$\theta_1 = -\frac{q_n(1,z)\,dz}{h_{n+1}(1,z)}. \tag{10.7}$$

The equation (10.6) can be rewritten in the other chart (y,w) on the complex Möbius band. After the change of variables $z=1/w$, $x=yw$ we obtain an analogous Pfaffian system

$$dy = y\vartheta_1 + y^2\vartheta_2 + \cdots, \tag{10.8}$$

with the meromorphic coefficients $\vartheta_i \in \Lambda^1(\mathbb{E})\otimes\mathfrak{M}(\mathbb{E})$ related to the coefficients θ_i of the system (10.6) by the formulas

$$\vartheta_1 = \theta_1 - \frac{dw}{w}, \qquad \vartheta_k = w^{k-1}\theta_k,\ k\geqslant 2. \tag{10.9}$$

The nontrivial formula for transition from θ_1 to ϑ_1 is the consequence of the fact that the complex Möbius band $\mathbb{M}$ on which the blow-up is defined, *is not* the Cartesian product $\mathbb{E}\times\mathbb{C}$. The *linearization form* θ_1 should rather be considered as a meromorphic connexion on the nontrivial *normal line bundle* over $\mathbb{E}$ (cf. with Remark 14.8 and especially §17**G**).

Remark 10.14. Conversely, a foliation $\mathcal{F}'$ on the complex Möbius band defined by the holomorphic (convergent) Pfaffian equation (10.6) and symmetric by the complex conjugacy $(z,x)\mapsto(\bar z,\bar x)$, always blows down to a singular real analytic foliation $\mathcal{F}$ on $(\mathbb{R}^2,0)$ defined by a real analytic form ω, *provided that* the point at infinity $z=\infty$ is a nonsingular or at worst a finite order pole for *all* forms ϑ_k. The latter assumption means that $\sup_k \operatorname{ord}_{w=0}\vartheta_k < +\infty$.

In particular, assume that $\Sigma\subset\mathbb{C}$ is a finite set (necessarily symmetric with respect to the involution $z\mapsto\bar z$), disjoint with the real axis, $\Sigma\cap\mathbb{R}=\varnothing$, and θ_k are rational forms whose singularities always belong to Σ. Then the equation (10.6) corresponds to a generalized elliptic singularity, if the point $w=0$ is *nonsingular* for all forms ϑ_k, i.e., when

$$\theta_1 + \frac{dz}{z},\ z^{-1}\theta_2,\ \dots,\ z^{-k}\theta_k,\ \dots \text{ are holomorphic at } z=\infty \tag{10.10}$$

as 1-forms on $\mathbb{E}$ at the point $z=\infty$ (recall that this holomorphy for $\theta = R(z)\,dz$ means that $R(z)=O(z^{-2})$). In this case the conditions (10.10) imply that

$$\sum_{\Sigma} \operatorname{res}\theta_1 = -1 \tag{10.11}$$

where the summation is extended on all finite singularities of the form θ_1.

The rational 1-forms $\theta_i \in \Lambda^1(\mathbb{E}) \otimes \mathfrak{M}(\mathbb{E})$ depend on the homogeneous components of the 1-form $\omega \in \Lambda^1(\mathbb{C}^2, 0)$ in a rather simple way.

Lemma 10.15. *Assume that the blow-up of the real analytic form $\omega = \omega_n + \omega_{n+1} + \cdots$ is nondicritical. Then:*

(1) *The coefficients of the rational forms θ_k depend rationally on the coefficients of the initial form ω.*

(2) *The form θ_k* does not *depend on the coefficients of the homogeneous components of order $n+k$ and higher.*

(3) *If the principal homogeneous part ω_n is fixed, the first form θ_1 is uniquely determined and all other forms θ_k, $k \geqslant 2$, depend* polynomially *on the remaining coefficients of higher order terms $\omega_{n+1}, \omega_{n+2}, \ldots$ of the form ω.*

Proof. Everything follows immediately from (10.5) and computation of the reciprocal

$$\frac{1}{h_{n+1}(z) + x h_{n+2}(z) + \cdots} = \frac{1}{h_{n+1}(z)} \left(1 - x \cdot \frac{h_{n+2}(z)}{h_{n+1}(z)} + \cdots\right)$$

on any compact set $K \times (\mathbb{C}, 0)$, $K \Subset \mathbb{C} \smallsetminus \Sigma$. □

Remark 10.16. It would be wrong to assume that, conversely, the principal homogeneous part ω_n is determined by the linearization form θ_1 only. For instance, the form θ_1 may be nonsingular at some points of Σ (when p_n and q_n have common factor), whereas some of the higher forms θ_k, $k \geqslant 2$, may have poles and therefore necessarily contribute to ω_n.

To compute the coefficients of the semi-monodromy map associated with the cross-section $z = 0$ with the chart $u \in (\mathbb{C}^1, 0)$ on it, we will integrate the equation (10.6) in the form $x = X(z, u)$ subject to the initial condition $X(0, u) = u$. Expanding this solution in the series $X(z, u) = \sum_{k \geqslant 1} u^k X_k(z)$ and substituting this expansion into (10.6), we obtain a triangular (infinite) system of ordinary differential equations in the Pfaffian form with the initial conditions

$$\begin{aligned} dX_1 &= X_1\theta_1, & X_1(0) &= 1, \\ dX_2 &= X_2\theta_1 + X_1^2\theta_2, & X_2(0) &= 0, \\ dX_3 &= X_3\theta_1 + 2X_1X_2\theta_2 + X_1^3\theta_3, & X_3(0) &= 0, \\ &\vdots & &\vdots \end{aligned} \tag{10.12}$$

This system can be recursively solved in quadratures, since on each step the equation for X_k is linear nonhomogeneous with the same linear part and variable yet known nonhomogeneity.

The coefficients of the semi-monodromy map are obtained as the result of analytic continuation of solutions of the system (10.12) along the loop $\mathbb{R}P^1$ (i.e., along the real line across infinity),

$$\Delta_{\mathbb{R}}(x) = \sum_{k\geqslant 1} a_k x^k, \qquad a_k = (\varDelta_{\mathbb{R}P^1} X_k)(0) \in \mathbb{R}, \qquad k = 1, 2, \dots, \tag{10.13}$$

where $\varDelta_{\mathbb{R}P^1}$ denotes the operator of analytic continuation of the function $X_k(\cdot)$ along $\mathbb{R}P^1$, not to be confused with the *map* $\Delta_{\mathbb{R}}$. Clearly, each coefficient a_k depends only on the forms $\theta_1, \dots, \theta_k$ and does not depend on the remaining forms $\theta_{k+1}, \theta_{k+2}, \dots$.

Remark 10.17. The algorithm of computation of the semi-monodromy and monodromy maps for generalized elliptic points, provides also a tool for *definition* of the (semi-)monodromy for *formal vector fields* or *formal* Pfaffian forms. Indeed, consider a formal Pfaffian form ω as in (10.3) but *without* assumption that the series converges. The condition (10.4) makes sense since it involves only the lowest order homogeneous terms ω_n of ω.

The "formal blow-up" of this form is well defined and gives a Pfaffian equation (10.6) with the forms θ_i still rational in z, but the series in the powers of x in the right hand side will be only formal.

It remains to notice now that the infinite triangular system of Pfaffian equations (10.12) remains exactly the same (no changes are required) and solving any finite number of equations from this system *determines uniquely* the infinite formal series (10.13) for the holonomy $\Delta_{\mathbb{R}}$. Thus the map $\Delta_{\mathbb{R}} \in \operatorname{Diff}[[\mathbb{R}, 0]]$ is consistently defined for the specific choice of the cross-section $\tau = \{z = 0\}$. Choosing any other cross-section $\{z = \varphi(x)\}$, even formal so that $\varphi \in \mathbb{C}[[x]]$, may change $\Delta_{\mathbb{R}}$ by the formal conjugacy: the arguments remain the same.

Finally, we remark that if the homogeneous forms $\omega_n, \omega_{n+1}, \dots$ depend analytically on any additional parameters $\lambda_1, \dots, \lambda_m$ in the sense of Definition 4.17, then the coefficients of the formal holonomy (semi-monodromy) will depend analytically on λ as far as the form remains generalized elliptic, that is, the roots of the homogeneous polynomial h_{n+1} in (10.4) remain off the real axis.

10E. Relative decidability of alternative in the generalized elliptic case. The established structure of the map $\Delta_{\mathbb{R}}$ allows us to prove *relative* decidability of the alternative for generalized elliptic singularities with fixed principal part. Denote by $\mathcal{B}(\omega_n) = (j^n)^{-1}(\omega_n) = \{\omega = \omega_n + \omega_{n+1} + \cdots\} \subseteq \varLambda^1(\mathbb{R}^2, 0)$ the space of all holomorphic forms *with the fixed principal homogeneous part* ω_n.

Theorem 10.18 (see [**Ily72a**]). *For generalized elliptic foliations with the principal part ω_n the center-focus alternative is ultimately decidable within the class $\mathcal{B}(\omega_n)$.*

Proof. We show that in the assumptions of the theorem, the coefficients $a_k = a_k(\omega) = a_j(\omega_{n+1}, a_{n+2}, \dots)$ of the semi-monodromy map $\Delta_{\mathbb{R}}(x) = a_1 x + a_2 x^2 + \cdots$ are *quasi*homogeneous polynomials in the *nonprincipal* Taylor coefficients $\omega_{n+1}, \omega_{n+2}, \cdots$ of ω. When written as an argument, each

ω_k is identified with the string of its coefficients which are in turn natural coordinates on the jet space.

By Lemma 10.15, each coefficient a_k depends only on the components $\omega_n, \dots, \omega_{n+k-1}$ and this dependence is real analytic.

Consider an arbitrary real number $0 \neq \mu \in \mathbb{R}$ and the linear transformation $D_\mu = (x, y) \mapsto (\mu x, \mu y)$. This transformation acts diagonally on 1-forms if we choose a monomial basis. After the appropriate rescaling the 1-form

$$\mu^{-n-1} D_\mu^* \omega = \omega_n + \mu\omega_{n+1} + \mu^2\omega_{n+2} + \cdots$$

again belongs to $\mathcal{B}(\omega_n)$.

On the other hand, D_μ changes the chart on the x-axis by the linear transformation $x \mapsto \mu x$ and hence transforms the semi-monodromy map $\Delta_\mathbb{R}$ into

$$\mu^{-1}\Delta_\mathbb{R}(\mu x) = a_1 x + \mu a_2 x^2 + \mu^2 a_3 x + \cdots .$$

Since the coefficients of the semi-monodromy are uniquely defined, we conclude that

$$a_k(\mu\omega_{n+1}, \mu^2\omega_{n+2}, \dots, \mu^{k-1}\omega_{n+k-1}) = \mu^{k-1} a_k(\omega_{n+1}, \omega_{n+2}, \dots, \omega_{n+k-1}).$$

In other words, each a_k is a quasihomogeneous analytic function of its arguments. Such a function is necessarily a quasihomogeneous polynomial.

The ultimate algebraic decidability of the center-focus alternative now follows immediately from ultimate decidability of the identity $\Delta_\mathbb{R} \circ \Delta_\mathbb{R} = \mathrm{id}$ for holomorphic self-maps (Problem 10.4). Indeed, since a_j are polynomial functions on $\mathcal{B}(\omega_n)$, vanishing of any finite number of coefficients of $\Delta_\mathbb{R} \circ \Delta_\mathbb{R}$ is an algebraic condition on a finite jet of ω. If *all* nonlinear coefficients of $\Delta_\mathbb{R} \circ \Delta_\mathbb{R}$ vanish, then the singularity is a center. □

10F. Decidability to codimension 1. Analyzing system (10.12), we immediately see that the first coefficient $a_1(\omega)$ depends *nonalgebraically* on (the Taylor coefficients of) θ_1. Yet despite this nonalgebraicity, the *neutrality condition* $a_1(\omega) = -1$ on 1-jet of the self-map $\Delta_\mathbb{R} \in \mathrm{Diff}(\mathbb{R}^1, 0)$ to be 2-periodic, turns out to be algebraically decidable. This assertion is not completely trivial, since its complex counterpart fails (Problem 10.9).

Theorem 10.19.

1. *The multiplicator* $a_1 = a_1(\omega_n)$ *of the semi-monodromy map* $\Delta_\mathbb{R}$ *of a generalized elliptic singular point is equal to* -1, *if and only if*

$$\sum_{\operatorname{Re} z > 0} \operatorname{res}_z \theta_1 = -\frac{1}{2}. \tag{10.14}$$

2. *The center-focus alternative for generalized elliptic singularities is algebraically decidable to codimension* 1.

Proof. The first equation of (10.12) can be immediately integrated, yielding for the solution $X_1(z)$ and the multiplicator a_1 of its continuation along $\mathbb{R}P^1$ the *transcendental* expressions

$$X_1(z) = \exp \int_0^z \theta_1, \qquad a_1 = \exp \oint_{\mathbb{R}P^1} \theta_1.$$

The *neutrality condition* $a_1 = -1$ holds if and only if $\oint_{\mathbb{R}P^1} \theta_1 = \pi i(2m+1)$, $m \in \mathbb{Z}$, i.e.,

$$\sum_{\operatorname{Re} z>0} \operatorname{res}_z \theta_1 = \frac{1}{2} + m, \qquad m = 0, \pm 1, \pm 2, \dots. \tag{10.15}$$

This equality is not yet an algebraic condition, since it is the union of *infinitely many* conditions for different values of $m \in \mathbb{Z}$. However, since ω is *real* on the real axis, its singular locus Σ is symmetric by the reflection $z \mapsto \bar{z}$, and the residues at symmetric points are complex conjugate. The total of *all* residues of θ_1 on the whole plane $\mathbb{C}$ is -1 by (10.11). Therefore, the real part of the expression in the left hand side of (10.15) is $-\frac{1}{2}$, which is compatible with the right hand side only when $m = -1$, thus proving (10.14).

The second assertion of the theorem immediately follows from the first one, since (10.14) is an algebraic condition on the form θ_1. □

10G. Nondecidability of the weak focus stability alternative. Inspection of the next nontrivial equation in (10.12) already suggests nonalgebraicity of the second nontrivial condition $a_3(\omega) = 0$.

The real topological type of vector fields with $a_1 = -1$ and $a_3 \neq 0$ is called a *weak focus*: the weakness means that the convergence of integral trajectories to the origin is slower than that of logarithmic spirals. Weak foci can be stable (if $a_3 < 0$) and unstable if $a_3 > 0$; see Problem 10.12.

Remark 10.20. Note that if $a_1 = -1$, then for any choice of $a_2(\omega)$ the square $\Delta_{\mathbb{R}} \circ \Delta_{\mathbb{R}}$ starts from cubic terms, hence nonalgebraicity of the condition $a_3(\omega) = 0$ would mean that the unrestricted center-focus alternative is not decidable to codimension 2. In other words, Theorem 10.19 establishes a *sharp* bound on the codimension to which the unrestricted center-focus alternative is algebraically decidable (Problem 10.14).

To prove the nonalgebraicity of the condition $a_3(\omega) = 0$, we consider a polynomial foliation which *after blow-up* is defined by a rational Pfaffian equation in the affine chart (x, z), $z = y/x$, of the following form:

$$\begin{aligned} dx &= x\theta_1 + x^3\theta_3, \qquad \theta_1 = \left(\frac{A}{z^2+4} - \frac{A+1}{z^2+1}\right) z\,dz, \\ \theta_3 &= \mu dz + \frac{zdz}{z^2+\lambda^2}, \qquad \lambda, \mu \in \mathbb{R}, \quad \lambda \neq 0. \end{aligned} \tag{10.16}$$

Here $A \in \mathbb{R} \smallsetminus \mathbb{Z}$ is any fixed *noninteger* number. The foliation defined by (10.16) on the complex Möbius band $\mathbb{M}$ tangent to the exceptional divisor $\mathbb{E}$ can be blown down to a polynomial foliation defined by a polynomial form $\omega = 0$ in $\mathbb{C}^2$ by Remark 10.14 (see also Problem 10.13). Note that the forms $\theta_1 + \frac{dz}{z}$ and $z^{-2}\theta_3$ are both holomorphic at $z = \infty$, therefore the infinite point $z = \infty$ on $\mathbb{E}$ is *nonsingular* in the other affine chart on $\mathbb{M}$. In particular, the holonomy operator along the real equator can be replaced by the holonomy along the contour Γ which consists of a real segment $[-R, R]$ and the large semicircle $\{|z| = R,\ \operatorname{Im} z \geqslant 0\}$ in the upper half-plane,

$$\Gamma = [-R, R] \cup \{|z| = R,\ \operatorname{Im} z \geqslant 0\} \subset \mathbb{C}, \qquad R \gg 2, \tag{10.17}$$

with a standard orientation inherited from $\mathbb{C}$.

The conditions (10.10) for this system are obviously verified, meaning that in the semialgebraic domain $\lambda \neq 0$ the equation (10.16) is generalized elliptic. The total residue of the form θ_1 at the singular points $i, i\sqrt{2}$ in the upper half-plane is exactly $-\frac{1}{2}$, so the condition (10.14) is automatically verified for all values of λ, μ.

Obviously, the second coefficient $a_2 = a_2(\lambda, \mu)$ of $\Delta_{\mathbb{R}}$ is zero, since the term $x^2\theta_2$ is absent in (10.16). The third coefficient, $a_3 = a_3(\lambda, \mu)$ is a real analytic function of λ, μ in the domain $\lambda \neq 0, 1, 2$ where (10.16) is generalized elliptic. The generalized ellipticity holds also for the values $\lambda = 1, 2$, yet the holomorphic dependence on parameters fails at these points, as will be seen from the proof of the following result.

Theorem 10.21. *The second integrability condition $a_3(\lambda, \mu) = 0$ for the family* (10.16) *defines a nonalgebraic real curve on the plane of parameters $\{\lambda > 0,\ \mu \in \mathbb{R}\}$.*

The complement of this curve $\{a_3(\lambda, \mu) = 0\}$ consists of sufficient jets (foci), thus Theorem 10.21 indeed proves undecidability of the center-focus problem.

Proof of Theorem 10.21. Consider the holonomy Δ_Γ of the system (10.16) along the contour Γ; see Fig. II.7. This holonomy depends on the parameters λ, μ, and we will show that the condition $\Delta_\Gamma \circ \Delta_\Gamma = \mathrm{id}$ is *non-algebraic* with respect to these parameters.

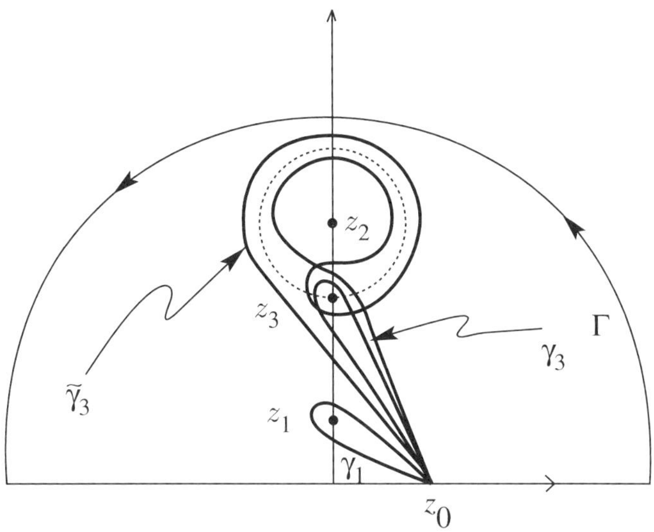

Figure II.7. Holonomy group and continuous deformation of the loops generating this group, when parameters change

To do this, we first transform the first three (in fact, only two, since $\theta_2 = 0$ is absent) equations from the corresponding system (10.12) into a linear form. This is achieved by the substitution $y_1 = X_1^2$, $y_2 = X_3/X_1^2$. Such a substitution naturally arises when solving the first three equations from (10.12) by the method of variation of constants. The resulting equations have the form

$$\begin{cases} dy_1 = 2\theta_1 y_1, & y_1 = X_1^2, \\ dy_2 = \quad \theta_3 y_1, & y_2 = X_3/X_1^2. \end{cases} \tag{10.18}$$

This system is linear, and it obviously has *identical* holonomy associated with the contour Γ, if and only if the holonomy of the initial nonlinear system is 2-periodic. We will show that this condition is nonalgebraic in (λ, μ).

Because of the reducibility of this system, its holonomy group can be easily computed. The linearity of the system implies that this group is linear. In what follows we perform computations with linear systems, that will be described in more details in §15**C**.

The holonomy group is generated by three linear operators corresponding to circulation around the three singular points. Fix a base point $z_0 \in \mathbb{R}_+$ somewhere on the positive semiaxis and let γ_i, $i = 1, 2, 3$, be the standard small loops going around the three singular points $z_1 = i$, $z_2 = 2i$ and $z_3(\lambda) = i\lambda$ back and forth along the line segments. Fix a solution $Y = Y(z; \lambda, \mu)$ of the system (10.18) with the initial data $y_1(z_0) = 1$, $y_2(z_0) = 0$.

Note that the form θ_1 does not depend on the parameters, and the first coordinate subspace is invariant (i.e., the result of analytic continuation of y_1 does not depend on y_2). Therefore the function y_1 admits a holomorphic branch along the loop γ_3. On the other hand, the form θ_1 admits a holomorphic branch along the paths γ_1, γ_2. This allows us to compute the integrals $\int_{\gamma_i} y_1\theta_3$. These integrals allow us to represent the linear transformations

$$\Delta_{\gamma_i} Y = YM_i, \qquad M_i \in \mathrm{GL}(2, \mathbb{C}),$$

by the three 2×2-matrices M_1, M_2, M_3. These matrices, with entries depending on the parameters λ, μ, are as follows:

$$\begin{gathered} M_1 = \begin{pmatrix} \alpha & \beta_1 \\ & 1 \end{pmatrix} \qquad M_2 = \begin{pmatrix} \alpha^{-1} & \beta_2 \\ & 1 \end{pmatrix} \qquad M_3 = \begin{pmatrix} 1 & \beta_3 \\ & 1 \end{pmatrix} \\ \alpha = \exp 2\pi i A, \qquad A = 2\operatorname{res}_{z_1}\theta_1 = -2\operatorname{res}_{z_2}\theta_1, \\ \beta_1 = (1-\alpha)\int_{\ell_1} y_1\theta_3, \qquad \beta_2 = (1-\alpha^{-1})\int_{\ell_2} y_1\theta_3, \\ \beta_3 = 2\pi i\, y_1(z_3)\operatorname{res}_{z_3}\theta_3 = \pi i\, y_1(z_3). \end{gathered} \tag{10.19}$$

In all formulas ℓ_i, $i = 1, 2, 3$, denote the line segments connecting the base point z_0 with the points z_i. The integrals involving the multivalued solution along these segments, as well as the value $y_1(z_3)$, are obtained by continuing the branch with $y_1(z_0) = 1$ along these segments.

The values of $\beta_1, \beta_2, \beta_3$ depend on the parameters λ, μ in a locally holomorphic (as soon as $\lambda \notin \{0, 1, 2\}$) yet rather complicated way.

First since the form θ_3 itself depends on these parameters, the integrals β_1, β_2 are linear functions of μ and rational functions of λ. Indeed, the form θ_1 has no singularities at z_3 and hence such a variation does not affect the result of integration along ℓ_1 and ℓ_2: the branch of y_1 on these segments remains the same. Since θ_3 depends linearly on μ and rationally on λ, the matrix entries $\beta_i(\lambda, \mu)$, $i = 1, 2$ are also single-valued functions of the parameters λ, μ, linear in μ.

On the contrary, if the singular point $z_3 = i\lambda$ makes a full turn along a circular path around, say, the point $z_2 = 2i$, then the corresponding path γ_3 will be replaced by the conjugate loop $\widetilde{\gamma}_3 = \gamma_2\gamma_3\gamma_2^{-1}$ in the fundamental group (this is called the *braid group action*) . Accordingly, the integral

$$\int_{\widetilde{\gamma}_3} y_1\theta_3 = 2\pi i\, \widetilde{y}_1(z_3)\operatorname{res}_{z_3}\theta_3 = i\pi\, \widetilde{y}_1(z_3) = \alpha^{-1}\int_{\gamma_3} y_1\theta_3$$

differs from its initial value by the multiplier α^{-1} as a result of the transition to the different branch of the function $\widetilde{y}_1 = \alpha^{-1}y_1$ (the first component). In the same way the result of the point z_3 circulating around the point $z_1 = i$ consists of multiplication of β_3 by α.

In other words, if α is not a root of unity, i.e., if A is an irrational number, then the matrix entry $\beta_3 = \beta_3(\lambda)$ has logarithmic ramification points at $\lambda = 1$ and $\lambda = 2$ in the half-plane $\{\operatorname{Re}\lambda > 0\}$: when λ goes around one of these points, the value $\beta_3(\lambda)$ gets multiplied by α or α^{-1} respectively. Thus for $A \notin \mathbb{Q}$ the function $\beta_3(\lambda)$ is a *nonalgebraic function of* λ.

The holonomy of the system (10.18) associated with the contour Γ, is the linear operator represented by the product of the three upper-triangular matrices $M = M_2M_3M_1$, itself an upper-triangular 2×2-matrix with units on the diagonal. From the above analysis it follows that the nonzero off-diagonal entry $\beta_* = \beta_*(\mu, \lambda)$ of the matrix product M is a sum of a linear form in μ, a rational (single-valued) function of λ and a transcendental function with nontrivial logarithmic ramification at the points $\lambda = 1, 2$.

We will show that the function β_* depends on μ in a nontrivial way. To verify this, we fix λ and let $|\mu|$ grow to infinity. The off-diagonal term of the operator Δ_Γ is equal to the integral

$$\oint_\Gamma y_1\theta_3 = \mu \oint_\Gamma y_1\,dz + O(1) = \mu \int_{\mathbb{R}} y_1\,dz + O(1) \qquad \text{as } |\mu| \to \infty.$$

Yet along the real equator $\mathbb{R}$ the function y_1 is single-valued and *everywhere positive* as a solution of the first equation in (10.18) with the form θ_1 *real on* $\mathbb{R}$ and the initial value $y_1(z_0) = 1$ which is *positive*. Thus we see that $\beta_*(\mu, \lambda) = C\mu + L(\lambda)$ with $C > 0$ and L a function with logarithmic singularities at $\lambda = 1, 2$.

Thus the trivial holonomy condition $M_2M_3M_1 = E$ occurs on the non-algebraic curve $\{\mu = -L(\lambda)/C\}$ (the graph of a transcendental function) on the (μ, λ)-plane of the parameters. □

Remark 10.22. Nonalgebraicity of the condition $\{a_3(\omega) = 0\}$ in fact does not mean that the alternative is not decidable in the sense of Definition 10.2. Indeed, the variety of centers is given by the infinite number of equations $\{a_j(\omega) = 0\}$, $j = 2, 3, \ldots$, imposed on all Taylor coefficients of the square $\Delta_{\mathbb{R}} \circ \Delta_{\mathbb{R}}$ of the monodromy map. Though the equations are nonalgebraic, the corresponding zero locus N_k of vanishing of the first k coefficients may be a proper analytic subset of a larger but *algebraic* variety $N'_k \subset J^k$. If the codimension of N'_k grows to infinity together with k, this would mean decidability of the alternative.

What the assertion of Theorem 10.21 implies is algebraic *nondecidability of the Lyapunov stability* for generalized elliptic singularities. Indeed, the stable foci correspond to the domain $S^+ = \{a_3 > 0\}$, while unstable foci are determined by the inequality $S^- = \{a_3 < 0\}$. These sets are relatively open in the locus $\{a_1 = 0\}$ in any jet space J^k, and the hypersurface separating them is real analytic but nonalgebraic in the intersections. Clearly, such a situation is incompatible with algebraic decidability to any codimension.

It turns out that the Lyapunov stability for generalized elliptic singularities is in fact *ultimately analytically decidable* in some rigorously defined sense (we do not give details here). This result was achieved by N. Medvedeva [**Med06**].

Exercises and Problems for §10.

Exercise 10.1. Prove that no finite jet of a planar real analytic vector field can be sufficient for a center.

Suggestion. Consider together with the field $F(z)$ (in the complex notation) its small perturbation $F(z) + if(z)F(z)$ where the scalar function $f(z)$ is flat and nonnegative.

Exercise 10.2. Prove that the property of having local (strict) minimum is ultimately algebraically decidable for germs of real analytic functions of one variable.

An isolated singularity of a vector field (resp., fixed point of a self-map) is *Lyapunov stable*, if for any open neighborhood U of this point one can find a (smaller) neighborhood V such that any trajectory of the field (resp., the orbit of the self-map), starting in V, never leaves U.

Exercise 10.3. Prove that Lyapunov stability is algebraically decidable for germs of real analytic vector fields on the real line.

Problem 10.4. Prove that the problem of deciding, whether a germ is periodic with period 2, is ultimately algebraically decidable for germs of real analytic self-maps tangent to the symmetry $x \mapsto -x$.

Problem 10.5. Prove that the periodicity/aperiodicity alternative for holomorphic self-maps $\operatorname{Diff}(\mathbb{C}, 0)$ is not algebraically decidable.

Suggestion. Prove that this problem is undecidable for *linear* self-maps.

Exercise 10.6. Prove that an elliptic singular point (in the sense of Definition 4.28) is generalized elliptic.

Exercise 10.7. Prove the assertions made in Remark 10.14.

Problem 10.8. Let F, F' be two formally orbitally equivalent generalized elliptic formal vector fields. Prove that their formal monodromy maps as defined in Remark 10.17, are formally conjugate in the group $\operatorname{Diff}[[\mathbb{R}^1, 0]]$.

Problem 10.9. Consider the set $\mathcal{B}$ of germs of holomorphic 1-forms, which have no singularities on the real axis after the first blow-up (an analog of generalized ellipticity condition). We will call a form $\omega \in \mathcal{B}$ a *pseudo-center*, if the holonomy operator $\Delta_{\mathbb{R}}$ associated with the loop $\mathbb{R}P^1 \subset \mathbb{P}^1$ of the blown-up foliation, is 2-periodic. Otherwise we call the germ *pseudo-focus*.

Prove that the set $\mathcal{B}$ is semialgebraic, but the alternative pseudo-center/pseudo-focus relative to $\mathcal{B}$ is not algebraically decidable to codimension 1.

Exercise 10.10. Determine, for which values of the real parameters a, b the analytic singularity

$$\dot{x} = -x^2 + axy + \cdots, \quad \dot{y} = -y + bx^2 + \cdots$$

(the dots denote cubic terms) is Lyapunov stable.

Problem 10.11. Give a necessary and sufficient condition of Lyapunov stability for *nonmonodromic* singularities with nondicritical complete desingularization.

Problem 10.12. Prove that the weak focus with the semi-monodromy map $\Delta_{\mathbb{R}}(x) = -x + ax^3 + \cdots$ is stable if $a > 0$ and unstable if $a < 0$.

Problem 10.13. What is the order of singularity obtained by blowing down the foliation $\mathcal{F}$ defined by the Pfaffian equation (10.16) in a neighborhood of the exceptional divisor $\mathbb{E}$?

Problem 10.14. Prove that Lyapunov stability for germs of planar vector fields is algebraically decidable to codimension 11, but not to codimension 12.

11. Holonomy and first integrals

In this section we study the inter-relations between analytic and topological properties of singular foliations. The main tool of this study is construction and analysis of some finitely generated subgroups of the group $\operatorname{Diff}(\mathbb{C}, 0)$, especially the *vanishing holonomy group*. Using properties of this group, we show, following [**Mou82**], that *integrability* of a real analytic foliation, existence of nontrivial *analytic* first integral, is equivalent to the topological property of being a center, for elliptic foliations (this result is known as the *Poincaré–Lyapunov theorem.*). On the other hand, the equivalence fails for generalized elliptic foliations with degenerate linear part: such systems can be centers without analytic integrals.

The second part of the section is devoted to generalizations of the Poincaré–Lyapunov theorem for arbitrary isolated singularities of *holomorphic* (nonreal) foliations on $(\mathbb{C}^2, 0)$. We introduce, following the seminal paper by J.-F. Mattei and R. Moussu [**MM80**], the class of (topologically) simple foliations and show that simplicity of a holomorphic foliation is necessary and sufficient for its analytic integrability.

11A. Integrability and its decidability. Thus far the term "integrability" was used in three different senses: for distributions (when integrability means existence of foliation tangent to the distribution), for differential equations (which sometimes admit complete solution in quadratures), and for groups of conformal germs (when integrability means existence of functions constant along all orbits). Integrability of foliations, introduced below, is very close to the last notion: a foliation is integrable if there exists a nontrivial holomorphic function constant along all leaves of the foliation.

Note that at any nonsingular point of a holomorphic foliation there exists a germ of holomorphic function with nonvanishing differential, which is constant along the local leaves (plaques) of the foliation. Thus an integrable distribution near a *nonsingular* point generates an integrable foliation.

Definition 11.1. A singular foliation $\mathcal{F} = \{\omega = 0\}$ on $(\mathbb{C}^2, 0)$ is said to be *integrable*, if there exists a nonconstant holomorphic function (germ) $u \in \mathcal{O}(\mathbb{C}^2, 0)$ such that $\omega \wedge du = 0$.

Equivalently, the germ of a vector field $F \in \mathcal{D}(\mathbb{C}^2, 0)$ is *integrable*, if there exists a nonconstant holomorphic function such that $Fu = 0$. In both cases the nonconstant function is called *first integral*, or simply *integral* of the foliation.

Every leaf of an integrable foliation entirely belongs to a level curve $\{u = \text{const}\}$ and hence is an *analytic curve* in $(\mathbb{C}^2, 0)$.

The first integral, if it exists, is by no means unique: any nonconstant function $f \in \mathcal{O}(\mathbb{C}, 0)$ applied to a first integral $u \in \mathcal{O}(\mathbb{C}^2, 0)$ produces another first integral $v = f \circ u$. Clearly, if the germ f is invertible in $\mathrm{Diff}(\mathbb{C}, 0)$, the two integrals can exchange their roles. All the way around, if f is noninvertible, the level curves of v necessarily consist of *several* leaves of the foliation: for all small c, the preimage $f^{-1}(c)$ consists of more than one point, hence $v^{-1}(c)$ cannot be connected.

Definition 11.2. A nonconstant holomorphic function $u \in \mathcal{O}(\mathbb{C}^2, 0)$ is called a *primitive* (first) integral of an integrable foliation, if the level curves $\{u = \text{const}\}$ are *all connected* (in a sufficiently small neighborhood of the origin).

Proposition 11.3. *If u is a primitive first integral of a foliation $\mathcal{F}$, then any other first integral v is an analytic function of u, $v = f(u)$ for some analytic nonconstant germ $f \in \mathcal{O}(\mathbb{C}, 0)$.*

Proof. If both u and v are first integrals, then $dv \wedge du \equiv 0$, which means that v takes constant values on each connected component of any level curve $\{u = \text{const}\}$. By the implicit function theorem, v is an analytic function of u outside the set of the critical values of the latter. If the neighborhood of the singularity is sufficiently small, there is only one critical value $u = 0$, even if the critical point of u were nonisolated, while the image by u of any such neighborhood is always an open set containing the origin. Thus $v = f(u)$, where f is analytic and bounded in a *punctured* neighborhood of the origin on the u-line. By the removable singularity theorem, f is analytic also at the origin, $f \in \mathcal{O}(\mathbb{C}, 0)$. □

The definition of integrability admits a formal counterpart: a formal Pfaffian form ω (resp., formal planar vector field F) is *formally integrable*, if there exists a nonconstant formal series $u \in \mathbb{C}[[x,y]]$ such that $\omega \wedge du = 0$ (resp., $Fu = 0$) on the level of formal series. However, we will show later in §11$\mathbf{G}_2$ that formal integrability for *analytic* 1-forms, vector fields and foliations coincides with analytic integrability.

Theorem 11.4. *Integrability of foliations is algebraically decidable.*

Proof. The formal identity $\omega \wedge du = 0$ involving the series $u = u_2 + u_3 + \cdots \in \mathbb{C}[[x,y]]$ and the Pfaffian 1-form $\omega = \omega_1 + \omega_2 + \cdots$ is equivalent to the infinite triangular system of polynomial identities involving the homogeneous components u_i, ω_j respectively,

$$\forall k = 1, 2, \dots \qquad \sum_{i+j=k+1} \omega_i \wedge du_j = 0. \tag{11.1}$$

Truncation of this system to any order $k \leqslant N$ is a linear homogeneous system of algebraic equations on the unknown components $u_2, \dots, u_N$ with the coefficients linearly depending on the forms $\omega_1, \dots, \omega_N$. Nontrivial solvability of the system is a semialgebraic condition on the N-jet of the form ω by the Tarski–Seidenberg theorem. We leave it to the reader to verify that the codimension of neutral jets tends to infinity with the order of the jet. □

11B. Integrability of real foliations. Integrability of real analytic foliations is closely related to having a singularity of the topological type "center".

Proposition 11.5. *A monodromic integrable singularity is a center.*

Proof. Without loss of generality we may choose the cross-section τ used in the construction of the monodromy map (see Definition 9.6) such that the restriction of the integral on τ is a nonconstant function. Being real analytic, this function gives one-to-one parametrization of all points on the positive semi-section in a sufficiently small neighborhood of the singularity. Thus a real leaf crossing τ at some point a, cannot cross it again at a different point, which means that all real leaves are closed. □

For elliptic singularities analytic integrability can be easily replaced by formal integrability.

Proposition 11.6. *Formally integrable* elliptic *singularity is a center.*

Proof. Formal integrability is obviously invariant by formal orbital classification. We prove first that for an elliptic vector field in the formal normal form (4.10), the formal integrability is equivalent to formal orbital linearizability.

Indeed, if the vector field is nonlinearizable, then in suitable "formal coordinates" (x, y) it takes the form (see Table I.1)

$$F = \mathbf{I} + (r^{2k} + ar^{4k+2})\mathbf{E}, \qquad r^2 = x^2 + y^2,$$

where $\mathbf{I}$ and $\mathbf{E}$ are the rotation field $-y\frac{\partial}{\partial x}+x\frac{\partial}{\partial y}$ and the Euler field $x\frac{\partial}{\partial x}+y\frac{\partial}{\partial y}$ respectively. If $u = u_m + u_{m+1} + \cdots$ is a nontrivial formal integral, then the functions $u_m, \dots, u_{m+2k-2}$ must be radial (depend only on the polar radius r), hence powers of r^2 with constant coefficients. In particular, $m = 2n$ must be even, so that u starts with the term $r^{2n} + \cdots$. Yet for the determination of the term u_{2n+2k} we obtain the equation $\mathbf{I}u_{2n+2k} = (2n + 2k)r^{2k+2n}$ which is nonsolvable, since the right hand side has nonzero integral on all circles $r = \text{const}$.

On the other hand, if the field is formally linearizable, $F = \mathbf{I}$, then it is obviously formally integrable, $u = r^2$ being the nontrivial integral.

To prove the proposition, it is sufficient to show that a formally orbitally linearizable real *analytic* foliation is necessarily a center. Indeed, formal orbital linearizability means that the monodromy map is formally equivalent to the identity in the group $\operatorname{Diff}[[\mathbb{R}^1, 0]]$ (Problem 10.8). But then the monodromy map is itself identity (cf. with Theorem 6.8), which means that the singularity is a center. □

In other words, we proved that out of the three conditions,

(AI) existence of analytic first integral,

(FI) existence of formal first integral,

(C) center (identical monodromy map),

the first obviously implies the second and the third, regardless of whether the monodromic singularity is elliptic or not.

The implication (FI) $\Longrightarrow$ (C) is asserted by Proposition 11.6 for elliptic singularities, yet in fact it is valid without this assumption; see §11$\mathbf{G}_2$. We will now discuss the remaining implication (C) $\Longrightarrow$ (AI) showing that for *elliptic* singularities, all three conditions are equivalent. This is the famous Poincaré–Lyapunov theorem, proved by Poincaré for polynomial differential equations and by Lyapunov for analytic singularities. The modern proof given below, is based on [**Mou82**].

Theorem 11.7 (Poincaré–Lyapunov). *A real analytic elliptic singularity which is a center, admits a real analytic first integral with the nondegenerate quadratic part.*

As a corollary to this result and Proposition 11.6, we have a result on "convergence of formal integrals".

Corollary 11.8. *An elliptic singularity which admits a formal first integral with nondegenerate quadratic part, also admits an analytic first integral with the same property.*

For elliptic singularities integrability is ultimately decidable. □

Remark 11.9. The stress in the assertion of Theorem 11.7 is on *analyticity* of the first integral. Indeed, existence of a first integral that is simply continuous at the origin $x = y = 0$ and real analytic outside, is obvious. Indeed, take the cross-section $\tau = \{y = 0,\ x > 0\}$ and the function x^2 on it, and extend this function on the entire neighborhood of the origin by a constant along each trajectory of the vector field. Since all trajectories are closed, this extension is unambiguous and real analytic outside the origin where its continuity is obvious. Applying this construction in the coordinates linearizing any finite order jet, we can in fact guarantee smoothness of the constructed integral to any finite order and even its C^∞-smoothness (Exercise 11.13).

Note that the isolated point where the analyticity break may eventually occur, is a *small* set of codimension 2. Thus, if all objects were defined in $(\mathbb{C}^2, 0)$ rather than in $(\mathbb{R}^2, 0)$, the analyticity would follow automatically unlike in the real context where no removable singularity theorems are available. In other words, the natural way to prove analyticity is to complexify the situation.

The proof of Theorem 11.7 is based on application of results on integrability of groups of conformal germs (see §6**C**) to the vanishing holonomy group introduced below. For an elliptic singularity this group is especially simple (cyclic).

11C. Vanishing holonomy of singularity. Vanishing holonomy already appeared implicitly in §10**C**.

Definition 11.10. The *vanishing holonomy group* of an isolated *nondicritical* singular foliation $\mathcal{F}$ is the holonomy group of the exceptional leaf $L = \mathbb{E} \smallsetminus \operatorname{Sing} \mathcal{F}'$ for the foliation $\mathcal{F}' = \sigma^*\mathcal{F}$ on the Möbius band $\mathbb{M}$ obtained by a simple blow-up $\sigma\colon (\mathbb{M}, \mathbb{E}) \to (\mathbb{C}^2, 0)$ of the singularity.

For real analytic foliations the vanishing holonomy requires preliminary complexification. By construction, the vanishing holonomy is a finitely generated subgroup of the group $\operatorname{Diff}(\mathbb{C}, 0)$ of conformal germs. For dicritical singularities the vanishing holonomy group is not defined.

Computations from §10**C** show that after a simple blow-up of the elliptic singularity the foliation $\mathcal{F}'$ has exactly two singular points on the exceptional divisor $\mathbb{E}$ at the points $z = \pm i$, both of them saddles with the ratio of eigenvalues equal to $-\frac{1}{2}$. By the Hadamard–Perron Theorem 7.1, each saddle has two holomorphic invariant curves. One of them is the common complex separatrix $\mathbb{E}$, the other are holomorphic curves W_+ and W_- transversal to $\mathbb{E}$ at $z = \pm i$ respectively.

The fundamental group of $L = \mathbb{E} \smallsetminus \Sigma$, $\Sigma = \{\pm i\}$, is cyclic and generated by the loop $\mathbb{R}P^1$ (the equator of the Riemann sphere). Therefore the vanishing holonomy group H of an elliptic singularity is cyclic, generated by the single germ of the semi-monodromy $f = \Delta_{\mathbb{R}}$,

$$f = \Delta_{\mathbb{R}}|_\tau \in \operatorname{Diff}(\mathbb{C}, 0), \quad \tau = \{z = 0\} \cong (\mathbb{C}, 0), \qquad f(x) = -x + \cdots. \tag{11.2}$$

As explained in §10**C**, the monodromy of a real elliptic singularity is the square of the vanishing holonomy generator f.

If the foliation $\mathcal{F}$ is integrable and u is an analytic integral, then this analytic integral pulls back on $\mathbb{M}$ as an analytic function on $(\mathbb{M}, S)$ constant along the leaves of $\mathcal{F}'$. By definition of the holonomy, this means that the restriction of u on the cross-section τ used to compute the holonomy maps, is invariant by the vanishing holonomy group H. In other words, we have the obvious implication.

Proposition 11.11. *The vanishing holonomy group of an integrable foliation is an integrable subgroup (in the sense of Definition 6.24) in* $\operatorname{Diff}(\mathbb{C}, 0)$. □

Integrability of a group of conformal germs is a very stringent condition, implying that the group is finite abelian and linearizable; see §6**C**. Together with Proposition 11.11 this gives necessary conditions of integrability of foliations.

For an elliptic center, the generator f of the vanishing holonomy group H is 2-periodic, hence integrable. Thus the necessary conditions dictated by Proposition 11.11 are fulfilled. The Poincaré–Lyapunov theorem asserts that for elliptic singularities they are also sufficient, i.e., if $H \cong \mathbb{Z}_2$, then there exists a real analytic function $u \in \mathcal{O}(\mathbb{R}^2, 0)$ constant along the integral curves.

The proof will indeed be organized by proper complexification of the insufficient "real" arguments mentioned in Remark 11.9. First, we construct an integral $\widetilde{u} \in \mathcal{O}(\tau, 0)$ of the holonomy group H, a function analytic on the cross-section τ and invariant by all holonomy maps. Using the construction of saturation (Lemma 2.18), we extend the integral of the group to the integral of the foliation $\mathcal{F}'$ in a neighborhood of the exceptional leaf $L = \mathbb{E} \smallsetminus \Sigma$.

Local analysis of the singular points $\pm i$ shows that for each of them the local integral defined near one of the separatrices, can be extended along leaves of the foliation onto the full neighborhood of the singular point except the other separatrix. By the removable singularity theorem, we achieve an integral of the foliation $\mathcal{F}'$ defined in a full neighborhood of the exceptional divisor $\mathbb{E}$ in the complex Möbius band $\mathbb{M}$. Blowing down this integral, we obtain an analytic integral of the initial elliptic foliation $\mathcal{F}$.

11D. Complex topology and (non)integrability of elementary singularities. In order to carry out the above program, we need to generalize Lemma 2.18 for the singular context.

We consider the following problem. Let $\mathcal{F}$ be a singular foliation on a neighborhood of the origin $U = (\mathbb{C}^2, 0)$, say, a bidisk, with an irreducible complex separatrix S (analytic curve) through the unique singularity of $\mathcal{F}$ at the origin. Let $a \in S \smallsetminus \{0\}$ be a *regular* point on S and $u_0 \in \mathcal{O}(\mathbb{C}^2, a)$ the germ of holomorphic function at a, which is a local integral of $\mathcal{F}$ near a. The *extension problem* is to find an integral $u \in \mathcal{O}(U)$ of $\mathcal{F}$ in the entire domain

U in such a way that the germ of u at a coincides with u_0. For simplicity we may assume that u_0 is defined on a small cross-section $\tau\colon (\mathbb{C},0) \to (\mathbb{C}^2,a)$ to S at a.

The obvious obstruction for existence of such an extension is the holonomy of $\mathcal{F}$ associated with a small loop $\gamma \in \pi_1(S \setminus \{0\}, a)$. Indeed, if $f \in \operatorname{Diff}(\tau,a)$ is the holonomy map associated with the loop γ, then the extension problem can be solved only if the germ u_0 is invariant by f, that is, the holonomy germ f must be integrable. Yet integrability of f may be insufficient.

Indeed, assume that there is an infinite number of leaves of $\mathcal{F}$ passing near a, such that each of these leaves contains the origin in its closure. Then the foliation $\mathcal{F}$ cannot be integrable: the restriction of u on each of these leaves must be equal to the common constant $0 = u_0|_S$. This scenario is compatible with the identical monodromy (Exercise 11.3) and hence is a genuine obstruction for the extension of analytic integrals.

On the other hand, assume that there exists a small neighborhood $D \subset (\tau, a)$ of a on τ such that the saturation $U' = \operatorname{Sat}(D, \mathcal{F})$ is dense in U and the difference $S' = U \setminus U'$ is an analytic set. Then any f-invariant germ, in particular, u_0, can be extended along the leaves of the foliation $\mathcal{F}$ to a function $u \in \mathcal{O}(U')$ analytic and bounded in the complement to an analytic set, though eventually multivalued. If for some reasons the extension u is single-valued in U', then by the removable singularity theorem, the holomorphic function u can be extended on the whole of U and by construction is an integral of $\mathcal{F}$ extending the germ u_0.

These two patterns are clearly distinct for real foliations. Foliations which are topological nodes (including foci) or saddle-nodes, cannot be integrable, since infinitely many trajectories accumulate to the origin. On the other hand, for topological saddles saturation of a small cross-section to a separatrix S by trajectories of the foliation, entirely fill one the half-planes into which the other separatrix S' cuts a small neighborhood of the origin. One can expect that after complexification the saturation will fill the whole complement to $U' = U \setminus S'$. Recall that these types of real elementary singular foliations are determined by the characteristic ratio λ (the ratio of eigenvalues). The nodes, foci and saddle-nodes correspond to the case λ positive, nonreal or zero. The saddles (and centers) correspond to negative values of λ.

In the complex domain the "saddle" and "nodal" features are not mutually exclusive, yet for all elementary singularities the extension problem can be easily solved in the negative or affirmative sense.

It will be convenient to extend the relations $<, >, \leqslant, \geqslant$ from real to complex numbers in the following intuitive way: for $a, b \in \mathbb{C}$ we will write

$$\begin{aligned} a \geqslant b &\iff a - b \in \mathbb{R} \text{ and } a - b \geqslant 0, \\ a > b &\iff a - b \in \mathbb{R} \text{ and } a - b > 0. \end{aligned} \tag{11.3}$$

The relations $\ngeqslant, \ngtr$ are formed by logical negations of the above, while $a \leqslant b$ or $a < b$ mean that $b \geqslant a$ and $b > a$ respectively. Note that in contrast with the real case the relations $\leqslant$ and $>$ define only *partial order* on $\mathbb{C}$, so $a \ngtr b$ does not mean $a \leqslant b$ and vice versa, $a \ngeqslant b$ is not equivalent to $a < b$.

Lemma 11.12 (Complex "nodal" case). *Each leaf of an elementary singular foliation with the characteristic ratio $\lambda \nleqslant 0$ contains the singularity in its closure.*

Corollary 11.13. *None of these singularities are integrable.* □

Lemma 11.14 (Complex "saddle" case). *For any hyperbolic singular foliation $\mathcal{F}$ with the characteristic ratio* $\operatorname{Re}\lambda < 0$, *saturation* $\operatorname{Sat}(\tau, \mathcal{F})$ *of any cross-section τ to each separatrix fills a complement to the other separatrix in a small neighborhood of the origin.*

Corollary 11.15. *If the holonomy map $f \in \operatorname{Diff}(\tau, a)$ of a separatrix S of a hyperbolic singularity is integrable (this is possible only if $\lambda < 0$), then any germ $u_0 \in \mathcal{O}(\tau, a)$ invariant by f extends to an analytic integral of the foliation $\mathcal{F}$.*

Proof of Lemmas 11.12 and 11.14. If $\lambda \nleqslant 0$, then the corresponding vector field belongs to the Poincaré domain and one can use the holomorphic normal forms established in Theorem 5.5; cf. with Table I.1.

In the nonresonant case the field is linearizable and solutions are graphs of the multivalued function $y = cx^\lambda$. If x tends to the origin $x = 0$ along a logarithmic spiral, $x = \exp \alpha t$, $t \in \mathbb{R}_+$, $t \to +\infty$, $\operatorname{Re}\alpha < 0$, then the value of y varies along the spiral $y = c \exp \lambda\alpha t$. If λ is not negative, one can always find α such that both α and $\alpha\lambda$ belong to the left half-plane, $\operatorname{Re}\lambda\alpha < 0$, $\operatorname{Re}\alpha < 0$, so that the origin is the limit point of the leaf.

In the resonant case where λ or $1/\lambda$ is a natural number r, one can show by direct computation that, say, for the differential equation

$$\frac{dy}{dx} = r\frac{y}{x} + ax^{r-1}, \qquad a \in \mathbb{C},$$

all solutions tend zero as $x \to 0$ along the real axis.

In the saddle-node case we choose local coordinates so that the x-axis is tangent to the formal center manifold, while the y-axis is the hyperbolic manifold (analytic invariant curve). The differential equation corresponding to the vector field in these coordinates takes the form

$$\frac{dy}{dx} = \pm x^{-n}\big(y + O(x^{n+1}) + O(y^2)\big), \qquad n \geqslant 2.$$

Again by direct inspection one can verify that solutions $y = y(x)$ of this equation tend to zero exponentially fast as $x \to 0$ along to a positive, negative or imaginary semiaxis on the x-axis, depending on the sign in the normal form and parity of n.

In the "saddle" case $\lambda \not> 0$, the Hadamard–Perron Theorem 7.1 always applies, and therefore the foliation $\mathcal{F}$ has two holomorphic smooth complex separatrices that can be normalized to become coordinate axes. The differential equation defining the foliation $\mathcal{F}$ in these coordinates will take the form

$$\frac{dy}{dx} = \frac{y}{x}\big(\lambda + a(x,y)\big), \qquad a(0,0) = 0. \tag{11.4}$$

Rescaling the variables if necessary, we assume that the equation (11.4) is defined in the bidisk $U = \{|x| < 1,\ |y| < 1\}$, the cross-section τ is a small disk, $\tau = \{x = 1,\ |y| < \delta\}$, and the holomorphic term $a(x, y)$ is bounded in U,

$$|a(x,y)| < \tfrac{1}{2}|\operatorname{Re}\lambda| \qquad \forall (x,y) \in U.$$

We will show that for each point $(x_0, y_0) \in U$ with $x_0 \neq 0$ there is a path on the punctured x-plane connecting x_0 with $x = 1$ and a positive finite constant C depending only on the equation (11.4) such that the solution $y = y(x)$ of this equation with the initial condition $y(x_0) = y_0$ admits continuation along this path without leaving U and the value $y(1) = y_1$ at the end satisfies the inequality $|y_1| < C|y_0|$.

For all points x on the circumference this assertion is obvious: any such point can be connected with $x = 1$ by an arc of the circle $\{|x| = 1\}$. Since the circle is compact and the right hand side of (11.4) is bounded on it, we can choose $C = \exp 2\pi(|\lambda| + A)$, $A = \max_U |a(x,y)|$.

A point inside the disk is connected first by the radial segment with a point on the boundary, and then by a circular arc with $x = 1$ as explained above. Along the radial segment we have the differential inequality

$$\frac{d|y|^2}{dx} = \frac{\bar y y}{x}(\lambda + a(x,y)) + \frac{y\bar y}{x}\overline{(\lambda + a(x,y))} = 2\frac{|y|^2}{x}\operatorname{Re}(\lambda + a(x,y)) < 0,$$

which means that the corresponding flow map is contracting and one can choose $C = 1$.

Thus we proved that every leaf passing through interior point (x_0, y_0) of the bidisk U with $|y_0| < \delta/C$, except for the vertical axis $x = 0$, crosses the section $\tau \subseteq \{x = 1\}$ at some point $(1, y_1)$ with $|y_1| < \delta$, i.e., belongs to the saturation of the cross-section by leaves of the foliation. $\square$

Proof of Corollary 11.15. Every point not on the other separatrix S' of the saddle singular point, can be connected by at least one path on the leaf with a point on the cross-section τ, which means that the germ u_0 extends as a multivalued analytic function on $U \smallsetminus S'$ constant along the leaves.

The fundamental group of $U \smallsetminus S'$ is cyclic, generated by the loop $\gamma \subset S$ going around the origin on the smooth separatrix. Since u_0 is invariant by the holonomy map $f = \Delta_\gamma$ associated with this very loop, the extension u is in fact a single-valued and bounded holomorphic integral of $\mathcal{F}$ on the complement $U \smallsetminus S'$ to the analytic curve S'. By the removable singularity theorem, u extends as an analytic integral of $\mathcal{F}$ on U. □

11E. Poincaré–Lyapunov theorem: proof and (counter)examples. Now everything is ready for the proof of Theorem 11.7.

Proof of Theorem 11.7. Assume that an elliptic singular real analytic foliation $\mathcal{F}$ is a center, and consider blow-up of its complexification, the foliation $\mathcal{F}'$ on the complex Möbius band $\mathbb{M}$ near the exceptional divisor $\mathbb{E}$. The foliation $\mathcal{F}'$ has two singular points $\Sigma = \{\pm i\}$, both off the real equator $\mathbb{R}P^1 \subset \mathbb{E}$.

The semi-monodromy map $f = \Delta_{\mathbb{R}} \in \operatorname{Diff}(\tau, 0)$ associated with the equator $\mathbb{R}P^1$ must be 2-periodic (an *involution*), as explained in §10**C**. It generates the vanishing holonomy group,

$$H = \{\mathrm{id}, f\} \cong \mathbb{Z}_2, \qquad f\colon x \mapsto -x + \cdots, \qquad f \circ f = \mathrm{id}. \tag{11.5}$$

By Proposition 6.25, the vanishing holonomy group H is integrable, hence there exists a germ $u_0 \in \mathcal{O}(\tau, 0)$, holomorphic on the cross-section τ and invariant by f. By Saturation Lemma 2.18, the germ u_0 can be extended as an integral of the foliation $\mathcal{F}'$ near the leaf $L = \mathbb{E} \smallsetminus \Sigma$, where Σ is the singular locus of $\mathcal{F}'$ consisting of two points $\pm i$. Each of these points is a "complex saddle" with the same characteristic ratio equal to the negative number $-\frac{1}{2}$. The restriction of u on any cross-section to $\mathbb{E}$ close to these points, is invariant by the respective local holonomy map. By Corollary 11.15, the integral u extends analytically on the full neighborhoods of both singular points, producing a holomorphic first integral of $\mathcal{F}'$ on $\mathbb{M}$.

The blow-down $(\sigma^*)^{-1}u$ of u is a holomorphic first integral of the initial elliptic foliation $\mathcal{F}$ on the punctured neighborhood of the origin in $(\mathbb{C}^2, 0)$. Again by the removable singularity theorem, the blow-down extends to a holomorphic first integral on the plane.

These arguments alone do not yet guarantee that the constructed integral u has nondegenerate quadratic part, but a closer inspection of Proposition 6.25 yields the principal part of a germ invariant by a periodic map of period p. In the case $p = 2$ the proof of Proposition 6.25 implies that the germ u_0 defined on the cross-section τ at the point $\{z = 0\} \in \mathbb{E}$ with the chart x on it, can be chosen with the 2-jet $u_0(x) = x^2 + \cdots$. The corresponding extension after blow-down is the analytic function whose 2-jet is rotationally symmetric and $u(x, 0) = x^2 + \cdots$. The only possibility is that $u(x, y) = x^2 + y^2 + \cdots$, i.e., the integral is nondegenerate. The proof of the Poincaré–Lyapunov theorem is complete. □

Literally the same proof applies to a much more general situation and gives a partial inversion of Proposition 11.11.

Theorem 11.16. *Assume that all singularities which appear after a single blow-up of a singular holomorphic foliation $\mathcal{F}$, are elementary. Then $\mathcal{F}$ is integrable if and only if the following two conditions are met:*

1. *the vanishing holonomy group of $\mathcal{F}$ is integrable, and*
2. *all singularities of $\mathcal{F}' = \sigma^*\mathcal{F}$, are complex saddles with negative rational characteristic ratios $\lambda_i < 0$.*

Proof. If the holonomy group is nonintegrable, then it must contain an aperiodic element. By Theorem 6.34, the foliation has uncountably many leaves which intersect an analytic cross-section by an infinite number of points, which is impossible in the integrable case.

If a holonomy map associated with a separatrix of an elementary singularity is periodic, then the corresponding multiplicator must be a root of unity, that is, the characteristic ratio λ_i is a rational number, $\lambda_i \in \mathbb{Q}$. The nodal ($\lambda_i > 0$) and saddle-nodal cases ($\lambda_i = 0$) are incompatible with the integrability by Lemma 11.12, which leaves only one possibility that all these singularities are saddles.

Conversely, a first integral of the vanishing holonomy group can be extended to a full neighborhood of each saddle by Corollary 11.15, producing a global integral of the foliation $\mathcal{F}'$ and hence of $\mathcal{F}$. □

On the other hand, the proof of Theorem 11.7 clarifies the role played by the assumption on the linear part of the real singular foliation $\mathcal{F}$. This assumption guarantees that the vanishing holonomy group is generated by a single map, the semi-monodromy. For generalized elliptic singularities, the vanishing holonomy map may have more than one generator, so periodicity of only one of them does not imply periodicity of the entire vanishing holonomy. In other words, one should expect existence of degenerate singular real analytic foliations with center but without analytic (first) integrals.

One of the earliest counterexamples of this sort was constructed in [**NS60**, §4.656, p. 122] explicitly.

Example 11.17. Consider the function

$$u(x, y) = (2x^2 + y^2)\exp\bigl[1/(x^2 + y^2)\bigr].$$

This function is real analytic on the punctured real plane, but has essential singularities on the imaginary cross $x = \pm iy$. The logarithmic derivative $\omega = \frac{du}{u}$ is a rational 1-form with a second order pole on this cross, and the Pfaffian equation $\omega = 0$ can be transformed to a polynomial vector field. This vector field has transcendental first integral, hence center on the real plane, but cannot have an analytic first integral.

Similar examples can be constructed from real analytic functions $f_j(x, y)$ which vanish at the origin $x = y = 0$ and are positive on a punctured neighborhood $(\mathbb{R}^2, 0) \smallsetminus \{(0, 0)\}$. For any collection of positive weights $\alpha_j > 0$ linearly independent over $\mathbb{Q}$, the function $f = \prod_j f_j^{\alpha_j}$ is a nonanalytic first integral for the corresponding foliation $\omega = 0$ with the rational 1-form $\omega = df/f$.

Another way to construct counterexamples is to start directly from the nonintegrable vanishing holonomy.

Example 11.18. Let $\theta = \theta_1$ be a real rational meromorphic 1-form on $\mathbb{E} = \mathbb{P}^1$ without real poles, satisfying the condition (10.10). Consider the corresponding Pfaffian equation (10.6) with $\theta_2 = \theta_3 = \cdots = 0$. By Remark 10.14, this equation can be blown down to a generalized elliptic singularity.

Being linear in x, the equation (10.6) is integrable, and all holonomy maps are linear in the natural chart x. By (10.11) and the symmetry of θ by the involution $z \mapsto \bar{z}$, the total residue of all singularities in each half-sphere $\pm \operatorname{Im} z > 0$ on $\mathbb{E}$ is $-\frac{1}{2}$. Thus the holonomy of the real (projective) line is 2-periodic (the linear symmetry $x \mapsto -x$), so the first return map of the real singularity would be a center.

On the other hand, if there is more than one pole of θ, the above constraints are compatible with the fact that some of the corresponding residues are *not negative rational numbers*. This means that the holonomy operators for small loops around these singularities cannot be periodic.

Clearly, this is impossible for an integrable singularity by Theorem 11.16.

In yet another example centrality follows from axial symmetry of the foliation on the real plane.

Example 11.19 (R. Moussu [**Mou82**]). The real polynomial 1-form

$$\omega = x^3\,dx + y^3\,dy - \tfrac{1}{2}x^2y^2\,dy \tag{11.6}$$

defines a real analytic singular foliation on $(\mathbb{R}^2, 0)$. The singular point at the origin is the center, being symmetric by the mirror symmetry (involution) $(x, y) \mapsto (-x, y)$.

The principal part (3-jet) of ω is integrable: $j^3\omega = \frac{1}{4}d(x^4+y^4)$. However, by direct inspection one can show that there is no 5-jet of the form $u = x^4 + y^4 + \cdots$ such that $j^5(\omega \wedge du) = 0$.

11F. Simple foliations on $(\mathbb{C}^2, 0)$. As was already noted, the Poincaré–Lyapunov Theorem 11.7 relates certain topological simplicity of a real analytic foliation with its integrability which is an analytic property. We will describe a generalization of this theorem for arbitrary singular holomorphic foliations on $(\mathbb{C}^2, 0)$, establishing necessary and sufficient topological conditions for integrability.

Topology of integrable foliations is necessarily simple in the following precise sense.

Definition 11.20. A singular foliation $\mathcal{F}$ on $(\mathbb{C}^2, 0)$ is *simple*, if

(1) all leaves are relatively closed in $(\mathbb{C}^2, 0) \smallsetminus \{0\}$, and
(2) at most countably many leaves contain the isolated singular point $\{0\}$ in their closure.

Obviously, any integrable foliation is simple in the sense of this definition. Moreover, the number of leaves adjacent to the singular point, is in fact at most finite: every such leaf must be an irreducible component of the germ of the analytic curve $\{u = 0\} \subset (\mathbb{C}^2, 0)$.

The inverse result generalizes Poincaré–Lyapunov Theorem 11.7 as well as Theorems 11.16 and 6.34.

Theorem 11.21 (J.-F. Mattei and R. Moussu [**MM80**], [**Mou98**]). *A simple singular holomorphic foliation is always integrable, moreover, it always admits a primitive first integral.*

Proof. The proof of Theorem 11.21 is achieved by induction in the number of blow-ups required to desingularize the foliation completely.

1. As a base of induction, we verify case by case the assertion of the theorem for all elementary singularities.

If the holonomy map associated with a separatrix of a saddle point is periodic, then the map is integrable and by Corollary 11.15 this integral extends as an integral of the foliation. If this integral is not primitive, a suitable root of it (in the chart linearizing the holonomy) is primitive.

On the other hand, aperiodicity of the holonomy means that the foliation is not simple: by Lemma 6.33, there are uncountably many leaves not relatively closed. Nonsaddle elementary singularities are also nonsimple by Lemma 11.12. The base of induction is verified.

2. To implement the induction step, let $\mathcal{F}$ be a simple foliation and $\mathcal{F}'$ its standard blow-up. The blow-up obviously must be nondicritical (otherwise $\mathcal{F}$ is not simple), thus $\mathcal{F}'$ has finitely many isolated singularities $\{a_1, \dots, a_m\} = \Sigma$ on the exceptional divisor $\mathbb{E} \subset \mathbb{M}$. Clearly, all these singularities must also be topologically simple (though by no means elementary).

By the induction assumption, near each singularity a_i the foliation $\mathcal{F}'$ is integrable and admits a primitive first integral $u_i \in \mathcal{O}(\mathbb{M}, a_i)$. Furthermore, the vanishing holonomy group H is also integrable, i.e., H admits an analytic "semi-global" integral $u_0 \in \mathcal{O}(\mathbb{M}, \mathbb{E} \smallsetminus \Sigma)$. Our goal will be achieved if we replace all these "partial integrals" by functions of the form $\varphi_i \circ u_i$, $i = 0, 1, \dots, m$ which will agree with each other and form a single holomorphic integral $u \in \mathcal{O}(\mathbb{M}, S)$ with connected level curves.

To do this, we will explicitly construct a finitely generated subgroup $G = G_{\mathcal{F}} \subset \operatorname{Diff}(\mathbb{C}, 0)$ of conformal germs such that the orbits of this subgroup (more precisely, of a pseudogroup obtained by specifying domains of the generators) will *coincide* with the intersections of leaves of $\mathcal{F}'$ with the cross-section: $G(b) = \{L_b \cap \tau\}$ for all sufficiently small points $b \in (\tau, a)$, where $L_b \in \mathcal{F}'$ is the leaf of the foliation $\mathcal{F}'$. Note that G must contain the vanishing holonomy group H, yet the latter group can be too small for that purpose (Problem 11.6).

If the foliation $\mathcal{F}$ is simple, the orbits of the group $G_{\mathcal{F}}$ must all be finite, that is, the group itself must be integrable by Theorem 6.34. The integral u of the foliation $\mathcal{F}'$ will be obtained away from the singular locus Σ by extending the primitive integral u_* of the group G, and we will show that this primitive integral, after continuation along leaves of $\mathcal{F}$ into a complement of the singular locus, extends further as a suitable function $\varphi_i(u_i)$ into a neighborhood of each singularity a_i for all $i = 1, \dots, m$.

The detailed description of the construction follows.

3. Without loss of generality, using continuation along leaves over the paths γ_i connecting the singular points a_i with the base point $a \in S \smallsetminus \Sigma$, we can assume that all local integrals are defined on the same cross-section $\tau \colon (\mathbb{C}, 0) \to (\mathbb{M}, a)$ at the base point.

By assumption, the integrals u_i were all primitive, hence the level sets $\{u_i = \varepsilon\} \subset \tau$ belong to intersection of the same leaf with τ. Let $p_i = \operatorname{ord}_a u_i$ be the respective orders, so that each level set consists of $p_i \geqslant 1$ points for $\varepsilon \neq 0$.

Let $f_i \in \operatorname{Diff}(\mathbb{C}, 0)$ be a holomorphic map which generates the complete symmetry group of the germ u_i (see Remark 6.27) for each $i = 0, 1, \dots, m$.

By construction, f_i takes any level set of u_i into itself and hence the orbit belongs to the same leaf.

The group $G \subset \operatorname{Diff}(\mathbb{C}, 0)$ generated by the germs $f_0, f_1, \dots, f_m$, has the same property: its orbits remain on the same leaves of the foliation $\mathcal{F}'$.

We claim that if the foliation $\mathcal{F}$ (and hence $\mathcal{F}'$) were integrable, the group G is also integrable. Indeed, otherwise by Theorem 6.34 there would coexist uncountably many leaves of $\mathcal{F}'$ crossing τ by infinite point sets.

4. By the structural Corollary 6.26, the group G is cyclic and generated by a germ that we denote by $f_* \in \operatorname{Diff}(\mathbb{C}, 0)$. Let $u_* \in \mathcal{O}(\mathbb{C}, 0)$ be the corresponding minimal f_*-invariant integral of the group G. Each map f_i, $i = 0, 1, \dots, m$, is an iterate of this generator, $f_i = f_*^{\circ q_i}$ for some $q_i \in \mathbb{N}$. We claim that there exist holomorphic functions φ_i of order q_i such that u_* coincides with $\varphi_i \circ u_i$.

To prove this claim (see also Problem 11.7), consider for every $i = 0, 1, \dots, m$ the quotient space $Q_i = (\mathbb{C}, 0)/f_i$ of a small neighborhood of the origin by the action of the periodic germ f_i. This space is the germ of a nonsingular analytic curve. By the construction of f_i, the natural chart on Q_i is given by the function u_i. The map f_* descends on the quotient space as a q_i-periodic self-map, and its integral u_* descends on Q_i as the integral of this periodic map. In a suitable holomorphic chart $w = w(u_i)$ such an integral is the monomial of degree q_i of w, $u_* = w^{q_i}$, that is, a function φ_i of order q_i of the natural coordinate u_i.

5. The standard construction of saturation allows us to extend u_* as the holomorphic function on $\mathbb{M}$ near the exceptional divisor $\mathbb{E}$ with the deleted singular points, while keeping the identities $u_0 = \varphi_i \circ u_i$ near these points (these identities are preserved by backwards continuation along leaves of $\mathcal{F}'$ over the paths γ_i).

Since the functions $\varphi_i \circ u_i$ are well defined and holomorphic in some full neighborhoods of the singular points $a_i \in \Sigma$, these identities allow us to extend the integral u_* to the neighborhood of each singular point. One can easy see that this integral is primitive, since its level sets coincide with the connected local leaves of $\mathcal{F}'$ near $\mathbb{E}$. The proof of Theorem 11.21 is complete. □

11G. Survey of further results. Here we briefly mention some of the results that link integrability with properties of the holonomy group.

11G$_1$. *Primitive* vs. *nonprimitive integrals.* The demonstration of Theorem 11.21 automatically produces a primitive first integral for any simple foliation. Any integrable foliation is automatically simple, therefore existence of *any* analytic integral implies existence of a *primitive* integral [**MM80**]. Yet it is instructive to have a direct construction of this integral.

Theorem 11.22. *Any integrable foliation admits a primitive integral.*

Sketch of the direct proof. Consider u as the map, $u\colon (\mathbb{C}^2, 0) \to (\mathbb{C}, 0)$ and choose a sufficiently small ball $B \subseteq (\mathbb{C}^2, 0)$ whose boundary is transversal to the zero level curve. Then one can choose a small disk D in the target such that $u = 0$ is the only critical value of the map in this disk. Denote by M the preimage $u^{-1}(D) \cap B$. Then the map $u\colon M \to D$ is a proper map, surjective outside the origin in the target space.

It can be shown (cf. with the arguments that appear in the more difficult context in §26**F**), that $u\colon M \to D^* = D \smallsetminus \{0\}$ is a topological bundle: preimages of points $X_t = u^{-1}(t) \subset M$ are continuously depending on $0 \neq t \in D$. Without loss of generality we may assume that D is the unit disk.

Assume that the first integral u is not primitive (otherwise one can take $w = u$ and use Proposition 11.3). Assume that some regular fiber, say, X_1, consists of p different connected components, denoted by $C_0, \dots, C_{p-1}$. Denote by γ the oriented unit circle bounding D and consider the operators $\Delta_{1,t}$ of continuation of the fibers X_t along γ, which is a homeomorphism between X_1 and X_t, continuously depending on t for t varying along γ. After making the full turn the fiber X_1 returns onto itself, but eventually with permuted connected components: $\Delta = \Delta_{1,\exp 2\pi i}$ maps X_1 into X_1 but $\Delta(C_j) = C_{\sigma(j)}$.

We claim that this permutation σ is cyclic. Indeed, consider any two components $C \neq C'$ and a path α in M connecting these components while avoiding the singular fiber $\Sigma = u^{-1}(0)$ (that is, a continuous map $\alpha\colon [0,1] \to M \smallsetminus \Sigma$ such that $\alpha(0) \in C$, $\alpha(1) \in C'$). The image $u \circ \alpha\colon [0,1] \to D^*$ is a closed loop in D^* which is homotopic to k simple turns around the origin. Then the kth iterate Δ^k maps C into C'. Since C, C' were arbitrary, we proved that permutation σ is transitive and hence a cycle of maximal length on p symbols. In other words, the components $C_0, \dots, C_{p-1}$ can be enumerated in such a way that $\Delta(C_j) = C_{j+1 \bmod p}$.

Consider the function $w\colon M \to \mathbb{C}$ which can be described as "the pth root of u with separately chosen branches". More precisely, define the function first on $X_1 = \bigcup_0^{p-1} C_j$ and then continue it analytically as follows.

Assign the value of w on each component C_j as the constant equal to $\exp(2\pi i j/p)$ so that passing from C_j to $C_{j+1 \bmod p}$ results in w being multiplied by the primitive root of unity of degree p. Then for each $t \in D^*$ and each point $x \in X_t$ we let $w(x)$ be the value of the root $t^{1/p}$ obtained by continuation along the path connecting t with the base point $t = 1$.

This function is well defined (single-valued) and continuous. Indeed, consider any path α in M, connecting C_j with $C_{j+1 \bmod p}$ and avoiding Σ. The corresponding loop $\beta = u \circ \alpha \in \pi_1(D^*, 1)$ has index 1 mod p in D^*, because of the ordering of the components. After continuation of the branch of the root $t^{1/p}$ along β the value $w(C_j) = \exp(2\pi i j/p)$ will be multiplied by $\exp(2\pi i/p)$ which coincides with the value of w assigned to the component $C_{j+1 \bmod p}$. Thus the definition of the function w is self-consistent in the complement $M \smallsetminus \Sigma$.

Moreover, w is also analytic outside Σ. Indeed, locally near any connected component of any fiber X_t this function is a lift of the appropriate branch of the analytic function $u^{1/p}$, hence analytic itself.

Being bounded, w extends analytically on the entire space M as a first integral with connected level sets. The proof of Theorem 11.22 is complete. □

11**G**$_2$. *Formal and true integrability.* Existence of a first integral is possible to establish via desingularization, if integrability of saddles at the end is known. Alternatively, one can look for a formal integral as a formal solution $u = u_2 + u_3 + \cdots$ for the triangular system of linear equations (11.1) (this solution may start with several zero terms but ultimately must be nontrivial).

Yet the formal series $u \in \mathbb{C}[[x, y]]$ found in such a way, does not necessarily have to converge, moreover, together with convergent solutions, if they exist, there will always be divergent ones of the form $g(u(x, y))$, where g is a *divergent* series in one variable.

However, existence of at least one nonzero formal solution implies existence of holomorphic first integrals. For elliptic singular points it was proved in Proposition 11.6. The general result, also due to J.-F. Mattei and R. Moussu [**MM80**], holds for all isolated singularities.

Theorem 11.23. *Assume that the holomorphic foliation $\mathcal{F} = \{\omega = 0\}$ in $(\mathbb{C}^2, 0)$ has a formal first integral $u \in \mathbb{C}[[x, y]]$. Then there exists a holomorphic first integral $0 \neq v \in \mathcal{O}(\mathbb{C}^2, 0)$.*

Sketch of the proof. We prove that formally integrable foliations are always simple by induction in the number of steps required for complete desingularization of the singularity.

Indeed, an elementary singularity is simple if and only if it has a negative characteristic ratio and the holonomy maps associated with holomorphic separatrices are periodic. A formally integrable elementary singularity has negative rational hyperbolicity ratio and admits formal first integral for the respective holonomy (Problem 11.8). But a formally integrable holomorphic self-map from $\operatorname{Diff}(\mathbb{C}, 0)$ is necessarily periodic by Problem 6.4. This observation establishes the base of induction.

Consider an arbitrary formally integrable foliation after a blow-up (necessarily nondicritical) produces a foliation with isolated singularities, all of them formally integrable. By the induction assumption, for each of the singularities there are only finitely many leaves that contain these singularities in their closure.

Thus if $\mathcal{F}'$ were not simple, then any leaf eventually accumulating to the exceptional divisor or not relatively closed, must intersect infinitely many times any cross-section to a nonsingular point $a \notin \Sigma$. This means that the vanishing holonomy group H contains an infinite nonperiodic orbit.

On the other hand, the group H is formally integrable (again by Problem 11.8). By Problem 6.4, H is analytically integrable, which contradicts existence of an infinite aperiodic orbit.

Thus a formally integrable holomorphic foliation is necessarily simple. By Theorem 11.21, it admits an analytic integral. □

In fact, both Theorems 11.21 and 11.23 are particular 2-dimensional cases of more general results concerning holomorphic singular foliations in $(\mathbb{C}^n, 0)$. We will not discuss these generalizations.

11**G**$_3$. *Meromorphic and Darboux integrability.* Besides holomorphic integrals, one may consider more general types of integrals, say, *meromorphic integrals.* The definition remains formally the same: the germ of a singular foliation $\mathcal{F}$ defined in $(\mathbb{C}^2, 0)$ by the Pfaffian equation $\omega = 0$ is said to be *meromorphically integrable*, if there exists a nonconstant meromorphic germ $u \in \mathcal{M}(\mathbb{C}^2, 0)$ such that $\omega \wedge du \equiv 0$.

Theorem 11.21 fails for meromorphically integrable foliations, as they can be nonsimple (Exercise 11.10). Still meromorphic integrability implies that the holonomy associated with any analytic separatrix, as well as the vanishing holonomy group (if it is defined) is periodic and linearizable. We prove a more general result.

Definition 11.24. A singular holomorphic foliation on $(\mathbb{C}^2, 0)$ is said to be *Darboux integrable*, if it can be defined by a *closed* meromorphic 1-form ω.

Definition 11.25. A closed meromorphic 1-form is called *logarithmic*, if all poles of this form are of the first order. A foliation generated by a logarithmic form, is called *logarithmic foliation*.

Both (truly) and meromorphically integrable foliations correspond to particular cases of Darboux integrable foliations.

Theorem 11.26. *The holonomy group associated with any analytic separatrix of a logarithmic foliation, as well as the vanishing holonomy (if the foliation is nondicritical), is abelian and linearizable (i.e., isomorphic to a subgroup of $\mathbb{C}^*$).*

The proof is based on the description of closed meromorphic forms on $(\mathbb{C}^2, 0)$. Let $\Sigma = \bigcup_{i=1}^n C_i$ be the germ an analytic curve in $(\mathbb{C}^2, 0)$, represented as the union of the irreducible components $C_i = \{f_i = 0\}$, $i = 1, \dots, n$ defined by square-free germs $f_i \in \mathcal{O}(\mathbb{C}^2, 0)$.

Lemma 11.27. *Any closed 1-form $\omega \in \Lambda^2(\mathbb{C}^2, 0)$ with the polar locus on $\Sigma = \bigcup_{i=1}^n \{f_i = 0\}$, admits the representation*

$$\omega = \sum_{j=1}^n \lambda_j \frac{df_j}{f_j} + d\left(\frac{g}{f_0}\right), \qquad f_0, g \in \mathcal{O}(\mathbb{C}^2, 0),\ \lambda_j \in \mathbb{C}, \tag{11.7}$$

where the holomorphic germ f_0 is nonvanishing outside Σ.

A logarithmic form ω is a linear combination of logarithmic derivatives modulo an exact holomorphic form,

$$\omega = dg + \sum_{j=1}^n \lambda_j \frac{df_j}{f_j}, \qquad \lambda_1, \dots, \lambda_n \in \mathbb{C},\ g \in \mathcal{O}(\mathbb{C}^2, 0). \tag{11.8}$$

Proof of the lemma. The primitive of a closed 1-form with the polar locus Σ is a multivalued function on the complement $(\mathbb{C}^2, 0) \smallsetminus \Sigma$, ramified over Σ. The fundamental group of the complement is generated by small loops δ_j around smooth points on the irreducible components C_j, defined modulo free homotopy. Let

$$\lambda_j = \frac{1}{2\pi i} \oint_{\delta_j} \omega, \qquad j = 1, \dots, n$$

be the *residues* of the form ω on the irreducible components C_j. (The fact that the integral remains unchanged when δ_j is replaced by another loop freely homotopic to it, follows from the closedness of the form ω).

The 1-form $\omega' = \omega - \sum_{j=1}^n \lambda_j \frac{df_j}{f_j}$ is closed and has zero integrals over all loops δ_j. Hence ω' is *exact* in $(\mathbb{C}^2, 0) \smallsetminus \Sigma$; its primitive has at most polynomial growth near Σ and hence ω' is the differential of a *meromorphic* function g/f_0. By construction, f_0 may vanish only on the union of the loci $\{f_j = 0\}$. Hence all irreducible factors of f_0 should be in the list $\{f_1, \dots, f_n\}$.

If ω has only first order poles, so has the exact form $\omega' = \omega - \sum \lambda_j \frac{df_j}{f_j} = d(g/f_0)$. But the differential of any nonconstant meromorphic function has poles of order $\geqslant 2$, hence the exact form ω' must be holomorphic. $\square$

Proof of Theorem 11.26. We prove that the vanishing holonomy group is commutative if the foliation is nondicritical, explicitly constructing the linearizing chart.

Blowing up a logarithmic 1-form $\omega = \sum_{j=1}^n \lambda_j \frac{df_j}{f_j} + dg$, $f_j, g \in \mathcal{O}(\mathbb{C}^2, 0)$, we obtain a meromorphic 1-form on a small neighborhood $(\mathbb{M}, \mathbb{E})$ of the exceptional divisor in the complex Möbius band. Everywhere outside $\mathbb{E}$ this form obviously has poles of order no more than one. One can immediately verify that the eventual pole on $\mathbb{E}$ has the order at most one. Indeed, passing to the coordinates $(x, z = y/x)$, we can write $f_j(x, zx) = x^{p_j} \Phi_j(x, z)$, $\Phi_j|_S \not\equiv 0$, where $p_j = \operatorname{ord} f_j$, so that $\frac{df_j}{f_j} = p_j \frac{dx}{x} + \frac{d\Phi_j}{\Phi_j}$ has the first order pole on $\mathbb{E} = \{x = 0\}$. The blow-up is nondicritical if and only if the residue $\lambda_0 = \sum \lambda_j p_j$ is nonzero. The blow-up foliation in the chart (x, z) is given by the 1-form

$$\Omega = \lambda_0 \frac{dx}{x} + \sum_{j=1}^n \lambda_j \frac{d\Phi_j}{\Phi_j} + dG, \qquad \lambda_0 - \sum \lambda_j p_j \neq 0. \tag{11.9}$$

with the functions Φ_j, G holomorphic in x and z. The form Ω is closed.

Consider the nondicritical case and denote by Σ the union of roots of all polynomials $\Phi_j(0, z)$. Without loss of generality we may assume that all f_j are irreducible, in which case for each j the corresponding term $\Phi_j(0, \cdot)$ vanishes at only one point $a_j \in S$, though not all these points may be distinct.

Exactly as in the proof of Lemma 11.27, we can find complex numbers μ_j such that the 1-form $\Omega - \sum \mu_j \frac{dz}{z - z_j} - \lambda_0 \frac{dx}{x}$ is *exact* in the narrow band with the deleted cylindrical neighborhoods of the singular points $M^* = (\mathbb{M}, S) \smallsetminus \bigcup_{j=1} \{|z - z_j| < \varepsilon\}$, that is, there exists a holomorphic function

$H(x,z)$ such that

$$\Omega = \lambda_0 \frac{dx}{x} + \sum \mu_j \frac{dz}{z - z_j} + dH, \qquad H \in \mathcal{O}(M^*).$$

This formula immediately implies that the Pfaffian equation for leaves of the logarithmic foliation after the blow-up can be written in M^* as follows:

$$\frac{dW}{W} = -\sum_j \frac{\mu_j}{\lambda_0} \frac{dz}{z - z_j}, \qquad W = x \exp\bigl(\lambda_0^{-1} H(x,z)\bigr)$$

Consider the cross-section $\tau = \{z = z_0\}$ at a nonsingular point z_0 and the holomorphic chart $w = W|_\tau = x \exp\bigl(\lambda_0^{-1} H(x, z_0)\bigr)$ on it. The holonomy transformation in this chart can be instantly computed, since the variables in the above equation are separated: the holonomy map associated with the loop around the point z_j is the linear rotation,

$$\Delta_j \colon w \mapsto w \exp(-2\pi i \mu_j), \qquad j = 1, 2, \dots.$$

Thus the vanishing holonomy group consists of linear maps.

The proof of the linearizability of the holonomy associated with an arbitrary separatrix, is achieved by the same arguments as before, after the separatrix is desingularized to a smooth analytic curve transversal to the exceptional divisor. The details are left to the reader as an exercise. □

$11\mathbf{G}_4$. *Reversibility.* Another reason for the appearance of centers of real analytic foliations is a certain symmetry as in Example 11.19.

Assume that a real analytic foliation $\mathcal{F}$ has an isolated monodromic singularity at the origin and in addition is symmetric by a nontrivial involution $S \in \operatorname{Diff}(\mathbb{R}^2, 0)$: $S \circ S = \mathrm{id}$, S reverses the orientation and $S^*\omega = \omega$. Such involution in suitable coordinates is an axial symmetry $(x, y) \mapsto (-x, y)$.

The corresponding vector field F will be *antisymmetric*: $S_*F = -F$, which explains why the corresponding singularities are called *reversible*.

Proposition 11.28. *A monodromic reversible singularity is a center.* □

It turns out that for some types of singularities reversibility is the only possible scenario of producing centers.

Theorem 11.29 (M. Berthier, R. Moussu [**BM94**]). *A singular real analytic foliation defined by 1-form $\omega \in \Lambda^1(\mathbb{R}^2, 0)$ with the linear part $\omega = y\,dy + \cdots$ is a center if and only if it is reversible, i.e., there exists an analytic involution $S \in \operatorname{Diff}(\mathbb{R}^2, 0)$ such that $S^*\omega = \omega$.* □

One can generalize reversibility by considering *foldable* (*equivariant*) foliations generated by *generalized folds*, real analytic maps $\Phi \colon (\mathbb{R}^2, 0) \to (\mathbb{R}^2, 0)$ which are many-to-one proper maps. We say that a foliation is *foldable by Φ*, if it is a Φ-preimage $\Phi^{-1}(\mathcal{G})$ of another real analytic (eventually, nonsingular) foliation $\mathcal{G}$ on $(\mathbb{R}^2, 0)$. The folding map associated with the standard axial symmetry has the form $(x, y) \mapsto (x^2, y)$. Monodromic foldable foliations also are centers.

Yet it would be wrong to conclude that generalized reversibility and Darboux integrability are the only reasons why a real foliation may be a center. In the illuminating paper [**BCLN96**] an example of a real analytic foliation is constructed, which has a center but is not foldable by any nontrivial proper map and does not admit even Liouvillean (multivalued) integral[2].

Exercises and Problems for §11.

Problem 11.1. Prove that a singular foliation is integrable if and only if its blow-up, the foliation $\mathcal{F}' = \sigma^*\mathcal{F}$ on the complex Möbius band $\mathbb{M}$, is integrable in a neighborhood of the exceptional divisor.

Exercise 11.2. Give a proof of Proposition 11.6 based on Problem 6.4.

Exercise 11.3. Construct a singular foliation with a separatrix S, such that the holonomy of S is identical but the saturation of a small cross-section τ to S by leaves of $\mathcal{F}$ is not dense in any neighborhood of the singularity.

Problem 11.4. Prove the assertion of Lemma 11.14 under the relaxed assertion on the characteristic ratio $\lambda \not\geq 0$.

Problem 11.5. Compute the vanishing holonomy group for foliations described in Examples 11.17 and 11.19 and *prove* that these examples are indeed nonintegrable.

Problem 11.6. Consider the integrable foliation in $(\mathbb{C}^2, 0)$ with the first integral $x^2(x+y)^3y^4$. Compute the holonomy of each separatrix and compare the orbits of each holonomy map with the intersections of leaves with small cross-sections to the respective separatrix.

Problem 11.7. Prove the claim from part **4** of the proof of Theorem 11.21 (see p. 193) by direct reasoning in the chart linearizing the map f_0.

Problem 11.8. Assume that a singular holomorphic foliation is formally integrable. Prove that any holonomy map associated with any separatrix, and the entire vanishing holonomy group are formally integrable.

Exercise 11.9. At which moment the proof of Theorem 11.21 fails if instead of the larger group G the vanishing holonomy group H were taken?

Exercise 11.10. Give an example of nonsimple meromorphically integrable foliation.

Exercise 11.11. Compute the holonomy of each separatrix for a foliation with the meromorphic integral $u = x^p/y^q$, $p, q \in \mathbb{N}$.

[2]A function is called Liouvillean, if it can be obtained by finite differential extension of the field of rational functions by algebraic functions, exponents and primitives of exponents. Existence of a Liouvillean first integral is closely related to the solvability of the vanishing holonomy group; see [**BCLN96**] and references therein.

Exercise 11.12. Prove Proposition 11.28.

Exercise 11.13. Construct a real analytic foliation with C^∞-smooth integral but without formal or analytic integrals.

Suggestion. Modify Example 11.17.

Exercise 11.14. Show that the vanishing holonomy group is an analytic invariant of the nondicritical foliations: if two such foliations are analytically equivalent, the respective groups are analytically conjugated.

Problem 11.15. Is the inverse assertion true?

Suggestion. Solve the next two problems.

Problem 11.16. Consider nondicritical foliations having at most three hyperbolic singularities on the exceptional divisor after blow-up.

Prove that for such singularities analytic conjugacy of the vanishing holonomy groups implies analytic equivalence of the foliations themselves.

Problem 11.17. How many smooth pairwise transversal holomorphic curves through the origin can be *simultaneously holomorphically rectified* (transformed to straight lines by a suitable biholomorphism)?

Exercise 11.18. Prove that integrability of planar analytic foliations is *ultimately* decidable.

12. Zeros of parametric families of analytic functions and small amplitude limit cycles

This section, somewhat aside from the mainstream, deals with analytic local multiparametric families (deformations) of functions of one variable (real or complex). If a function has an isolated root of multiplicity $\mu < \infty$, then by the Weierstrass preparation theorem any deformation of this function has no more than μ zeros nearby (exactly μ in the complex analytic settings). We describe an object, called *Bautin ideal*, that determines the bound for the number of isolated zeros in the case where deformations of an *identically zero* function are considered. This ideal was introduced by R. Roussarie [**Rou89**]; in our exposition we focus on the additional structure (filtration) on the Bautin ideal and discuss its functoriality.

This subject is traditionally linked to the problem of describing bifurcations of limit cycles from an elliptic center. The problem was studied first by Poincaré and H. Hopf and later by A. Andronov and L. Pontryagin. In the least degenerate case it is customarily referred to as the *Andronov–Hopf bifurcation*. N. Bautin formulated the problem in full generality, including cases of infinite degeneracy (centers), and gave a complete solution for quadratic vector fields in 1939; see [**Bau54**]. We give in §13**A** the modern exposition of this work, based on [**Żoł94**].

12A. Poincaré–Andronov–Hopf–Takens bifurcation: small limit cycles bifurcating from elliptic points. Consider a real analytic *local family* of planar vector fields $F_\lambda = F(x, y; \lambda)$ defined in a small neighborhood $(\mathbb{R}^2, 0)$ of the origin on the real plane and depending analytically on a number of real parameters $\lambda = (\lambda_1, \dots, \lambda_n) \in (\mathbb{R}^n, 0)$. Suppose that this family is *elliptic*, i.e., the eigenvalues of the linearization matrix $A(0)$ form a pair of nonzero complex conjugate numbers.

This assumption immediately implies that the singular point itself depends analytically on the parameters (by the implicit function theorem). Moreover, the local coordinates (x, y) can be chosen such that linear part $\mathbf{A}$ of F has the form

$$\mathbf{A} = \alpha(\lambda)\mathbf{E} + \beta(\lambda)\mathbf{I}, \qquad \mathbf{E} = x\frac{\partial}{\partial x} + y\frac{\partial}{\partial y}, \quad \mathbf{I} = y\frac{\partial}{\partial x} - x\frac{\partial}{\partial y}, \tag{12.1}$$

with real analytic coefficients (germs) $\alpha(\lambda)$ and $\beta(\lambda)$ before the radial (Euler) vector field $\mathbf{E}$ and the rotation field $\mathbf{I}$. The ellipticity assumption means that the real analytic function $\beta(\lambda)$ is nonvanishing.

The monodromy (first return) map $P(\cdot, \lambda)$ for any elliptic family is real analytic and depends analytically on the parameters by Theorem 10.12. Denote by $f(x, \lambda)$ the *displacement function* $f = P - \mathrm{id}$ for some choice of a cross-section, say, the semiaxis $\tau_+ = \{y = 0, x > 0\}$, and an analytic chart x on this cross-section. By definition, sufficiently small limit cycles of the field F_λ intersect τ_+ at *isolated* zeros of f.

The number of small limit cycles born by small perturbations from a singular point, is usually referred to as the *cyclicity* of this singular point relative to the family $F = \{F_\lambda\}$.

Cyclicity can be relatively easily determined if the field $F(\cdot, 0)$ is *not a center*. In this case the real analytic displacement function $f(\cdot, 0)$ is different from the identical zero and hence there exists a finite natural number μ such that $f(x, 0) = cx^\mu + O(x^{\mu+1})$ with some $c \neq 0$.

In this case there exist $\varepsilon > 0$ and $\delta > 0$ such that for all $|\lambda| < \varepsilon$ the function $f(\cdot, \lambda)$ has *no more than μ roots* in the interval $(0, \delta)$, necessarily isolated. In fact, in the analytic case we are dealing with, the number of zeros of the complexified function is exactly equal to μ in the small complex *disk* $\{|x| < \delta\} \subseteq (\mathbb{C}^1, 0)$, if the zeros are counted with multiplicities.

The proof follows from the standard Rouché-type argument. The holomorphic function $f(x, 0) = cx^\mu(1 + o(1))$ is nonvanishing on a sufficiently small circle $\{|x| = \delta\}$ and its *variation of argument* (index) along this circle is equal to $2\pi\mu$. By continuity in the parameters, the variation of argument of $f(\cdot, \lambda)$ along $\{|x| = \delta\}$ remains the same for all $|\lambda| < \varepsilon$ if $\varepsilon > 0$ is sufficiently small. By the argument principle, the number of complex roots of $f(\cdot, \lambda)$ in the disk $\{|x| < \delta\}$ is equal to μ.

The bound for cyclicity established by this simple argument, does not depend on the family, only on the field $F(\cdot, 0)$. On the other hand, these arguments break almost completely if the field $F(\cdot, 0)$ is integrable (center). In this case the bound necessarily depends on the family. This section describes the algebraic procedure that allows us to produce an upper bound for the cyclicity of an elliptic family of real analytic planar vector fields.

12B. Bautin ideal and generating functions. The initial steps of the construction exposed below, refer to *semi*formal series introduced in Definition 4.17.

Let $\mathfrak{A}$ be a Noetherian ring of functions. The most important are the particular cases when $\mathfrak{A}$ is:

(1) the rings of germs $\mathcal{O}(\mathbb{C}^m, 0)$ or $\mathcal{O}(\mathbb{R}^n, 0)$, complex or real analytic respectively,

(2) the ring $\mathcal{O}(U)$ of analytic functions in a domain $U \subseteq \mathbb{R}^n$, or $U \subseteq \mathbb{C}^n$,

(3) the ring of polynomials in m variables $\lambda_1, \dots, \lambda_m$ (again, real or complex).

However, sometimes we will need combinations of these types, e.g., investigation of quadratic planar vector fields in §13**A** requires working in the ring $\mathfrak{A} = \mathcal{O}(\mathbb{R}^1, 0) \otimes \mathbb{R}[\lambda_1, \dots, \lambda_5] \subset \mathcal{O}(\mathbb{R}^6, 0)$ of analytic germs polynomially depending on all variables except for the first one.

In this section we refer to the variables $\lambda_1, \dots, \lambda_m$ as the *parameters* and U as the *parameter space*. Using anyone of these rings, we can construct the rings $\mathfrak{A}[[x, y, \dots]]$ of semiformal series, formal in the variables $x, y, \dots$ with coefficients from the algebra $\mathfrak{A}$ of one of the above types.

With any sequence of functions

$$a_0(\lambda), a_1(\lambda), \dots, a_k(\lambda), \dots, \qquad a_k \in \mathfrak{A}, \tag{12.2}$$

we can associate a growing chain of ideals,

$$\begin{gathered} B_0 \subseteq B_1 \subseteq \cdots \subseteq B_k \subseteq \cdots \subseteq (1) = \mathfrak{A}, \\ B_k = \langle a_0, a_1, \dots, a_k \rangle . \end{gathered} \tag{12.3}$$

Since the ring $\mathfrak{A}$ is Noetherian, the chain (12.3) stabilizes at some moment $k = \nu$, so that $B_{\nu-1} \neq B_\nu = B_{\nu+1} = \cdots$.

With the sequence (12.2) we will associate the *generating function*, the semiformal series in one formal variable

$$a(x, \lambda) = \sum_{k \geqslant 0} a_k(\lambda)\, x^k \in \mathfrak{A}[[x]]. \tag{12.4}$$

Conversely, with any formal or converging series $a(x, \lambda)$ of the form (12.4) we can associate the sequence of its coefficients (12.2), the ascending chain of ideals (12.3), denoted by $B_k(a)$, and the limit ideal

$$B(a) = \lim_{k\to\infty} B_k(a) = B_\nu(a). \tag{12.5}$$

Definition 12.1. The ideal $B(a)$ is called the *Bautin ideal* of the semiformal series $a(x, \lambda)$. The chain of ideals (12.3) will be referred to as the *Bautin chain* and denoted $\mathfrak{B}(a)$. The stabilization moment ν is the *Bautin index*.

We stress that the enumeration of ideals in the Bautin chain begins with B_0 which, however, may be zero ideal. For application to *real analytic* problems instead of the Bautin index we will use another number, the Bautin depth that is by one less the number of *nonzero different* ideals in the chain (12.3).

Definition 12.2. The *Bautin depth* of the chain (12.3) is the number of instances in which the inclusion is *strict* and *nontrivial*,

$$\mu = \#\{k \in \mathbb{N} : 0 \neq B_{k-1} \neq B_k\} \geqslant 0.$$

Obviously, $\mu \leqslant \nu$, with the equality possible only if $0 \neq B_0 \subsetneq \cdots \subsetneq B_\nu = B_{\nu+1} = \cdots$.

For two Bautin chains of ideals $\mathfrak{B} = \{B_k\}$ and $\mathfrak{B}' = \{B'_k\}$ in the same ring $\mathfrak{A}[[x]]$ we will write $\mathfrak{B} = \mathfrak{B}'$ if all ideals in the two chains coincide, and $\mathfrak{B} \subseteq \mathfrak{B}'$ when $B_k \subseteq B'_k$ for all $k = 0, 1, 2, \dots$.

Remark 12.3 (terminological)**.** The term "Bautin ideal" is rather standard and widely used [**Rou98, Yom99**], whereas the combination "Bautin chain" is not. In algebraic terms the Bautin chain $\mathfrak{B}(a)$ defines a *filtration* on the Bautin ideal $B(a)$. In order to be consistent with the accepted terminology, we will speak mostly about Bautin ideals, while always bearing in mind that they are in fact filtered. We will use the notation $\mathfrak{B}(a)$ for the Bautin ideal in order to stress the fact that it is considered together with the filtration, whereas $B(a)$ usually denotes the unfiltered ideal.

On the contrary, the term "Bautin depth" seems to be new. The reason why the Bautin depth is introduced, is closely related to the so-called *fewnomial theory* developed by A. Khovanskii [**Kho91**]. Its usefulness will be clear from Example 12.10 and Theorem 12.25.

Recall that the *radical* of an ideal $B \subseteq \mathfrak{A}$ is

$$\sqrt{B} = \{f \in \mathfrak{A} : f^k \in B \text{ for some } k \in \mathbb{N}\}. \tag{12.6}$$

Obviously, $B \subseteq \sqrt{B}$. The ideal is *radical* (adjective), if $B = \sqrt{B}$.

For *polynomial* ideals in $\mathfrak{A} = \mathbb{C}[\lambda_1, \dots, \lambda_n]$ over the algebraically closed field $\mathbb{C}$, the radical consists of all polynomials vanishing on the *complex null locus* $X_B = \{\lambda \in \mathbb{C}^n : f(\lambda) = 0 \, \forall f \in B\}$ of the ideal $B \subseteq \mathbb{C}[\lambda_1, \dots, \lambda_n]$. This assertion is known under the name *Nullstellensatz* introduced by D. Hilbert. By the Nullstellensatz, the radical polynomial ideals over $\mathbb{C}$ are in one-to-one correspondence with their null loci: any radical polynomial ideal can be characterized as the *biggest* ideal with the same null locus.

The *null locus* X_B (real or complex) of the Bautin ideal B corresponds to the parameter values when the series $a(\cdot, \lambda)$ vanishes identically.

The Bautin ideal (and more generally, the Bautin chain) describes parametric deformations of the identically zero functions (series). In a similar way, we can introduce ideals describing deformations of "maximally degenerate" objects of other types, that can be translated into univariate series.

(1) Families of formal self-maps, defined as automorphisms $\operatorname{Aut}\mathfrak{A}[[x]]$ preserving the ring $\mathfrak{A} = \mathbb{C}[\lambda]$ (we will be only interested in the case of one formal variable x);

(2) Families of formal vector fields, defined as derivations $\operatorname{Der}\mathfrak{A}[[x]]$ (one-dimensional) or $\operatorname{Der}\mathfrak{A}[[x, y]]$ (planar families), with the field of constants $\mathbb{C}(\lambda)$, the field of fractions of the ring $\mathfrak{A}$;

Sometimes we will use the notation $\mathbb{C}[[x]] \otimes \mathfrak{A}$ for the algebra of semiformal series (the tensor product is over the ground field $\mathbb{R}$ or $\mathbb{C}$ depending on the type of the ring $\mathfrak{A}$). This tensorial notation will be extended to other types of semiformal objects: $\operatorname{Diff}[[\mathbb{C}^1, 0]] \otimes \mathfrak{A}$ denotes semiformal self-maps, $\mathcal{D}[[\mathbb{R}^2, 0]] \otimes \mathfrak{A}$ stands for formal real planar vector fields, *etc.*

12C. Basics of formal theory. We begin by pointing out several almost obvious properties of the Bautin ideals of "univariate" objects. These properties reflect simple combinatorics of coefficients of product and composition of formal series in one independent variable.

We start by observing that a semiformal series $f = \sum a_k x^k \in \mathfrak{A}[[x]]$ is invertible, i.e., $1/f \in \mathfrak{A}[[x]]$, if and only if $a_0 \in \mathfrak{A}$ is invertible in $\mathfrak{A}$, in particular, a_0 is a nonzero constant if $\mathfrak{A}$ is a polynomial ring. In a similar way a semiformal self-map $H \colon x \mapsto y = \sum c_k x^k$ is well defined if and only if $c_0 \equiv 0$ and invertible (i.e., $H^{-1} \in \operatorname{Aut}\mathfrak{A}[[x]]$) if and only if c_1 is invertible in $\mathfrak{A}$.

Proposition 12.4. *If $f, g \in \mathfrak{A}[[x]]$, then $\mathfrak{B}(fg) \subseteq \mathfrak{B}(f)$. If g is invertible in $\mathfrak{A}[[x]]$, then $\mathfrak{B}(fg) = \mathfrak{B}(f)$.*

Proof. Denote by $f_k, g_k \in \mathfrak{A}$ the Taylor coefficients of f and g respectively, and by f'_k the coefficients of their product fg. Then, obviously,

$$f'_k = g_0 f_k \bmod \langle f_0, f_1, \dots, f_{k-1} \rangle ,$$

which means that $B_k(fg) \subseteq B_k(f)$ for all $k = 0, 1, \dots$. The first assertion is thus proved; the second assertion is obvious by the invertibility criterion in $\mathfrak{A}[[x]]$. $\square$

The Bautin ideal is in fact independent of the choice of the coordinate x, or, in algebraic terms, of the generator of the ring $\mathfrak{A}[[x]]$.

Proposition 12.5. *The Bautin ideal is invariant by formal conjugacy: for any $f \in \mathfrak{A}[[x]]$ and any semiformal automorphism $H\colon x \mapsto y = \sum_1^\infty c_k x^k$ of $\mathfrak{A}[[x]]$, the Bautin ideals of f and $Hf = f \circ y$ coincide.*

Proof. Denote the Taylor coefficients of f and $f' = f \circ y$ by f_k and f'_k respectively. Re-expanding $f' = \sum f_k y^k$, we obtain

$$f'_k = c_1^k f_k \bmod \langle f_0, f_1, \dots, f_{k-1} \rangle .$$

The morphism H is invertible in $\operatorname{Aut}\mathfrak{A}[[x]]$ if and only if c_1 is invertible in $\mathfrak{A}$. This immediately means that $B_k(f') = B_k(f)$. $\square$

A semiformal family of vector fields $F \in \mathfrak{A} \otimes \mathfrak{D}[[\mathbb{C}^1, 0]]$ on the line having a singularity at the origin, preserves the maximal ideal $\mathfrak{m} = \mathfrak{A} \otimes \langle x \rangle$: $F\mathfrak{m} \subseteq \mathfrak{m}$.

Definition 12.6. The Bautin ideal of the semiformal vector field $F \in \operatorname{Der}\mathfrak{A}[[x]]$ is the Bautin ideal of the semiformal series Fx.

Definition 12.7. The Bautin ideal of an endomorphism H is the Bautin ideal of the difference $H - \mathrm{id}$, i.e., the Bautin ideal of the series $Hx - x$.

These definitions in fact do not depend on the choice of the generator.

Proposition 12.8. *(Semi)formally equivalent vector fields or self-maps have the same Bautin ideals.*

Proof. By Proposition 12.4, if F is a derivation, then for any automorphism $H\colon x \mapsto y = Hx \in \mathfrak{A}[x]$, we have

$$\mathfrak{B}(FHx) = \mathfrak{B}(g \cdot Fx) = \mathfrak{B}(Fx), \qquad g = \frac{dH}{dx} \text{ invertible series.}$$

If $G \in \operatorname{Aut}\mathfrak{A}[[x]]$, then for any automorphism H we have $\mathfrak{B}(H^{-1}GHx - x) = \mathfrak{B}(H^{-1}(G - \mathrm{id})Hx)$ by Proposition 12.5. $\square$

In coordinates, the Bautin ideal of the semiformal vector field $F = f(x, \lambda)\frac{\partial}{\partial x}$ is the Bautin ideal of the coefficient (series) f. If $g = \sum_{k \geqslant 0} g_k\, x^k$, $F = \sum_{k \geqslant 1} f_k\, x^k \frac{\partial}{\partial x}$ and $Fg = g' = \sum_{k \geqslant 0} g'_k\, x^k$, then

$$g'_0 = 0, \quad g'_k = k f_1 g_k \bmod \langle g_0, \dots, g_{k-1} \rangle, \qquad k = 1, 2, \dots. \tag{12.7}$$

Remark 12.9. Note that since a formal derivation F must have zero "free terms", the Bautin chain $\mathfrak{B}(F)$ always starts with the zero ideal $B_0(F) = 0$.

Noninvertible but not identically zero transformations of the formal variable may change the Bautin chain (i.e., the filtration on the Bautin ideal) without changing its limit (the ideal itself).

Example 12.10. Consider a semiformal vector field $F = f(z, \lambda)\frac{\partial}{\partial z}$ with $f(z, \lambda) = a_1(\lambda)\, z + a_2(\lambda)\, z^2 + a_3(\lambda)\, z^3 + \cdots$. The substitution $z = x^2$ brings this vector field to the field $f'(x, \lambda)\frac{\partial}{\partial x}$ with $f'(x, \lambda) = \frac{1}{2}x^{-1} f(x^2, \lambda) = \frac{1}{2}[a_1(\lambda)\, x + a_2(\lambda)\, x^3 + a_3(\lambda)\, x^5 + \cdots]$.

The Bautin chain $\mathfrak{B}'$ for the transformed vector field is obtained by "shearing transformation" of the chain $\mathfrak{B}$:

$$B_1' = B_2' = B_1, \quad B_3' = B_4' = B_2, \quad \dots \quad B_{2k-1}' = B_{2k}' = B_k.$$

Clearly, this transformation does not affect the Bautin ideal as the limit of the Bautin chain, and changes the Bautin index. Yet the Bautin depth remains the same.

Remark 12.11. More generally, let $d \geqslant 2$ be a natural number. Then one can introduce the dth *periodic Bautin ideal* of semiformal families of self-maps as the Bautin (filtered) ideal of the displacement of the dth *iterated* power $H^{\circ d} = \underbrace{H \circ \cdots \circ H}_{d \text{ times}}$ of the formal map H, $\mathfrak{B}^{\circ d}(H) = \mathfrak{B}(H^{\circ d})$. This iterated ideal describes analytic perturbations of *periodic* formal maps.

The main (though still very simple) result of this section compares the Bautin ideals of a (semi)formal vector field F and that of its (semi)formal flow $\exp tF$.

Proposition 12.12. *The Bautin ideals $\mathfrak{B}(F)$ and $\mathfrak{B}(\exp tF)$ of a vector field F and its formal flow map respectively, coincide for $t \neq 0$.*

Proof. We use the exponential series (3.8), extending it on the algebra $\mathfrak{A}[[x]]$ literally,

$$\exp tF = \operatorname{id} + tF + \frac{t^2}{2!} F^2 + \cdots + \frac{t^k}{k!} F^k + \cdots. \tag{12.8}$$

The "matrix" of the operator F in the basis $1, x, x^2, x^3, \dots$ of $\mathfrak{A}[[x]]$ is the infinite matrix

$$M_F = \begin{pmatrix} 0 & & & & & \\ & a_1 & & & & \\ & a_2 & 2a_1 & & & \\ & a_3 & 2a_2 & 3a_1 & & \\ & a_4 & 2a_3 & 3a_2 & 4a_1 & \\ & \vdots & & & & \ddots \end{pmatrix}.$$

The proposition can be proved by inspection of the structure of the powers F^k and hence of the entire sum (12.8). Looking at the first coefficient, we see that $F^k x = a_1^k\, x + O(x^2)$, so that $(\exp tF)x = x + \sum_{k\geqslant 1} \frac{t^k}{k!}\, a_1^k\, x + O(x^2)$, and therefore the first Bautin ideal $B_1(\exp tF)$ is the ideal

$$B_1(\exp tF) = \langle \exp(ta_1) - 1 \rangle = \langle ta_1(1 + \cdots) \rangle = \langle ta_1 \rangle .$$

Assume by induction that the equalities $B_i(\exp tF) = B_i(F) = \langle a_1, \dots, a_i \rangle$ are proved for all $i = 1, 2, \dots, k-1$. To prove that $B_k(\exp tF) = B_k(F)$, note that modulo the ideal $\langle a_1, \dots, a_{k-1} \rangle\, [[x]] \subseteq \mathfrak{A}[[x]]$, the derivation F coincides with the derivation $[a_k\, x^k + O(x^{k+1})]\frac{\partial}{\partial x}$. Substituting it into the exponential series, we obtain

$$(\exp tF)x = x + t\, a_k\, x^k + O(x^{k+1}) + \frac{t^2}{2!} O(x^{2k-1}) + \cdots \quad \bmod \langle a_1, \dots, a_{k-1} \rangle .$$

By induction, the coincidence of the ideals is proved. □

Remark 12.13. All assertions of this section on coincidence of Bautin ideals become completely transparent if the (filtered) Bautin ideal were replaced by the respective zero locus $X = X_B \in (\mathbb{C}^n, 0)$. This locus corresponds to trivial objects (identically zero formal vector fields and identical formal self-maps). Obviously, such objects remain trivial by any conjugacy.

12D. Bautin ideal of a convergent series. It was already noted (see Proposition 12.5), that formal changes of variables leaving the origin fixed, preserve the Bautin ideals of various "one-dimensional" objects.

For *convergent* (analytic) families of functions the *translation* (shift) of the variable x also keeps the Bautin ideal.

Theorem 12.14. *Assume that the series $\sum_{k\geqslant 0} a_k(\lambda)\, x^k$ is convergent in some small neighborhood of the origin $(x, \lambda) \in (\mathbb{C}^1, 0) \times (\mathbb{C}^n, 0)$.*

Then for any analytic germ $t\colon (\mathbb{C}^n, 0) \to (\mathbb{C}, 0)$ the nonfiltered Bautin ideal of the shifted function $S_t(a)$, $S_t(a)(x, \lambda) = a(x + t(\lambda), \lambda)$ does not depend on the germ t,

$$B(S_t(a)) = \lim_k B_k(S_t(a)) = \lim_k B_k(a) = B(a).$$

In other words, the ideal

$$B(a;y) = \big\langle a(y,\lambda), \frac{\partial a}{\partial x}(y,\lambda), \frac{\partial^2 a}{\partial x^2}(y,\lambda), \dots, \frac{\partial^k a}{\partial x^k}(y,\lambda), \dots \big\rangle \subseteq \mathfrak{A} \qquad (12.9)$$

generated by the derivatives at a variable point $y \in (\mathbb{C}^1, 0)$, *is independent of this point as long as it remains in the domain of analyticity of* a.

Remark 12.15 (important). The Bautin chains (*filtrations*) induced on the limit ideal, are *not preserved* by the shift. In other words, the ideal $B(a;y)$ depends on the point y if considered as a filtered ideal.

The following corollary restores the complete invariance of the Bautin ideals by *arbitrary* analytic changes of variables.

Corollary 12.16. *The Bautin ideal is invariant by analytic conjugacy fixing the origin: for any analytic family of functions* $f(x,\lambda)$ *and any family* $H = H(x,\lambda)$ *of invertible analytic transformations depending on parameters with* $H(0,0) = (0,0)$, *the Bautin ideals of* f *and* $Hf = f \circ y$ *coincide.*

Proof of the corollary. An arbitrary family H can be represented as a composition of a translation (shift) $(x,\lambda) \mapsto (x+t(\lambda),\lambda)$, and a holomorphic transformation H' preserving the origin, $H'(0,\lambda) \equiv 0$ for all λ. The germ $t(\lambda)$ is holomorphic and $t(0) = 0$. □

The proof of Theorem 12.14 is based on a rather nontrivial fact, the *closedness* of analytic ideals, which in turn is a consequence of the fact that the operator of expansion in generators of an analytic ideal is bounded.

Let $I \subseteq \mathcal{O}(C^n, 0)$ be an ideal generated by the germs of analytic functions $a_1(\lambda), \dots, a_n(\lambda)$. Denote by $D \cong (\mathbb{C}^n, 0)$ a small polydisk D centered at the origin, on which all germs a_k extend as holomorphic functions. Recall that $\|f\|_D = \sup_{\lambda \in D} |f(\lambda)|$ denotes the norm on the space of holomorphic functions $\mathcal{O}(D)$.

Theorem 12.17 (Division theorem for germs [**Her63**]). *For any polydisk* $D' \Subset D$ *there exist a constant* K *depending, in general, on* D', *such that any holomorphic function* $f \in \mathcal{O}(D')$ *whose germ at the origin belongs to* I, *admits expansion* $f = \sum_1^m h_i a_i$ *with* h_i *also holomorphic in* D' *and*

$$\|h_i\|_{D'} \leqslant K \|f\|_{D'}.$$

This theorem implies that ideals in the ring of germs are *closed* in the following sense.

Corollary 12.18 (closedness of ideals). *If a sequence of functions* $\{f_i\}_{i=1}^{\infty}$ *is defined in a common open neighborhood of the origin, converges uniformly on a smaller set, and their germs at the origin belong to an arbitrary ideal* $I \subseteq \mathcal{O}(\mathbb{C}^n, 0)$, *then the germ of the limit function also lies in this ideal.* □

Remark 12.19. Formulation of Theorem 12.17 is somewhat technical because of the interplay between germs and representing them as holomorphic functions: the ring of germs cannot be equipped by a single norm with respect to which the ideals are closed.

There exists a parallel assertion for polynomials that is free of this drawback. For a (multivariate) polynomial $p = \sum c_\alpha \lambda^\alpha \in \mathbb{C}[\lambda]$ denote by $|p|$ the sum of absolute values of all its coefficients, $|p| = \sum_\alpha |c_\alpha|$. The correspondence $p \mapsto |p|$ is a multiplicative norm on the algebra of the complex polynomials, $|p+q| \leqslant |p| + |q|$, $|pq| = |p|\,|q|$.

Consider an arbitrary *polynomial* ideal $I = \langle a_1, \dots, a_m \rangle \subset \mathbb{C}[\lambda_1, \dots, \lambda_n]$. By definition of the basis, any other polynomial $q \in I$ from this ideal can be expanded as $q = \sum_1^m h_i a_i$ with some polynomial coefficients $h_1, \dots, h_m \in \mathbb{C}[\lambda]$. This expansion is by no means unique, however, it is well-posed in the following precise sense.

Theorem 12.20 (Hironaka division theorem for polynomial ideals)**.** *For some (hence, for any) basis $a_1, \dots, a_m$ of an arbitrary polynomial ideal $I \subseteq \mathbb{C}[\lambda_1, \dots, \lambda_n]$ there exist two finite constants K_1, K_2, depending in general on the choice of the basis, such that any member $q \in I$ admits expansion $q = \sum_1^m h_i a_i$ with*

$$\deg h_i \leqslant \deg q + K_1, \qquad |h_i| \leqslant K_2^{\deg q}\,|q|.$$

This result can be proved by thorough inspection of the division algorithm involving Gröbner bases of ideals [**CLO97**]. In this form the result appears in [**Yom99**].

Proof of Theorem 12.14. Consider a series $\sum a_k(\lambda)\, x^k$ converging to a function $a(x, \lambda)$ holomorphic in some polydisk $U \times D \subseteq (\mathbb{C}^{n+1}, 0)$. Consider first the case where $t \in \mathbb{C}$ is an independent variable parameter. The coefficients $a_{k,t} \in \mathfrak{A}$ of the expansion of $S_t(a)(x, \lambda) = a(t + x, \lambda)$ with the center t, i.e., the derivatives of $a(\cdot, \lambda)$ at the point t, coincide (modulo the factorial coefficients) with the derivatives of the shifted function at t. In particular,

$$a_{0,t}(\lambda) = a(t, \lambda) = \sum_0^\infty a_k(\lambda)\, t^k.$$

This series converges if $|t|$ is sufficiently small and its kth partial sums belong to $B_k(a) \subseteq B(a)$. By Corollary 12.18, the limit belongs to $B(a)$. Differentiating this converging series termwise in t proves that the kth partial sum for $k!\, a_{j,t}(\lambda) = \partial^k a(t, \lambda)/\partial t^k$ belongs to $B_{k+j}(a) \subseteq B(a)$ for all $j = 1, 2, \dots$. Thus the ideal generated by $a_{j,t}$ belongs to $B(a)$,

$$B(S_t(a)) = \langle a_{0,t}, a_{1,t}, \dots, a_{j,t}, \dots, \rangle \subseteq B(a).$$

The inclusion remains valid after substitution of a holomorphic germ $t = t(\lambda)$ instead of the formal parameter t. The arguments being symmetric (reversible), we conclude that the two ideals in fact coincide. □

Another very important corollary of the closedness of the ideals is the possibility of grouping their terms. Consider a convergent series $a(x, \lambda) = \sum a_k(\lambda)x^k$ and its filtered Bautin ideal $\mathfrak{B}(a)$ in the ring $\mathfrak{A} = \mathcal{O}(\mathbb{C}^n, 0)$.

Lemma 12.21. *If the Bautin depth of the Bautin ideal $\mathfrak{B}(a)$ is equal to μ, then the germ a can be represented as the finite sum*

$$a(x,\lambda) = \sum_{j=0}^{\mu} a_{k_j}(\lambda)\, x^{k_j} h_j(x,\lambda), \qquad 0 \leqslant k_0 < k_1 < \cdots < k_\mu, \qquad h_j(0,0) = 1, \qquad j = 0,1,\ldots,\mu. \tag{12.10}$$

Here k_j are the instances where the strict inclusions in the chain (12.3) *occur, $B_{k_j - 1} \neq B_{k_j}$.*

Proof. If the series a converges, then $\|a_k\|_U \leqslant Cr^{-k}$ for some positive constants $0 < r, C < +\infty$.

By definition of the Bautin depth, the functions $\{a_{k_0}, a_{k_1}, \ldots, a_{k_\mu}\} = \{a_j : j \in S\}$, $S \subset \mathbb{N}$, which will be referred to as *basic elements*, generate the limit Bautin ideal $B(a)$. Therefore all coefficients a_k can be expressed as combinations of the basic elements,

$$a_k = \sum_{j \in S, j \geqslant k} h_{kj} a_j, \qquad h_{kj} \in \mathfrak{A}, \quad k = 0,1,\ldots, \tag{12.11}$$

(we expand the basic elements in a trivial way so that $h_{jj} \equiv 1$ for all $j \in S$). By Theorem 12.17, the representation can be chosen so that $\|h_{kj}\|_U \leqslant C' r^{-k}$ with another constant C'. But this means that the series

$$h'_j(x,\lambda) = \sum_k h_{kj}(\lambda)\, x^k = x^{k_j} h_j(x,\lambda),$$
$$h_{kj}(x,\lambda) = 1, \qquad j = 0,1,\ldots,\mu,$$

is convergent and begins with the term x^{k_j}. Multiplying the identities (12.11) by x^k and rearranging the terms of the converging series, we obtain the required representation. □

12E. Bautin index and cyclicity. Let $f = f(x, \lambda) \in \mathcal{O}(\mathbb{C}^{n+1}, 0)$ be a holomorphic (or real analytic) germ represented by a function holomorphic in a small polydisk $D \times U$. This function can be considered as an analytic *local family* of functions in $\mathfrak{A} \otimes \mathcal{O}(D)$, $\mathfrak{A} = \mathcal{O}(U)$.

Definition 12.22. The *complex cyclicity* (sometimes referred to as *local valency*) of the complex analytic local family of functions $f(x, \lambda)$ is the smallest integer number $\mu \in \mathbb{N}$ such that the number of *isolated* zeros of the function $f(\cdot, \lambda)$ in a sufficiently small polydisk $\{|x| < \delta, |\lambda| < \varepsilon\}$ does not exceed μ,

$$\exists \varepsilon > 0, \delta > 0 \quad \forall |\lambda| < \varepsilon, \quad \#\{x \colon |x| < \delta,\ f(x, \lambda) = 0\} \leqslant \mu. \tag{12.12}$$

Here and below by $\#M$ we will denote the number of *isolated* points in a real or complex analytic set $M \subseteq U$.

Remark 12.23 (terminological). The term *cyclicity* is related to bifurcations of limit cycles, as explained in §12**A**. Assume that L is a limit cycle of a planar real analytic vector field analytically depending on parameters $\lambda_1, \dots, \lambda_n$ varying near the origin in $\mathbb{R}^n$. Let $f(x, \lambda)$ be the displacement function for the first return (real holonomy) map associated with any choice of the cross-section to L. Then cyclicity of the germ f is equal to the maximal number of limit cycles that can be observed in a small annulus around L for any sufficiently small value of the parameters.

For applications to the study of small limit cycles of elliptic vector fields, we need a modification of the construction.

Definition 12.24. The *real cyclicity* of a real analytic local family of functions $f(x, \lambda)$ is the maximal number of *positive* isolated roots of $f(\cdot, \lambda)$ in a sufficiently small semi-interval $(\mathbb{R}^1_+, 0)$, the maximum taken over all small values of the parameters $\lambda \in (\mathbb{R}^n, 0)$.

The formal definition with quantifiers coincides with (12.12) except that instead of the disk $\{|x| < \delta\}$ one has to take the real interval $\{0 < x < \delta\}$.

By definition, cyclicity is defined for a family, i.e., for a deformation, though if $f_0 = f(\cdot, 0)$ is not identically zero, it can be majorized uniformly over all analytic families containing f_0, as explained in §12**A**.

Theorem 12.25.

1. *If f is a real analytic germ and the associated Bautin ideal $\mathfrak{B}(f) \subseteq \mathcal{O}(\mathbb{R}^n, 0)$ has the depth μ, then the real cyclicity of the family on the real semiaxis is $\leqslant \mu$.*

2. *If $f(x, \lambda) = \sum_0^\infty a_k(\lambda)\, x^k$ is an holomorphic germ and the associated Bautin ideal $\mathfrak{B}(f) \subseteq \mathcal{O}(\mathbb{C}^n, 0)$ has index ν, then the complex cyclicity of the family is $\leqslant \nu$.*

Proof. The real assertion is proved by the classical derivation-division process which is one of ingredients of the much broader *fewnomial theory*

[**Kho91**]. The complex counterpart is treated using the Cartan inequality and the perturbation technique following [**Yak00**].

1. By Lemma 12.21, the germ f can be represented as the finite sum $f(x,\lambda) = \sum_{j\in S} a_j(\mu)\, x^{k_j} h_j(x,\lambda)$, $S = \{k_0,\dots,k_\mu\} \subset \mathbb{N}$ (see (12.10)) with $k_0 < k_1 < \cdots < k_\mu$.

The neighborhood $U = (\mathbb{R}^n, 0)$ of the origin in the parameter space can be represented as the union of the domains where the jth coefficient a_j is not too small compared to the other coefficients a_i, $i \neq j$,

$$U = Z \cup U_0 \cup \cdots \cup U_\mu, \qquad Z = \{\lambda\colon a_0 = \cdots = a_\mu = 0\},$$
$$U_j = \{\lambda\colon 2(\mu+1)\,|a_j| > \sum_{i\neq j} |a_i|\}, \qquad j = 0,\dots,\mu.$$

For $\lambda \in Z$ there is nothing to prove since $f(x,\lambda) \equiv 0$ there. It remains to show that $f(x,\lambda)$ has no more than μ zeros in some interval $(0,\varepsilon)$ uniformly over λ restricted to each U_j.

Consider the following *derivation-division process*. The sum involving $\mu+1$ terms $f(x,\lambda) = f_0(x,\lambda) = \sum_{j\in S} a_j(\lambda)\, x^{k_j} h_j(x,\lambda)$ is divided by the function $x^{k_0} h_0(x,\lambda)$ and then the derivative in x is taken. This division leaves the sum real analytic since the exponents k_j increase and $h_0(0,0) \neq 0$. As a result, the first term disappears completely and the remainder $f_1(x,\lambda)$ has the same structure, $f_1(x,\lambda) = \sum_{j\in S\smallsetminus\{k_0\}} a_j(\lambda) x^{k_j-k_0} h_{j,1}(x,\lambda)$, but with different exponents $k_j - k_0 > 0$ and some analytic invertible coefficients, $h_{j,1}(0,0) \neq 0$, $j \in S \smallsetminus \{k_0\}$.

Fix one of the domains U_ν. After ν "division+derivation" steps described above, we arrive at the function

$$f_\nu(x,\lambda) = a_\nu(\lambda)\, x^{k_\nu - k_{\nu-1}} + \sum_{j\in S\smallsetminus\{k_0,\dots,k_{\nu-1}\}} a_j(\lambda) x^{k_j - k_{\nu-1}} h_{j,\nu}(x,\lambda).$$

This function is nonvanishing for all values of $\lambda \in U_\nu$ on a sufficiently small real interval $(0,\varepsilon)$. Indeed, the exponents $k_j - k_{\nu-1}$ are all bigger than $k_\nu - k_{\nu-1}$ because of the monotonicity of the indices k_i, and the ratios $|a_j(\lambda)|/|a_\nu(\lambda)|$ do not exceed $\frac{1}{2(\mu+1)}$ by construction of U_ν. Thus the first term in f_ν dominates on a sufficiently small interval $(0,\varepsilon)$ the rest of the sum, therefore f_ν has the same sign as $a_\nu(\lambda) \neq 0$ in U_ν.

It remains to notice that each step "division+derivation" may decrease the number of isolated zeros on $(0,\varepsilon)$ at most by 1:

$$\#\{x \in (0,\varepsilon)\colon f_\nu(x,\lambda) = 0\} \geqslant \#\{x \in (0,\varepsilon)\colon f_{\nu-1}(x,\lambda) = 0\}$$

for any $\nu = 1,2,\dots,\mu$. Indeed, multiplication by any power of x does not affect the number of roots on any positive interval, while derivation can decrease the number of roots by 1 at worst. This follows from the

Rolle lemma, since (i) between any two *distinct* roots of f there must be at least one root of the derivative, and (ii) the *multiplicity* of a multiple root decreases after derivation exactly by 1.

Since $f_\nu(x,\lambda)$ is nonvanishing on $(0,\varepsilon)$ for $\lambda \in U_\nu$, the function $f = f_0$ has no more than ν isolated zeros there. On the union $\bigcup_{\nu=0}^{\mu} U_\nu$ the function f has no more than μ real roots. The statement on real zeros is proved.

2. To prove the assertion on complex zeros, we use the same representation $f(x,\lambda) = \sum_0^\mu a_j(\lambda)x^{k_j}\, h_j(x,\lambda)$ (see (12.10)) which should be further prepared as follows. Let $D = \{|x| < \varepsilon\} \subset \mathbb{C}$ be a small disk on which the functions h_j are explicitly bounded, say, by 2 uniformly over λ. Restricting the parameters on the domain U_j and dividing the function f by a_j there, we obtain

$$a_j^{-1}(\lambda)\, f(x,\lambda) = p_j(x,\lambda) + x^{k_j+1} q_j(x,\lambda), \tag{12.13}$$

where p_j are monic polynomials of degree k_j, while the remainders q_j are explicitly bounded,

$$\begin{aligned} p_j(x,\lambda) &= x^{k_j} + \sum_{k<k_j} b_{kj}(\lambda)\, x^k, \qquad b_j \in \mathcal{O}(U_j), \\ |q_j(x,\lambda)| &\leqslant C = 4(\mu+1), \qquad (x,\lambda) \in D \times U_j. \end{aligned} \tag{12.14}$$

The rest of the proof goes independently for each domain U_j. We show that a function (12.13) constrained by the inequality (12.14) may have at most k_j complex zeros in a disk of radius

$$r_0 = \frac{1}{2}\bigl((8e)^{k_j}(C+1)\bigr)^{-1} \geqslant \frac{1}{2}\bigl((8e)^{\nu}(C+1)\bigr)^{-1}. \tag{12.15}$$

This will prove the theorem since $k_0 < \cdots < k_\mu = \nu$. To simplify the notation, we omit explicit dependence on λ.

Let r be a positive number between 0 and ε to be chosen later. As the polynomial p_j is monic, by Cartan inequality [**Lev80**] there exists a finite number of *exceptional disks* with the sum of their diameters less than r such that *outside* their union p_j admits the *lower* bound $|p_j(x)| \geqslant (r/4e)^{k_j}$, where $e \approx 2.71828\ldots$ is the Euler number.

Consider the annulus $\{r \leqslant |x| \leqslant 2r\}$ foliated by concentric circumferences $\{|x| = \rho\}$, $r \leqslant \rho \leqslant 2r$. As the sum of diameters of the exceptional disks is less than r, at least one such circumference is disjoint with their union and hence p_j is bounded from below on it by $(r/4e)^{k_j}$.

On the other hand, on any such circumference the term $x^{k_j+1}q_j(x)$ admits an explicit upper bound using (12.14):

$$|x^{k_j+1}q_j(x)|_{|x|=\rho} \leqslant C\frac{\rho^{k_j+1}}{1-\rho} \leqslant C\frac{(2r)^{k_j+1}}{1-2r}.$$

The *domination inequality* $(r/4e)^{k_j} > (2r)^{k_j+1}C/(1-2r)$ ensures that the Rouché theorem applies to the circumference $\{|x| = \rho\}$ and guarantees that the number of roots of p_j and $a_j^{-1}f$ (the former being at most k_j) in the disk $\{|x| \leqslant r\}$ coincide. Resolving the domination inequality with respect to r gives $r < \frac{1}{2}\big((8e)^{k_j}(C+1)\big)^{-1}$. □

Remark 12.26. The proof of Theorem 12.25 is constructive in the sense that, knowing the parameter K characterizing the ideal in Theorem 12.17, one can produce explicitly the lower bound for the size of the interval or disk containing no more than the asserted number of roots (in the complex case this was done explicitly).

The simple bound of this type asserted by Theorem 12.25, is not the best known one. In [**RY97**] N. Roytwarf and Y. Yomdin considered the general problem of uniform localization of zeros of an analytic family of functions with the specified Bautin ideal and explicit constraints on the growth of Taylor coefficients, the so-called *Bernstein classes*. Using a dual description of the Bernstein classes in terms of the growth rate of the functions represented by the series, they obtain a lower bound for the radius of the disk in which at most ν zeros can occur. This bound was achieved in the form $r_0 = \big(8^\nu \max(C,2)\big)^{-1}$ (in the equivalent settings). These results are generalized in [**FY97**] for A_0-series with polynomial coefficients in $\mathfrak{A} = \mathbb{C}[\lambda_1, \dots, \lambda_n]$ of degree growing at most linearly and the norms at most exponentially.

Yet somewhat surprisingly, the best result can be obtained by properly "complexifying" the derivation-division process, based on the complex analog of the Rolle lemma [**KY96**]. In this way one can prove that the number of small complex isolated roots in the family (12.13)–(12.14) does not exceed ν in the disk of radius $r_s = \frac{1}{2}(1-s^{-1})\big(s^{\nu+1}C+1\big)^{-1}$ for any value of $s > 1$. All details can be found in [**Yak00**].

The assertion of Theorem 12.25 can be improved in another direction. An *integral closure* of an ideal $I \subset \mathfrak{A}$ is the collection of all roots $y \in \mathfrak{A}$ of all equations of the form $y^n + q_1 y^{n-1} + \cdots + q_{n-1}y + q_n = 0$ with the coefficients q_k belonging to the kth powers of I, $q_k \in I^k$. If $B = \langle a_0, a_1, \dots, a_n, \dots \rangle$ is the filtered Bautin ideal, its *reduced Bautin index* is defined in [**HRT99**] as the minimal number $r \in \mathbb{N}$ such that the integral closure of $\langle a_0, \dots, a_r \rangle$ coincides with B. Obviously, the reduced Bautin index does not exceed its (usual) Bautin index. In [**HRT99**] an analog (also constructive) of the second assertion of Theorem 12.25 is proved for the reduced Bautin index rather than ν.

Theorem 12.25 is a general tool linking cyclicity of analytic families of functions of one variable with the depth of the corresponding Bautin chain of ideals generated by the coefficients. In the next sections this tool will be applied to the study of bifurcations of limit cycles in analytic vector fields on the plane.

12F. Elliptic vector fields on the plane: Bautin and Dulac ideals. Consider a real analytic family of vector fields on the plane,

$$
\begin{gathered}
F = \mathbf{A} + \text{nonlinear terms}, \qquad \mathbf{A} = \alpha(\lambda)\mathbf{E} + \beta(\lambda)\mathbf{I}, \\
\mathbf{E} = x\frac{\partial}{\partial x} + y\frac{\partial}{\partial y}, \qquad \mathbf{I} = y\frac{\partial}{\partial x} - x\frac{\partial}{\partial y}, \\
\mathfrak{A} = \mathcal{O}(\mathbb{R}^n, 0), \quad \alpha, \beta \in \mathfrak{A}, \quad F \in \mathfrak{A} \otimes \mathcal{D}(\mathbb{R}^2, 0),
\end{gathered}
\tag{12.16}
$$

with the linear part $\mathbf{A}$ normalized as in (12.1) and elliptic, i.e., $\alpha(0) = 0$, $\beta(0) \neq 0$.

There are *several* univariate semiformal series that are more or less naturally associated with the family F. One series is the first return map $P \in \mathfrak{A} \otimes \text{Diff}(\mathbb{R}^1, 0)$; this series is always convergent. The first return map depends on the choice of the cross-section and a local chart on it, but the corresponding Bautin ideal denoted by $\mathfrak{B}(F)$, is well defined by Proposition 12.8. The Bautin ideal is obviously invariant by the action of *real analytic orbital transformations* on the family (12.16).

As was already remarked before, the Poincaré return map for the elliptic family is the square of the holonomy operator $\Delta \in \mathfrak{A} \otimes \text{Diff}(\mathbb{R}^1, 0)$ associated with the real equator on the Möbius band after blow-up of the corresponding foliation. The 2-periodic Bautin ideal of the holonomy map represented by a convergent series, by definition is the Bautin ideal of the Poincaré return map (see Remark 12.11).

The third series, sometimes referred to as the Poincaré–Dulac series, appears as the generating function of the coefficients of the Poincaré–Dulac orbital formal normal form. By Theorem 4.18, there exists a semiformal transformation bringing the family (12.16) to the rotationally invariant normal form (4.9). After division by the nonvanishing formal series the normal form can be further transformed to the semiformal vector field

$$F' = f(r^2, \lambda)\mathbf{E} + \mathbf{I}, \qquad f(u, \lambda) = \sum_{k=1}^{\infty} f_k(\lambda)\, u^k \in \mathfrak{A}[[u]]. \tag{12.17}$$

The semiformal series f occurring the orbital formal normal form (12.17), will be referred as the Poincaré–Dulac series. By construction, the Poincaré–Dulac series can apriori be divergent and is not uniquely defined (because of the freedom in the choice of resonant coefficients during the Poincaré–Dulac normalization). Later on we explain an invariant construction for this series and introduce in Definition 12.31 the corresponding *Dulac ideal* $\mathfrak{D}(F)$ which will also be a formal orbital invariant of the family (12.16).

In addition to these series and corresponding chains of ideals, there are some other univariate semiformal series, usually associated with certain methods of formal integration. Yet these series may produce filtered ideals $B_0 \subseteq B_1 \subseteq \cdots \subseteq \mathfrak{A}$ that are noninvariantly related to the family (12.16): only the corresponding filtered zero loci $X_k = \{\lambda \colon a(\lambda) = 0\, \forall a \in B_k\}$ in the parameter space $(\mathbb{R}^n, 0)$ usually have invariant meaning.

Vanishing of all coefficients of the return map P means that the field F exhibits a center for the corresponding values of the parameters. Vanishing of all coefficients of the Poincaré–Dulac series f in (12.17) means that the field F is formally orbitally linearizable and hence admits a formal first integral. By Proposition 11.6, the two properties are equivalent for elliptic vector fields, which means that the respective zero loci of the two (unfiltered) ideals $\mathfrak{B}(F)$ and $\mathfrak{D}(F)$ coincide.

This observation suggests a conjecture that the ideals generated by coefficients of these two series, should also coincide. This assertion, if true, can be considered as a parametric generalization of Proposition 11.6 and the Poincaré–Lyapunov Theorem 11.7 with appropriate implications for cyclicity of the elliptic families.

This conjecture turns out, broadly speaking, true. However, in order to make its formulation precise, one has to overcome several technical obstacles arising since the normal form can be divergent. Furthermore, we give an alternative construction for the Dulac ideal that will be invariant by formal transformations (Definition 12.31).

12$\mathbf{F}_1$. *Formal first return map for semiformal families.* The monodromy and vanishing holonomy maps for an elliptic vector field can be consistently defined in the (semi)formal category. Indeed, one can blow up the origin on the plane and transform the Pfaffian equation associated with the *semiformal* vector field on the plane,

$$
\begin{gathered}
F = \alpha(\lambda)\mathbf{E} + \beta(\lambda)\mathbf{I} + (\text{nonlinear terms}) \in \mathfrak{A} \otimes \mathcal{D}[[\mathbb{R}^2, 0]], \\
\alpha, \beta \in \mathfrak{A} = \mathcal{O}(\mathbb{R}^n, 0), \quad \beta(0) \neq 0, \ \alpha(0) = 0
\end{gathered} \tag{12.18}
$$

to the form (10.6). In this formal Pfaffian equation the 1-forms $\theta_k \in \mathfrak{A} \otimes \Lambda^1(\mathbb{E})$ on the projective line $\mathbb{E}$ continue to be meromorphic, nonsingular on the real equator $\mathbb{R}P^1 \subset \mathbb{E}$ and analytically depending on the parameters $\lambda \in (\mathbb{R}^n, 0)$ by virtue of the formulas (10.5). The only difference with the analytic case is that the series in powers of the variable x in the right hand side of (10.6) is in general divergent. Yet despite this divergence integration of the triangular system of linear ODE's (10.12) yields the "formal holonomy map" $\Delta_{\mathbb{R}}\mathfrak{A} \otimes \operatorname{Diff}[[\mathbb{R}, 0]]$ as a well-defined semiformal series (10.13); see Remark 10.17. The formal square $\Delta_{\mathbb{R}} \circ \Delta_{\mathbb{R}} \in \mathfrak{A} \otimes \operatorname{Diff}[[\mathbb{R}, 0]]$ defines the semiformal first return map.

Definition 12.27. The Bautin ideal $\mathfrak{B}(F)$ (with the corresponding filtration) of a semiformal elliptic family of vector fields $F \in \mathfrak{A} \otimes \mathcal{D}[[\mathbb{R}^2, 0]]$, is the Bautin ideal $\mathfrak{B}(P)$ of its semiformal first return map P as defined above.

However, instead of the algebraic blow-up, for computations of the formal return map we can use the trigonometric blow-up passing to the polar coordinates (r, φ) on the real plane $\mathbb{R}^2$; see Exercise 12.4.

12$\mathbf{F}_2$. *Quotient equation.* In this section we give an invariant definition of the second ideal associated with a semiformal elliptic family (12.18). This definition is based on *integrability* of the Poincaré–Dulac formal normal form for all planar singularities.

The integrability, usually understood as a possibility to express solutions of a differential equation in elementary functions, quadratures and their inverse (implicit) functions, implies that a given planar vector field can be reduced to a one-dimensional vector field. In the particular context we say that the planar semiformal vector field $F \in \mathfrak{A} \otimes \mathcal{D}[[\mathbb{R}^2, 0]]$ *admits projection on the line*, if there exists another semiformal vector field $G \in \mathfrak{A} \otimes \mathcal{D}[[\mathbb{R}^1, 0]]$ on the real line and a projection, a semiformal "map" $u\colon \mathbb{R}^2 \to \mathbb{R}^1$ analytically depending on the parameters λ, which conjugates the fields F and G. Interpreting u as a semiformal series from $\mathfrak{A}[[x, y]]$, i.e., choosing a fixed chart on the target space, and writing the vector field in this chart as $G = g(u)\frac{\partial}{\partial u}$ with $g \in \mathfrak{A}[[u]]$, we derive the necessary and sufficient condition of the conjugacy between F and G under the form

$$Fu(x, y) = g(u(x, y)), \qquad u \in \mathfrak{A}[[x, y]], \quad g \in \mathfrak{A}[[u]]. \tag{12.19}$$

The solution of this equation, if it exists, can be found by the method of indeterminate coefficients, using the ansatz $g(v) = \sum_{k=1}^{\infty} g_k v^k$ and finding consecutively the unknown homogeneous components of the series $u = u_p + u_{p+1} + \cdots$ starting from the leading order $p \geqslant 1$.

Example 12.28. The leading component $u_p \in \mathbb{R}[x, y]$ must be an eigenvector of the linear differential operator associated with the linear part $\mathbf{A} = \alpha(\lambda)\mathbf{E} + \beta(\lambda)\mathbf{I}$ of the vector field F.

The operator $\mathbf{A}$ restricted on the space of homogeneous polynomials of degree p, is diagonalizable over the complex field: a monomial $z^k \bar{z}^{p-k}$ is an eigenvector with the eigenvalue $k\mu + (p-k)\bar{\mu}$, where $\mu = \alpha + i\beta$ is the eigenvalue of the 2×2-matrix A of the linear vector field $\mathbf{A}$. Yet for an elliptic family all these eigenvalues and eigenvectors are nonreal, the only exception being the middle eigenvalue associated with the eigenvector $(z\bar{z})^{p/2} = r^p$ for even values of p, starting from $p = 2$.

Having this example in mind, it makes sense to normalize solutions of the equation (12.19) by the additional requirement that

$$u_2(x, y) = x^2 + y^2. \tag{12.20}$$

Definition 12.29. The *quotient equation* for a semiformal elliptic family F of real planar vector fields is the equation (12.19) on the unknown semiformal series $u \in \mathfrak{A}[[x, y]]$, $g \in \mathfrak{A}[[u]]$, normalized by the condition (12.20). The *quotient field* is the semiformal vector field $G_F = g(u)\frac{\partial}{\partial u} \in \mathfrak{A} \otimes \mathcal{D}[[\mathbb{R}^1, 0]]$.

In [**Arn69**] the quotient equation is introduced under the name *cocycle*, but this term is too overburdened and will never be used in this sense.

Existence of the projection is obviously invariant by the formal conjugacy, but in general not invariant by the *orbital* formal conjugacy. On the other hand, the semiformal projection u is certainly not uniquely defined, as any semiformal transformation $u \mapsto u' \in \mathfrak{A}[[u]]$ tangent to the identity, produces another projection and another quotient field G', albeit semiformally conjugate to the initial field G. It turns out that this is the only freedom, and hence the Bautin ideal of the quotient field is an invariant of elliptic semiformal fields.

Lemma 12.30. *Any semiformal elliptic planar vector field $F \in \mathfrak{A} \otimes \mathcal{D}[[\mathbb{R}^2, 0]]$ admits a semiformal projection on the real line.*

The quotient field $G = G_F \in \mathfrak{A} \otimes \mathcal{D}[[\mathbb{R}^1, 0]]$ normalized by condition (12.20) is defined by F uniquely modulo a semiformal conjugacy from $\mathfrak{A} \otimes \operatorname{Diff}[[\mathbb{R}^1, 0]]$. The Bautin filtered ideal $\mathfrak{B}(G_F)$ of the quotient field is uniquely defined by F.

Proof. Solvability of the quotient equation does not depend on the semiformal conjugacy class of the elliptic field: the transformation conjugating two fields F, F' simultaneously conjugates the respective series u, u'.

Thus one can assume without loss of generality that the elliptic semiformal family is already in the normal form given by the first assertion of Theorem 4.18:

$$F' = a(r^2, \lambda)\mathbf{E} + b(r^2, \lambda)\mathbf{I}, \qquad a, b \in \mathfrak{A}[[r^2]], \quad r^2 = x^2 + y^2; \tag{12.21}$$

cf. with (4.9). For the field F' in the normal form (12.21) all assertions of the lemma are immediate. Indeed, the two series

$$u(x, y) = x^2 + y^2, \qquad g(u) = 2u\, a(u) \tag{12.22}$$

give a solution of (12.19) by the Euler identity $\mathbf{E}u = 2u$ and the rotational symmetry $\mathbf{I}u = 0$.

Next we show that *any* solution $u = u_2 + u_3 + \cdots$ of the quotient equation starting with $u_2 = x^2 + y^2$, is in fact rotationally symmetric, i.e., depends on r^2 only. Indeed, apply the differential operator $\mathbf{I}$ to both parts of the quotient equation: using the fact that $\mathbf{I}$ and F' commute (Theorem 4.18), we conclude that $s = \mathbf{I}u$ is a solution of the equation

$$F's = \varphi(u) \cdot s, \qquad s \in \mathfrak{A}[[x, y]], \quad \varphi = 2\tfrac{d}{du}\big(ua(u)\big) \in \mathfrak{A}[[u]].$$

In particular, if $s \not\equiv 0$, then the principal homogeneous polynomial term $s_q \in \mathbb{R}[x, y]$ of degree $q \geqslant 3$ is an eigenvector of the operator $\mathbf{A} = a(0)\mathbf{E} + b(0)\mathbf{I}$. As follows from Example 12.28, the only possibility for a nonzero polynomial to be such an eigenvector occurs when the order q is even and $s_q = (x^2 + y^2)^{q/2}$. Clearly, such a polynomial has a nonzero average on any circle $r = \text{const}$ and hence cannot be a derivative of the form $\mathbf{I}u_{q+1}$ of any polynomial u_{q+1}. Thus any solution of the quotient equation is rotationally symmetric, $\mathbf{I}u = 0$, and hence a semiformal series in r^2. □

12F$_3$. *Dulac ideal.* Lemma 12.30 asserts that the Bautin ideal of the quotient vector field is an invariant object.

Definition 12.31. The *Dulac ideal* $\mathfrak{D}(F) \subseteq \mathfrak{A}$ of the semiformal planar elliptic vector field F is the (filtered) Bautin ideal $\mathfrak{B}(G_F)$ of the quotient semiformal vector field G_F on the real line.

Theorem 12.32. *The Dulac chain (filtered ideal) $\mathfrak{D}(F) = \{D_k\}$ of any semiformal elliptic family F is obtained by shearing the Bautin chain $\mathfrak{B}(F) = \{B_k\}$ of this family, i.e.,*

$$D_1 = D_2 = B_1, \ D_3 = D_4 = B_2, \ \dots \ D_{2k-1} = D_{2k} = B_k, \ \dots. \tag{12.23}$$

In particular,

$$D(F) = \lim D_k(F) = \lim B_k(F) = B(F).$$

In other words, the Dulac and Bautin ideals coincide as unfiltered ideals in $\mathfrak{A}$, and have the same depth as filtered ideals.

Proof. For the elliptic family in the formal normal form (12.21) the Dulac ideal is the ideal of the univariate series $ua(u, \lambda)$ in the powers of the formal variable u.

To compute the monodromy map, we transform the normal form to the polar coordinates $(r, \varphi) \in (\mathbb{R}^1_+, \mathbb{S}^1)$; note that the restriction of r on the positive cross-section coincides with the chart x. In the polar coordinates (12.21) takes the form

$$\frac{dr}{dt} = r\,a(r^2, \lambda), \qquad \frac{d\varphi}{dt} = b(r^2, \lambda). \tag{12.24}$$

The first equation is obtained from the quotient equation (12.22), the second follows from the formulas $\mathbf{E}\varphi = 0$, $\mathbf{I}\varphi = 1$. The field (12.24) is orbitally equivalent to the vector field

$$\frac{dr}{dt} = \frac{r\,a(r^2, \lambda)}{b(r^2 m\lambda)}, \qquad \frac{d\varphi}{dt} = 1, \tag{12.25}$$

since the series $b(r^2, \lambda)$ is invertible in $\mathfrak{A}[[r^2]]$ (recall that $b(0,0) \neq 0$ is the ellipticity condition). The monodromy map for the field (12.25) is the time 2π flow $\exp 2\pi G'$ of the semiformal vector field

$$G'(r) = b(r^2, \lambda)^{-1} \cdot r\,a(r^2, \lambda)\tfrac{\partial}{\partial r}, \tag{12.26}$$

which differs from the quotient field $G = ua(u, \lambda)\frac{\partial}{\partial u}$ by the "folding transformation" $u = r^2$ (cf. with Example 12.10) and division by the invertible series b. The latter transformation does not affect the Bautin ideal, while the former shears the chain $\mathfrak{B}(G_F)$, as explained in Example 12.10. This proves the theorem. □

This theorem immediately implies a number of corollaries.

Corollary 12.33. *The Dulac ideal is invariant by* orbital *semiformal conjugacy of the elliptic family.*

Proof. The Bautin ideal is invariant by the orbital transformation $F \mapsto sF$ with an invertible series $s \in \mathfrak{A}[[x,y]]$. $\square$

Corollary 12.34. *Cyclicity of an elliptic family of real analytic vector fields is equal to the depth of the corresponding Dulac chain (filtered ideal).*

Proof. Each small limit cycle appearing in the elliptic analytic family of planar vector fields (12.18) corresponds to a unique isolated *positive* root of the Poincaré displacement function $\delta_F(r) = P(r) - r$, where P is the monodromy map. The Bautin ideal of the function δ_F is by definition the Bautin ideal of the family F, $\mathfrak{B}(\delta_F) = \mathfrak{B}(F)$. By the first assertion of Theorem 12.25, cyclicity of the analytic family of real functions δ_F is equal to the depth of this ideal. By Theorem 12.32, the depths of $\mathfrak{B}(F)$ and $\mathfrak{D}(F)$ coincide. $\square$

12G. Universal polynomial families, cyclicity and localized Hilbert problem. Consider the *universal family* of elliptic polynomial vector fields of a given degree d,

$$F = \alpha\mathbf{E} + \beta\mathbf{I} + \sum_{2 \leqslant i+j \leqslant d} \lambda'_{ij} x^i y^j \frac{\partial}{\partial x} + \lambda''_{ij} x^i y^j \frac{\partial}{\partial y}. \tag{12.27}$$

parameterized by the real parameters

$$\alpha \in \mathbb{R}^1, \quad \beta \in \mathbb{R}^1 \smallsetminus \{0\}, \quad \lambda = \{\lambda'_{ij}, \lambda''_{ij}\} \in \mathbb{R}^n, \qquad n = n(d).$$

Cyclicity of the origin in the family (12.27) is closely related to Hilbert's sixteenth problem about the number and location of limit cycles of a polynomial vector field of degree d; see §24**A**. Knowing this cyclicity would answer the question about the maximal number of *small* limit cycles near the origin, at least for vector fields close to linear centers.

12**G**$_1$. *Reduced universal family and chains of polynomial ideals.* As follows from Corollary 12.34, cyclicity of the elliptic family (12.27) is equal to the depth of the Dulac chain (filtered ideal). Despite the fact that the universal family is polynomial, Dulac and Bautin ideals for the universal family (12.27) apriori belong only to the ring $\mathcal{O}(\mathbb{R}^{n+2}, 0)$ of real analytic germs of functions of $n+2$ variables α, β, λ.

However, the question about cyclicity of the family (12.27) can be reduced to computation of the depth of some *polynomial* filtered ideal, the Dulac ideal of the auxiliary *reduced family* with fixed linear part $\mathbf{I} = y\frac{\partial}{\partial x} - x\frac{\partial}{\partial y}$

(pure rotation),

$$F' = \mathbf{I} + \sum_{2\leqslant i+j\leqslant d} \lambda'_{ij} x^i y^j \tfrac{\partial}{\partial x} + \lambda''_{ij} x^i y^j \tfrac{\partial}{\partial y}. \qquad (12.28)$$

Denote by $\mathfrak{D} = \{D_k\}$ and $\mathfrak{D}' = \{D'_k\}$ the Dulac chain of ideals for the corresponding families (12.27) and (12.28),

$$\mathfrak{D} = \{D_k\},\ D_k \subseteq \mathcal{O}(\mathbb{R}^{n+2}, 0), \qquad \mathfrak{D}' = \{D'_k\},\ D'_k \subseteq \mathcal{O}(\mathbb{R}^n, 0).$$

Denote the depths of these chains by μ and μ' respectively.

Proposition 12.35. *The auxiliary chain of ideals $\mathfrak{D}'$ is generated by polynomials in λ and $D'_1 = 0$. The depths of the two chains differ by* 1,

$$\mu = \mu' + 1.$$

Proof. To compute the first Dulac ideal of a vector field, one has to solve the linearization of the quotient equation: since $\mathbf{I}r^2 = 0$, we have for the leading coefficient g_1 the equality $g_1 r^2 = \alpha \mathbf{E} r^2$ so that $D_1 = \langle \alpha \rangle$, $D'_1 = \{0\}$.

Since the ideal D_1 is radical, the quotient ideals $D_k \bmod D_1$ are isomorphic to the ideals obtained by fixing $\alpha = 0$ in the universal family. Moreover, since the Dulac ideal(s) are invariant by orbital transformation, one can also fix the second parameter, setting $\beta = 1$. Then the universal family takes the form (12.28), and we conclude that

$$D'_k = D_k \bmod D_1, \qquad \text{for all } k \geqslant 2.$$

This instantly proves the proposition. The fact that ideals D'_k are polynomial (i.e., belong to $\mathbb{R}[\lambda]$) follows from the fact that the nonlinear coefficients of the monodromy map are quasihomogeneous functions of the parameters λ, as explained in Theorem 10.18. □

Proposition 12.35 reduces the transcendental problem on the number of small limit cycles that can appear near an elliptic singular point of a polynomial vector field of degree d, to a completely algebraic problem of determination of the depth of a growing chain of *polynomial* ideals $D_i \subseteq \mathbb{R}[\lambda]$. This chain of ideals in "universal" in the sense that it serves all deformations of planar polynomial vector fields of the given degree d.

Computing any finite number of ideals in the Dulac chain $\mathfrak{D}$ is theoretically feasible and can be relegated to one of many existing symbolic computation programs. Yet computation of the Bautin index (or depth) of the chain is a problem beyond the reach of any computer algebra system, even if we ignore the practical limitations on memory and time. Indeed, after observing that the chain $\mathfrak{D}$ stops growing at some moment μ, one has to *prove* that all *infinitely many* remaining coefficients of, say, the series $g(u, \lambda)$, belong to the ideal generated by the first μ of them.

In practice the only case for which the construction of Dulac ideal and computation of its depth is fully implemented, is that of *quadratic* vector fields corresponding to $d = 2$. The corresponding results are described in §13.

12**G**$_2$. *Practical computation of the Dulac chain.* The most important advantage of working with the Dulac chain (ideal) rather than with the Bautin ideal, is practical: computation of $\mathfrak{D}$ does not require solving differential equations which is a necessary step when computing the first return map (cf. with §10**E**). From the outset we can work with the reduced universal polynomial family (12.28) with the linear part $\mathbf{I} = y\frac{\partial}{\partial x} - x\frac{\partial}{\partial y}$.

Assume that the Taylor polynomials U_{p-1} and $G_{(p-1)/2}$ of degrees $p-1$ and $\lfloor (p-1)/2 \rfloor$ of the respective series $u \in \mathbb{R}[\lambda][[x, y]]$ and $g \in \mathbb{R}[\lambda][[u]]$ are already found (for p odd we take the integer part of the ratio $(p-1)/2$). Recall that, for the reduced family, $g_1 = 0$.

The homogeneous component u_p and the next coefficient $g_{p/2}$ of the series g will be a solution of the equation

$$\begin{gathered} \mathbf{I}u_p = v_p + g_{p/2} r^p, \\ v_p = p^{\text{th}} \text{ degree terms of } G_{(p-1)/2}(U_{p-1}) - \mathbf{N}U_{p-1} \end{gathered} \tag{12.29}$$

where $\mathbf{N} = F' - \mathbf{I}$ is the nonlinear part of the vector field F' considered as a differential operator with polynomial coefficients and polynomially depending on the parameters λ. The term $g_{p/2} r^p$ is absent when p is odd.

This equation is always solvable. If p is odd, then any homogeneous polynomial occurring in the right hand side of (12.29) has zero average on any circle $r = \text{const}$ and hence admits a unique homogeneous primitive $u_p = \mathbf{I}^{-1} v_p$.

If p is even, then the average of v_p on the circles may be nonzero, but it will be always of the form cr^p for some constant $c \in \mathbb{R}$. If we set $g_{p/2} = -c$, the right hand side will have zero average and hence a polynomial primitive $u_p = \mathbf{I}^{-1}(v_p + g_{p/2} r^p)$, which is defined uniquely modulo addition of $c' r^p$ (this nonuniqueness creates nonuniqueness of solution of the quotient equation).

In both cases we can determine uniquely the next Taylor coefficients u_p and $g_{p/2}$ so that the process continues by induction. Inspection of this process yields an independent proof of polynomial dependence of all coefficients on the parameters λ, the nonlinear coefficients of the universal polynomial family.

12**G**$_3$. *Dulac ideal and Poincaré–Lyapunov constants.* Besides the quotient equation (12.19), there are other constructions which associate a univariate semiformal series with an elliptic family. For instance, in [**Sch93**] and in some other sources the following equation appears,

$$Fv = b(r^2, \lambda), \qquad b(r^2, \lambda) = \sum_{j \geqslant 1} b_j(\lambda)\, r^{2j}, \tag{12.30}$$

where $v \in \mathfrak{A}[[x, y]]$ is a semiformal series in two variables, and F is an elliptic family with the fixed linear part $\mathbf{I}$. The equation (12.30) also always admits a formal solution (v, b) for the same reasons as explained in §12**G**$_2$. The coefficients $b_j \in \mathfrak{A}$ are called by different names as *Poincaré–Lyapunov constants*, *Lyapunov values*, *focal values*, *etc.* Starting from the generating series b, in the standard way arrive at a growing chain of ideals

$$\langle b_1 \rangle \subseteq \langle b_1, b_2 \rangle \subseteq \langle b_1, b_2, b_3 \rangle \subseteq \cdots \tag{12.31}$$

in the ring of polynomials $\mathbb{R}[\lambda]$.

The construction of Poincaré–Lyapunov constants and the chain of ideals (12.31) is not intrinsically invariant (unlike the definition of the Dulac ideal). Nevertheless, the common zero locus of the first k polynomials, $\{b_1 = \cdots = b_k = 0\} \in \mathbb{R}^n$ corresponds to parameter values for which the elliptic field admits a jet of order $2k$ of the first integral. The same condition in terms of the Dulac ideal translates as a vanishing of the first k coefficients of the vector fields G. Thus at least as far as the Dulac ideals remain radical, the chains of ideals $\mathfrak{D}$ and (12.31) coincide. We will not explore this direction.

Exercises and Problems for §12.

Exercise 12.1. Assume that $\mathfrak{A} = \mathcal{O}(\mathbb{C}^n, 0)$, $\mathfrak{m} \subseteq \mathfrak{A}$ is the maximal ideal, and the coefficients of a semiformal series $f = \sum_1^\infty a_k x^k$ have the following properties:

(1) the first m coefficients $a_1, \dots, a_m$ belong to $\mathfrak{m}^2$,

(2) the next n coefficients $a_{m+1}, \dots, a_{m+n}$ span the linear space $\mathfrak{m}/\mathfrak{m}^2 \cong \mathbb{C}^n$.

Compute the index of the corresponding Bautin chain.

Exercise 12.2. Let X_k be the zero locus of the kth Bautin ideal of a semiformal vector field (or a self-map). Give a direct proof that this locus is the same for any two formally equivalent vector fields (resp., self-maps); cf. with Remark 12.13.

Problem 12.3. Formulate and prove an analogous statement for zero loci of Bautin and Dulac ideals of an elliptic family of planar vector fields.

Exercise 12.4. Compute the formal holonomy map for the semiformal elliptic field (12.17) in the Poincaré–Dulac normal form.

Exercise 12.5. Consider the planar elliptic polynomial family given in the complex notation by the differential equation

$$\dot{z} = z(i + \lambda z^p \bar{z}^p), \qquad \lambda \in \mathbb{R},\ p \geqslant 1.$$

Compute the Bautin and Dulac ideals and Lyapunov "constants" for this field.

Exercise 12.6. Let f be a complex analytic family satisfying the assumptions of Example 12.1.

Prove that the depth of the translated ideal $\mathfrak{B}(f(x+t))$ is the same for all sufficiently small $t \in (\mathbb{C}^1, 0)$ except for a possible discrete set.

13. Quadratic vector fields and the Bautin theorem

13A. Quadratic vector fields. The only universal polynomial family for which the depth of the Bautin ideal was computed, is the family of *quadratic vector fields* corresponding to $d = 2$. In this section we prove the following famous theorem.

Theorem 13.1 (N. Bautin [**Bau39, Bau54**]). *Cyclicity of an elliptic singular point in the family of quadratic vector fields is equal to* 3.

The Bautin theorem inspired the conjecture that the number of *all* limit cycles of a planar quadratic vector field can be at most 3. This conjecture was believed to be true until in 1980 Shi Songling discovered an example of a quadratic vector field in which 3 small limit cycles coexist with one "large" limit cycle away from the elliptic singularity [**Shi80a**].

Consider the reduced family F' of quadratic vector fields with the fixed linear part, which we write as a system of differential equations

$$\begin{aligned} \dot{x} &= \quad y + \lambda_1 x^2 + \lambda_2 xy + \lambda_3 y^2, \\ \dot{y} &= -x + \lambda_4 x^2 + \lambda_5 xy + \lambda_6 y^2. \end{aligned} \tag{13.1}$$

By Corollary 12.34 and Proposition 12.35, the Bautin theorem follows from the following purely algebraic fact.

Theorem 13.2. *The reduced Bautin chain of ideals $\mathfrak{B}' = \{\mathbf{B}'_k\}$ for the family* (13.1) *of quadratic vector fields with the rotation linear part* $\mathbf{I}$, *has depth* 2, *specifically,*

$$0 \neq \mathbf{B}'_2 \subsetneq \mathbf{B}'_3 \subsetneq \mathbf{B}'_4 = \mathbf{B}'_5 = \mathbf{B}'_6 = \cdots. \tag{13.2}$$

The proof of Theorem 13.2 occupies the rest of §13**A** and all of §13**B**.

It was already noted on several occasions that many assertions concerning Bautin ideals admit counterparts concerning the respective zero loci in the space of the parameters, and almost always these assertions are much simpler. The Bautin theorem is not an exception: its proof is based on a no less remarkable theorem proved by H. Dulac in 1908.

Together with the chain of *real* polynomial ideals $\mathfrak{B}' \subseteq \mathbb{R}[\lambda]$ consider the chain of their *complexified* zero loci

$$\begin{aligned} \mathbb{C}^6 \supseteq X_2 \supseteq X_3 \supseteq X_4 \supseteq \cdots \supseteq X_k \supseteq \cdots, \\ X_k = \{\lambda \in \mathbb{C}^n : p(\lambda) = 0 \, \forall p \in \mathbf{B}'_k\}. \end{aligned} \tag{13.3}$$

The limit $X = \lim_{k\to\infty} X_k$ of the chain (13.3) consists of the complex values of the parameters λ for which the *complex* vector field is formally integrable, i.e., there exists a formal solution $u = (x^2 + y^2) + \cdots$ of the quotient equation $F'u \equiv 0$ corresponding to $g \equiv 0$. By Proposition 11.6, in this case there exists another, convergent formal integral.

Theorem 13.3 (H. Dulac [**Dul08**]). *The complex variety $X_4 \subseteq \mathbb{C}^6$ corresponds to integrable quadratic systems.*

In other words, the chain of complex algebraic varieties (13.3) stabilizes on the fourth term, $X_4 = X_5 = \cdots = X$.

The chain of ideals (13.2), starting from the term $\mathbf{B}'_4$, in principle may exhibit nontrivial growth, but only in such a way that the zero loci of all subsequent ideals $\mathbf{B}'_4, \mathbf{B}'_5, \ldots$ remain constant. This is, however, impossible. The following theorem asserts that $\mathbf{B}'_4$ is *the biggest* ideal with the null locus X_4, so that further growth of the Bautin chain $\mathfrak{B}'$ is impossible.

Theorem 13.4 (H. Żołądek [**Żoł94**]). *The ideal* $\mathbf{B}_4'$ *from the Bautin chain* (13.2) *is radical: any polynomial* $p \in \mathbb{C}[\lambda]$ *vanishing on* X_4*, belongs to* $\mathbf{B}_4'$.

Theorem 13.2 obviously follows from Theorems 13.3 and 13.4, whose complete proofs are postponed until §13**B**. Here we outline the general structure of these proofs in a brief historical discourse. From the outset it should be stressed that heavy computations cannot be avoided, though almost all of them can now be done by computers.

The first step is to compute the initial segment of the Bautin chain. On the level of null loci this computation was done by Dulac in [**Dul08**]. To minimize the number of independent parameters, Dulac used rotation of the coordinates (x, y) on the real plane to reduce the vector field to the so-called *Kapteyn normal form* involving only 5 parameters $\widetilde{\lambda}_2, \dots, \widetilde{\lambda}_6$ (different from the initial parameters $\lambda_1, \dots, \lambda_6$),

$$\begin{aligned} \dot{x} &= -y - \widetilde{\lambda}_3\, x^2 + (2\widetilde{\lambda}_2 + \widetilde{\lambda}_5)\, xy + \widetilde{\lambda}_6\, y^2, \\ \dot{y} &= x + \widetilde{\lambda}_2\, x^2 + (2\widetilde{\lambda}_3 + \widetilde{\lambda}_4)\, xy - \widetilde{\lambda}_2\, y^2. \end{aligned} \tag{13.4}$$

For this family Dulac derived the polynomial conditions over $\mathbb{R}[\widetilde{\lambda}]$ necessary for existence of a 7-jet of a first integral $u = (x^2 + y^2) + \cdots$, and discovered that under these conditions the vector field is integrable.

Bautin recycled the computations of Dulac to compute (by hand!) the coefficients of the return map and discovered that the corresponding Dulac ideal $\widetilde{D}_7 = \langle \widetilde{a}_3, \widetilde{a}_5, \widetilde{a}_7 \rangle \subseteq \mathbb{R}[\widetilde{\lambda}]$ of the quotient equation is *not radical*. The main lemma of the paper [**Bau54**], proved by lengthy calculations (partially explained in [**Yak95**]), claims that despite this nonradicality, all higher coefficients of the quotient equation in fact belong to $\widetilde{D}_7$.

This circumstance remained completely mysterious until H. Żołądek in 1994 realized that both nonradicality of the ideal $\widetilde{D}_7$ in the Bautin chain and the fact that this chain stabilizes *despite* this nonradicality, are aberrations caused by the Kapteyn form, since transformation of the general equation (13.1) to the Kapteyn form (13.4) is singular (discontinuous). When written with respect to the original parameters λ, the respective (Bautin or Dulac) ideals $\mathbf{B}_4' = D_7$ become radical. Żołądek himself in [**Żoł94**] gave an elementary (though long and technical) proof of this radicality with respect to the ring of polynomials equivariant by a natural circle action (see Remark 13.6 below) and noted in passing that the equivariance is irrelevant and the fact remains true in the full ring $\mathbb{C}[\lambda]$, though the proof of this is "much more complicated" [**Żoł94**, Remark 1, p. 236].

However, unlike the claim on effective termination of the infinite chain of ideals which amounts to the *infinite* number of equalities between individual ideals in the chain, the claim on radicality of a given *single* ideal admits verification in finite time. Moreover, algorithms for computing the radical of a polynomial ideal given by its generators, as well as the coincidence test for two such ideals are well developed and efficient computer algebra systems exist for implementing them. Proving Theorem 13.4 can be completely delegated to the computer in the same way as computation of the initial coefficients of formal integrals, normal forms, *etc.* This observation in some sense "downgrades" Theorem 13.4 to the level of a polynomial identity which for the moment cannot be proved by any method other than direct tedious computation. Below we give a five-line script for `CoCoA` (Commutative Computer Algebra, [**CNR00**]), which computes the radical $\sqrt{\mathbf{B}_4'}$ and verifies that it coincides with $\mathbf{B}_4'$.

Unlike Theorem 13.4, Dulac Theorem 13.3 is a claim that requires human intervention and ingenuity (together with unavoidable computations).

13B. Dulac center conditions. It was another observation of H. Żołądek that using the "complex notation" greatly simplifies computations. If we identify a point (x, y) on the real plane $\mathbb{R}^2$ with the complex number $z = x + iy \in \mathbb{C}$, then any quadratic vector field with the linear part $\mathbf{I}$ can be written as

$$\dot{z} = iz + A\,z^2 + B\,z\bar{z} + C\,\bar{z}^2, \qquad A, B, C \in \mathbb{C}, \tag{13.5}$$

with *complex* coefficients A, B, C. This observation can be explained by the fact that after complexification (allowing the coefficients λ to take complex values) the linear part can be diagonalized by passing to the coordinates $z = x + iy$, $w = x - iy$. The complex quadratic vector field $F \in \mathcal{D}[\mathbb{C}^2]$ acquires then the form given by the differential equations

$$\begin{aligned} \dot{z} &= iz + Az^2 + Bzw + Cw^2, \\ \dot{w} &= -iw + C'z^2 + B'zw + A'w^2, \end{aligned} \qquad A, \dots, C' \in \mathbb{C}. \tag{13.6}$$

The real vector fields (with real values of the parameters λ) correspond to systems of the form (13.6) with the complex parameters $A, \dots, C'$ meeting the conditions

$$A' = \bar{A}, \quad B' = \bar{B}, \quad C' = \bar{C} \tag{13.7}$$

(the bar denotes the complex conjugation), after restriction on the real subspace $\mathbb{R}^2 \cong \{w = \bar{z}\} \subseteq \mathbb{C}^2$. Clearly, solving the quotient system (12.19) when the vector field F has diagonal linear part, is much easier; see §12$\mathbf{G}_2$.

The first several steps of formal solution of the quotient equation for the equation (13.6) yield the following results for coefficients of the series $g(u) = g_1 u + g_2 u^2 + \cdots$,

$$\begin{aligned} g_1 &= 0, \\ g_2 &= c_2\,(AB - A'B'), \\ g_3 &= c_3\,[(2A + B')(A - 2B')CB' - (2A' + B)(A' - 2B)C'B], \\ g_4 &= c_4\,(BB' - CC')[(2A + B')B'^2C - (2A' + B)B^2C'], \end{aligned} \tag{13.8}$$

where $c_i \neq 0$ are *nonzero* constants, $i = 2, 3, 4$. Under the "reality" assumptions (13.7) these conditions take the form

$$\begin{aligned} g_1 &= 0, \\ g_2 &= c_2\,\operatorname{Im}(AB), \\ g_3 &= c_3\,\operatorname{Im}[(2A + \bar{B})(A - 2\bar{B})\bar{B}C], \\ g_4 &= c_4\,\operatorname{Im}[(|B|^2 - |C|^2)(2A + \bar{B})\bar{B}^2C], \end{aligned} \tag{13.9}$$

as they appear in [**Żoł94**]. Clearly, cancellation of the nonzero constants does not change the chains of ideals, so from now on we will omit them.

In §12**G**$_2$ we explained how the computations of the polynomials $g_{2,3,4}$ should be organized; the algorithm described there, can be easily made into a code for `Mathematica`.

Remark 13.5. Computation of the coefficients of the first return map is considerably more resource-consuming than that of the quotient equation. Bautin in [**Bau54**] reveals no details, only the ultimate results. This computation was reproduced using computers (see [**FLLL89**]) confirming Bautin's formulas modulo an inessential error in the numeric coefficient c_4. Żołądek in [**Żoł94**] double-checked part of the results using perturbations technique. All existing methods corroborate the formulas (13.9).

13C. Irreducible components of the Dulac variety. The Dulac variety $X_4 = \{g_2 = g_3 = g_4 = 0\} \subseteq \mathbb{C}^6$ is reducible and consists of 4 components (their names will be later explained by the different mechanisms of integrability),

$$\begin{aligned} V_\triangle &= \{B = B' = 0\} && \text{(Darbouxian)},\\ V_H &= \{2A + B' = 2A' + B = 0\} && \text{(Hamiltonian)},\\ V_\ominus &= \{AB - A'B' = B'^3C - B^3C' = 0\} && \text{(symmetric)},\\ V_\between &= \{A - 2B' = A' - 2B = BB' - CC' = 0\} && \text{(meromorphic)}. \end{aligned} \tag{13.10}$$

Indeed, the locus $B = B' = 0$ of codimension 2 satisfies all equations (13.8) and gives the component of X_4 denoted by $V_\triangle$. Outside $V_\triangle$ the equation $g_2 = 0$ yields $A/B' = A'/B$; denoting this common value by R, we transform the remaining equations $g_3 = 0$, $g_4 = 0$ respectively to

$$(2R+1)(R-2)(B'^3C - B^3C') = 0,$$
$$(BB' - CC')(2R+1)(B'^3C - B^3C') = 0.$$

Two more components are given by the equations $2R+1 = 0$ which (together with $g_2 = 0$) corresponds to the locus V_H, and the equation $B'^3C - B^3C = 0$ that defines $V_\ominus$. Outside all these components of codimension 2 the last remaining component is defined by the equations $R = 2$, $BB' - CC' = 0$ which gives us $V_\between$.

13D. Proof of the Dulac Theorem 13.3. We begin the proof by noting that the linear part of normal form (13.6) is invariant by diagonal transformations $(z, w) \mapsto (\gamma z, \gamma' w)$, $\gamma, \gamma' \in \mathbb{C} \smallsetminus \{0\}$, in particular, by the transformations $(z, w) \mapsto (\gamma z, \gamma^{-1} w)$. These transformations, however, change the coefficients $A, \dots, C'$ of the field as follows:

$$\begin{aligned} (z, w) &\mapsto (\gamma z, \gamma^{-1} w),\\ (A, B, C, A', B', C') &\mapsto (\gamma A, \gamma^{-1} B, \gamma^{-3} C, \gamma^3 A', \gamma B', \gamma^{-1} C'). \end{aligned} \tag{13.11}$$

These formulas define an action of $\mathbb{C} \smallsetminus \{0\}$ on the space of the coefficients; all components of the loci (13.10) are invariant by this action.

Remark 13.6. Though the action of $(\mathbb{C}\smallsetminus\{0\})^2$ or $\mathbb{C}\smallsetminus\{0\}$ does not preserve the subset of real systems (13.7), the restriction of this action on the circle $\mathbb{S}^1 = \{|\gamma| = 1,\ \gamma' = \gamma^{-1} = \bar{\gamma}\}$, corresponding to the rigid rotation of the real plane $z \mapsto \gamma z$, induces the circle action on the space of real quadratic vector fields with an elliptic singular point at the origin. It is this circle action that was used by Żołądek in [**Żoł94**] to simplify the proof of radicality.

We prove Theorem 13.3 by proving separately that each of the four components (13.10) corresponds to integrable systems.

1. V_H: *Hamiltonian case.* The divergence of the vector field (13.6) is

$$i + 2Az + Bw + (-i) + B'z + 2A'w = z(2A + B') + w(2A' + B)$$

and vanishes identically along the component V_H. The corresponding Hamiltonian is a cubic polynomial $\frac{1}{2}zw + \cdots$.

When establishing integrability of vector fields for the three remaining components of the locus (13.10), we will first establish it for a particular combination of parameters in the corresponding component and then show that by a suitable action (13.11) any other point on this component can be brought to this particular form.

2. $V_\ominus$: *Symmetric*, or *reversible case.* The component $V_\ominus$ parameterizes systems whose phase portrait is symmetric by a line passing through the origin.

Indeed, if

$$A' = -A, \quad B' = -B, \quad C' = -C, \tag{13.12}$$

then the vector field (13.6) is *anti*-invariant by the symmetry $\sigma\colon (z, w) \mapsto (w, z)$: this symmetry preserves the field modulo the constant factor -1, $\sigma_* F = -F$. Therefore the complex holomorphic foliation $\mathcal{F}$ is symmetric (σ sends leaves into leaves). We claim that this symmetry implies integrability.

Indeed, denote by $\Delta_\mathbb{R}$ the holonomy (semi-monodromy) map of $\mathcal{F}$ after blow-up, corresponding to the symmetric cross-section $\tau = \{z + w = 0\}$; see Definition 10.11. The symmetry σ changes the orientation of the loop (equator) $\mathbb{R} \subset \mathbb{E}$ on the exceptional divisor $\mathbb{E}$, on the other hand, it does not change the intersection points between the leaves and the cross-section. Therefore

$$\Delta_\mathbb{R}^{-1} = \Delta_{\sigma(\mathbb{R})} = \Delta_\mathbb{R},$$

which means that $\Delta_\mathbb{R}$ is 2-periodic, $\Delta_\mathbb{R}^2 = \mathrm{id}$, and the field is a center.

Now we claim that any other combination of parameters on $V_\ominus$ can be brought to the special form (13.12) by a suitable action (13.11). Indeed, the equations of $V_\ominus$ can be reduced to the form

$$A/A' = B'/B, \qquad (B'/B)^3 = C'/C. \tag{13.13}$$

By a suitable choice of γ one can make the ratio A/A' equal to -1. The equations (13.13) imply then that the other two ratios B'/B and C'/C are automatically equal to -1, i.e., the conditions (13.12) are achieved. Thus any combination of the parameters on $V_\ominus$ corresponds to a field having a symmetry axis and hence integrable.

Darbouxian cases. In both the two remaining cases the vector field has several (real algebraic) invariant curves $p_i(z, w) = 0$. Starting from the functions p_i one can construct Darbouxian integrals of the form $\Phi = \prod p_i^{\alpha_i}$ with suitable (in general, noninteger or even nonreal) exponents $\alpha_i \in \mathbb{C}$.

3. $V_\triangle$: *Darbouxian triangle.* The component $V_\triangle$ defined by the condition $B = B' = 0$, corresponds to vector fields having (generically) three invariant lines. To see them, note that the straight line $\{w - z = \alpha\}$, $\alpha \in \mathbb{C}$ is invariant by the field (13.6) with $B = 0$, if and only if

$$C' + A' = C + A, \quad 2\alpha(C - A') + 2i = 0, \quad \alpha^2(C - A') + i\alpha = 0. \tag{13.14}$$

To see this, it is sufficient to differentiate $z - w + \alpha$ along the field (13.6) and restrict the result $iz + Az^2 + Cw^2 + iw - C'z^2 - A'w^2$ on the line $w = z + \alpha$; the corresponding quadratic polynomial must vanish identically, which yields the three equations (13.14).

This system (13.14) admits solution α only if

$$C' + A' = C + A; \tag{13.15}$$

moreover, if $C \neq A'$ (i.e., generically), this solution indeed exists. For an arbitrary combination A, C, A', C' the condition (13.15) can be achieved by a suitable diagonal action (13.11): one should resolve the equation

$$\gamma^{-3}C + \gamma A = \gamma^3 C' + \gamma^{-1} A' \tag{13.16}$$

with respect to $\gamma \in \mathbb{C} \setminus \{0\}$. This equation of degree 6, cubic with respect to γ^2, generically has three pairs of roots differing by a sign in each pair; each pair of roots corresponds to an invariant line.

Thus we conclude that for the parameter values in the component $V_\triangle$, the vector field F has (generically) three invariant straight lines $p_i = 0$, $i = 1, 2, 3$, two of them eventually conjugate. The invariance means that the derivatives Fp_i are divisible by p_i in the ring of polynomials in z, w. Denote by q_i the corresponding *cofactors*, the polynomials such that

$$Fp_i = q_i p_i, \qquad i = 1, 2, 3, \quad \deg q_i = 1.$$

Clearly, $q_i(0, 0) = 0$. Since any three homogeneous linear forms on $\mathbb{C}^2$ are linearly dependent, there exist three nonzero complex numbers $\alpha_1, \alpha_2, \alpha_3$

such that $\sum \alpha_i q_i = 0$. The direct computation shows (cf. with §25**G** below) that the function $\varPhi = \prod_1^3 p_i^{\alpha_i}$ is the first integral:

$$F\varPhi = \varPhi \cdot \sum_1^3 \frac{F p_i^{\alpha_i}}{p_i^{\alpha_i}} = \varPhi \cdot \sum_1^3 \alpha_i q_i = 0.$$

Since $p_i(0,0) \neq 0$, every branch of $\varPhi$ is analytic at the singular point. Thus the component $V_\triangle$ corresponds to the Darbouxian integrable vector fields having an invariant triangle $p_1 p_2 p_3 = 0$.

Thus a generic vector field corresponding to the component $V_\triangle$ is a center. Yet since being center is a closed property, the entire component $V_\triangle$ consists of centers.

4. $V_{\between}$: *Meromorphic integrable systems.* In the last remaining case where the parameters belong to the component $V_{\between}$, we show that one can find a meromorphic (rational) first integral as a ratio of two degree 6 polynomials, both nonzero at the singular point.

By a suitable action $(z,w) \mapsto (\gamma z, \gamma' w)$ multiplying B by γ and B' by γ', the vector field can be brought to the form with $B = B' = 1$. The remaining equations of $V_{\between}$ imply then that

$$B = B' = 1, \qquad A = A' = 2, \qquad CC' = 1, \tag{13.17}$$

so that the vector field has the form

$$\begin{aligned} \dot z &= \quad iz + 2z^2 + zw + Cw^2, \\ \dot w &= -iw + (1/C)\, z^2 + zw + 2w^2. \end{aligned} \tag{13.18}$$

We show that this vector field has two invariant curves, a quadric $\{p_2(z,w) = 0\}$ and a cubic $\{p_3(z,w) = 0\}$, with the corresponding cofactors *coinciding* modulo the rational coefficient,

$$Fp_2 = 2(z+w)p_2, \quad Fp_3 = 3(z+w)p_3. \tag{13.19}$$

Consequently, the *rational* first integral of the field F has the form $\varPhi = p_2^3 p_3^{-2}$. The polynomials $p_2(z,w) = \sum_{i+j \leqslant 3} (P_2)_{ij}\, z^{i-1} w^{j-1}$ and $p_3(z,w) = \sum_{i+j \leqslant 4} (P_3)_{ij}\, z^{i-1} w^{j-1}$ have the following coefficient matrices,

$$P_2 = \begin{pmatrix} -1 & -2\,i & C \\ 2\,i & -2 & \\ \frac{1}{C} & & \end{pmatrix}, \quad P_3 = \begin{pmatrix} \frac{2\,i}{1+C} & \frac{-6}{1+C} & -3\,i & C \\ \frac{6}{1+C} & \frac{3\,i\,(1+C)}{C} & -3 & \\ \frac{-3\,i}{C} & \frac{3}{C} & & \\ -\frac{1}{C^2} & & & \end{pmatrix}$$

and the fact that they satisfy condition (13.19), can be verified by a direct (though tedious) computation. Actually, they were found by `Mathematica` [**Wol96**] as solutions of (13.19) using the indeterminate coefficients method.

Thus all four components (13.10) correspond to nonlinear centers, which completes the proof of Theorem 13.3. □

13E. Symbolic computations and the "proof" of the Żołądek Theorem 13.4. We have to prove that the ideal generated in the polynomial ring in 6 variables $\mathbb{C}[A, B, C, A', B', C']$ by the three polynomials g_2, g_3, g_4 from (13.8), is radical. Checking radicality is a task that is well algorithmized. The computer system CoCoA includes both the computation of the complex radical and the coincidence test for two ideals defined by their generators, as the standard functions; see [**CNR00**].

```
Use R::=Q[axbycz];

   G2:=ab-xy;
   G3:=(2a+y)(a-2y)cy-(2x+b)(x-2b)zb;
   G4:=(by-cz)((2a+y)y^2c-(2x+b)b^2z);

D:=Ideal(G2,G3,G4);
D=Radical(D);
```

Figure II.8. The CoCoA code verifying radicality of the Bautin ideal

The code checking radicality, is given in Fig. II.8. Due to the technical constraints (independent variables should be denoted by lowercase letters) we denoted by `a,b,c,x,y,z` the variables A, B, C, A', B', C' respectively. The first line instructs the computer to use the ring of characteristic zero in the six indeterminates, then `D` is defined as the ideal generated by the polynomials `G2,G3,G4` encoding respectively g_2, g_3, g_4. Finally, the last line is the logical command checking equality between the ideal `D` and its radical `Radical(D)`. After 2 seconds of computations on a laptop, the program prints `TRUE`. This proves the Żołądek theorem. □

13F. Concluding remarks. We conclude the proof of the Bautin theorem by two technical remarks.

Remark 13.7. The "complex notation" (i.e., writing the quadratic vector field so that its linear part is diagonal) simplifies computations not only for humans, but also for computers. An attempt to compute the radical of the Bautin ideal $\mathbf{B}'_4$ written for the real system (13.1) fails miserably, apparently because the corresponding polynomials g_i have too many monomial terms for the standard algorithms to cope with (recall that we are dealing with polynomials of degree 6 in 6 independent variables!).

Remark 13.8. One may recycle information already stored in the equations of the four Dulac loci (13.10) to simplify computation of the radical $\sqrt{\mathbf{B}'_4}$. Indeed, this radical is the *intersection* of the four ideals $J_\triangle, J_H, J_\ominus$ and $J_\between$ in $\mathbb{C}[A, \dots, C']$ which consist of polynomials vanishing on the respective components.

However, one has to bear in mind that while the three ideals,

$$\begin{aligned} J_{\triangle} &= \langle B, B' \rangle, \\ J_H &= \langle 2A + B', 2A' + B \rangle, \\ J_{\Diamond} &= \langle A - 2B', A' - 2B, BB' - CC' \rangle, \end{aligned}$$

are all radical, the polynomial equations defining $J_\ominus$ in (13.10), span a *non*-radical ideal,

$$\begin{aligned} J_\ominus &= \sqrt{\langle AB - A'B', B'^3C - B^3C' \rangle} \\ &= \langle AB - A'B', B'^3C - B^3C', AB'^2C - A'B^2C', A^2B'C - A'^2BC' \rangle. \end{aligned}$$

In any case, computing intersection of ideals (i.e., computing a basis for the intersection) is in general a tedious task which amounts to computing resultants and elimination of variables. On top of that one should solve the membership problem, checking that all elements of the constructed basis again belong to $\mathbf{B}'_4$. To double-check the above described `CoCoA`-proof of Theorem 13.4, these computations were also implemented (by another `CoCoA` script) and gave the same answer, thus further reducing the chances of computer- or human-generated errors.

Exercises and Problems for §13.

Any line on the plane cannot have more than two isolated tangencies with a quadratic vector field. This obvious observation immediately implies a number of simple geometric properties of real quadratic foliations. *All problems below are formulated for quadratic foliations* (vector fields).

Problem 13.1. Prove that any periodic orbit is convex.

Problem 13.2. Prove that a limit cycle necessarily contains a singular point inside. Prove that this point cannot be neither saddle nor node. Prove that this point must be unique and is necessarily a focus.

Problem 13.3. A *Lotka–Volterra system* is a quadratic vector field tangent to the coordinate axes and having a saddle point at the origin.

Write explicitly this vector field and prove that it has a unique equilibrium in the positive quadrant $\{x, y > 0\}$. Find the conditions guaranteeing that the Lotka–Volterra system has a center. List the Dulac components to which integrable Lotka–Volterra systems may belong.

Problem 13.4. Determine cyclicity of the origin for an analytic unfolding of the quadratic center

$$\dot{z} = iz + z^2 + z\bar{z} + i\bar{z}^2$$

in the class of quadratic fields.

14. Complex separatrices of holomorphic foliations

In this section we generalize the result on existence of holomorphic invariant curves from the hyperbolic or semi-hyperbolic context of §7**A** to arbitrary isolated *planar* singularities. The invariant curves will be *analytic*, but in general nonsmooth. Their order at the singular point in many cases can be majorized by the order of the vector field generating the foliation: the corresponding result can be considered as a solution of a *local version of the Poincaré problem* (on degree of algebraic solutions of polynomial differential equations, which will be treated in §25).

14A. Invariant curves. Consider the germ of a holomorphic foliation $\mathcal{F}$ on $(\mathbb{C}^2, 0)$, defined by a Pfaffian equation $\{\omega = 0\}$, with an isolated singularity of multiplicity μ and order n at the origin.

Recall (see Definition 2.27) that a complex separatrix of $\mathcal{F}$ is a leaf $L \in \mathcal{F}$ whose closure $L \cup \{0\}$ is an analytic curve $\gamma = \{f = 0\} \subset (\mathbb{C}^2, 0)$.

For an elementary singular point, there always exists at least one *smooth* complex separatrix. More precisely, there are two smooth complex separatrices if the singular point is *not* a saddle-node or a resonant node, and one or two smooth separatrices in the latter cases. The question on existence of complex separatrices for more degenerate singular points was first discussed by C. Briot and J. Bouquet in 1856. However, the complete solution was achieved only in 1982 by C. Camacho and P. Sad [**CS82**].

Theorem 14.1 (C. Camacho–P. Sad, 1982). *Every isolated singularity of a planar holomorphic vector field admits a complex separatrix.*

Remark 14.2. If $\mathcal{F}$ is a real analytic foliation on $(\mathbb{R}^2, 0)$, then a real separatrix, if it exists, is necessarily a characteristic trajectory. Thus nondegenerate foci and centers do not have real separatrices, though they have a pair of complex separatrices. The inverse claim is false: a characteristic trajectory is not necessarily a separatrix. For instance, among all real trajectories of a nonresonant node with an irrational characteristic ratio, only two are separatrices, the rest being nonanalytic at the origin.

The idea of the proof of Theorem 14.1 is to blow up the foliation until it has only elementary singularities. Each such singularity has at least one complex separatrix. If this separatrix is *not contained* in the vanishing divisor D (preimage of the singular point), then the image of this separatrix will be a nonconstant analytic curve and hence a complex separatrix. To prove the theorem, one has to show that after complete desingularization, at least one elementary singularity always has an invariant curve (it will be always a hyperbolic invariant curve) transversal to D. This is achieved by careful study of characteristic ratios of hyperbolic singularities that appear by blow-up. The most difficult combinatorial part of the original proof from [**CS82**] was recently simplified by J. Cano [**Can97**], whose proof we largely follow.

14B. Linearization along invariant curves and index of a complex separatrix. We start by introducing in invariant terms the notion which generalizes the characteristic ratio of nondegenerate singularities. The

Camacho–Sad index (or simply *index* of a smooth complex separatrix of a foliation is defined in terms of residue of the linearization of the foliation along this separatrix.

The construction of linearization along a smooth invariant curve S (leaf or separatrix) is intuitively rather clear. Assume that S is locally given by the equation $\{y = 0\}$ and a holomorphic Pfaffian form $\omega = f\,dx + g\,dy \in \Lambda^1(\mathbb{C}^2, 0)$ with isolated singularities vanishes on the tangent direction to S. This condition means that $f(x, 0) \equiv 0$. Keeping only the terms of first order in y and dy, we have

$$f(x, y) = a(x)y + O(y^2), \qquad g(x, y) = b(x) + O(y),$$

so that the "linearized" Pfaffian equation (truncated to the terms linear in y and dy) takes the form

$$y\,a(x)\,dx + b(x)\,dy = 0. \tag{14.1}$$

Denote by θ the meromorphic 1-form on the curve S,

$$\theta = -\frac{a(x)}{b(x)}\,dx, \qquad \theta \in \Lambda^1(S) \otimes \mathcal{M}(S). \tag{14.2}$$

For reasons to be explained in Chapter III, which is entirely devoted to linear systems, the form θ will be called the *connexion form* of the foliation $\mathcal{F}$ along the smooth invariant curve S. Using the connexion form, we can rewrite the linearized equation (14.1) as follows:

$$dy = y\theta, \qquad y \in \mathbb{C}, \quad \theta \in \Lambda^1(S) \otimes \mathcal{M}(S);$$

cf. with the nonlinear equations (10.6).

The form θ is only meromorphic on S: from its definition it follows immediately that it is holomorphic at all *nonsingular* points of S. Singularities of the foliation, corresponding to isolated roots of the holomorphic function $b \in \mathcal{O}(S)$, are *poles* of the connexion form.

Definition 14.3. The *index* $i(p, S, \mathcal{F})$ of the smooth analytic invariant curve (separatrix) S passing through a singular point $p \in S$ of a singular foliation $\mathcal{F}$ is the residue $\operatorname{res}_p \theta$ of the connexion form (14.2) along S.

In the notation below we will sometimes omit one or more arguments from the list $i(p, S, \mathcal{F})$, when they are unambiguously determined by the context.

The construction of linearization and the connexion form θ via special local coordinates leaves open the question, to what extent different parts of this construction, in particular, the definition of index, are invariant.

To show that the index in fact does not depend on either the coordinates used for the linearization, or on the choice of ω (i.e., remains the same if ω

is replaced by a multiple $u\omega$, $u \neq 0$), we re-expose *the same* construction in more invariant terms as follows.

Consider a holomorphic 2-dimensional manifold M covered by an atlas of charts U_α, and assume that this manifold carries a singular holomorphic foliation $\mathcal{F}$ with a globally defined smooth separatrix, a complex curve $S \subset M$. In each chart U_α the foliation is defined by a different Pfaffian equation $\{\omega_\alpha = 0\}$, and the separatrix by a different holomorphic equation $\{h_\alpha = 0\}$ with nonvanishing differential dh_α. On the pairwise intersections $U_{\alpha\beta} = U_\alpha \cap U_\beta$ the corresponding forms and functions differ by invertible holomorphic factors:

$$\begin{aligned} \omega_\alpha &= u_{\alpha\beta}\omega_\beta, \qquad h_\alpha = v_{\alpha\beta}h_\beta, \\ u_{\beta\alpha} &= 1/u_{\alpha\beta}, \quad v_{\beta\alpha} = 1/v_{\alpha\beta}, \qquad u_{\alpha\beta}, v_{\alpha\beta} \in \mathcal{O}(U_{\alpha\beta}). \end{aligned} \tag{14.3}$$

We show first that in each neighborhood U_α, if it is sufficiently small, the Pfaffian equation $\omega_\alpha = 0$ is equivalent to the equation $dh_\alpha - h_\alpha\theta_\alpha = 0$, where θ_α is a suitable 1-form on U_α whose restriction of S is uniquely defined.

Proposition 14.4. *Assume that $U \cong (\mathbb{C}^2, 0)$ is a small neighborhood and a smooth curve $S \subset U$ is given by the equation $\{h = 0\}$, where h is a holomorphic function on M with the differential dh not vanishing on S.*

Then any holomorphic 1-form ω tangent to S can be represented as

$$\omega = g(dh - h\,\theta), \tag{14.4}$$

where g is a holomorphic function and θ a meromorphic 1-form whose poles can be only at singular points of ω.

The restrictions of the function g and the form θ on S and the tangent bundle $\mathbf{T}S = \bigcup_{a\in S} \mathbf{T}_a S$ respectively, are uniquely defined by ω and h.

Proof. Since ω vanishes on vectors tangent to S, we have $\omega = g\,dh$ at all points of S (two forms with the same null space must be proportional). The holomorphic function $g\colon (S, 0) \to \mathbb{C}$, originally defined only on S, can be extended on the neighborhood of S in M; this extension (denoted again by g) is vanishing only at singular points of ω on S.

The difference $\omega - g\,dh$ is a 1-form vanishing identically at all points of S and hence divisible by h: $\omega - g\,dh = h\vartheta$, where ϑ is a holomorphic 1-form. Denote by θ the *meromorphic* 1-form $\theta = g^{-1}\vartheta$: this yields the representation (14.4).

The extension of g from S on M is nonunique, hence θ is nonunique. However, if $\omega = g'(dh - h\,\theta')$ is an alternative representation with a different choice of g', θ', then g and g' must coincide on S and hence their difference is divisible by h, $g - g' = uh$. From equality between the two representations $g(dh - h\,\theta) = (g + uh)(dh - h\,\theta')$ of the same form ω it follows that $g(\theta' - \theta) = u(dh - h\theta')$. Both terms dh and $h\theta'$ in the right hand side vanish on vectors tangent to S, hence the restrictions of θ and θ' on TS coincide. □

The restriction of the 1-form θ on S, the meromorphic 1-form $\theta \in \Lambda^1(S, 0) \otimes \mathfrak{M}(S, 0)$, in the local coordinates coincides with the expression (14.2) obtained by the straightforward computation in local coordinates.

Corollary 14.5. *The connexion form θ is not changed when ω is replaced by a proportional form $u\omega$ with $u|_S \neq 0$.*

If the function h is replaced by a proportional function $h' = vh$, $v|_S \neq 0$, then θ is replaced by the form

$$\theta' = \theta + v^{-1}\,dv, \qquad v|_S \neq 0. \tag{14.5}$$

Consequently, the residue $\operatorname{res}_a \theta$ of the form (14.4) does not depend either on the choice of ω or on the choice of the holomorphic function h defining the local equation of S. □

As a result of this local analysis, we conclude that the collection of local data (14.3) defining a global separatrix $S \subset M$ of a foliation $\mathcal{F}$ on M, defines a collection of meromorphic 1-forms θ_α with the following properties:

$$\theta_\alpha \in \Lambda^1(S \cap U_\alpha) \otimes \mathbf{M}(S \cap U_\alpha), \qquad \theta_\alpha = \theta_\beta + \frac{dv_{\alpha\beta}}{v_{\alpha\beta}} \qquad \text{on } U_{\alpha\beta}. \tag{14.6}$$

Such a collection will be identified later in §17 with a *meromorphic connexion on the normal line bundle* over S.

For convenience we will assume that the index of a holomorphic curve at a *nonsingular point* of a foliation is always zero.

The following proposition explains why the Camacho–Sad index is a proper geometric generalization of the characteristic ratio.

Proposition 14.6. *Let $S = S_1$ be a smooth invariant curve through an elementary singular point of a planar foliation $\mathcal{F}$.*

If the eigenvalue λ_1 of the linearization matrix, associated with the eigenvector tangent to S, is nonzero, then the index of the singularity is equal to the characteristic ratio λ_2/λ_1, where λ_2 is the other eigenvalue, zero or nonzero,

$$i(0, S_1, \mathcal{F}) = \lambda_2/\lambda_1.$$

Proof. The proof immediately follows from the computation in a coordinate system which normalizes the 2-jet of the field to the form given in Table I.1. □

As an immediate corollary, we conclude that for a foliation with two transversal smooth separatrices S_1, S_2 the corresponding indices are reciprocal,

$$i(0, S_1, \mathcal{F}) = \lambda_2/\lambda_1 = [i(0, S_2, \mathcal{F})]^{-1}. \tag{14.7}$$

The index of a hyperbolic invariant curve of a saddle-node is zero. Note, however, that if a saddle-node has a *holomorphic center* manifold, then its index may well be nonzero: for the normal form $\omega = y\,dx - (x^n + ax^{2n-1})\,dy$ the index of the x-axis is equal to

$$\operatorname{res}_{x=0} \frac{dx}{x^n + ax^{2n-1}} = \operatorname{res}_0[x^{-n}(1 - ax^{n-1} + \cdots)] = -a.$$

14C. Total index along a smooth compact invariant curve. Consider a singular foliation $\mathcal{F}$ on a complex 2-dimensional surface M and assume that a smooth compact holomorphic curve S becomes a leaf of $\mathcal{F}$ after deleting from it the singular points $a_1, \dots, a_n$ of the latter.

Theorem 14.7. *Assume that S is a smooth compact holomorphic curve on a complex 2-dimensional manifold M.*

Then for all foliations $\mathcal{F}$ on M which are tangent to S, the sum of indices of $\mathcal{F}$ at all singular points $\operatorname{Sing}\mathcal{F} \cap S$ is the same and depends only on S and M:

$$\sum_{a \in S} i(a, S, \mathcal{F}) = c(S, M). \tag{14.8}$$

Proof. Consider a covering of M by open neighborhoods U_α, the corresponding local equations $\{h_\alpha = 0\}$ for S and two singular holomorphic foliations $\mathcal{F}, \mathcal{F}'$ tangent to S.

These foliations define two collections of the connexion forms, denoted respectively by θ_α and θ'_α, on the open covering of S by the (relatively) open domains $U_\alpha \cap S$.

On the pairwise intersections $U_{\alpha\beta} \cap S$ we have the formulas (14.6) for each collection $\{\theta_\alpha\}$ and $\{\theta'_\alpha\}$ separately, but *with the same terms* $v_{\alpha\beta}$ which are determined solely by the choice of the local equations for S.

Subtracting one representation from the other, we see that the differences $\xi_\alpha = \theta_\alpha - \theta'_\alpha \in \Lambda^1(U_\alpha) \otimes \mathcal{M}(U_\alpha)$ satisfy the identity

$$\xi_\alpha = \xi_\beta + \frac{dv_{\alpha\beta}}{v_{\alpha\beta}} - \frac{dv_{\alpha\beta}}{v_{\alpha\beta}} = \xi_\beta \qquad \text{on } S \cap U_{\alpha\beta}.$$

In other words, the 1-forms ξ_α together correctly define a global meromorphic 1-form $\xi \in \Lambda^1(S) \otimes \mathcal{M}(S)$.

It remains to notice that the sum of residues of any such form is zero, $\sum \operatorname{res}_a \xi = 0$, if S is compact without boundary; see [**For91**]. On the other hand, singularities of ξ are all in the union of the singular loci of the foliations $\mathcal{F}$ and $\mathcal{F}'$, and in each neighborhood U_α,

$$\sum_{a \in U_\alpha} \operatorname{res}_a \xi = \sum_{a \in U_\alpha} \operatorname{res}_a \theta - \sum_{a \in U_\alpha} \operatorname{res}_a \theta' = \sum_{a \in U_\alpha} i(a, S, \mathcal{F}) - i(a, S, \mathcal{F}').$$

Adding these equalities over all singular points, we conclude that

$$0 = \sum_{a \in S} i(a, S, \mathcal{F}) - \sum_{a \in S} i(a, S, \mathcal{F}'),$$

as asserted by the theorem. □

Remark 14.8 (forward reference). This elementary proof is a particular case of the general argument explained in full details in Chapter III (cf. with Theorem 17.33). In geometric terms introduced there, Corollary 14.5 means that the sum of residues of any meromorphic connexion on a line bundle does not depend on the connexion, only on the bundle.

The common number $c(S, M)$ is the degree of this (normal) bundle, a topological invariant of the embedding of S in M.

Theorem 14.7 provides an easy way of computing the total index of any invariant curve on M: it is sufficient to find any foliation tangent to S. For instance, when S can be defined by *one global* equation, $S = \{h = 0\}$ on M, the total index of S at all singularities is zero for any foliation tangent to S. Indeed, the Pfaffian form dh is (by assumption on the smoothness) nonsingular at all points of the level curve S, hence has the total index equal to zero.

Another application of this sort is the following result. Consider the exceptional divisor $\mathbb{E}$ in the complex Möbius band.

Lemma 14.9.

$$c(\mathbb{E}, \mathbb{M}) = -1. \tag{14.9}$$

Corollary 14.10. *The simple blow-up $\mathcal{F}'$ of any nondicritical foliation $\mathcal{F}$ satisfies the identity*

$$\sum_{b \in \mathbb{E}} i(b, \mathbb{E}, \mathcal{F}') = -1. \tag{14.10}$$

Proof of the lemma. Consider the *nonsingular* foliation $dy = 0$ on $(\mathbb{C}^2, 0)$. After the monoidal blow-up $y = xz$ we obtain the foliation $\mathcal{F}'$ on the Möbius band $\mathbb{M}$, defined by the Pfaffian equation $\omega' = z\,dx + x\,dz$. This foliation has a unique nondegenerate saddle at $\{z = x = 0\}$ with the characteristic ratio -1 on the exceptional divisor $\mathbb{E} \subset \mathbb{M}$. By Theorem 14.7, $c(\mathbb{E}, \mathbb{M}) = -1$. □

14D. Index and blow-up. Let S be an integral curve through a singular point $a = 0$ of a singular foliation $\mathcal{F}$ on $(\mathbb{C}^2, 0)$. Consider the blow-up $\mathcal{F}'$ of $\mathcal{F}$ on the complex Möbius band $\mathbb{M}$. Denote by S' the blow-up of the curve S and let $a' = S' \cap \mathbb{E}$.

Lemma 14.11.

$$i(a', S', \mathcal{F}') = i(a, S, \mathcal{F}) - 1. \tag{14.11}$$

Proof. Consider the local coordinates such that $S = \{y = 0\}$. In these local coordinates the linearized Pfaffian equation of $\mathcal{F}$ has the form $dy - y\theta = 0$, where $\theta \in \Lambda^1(S, 0) \otimes \mathcal{M}(S, 0)$ is the connexion form.

Blowing up corresponds to the change of variables $y = xz$; the curve S' in the chart (x, z) is given by the equation $z = 0$. After the change we obtain the Pfaffian equation $x\,dz + z\,dx - xz\theta = 0$, which after division by x takes the form

$$dz - z\theta' = 0, \qquad \theta' = \theta - \frac{dx}{x}.$$

Since the change of variables is linear in z, no additional linearization is required. Computation of residues of θ and θ' at the origin yields (14.11). □

14E. Cano points. Consider a divisor with normal crossings D on a complex 2-dimensional holomorphic manifold M, and a singular foliation $\mathcal{F}$ tangent to D. As before, this means that $D \setminus \operatorname{Sing}\mathcal{F}$ is the union of leaves of the foliation $\mathcal{F}$. The following definition is given in terms of the index of one or two separatrices through a singular point and the partial order (11.3) on the complex numbers.

Definition 14.12. A singular middle point a on the divisor D is called the *Cano middle point* for the foliation $\mathcal{F}$, if

$$i(a, D) \not\geq 0. \tag{14.12}$$

A (singular) corner point $a \in D_+ \cap D_-$ on the intersection of two smooth components is called the *Cano corner point*, if two conditions

$$i(a, D_-) < 0, \tag{14.13}$$

$$i(a, D_+) \not\geq [i(a, D_-)]^{-1} \tag{14.14}$$

hold *simultaneously*. A *Cano point* is a Cano middle point or a Cano corner point.

Note that the two curves $D_\pm$ play asymmetric roles in (14.13)–(14.14), thus being a Cano corner point is the property of the triplet (a, D_+, D_-) *or* the triplet (a, D_-, D_+).

Proposition 14.13.

(1) *A Cano middle point which is elementary, must have an holomorphic separatrix passing through it and not contained in D;*

(2) *a Cano corner point cannot be elementary.*

Proof. Both assertions follow from Proposition 14.6.

1. If the Cano middle point is a saddle-node, then its hyperbolic invariant manifold (curve) cannot locally coincide with D, since in this case the index would be zero.

2. A nondegenerate elementary Cano point must have *two* hyperbolic invariant curves (complex separatrices). Indeed, as soon as the ratio of the two eigenvalues is not a positive real, this is asserted by the Hadamard–Perron Theorem 7.1.

But two transversal separatrices of a *middle* point cannot simultaneously belong to the vanishing divisor.

3. A Cano corner point cannot have zero index along any smooth component, since then the other index $i(D_-)$ must be negative and the inequality $0 = i(D_+) \geqslant 1/i(D_-)$ means violation of the Cano property (14.14).

Thus a saddle-node cannot be a Cano corner point. Similarly, a nondegenerate singularity cannot be a Cano corner point since in this case $i(D_+) = 1/i(D_-)$ in contradiction with (14.14) even if both are negative and (14.13) holds. □

The *raison d'étre* of the Cano points is their persistence under nondicritical blow-up. Consider a singular foliation $\mathcal{F}$ tangent to a divisor D with normal crossings, and let $a \in D \cap \operatorname{Sing} \mathcal{F}$ be a singular point, either corner or a middle with respect to D. After the simple blow-up $\sigma\colon (\mathbb{M}, \mathbb{E}) \to (\mathbb{C}^2, a)$ we obtain the new foliation $\mathcal{F}'$ defined on the neighborhood of $\sigma^{-1}(D) = D' \cup \mathbb{E}$, where the D' is the blow-up of the components of D and $\mathbb{E}$ the exceptional divisor.

Lemma 14.14 (J. Cano [**Can97**]). *If $a \in D$ is a Cano point, then at least one of the singularities that appear by the nondicritical blow-up of a on $\mathbb{E}$, is again a Cano point with respect to the divisor $\sigma^{-1}(D)$.*

Proof. 1. Consider first the case where a is a middle Cano point, i.e., D consists of a single smooth curve through a. In this case the blow-up D' is a single smooth curve on $\mathbb{M}$ which intersects the exceptional divisor $\mathbb{E}$ transversally at the corner point $a' = \mathbb{E} \cap D' \in \sigma^{-1}(D)$. The singular locus for the blow-up foliation $\mathcal{F}'$ on $\mathbb{M}$ consists of the corner point a' and, eventually, several middle points $m_1, \dots, m_k$ on $\mathbb{E}$.

Assume that all singularities of $\mathcal{F}'$ are non-Cano.

Since the points m_j are middle, the negation of (14.12) yields $i(m_j, \mathbb{E}) \geqslant 0$ for all of them. By Corollary 14.10,

$$i(a', \mathbb{E}) = -1 - \sum_j i(m_j, \mathbb{E}) \leqslant -1 < 0.$$

Therefore (14.13) holds for a' and $\mathbb{E}$. Since the corner point a' is also non-Cano, the negation of (14.14) yields $i(a', D') \geqslant 1/i(a', \mathbb{E}) \geqslant -1$. By (14.11),

$$i(a, D) = 1 + i(a', D') \geqslant 1 + 1/i(a', \mathbb{E}) \geqslant 1 - 1 = 0,$$

and we arrive at a contradiction with the assumption that a was a middle Cano point. The contradiction proves that among all singularities $\operatorname{Sing} \mathcal{F}' = \{a', m_1, \dots, m_k\}$ on $\mathbb{E}$ there must be at least one Cano point.

2. Consider the case where $a \in D_- \cap D_+$ is a Cano corner point with $I = i(a, D_-) < 0$. After the blow-up $\sigma^{-1}(D)$ has *two* corner points $a'_\pm = D'_\pm \cap \mathbb{E}$

on the intersection of $\mathbb{E}$ with the blow-ups $D'_\pm$ of the smooth components $D_\pm$ both singular for $\mathcal{F}'$. The foliation $\mathcal{F}'$ may also have one or more middle singular points $m_1, \dots, m_k \in \mathbb{E} \smallsetminus D'$.

Assume that all these singularities are non-Cano points.

Then $i(a'_-, D') = i(a, D_-) - 1 < 0$ and (14.13) holds for a'_- and D'_-. Since a'_- is non-Cano, then by negation of (14.14) we obtain

$$i(a'_-, \mathbb{E}) \geqslant 1/i(a'_-, D'_-) = 1/[i(a, D_-) - 1] = 1/(I - 1).$$

If all middle points are non-Cano, their indices $i(m_j, \mathbb{E})$ are nonnegative and

$$i(a'_+, \mathbb{E}) = -1 - i(a'_-, \mathbb{E}) - \sum i(m_j, \mathbb{E}) \leqslant -1 - 1/(I - 1) = I/(1 - I).$$

This last quantity is negative so (14.13) holds for a'_+ and $\mathbb{E}$. Since a'_+ is non-Cano, the negated (14.14) implies that $i(a'_+, D'_+) \geqslant 1/i(a'_+, \mathbb{E})$. Again by (14.11),

$$i(a, D_+) = 1 + i(a'_+, D'_+) \geqslant 1 + 1/i(a'_+, \mathbb{E}) \geqslant 1 + (1 - I)/I = 1/I.$$

As a result we conclude that $i(a, D_+) \geqslant 1/I = 1/i(a, D_-)$ in contradiction with the assumption that a was a corner Cano point. This contradiction proves that among $\operatorname{Sing} \mathcal{F}' = \{a'_\pm, m_1, \dots, m_k\}$ there must be at least one Cano point. □

14F. Proof of the Camacho–Sad theorem. Consider a singular foliation $\mathcal{F}_0$ at an isolated singular point. By Theorem 8.14, there exists a map $\pi\colon (M, D) \to (\mathbb{C}^2, 0)$ resolving all singularities of $\mathcal{F}$. Expanding π as a composition of simple blow-ups, we obtain a chain of holomorphic 2-dimensional surfaces M_k carrying singular foliations $\mathcal{F}_k$ and simple blow-down maps $\pi_k\colon M_{k+1} \to M_k$ such that the preimage of the origin by any composition $\pi_k \circ \cdots \circ \pi_1$ is a vanishing divisor D_k with normal crossings. At the end we obtain the foliation $\mathcal{F}_n = \pi^*\mathcal{F}$ which has only elementary singularities on $D_n = \pi^{-1}(0)$.

If one of the blow-ups π_k were dicritical, there would be infinitely many leaves of $\mathcal{F}_k$ transversal to D_k, which after blow-down produce complex separatrices. Thus we may consider only the case where all blow-ups π_k are nondicritical.

We claim that in this case at least one singularity of each $\mathcal{F}_k$ is a Cano point. Indeed, the first vanishing divisor $D_1 = \mathbb{E}$ is smooth. If all singularities from $\operatorname{Sing} \mathcal{F}_1$ are non-Cano, then by negation of (14.12) we would have $i(a, \mathbb{E}) \geqslant 0$ for each $a \in \operatorname{Sing} \mathcal{F}_1$. Adding these inequalities, we see that $\sum_{a \in \mathbb{E}} i(a, \mathbb{E}, \mathcal{F}_1) \geqslant 0$ in contradiction with Corollary 14.10. The contradiction proves presence of at least one Cano (middle) point on D_1.

By Lemma 14.14, the π_2-preimage of the Cano point p_1 on D_1 must contain a Cano point $p_2 \in D_2$, either corner or middle. For the same reason the preimage $\pi_3^{-1}(p_2)$ must contain a Cano point $p_3 \in D_3$, *etc.*, until we find a Cano point $p_n \in D_n$. By the assumption on the resolution, $\mathcal{F}_n$ has only elementary points.

By Proposition 14.13, an elementary Cano point has a complex separatrix not contained in D_n. Its blow-down is the complex separatrix of the initial singularity. The proof of the theorem is complete. □

* * *

14G. Local Poincaré problem. The next natural question is the *number* of "different" complex separatrices of a singular foliation $\mathcal{F}$ generated by a holomorphic vector field $F \in \mathcal{D}(\mathbb{C}^2, 0)$. Note that if $C_1, \dots, C_r$ are different irreducible analytic separatrices of $\mathcal{F}$ with the local *reduced* (square-free) equations $\{f_k = 0\}$, $f_k \in \mathcal{O}(\mathbb{C}^2, 0)$, then the union $C = \bigcup_k C_k$ is a reducible analytic separatrix with the square-free local equation $\{f = 0\}$, $f = \prod_k f_k$. Thus the problem of "counting" different separatrices can be transformed into the question about the "maximal" square-free germ f such that the Lie derivative Ff is divisible by f in $\mathcal{O}(\mathbb{C}^2, 0)$, $Ff = 0 \bmod \langle f \rangle$.

In several cases this problem obviously admits no meaningful solution. Assume that the foliation $\mathcal{F}$ is dicritical. Then there is a continuum of smooth irreducible analytic separatrices, thus any finite union of them is also an analytic separatrix, but none such union is maximal.

Definition 14.15. A singular point of a *holomorphic* foliation is called *generalized dicritical*, if it has infinitely many analytic separatrices.

Obviously, the singularity is generalized dicritical if and only if its complete desingularization as described in §8, involves at least one dicritical blow-up (or contains a rational node which after subsequent desingularization involves a dicritical blow-up; see Problem 8.7). Then all leaves that cross transversally the corresponding exceptional divisor, will become analytic separatrices after blowing down.

In all other cases the number of smooth separatrices of the desingularized foliation, and hence that of the initial foliation, is finite, and one may ask about their number. Yet a more appropriate characteristic is not the number, but rather the *order* of the "maximal" invariant curve. It turns out that in the situation when the maximal separatrix exists, its order admits an upper bound in terms of the *order* of the singular foliation.

Remark 14.16 (forward reference). A similar problem arises when F is a polynomial vector field on the plane $\mathbb{C}^2$ and $C = \{f = 0\}$ an invariant algebraic curve. In this case $Ff = fg$, where the *cofactor* $g \in \mathbb{C}[x, y]$ is also polynomial, and the question, usually referred to as the *Poincaré problem*, arises: determine an apriori bound on the *degree* $\deg f$ in terms of $\deg F$. This problem will be discussed in detail in §25**B**.

Definition 14.17. The order of a planar analytic curve $C \subset (\mathbb{C}^2, 0)$ at the point $0 \in C$, denoted by $\operatorname{ord}_0 C$, is the degree of the principal homogeneous terms of any reduced (square-free) convergent Taylor series f which locally defines the germ of the curve.

Though the local equation of a curve *is not* uniquely defined, the order is (Problems 14.1 and 14.3). The order of a curve at its smooth point is 1. Curves of order $\geqslant 2$ exhibit a singularity: a generic curve of order 2 is the transversal intersection of two smooth branches.

The reason why the order is convenient for "counting" analytic curves, is its additivity.

Proposition 14.18. *The order of a finite union of germs of pairwise different analytic curves $C = \bigcup_k C_k$ with isolated pairwise intersections $C_j \cap C_k = \{0\}$ for $j \neq k$, is the sum of their respective orders.* □

A parallel construction can be used to define the order of a singular foliation.

Definition 14.19. The order of a holomorphic foliation $\mathcal{F}$ at an isolated singular point $(\mathbb{C}^2, 0)$ is defined as the order of any holomorphic 1-form ω with an isolated singularity, defining $\mathcal{F}$ by the Pfaffian equation $\omega = 0$,

$$\operatorname{ord}_0 \mathcal{F} = \operatorname{ord}_0 \omega, \qquad \omega \in \Lambda^1(\mathbb{C}^2, 0), \tag{14.15}$$

where the order $\operatorname{ord}_0 \omega$ is the degree ν of the first nonzero homogeneous component of the Taylor expansion $\omega = \omega_\nu + \omega_{\nu+1} + \cdots$.

This order is also defined independently of the choice of local coordinates and the Pfaffian form; it can be equally defined as the order of a holomorphic vector field defining $\mathcal{F}$ locally.

The following result by C. Camacho, A. Lins Neto and P. Sad in [**CLNS84**] gives a sharp inequality between the orders of a foliation and its *maximal* analytic invariant curve.

Theorem 14.20. *Assume that a singular holomorphic foliation $\mathcal{F}$ on $(\mathbb{C}^2, 0)$ is not generalized dicritical, i.e., has at most finitely many separatrices.*

Then the order of any local separatrix $C \subset (\mathbb{C}^2, 0)$ satisfies the inequality

$$\operatorname{ord}_0 C \leqslant \operatorname{ord}_0 \mathcal{F} + 1. \tag{14.16}$$

If the complete desingularization of $\mathcal{F}$ has no saddle-nodes and C is the union of all *separatrices passing through the singular point, then the inequality becomes the equality,*

$$\operatorname{ord}_0 C = \operatorname{ord}_0 \mathcal{F} + 1. \tag{14.17}$$

The proof of Theorem 14.20 is based on thorough investigation of the desingularization process.

14H. Weight of a component of the vanishing divisor. Consider an arbitrary desingularization, a nonconstant holomorphic map $\pi\colon (M,S) \to (\mathbb{C}^2,0)$ holomorphically invertible outside the origin. Denote by $S = \bigcup_{j=1}^m L_j$ the vanishing divisor, $S = \pi^{-1}(0)$, which is the union of projective lines, $S = \bigcup L_j$, $L_i \pitchfork L_j$ for $i \neq j$. We will associate with each component L_j its *weight*, a natural number, which measures the topological complexity of the map π near L_j. As before, we will distinguish between middle (smooth) points of the divisor S, which belong to only one component L_j, and corner points which belong to the intersection of two smooth components.

The construction of the weight starts with the following observation.

Lemma 14.21. *For any holomorphic cross-section $\tau\colon (\mathbb{C}^1,0) \to (M,a)$ to the exceptional divisor S at a noncorner point $a \in L$ of a component $L \subseteq S$, the order of its blow-down curve $\gamma = \pi \circ \tau\colon (\mathbb{C}^1,0) \to (\mathbb{C}^2,0)$, does not depend on the choice of the cross-section as soon as the point a remains noncorner on the same component L.*

This lemma makes the following definition self-consistent.

Definition 14.22. The *weight* $w(L)$ of a component $L \subseteq S = \pi^{-1}(0)$ with respect to the blow-up $\pi\colon (M,S) \to (\mathbb{C}^2,0)$ is the order of any blow-down $\pi \circ \tau$ for an arbitrary cross-section τ to L in M.

Proof of Lemma 14.21. Let τ,τ' be two cross-sections to the same component L at two different nonsingular points a,a', and let $w,w' \in \mathbb{N}$ be the respective orders of the curves,

$$w = \operatorname{ord}_0 \gamma, \quad w' = \operatorname{ord}_0 \gamma', \qquad \gamma = \pi \circ \tau, \quad \gamma' = \pi \circ \tau'.$$

These orders can be described as the numbers of intersection points between the curves γ,γ' and the affine line $\ell_\varepsilon = \{l = \varepsilon\}$ for a generic linear function $l\colon \mathbb{C}^2 \to \mathbb{C}$ and all sufficiently small $0 \neq \varepsilon \in (\mathbb{C},0)$ (cf. with Problem 14.3).

Consider the *nonsingular* foliation $\mathcal{G}$ on $(\mathbb{C}^2,0)$ defined by the Pfaffian equation $dl = 0$. The foliation $\mathcal{G}$ is *integrable* and not generalized dicritical: there is only one leaf ℓ_0 whose closure passes through the "singular" point at the origin. The blow-up $\mathcal{G}' = \pi^*\mathcal{G}$ of this foliation is a singular foliation on M. The singular locus of this foliation can be easily described: after the first simple blow-up there will be a unique hyperbolic singularity $s \in \mathbb{E}$, and by choosing a generic direction l one may without loss of generality assume that this point will never be blown further by π (we will say that the direction l is nonexceptional for π). Thus $\operatorname{Sing}\mathcal{G}'$ consists, besides s, of only corner singularities on the intersections of smooth components of the exceptional divisor.

Both τ and τ' are cross-sections for $\mathcal{G}'$: indeed, by construction both are transversal to *the same leaf* $L \in \mathcal{G}'$ of the latter. Hence the holonomy correspondence (map) between these cross-sections along leaves of $\mathcal{G}'$ is well defined: in particular, any other leaf of $\mathcal{G}'$ that crosses τ, crosses also τ' and vice versa. Hence the number of intersections between $C_\varepsilon = \pi^{-1}(\ell_\varepsilon)$ with each of the cross-sections τ, τ' is the same for all small $\varepsilon \neq 0$. $\square$

Remark 14.23. From Lemma 14.21 it follows that the weight $w(L)$ of a component $L \subset \pi^{-1}(0)$ can be alternatively defined as the order of the restriction of $\pi^* l$ on a cross-section to L at any middle point $a \in L$ for a generic linear function $l\colon \mathbb{C}^2 \to \mathbb{C}$. The genericity condition is the same as in the proof of Lemma 14.21: l must be nonexceptional for π.

The weights of components can be computed recursively if π is represented as a composition of simple blow-ups. Assume that S is a divisor with normal crossings in a holomorphic surface M and $a \in S$ is a smooth or corner point.

After an extra simple blow-up σ with the center at a we obtain another manifold M' with a new vanishing divisor $S' \subset M'$ and the chain of maps

$$(M', S', \mathbb{E}) \xrightarrow{\sigma} (M, S, a) \xrightarrow{\pi} (\mathbb{C}^2, 0, 0).$$

The exceptional divisor $\mathbb{E} \subset M'$ is a "newly created" component of S', while all other components of S' are blow-ups of "old" smooth components of S.

The composition $\pi' = \sigma \circ \pi\colon (M', S') \to (\mathbb{C}^2, 0)$ is a blow-up. Denote by $w'(L')$ the weights of the smooth components of the new vanishing divisor S', to distinguish them from the weights associated with the blow-up map $\pi\colon (M, S) \to (\mathbb{C}^2, 0)$.

The center of the blow-up a can be middle (belonging to only one smooth component of S) or corner (on the intersection of two components). In both cases the weight of its preimage $\mathbb{E} = \sigma^{-1}(a)$ is the sum of the weights of these components. More accurately, we have the following assertion.

Lemma 14.24. *The weights of components $L_j' \subseteq S'$ which are blow-up of the respective components $L_j \subseteq S$, are unchanged, $w'(L_j') = w(L_j)$.*

The weight $w'(\mathbb{E})$ of the exceptional divisor $\mathbb{E} = \sigma^{-1}(a) \subseteq S'$ is equal to the sum of the weights of the components of S passing through a,

$$\text{if} \quad \sigma(\mathbb{E}) = a, \quad \text{then} \quad w'(\mathbb{E}) = \sum_{L\colon a \in L \subseteq S} w(L). \tag{14.18}$$

Proof. The first assertion is obvious, since π' and π are biholomorphically equivalent outside a by definition of the simple blow-up σ.

To prove the second assertion, assume that the local coordinates are chosen so that one or two components of S that contain a, are coordinate axes of suitable local coordinates $(x,y) \in (\mathbb{C}^2,0)$. Consider the pullback π^* of a generic linear function l. We claim that in these coordinates

$$\pi^* l(x,y) = \begin{cases} y^w h(x,y), & a \text{ a middle point}, \\ x^v y^w h(x,y), & a \text{ a corner point}, \end{cases} \qquad h(0,0) \neq 0,$$

where v and w are the weights of the respective components $\{x=0\}$ and $\{y=0\}$ of the vanishing divisor S.

Indeed, divisibility of $\pi^* l$ by the appropriate powers of x and y follows from Remark 14.23. It remains to prove that the corresponding quotient h is in both cases nonvanishing. Yet if $h(0,0)=0$, then the level curve $\pi^* l = 0$ will have a branch not in the vanishing divisor, which is therefore a complex separatrix of the trivial foliation $dl = 0$, which contradicts the choice of l.

To complete the proof of the lemma, note that a generic cross-section $\tau\colon (\mathbb{C},0) \to (M',a')$ to the exceptional divisor $\mathbb{E} \subset M'$ at the point $a' \in \mathbb{E}$ is mapped by σ to a smooth analytic curve through the origin in the (x,y)-plane, transversally crossing one or both axes. Such a curve can be parameterized as $t \mapsto (\alpha t + \cdots, \beta t + \cdots)$, $\alpha\beta \neq 0$. Restricting the function $\pi^* l$ on this curve, we obtain a holomorphic germ of order w or $w+v$ respectively. The lemma follows now from Remark 14.23. □

The simple combinatorial law (14.18) allows us to compute immediately the order of separatrices of a foliation if the complete desingularization is known. If a hyperbolic saddle (with nonpositive characteristic ratio) occurs as a middle singularity on a component L of weight w of the vanishing divisor, then the smooth holomorphic invariant curve transversal to L at this point produces (after blow-down) a local separatrix of order w for the initial foliation.

14I. Weighted sum of vanishing orders. Let $\mathcal{F}$ be the germ of a singular holomorphic foliation on $(\mathbb{C}^2,0)$ generated by a holomorphic vector field $F \in \mathcal{D}(\mathbb{C}^2,0)$ with an isolated singular point at the origin, and γ the germ (at the origin) of an *irreducible* invariant curve (a separatrix) $C = \{f=0\} \subseteq (\mathbb{C}^2,0)$ for $\mathcal{F}$ (as usual, f is assumed to be square-free). Denote by $\gamma\colon (\mathbb{C}^1,0) \to (\mathbb{C}^2,0)$ also the local parametrization established in Theorem 2.26.

Definition 14.25. The *vanishing order* of the foliation $\mathcal{F}$ along an irreducible separatrix γ is the order of the holomorphic vector field $\gamma^* F \in \mathcal{D}(\mathbb{C}^1,0)$, the pullback of F,

$$\varkappa_0(\mathcal{F},\gamma) = \operatorname{ord}_0 \gamma^* F, \qquad F \in \mathcal{D}(\mathbb{C}^2,0). \tag{14.19}$$

Clearly, this definition does not depend either on the arbitrariness of the choices of the field F generating $\mathcal{F}$ or on the parametrization of γ. If γ is a smooth curve, then the vanishing order is equal to the order of zero of the restriction of F on γ. Note that the Pfaffian form ω defining the foliation, cannot be directly used for computation of the vanishing order, since $\gamma^*\omega \equiv 0$.

The vanishing order is obviously invariant by biholomorphisms, yet non-invertible maps (blow-ups) can change it.

Proposition 14.26. *Let $\mathcal{F}$ be the germ of a nondicritical singular holomorphic foliation on $(\mathbb{C}^2, 0)$ and $\mathcal{F}'$ the blow-up of $\mathcal{F}$ by the standard monoidal map $\sigma\colon (\mathbb{M}, \mathbb{E}) \to (\mathbb{C}^2, 0)$. Then*

$$\sum_{a\in\mathbb{E}} \varkappa_a(\mathcal{F}', \mathbb{E}) = \operatorname{ord}_0 \mathcal{F} + 1. \tag{14.20}$$

If $\gamma \subset (\mathbb{C}^2, 0)$ is an irreducible separatrix of $\mathcal{F}$ and γ' its blow-up which intersects $\mathbb{E}$ at a point $a = \gamma' \cap \mathbb{E}$, then

$$\varkappa_a(\mathcal{F}', \gamma') = \varkappa_0(\mathcal{F}, \gamma) - \operatorname{ord}_0 \gamma \cdot (\operatorname{ord}_0 \mathcal{F} - 1). \tag{14.21}$$

Proof of the proposition. Let F be a holomorphic vector field which generates $\mathcal{F}$. Denote $F_\nu = p_\nu(x, y)\frac{\partial}{\partial x} + q_\nu(x, y)\frac{\partial}{\partial y}$ the principal homogeneous component of order $\nu = \operatorname{ord}_0 \mathcal{F}$ of the field F. Without loss of generality we may assume that all singularities of the foliation $\mathcal{F}'$ belong to the affine chart (x, z), $z = y/x$ on the complex Möbius band $\mathbb{M}$. In this chart the vector field F' defining $\mathcal{F}'$, takes the form

$$F' = [q_\nu(1, z) - zp_\nu(1, z) + x(\cdots)]\frac{\partial}{\partial z} + x[p_\nu(1, z) + x(\cdots)]\frac{\partial}{\partial x}.$$

After restriction on the smooth exceptional divisor $\mathbb{E} = \{x = 0\}$ we obtain the polynomial field $Z = h_{\nu+1}(z)\frac{\partial}{\partial z}$, $h_{\nu+1} = q_\nu(1, z) - zp_\nu(1, z)$. The vanishing order of the field Z at each singularity corresponding to a root of the polynomial $h_{\nu+1}$, is equal to the multiplicity of this root. The total order of all roots is equal to $\nu + 1 = \deg h_{\nu+1}$. This proves (14.20).

To prove (14.21), we compare the vector field F' generating its blow-up $\mathcal{F}' = \sigma^*\mathcal{F}$ near $a \in S \cap \gamma'$ with the vector field $F'' = \sigma_*^{-1}F$ obtained by the pullback of F on $\mathbb{M}$. These two fields are tangent to the same foliation $\mathcal{F}'$ and differ by the factor $x^{\nu-1}$, i.e., $F'' = x^{\nu-1}F'$. Thus the restriction of the field F' on the curve γ' differs from that of F'' by the scalar factor $(x \circ \gamma'(t))^{\nu-1}$. It remains to notice that the function $x \circ \gamma'(t)$ itself has the vanishing order at $t = 0$ equal to $\operatorname{ord}_0 \gamma$ by Remark 14.23. □

The first assertion of Proposition 14.26 suggests that the sum of vanishing orders of all singularities obtained by a blow-up $\pi\colon (M, S) \to (\mathbb{C}^2, 0)$ of an isolated singular foliation $\mathcal{F}$, whether complete or not, could be expressed via the order $\operatorname{ord}_0 \mathcal{F}$ of the singularity. This is indeed the case, though the definition of the vanishing order must be modified because of the corner (nonsmooth) points of the vanishing divisor.

Let $S = \bigcup L_j$ be decomposition of the vanishing divisor of the blow-up π and $\mathcal{F}'$ the blow-up of the foliation $\mathcal{F}$ on M. Let $a \in S$ be a point on the vanishing divisor, either "middle" (smooth) or corner (the transversal intersection of two different smooth components). Let $L \ni a$ be a smooth component of the vanishing divisor through a. Denote

$$\kappa_a(\mathcal{F}', L) = \begin{cases} \varkappa_a(\mathcal{F}', L) & \text{if } a \text{ is a middle point,} \\ \varkappa_a(\mathcal{F}', L) - 1 & \text{if } a \text{ is a corner point,} \end{cases} \tag{14.22}$$

where $\varkappa_a(\mathcal{F}', L)$ is the vanishing order introduced in Definition 14.25. In what follows we refer to κ as the vanishing order along a *component of the vanishing divisor*: this should not lead to confusion with the vanishing order along a *separatrix*.

The definition (14.22) seems to be somewhat artificial, yet it leads to the elegant formula for the sum of weighted vanishing orders, due to the same authors [**CLNS84**].

Theorem 14.27. *Assume that $\mathcal{F}$ is a singular holomorphic foliation on $(\mathbb{C}^2, 0)$ and $\pi\colon (M, S) \to (\mathbb{C}^2, 0)$ a blow-up without dicritical components, i.e., the corresponding foliation $\mathcal{F}'$ is everywhere tangent to the vanishing divisor $S = \pi^{-1}(0) = \bigcup L_j$.*

Then the weighted sum of vanishing orders of $\mathcal{F}'$ along all components of the vanishing divisor is by one greater than the order of the initial singularity:

$$\sum_{L_j} \sum_{a \in L_j} w(L_j)\kappa_a(\mathcal{F}', L_j) = \operatorname{ord}_0 \mathcal{F} + 1. \tag{14.23}$$

Though the summation in (14.23) is formally extended over all points of each component L_j, only singularities of $\mathcal{F}'$ may contribute nonzero terms. On the other hand, all corner points appear in this sum twice, contributing the vanishing order along each of the two smooth components passing through them.

Demonstration of Theorem 14.27. The proof goes by induction in the number of simple blow-ups necessary to obtain π.

For the standard monoidal map $\sigma\colon (\mathbb{M}, \mathbb{E}) \to (\mathbb{C}^2, 0)$ the identity (14.23) coincides with (14.20), since the weight of the unique component of the exceptional divisor is equal to 1.

Assume now that for a map $\pi\colon (M, S) \to (\mathbb{C}^2, 0)$ the formula (14.23) is true. Consider an arbitrary point a, a simple blow-up σ of M at a and the composition $\pi'\colon (M',S') \to (\mathbb{C}^2,0)$, $\pi' = \sigma \circ \pi$. The proof will be achieved if we establish the equality (14.23) for the blow-up π'. This will be achieved by showing that the weighted sum which occurs in the left hand side of the equality (14.23), computed for the foliation $\mathcal{G} = \pi^*\mathcal{F}$ and its simple blow-up $\mathcal{G}' = \sigma^*\mathcal{G} = \pi'^*\mathcal{F}$, is the same.

As usual, we have to consider two cases, when a is a middle point on some smooth component, and when it is a corner formed by two components.

1. Assume that $a = (0,0)$ is a middle point of order ν on the separatrix $L = \{y = 0\}$ of the foliation $\mathcal{G} = \pi^*\mathcal{F}$, and assume that the weight of this component is equal to w. Then the contribution of this point to the sum (14.23) for the foliation $\mathcal{G}$ on M is just $w\kappa$, where $\kappa = \varkappa$ is the vanishing order of $\mathcal{F}$ along L.

After the blow-up σ we obtain the foliation $\mathcal{G}' = \sigma^*\mathcal{G}$; there will appear a new component of S', the vanishing divisor $\mathbb{E} = \sigma^{-1}(0)$, and the curve L', the blow-up of L, will cross $\mathbb{E}$ at the new corner singularity of the foliation $\mathcal{G}'$.

The contribution of the new singularities to the sum (14.23) for the blow-up foliation $\mathcal{G}'$ consists of the total sum of weighted vanishing orders along $\mathbb{E}$, plus the vanishing order along L' at a'. The new weights can be immediately computed by Lemma 14.24:

$$w(\mathbb{E}) = w(L) = w, \qquad w(L') = w(L) = w.$$

Note that all singularities of $\mathcal{G}'$ on $\mathbb{E}$, except for a', are middle points, while a' is a corner. Thus the (nonweighted) sum of vanishing orders is equal to

$$\left[\sum_{b\in\mathbb{E}} \varkappa_b(\mathcal{G}',\mathbb{E}) - 1\right] + [(\varkappa_a(\mathcal{G}',a') - 1)] = [\nu + 1 - 1] + [\varkappa - (\nu - 1) - 1] = \varkappa$$

by (14.20), (14.21), (14.22) since all invariant curves are smooth of order 1. Thus the weighted contribution from all singular points on $\mathbb{E}$ is equal to $w\varkappa = w\kappa$, the same as before, that is, the total weighted sum of vanishing orders for $\mathcal{G}$ and $\mathcal{G}' = \sigma^*\mathcal{G}$ remains the same.

2. If the point a is a corner formed by two smooth components L_1 and L_2 of the weights w_1 and w_2 with the vanishing orders $\varkappa_1$ and $\varkappa_2$ respectively, then the corresponding contribution from a to the weighted sum for the foliation $\mathcal{G}$ is equal to

$$w_k\kappa_1 + w_2\kappa_2 = w_k(\varkappa_1 - 1) + w_2(\varkappa_2 - 1). \tag{14.24}$$

This is to be compared with the contribution from all singularities on the newly created exceptional divisor $\mathbb{E}$ which now carries not one, but *two* corner points $a'_{1,2}$ on the intersection of $\mathbb{E}$ with the blow-ups $L'_{1,2}$ of $L_{1,2}$. The weight of $\mathbb{E}$ is equal to $w_1 + w_2$, thus, denoting again by $\nu = \operatorname{ord}_a \mathcal{G}$ the order of the corner singularity, we obtain for the contribution

$$w(\mathbb{E})\sum_{b\in\mathbb{E}}\kappa_b(\mathcal{G}',\mathbb{E})+w(L_1')\kappa_{a_1'}(\mathcal{G}',L_1')+w(L_2')\kappa_{a_2'}(\mathcal{G}',L_2')$$

to the sum (14.23) from all singularities on $\mathbb{E}$ the value

$$(w_1+w_2)[\nu+1-2]+w_1[\varkappa_1-(\nu-1)-1]+w_2[\varkappa_2-(\nu-1)-1], \tag{14.25}$$

again by the same formulas (14.20)–(14.22). One can immediately verify that the expressions (14.24) and (14.25) coincide, which means that the total sum (14.23) remains the same for the foliations $\mathcal{G}$ and $\mathcal{G}'$. This completes the inductive proof of the theorem. □

14J. Minimality of integrable foliations. Theorem 14.20 is proved by comparing the foliation $\mathcal{F}$ with an integrable foliation $\mathcal{H}$ with the same separatrices.

Demonstration of Theorem 14.20. Let $\mathcal{F}$ be a singular holomorphic foliation on $(\mathbb{C}^2,0)$ having only finitely many analytic separatrices, and C one of these separatrices defined by a square-free equation $\{f=0\}$, $f\in\mathcal{O}(\mathbb{C}^2,0)$. Since f is square-free, the Pfaffian form df vanishes only at the origin.

Consider the *integrable* foliation $\mathcal{H}$ defined by the Pfaffian equation $\{df=0\}$ on $(\mathbb{C}^2,0)$. By construction, the curve C is a common separatrix for both $\mathcal{F}$ and $\mathcal{H}$. By definition of the order,

$$\operatorname{ord}_0\mathcal{H}=\operatorname{ord}_0 f-1=\operatorname{ord}_0 C-1, \tag{14.26}$$

since $\mathcal{H}$ can be generated by the Hamiltonian vector field $H=-\frac{\partial f}{\partial y}\frac{\partial}{\partial x}+\frac{\partial f}{\partial x}\frac{\partial}{\partial y}$ with an isolated singularity. The foliation $\mathcal{H}$ is automatically not generalized dicritical: the curve C is its maximal separatrix.

The proof of the theorem will be obtained by comparing the weighted sums of the vanishing orders for $\mathcal{F}$ and $\mathcal{H}$ after blow-up by a map $\pi\colon(M,S)\to(\mathbb{C}^2,0)$ which desingularizes completely *both foliations* $\mathcal{F},\mathcal{H}$. Existence of such a simultaneous desingularization is obvious: first one has to desingularize $\mathcal{F}$ and then to continue by blowing up singularities of $\mathcal{H}$ regardless of the nature of singularities of $\mathcal{F}$ (and their mere presence) at the singularities of $\mathcal{H}$. Since blow-up of an elementary singularity (or a nonsingular point) is again elementary, we end up constructing a map $\pi\colon(M,S)\to(\mathbb{C}^2,0)$ such that blow-up by π of both foliations, denoted $\mathcal{F}'$ and $\mathcal{H}'$ respectively, has only elementary singularities on S. The foliation $\mathcal{H}'$ remains to be analytically integrable, since this property is stable by blow-ups.

An elementary integrable singularity can be only a saddle which has two smooth separatrices with the vanishing order along each one equal to

1. Thus we have for any smooth component $L \subset S$ of the vanishing divisor, the following values:

$$\forall a \in L \subseteq S, \qquad \kappa_a(\mathcal{H}', L) = \begin{cases} 1, & a \text{ a middle } \textit{singularity} \text{ for } \mathcal{H}', \\ 0, & a \text{ a corner point.} \end{cases} \tag{14.27}$$

As for the foliation $\mathcal{F}'$, it must have singularities at every corner point of S and also at any middle singularity of $\mathcal{H}'$. Indeed, through any such middle singularity passes a smooth analytic separatrix common for $\mathcal{H}$ and $\mathcal{F}'$ and transversal to S.

Since the vanishing order at a singular point is at least one, we have the following inequalities:

$$\forall a \in L \subseteq S, \qquad \kappa_a(\mathcal{F}', L) \geqslant \begin{cases} 1, & a \text{ a middle } \textit{singularity} \text{ for } \mathcal{H}', \\ 0, & a \text{ any other point.} \end{cases} \tag{14.28}$$

These inequalities become equalities if the foliation $\mathcal{F}'$ has only hyperbolic singular points on S and the separatrix C used to construct the Hamiltonian foliation, is maximal. Indeed, under these additional assumptions any middle singularity a of $\mathcal{F}'$ on S has another separatrix transversal to S; because of the maximality, it is also a separatrix for $\mathcal{H}'$, which means that a is also singular for $\mathcal{H}'$.

By Theorem 14.27, the weighted sum of vanishing orders is equal to the order of the initial foliations, related to the order of the separatrix C by (14.26). Adding together the equalities (14.27) and inequalities (14.28) with the corresponding weights $w(L)$ over all smooth components L of the vanishing divisor S, we obtain the inequality

$$\operatorname{ord}_0 C = 1 + \operatorname{ord}_0 \mathcal{H} = \sum_L w(L)\kappa_a(\mathcal{H}', L) \leqslant \sum_L w(L)\kappa_a(\mathcal{F}', L) = 1 + \operatorname{ord}_0 \mathcal{F},$$

which becomes equality if the complete desingularization of $\mathcal{F}$ has no saddle-nodes and C is maximal. This proves the theorem. □

The main idea of the proof of Theorem 14.20 is the comparison of two foliations sharing a common separatrix: one arbitrary and another integrable.

Let $\mathcal{F}$ and $\mathcal{H}$ be two singular holomorphic foliations on $(\mathbb{C}^2, 0)$ having nontrivial common leaves. If the foliations are different, then these common leaves can be only analytic separatrices, as was observed by R. Moussu. Indeed, if the wedge product of the two Pfaffian forms defining these foliations is not identically zero, then it can vanish only on an analytic curve which should contain all leaves common for $\mathcal{F}$ and $\mathcal{H}$.

Denote by $C \subseteq (\mathbb{C}^2, 0)$ the common separatrix (in general, reducible) for $\mathcal{F}$ and $\mathcal{H}$. Then for each connected (irreducible) component $\gamma \subseteq C$ one can compare the vanishing orders of $\mathcal{F}$ and $\mathcal{H}$ on this component. It turns out, that the integrable foliations possess the following minimality property: the vanishing order is minimal compared to any other holomorphic foliation with the same separatrix.

Theorem 14.28. *Let $C \subset (\mathbb{C}^2, 0)$ be an analytic curve which is a common separatrix for two singular holomorphic foliations $\mathcal{F}$ and $\mathcal{H}$. Assume that:*

(1) *$\mathcal{F}$ is not generalized dicritical,*

(2) *$\mathcal{H}$ is holomorphically integrable,*

(3) *C is maximal for $\mathcal{H}$.*

Then for any irreducible component $\gamma \subseteq C$ the vanishing orders of $\mathcal{F}$ and $\mathcal{H}$ satisfy the inequality

$$\varkappa_0(\mathcal{F}, \gamma) \geqslant \varkappa_0(\mathcal{H}, \gamma). \tag{14.29}$$

This assertion for elementary singularities (Problem 14.8), is equivalent to the inequalities (14.27)–(14.28) which are the key ingredient in the proof of Theorem 14.20. In turn, the latter theorem is a key tool for demonstration of Theorem 14.28 in the general case. We stress that here and below we deal with the "true" vanishing order $\varkappa_0(\cdot, \gamma)$ and *not* the "modified" vanishing order $\kappa_0(\cdot, \gamma)$ as in (14.22), even for separatrices γ which are "partial" (nonmaximal, e.g., separate sides of corner points).

Proof of Theorem 14.28. Consider a branch of the *simultaneous* desingularization of the foliations $\mathcal{F}$ and $\mathcal{H}$ on the common irreducible curve. This means that we consider a sequence of blow-ups $\pi_1, \dots, \pi_k$, obtained as follows:

(1) $\pi_1 = \sigma_1$ is a standard blow-up at the origin $a_0 = (0) \in \mathbb{C}^2$;

(2) $\pi_{i+1} = \sigma_{i+1} \circ \pi_i$ is the composition of π_i and the standard blow-up σ_{i+1} of a point $a_i \in S_i = \pi_i^{-1}(0)$ on the vanishing divisor S_i.

Each subsequent center of desingularization, the point $a_i \in S_i$ is at the intersection between S_i and the (strict) blow-up γ_i of the curve $\gamma = \gamma_0$ by π_i: since the curves $\gamma_0 = \gamma$, $\gamma_1, \dots, \gamma_k$ are all irreducible, this intersection point is uniquely determined.

Denote by $\mathcal{F}_i$ and $\mathcal{H}_i$ the blow-ups of $\mathcal{F}$ and $\mathcal{H}$ respectively by the maps π_i, $i = 1, \dots, k$. Each γ_i is a common irreducible separatrix through a_i for both foliations $\mathcal{F}_i$ and $\mathcal{H}_i$.

In the assumptions of the lemma, all singularities $\mathcal{F}_i$ at a_i are not generalized dicritical and all $\mathcal{H}_i$ are integrable foliations. Denote

$$\nu_i = \operatorname{ord}_{a_i} \mathcal{F}_i, \quad \mu_i = \operatorname{ord}_{a_i} \mathcal{H}_i, \quad \rho_i = \operatorname{ord}_{a_i} \gamma_i, \qquad i = 1, \dots, k.$$

Iterating the equalities (14.21), we obtain the explicit expressions for the tangency orders

$$\varkappa_0(\mathcal{F},\gamma) = \rho_1(\nu_1 - 1) + \cdots + \rho_k(\nu_k - 1) + \varkappa_{a_k}(\mathcal{F}_k, \gamma_k), \tag{14.30}$$

$$\varkappa_0(\mathcal{H},\gamma) = \rho_1(\mu_1 - 1) + \cdots + \rho_k(\mu_k - 1) + \varkappa_{a_k}(\mathcal{H}_k, \gamma_k). \tag{14.31}$$

By Theorem 14.20, we have the inequalities between all orders,

$$\nu_i \geqslant \mu_i, \qquad i = 0, 1, \dots, k. \tag{14.32}$$

By construction, the terminal point a_k is elementary for both $\mathcal{F}_k$ and $\mathcal{H}_k$. The curve γ_k obtained on the last step, is smooth and transversal to the vanishing divisor S_k, since all analytic separatrices of *elementary* integrable singularities are smooth and transversal to each other. All these properties imply that

$$\varkappa_{a_k}(\mathcal{F}_k, \gamma_k) \geqslant 1 = \varkappa_{a_k}(\mathcal{H}_k, \gamma_k). \tag{14.33}$$

The relations (14.30)–(14.33) taken together prove (14.29). □

Exercises and Problems for §14.

Problem 14.1. Show that the definitions of the order of an analytic curve and of a singular foliation at an isolated singularity are self-consistent.

Exercise 14.2. Let $u \in \mathcal{O}(\mathbb{C}^2, 0)$ be a primitive integral of an holomorphic foliation $\mathcal{F}$. Is it true that $\operatorname{ord}_0 u = \operatorname{ord}_0 \mathcal{F} + 1$?

Problem 14.3. Prove that an order of a planar analytic curve C at a point a is equal to the multiplicity of intersections between C and a *generic* line passing through a. Formulate and prove an analogous result for the order of foliation at an isolated singular point.

Exercise 14.4. Compare the vanishing order of a holomorphic foliation on a local separatrix before and after a *dicritical* blow-up; cf. with Proposition 14.26.

Problem 14.5. Prove that resonant nodes (with integer or inverse integer characteristic ratio) can be desingularized by suitable blow-up.

Problem 14.6. Describe all elementary singularities which are generalized dicritical.

Problem 14.7. Assume that C, the *maximal* separatrix of an integrable foliation $\mathcal{H}$, is the union of pairwise transversal smooth curves. Compute the vanishing order of $\mathcal{H}$ along any such curve.

Problem 14.8. Prove Theorem 14.28 for the case where $\mathcal{H}$ is an elementary singularity.

Problem 14.9. Prove that any derivation $F \in \operatorname{Der} \mathfrak{A}$ always possesses an "eigenvector" $a \in \mathfrak{A}$ such that $Fa = ab$, $b \in \mathfrak{A}$, for the following two algebras,

(1) $\mathfrak{A} = \mathcal{O}(\mathbb{C}^2, 0)$, the algebra of holomorphic germs;

(2) $\mathfrak{A} = \mathbb{C}[[x, y]]$, the algebra of formal series in two variables.

Is this assertion true when $\mathfrak{A} = \mathbb{C}[x, y]$ is the algebra of polynomials? (see §25 for a hint).

Problem 14.10. A separatrix C of a holomorphic foliation $\mathcal{F}$ on $U = (\mathbb{C}^2, 0)$ is called *isolated*, if for some regular point $a \in C \smallsetminus \{0\}$ there exists a small open neighborhood V, $a \in V \subseteq U$, such that the only separatrix of $\mathcal{F}$ which crosses V, is C itself. A separatrix is called *identical*, if any regular point $a \in C \smallsetminus \{0\}$ has a neighborhood V such that any point $b \in V$ belongs to a separatrix of the foliation.

Prove that any separatrix of a singular holomorphic foliation is either isolated, or identical.

Problem 14.11. Prove that the number of isolated separatrices of an isolated singularity of holomorphic foliation is always finite.

Chapter III

Local and global theory of linear systems

Analysis of holomorphic vector fields and analytic foliations beyond the local theory exposed in Chapter I, is very difficult in more than two dimensions. Perhaps the only case where such a study is possible, both locally and globally, is that of (nonautonomous) *linear systems.* These systems exist on a rather special type of holomorphic manifolds, *holomorphic vector bundles.* The latter are "locally cylindric manifolds" made of cylinders (Cartesian products) $U \times \mathbb{C}^n$, $U \subseteq \mathbb{C}$ in the same way the manifolds are made of locally Euclidean charts. In this section we develop local and global theory of linear systems and their singularities.

15. General facts about linear systems

15A. Linear differential equations: Pfaffian, ordinary, matrix. Let T be a Riemann surface, a complex one-dimensional (connected) manifold, which will play the role of the "complex time axis". The particular cases most important for the following are an arbitrarily small neighborhood of a point (punctured or not), subdomains of the complex line $\mathbb{C}$ and the Riemann sphere (the projective line $\mathbb{P}^1$ denoted for brevity by $\mathbb{P}$). We completely ignore in this book the linear systems defined over Riemann surfaces of positive genus.

Let n be a natural number and $\omega_{ij} \in \Lambda^1(T)$, $i,j = 1,\dots,n$ a collection of n^2 holomorphic differential forms on T, arranged as an $n\times n$-*matrix* 1-*form*

$$\Omega = \begin{pmatrix} \omega_{11} & \cdots & \omega_{1n} \\ \vdots & \ddots & \vdots \\ \omega_{n1} & \cdots & \omega_{nn} \end{pmatrix} \in \operatorname{Mat}(n, \Lambda^1(T)).$$

Consider the complex n-space $\mathbb{C}^n$ equipped with the coordinates $x = (x_1,\dots,x_n)$ and the Cartesian product $T\times\mathbb{C}^n$. The one-dimensional distribution on $T\times\mathbb{C}^n$ defined by the common null space of the n holomorphic 1-forms $\theta_i = dx_i - \sum_{j=1}^n \omega_{ij}x_j \in \Lambda^1(T\times\mathbb{C}^n)$, $i = 1,\dots,n$, defines a holomorphic foliation. Its leaves, considered as graphs of holomorphic vector functions $x(\cdot)\colon T\to\mathbb{C}^n$, are *solutions* of the *system of linear Pfaffian differential equations*

$$dx = \Omega x, \qquad \text{or, after expansion,} \qquad dx_i = \sum_{j=1}^n \omega_{ij}x_j. \tag{15.1}$$

Note that in general linear systems have only *multivalued* holomorphic solutions, since the leaves may cross many times the "vertical" lines $\{t = \text{const}\}$ which will be called *fibers* (having in mind future generalizations of the theory).

If $\widetilde{U}\subseteq T$ is a chart on T with a coordinate function $t\colon \widetilde{U}\to\mathbb{C}$ on it with the range $U = t(\widetilde{U})$, then the 1-forms ω_{ij} and the respective matrix Ω can be represented as

$$\omega_{ij} = a_{ij}(t)\,dt, \qquad \text{resp.,} \qquad \Omega = A(t)\,dt,$$

where $a_{ij}(t)$ are holomorphic functions on U together forming the holomorphic matrix function $A(t)\in\operatorname{Mat}(n,\mathcal{O}(U))$. In the chart t the system of Pfaffian equations (15.1) takes the form of a system of n *ordinary* linear differential equations

$$\dot{x}(t) = A(t)x(t), \qquad t\in U\subseteq\mathbb{C}, \quad x = (x_1\dots,x_n)^\top\in\mathbb{C}^n. \tag{15.2}$$

Together with vector solutions of the equations (15.1) or (15.2), it is very useful to consider also their *matrix* solutions. While any rectangular matrix solution with n rows can be considered, the most important is the case of *square* $n\times n$-matrices. To distinguish the matrix equation from the vector one, we will choose the capital letters, writing

$$\begin{aligned} &dX = \Omega X, && && \dot{X}(t) = A(t)X(t), \\ &\Omega\in\operatorname{Mat}(n,\Lambda^1(T)), && \text{or} && A(t)\in\operatorname{Mat}(n,\mathcal{O}(T)), \\ &\text{(Pfaffian)}, && && \text{(ordinary)}, \end{aligned} \tag{15.3}$$

$$X = X(t)\in\operatorname{Mat}(n,\mathcal{O}(t)), \quad \Omega = A(t)\,dt, \quad t\in T.$$

Such matrices represent n-tuples of vector solutions of (15.2) or (15.1).

15B. Fundamental solutions. In general not all solutions of differential equations can be continued over all paths. Linear systems are exceptionally well behaved in this respect.

Proposition 15.1. *Any solution of a differential system* (15.1) *can be continued as an analytic vector function along any simple path* $\gamma \subset T$. *The result of this continuation is a linear map* Δ_γ *between the fibers* τ_a *and* τ_b *at the endpoints* $a, b \in T$ *of the path* γ.

Proof. The zero solution obviously can be continued along any path, thus all sufficiently close solutions can also be continued. Yet because of the linearity, any solution admits continuation.

More precise arguments run as follow. The *null leaf*, the curve $T \times \{0\} \subset T \times \mathbb{C}^n$, is always the leaf of *any* foliation defined by a linear system (15.1). This curve is transversal to each fiber $\tau_a = \{a\} \times \mathbb{C}^n \subset T \times \mathbb{C}^n$.

Let $\gamma \subset T$ be any path connecting two points $a, b \in T$. Then for any foliation $\mathcal{F}$ defined by a system (15.1) the holonomy (correspondence) map $\Delta_\gamma \colon (\tau_a, a) \to (\tau_b, b)$ is always defined between *sufficiently small* neighborhoods of the points a and b on the respective cross-sections τ_a, τ_b, as explained in §2**C**. Yet because of the linearity of the system, the holonomy map is in fact *linear* and hence can be defined between the entire transversals $\tau_a, \tau_b \cong \mathbb{C}^n$.

Indeed, let $x'(t), x''(t)$ be solutions of (15.1) corresponding to the initial conditions $v', v'' \in \tau_a$ and small enough so that both their graphs and the graph of their sum $x(t) = x'(t) + x''(t)$ belong to a sufficiently small tubular neighborhood of the curve $\gamma \subset T \times \{0\} \subset T \times \mathbb{C}^n$. Then the sum $x(t)$ also satisfies (15.1)

$$dx = d(x' + x'') = dx' + dx'' = \Omega x' + \Omega x'' = \Omega(x' + x'') = \Omega x,$$

with the initial condition $v = v' + v''$ and takes the terminal value $\Delta_\gamma(v) = \Delta_\gamma(v') + \Delta_\gamma(v'')$. A similar argument shows also that $\Delta_\gamma(cv) = c\Delta_\gamma(v)$ for all sufficiently small v and cv. □

From now on we will always consider the holonomy maps Δ_γ as globally defined (automatically invertible) linear maps between different copies of $\mathbb{C}^n$ if $a \neq b$ or linear self-maps from $\mathrm{GL}(n, \mathbb{C})$ if $a = b$ and the path γ is a closed loop. As follows from the general properties of the holonomy, the map Δ_γ depends only on the homotopy class of the path γ with the fixed endpoints. In particular, if T is simply connected, then the correspondence map Δ_a^b is well defined for any two endpoints $a, b \in T$. In this case solutions of the linear system obviously form a *linear space* over $\mathbb{C}$.

Definition 15.2. A tuple of solutions is called a *fundamental system of solutions* of the systems (15.1) or (15.2) on a simply connected base T, if it is a basis in the linear space of all such solutions. A *fundamental matrix solution* of the equation (15.3) is a holomorphic matrix function $X\colon T \mapsto \operatorname{Mat}(n,\mathbb{C})$ which is everywhere *nondegenerate*, $\det X(t) \neq 0$ for all $t \in T$.

The following basic result describes the structure of the linear space of solutions of a linear system.

Theorem 15.3. *Any linear system* (15.1) *of order* n *over a simply connected Riemann surface* T *admits an* n*-dimensional linear space of solutions. The "evaluation map"* $x(\cdot) \mapsto x(a)$ *assigning to any solution* $x(\cdot)$ *its value* $x(a)$ *is an isomorphism between this space and the vertical cross-section* $\tau_a = \{a\} \times \mathbb{C}^n$ *for any choice of the point* $a \in T$.

Any n *solutions with linearly independent initial conditions in* τ_a *form a fundamental matrix solution defined on the entire surface* T. *Any two fundamental matrix solutions* $X(t), X'(t)$, *differ by a constant right matrix factor,* $X'(t) = X(t)C$, *where* $C \in \operatorname{Mat}(n,\mathbb{C})$.

Proof. Let $a \in T$ be a fixed base point. Then for any $s \in T$ the holonomy map Δ_a^s between the cross-sections $\tau_a = \{t = a\}$ and $\tau_s = \{t = s\}$ is a uniquely defined linear operator by Proposition 15.1, and for any initial value $v \in \tau_a$ the function

$$s \longmapsto x_v(s) = \Delta_a^s(v), \qquad s \in T,$$

is a globally defined solution to the linear system (15.1). The map $v \mapsto x_v(\cdot)$, inverse to the evaluation map $x(\cdot) \mapsto x(a)$, is a linear operator between two linear spaces: it is surjective by the existence part and injective by the uniqueness part of the assertion of Theorem 1.1.

Choosing any n linearly independent initial values $v_1, \dots, v_n \in \tau_a$ and arranging them into the nondegenerate square matrix V, we may as before use the holonomy Δ_a^s to construct the matrix solution $s \mapsto X_V(s)$. This solution is automatically nondegenerate at every point, since all holonomy operators Δ_a^s are invertible.

To prove the last remaining assertion, consider the quotient $C(t) = X^{-1}(t)X'(t)$ of two fundamental matrix solutions for the same system (15.3). Differentiating this quotient, we obtain

$$\begin{aligned} dC = d(X^{-1}X') &= -X^{-1}\,dX \cdot X^{-1}X' + X^{-1}\,dX' \\ &= -X^{-1}\Omega X' + X^{-1}\Omega X' = 0, \end{aligned}$$

which means that this quotient is a constant invertible matrix C. □

Remark 15.4. An alternative proof of the fact that any solution of a linear system can be continued along any path, can be achieved by purely real arguments. We start with a general a priori growth rate bound characteristic for linear systems.

Lemma 15.5 (Gronwall inequality). *Let $A(\cdot)$ be a continuous matrix function on the real interval $t \in [t_0, t_1] \subset \mathbb{R}$ of explicitly bounded norm,*

$$\forall t \in [t_0, t_1] \qquad A(t) \in \operatorname{Mat}(n, \mathbb{C}), \qquad \|A(t)\| \leqslant c.$$

Then any solution $x(t)$ of the linear system (15.2) *satisfies the inequality*

$$\|x(t)\| \leqslant \|x(t_0)\| \exp(c\,|t - t_0|).$$

Proof. By the limit triangle inequality, $\frac{d}{dt}\|x(t)\| \leqslant \|\frac{d}{dt}x(t)\|$, therefore

$$\tfrac{d}{dt}\|x(t)\| \leqslant \|A(t)\|\,\|x(t)\| \leqslant c\,\|x(t)\|.$$

Therefore the logarithmic derivative $\frac{d}{dt} \ln \|x(t)\|$ is bounded by c everywhere on $[t_0, t_1]$, so that its growth between t_0 and an arbitrary t is no greater than $C\,|t - t_0|$. This immediately implies the inequality for the norm $\|x(t)\|$ itself. □

By the Gronwall inequality, any solution with the initial condition $x_0 \in \mathbb{R}^n$ cannot leave the compact set $K = [t_0, t_1] \times \{\|x\| \leqslant R'\} \subset \mathbb{R}^{1+n}$, $R' = \|x_0\| \exp(R\,|t_1 - t_0|)$, except for the right section $K \cap \{t_1\} \times \mathbb{R}^n$. On the other hand, by one of the fundamental theorems for real ordinary differential equations [**Arn92**], any solution beginning in any compact $K \subset \mathbb{R} \times \mathbb{R}^n$ in the "space-time" can be continued until it reaches the boundary of K. Together with the above argument, this implies that solutions of linear systems on real intervals are always globally defined.

One can use this real theorem to continue solutions along arbitrary parameterized curves in T. It remains to prove that these restricted solutions are in fact holomorphic on T and prove (in the same way as before) that the results are independent of the choice of the curves in case the domain is simply connected.

Remark 15.6 (variation of constants). Solution of nonhomogeneous systems can be reduced to that of homogeneous systems using the method of *variation of constants.* If $X(t)$ is a fundamental matrix solution of the linear system $dX = \Omega X$, then a particular solution of the nonhomogeneous system $dY = \Omega Y + \Theta$, where Θ is a known matrix 1-form on T, is given by the formulas

$$Y(t) = X(t)C(t), \qquad dC = X^{-1}\Theta, \tag{15.4}$$

where solutions of the second equation can be found by immediate integration, $C = \int X^{-1}\Theta$, since any holomorphic 1-form on a simply connected Riemann surface T is exact. Any other solution of the nonhomogeneous system can be obtained as the sum of the particular solution $Y(t)$ and a general solution of the homogeneous system.

15C. Monodromy and holonomy. If the Riemann surface T is not simply connected, the leaves of the foliation $\mathcal{F}$ tangent to the distribution $dx - \Omega x = 0$ on $T \times \mathbb{C}^n$ in general are not graphs of vector functions: they may intersect the fibers $\tau_a = \{t = a\} \times \mathbb{C}^n \subset T \times \mathbb{C}^n$ by more than one point. In the classical language it is said that solutions of the system (15.3) are *multivalued* functions of t. Speaking geometrically, the foliation $\mathfrak{F}$ may have a nontrivial *holonomy group*.

As it was defined in §2**C**, this group associates with any loop $\gamma \in \pi_1(T, a)$ on the null leaf $T \cong L_0 = \{x = 0\} \in \mathfrak{F}$ a linear invertible self-map Δ_γ of the cross-section τ_a. If a coordinate system is fixed on the section τ_a, then Δ_γ becomes a square matrix denoted by F_γ. By construction, for any fundamental matrix solution $X(t)$ of (15.3) the result of its analytic continuation over the loop γ is

$$\Delta_\gamma X(a) = F_\gamma X(a), \qquad F_\gamma \in \mathrm{GL}(\tau_a) \cong \mathrm{GL}(n, \mathbb{C}). \tag{15.5}$$

Note that the linear operators F_γ depend on the choice of the base point a.

A different construction requires a choice of fundamental matrix solution. If $X(t)$ is such a solution, then the result of its analytic continuation along a loop $\gamma \in \pi_1(T, a)$ is another fundamental matrix solution. By Theorem 15.3, there exists a constant matrix $M = M_\gamma$, called the *monodromy matrix*, such that

$$\Delta_\gamma X(t) = X(t) \cdot M_\gamma, \qquad M_\gamma \in \mathrm{GL}(n, \mathbb{C}). \tag{15.6}$$

The monodromy matrices *do not depend on the choice of the base point $a \in T$* in the following sense: the identity (15.6) holds for all points t sufficiently close to a, if we identify the loops $\gamma \in \pi_1(T, t)$ for different base points t sufficiently close to a. On the other hand, the monodromy matrices depend on the choice of a fundamental solution $X(t)$: choosing a different fundamental solution $X'(t) = X(t)C$ results in replacing M_γ by $M'_\gamma = C^{-1} M_\gamma C$.

Both correspondences, the holonomy $\gamma \mapsto F_\gamma$ and the monodromy $\gamma \mapsto M_\gamma$, are linear representations of the fundamental group:

$$\forall \gamma_1, \gamma_2 \in \pi_1(T, a) \qquad M_{\gamma_1 \cdot \gamma_2} = M_{\gamma_2} M_{\gamma_1}, \qquad F_{\gamma_1 \cdot \gamma_2} = F_{\gamma_2} F_{\gamma_1}, \tag{15.7}$$

where $\gamma_1 \cdot \gamma_2$ is the composite loop circumscribing first γ_1 and then γ_2. The two representations are equivalent: as follows from their definitions, the monodromy matrices M_γ *numerically coincide* with the holonomy matrices F_γ for the standard choice of coordinates on $\mathbb{C}^n$ and a special choice of the fundamental solution $X(t)$, normalized by the condition $X(a) = E$. The image of these representations in $\mathrm{GL}(n, \mathbb{C})$ will be referred to as the *monodromy group* of the linear system (15.3).

15D. Gauge transform and gauge equivalence. The special structure of the phase space on which linear systems are defined, restricts the class of admissible transformations. Instead of arbitrary biholomorphisms of the Cartesian product $T \times \mathbb{C}^n$, only maps linear in the "vertical" ("linear") variables are allowed.

More precisely, consider two cylinders $S = T \times \mathbb{C}^n$ and $S' = T' \times \mathbb{C}^n$ over two Riemann surfaces T and T' respectively. Each cylinder is naturally equipped by the projection $\pi\colon S \to T$ (resp., $\pi'\colon S' \to T'$) on the base. A *gauge map*, or *gauge transform* between these two cylinders is a holomorphism $H\colon S \to S'$ which respects these projections and is linear on each fiber $\tau_a = \pi^{-1}(a) = \{a\} \times \mathbb{C}^n$, for any $a \in T$. This means that there exists a holomorphic map $h\colon T \to T'$ such that

$$\begin{aligned} \pi' \circ H = h \circ \pi, &\qquad H|_{\tau_a}\colon \tau_a \to \tau_{h(a)} \quad \text{is linear,} \\ \tau_a = \pi^{-1}(a), &\qquad \tau_{h(a)} = \pi'^{-1}(h(a)). \end{aligned} \tag{15.8}$$

In coordinates a gauge map takes the form

$$(t, x) \mapsto \bigl(h(t), H(t)x\bigr), \qquad H \in \mathrm{GL}(n, \mathcal{O}(T)), \tag{15.9}$$

where $H(\cdot)$ is a holomorphic matrix function ("linear change of the dependent variables"). If necessary, we will specify explicitly that the gauge transform is fibered over the base map h. A gauge map is invertible if and only if h is a biholomorphism and $H(t)$ is invertible for any $t \in T$. In practice we will almost always consider cylinders over the same Riemann surface and use maps fibered over the identity map $h = \mathrm{id}$. The holomorphic invertible matrix function $H = H(t) \in \mathrm{GL}(n, \mathcal{O}(T))$ is called the *conjugacy matrix*.

Gauge equivalence naturally acts on linear systems defined on the respective cylinders. If $X(t)$ is a fundamental matrix solution to a system $dX = \Omega X$ and $H\colon (t, x) \mapsto \bigl(t, H(t)x\bigr)$ a gauge map, then the image $X'(t) = H(t)X(t)$ is a fundamental matrix solution to another linear system $dX' = \Omega' X'$. One can immediately see by expanding the expression for the derivative $dX' \cdot X'^{-1}$, that

$$\Omega' = dH \cdot H^{-1} + H \cdot \Omega \cdot H^{-1}. \tag{15.10}$$

Two linear systems of the same order defined on the same Riemann surface T, are said to be *gauge equivalent* (more precisely, holomorphically gauge equivalent) if they can be transformed into each other by an invertible gauge map.

Clearly, *gauge equivalent systems have isomorphic monodromy and holonomy groups.* The corresponding matrix representations are equivalent. If the two fundamental solutions used to compute the monodromy group are $X(t)$ and $X'(t) = H(t)X(t)$, then the monodromy matrices *coincide* identically. This explains why in many cases the monodromy matrices are more convenient to deal with than the holonomy operators. The holonomy groups for two gauge equivalent systems, if associated with the same base point $a \in T$, are linearly conjugate by the map $H(a) \in \mathrm{GL}(\tau_a)$.

15E. Systems with isolated singularities. A *linear system with singularities* over a Riemann surface T is a singular holomorphic foliation $\mathcal{F}$ on $T \times \mathbb{C}^n$ which coincides with a foliation defined by some linear system (15.3) outside a "small" exceptional set. The exception, nonanalyticity locus Σ of the matrix 1-form Ω, is a subset of the complex one-dimensional base T. It is reasonable to assume that this set is discrete (zero-dimensional), so that $\mathcal{F}$ is defined on the complement $(T \smallsetminus \Sigma) \times \mathbb{C}^n$ to an analytic subset of complex codimension 1.

One can show that in order to be extendable to the complement of an analytic subset of codimension $\geqslant 2$, the matrix elements of the form must be *meromorphic*, i.e., have at worst a finite order pole at every point of the singular locus Σ; cf. with Problem 15.5. In this case, assuming that the singular point is at the origin $t = 0$, the foliation can be generated by a holomorphic vector field

$$F = t^{r+1}\frac{\partial}{\partial t} + \sum_{j=1}^{n} a_{ij}(t)x_j\frac{\partial}{\partial x_j}, \qquad r \in \mathbb{Z}_+, \quad (t,x) \in (\mathbb{C}^1, 0) \times \mathbb{C}^n. \tag{15.11}$$

One can see immediately that the singularities of the foliation after maximal extension are all *isolated* (apriori they should only form a locus of complex codimension $\geqslant 2$) and belong to the closure of the null leaf $T \times \{0\}$ which thus becomes a *common separatrix for all singularities*. This observation explains the special role that the null leaf plays in investigation of the linear systems.

Definition 15.7. A *linear system with singularities* on a Riemann surface T is the singular holomorphic foliation defined by a Pfaffian system (15.3) with a *meromorphic* matrix 1-form $\Omega \in \operatorname{Mat}(n, \mathfrak{M}(T))$ with the polar locus $\Sigma \subset T$. Points of this locus are called *singular points*, or simply *singularities* of the linear system.

Clearly, a linear system with singularities on Σ is a holomorphic (nonsingular) linear system on the Riemann surface $T' = T \smallsetminus \Sigma$. Since T' is not simply connected when $\Sigma \neq \varnothing$, the holonomy (monodromy) group of this restricted system is usually nontrivial.

Definition 15.8. The monodromy group of a linear system with singularities on the Riemann surface is the monodromy group of its restriction on $(T \smallsetminus \Sigma) \times \mathbb{C}^n$.

Example 15.9 (Euler system). A linear system with constant coefficients, $\Omega = A\,dt$, has no singularities on $\mathbb{C}$ but when considered on $\mathbb{P}$, it has a pole of second order at infinity: in the chart $w = 1/t$, it has the Pfaffian matrix $\Omega = -Aw^{-2}dw$.

A simplest nontrivial example of a linear system on $\mathbb{P}$ having the minimal number of *simple* poles, is the *Euler system*,

$$dX = \Omega X, \qquad \Omega = A\,t^{-1}dt, \qquad A \in \operatorname{Mat}(n, \mathbb{C}), \tag{15.12}$$

defined by a single constant matrix A called the *residue*. The singular locus of the system (15.12) on $\mathbb{P}$ consists of two points $\{0, \infty\}$.

The Euler system can be immediately integrated. Consider the logarithmic chart $z = \ln t$ on the universal covering $\mathbb{C}$ of $\mathbb{P} \smallsetminus \Sigma$. In this chart (15.12) becomes a system with constant coefficients $\Omega = A\,dz$, whose fundamental solution is given by the matrix exponent. In the initial chart the exponential solution takes the form

$$X(t) = t^A = \exp(A \ln t), \qquad t \neq 0 \tag{15.13}$$

which is indeed ramified over Σ.

The fundamental group of $\mathbb{P} \smallsetminus \Sigma = \mathbb{C} \smallsetminus \{0\}$ is cyclic, generated by the loop $s \mapsto \exp 2\pi i s$, $s \in [0, 1]$, around the origin. The monodromy matrix of the Euler equation, corresponding to the above constructed fundamental solution, can be easily computed:

$$M = \exp 2\pi i A \tag{15.14}$$

(going around the origin corresponds to choosing a different branch of the logarithm, shifted by $2\pi i$ from the initial one).

The integer index $r \geqslant 0$ determining the order of pole of the matrix Ω at a singular point, is called the *Poincaré rank* of the corresponding singularity. The holomorphic gauge transformations act in a natural way on meromorphic linear systems as well. Obviously, the Poincaré rank is invariant by the gauge equivalence.

Remark 15.10. Once the class of holomorphic linear linear systems is extended to the class of meromorphic linear systems (with singularities), it is natural to extend also the class of admissible gauge transformations, relaxing holomorphy of the matrix function $H(t)$ in (15.9) to meromorphy of $H(t)$ together with its inverse $H^{-1}(t)$.

However, the *meromorphic gauge equivalence* introduced this way, is too strong. In particular, any two systems with poles of first order (i.e., of Poincaré rank zero) are meromorphically gauge equivalent if and only if their monodromy groups are isomorphic, both locally and globally (Problem 16.2). On the other hand, the Poincaré rank is not necessarily preserved by meromorphic gauge transformations.

Exercises and Problems for §15.

Problem 15.1. Prove directly, using Theorem 1.1, that for any point $a \in T$ there exists a small neighborhood $U_\alpha \subset T$ of a such that the linear system (15.3) has a fundamental matrix solution X_α in U, normalized by the condition $X_\alpha(a) = E$.

Exercise 15.2. Assume that the covering $\{U_\alpha\}$, constructed in the previous problem, is finite. Prove that the constant matrix factors $C_{\alpha\beta} = X_\alpha^{-1} X_\beta$ defined on the pairwise intersections $U_{\alpha\beta} = U_\alpha \cap U_\beta$, satisfy the identities

$$C_{\alpha\beta} C_{\beta\alpha} = \mathrm{id}, \qquad C_{\alpha\beta} C_{\beta\gamma} C_{\gamma\alpha} = \mathrm{id} \tag{15.15}$$

on $U_\alpha \cap U_\beta$ and $U_\alpha \cap U_\beta \cap U_\gamma$ respectively (whenever the latter intersections are nonempty).

Problem 15.3. Consider a *simply connected* Riemann surface T and its covering U_α by open domains such that all nonempty pairwise and triple intersections are connected.

Prove that for any collection of matrices $C_{\alpha\beta}$ satisfying the identities (15.15), one can find constant matrices C_α so that $C_{\alpha\beta} = C_\alpha^{-1} C_\beta$ whenever the intersection $U_\alpha \cap U_\beta$ is nonempty.

What happens if T is not simply connected?

Problem 15.4. Derive from Exercise 15.2 and Problem 15.3 that any linear system on a simply connected Riemann surface admits a globally defined fundamental matrix solution.

Problem 15.5. Let $\mathcal{F}$ be a holomorphic foliation generated by a linear system $dx - \Omega x = 0$ on the cylinder $T \times \mathbb{C}^n$ outside the locus $\Sigma \times \mathbb{C}^n$, where $\Sigma \subset T$ is a discrete set.

Prove that this foliation extends as a singular holomorphic foliation with a singular locus of codimension $\geqslant 2$ on $T \times \mathbb{C}^n$ if and only if Ω has a finite order pole at every point of Σ.

Problem 15.6. Prove that any linear system on $\mathbb{P}$ with two simple poles is gauge equivalent to the Euler system (15.12).

Exercise 15.7. Prove that *any* nondegenerate matrix M can be realized as the monodromy of an appropriate Euler system.

Problem 15.8. Let $A_1, \dots, A_m \in \operatorname{Mat}(n, \mathbb{C})$ be *commuting* constant matrices with $A_1 + \cdots + A_m = 0$, and $t_1, \dots, t_m \in \mathbb{C}$ different points.

Prove that the rational 1-form $\Omega = \sum_1^m A_j \frac{dt}{t-t_j}$ defines a singular linear system on $\mathbb{P}^1$. Describe the singular locus and the monodromy group of this system.

Problem 15.9. Prove that two holomorphically or meromorphically gauge equivalent linear systems have isomorphic monodromy groups.

Problem 15.10 (Liouville–Ostrogradskii formula). Let $X(t)$ be a meromorphic matrix function in a domain U with $\det X \not\equiv 0$, and $\Omega = dX \cdot X^{-1}$ the meromorphic matrix 1-form (the "logarithmic derivative" of X). Prove that the scalar 1-form $\operatorname{tr}\Omega$ is the logarithmic derivative of $\det X$, i.e., it satisfies the identity $\operatorname{tr}\Omega = d(\det X) \cdot (\det X)^{-1}$ in U; cf. with (1.16).

16. Local theory of regular singular points and applications

In this section we consider linear systems defined by the *germs* of meromorphic 1-forms $\Omega = A(t)\,dt$ at the origin which is a singular point $t \in (\mathbb{C}, 0)$ of finite order $r+1$. Such a germ will be referred to as a *singular point* of a linear system or simply a *singularity*.

The fundamental group of the punctured neighborhood $(\mathbb{C}, 0) \smallsetminus \{0\}$ is infinite cyclic, generated by a single loop γ_0 going counterclockwise around the origin. The corresponding operator of analytic continuation will be denoted by Δ. In a similar way indication of the loop will be omitted in the notations for the monodromy matrix

$$\Delta X(t) = X(t)M, \qquad M \in \mathrm{GL}(n, \mathbb{C}). \tag{16.1}$$

The notion of gauge equivalence (holomorphic or meromorphic) can be easily localized so that one can speak about (locally) *holomorphically* (*meromorphically*) *equivalent singularities* of linear systems. More precisely, this means that we consider gauge maps of the "infinitely narrow cylinder" $(\mathbb{C}, 0) \times \mathbb{C}^n$ into itself, having the form (15.9), where $h\colon (\mathbb{C}, 0) \to (\mathbb{C}, 0)$ is the identical germ and the invertible matrix function $H(t)$ belongs to $\mathrm{GL}(n, \mathcal{O}(\mathbb{C}, 0))$ or $\mathrm{GL}(n, \mathcal{M}(\mathbb{C}, 0))$ respectively. Furthermore, we can consider *formal* gauge transforms, when H is a *formal matrix series*, $H \in \mathrm{GL}(n, \mathbb{C}[[t]])$. Such formal gauge transforms naturally act on *formal linear systems* defined by formal Pfaffian equations

$$t^{r+1}\,dx = \Omega x, \qquad \Omega = A(t)\,dt, \qquad A \in \mathrm{Mat}(n, \mathbb{C}[[t]]).$$

Our immediate goal in this section is to give a local classification (holomorphic, meromorphic or formal) of singularities of linear systems. It turns out that for a special class of singularities, so-called *regular* singularities, the problem admits complete solution.

16A. Regular singularities. A pole of an analytic function $f(t)$ can be described as an isolated singular point at which the absolute value $|f(t)|$ grows at most polynomially in $|t|^{-1}$ (assuming the singular point at the origin). This moderate growth condition ensures numerous important properties, the most important of them being finiteness of the number of Laurent terms for f. A parallel notion can be defined for singularities of linear systems, but special care has to be exercised because of the multivaluedness of their solutions.

Definition 16.1. A vector or matrix function $X(t)$, eventually ramified at the origin, is said to be of *moderate growth* there if its norm grows at most polynomially in $|t|^{-1}$ as t tends to the origin in any sector $\alpha < \operatorname{Arg} t < \beta$ of opening less than 2π.

$$\|X(t)\| \leqslant C|t|^{-d}, \qquad \text{as } |t| \to 0^+,\ \alpha < \operatorname{Arg} t < \beta, \tag{16.2}$$

for some finite d and C (which a priori may depend on the sector).

Definition 16.2. A singular point of a linear system is called *regular*, if some (hence any) fundamental matrix solution $X(t)$ of the system has moderate growth at this point.

Differentiating the formula $dX = \Omega X$, we see that all derivatives of components of a fundamental solution also grow moderately at a regular singularity, since the meromorphic matrix form has at worst a pole at the singular point. This observation also remains valid for the higher derivatives of any finite order.

Remark 16.3. This terminology is counterintuitive, since "regular" does not mean "nonsingular". However, it is too firmly established to replace the adjective "regular" by "tame" or "moderate" which would be less confusing.

Lemma 16.4. *For a regular singularity, the inverse $X^{-1}(t)$ of any fundamental solution also grows moderately.*

Proof. From the monodromy property (16.1), the determinant $h(t) = \det X(t)$ of any solution, is ramified over the origin:

$$\Delta h(t) = \mu h(t), \qquad \mu = \det M \in \mathbb{C}^*.$$

The function $t^{-\lambda}h(t)$, $\lambda = (2\pi i)^{-1} \ln \mu$, is therefore single-valued, not identically zero and growing moderately as $t \to 0$. Hence it must have a zero or pole of some finite order $k \in \mathbb{Z}$,

$$\det X(t) = t^{k-\lambda}u(t), \qquad u \in \mathcal{O}(\mathbb{C}, 0),\ u(0) \neq 0.$$

Therefore the reciprocal $1/h(t)$ is a function of moderate growth. The inverse X^{-1} can be expressed as $(\det X)^{-1}$ times the adjugate matrix formed by all $(n-1) \times (n-1)$-minors of $X(t)$. Hence $X^{-1}(t)$ also grows moderately. □

Corollary 16.5. *Let $X(t)$ be a* monodromic *matrix function, such that $\Delta X(t) = X(t)M$ for some nondegenerate matrix M. If $X(t)$ has moderate growth, then the "logarithmic derivative" $\Omega = dX \cdot X^{-1}$ is a* meromorphic *matrix 1-form.*

Proof of the corollary. The form Ω is single-valued in the punctured neighborhood of the singular point: $\Delta\Omega = dX \cdot MM^{-1}X^{-1} = \Omega$. Because of the moderate growth, Ω has at worst a pole at this point. □

Lemma 16.6. *If the homogeneous linear system* (15.3) *is regular at the origin and $b(t)$ is a vector function of moderate growth at $t = 0$, then solutions of the nonhomogeneous system $\dot{x} = A(t)x + b(t)$ also have moderate growth.*

Proof. This follows from the explicit formula (15.4). □

Meromorphic classification of regular singularities is very simple. Recall that cyclic matrix groups are isomorphic if their generators M, M' are conjugated by an invertible matrix, $M' = C^{-1}MC$. Isomorphism of the monodromy groups is a necessary condition of any gauge equivalence (Problem 15.9). For meromorphic gauge equivalence there are no other obstructions.

Theorem 16.7 (meromorphic classification of regular singularities). *Any two regular singularities with the same monodromy are meromorphically gauge equivalent to each other.*

In particular, any regular singularity is meromorphically equivalent to an Euler system.

Proof. Without loss of generality we can find two fundamental matrix solutions $X(t)$ and $X'(t)$ for the two systems, which have *the same* monodromy matrix $M \in \mathrm{GL}(n, \mathbb{C})$:

$$\Delta X(t) = X(t)M, \qquad \Delta X'(t) = X'(t)M.$$

Then the matrix ratio $H(t) = X'(t)X^{-1}(t)$ is single-valued in the punctured neighborhood of the singular point, since

$$\Delta H = X'M \cdot M^{-1}X^{-1} = H.$$

Since H has (together with X', X and X^{-1}) moderate growth, we conclude that H is a meromorphic matrix function, holomorphically invertible everywhere outside the singular point. By construction, H as a gauge map conjugates X with $X' = HX$.

Any monodromy matrix M has a matrix logarithm, thus there exists a complex matrix A such that $\exp 2\pi i A = M$. The corresponding Euler system $dX = AX$ with the fundamental matrix solution $X(t) = t^A$ has an arbitrary specified monodromy (Exercise 15.7). □

The explicit formula (15.13) for solutions of the Euler system implies the following corollary.

Corollary 16.8. *Any fundamental matrix solution of a linear system with a regular singularity at the origin, can be represented as*

$$X(t) = H(t)\, t^A, \qquad H \in \mathrm{GL}(n, \mathfrak{M}(\mathbb{C}, 0)),\ A \in \mathrm{Mat}(n, \mathbb{C}) \tag{16.3}$$

with some constant matrix A and meromorphic invertible matrix function (germ) $H(t)$. □

16B. Fuchsian singularities. The problem of detecting regular singularities is in general rather difficult. For instance, Exercise 16.3 shows that no necessary condition of regularity can be given in terms of the Poincaré rank. However, there exists a simple *sufficient* condition of regularity.

Definition 16.9. A singularity is called *Fuchsian*, if its Pfaffian matrix has a simple pole, i.e., if its Poincaré rank r is equal to zero:

$$\Omega = (A_0 + A_1 t + \cdots)\, t^{-1} dt, \qquad A_0, A_1, \cdots \in \operatorname{Mat}(n, \mathbb{C}).$$

The matrix coefficient A_0 before the term t^{-1} is called the *residue* of the Fuchsian singularity.

Theorem 16.10 (L. Sauvage, 1886). *Any Fuchsian singularity is regular.*

Proof. In the logarithmic chart $z = \ln t$ the Fuchsian system defined by a matrix 1-form $\Omega = A(t) \cdot t^{-1}\, dt$ with the first order pole, becomes the linear system defined in some "sufficiently left" half-plane $\{\operatorname{Re} z < -B\}$, $B \gg 0$ by a *bounded* $2\pi i$-periodic matrix 1-form $\Omega' = A(\exp z)\, dz$.

By the Gronwall inequality (Lemma 15.5), in any horizontal semi-strip $\{\alpha < \operatorname{Im} z < \beta,\ \operatorname{Re} z < -B\}$ the norm of the fundamental matrix solution $\|X(z)\|$ grows no faster than $\|X(a)\| \cdot \exp K|z - a|$, where a is a point on the right boundary of the strip and $K = \sup \|A(z)\| < +\infty$. Since the semi-strip is horizontal, $|z - a| \leqslant |\beta - \alpha| + |\operatorname{Re} z - B|$ on it. Combining these estimates and returning to the initial chart $t = \exp z$, we obtain the bound $\|X(t)\| \leqslant \operatorname{const} |t|^{-K}$ in the sector bounded by the rays $\operatorname{Arg} t = \alpha$ and $\operatorname{Arg} t = \beta$. □

Corollary 16.11. *Any Fuchsian singularity is meromorphically gauge equivalent to an Euler system.* □

However, it would be wrong to assume that a Fuchsian system with the residue matrix A_0 is always meromorphically equivalent to the Euler system $t\dot{x} = A_0 x$ with the same matrix A_0 (cf. with Problem 16.6). In the next several subsections we establish a polynomial integrable normal form for the local *holomorphic* classification of Fuchsian systems and prove its integrability, computing explicitly the fundamental solution and the monodromy.

16C. Formal classification of Fuchsian singularities. The first step in the local holomorphic classification of Fuchsian singularities consists of studying *formal equivalence.* Recall that two singularities Ω, Ω' are formally (gauge) equivalent, if there exists a *formal gauge transformation* defined by a formal series $H \in \operatorname{GL}(n, \mathbb{C}[[t]])$ such that the identity (15.10) holds on the level of formal power series.

As was observed by V. I. Arnold, the formal classification of Fuchsian singularities of linear systems can be reduced to the formal classification of *nonlinear* vector fields. Indeed, consider a system of linear equations

$$\dot{x} = t^{-1}(A_0 + tA_1 + t^2 A_2 + \cdots)\,x,$$

and the corresponding meromorphic vector field (15.11) with $r = 0$ in $(\mathbb{C}, 0) \times \mathbb{C}^n$. This analytic field is associated with the system of holomorphic *nonlinear* ordinary differential equations

$$\begin{aligned} \dot{x} &= A_0 x + tA_1 x + \cdots, \\ \dot{t} &= t, \end{aligned} \tag{16.4}$$

having an isolated singular point at the origin $(0, 0)$.

The linearization matrix that is block diagonal with two blocks, one being the residue matrix A_0 of size $n \times n$ and another 1×1-block consisting of the single entry 1. Without loss of generality we can assume that the matrix A_0 is already in the upper-triangular Jordan normal form; its eigenvalues will be denoted $\lambda_1, \dots, \lambda_n$.

By the Poincaré–Dulac theorem, after an appropriate formal transformation one can remove from the system (16.4) all nonresonant terms. Yet the system (16.4), linear in all variables but one, has its specifics. On one hand, only the formal transformations from $\operatorname{Diff}[[\mathbb{C}^{n+1}, 0]]$ preserving the t-variable and linear in x-variables, are allowed by definition of the formal gauge equivalence. On the other hand, all resonant monomials are linear in $x_1, \dots, x_n$ and have the form $t^k x_j \frac{\partial}{\partial x_i}$. Thus the only resonances between the eigenvalues $\lambda_1, \dots, \lambda_n, 1$ that can prevent these monomials to be eliminated from (16.4), should have the form $\lambda_i = \lambda_j + k$ with $k \in \mathbb{Z}_+$; all other eventual resonances correspond to monomials that do not appear in (16.4) from the outset.

Definition 16.12. A Fuchsian singularity with the residue matrix A_0 is *resonant*, if there are two eigenvalues of A_0 that differ by a natural number. Otherwise the Fuchsian singularity is *nonresonant.*

In the resonant case one can immediately describe all resonant monomials linear in x. If $A(t) = \sum_{k=0}^{\infty} t^k A_k$ is the matrix function containing only monomials resonant in the sense of Poincaré–Dulac, then *the matrix coefficient A_k may have nonzero entry at the (i, j)th position only if $\lambda_i - \lambda_j = k$.* If the eigenvalues are arranged in the nonincreasing order in the sense of the partial order (11.3),

$$\lambda_i > \lambda_j \text{ in the sense (11.3)} \implies i < j \qquad \forall i, j, \tag{16.5}$$

then the matrix $A(t)$ is upper-triangular.

This condition formulated in terms of matrix elements, can be reformulated in terms of commutation of special matrices, i.e., as identity in $\mathrm{GL}(n,\mathbb{C})$. Denote by $\Lambda = \operatorname{diag}\{\lambda_1,\dots,\lambda_n\}$ the diagonal part of the residue matrix A_0. For any constant matrix C the conjugacy $C \mapsto t^\Lambda C t^{-\Lambda}$ by the power matrix function t^Λ multiplies (i,j)th element of C by $t^{\lambda_i-\lambda_j}$. Therefore the resonant terms $A_k t^k$ can be described via their commutator with t^Λ as follows:

$$t^\Lambda A_k t^{-\Lambda} = t^k A_k, \qquad k = 1,2,\cdots. \tag{16.6}$$

Definition 16.13. A linear system of equations

$$\dot{x} = t^{-1}(A_0 + tA_1 + \cdots + t^k A_k + \cdots)x, \qquad A_k \in \operatorname{Mat}(n,\mathbb{C}), \tag{16.7}$$

with the residue matrix A_0 is said to be in the *Poincaré–Dulac–Levelt normal form*, if

(1) the residue matrix A_0 is in the upper-triangular block diagonal Jordan form with the diagonal part $\Lambda = \operatorname{diag}\{\lambda_1,\dots,\lambda_n\}$,
(2) the eigenvalues are enumerated in the nonincreasing order: if $\lambda_i - \lambda_j = k \in \mathbb{N}$, then $i < j$, i.e., λ_j, *follows after* λ_i,
(3) the higher order matrix coefficients A_k satisfy the condition (16.6).

Remark 16.14. It is convenient to arrange the eigenvalues $\lambda_1,\dots,\lambda_n$ in such a way that all eigenvalues with integer differences stay together and form a *resonant group* (the order of different, "incomparable" resonant group is not essential). If inside each group the eigenvalues are arranged in the decreasing order, then the matrix $A(t)$ will have block-diagonal form with upper-triangular blocks corresponding to each resonant group. Such an arrangement will be convenient for reasoning.

Note also that the condition (16.6) for systems in the normal form is automatically satisfied also for $k = 0$: the matrix in the Jordan form commutes with its diagonal part. The requirement that A_0 is (nonstrictly) upper-triangular is explicitly stated in Definition 16.13.

In the nonresonant case the Poincaré–Dulac–Levelt form is especially simple: it must be an Euler system with all A_k absent for $k \geqslant 1$. As there can be only finitely many differences between the eigenvalues, the Poincaré–Dulac–Levelt normal form is necessarily *polynomial.*

Theorem 16.15 (Poincaré–Dulac theorem for Fuchsian singularities). *A Fuchsian singularity is formally equivalent to an upper-triangular system in the Poincaré–Dulac–Levelt normal form* (16.7)–(16.6).

In particular, a Fuchsian system with a nonresonant residue matrix A_0 *is formally equivalent to the Euler system* $t\dot{x} = A_0 x$.

The proof of this theorem follows immediately from the Poincaré–Dulac Theorem 4.10. Indeed, Definition 16.13 is specifically designed so that the normal form contains all resonant terms and only them. All other (nonresonant) monomials can be eliminated from the system (16.4).

It remains only to check whether the resulting formal transformation will be linear in x_i and preserving the t-coordinate identically. This can be seen by inspection of the Poincaré–Dulac method: the normalizing map is constructed as an infinite composition of polynomial maps, each preserving the t-coordinate and linear in the x-coordinates, since only monomials of this form may need to be eliminated on each step.

However, the direct proof, largely parallel to the proof of Theorem 4.10, is shorter.

Direct proof of the theorem. To remove nonresonant terms of order $k-1$ from the Fuchsian system whose matrix $A(t) = t^{-1}\sum_{j\geqslant 0} t^j A_j$ has all lower order terms already normalized, consider a gauge equivalence with the conjugacy matrix $H(t) = E+t^k H_k$, whose inverse is $H^{-1}(t) = E-t^k H_k+\cdots$. The transformed system will have the terms of order $(k-1)$ as follows:

$$\begin{aligned} A'(t) &= kt^{k-1}H_k + t^{-1}(E+t^kH_k)A(t)(E-t^kH_k+\cdots) \\ &= A(t) + t^{k-1}(kH_k + H_kA_0 - A_0H_k) + \cdots . \end{aligned}$$

This computation shows that all matrix coefficients $A'_0, \dots, A'_{k-1}$ of $A'(t)$ will remain the same as the matrix coefficients of $A(t)$, while the last matrix coefficient A'_k can be modified by subtracting (or adding) any matrix B representable as $kH + [H, A_0]$ for some $H \in \operatorname{Mat}(n, \mathbb{C})$.

The operator of twisted commutation $\mathbf{T}_k = k + \operatorname{ad}_{A_0} \colon \mathcal{D}_1 \to \mathcal{D}_1$ on the space $\mathcal{D}_1$ of linear vector fields (matrices) is lower triangular[1] in the basis $\{x_i\frac{\partial}{\partial x_j} : 1 \leqslant i, j \leqslant n\}$ by Lemma 4.5 with the eigenvalues $\lambda_i - \lambda_j - k$ on the diagonal. All nonresonant vector monomials $x_i\frac{\partial}{\partial x_j}$ belong to the image of $\mathbf{T}_k$ and hence can be eliminated, as explained in §4**C**.

In other words, the matrices A'_k can be brought into the resonant normal form containing nonzero entries only on (i, j) such that $\lambda_i - \lambda_j = k$. This entails the condition $t^\Lambda A'_k t^{-\Lambda} = t^k A'_k$. The process continues further by induction in k. □

16D. Holomorphic classification of Fuchsian singularities. As we have seen before, convergence of formal normalizing transformations for arbitrary nonlinear vector fields can be a rather delicate issue. However, for Fuchsian systems the situation is ideal.

Theorem 16.16 (holomorphic classification of Fuchsian singularities). *Any formal gauge transformation conjugating two Fuchsian singularities, always converges.*

[1]Triangularity occurs with respect to the order of vector monomials chosen as the Lemma 4.5, regardless of the order of the variables $x_1, \dots, x_n$ themselves.

In particular, any Fuchsian singularity is locally holomorphically equivalent to a polynomial Fuchsian system in the upper-triangular normal form (16.7)–(16.6). *A nonresonant Fuchsian system is holomorphically equivalent to an Euler system.*

The proof of this result can be obtained by several arguments. First, one can modify the proof of the Poincaré normalization Theorem 5.5 to show that the series converges; this is possible since all nonzero "small" denominators $\lambda_i - \lambda_j - k$ are in fact bounded away from zero, exactly as in the Poincaré domain. However, there is an alternative simple proof avoiding all technical difficulties.

We start with a lemma concerning convergence of *formally meromorphic* solutions of Fuchsian systems. By definition, a formally meromorphic solution of a linear system (15.2) is a formal vector Laurent series

$$x(t) = \sum_{t=-d}^{+\infty} t^k x_k, \qquad x_{-d}, \dots, x_0, x_1, \dots \in \mathbb{C}^n, \tag{16.8}$$

satisfying formally the equation (15.2).

Lemma 16.17. *Any formal meromorphic solution of a regular system is convergent and hence truly meromorphic.*

Proof. The property of having only convergent formally meromorphic solutions, is obviously invariant by (truly) meromorphic gauge equivalence of linear systems. As any regular system is meromorphically equivalent to an Euler system (Theorem 16.7), the assertion of the lemma is sufficient to prove only in this particular case.

For an Euler system $t\dot{x} = Ax$, $A \in \operatorname{Mat}(n, \mathbb{C})$, any formal solution (16.8) after substitution gives an infinite number of conditions

$$kx_k = Ax_k, \qquad k = -d, \dots, 0, 1, \dots.$$

Each of these conditions means that the vector coefficient x_k must be either zero or an eigenvector of A with the eigenvalue $k \in \mathbb{Z}$. But as soon as $|k|$ exceeds the spectral radius of A, the second variant becomes impossible and hence all formal meromorphic solutions of the Euler system must be Laurent (vector) polynomials, thus converging. □

Proof of Theorem 16.16. Let $H(t)$ be a formal matrix Taylor series conjugating two Fuchsian singularities $\Omega_i = A_i(t)\, t^{-1} dt$, $i = 1, 2$. By (15.10), it means that

$$t^{-1}A_2 = \dot{H} \cdot H^{-1} + t^{-1} H A_1 H^{-1},$$

implying the "matrix differential equation" for the matrix function $H(t)$,

$$t\dot{H} = A_1 H - H A_2.$$

This is *not* the equation in the form (15.3) with respect to the unknown matrix function H, since both left and right matrix multiplication occurs in the right hand side of this equation. However, it can be expanded to a *system* of n^2 linear ordinary differential equations with respect to all n^2 *entries* of the matrix H. The coefficients of this large $(n^2 \times n^2)$-system are picked from among the entries of $t^{-1}A_i(t)$ and hence exhibit at most a simple pole at the origin.

All this means that $H(t)$ is a formal vector solution to a Fuchsian system of order n^2. By Lemma 16.17, it converges. □

16E. Integrability of the normal form. Similarly to the nonlinear resonant Poincaré–Dulac normal forms, the Poincaré–Dulac–Levelt form is integrable even in the resonant case. This allows us to compute explicitly the corresponding monodromy operator.

Consider the matrix polynomial $A(t) = A_0 + A_1 t + A_2 t^2 + \cdots + A_d t^d \in \mathrm{Mat}(n, \mathbb{C}[t])$ in the Poincaré–Dulac–Levelt normal form, i.e., with the matrix coefficients A_k satisfying the conditions (16.6). The constant matrix difference

$$I = A(1) - \Lambda = (A_0 - \Lambda) + A_1 + \cdots + A_d, \tag{16.9}$$

is called the *characteristic matrix* of the corresponding Poincaré–Dulac–Levelt normal form.

The characteristic matrix I is nilpotent. Indeed, by Remark 16.14 all matrices $A_1, \dots, A_d$ are strictly upper-triangular, and so is $A_0 - \Lambda$. Thus I is a strictly upper-triangular matrix involving contributions from both off-diagonal terms of the Jordan form of the residue A_0 and also from the higher order terms of $A(t)$. Notice that in general Λ and I *do not commute.*

The characteristic matrix I allows us to write explicitly the fundamental matrix solution of a linear system in the normal form.

Lemma 16.18. *The system in the Poincaré–Dulac–Levelt normal form with the characteristic matrix I and the diagonal part of the residue Λ admits the fundamental matrix solution*

$$X(t) = t^{\Lambda} t^{I}. \tag{16.10}$$

Proof. Direct computation yields

$$\begin{aligned} t\dot{X}X^{-1} &= \Lambda + t^{\Lambda} I t^{-\Lambda} = t^{\Lambda}(\Lambda + I)t^{-\Lambda} \\ &= t^{\Lambda}(\Lambda + A_0 - \Lambda + A_1 + \cdots + A_d)t^{-\Lambda} \\ &= (\Lambda + A_0 - \Lambda) + tA_1 + \cdots + t^d A_d \\ &= A(t) \end{aligned}$$

by virtue of (16.6). □

If the matrices t^I and t^Λ were commuting, the monodromy of the system would be equal to the product $\exp(2\pi i\Lambda)\exp(2\pi iI)$ (in any order). It turns out that the formula still holds even if $[t^I, t^\Lambda] \neq 0$.

Corollary 16.19. *The monodromy matrix M of the Poincaré–Levelt normal form is the product of two commuting matrices,*

$$M = \exp(2\pi i\Lambda)\exp(2\pi iI) = \exp(2\pi iI)\exp(2\pi i\Lambda). \tag{16.11}$$

Proof. Recall that a root subspace of an operator A_0 corresponding to an eigenvalue λ is the maximal invariant subspace in $\mathbb{C}^n$, on which $A_0 - \lambda E$ is nilpotent.

The space $\mathbb{C}^n$ is the direct sum of *resonant subspaces*: by definition, each such subspace is the union of the *root subspaces* of all eigenvalues whose difference is an integer number. By construction, each resonant subspace is invariant by A_0. The conditions (16.6) guarantee also that the resonant space is invariant by all higher matrix coefficients A_k, $k = 1, 2, \ldots$.

The exponent of the diagonal term

$$\exp(2\pi i\Lambda) = \operatorname{diag}\{\exp 2\pi i\lambda_1, \ldots, \exp 2\pi i\lambda_n\}$$

is a *scalar* matrix on each resonant subspace of A, because all eigenvalues corresponding to this subspace have integer differences. Hence on each resonant subspace $\exp(2\pi i\Lambda)$ commutes with I, thus also with t^I and $\exp(2\pi iI)$. Ultimately the monodromy operator Δ around the singularity can be expressed as follows:

$$\begin{aligned}\Delta X(t) &= t^\Lambda \exp(2\pi i\Lambda)\, t^I \exp(2\pi iI)\\ &= t^\Lambda t^I \exp(2\pi i\Lambda)\exp(2\pi iI)\\ &= X(t)M,\end{aligned}$$

where M is given by the commuting product (16.11). □

For a nilpotent matrix I the matrix power $t^I = \exp(\ln t I)$ is a matrix polynomial in $\ln t$ of degree $\leqslant n$, hence Lemma 16.18 indeed yields a solution of the system in a closed form. Yet the true power of this result is a description of *invariant subspaces*, coordinate subspaces in $\mathbb{C}^n$ of different dimensions, which are invariant by the flow of the Fuchsian system (15.2).

Corollary 16.20. *Eigenvalues ν_j of the monodromy operator of a Fuchsian singular point may be put in one-to-one correspondence with the eigenvalues λ_j of the residue matrix in such a way that $\nu_j = e^{2\pi i\lambda_j}$.*

Proof. This is an immediate consequence of Lemma 16.18. It may be checked directly for the fundamental matrix (16.10). Choice of another fundamental matrix results in conjugacy of the monodromy operator, hence, leaves the eigenvalues unchanged. □

16F. Further simplification of the normal form for Fuchsian systems. Different Poincaré–Dulac–Levelt normal forms may still be holomorphically equivalent to each other. The problem of *complete* holomorphic classification, including recognition of pairwise nonequivalent normal forms, was only very recently reduced to a purely algebraic problem of classification of upper-triangular matrices by the Heisenberg group.

More precisely, consider the splitting of $\mathbb{C}^n$ into the resonant subspaces as described in Remark 16.14, with eigenvalues in each resonant group following in the nonincreasing order.

The Poincaré–Dulac–Levelt normal form implies that the characteristic matrix I (see (16.9)) of the system (16.7)–(16.6), is block-diagonal with respect to this resonant splitting and each block is upper-triangular.

Theorem 16.21 (Complete holomorphic classification of Fuchsian singularities [**VR04**]). *Two different systems in the Poincaré–Dulac–Levelt normal form are holomorphically equivalent if and only if their characteristic matrices* (16.9) *are conjugated by a constant matrix which is block-diagonal with upper-triangular blocks.*

Proof. Since the residue matrices are invariant, we can assume by (16.10) that both systems are in the normal form with the fundamental matrix solutions

$$X_1(t) = t^{\Lambda} t^{I_1} \qquad \text{and} \qquad X_2(t) = t^{\Lambda} t^{I_2} \tag{16.12}$$

with common diagonal matrix Λ. If these systems are holomorphically conjugate, then for some analytic matrix-function $H(t) \in \mathrm{GL}(n, \mathcal{O}(t))$ and a constant matrix $U \in \mathrm{GL}(n, \mathbb{C})$ we have $H(t)X_1(t) = X_2(t)U$, i.e.,

$$t^{-\Lambda} H(t)\, t^{-\Lambda} = t^{I_2} U\, t^{-I_1}. \tag{16.13}$$

Since I_1, I_2 are nilpotent matrices, the right hand side is a matrix polynomial in $\ln t$, while the left hand side is a converging matrix series involving only different powers of t. The equality is possible only if both parts are in fact *constant*. This constant is necessarily equal to U, as follows from the right hand side of (16.13) computed at $t = 1$:

$$H(t) = t^{\Lambda} U\, t^{-\Lambda}, \qquad t^{I_2} U = U\, t^{I_1}. \tag{16.14}$$

The fact that $H(t)$ involves only nonnegative powers of t, implies that U has the specified block-triangular structure (note that the matrix t^{Λ} is diagonal with entries t^{λ_i}, so that the matrix elements $h_{ij}(t)$ are of the form $u_{ij}\, t^{\lambda_i - \lambda_j}$). The second condition in (16.14) after derivation in t at $t = 0$ yields $I_2 U = U I_1$ which proves that the characteristic matrices I_1 and I_2 are conjugated by U as required.

Conversely, if U is the block-triangular matrix conjugating I_1 with I_2, then it also conjugates t^{I_1} with t^{I_2}. By assumption, $H(t) = t^{\Lambda} U\, t^{-\Lambda}$ is a matrix polynomial (involves only integer nonnegative powers of t), and, inverting the above computations, we conclude that the two Fuchsian systems in the Poincaré–Dulac–Levelt normal forms are holomorphically (in fact, polynomially) conjugated: $H(t)\, t^{\Lambda} t^{I_1} = t^{\Lambda} t^{I_2} U$. □

16G. Nonlocal theory of linear systems on $\mathbb{P}$: the Riemann-Fuchs theorem and the Riemann–Hilbert problem.

At the end of his short life, Riemann asked the following question: *How to describe all the functions that may occur as solutions of a linear differential equations of order n with regular singular points only?* He gave an answer in a short manuscript that was found ten years after his death. By that time Lazarus Fuchs developed the theory of complex linear equations and obtained the same answer.

In modern terms, his answer may be given in the language of systems.

Definition 16.22. A linear system (15.1) on $\mathbb{P}$ is called *regular* if Ω is meromorphic on $\mathbb{P}$, and all the poles of Ω are regular singular points for the system (15.1).

In the affine chart $t \in \mathbb{C}$, a regular system may be written as a system of differential equations

$$\dot{x} = A(t)x \tag{16.15}$$

with the matrix function $A(t)$ meromorphic on $\mathbb{P}$. Any meromorphic (matrix) function on $\mathbb{P}$ is rational. By definition, all the solutions of (16.15) have moderate growth at all the singular points. Moreover, the system has the *monodromy property*: for any fundamental matrix X, the circuit of a singular point a_j results in the right multiplication of X by a nonsingular matrix M_j.

Theorem 16.23 (Riemann-Fuchs theorem). *Any matrix function with a finite number of ramification points or logarithmic singularities which has the monodromy property and moderate growth is a fundamental matrix of some regular system.*

Proof. Let X be the matrix function from the statement of the theorem. Consider the matrix function

$$A = \dot{X}X^{-1}$$

This matrix function is meromorphic on $\mathbb{P}$ by Corollary 16.5. Hence, it is rational. The corresponding linear system is regular because the matrix X has moderate growth at all singular points by assumption. □

The Riemann-Fuchs theorem is very close to Theorem 16.7. The latter theorem, in turn, is based on Riemann's ideas.

16**G**$_1$. *Fuchsian systems and the Riemann-Hilbert problem.*

Definition 16.24. A linear system on the Riemann sphere is *Fuchsian* provided that all its singular points are Fuchsian in sense of Definition 16.9, that is, the matrix coefficient in the right hand side has simple poles only.

The following statement may be checked by the coordinate change $\tau = \frac{1}{t}$.

Proposition 16.25. *The system* (16.15) *has a regular* (*i.e., nonsingular*), *point at infinity if and only if* $\lim_{t\to\infty} tA(t) = 0$. *The same system has a Fuchsian singular point at infinity, provided that the previous relation fails, but* $\lim_{t\to\infty} A(t) = 0$. □

Corollary 16.26. *Any Fuchsian system on the Riemann sphere has the form*

$$\dot{x} = \sum_1^m \frac{A_j}{t - a_j} x.$$

Definition 16.27. The matrices A_j are called the *residue matrices* of the corresponding Fuchsian system.

This definition is a particular case of Definition 16.9:

Denote by M_j the monodromy operator of a_j. Under a proper choice of the loops γ_j around a_j, we have:

$$\gamma_1 \cdots \gamma_m = e,$$

where e is a trivial (contractible) loop on the punctured Riemann sphere $\mathbb{P} \smallsetminus (a_1, \ldots, a_m)$. Then

$$M_m \cdots M_2 M_1 = E. \tag{16.16}$$

Definition 16.28. The *monodromy data* is a collection of m points $a_1, \ldots, a_m$ as above and invertible linear operators $M_1, \ldots, M_m \in \mathrm{GL}(n, \mathbb{C})$ whose product in the specified order is the identity; see (16.16). The monodromy data is *realized* by a Fuchsian system if the monodromy map associated with each loop γ_j coincides with M_j for all $j = 1, \ldots, m$.

When the position of the singular points is irrelevant, we will consider the tuple $(M_1, \ldots, M_m)$ as the monodromy data.

The following problem stayed open for more than hundred years.

Riemann–Hilbert problem: *Is it possible to realize any tuple of invertible operators with the relation* (16.16) *as the monodromy data for some Fuchsian system?*

Hilbert (1900) conjectured the positive answer. Bolibruch (1989) constructed a counterexample.

The history, together with different statements of the problem, is presented in 18, which contains different positive and negative results on the problem. The exposition in 18 is geometrical. Below we present the counterexample of Bolibruch in a purely analytic setting. The presentation is based on [**Ily04**].

16**G**$_2$. *Equations of class B.* Nonrealizable monodromy data will be constructed in a special class defined below.

Definition 16.29. An ordered tuple of nonsingular linear operators is *of class B* provided that their product equals the identity and the following holds:

1. Any of the operators is equivalent to one Jordan cell.

2. The tuple is reducible, that is, all the operators have a common invariant subspace different from zero and the whole space.

Definition 16.30. A Fuchsian system is *of class B* (or a *Bolibruch system*) provided that its monodromy data is of class B.

Theorem 16.31. *The spectrum of any residue matrix of a Bolibruch system is a singleton. In other words, all the eigenvalues of such a matrix are equal.*

Without loss of generality, assume that infinity is not a singular point of the Fuchsian system under consideration. Then the system has the form

$$\dot{z} = \sum \frac{A_j}{t - a_j} z, \qquad \sum A_j = 0. \tag{16.17}$$

The following theorem provides a necessary condition for realizability of a monodromy data of class B by a Fuchsian system.

Theorem 16.32. *Suppose that the monodromy data of class B is realized by a Fuchsian system. Then the product of the eigenvalues of the operators of the tuple equals one (the unique eigenvalue of each operator is taken just once).*

Theorem 16.32 is an immediate consequence of Theorem 16.31. Indeed, in assumptions of Theorem 16.31, the unique eigenvalue ν_j of the monodromy operator M_j equals $e^{2\pi i \lambda_j}$, where λ_j is the unique eigenvalue of the corresponding residue matrix; see Corollary 16.20. Equation (16.17) implies that the sum of traces of the residue matrices vanishes. Theorem 16.31 implies that the sum of the (unique) eigenvalues of these matrices vanishes as well. Hence the product of the eigenvalues of the monodromy operators equals one.

Theorem 16.32 allows us to construct a tuple of three operators, nonrealizable as a monodromy data, the famous Bolibruch counterexample.

Theorem 16.33. *The following three matrices $M_1, M_2, M_3 \in GL(4, \mathbb{C})$,*

$$\begin{pmatrix} 1 & 1 & 0 & 0 \\ 0 & 1 & 1 & 0 \\ 0 & 0 & 1 & 1 \\ 0 & 0 & 0 & 1 \end{pmatrix} \quad \begin{pmatrix} 3 & 1 & 1 & -1 \\ -4 & -1 & 1 & 2 \\ 0 & 0 & 3 & 1 \\ 0 & 0 & -4 & -1 \end{pmatrix} \quad \begin{pmatrix} -1 & 0 & 2 & -1 \\ 4 & -1 & 0 & 1 \\ 0 & 0 & -1 & 0 \\ 0 & 0 & 4 & -1 \end{pmatrix} \tag{16.18}$$

cannot be realized as a monodromy data for a Fuchsian system.

Proof. It is easy to check that the three operators above form a data of class B. Indeed, their product is the identity and the Jordan normal form of each operator consists of one cell. On the other hand, the corresponding eigenvalues are $1, 1, -1$. Hence, these operators cannot be realized as the monodromy of a Fuchsian system by Theorem 16.32. □

In order to complete the justification of the counterexample, we need to prove Theorem 16.31.

16$\mathbf{G}_3$. *Invariant subsystems of the normalized system.*

Proof of Theorem 16.31. Theorem 16.31 is proved below for $n = 4$ under the assumption that the mutual invariant subspace of the operators has dimension 2, as in the example above. In the general case, 4 and 2 should be replaced by n and k, and the ratio $\frac{1}{2}$ below by the ratio $\frac{k}{n}$. The geometric proof of an improved version of Theorem 16.31 may be found at the end of Section 18.

Consider two solutions $x^1(t)$, $x^2(t)$ that span a plane preserved by all the monodromy operators. Take a base point $t = a$ and choose the coordinates in the phase space in such a way that the initial values are $x^1(a) = e_1$, $x^2(a) = e_2$, the first two columns of the identity matrix E. Let X be the fundamental matrix of the system (16.17) with the initial condition $X(a) = E$. In a neighborhood of any singular point the coordinate change $x_j = H_j(t)x$ bringing the original equation to the Poincaré–Dulac–Levelt normal form is defined. By (16.10), the fundamental matrix of the normalized system has the form

$$X_j = (t - a_j)^{\Lambda_j}(t - a_j)^{I_j}, \tag{16.19}$$

where Λ_j is a diagonal matrix. We assume that the eigenvalues λ_{ij} of Λ_j decrease in the sense that $\lambda_{ij} - \lambda_{i+1,j} \in \mathbb{Z}_+$. In this case I_j is an upper-triangular nilpotent matrix. The gauge transformation H_j may be extended to an arbitrary simply connected domain on the punctured Riemann sphere $\mathbb{P} \setminus (a_1, ..., a_m)$. Therefore, we may assume that the domain of any H_j contains a. The following natural question arises: *What is the image of the solutions x^1, x^2 under the gauge transformation H_j?*

The answer is the key point in the proof of Theorem 16.31: *the images of the solutions x^1, x^2 under the gauge transformation H_j belong to the plane spanned by the first two columns of the matrix* (16.19). The reason is that *a linear operator equivalent to one Jordan cell has one and only one invariant space in any dimension*; see Lemma 18.16 below.

In more detail, consider the first two columns of the fundamental matrix X_j. They form a 4×2 matrix $\tilde{V}_j = \begin{pmatrix} V_j \\ 0 \end{pmatrix}$, where V_j and 0 are 2×2 matrices, and

$$V_j(t) = \begin{pmatrix} (t-a_j)^{\lambda_{1,j}} & c_j(t-a_j)^{\lambda_{2,j}} \ln(t-a_j) \\ 0 & (t-a_j)^{\lambda_{2,j}} \end{pmatrix}, \quad \lambda_{1,j} - \lambda_{2,j} = k \in \mathbb{Z}_+. \tag{16.20}$$

When t circuits around a_j counterclockwise, this matrix is multiplied from the right by $m_j = \begin{pmatrix} \nu_j & \alpha_j \\ 0 & \nu_j \end{pmatrix}$, where $\nu_j = e^{2\pi i \lambda_j}$, $\alpha_j = 2\pi i c_j \nu_j a^{-k}$.

Here we do not need an explicit expression for the first two columns of X_j, but only the fact that they span an invariant plane of the monodromy operator. The solutions H_jx^1, H_jx^2 have the same property. But the monodromy operator M_j is equivalent to one Jordan cell. Hence, *these two planes coincide*. Therefore, for any j the equality

$$(x^1(t), x^2(t)) = H_j^{-1}(t)\tilde{V}_j$$

holds in some simply connected domain U_j that contains a and a_j. Denote by $Y(t)$ the upper left 2×2 minor of the matrix $Z(t)$. In the domain U_j it has the form:

$$Y(t) = h_j(t)V_j(t).$$

As $Y(a) = E,\ \det h_j \not\equiv 0$.

This representation of $Y(t)$ completes the main part of the proof of Theorem 16.31. The remaining part of the proof is based on the Liouville–Ostrogradskii formula and on the theorem about the sum of the residues of a meromorphic function. It is presented in the next subsection.

16$\mathbf{G}_4$. *The Wronski determinant of the invariant subsystem.* Consider the Wronskian

$$\det Y(t) = w(t).$$

The matrix-valued function Y on the punctured Riemann sphere has the properties of monodromy and regularity. Namely, under a circuit of a_j it is multiplied by m_j. On the other hand, it has moderate growth at any singular point a_j. Therefore, by the Riemann–Fuchs theorem, Y satisfies a regular linear system of the form $\dot{Y} = PY$. The matrix–valued function P is meromorphic on the Riemann sphere. Hence,

$$\sum_{b\in\mathbb{P}} \text{res}\ {}_b\text{tr}\ P = 0. \tag{16.21}$$

On the other hand, by the Liouville–Ostrogradskii formula,

$$\frac{d}{dt}(\ln w) = \text{tr}\ P. \tag{16.22}$$

Formula (16.20) implies that in U_j we have:

$$w(t) = (t - a_j)^{\lambda_{1,j}+\lambda_{2,j}} \det h_j(t).$$

Hence,

$$\text{res}_{a_j}\frac{d}{dt}(\ln w) \geqslant \lambda_{1,j} + \lambda_{2,j}. \tag{16.23}$$

At all the other points different from the singular ones, the residue of the logarithmic derivative of w is nonnegative because at these points w is holomorphic. Because of the ordering of the eigenvalues of the residue matrices, we have:

$$\lambda_{1,j} + \lambda_{2,j} \geqslant \frac{1}{2}\mathrm{tr}\ A_j \qquad \forall j \in \{1, \dots, m\}.$$

Theorem 16.31 is equivalent to the statement that all the inequalities above are equalities. Suppose, on the contrary, that at least one of these inequalities is strict. Then equations (16.21), (16.22), (16.23) imply:

$$0 = \sum_{b \in \mathbb{P}} \mathrm{res}_b \mathrm{tr}\ P > \frac{1}{2}\mathrm{tr}\ A_j = 0.$$

This contradiction proves Theorem 16.31. □

* * *

We conclude this section by the *Permutation Lemma* due to Bolibruch. It is a powerful tool in the study of linear systems.

16H. Monopoles. For linear systems defined on a compact Riemann curve, in particular, on $\mathbb{P}$, the notion of holomorphic equivalence is meaningless, since there are no holomorphic gauge transformations globally defined over $\mathbb{P}$. However, there is a sufficiently rich class of *meromorphic* gauge transformations holomorphic everywhere except for a single point.

Definition 16.34. A *monopole* is a rational matrix function on the Riemann sphere, holomorphic and holomorphically invertible everywhere except for one point.

If the pole is set at $t = \infty$, then the monopole is a *polynomial* matrix function $\Pi(t)$. Since it must be invertible everywhere except for $t = \infty$, $\det \Pi(t)$ is a polynomial without roots, e.g., a constant. Therefore $\Pi^{-1}(t)$ is also a polynomial matrix. Conversely, if both Π and Π^{-1} are polynomial, then both are monopoles with the pole at infinity:

$$\Pi \in \mathrm{GL}(n, \mathbb{C}[t]) \overset{\text{def}}{\iff} \Pi, \Pi^{-1} \in \mathrm{Mat}(n, \mathbb{C}[t]) \iff \Pi, \Pi^{-1} \text{ are monopoles.}$$

Example 16.35. If $D = \mathrm{diag}\{d_1, \dots, d_n\}$ is a diagonal matrix with nonincreasing integer entries $d_1 \geqslant \cdots \geqslant d_n$ and $\Pi(t)$ a constant or polynomial *upper-triangular* matrix function, then the conjugated matrix $t^D \Pi(t)\, t^{-D}$ will again be an upper-triangular matrix polynomial.

Indeed, after the conjugacy, every nonzero (i, j)th entry of $\Pi(t)$ will be multiplied by $t^{d_i - d_j}$ which is a Taylor monomial for all $i \leqslant j$.

In particular, if C is a constant upper-triangular matrix and D as above, then $t^D C\, t^{-D}$ is a monopole, since its determinant is a nonzero constant.

If a point $t = \infty$ is a singularity for a linear system (15.3), then after a monopole transform the singularity remains regular but may cease to be Fuchsian. Yet the same argument suggests that a regular non-Fuchsian singularity can sometimes be made Fuchsian by a monopole gauge transform. This modification cannot affect the other singularities, since the monopole map is holomorphically invertible there.

For instance, consider a linear system with a *nonsingular* point at infinity. Let $H = H(t)$ be the germ of a fundamental matrix solution near this point. The matrix function $t^D H(t)$ is a fundamental matrix solution to another system which in general have a singularity at infinity. This singularity will even be non-Fuchsian (though obviously regular). Indeed, the matrix form Ω' of this system is obtained from the nonsingular matrix form $\Omega = dH \cdot H^{-1}$ by the gauge transform

$$\Omega' = t^{-1} D\, dt + t^D \Omega t^{-D}. \tag{16.24}$$

The first term is always Fuchsian, yet the second will in general be non-Fuchsian at infinity unless Ω has very special properties depending on D. For instance, if the sequence of the integers $d_1, \dots, d_n$ is nondecreasing and Ω is upper-triangular, then Ω' is obviously Fuchsian.

However, it turns out that in this special case the regular non-Fuchsian singularity Ω' can be brought back into the Fuchsian form by a suitable monopole transform. The following result appears in [**Bol92**].

Lemma 16.36 (Permutation Lemma). *Any matrix germ at $t = \infty$ of the form $t^D H(t)$ with a holomorphically invertible factor $H(t) \in \mathrm{GL}(n, \mathcal{O}(\mathbb{P}, \infty))$ and an integer diagonal matrix D is monopole equivalent to a germ of the form $H'(t)\, t^{D'}$ with $H'(t)$ also holomorphic and invertible at infinity and D' a diagonal matrix with the same diagonal entries d_i, eventually in a permuted order.*

In other words, there exists a monopole $\Pi \in \mathrm{GL}(n, \mathbb{C}[t])$ such that

$$\Pi \cdot t^D \cdot H = H' \cdot t^{D'}. \tag{16.25}$$

Proof of Lemma 16.36. We start by proving the lemma in a simple particular case, and then reduce the general case to the former one by a series of suitable gauge transformations.

1. Consider first the case where the (constant) matrix $H(\infty)$ has all nonzero principal (upper-left) minors, while the diagonal matrix D is of the form $\left(\begin{smallmatrix} 0 & \\ & \nu E \end{smallmatrix}\right) = \operatorname{diag}\{0, \dots, 0, \nu, \dots, \nu\}$, $\nu > 0$. This means that D is block diagonal with only two distinct eigenvalues and they are arranged in the ascending order. We show that in this case the meromorphic germ $R(t) = t^D H(t)\, t^{-D}$ is monopole equivalent to a holomorphic germ $H'(t)$ that is automatically nondegenerate at infinity. This is a particular case of the lemma, when $D' = D$.

More precisely, we will show that in this case the monopole transformation can be chosen lower triangular with the block structure $\left(\begin{smallmatrix} E & 0 \\ * & E \end{smallmatrix}\right)$, so that the upper left blocks of

$H'(t)$ and $H(t)$ are the same. Denoting the appropriate blocks of $H(t)$ as follows yields:

$$H(t) = \begin{pmatrix} M(t) & N(t) \\ P(t) & Q(t) \end{pmatrix}, \qquad R(t) = t^D H(t)\, t^{-D} = \begin{pmatrix} M(t) & t^{-\nu} N(t) \\ t^{\nu} P(t) & Q(t) \end{pmatrix}.$$

The upper left block $M(t)$ is nondegenerate by assumption. The only elements that may have poles at infinity, are these of the lower right block $t^{\nu} P$. We show how these poles can be removed by lower triangular monopole transformations.

The principal Laurent part of the matrix $t^{\nu} P(t)$ at infinity can be expanded as

$$t^{\nu} P(t) = t^{\nu} P_{\nu} + t^{\nu-1} P_{\nu-1} + \cdots + t\, P_1 + P_0,$$

with constant rectangular matrices P_i. Linear combinations of rows of the nondegenerate matrix $M(0)$ generate any row of the appropriate length, in particular, any row of the constant matrix P_{ν}. Subtracting these combinations with the rational factor t^{ν} allows us to eliminate from $t^{\nu} P(t)$ all terms with poles of order ν at infinity. Being an elementary row operation, this corresponds to the left multiplication by an appropriate lower triangular monopole matrix $\Pi^{\nu}(t)$, polynomial in t and with determinant 1. Since elements of the upper right block of $R(t)$ were all divisible by $t^{-\nu}$, the lower right block of $R(t)$ will remain holomorphic after multiplication by $\Pi^{\nu}(t)$: one can easily see that

$$\Pi^{\nu} = \begin{pmatrix} E & 0 \\ -t^{\nu} P_{\nu} M^{-1} & E \end{pmatrix}, \qquad \Pi^{\nu} R = \begin{pmatrix} M & t^{-\nu} N \\ t^{\nu-1} P_{\nu-1} + \cdots & Q + P_{\nu} M^{-1} N \end{pmatrix}$$

with holomorphic matrices M, N, Q (the first of them invertible) and constant matrix P_{ν}. Note that the order of pole in the lower left corner is at most $\nu - 1$.

Iterating this step, by suitable left multiplications one can eliminate consecutively all terms with poles of order $\nu-1$, $\nu-2$ and so on until the constant terms will be eliminated. The overall product $\Pi^0(t)\Pi^1(t) \cdots \Pi^{\nu}(t)$ of all monopoles used in the process, will again be a monopole at infinity (polynomial in t), also lower triangular. This completes the proof in the particular case where the matrix D has only two distinct eigenvalues $0 < \nu$ ordered in the ascending (nondecreasing) order.

2. Any diagonal matrix D with ascending integer eigenvalues $d_1 \leqslant \cdots \leqslant d_n$ can be represented as a sum of several matrices of the type considered above. More precisely, we can always represent such D as the sum

$$D = D_0 + D_1 + \cdots + D_m, \qquad m \leqslant n-1, \tag{16.26}$$

so that D_0 is scalar (diagonal with a single eigenvalue) and each D_i with $i > 1$ is block diagonal with two eigenvalues 0 and $\nu_i > 0$ arranged in the ascending order. To see this, consider the monotonous integer function $i \mapsto d_i$, $i \in \{1, \dots, n\}$. This function can be represented as a sum of $m-1$ "step functions" (nonincreasing integer functions assuming only two values, one of them zero) plus a constant term. Indeed, the first difference $i \mapsto d_{i+1} - d_i$ is a nonnegative integer function which can be represented as the sum of $\leqslant m-1$ "delta-functions" taking a positive nonzero value only once. Taking "primitives" of these "delta-functions" (the sums restoring integer functions from their differences) and adding the "constant of integration" proves the claim: each step function can be considered as a diagonal matrix D_i with one zero and one positive eigenvalue.

Since the powers t^{D_i} commute between themselves, the terms in the representation (16.26) can be arranged so that the matrices with biggest-size upper-left (zero) block come last.

3. Splitting (16.26) permits us to prove the assertion of the lemma for every product $t^D H(t)$ where the diagonal matrix is ascending (its eigenvalues nondecreasing) and $H(t)$ having nonzero principal minors. In this case one can also choose $D = D'$. Indeed, in the representation

$$t^{D_0} t^{D_1} \cdots t^{D_m} H(t)$$

the term t^{D_m} can be permuted with $H(t)$ if the appropriate monopole $\Pi(t)$ is inserted between $t^{D_{m-1}}$ and t^{D_m}, as shown on Step 1. To do this, the whole product must be multiplied from the left by the matrix function

$$\Pi'(t) = t^{D_0+\cdots+D_{m-1}}\,\Pi(t)\,t^{-(D_0+\cdots+D_{m-1})}.$$

But since both D and all matrices D_i were ascending and $\Pi(t)$ lower triangular, the matrix $\Pi'(t)$ will again be a monopole; cf. with Example 16.35. By construction,

$$\Pi'(t)\,t^D H(t) = t^{D_0+\cdots+D_{m-1}}\,H'(t)\,t^{D_m},$$

and the upper-left corner of $H'(t)$ will coincide with that of $H(t)$. The process can be clearly continued by induction, since on the next step one may require nondegeneracy of only smaller or same size upper-left minors of $H(t)$, thus preserving inductively the assumptions required in Step 1. After m permutations all terms t^{D_i} will appear to the right from the holomorphically invertible term, while the scalar term t^{D_0} commutes with everything.

4. For an arbitrary nondegenerate $H(\infty)$, the required condition on principal minors can always be achieved by a suitable permutation of columns, that is, multiplying $t^D H$ from the right by a suitable constant permutation matrix P. By Step 3, $t^D H(t)P$ is monopole equivalent to $H'(t)\,t^D$ for any *ascending* matrix D. Therefore $t^D H(t)$ is monopole equivalent to $H'(t)P^{-1}\cdot P t^D P^{-1} = H''(t)t^{D'}$, where $D' = PDP^{-1}$ is a diagonal matrix with entries obtained by the permutation of entries of D.

5. The last remaining assumption that D is ascending, can also be removed by a suitable permutation of rows. Indeed, if P is a permutation matrix such that the entries of $D' = PDP^{-1}$ are ascending, then $t^D H$ is monopole equivalent to $t^{D'}H'$ with H' holomorphically invertible at infinity:

$$P\cdot t^D H = P t^D P^{-1}\cdot PH = t^{D'}H'.$$

By Step 4, $t^{D'}H'$ is monopole equivalent to $H''\,t^{D''}$ as required.

This proves Lemma 16.36 in full generality. □

Exercises and Problems for §16.

Exercise 16.1. Prove that any linear system at a *nonsingular* point is holomorphically gauge equivalent to the trivial (identically zero) system defined by the matrix 1-form $\Omega' \equiv 0$.

Problem 16.2. Prove that any two regular (in particular, Fuchsian) linear systems on $\mathbb{P}$ with the same singular loci are meromorphically gauge equivalent if and only if their monodromy groups are isomorphic.

Exercise 16.3. Compare the Poincaré ranks of a nonsingular point 0 and its meromorphic gauge transform by a diagonal matrix $H(t) = t^D = \operatorname{diag}\{t^{d_1},\dots,t^{d_n}\}$.

Exercise 16.4. Show that the definition of the residue matrix of a Fuchsian singularity does not depend on the choice of the chart t.

Problem 16.5. Prove that any algebraic function $x = x(t)$ of one complex variable t, defined by a polynomial equation $P(x,t) = 0$, satisfies (say, as the first component) a regular linear system over $\mathbb{P}$ of rank at most n.

Problem 16.6. Let $\Delta_a\colon \tau_a \to \tau_a$ be the holonomy operator corresponding to a simple positive loop around the origin beginning and ending at a nonsingular point $a \neq 0$ for a Fuchsian system $t\dot{x} = (A_0 + tA_1 + \cdots)x$. Prove that Δ_a depends analytically on a as $a \neq 0$, extends (as an analytic matrix function) at the origin $a = 0$ and the limit Δ_0 is equal to $\exp 2\pi i A_0$. Show that the operators Δ_a are conjugate to each other for all $a \neq 0$, but not necessarily to Δ_0.

Problem 16.7. Bring to the Poincaré–Dulac–Levelt normal form the linear systems with the matrix 1-form $\Omega = A(t)\frac{dt}{t}$, where $A(t)$ is one of the following matrix functions,

$$\begin{pmatrix} 1 & \sin 2t \\ & 2 \end{pmatrix}, \qquad \begin{pmatrix} 1 & e^t - 1 & t^3 \\ & 2 & t^2 \\ & & 3 \end{pmatrix}.$$

Problem 16.8. Prove that for any resonant tuple of the form $\boldsymbol{\lambda}^1 = (\lambda, \lambda + k)$ or $\boldsymbol{\lambda}^2 = (\lambda, \lambda + k, \lambda + k + m)$ there exists but a finite number of normal forms of equations with a Fuchsian singular point, for which the residue matrix has the spectrum $\boldsymbol{\lambda}^1$ or $\boldsymbol{\lambda}^2$.

17. Global theory of linear systems: holomorphic vector bundles and meromorphic connexions

Linear systems appear in a natural way by *linearization* of arbitrary complex one-dimensional holomorphic foliations along particular leaves (usually, separatrices). Example of such linearization for foliations on complex surfaces already appeared in the computation of the vanishing holonomy group in §10**D** and in slightly more general context in §14**B**. Both these examples suggest that, while locally a linear system "lives" on cylinders which are Cartesian products of the base leaf L by a complex linear space of the complementary dimension, globally the situation may be nontrivial. In particular, it may be impossible to define the linearized system globally over L by a single meromorphic 1-form (matrix or even scalar): the nontrivial relationship between 1-forms θ_1 and ϑ_1 in (10.9) shows that the linearized system is defined on a more general object than the "simple" Cartesian product $\mathbb{E} \times \mathbb{C}$. This object is called (holomorphic) *vector bundle*.

The material exposed in this section is rather standard and can be found in numerous sources, of which we recommend the books [**For91**, §29, §30] and [**Bol00**], but also [**GH78**, §0.5] and [**Wel80**, §2].

17A. Holomorphic vector bundles. A real or complex vector bundle of rank n over a topological manifold T (the "horizontal" base) is a topological manifold which is "built" from Cartesian cylinders $U_\alpha \times \mathbb{R}^n$ or $U_\alpha \times \mathbb{C}^n$ respectively, where U_α is a chart on T, in the same way as the base manifold is built from the charts U_α themselves. The added value is the linear structure along the "vertical" fibers $\{a\} \times \mathbb{R}^n$, resp., $\{a\} \times \mathbb{C}^n$. We will be interested only in the complex case. The formal definition looks as follows.

Definition 17.1. Let $\pi \colon S \to T$ be a continuous map between two topological spaces. A map Φ is called a *local trivialization* (sometimes *trivializing chart*, or simply *trivialization*) of π over an open subset $U \subseteq T$, if $\Phi \colon \pi^{-1}(U) \to U \times \mathbb{C}^n$ is a homeomorphism which conjugates π with the projection of the Cartesian product (cylinder) $\pi_0 \colon U \times \mathbb{C}^n \to U$ on the first component, so that $\pi_0 \circ \Phi = \pi$.

Trivializations play the role of special coordinate charts keeping track of the linear structure on the fibers.

Definition 17.2. The topological space S together with a continuous map (projection) $\pi \colon S \to T$ is called a *topological complex vector bundle* or rank n over a topological space T (called the *base*), if:

(1) for any point $a \in T$ of the base there exists an open neighborhood $U_\alpha \ni a$ and a trivialization Φ_α of π over U_α,

(2) the family of trivializations $\{\Phi_\alpha\}$ respects the linear structure of the *fibers* $\pi^{-1}(a)$: if Φ_α, Φ_β are two trivializations of π over two open domains with the nonempty intersection $U_{\alpha\beta} = U_\alpha \cap U_\beta$, then the *transition map* between them is a gauge transform fibered over the identity map as in §15**D**, i.e.,

$$\begin{gathered}\Phi_\beta \circ \Phi_\alpha^{-1} \colon U_{\alpha\beta} \times \mathbb{C}^n \to U_{\alpha\beta} \times \mathbb{C}^n, \\ \Phi_\beta \circ \Phi_\alpha^{-1}(a, x) = \big(a, H_{\beta\alpha}(a)\, x\big), \qquad H_{\beta\alpha}(a) \in \mathrm{GL}(n, \mathbb{C}),\ a \in U_{\alpha\beta}.\end{gathered} \tag{17.1}$$

The triplet $\pi \colon S \to T$ is called a *holomorphic* complex vector bundle, if both S and T are holomorphic manifolds, π is a holomorphic projection which admits *biholomorphic* trivialization near each point of T. In this case the transition maps are biholomorphic gauge transformations.

Preimages of points $\tau_a = \pi^{-1}(a)$ are called *fibers* of the vector bundle. The space S is called the *total space* of the vector bundle.

The bundles will usually be denoted by the symbols for the corresponding projections, provided that the two other components of the triplet (the total space and bundle) are clearly defined by the context.

Geometry provides a vast source of bundles. For any holomorphic manifold M of complex dimension n the collection of tangent vectors attached to different points of M has a natural structure of a holomorphic vector bundle of rank n over the base M, called the *tangent bundle*. Indeed, if $U \subset \mathbb{C}^n$ is a domain in the affine space, then vectors tangent to different points of U can be identified with elements of the vector space $\mathbb{C}^n$ itself. Thus every chart on M, defined in a domain $U \subset M$ provides a local trivialization of the tangent bundle. The tangent bundle is usually denoted $\mathbf{T}M$. In a similar way the cotangent bundle $\mathbf{T}^*M$ is defined as the collection of covectors (linear functionals on tangent spaces) at all points of M (see Problem 17.1).

17B. Cocycles. Obviously, if $\pi\colon S \to T$ is a topological (resp., holomorphic) vector bundle, then for each two local trivializations over overlapping domains the matrix functions

$$H_{\beta\alpha}\colon U_{\alpha\beta} \to \mathrm{GL}(n,\mathbb{C}), \qquad U_{\alpha\beta} = U_\alpha \cap U_\beta, \tag{17.2}$$

is continuous (resp., holomorphic) together with its inverse $H_{\beta\alpha}^{-1}$. Since the construction is symmetric with respect to the two trivializations, this inverse is the transition matrix $H_{\alpha\beta}$, i.e., we have the identities

$$H_{\alpha\beta} \cdot H_{\beta\alpha} \equiv E \qquad \text{on} \quad U_{\alpha\beta}. \tag{17.3}$$

Besides, if U_α, U_β and U_γ are *three* domains with the pairwise intersections $U_{\alpha\beta}, U_{\beta\gamma}, U_{\alpha\gamma}$ and a nonvoid triple intersection $U_{\alpha\beta\gamma}$, then

$$H_{\alpha\beta} \cdot H_{\beta\gamma} \cdot H_{\gamma\alpha} \equiv E \qquad \text{on} \quad U_{\alpha\beta\gamma}. \tag{17.4}$$

Indeed, this composition corresponds to the transition between the trivializations Φ_α, Φ_γ and Φ_β (in the specified order) back to Φ_α.

Definition 17.3. Let $\mathfrak{U} = \{U_\alpha\}$ be an open covering of the base T. A *holomorphic matrix cocycle* inscribed in this covering (or subordinated to this covering) is a collection of holomorphic matrix functions $\mathcal{H} = \{H_{\alpha\beta}\}$ defined in all nonempty pairwise intersections $U_{\alpha\beta}$ and satisfying the identities (17.3) and (17.4) on all nonempty double (resp., triple) intersections.

Definition 17.4. A *holomorphic matrix cochain* $\mathcal{G}$ subordinated to the covering $\mathfrak{U}$, is a collection of holomorphic matrix functions $G_\alpha \in \mathrm{Mat}(n, U_\alpha)$ defined and holomorphic in the domains of the covering. In a similar way meromorphic, vector and other types of cochains are defined with obvious modifications.[2]

Definition 17.5. The operator transforming a cochain $\mathcal{G} = \{G_\alpha\}$ into the cocycle $\mathcal{H} = \{H_{\alpha\beta}\}$ with $H_{\alpha\beta} = G_\alpha G_\beta^{-1}$, is called the *coboundary* (or *multiplicative matrix coboundary*, if necessary to distinguish it from similar operators).

Any family of trivializations of a holomorphic vector bundle defines a holomorphic matrix cocycle. Conversely, any holomorphic matrix cocycle inscribed in an open covering of T determines a holomorphic vector bundle over T.

[2]The notions of cocycle and cochain belong to algebraic topology which defines cohomology with coefficients in different sheaves. It would be more appropriate to use the terms 1-cocycle and 0-cochain rather than simply cocycle and cochain, yet we will never need the general case of k-cochains or k-cocycles in this book.

Theorem 17.6. *Any matrix cocycle inscribed in a covering of a holomorphic manifold T, can be realized as the collection of transition gauge maps between local trivializations of a suitable holomorphic vector bundle over the base T.*

Proof. Consider the disjoint union of cylinders $\widetilde{S} = \bigsqcup_\alpha U_\alpha \times \mathbb{C}^n$ together with the equivalence relation on it, identifying the points

$$U_\alpha \times \mathbb{C}^n \ni (a, x) \sim (a', x') \in U_\beta \times \mathbb{C}^n \quad \Longleftrightarrow \quad a = a' \in U_{\alpha\beta} \text{ and } x' = H_{\beta\alpha} x.$$

This relation is indeed symmetric by (17.3) and transitive by (17.4). The quotient space $S = \widetilde{S}/\sim$ by this relation obtains thus a natural structure of a holomorphic manifold with the charts $U_\alpha \times \mathbb{C}^n$. The Cartesian projections $\pi_\alpha \colon U_\alpha \times \mathbb{C}^n \to U_\alpha$ respect the equivalence and hence together define an analytic map $\pi \colon S \to T$. The cylinders $U_\alpha \times \mathbb{C}^n$ provide trivializations of the map π over U_α, and the transition maps between these trivializations tautologically coincide with the gauge transforms defined by the specified matrix functions from the cocycle. □

Description of vector bundles by matrix cocycles provides analytic tools (methods of theory of analytic matrix functions) for working in the geometric category of vector bundles.

Example 17.7. The trivial vector bundle $\pi \colon T \times \mathbb{C}^n \to T$, $\pi(a, x) = a$, of any rank n exists over any base T and is associated with the trivial cocycle $\{H_{\alpha\beta} = E\}$ inscribed in an arbitrary covering of T.

The definition of a vector bundle does not specify any particular choice of the trivializations (there mere existence is required). Clearly, if Φ_α is a trivialization of a vector bundle π over a domain $U_\alpha \subseteq T$ and $G_\alpha \colon U_\alpha \times \mathbb{C}^n \to U_\alpha \times \mathbb{C}^n$ a collection of invertible gauge map fibered over the identity, then $\Phi'_\alpha = G_\alpha \circ \Phi_\alpha$ is another trivialization over the same domain U_α. The cocycle $\mathcal{H}' = \{H'_{\alpha\beta}\}$ of the transition maps associated with the new collection of trivializations, is related to the initial cocycle as follows:

$$H'_{\alpha\beta} G_\beta = G_\alpha H_{\alpha\beta} \qquad \text{on} \quad U_{\alpha\beta}. \tag{17.5}$$

Definition 17.8. Two cocycles $\mathcal{H} = \{H_{\alpha\beta}\}$ and $\mathcal{H}' = \{H'_{\alpha\beta}\}$ inscribed in the same covering $\mathfrak{U} = \{U_\alpha\}$ are *equivalent*, if there exists a *holomorphic matrix cochain* $\mathcal{G} = \{G_\alpha\}$, such that (17.5) holds.

Summarizing, we conclude that each holomorphic vector bundle over the base T is associated with a family of equivalent holomorphic matrix cocycle inscribed in some open covering $\mathfrak{U} = \{U_\alpha\}$ of T. Conversely, any matrix cocycle can be realized by a suitable bundle.

The question that was not yet addressed is equivalence of bundles obtained from *different* coverings. Clearly, if a covering $\mathfrak{U} = \{U_\alpha\}$ is a refinement of another, more coarse covering $\mathfrak{U}' = \{U'_i\}$, i.e., if each U_α entirely belongs to one of the larger domains U'_i, then any cocycle inscribed in $\mathfrak{U}'$ can be refined to a cocycle inscribed in $\mathfrak{U}$, by restriction (postulating the identical transitions $H_{\alpha\beta} = \mathrm{id}$, if both U_α and U_β belong to the same larger domain U'_i). This allows us to define equivalence of two cocycles $\mathcal{H}, \mathcal{H}'$ inscribed in two different coverings $\mathfrak{U}, \mathfrak{U}'$, by passing to cocycles inscribed in the common refinement $\mathfrak{U}'' = \{U_\alpha \cap U'_i\}$.

Replacing the domains U_α by smaller ones, we can (and will always) assume that each of them are topological disks with smooth boundaries.

A difficult problem is to pass from fine to more coarse coverings. To that end one has to combine two trivializations over overlapping domains U_α, U_β into a trivialization over the union $U_\alpha \cup U_\beta$. This problem will be discussed in detail later, in §**17J**.

17C. Operations on bundles. Speaking informally, a holomorphic bundle is a union of linear spaces (fibers) parameterized by points of the base T in a locally trivial way. Most constructions of linear algebra can be translated into the category of vector bundles by implementing these constructions "fiberwise". We provide a brief glossary of the most basic terms.

Definition 17.9. A (holomorphic) *bundle map between two vector bundles* $\pi\colon S \to T$ and $\pi'\colon S' \to T'$ is a holomorphic map $F\colon S \to S'$ between the total spaces, which maps fibers of π linearly to fibers of π'.

Formally this means that there exists a map $f\colon T \to T'$ between the bases, such that $\pi' \circ F = f \circ \pi$. We say that the map F is *fibered over* f. Two vector bundles are equivalent, if there exists an invertible holomorphic bundle map between them.

To write bundle maps "in coordinates" we need to choose a pair of trivializations near a given point $a \in T$ and its image $a' = f(a)$. Consider a pair of domains $U_\alpha \subset T$ and $U'_i \subset T'$, containing a and a' respectively, and let Φ_α, Φ'_i respectively be two collections of trivializations of these two bundles. Then a bundle map becomes a gauge map between $U_\alpha \times \mathbb{C}^n$ and $U'_i \times \mathbb{C}^m$ (we do not assume that the two bundles have the same rank). In other words, in the trivializing charts the map $\Phi'_i \circ F \circ \Phi_\alpha^{-1}$ takes the form

$$U_\alpha \times \mathbb{C}^n \to U'_i \times \mathbb{C}^m, \qquad (a, x) \mapsto \bigl(f(a), F_{\alpha,i}(a) \cdot x\bigr),$$

with a $(n \times m)$-holomorphic matrix function $F_{\alpha,i}$. If instead of the trivialization Φ_α another trivialization Φ_β of the total space at the source is chosen, the matrix function $F_{\alpha,i}$ will be replaced by the matrix function $F_{\beta,i}$ which on the intersection $U_{\alpha\beta}$ satisfies the identity $F_{\beta,i}(a) = F_{\alpha,i}(a) \cdot H_{\alpha\beta}(a)$. A similar rule applies when changing the trivialization of the target total space.

Example 17.10. If the bundle S' is trivial (of some dimension m), then a bundle map between S and S' is defined by a cochain $\mathcal{F} = \{F_\alpha\}$ such that $F_\alpha \cdot H_{\alpha\beta} = F_\beta$.

Conversely, a map from the trivial bundle S' to S is defined by a cochain $\mathcal{G} = \{G_\alpha\}$ such that $H_{\alpha\beta} \cdot G_\beta = G_\alpha$.

Definition 17.11. A holomorphic cocycle $\mathcal{H} = \{H_{\alpha\beta}\}$ is called *solvable*, if there exists a holomorphic matrix cochain $\mathcal{G} = \{G_\alpha\}$ such that

$$H_{\alpha\beta} = G_\alpha G_\beta^{-1}. \tag{17.6}$$

By this definition, solvable cochains correspond to bundles which are holomorphically equivalent to the trivial bundle. In analytic terms cocycle is solvable if and only if it is equivalent to the trivial cocycle.

The general construction of a bundle map becomes more transparent if both the source and the tangent bundle π, π' are over the same base and the map is fibered over the identity. In this case it is natural to use trivializations $\Phi_\alpha, \Phi'_\alpha$ inscribed in the same covering. In each pair of trivializations the map $F \colon S \to S'$ is associated with a holomorphic matrix function

$$\Phi'_\alpha \circ F \circ \Phi_\alpha^{-1} \colon U_\alpha \times \mathbb{C}^n \to U_\alpha \times \mathbb{C}^n, \qquad (a, x) \mapsto \bigl(a, F_\alpha(a)x\bigr).$$

In other words, a bundle map is associated with a holomorphic matrix cochain (the matrices can be nonsquare, if the ranks of π, π' are different.

On the overlapping of two domains the two matrix functions F_α, F_β are related by the identity

$$F_\beta = H'_{\beta\alpha} F_\alpha H_{\alpha\beta}, \qquad \text{i.e.,} \qquad H'_{\alpha\beta} F_\beta = F_\alpha H_{\alpha\beta} \qquad \text{on} \quad U_{\alpha\beta}, \tag{17.7}$$

where $\{H_{\alpha\beta}\}, \{H'_{\alpha\beta}\}$ are two cocycles defining the bundles π, π' respectively. This identity coincides with (17.5) if the matrices F_α are holomorphic invertible, which again illustrates the notion of equivalence of cocycles as equivalence of the corresponding bundles.

Other linear algebraic constructions are introduced in a similar way. A *subbundle* S' of a holomorphic bundle $\pi \colon S \to T$ is a holomorphic submanifold $S' \subseteq S$ such that the restriction of π on S' is itself a vector bundle of some rank k less or equal to the rank of S. If S' is a subbundle, then one can define the *quotient bundle* S/S', whose fibers are quotient spaces τ_a/τ'_a, $\tau_a = \pi^{-1}(a)$, $\tau'_a = \tau_a \cap S' \subseteq \tau_a$. Given any two bundles π, π' over the same base, one can construct their direct sum $\pi \oplus \pi'$, the tensor product $\pi \otimes \pi'$, dual bundle π^*, *etc.*

For instance, the tangent and cotangent bundles $\pi = \mathbf{T}M$ and $\pi^* = \mathbf{T}^*M$ over any holomorphic manifold M are dual to each other: for every point $a \in M$ there is a bilinear pairing $\pi^{-1}(a) \times {\pi^*}^{-1}(a) \to \mathbb{C}$ between the fibers of these bundles.

The notions of holomorphic vector bundle, cocycle, cochain make perfect sense in the case of minimal rank $n = 1$. This case is especially important, since 1×1-matrices commute, and hence it is much easier to study cocycles and equivalence. To distinguish this case, bundles of rank 1 are called *line bundles*.

One construction very important for future applications, allows us to associate with a vector bundle of any rank a line bundle called *determinant*, though a more appropriate name would be the maximal wedge product.

Note that for any linear space of dimension n its n-times wedge power (the wedge product of n copies of the space) is one-dimensional. Thus for any bundle π of rank n the wedge product

$$\det \pi = \underbrace{\pi \wedge \cdots \wedge \pi}_{n \text{ times}}$$

is a line bundle. Every linear map $H \in \mathrm{GL}(n, \mathbb{C})$ induces a map $\det H \in \mathrm{GL}(1, \mathbb{C})$ between the wedge products,

$$x_1 \wedge \cdots \wedge x_n \mapsto Hx_1 \wedge \cdots \wedge Hx_n = (\det H) \cdot x_1 \wedge \cdots \wedge x_n.$$

This allows us to define the determinant of a bundle in terms of cocycles.

Definition 17.12. The *determinant* of a vector bundle $\pi \colon S \to T$ of rank n, associated with a cocycle $\mathcal{H}$, is the holomorphic vector bundle of rank 1, associated with the cocycle

$$\det \mathcal{H} = \{h_{\alpha\beta}\}, \qquad h_{\alpha\beta} = \det H_{\alpha\beta}. \tag{17.8}$$

One can instantly verify that $\det \mathcal{H}$ is indeed a (scalar) cocycle. From (17.5) it follows that equivalent cocycles produce *the same* determinant cocycle.

17D. Classification of line bundles over the Riemann sphere. As a first step towards classification of holomorphic vector bundles of arbitrary rank over the Riemann sphere $\mathbb{P}$ in §**17J**, we now give a complete classification of line bundles over $\mathbb{P}$.

Consider the *standard covering* of the Riemann sphere $\mathbb{P}$ by an atlas of two charts, $U_0 = \{|t| < r_0\} \subseteq \mathbb{C}$ (the disk in the affine part with the chart t inherited from the affine line) and $U_1 = \{|t| > r_1\} \cup \{\infty\}$ with the chart $z = 1/t$, in which it also becomes an open disk. The intersection $A = U_{01}$ of these two charts is the circular annulus $A = \{r_1 < t < r_0\}$. The exact choice of the parameters $r_1 < r_0$ is not important.

A (holomorphic matrix) cocycle inscribed in the standard covering consists of a single pair of matrix functions $H_{01}(t) = H_{10}^{-1}(t)$ holomorphic and invertible in the annulus A. Such cocycles will be called *Birkhoff–Grothendieck cocycles*. For instance, a Birkhoff–Grothendieck cocycle of rank 1 is just a nonvanishing function $h(t) = h_{01}(t) = 1/h_{10}(t)$ holomorphic in the annulus.

Denote by ξ_d the line bundle corresponding to the *standard Birkhoff–Grothendieck cocycle*

$$\mathfrak{L}_d = \{h_{01}, h_{10}\}, \quad h_{01}(t) = t^d\big|_{t\in A} = 1/h_{10}(t), \qquad d \in \mathbb{Z}, \tag{17.9}$$

in an annulus A. The integer number d will be referred to as the *degree* of the line bundle ξ_d and the corresponding standard cocycle.

Proposition 17.13. *Any scalar Birkhoff–Grothendieck cocycle $\mathfrak{L} = \{h_{01}(t), h_{10}(t)\}$ is equivalent to one of the standard cocycles* (17.9) *of some degree d. Standard cocycles of different degrees are not equivalent to each other.*

To prove the proposition, we need an additive (rather than multiplicative) analog of holomorphic solvability of cocycles.

Lemma 17.14. *Let $U, U' \subseteq \mathbb{P}$ be two domains such that both of them and their intersection $V = U \cap U'$ have piecewise-smooth boundary.*

Then any function $v \in \mathcal{A}(V)$ holomorphic in V and continuous on the closure $\overline{V} = V \cup \partial V$ can be represented as the difference, $v = u - u'$, with $u \in \mathcal{A}(U)$, $u' \in \mathcal{A}(U')$.

For continuous functions the corresponding claim is obvious: among other solutions, one can simply choose $u' = 0$ (such a function is defined everywhere) and construct u as an arbitrary continuation of the function v from a closed subset $\overline{V}$ to a larger set U. However, holomorphic functions are very rigid, and Lemma 17.14 is a nontrivial (though simple) fact.

Proof of Lemma 17.14. The function v can be represented by the Cauchy integral over the boundary ∂V. This boundary can be represented as the *disjoint* union of two parts, $\partial V = B \sqcup B'$, with $B \subset \partial U$ and $B' \subset \partial U'$. Thus we have

$$v(t) = \frac{1}{2\pi i}\oint_{\partial V} \frac{f(z)\,dz}{z-t} = \frac{1}{2\pi i}\oint_{B} \frac{f(z)\,dz}{z-t} - \frac{1}{2\pi i}\oint_{-B'} \frac{f(z)\,dz}{z-t}.$$

The integral over B (resp., B') is holomorphic in $U \subset \mathbb{P} \smallsetminus B$ (resp., $U' \subset \mathbb{P} \smallsetminus B'$), and both are continuous on the boundary. □

Example 17.15. The function u holomorphic in the annulus $A = U_{01}$ as above, can be expanded in a *converging* Laurent series. Collecting together nonnegative and negative powers of t, we obtain two series converging in the respective disks $U_0, U_1 \subset \mathbb{P}$.

Proof of Proposition 17.13. There exists a unique integer number d such that the argument of the function $t^{-d}h(t)$ is a well-defined function in the annulus $A = U_{01}$. This number is equal to the index (rotation number) of the loop $h(\mathbb{S}^1)$ around the origin, where $\mathbb{S}^1$ is the unit circle $\{|t| = 1\}$.

For such a choice of d the function $t^{-d}h(t)$ admits a well defined logarithm $u(t) = \ln\bigl(t^{-d}h(t)\bigr)$, a holomorphic function in A unique modulo $2\pi i\mathbb{Z}$. Expanding u as in Lemma 17.14, we obtain two functions u_0, u_1 holomorphic in the respective disks $D_i \subset \mathbb{P}$. The exponents $g_i = \exp u_i$ of these functions are holomorphic, nonvanishing in U_i and satisfy the identity $t^{-d}h(t) = g_0/g_1$ on U_{01}. Rewriting this identity in the form

$$h(t) \cdot g_1(t) = g_0(t)\, t^d,$$

we prove that the holomorphic cocycles $\mathfrak{L}$ and $\mathfrak{L}_d$ are equivalent; cf. with (17.5). The equality $t^{d'} g_1 = g_0\, t^d$ with $d \neq d'$ is impossible, since the variation of argument of each holomorphic nonvanishing function g_i along the circle is zero, while that of the ratio $t^{d-d'}$ is $2\pi(d - d')$. □

Proposition 17.13 gives classification of scalar cocycles inscribed in the standard covering of the Riemann sphere $\mathbb{P}$ by two charts. In fact, this particular case suffices to describe *all* scalar cocycles, hence *all* holomorphic line bundles over $\mathbb{P}$.

Theorem 17.16. *Any line bundle over the Riemann sphere is holomorphically equivalent to the standard bundle ξ_d of some degree $d \in \mathbb{Z}$.*

Proof. We first show that any line bundle π_0 over the unit disk $\mathbb{D} \subset \mathbb{C}$ is equivalent to the trivial bundle. Indeed, consider the cocycle $\mathfrak{L}$ which defines the bundle π_0. This cocycle is inscribed in a finite covering $\mathfrak{U}$. By further refinement of this covering we may assume that it is a "triangulation", i.e., the domains U_α are small ε-neighborhoods of triangles of some triangulation of $\mathbb{D}$ (one can also choose partition of the disk into small squares arranged in a grid). For our purposes it is important that the domains $U_1, \dots, U_N$ can be ordered in such a way that the intersections

$$U_{k+1} \cap (U_1 \cup \cdots \cup U_k), \qquad k = 1, 3, \dots, N-1,$$

are all connected and *simply connected*; see Fig. III.1.

Assume by induction that the cocycle $\mathfrak{L}$ is solvable over $U' = U_1 \cup \cdots \cup U_k$. Then, replacing the cocycle $\mathcal{H}$ by an equivalent cocycle, we may assume that all transitions h_{ij} between domains with numbers $\leqslant k$ are trivial. We claim that the cocycle can be trivialized also over $U' \cup U$, $U = U_{k+1}$. Indeed, in this case all we have to show is that any holomorphic invertible function h in the intersection $V = U \cap U'$ can be represented as the quotient of two functions, $h = g/g'$, holomorphic and invertible in U and U' respectively.

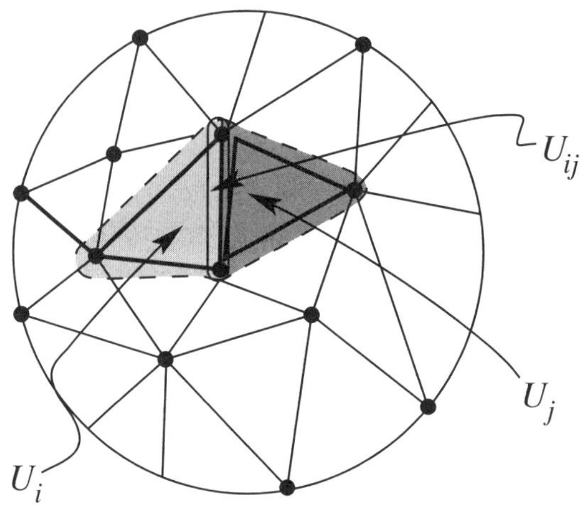

Figure III.1. Triangulation and "triangulated covering" of a disk

Since V by construction is simply connected, $\ln h$ is a well-defined holomorphic function which can be represented as a difference of two holomorphic functions by Lemma 17.14. After exponentiation we obtain the representation $h = g/g'$ and hence prove the solvability of the cocycle $\mathfrak{L}$ restricted on the union $U' \cup U = U_1 \cup \cdots \cup U_k \cup U_{k+1}$. By induction, the cocycle is solvable (and hence the corresponding line bundle π_0 is solvable).

Thus any holomorphic bundle over $\mathbb{P}$ can be trivialized over each of two charts of a standard Birkhoff–Grothendieck covering. This means that the problem of classification of *arbitrary* cocycles over $\mathbb{P}$ is reduced to classification of (scalar) Birkhoff–Grothendieck cocycles inscribed in a standard covering. By Proposition 17.13, each such cocycle is equivalent to a standard cocycle. □

17E. Sections of holomorphic vector bundles. Since the total space of a vector bundle is in general not a Cartesian product, we need a suitable generalization of the notion of vector functions.

Definition 17.17. A *section* of a holomorphic vector bundle $\pi \colon S \to T$ is a map $s \colon T \to S$, such that $\pi \circ s = \mathrm{id}$, i.e., such that the image of every point $a \in T$ belongs to the fiber $\pi^{-1}(a)$. We will specifically deal with continuous, *holomorphic* and *meromorphic* sections (the latter will be defined separately later).

Remark 17.18. Sometimes we will deal with "sections" defined only over some (open) subset U of the base T. In this case, to avoid confusion, we will say about *local sections*, explicitly specifying their domains.

Trivializations over domains U_α allow us to associate with every section s a holomorphic *vector cochain* $\{x_\alpha\}$, the collection of vector functions

$$x_\alpha \colon U_\alpha \to \mathbb{C}^n, \qquad x_\alpha = \Phi_\alpha \circ s\big|_{U_\alpha}.$$

Using a different trivialization Φ_β on the intersection of two domains of trivialization, replaces the function x_α by the function x_β,

$$x_\beta = H_{\beta\alpha} x_\alpha \qquad \text{on} \quad U_{\alpha\beta}. \tag{17.10}$$

Conversely, given a matrix cocycle $\mathcal{H} = \{H_{\alpha\beta}\}$ and a vector bundle defined by this cocycle, any holomorphic vector cochain $\{x_\alpha\}$ which satisfies the conditions (17.10) on the pairwise intersections, defines a section of the bundle.

However, not all bundles admit nontrivial (not identically zero) holomorphic sections (Problem 17.7).

Example 17.19. Sections of the tangent bundle $\mathbf{T}M$ are called (holomorphic) vector fields on M. Sections of the cotangent bundle are called (holomorphic) 1-forms. There are no globally defined holomorphic 1-forms without poles on the Riemann sphere $\mathbb{P}$ (otherwise their primitives would be globally defined holomorphic nonconstant functions), hence $\mathbf{T}^*\mathbb{P}$ does not admit holomorphic sections. Globally defined holomorphic vector fields on $\mathbb{P}$ do exist, but they must necessarily have zeros.

Absence of holomorphic sections motivates introduction of a slightly more general notion of a *meromorphic section* of a holomorphic bundle.

Definition 17.20. A *meromorphic* section of a holomorphic vector bundle defined by a holomorphic matrix cocycle $\mathcal{H} = \{H_{\alpha\beta}\}$, is a meromorphic vector cochain $\{x_\alpha\}$, $x_\alpha \in \mathcal{M}(U_\alpha) \otimes_{\mathbb{C}} \mathbb{C}^n$ which satisfies the identities (17.10) on the intersections $U_{\alpha\beta}$.

All meromorphic sections of a given bundle form an infinite-dimensional linear space over $\mathbb{C}$ and, moreover, a linear space over the field $\mathcal{M}(T)$ of meromorphic functions on the base T, since two sections can be added, and any meromorphic section can be multiplied by a meromorphic (scalar) function. The corresponding meromorphic vector cochains obey the obvious rules,

$$s = s' + s'' \iff x_\alpha = x'_\alpha + x''_\alpha, \qquad s' = \varphi \cdot s \iff x'_\alpha = \varphi x_\alpha.$$

The set of all meromorphic sections of a bundle $\pi \colon T \to S$ will be denoted by $\Gamma(\pi)$.

17F. Degree of a holomorphic bundle. Recall that the order of a meromorphic *scalar* function $\varphi \in \mathcal{M}(\mathbb{C}, 0)$ of a scalar argument $t \in (\mathbb{C}, 0)$ is the order (positive or negative) of its principal Laurent term, $\operatorname{ord}_0 \varphi = \nu$ if and only if $\varphi(t) = c_\nu t^\nu + c_{\nu+1} t^{\nu+1} + \cdots$, with $c_0 \neq 0$.

Definition 17.21. The order of a meromorphic *vector*-function $x(t) = \big(x_1(t), \dots, x_n(t)\big) \in \mathcal{M} \otimes \mathbb{C}^n$ is the minimal order of its components,

$$\operatorname{ord}_0 x = \min_{1 \leqslant j \leqslant n} \operatorname{ord}_0 .$$

One can instantly verify that $\operatorname{ord}_0 x(\cdot)$ is the unique integer number $d \in \mathbb{Z}$ such that $t^{-d} x(t)$ is holomorphic and nonvanishing at $t = 0$,

$$\operatorname{ord}_0 x(t) = d \iff t^{-d} x(t) \in \mathcal{O}(\mathbb{C}, 0) \otimes \mathbb{C}^n \text{ and } \lim_{t \to 0} t^{-d} x(t) \neq 0.$$

If π is a holomorphic vector bundle over a *one-dimensional* base (Riemann surface) T, then the *order* $\operatorname{ord}_a s$ of a meromorphic section $s \in \Gamma(\pi)$ at a given point $a \in T$ of the base can be defined as the order of the corresponding vector function x_α in any trivialization $U_\alpha \ni a$: since the transition cocycle consists of holomorphic matrix functions, this definition is self-consistent. The order is nonzero at all points except for a discrete set.

Proposition 17.22. *All nontrivial meromorphic sections of a line bundle over a compact Riemann surface T have the same total order*: *for any meromorphic section the sum*

$$\deg s = \sum_{a \in T} \operatorname{ord}_a s, \qquad s \in \Gamma(\pi), \tag{17.11}$$

is the same and depends only on the bundle π itself.

Proof. If the fibers are one-dimensional, then any two sections $s, s' \in \Gamma(\pi)$ are proportional, i.e., there exists a meromorphic *function* $\varphi \in \mathcal{M}(T)$ such that $s' = \varphi s$. Obviously, $\deg s' = \deg s + \sum_a \operatorname{ord}_a \varphi$, where the last term is the sum of orders of all poles and zeros of φ. Yet any meromorphic function φ considered as a map $\varphi \colon T \to \mathbb{P}$, assumes each value the same number of times (equal to the degree of this map). Applying this to the values 0 and ∞, we conclude that $\sum_a \operatorname{ord}_a \varphi = 0$, hence $\deg s = \deg s'$. □

Definition 17.23. The common degree of all meromorphic sections is called the *degree* of a line bundle π and denoted by $\deg \pi$.

The degree of arbitrary holomorphic vector bundle is *defined* as the degree of its determinant, the line bundle associated with the determinant cocycle (17.8),

$$\deg \pi = \deg(\det \pi). \tag{17.12}$$

We will need the following property of the degree. A holomorphic bundle map between bundles of the same dimension will be called *nondegenerate*, if it has a full rank at some point.

Lemma 17.24. *Let $\pi\colon S \to T$ and $\pi'\colon S' \to T$ be two bundles of the same rank over the same compact one-dimensional base T.*

If there exists a nondegenerate holomorphic bundle map $F\colon S \to S'$ fibered over the identity map of the base, then $\deg \pi \leqslant \deg \pi'$.

Proof. Consider first the case where S and S' are both *line* bundles defined by *scalar* cocycles $\mathcal{H} = \{h_{\alpha\beta}\}, \mathcal{H}' = \{h'_{\alpha\beta}\}$ on trivializations over the same covering $\mathfrak{U}$, then a bundle map between them is defined by a collection of holomorphic functions $f_\alpha \not\equiv 0$ related to the cocycles $\mathcal{H}, \mathcal{H}'$ by (17.7).

An arbitrary meromorphic section $s \in \Gamma(\pi)$ and its image $s' = Fs \in \Gamma(\pi')$ are defined by the meromorphic *scalar* cochains x_α and x'_α which satisfy the identity

$$x'_\alpha = f_\alpha x_\alpha. \tag{17.13}$$

Since $f_\alpha \not\equiv 0$, this implies that $\operatorname{ord}_a s'_\alpha = \operatorname{ord}_a s_\alpha + \operatorname{ord}_a f_\alpha \geqslant \operatorname{ord}_a s_\alpha$. Adding these inequalities over all points of T, we arrive at the inequality $\deg s' \geqslant \deg s$. By Proposition 17.22, this means that $\deg \pi' \leqslant \deg \pi$.

A general nondegenerate linear map $F\colon S \to S'$, represented by a matrix cochain $\{F_\alpha\}$, defines a nondegenerate map $\det F$ between the determinant bundles $\det \pi$ and $\det \pi'$. The map $\det F$ is defined by the nonzero scalar holomorphic cochain $f_\alpha = \det F_\alpha$: this follows (17.7) after passing to determinants and the definition of the determinant bundle (17.8). The lemma follows from the assertion for line bundles and the definition of degree of an arbitrary bundle. □

As a corollary, we may conclude that subbundles of a trivial bundle all have nonpositive degree.

Corollary 17.25. *Every subbundle of the trivial bundle over a compact Riemann curve has nonpositive degree.*

Proof of the corollary. Let $\pi\colon S \to T$ be a subbundle of rank n of the trivial bundle $\pi_0\colon T \times \mathbb{C}^{n+m} \to T$. We will prove that $\deg \pi \leqslant 0$. Indeed, one can always find a splitting of the fiber $\mathbb{C}^{n+m} = \mathbb{C}^n \oplus \mathbb{C}^m$ into two subspaces such that the fiber $\pi^{-1}(a)$ is transversal to $\mathbb{C}^m$ at some point $a \in T$. The projection on $\mathbb{C}^n$ parallel to $\mathbb{C}^m$ after restriction on the subbundle S becomes a holomorphic nondegenerate bundle map between π and the trivial subbundle $\pi' = T \times \mathbb{C}^n \to T$. By Lemma 17.24, $\deg \pi \geqslant \deg \pi' = 0$. □

17G. Holomorphic and meromorphic connexions. If $x\colon T \to \mathbb{C}^n$ is a holomorphic vector function of one or several variables, then its differential is a vector-valued 1-form on T. Once fibers over different points of the base T are different, as in the case of holomorphic vector bundles, the notion of *derivation of a section* needs to be appropriately modified. The result is the notion of a *connexion*, or in full *meromorphic connexion on a holomorphic vector bundle*.

Connexions can be described axiomatically by their geometric properties. Denote by $\Lambda^1(T) \otimes_{\mathfrak{M}(T)} \Gamma(\pi)$ the $\mathfrak{M}(T)$-module of meromorphic fiber-valued 1-forms on the base T of a holomorphic vector bundle π, the tensor product is taken over the field of meromorphic functions $\mathfrak{M}(T)$. By definition, a fiber-valued 1-form $\omega \otimes s$ can be evaluated on any meromorphic vector field $Z \in \mathfrak{D}(T)$, and the result will be the meromorphic section $\varphi \cdot s \in \Gamma(\pi)$, $\varphi = \omega(Z)$. This object generalizes the notion of a vector-valued 1-form. Now we give a generalization of the exterior derivative for vector-valued functions. This is a differential operator called a *connexion* on the bundle.

Definition 17.26. A *meromorphic connexion* on a holomorphic vector bundle π is a $\mathbb{C}$-linear operator

$$\nabla\colon \Gamma(\pi) \to \Lambda^1(T) \otimes \Gamma(\pi)$$

which satisfies the Leibnitz rule:

$$\begin{aligned} \nabla(\lambda s + \lambda' s') &= \lambda \nabla s + \lambda' \nabla s', \qquad && \forall s, s' \in \Gamma(\pi),\ \lambda, \lambda' \in \mathbb{C}, \\ \nabla(\varphi \cdot s) &= \varphi \cdot \nabla s + df \otimes s, && \forall s \in \Gamma(\pi),\ \varphi \in \mathfrak{M}(T). \end{aligned} \tag{17.14}$$

The result ∇s of a derivation is a fiber-valued 1-form on T.

Example 17.27. If π is a trivial bundle with $S = T \times \mathbb{C}^n$, then the standard (vector) exterior derivative

$$\nabla x = dx, \qquad \forall x\colon T \to \mathbb{C}^n$$

obviously satisfies the rules (17.14). In fact, for trivial bundles we can easily describe *all* differential operators satisfying the axioms (17.14). Indeed, if ∇, ∇' are two such operators, then their difference is a *linear operator on each fiber*: from (17.14) it immediately follows that

$$(\nabla - \nabla')(\varphi \cdot x) = \varphi \cdot [(\nabla - \nabla')x].$$

This means that the difference between the operators is defined by an $n \times n$-matrix-valued form: evaluated on a tangent vector at a point $a \in T$ of the base, it becomes a linear automorphism of the respective fiber $\pi^{-1}(a) \cong \mathbb{C}^n$ into itself.

In other words, any connexion ∇ *on the trivial bundle* can be represented using a suitable meromorphic matrix 1-form $\Omega \in \operatorname{Mat}(n, \Lambda^1(T) \otimes \mathfrak{M}(T))$ as the difference $\nabla = d - \Omega$, that is,

$$\nabla x = dx - \Omega x, \qquad \forall x \colon T \to \mathbb{C}^n, \tag{17.15}$$

The matrix 1-form Ω is called the *connexion form* of the connexion ∇.

For arbitrary (nontrivial) bundles such characterization is true only locally, in trivializing charts.

17H. Connexions vs. linear systems. If $F \colon S \to S'$ is an invertible holomorphic bundle map between two bundles π, π' over the same base, then this map allows us to carry any connexion on S to a connexion on S' and vice versa. Two connexions ∇, ∇' on the two bundles are called *F-related*, if $F(\nabla s) = \nabla'(Fs)$ for any section $s \in \Gamma(\pi)$. Here by Fs is denoted the section $s' \in \Gamma(\pi')$ obtained by application of F to the section s.

Assume that both S, S' are trivial bundles (of the same rank) and F is a gauge map defined by the matrix function $F(a) \in \mathrm{GL}(n, \mathbb{C})$ as in (15.9). It transforms a vector function $a \mapsto x(a)$ into the vector function $x'(a) = F(a)x(a)$. Thus two connexions, $\nabla = d - \Omega$ and $\nabla' = d - \Omega'$, defined by two matrix forms Ω, Ω', are F-related if and only if $F(dx - \Omega x) = d(Fx) - \Omega' F x$ for any vector-valued holomorphic function $x(\cdot)$. This condition is equivalent to the matrix identity

$$\Omega' = dF \cdot F^{-1} + F \Omega F^{-1} \tag{17.16}$$

which naturally coincides with the law of gauge transformation (15.10).

This observation allows us to represent any connexion on a holomorphic bundle by a collection of matrix 1-forms associated with different local trivializations of this bundle. Indeed, if Φ_α is a local trivialization of a holomorphic vector bundle with a meromorphic connexion ∇, then there exists a unique meromorphic connexion ∇_α on the trivial bundle $U_\alpha \times \mathbb{C}^n$, which is Φ_α-related to ∇. On the intersection of two charts $U_{\alpha\beta}$ two different trivializations lead to two different connexion 1-forms $\Omega_\alpha, \Omega_\beta$. By (17.16), these two matrix forms are related by the identity

$$dH_{\alpha\beta} = \Omega_\alpha H_{\alpha\beta} - H_{\alpha\beta} \Omega_\beta. \tag{17.17}$$

Conversely, given a collection of trivializations of a holomorphic vector bundle, related by a matrix cocycle $\mathcal{H} = \{H_{\alpha\beta}\}$ and an arbitrary collection of meromorphic matrix 1-form Ω_α satisfying the transition identities (17.16) on the pairwise intersections, we can define a meromorphic matrix connexion ∇ as the operator sending the vector cochain $\{x_\alpha\}$ defining an arbitrary section $s \in \Gamma(\pi)$ into the cochain $\{\theta_\alpha\}$ of vector-valued 1-forms $\theta_\alpha = dx_\alpha - \Omega x_\alpha$. It is a standard exercise to verify that if the initial cochain satisfies (17.10), then the cochain $\{\omega_\alpha\}$ defines a section of $\Lambda^1(T) \otimes \Gamma(\pi)$, i.e., satisfies the analogous identity $H_{\alpha\beta}\theta_\beta = \theta_\alpha$ on the pairwise intersections.

Describing a meromorphic connexion by its connexion (matrix) forms in suitable trivializations allows us to translate all notions and properties of theory of linear systems, which are invariant by holomorphic gauge equivalence, from the local theory of linear systems to the global context. We skip trivial checks.

Definition 17.28. The *singular locus* of a meromorphic connexion ∇ is defined as the collection of points at which the connexion matrix Ω_α in some (hence in any) trivialization has a pole. A meromorphic connexion is *holomorphic* if it has no poles.

Definition 17.29. A singular point of ∇ is *Fuchsian*, if it has a first order pole in some (hence in any) trivialization.

A singular point is *regular*, if it is regular for some (hence for any) linear system $dx = \Omega_\alpha x$.

For a Fuchsian connexion one can define its *residue* $\operatorname{res}_a \nabla$ at each Fuchsian singularity. This is a linear operator of the fiber $\pi^{-1}(a)$ into itself, defined in the local trivializing chart as the residue of the corresponding matrix connexion form:

$$
\begin{gathered}
\operatorname{res}_a \nabla \colon \pi^{-1}(a) \to \pi^{-1}(a), \\
\operatorname{res}_0(d-\Omega) = A \iff \Omega = (t^{-1}A + A_0 + A_1 t + \cdots)\, dt.
\end{gathered}
\tag{17.18}
$$

A vector function $x(\cdot)$ whose differential vanishes, $dx(\cdot) \equiv 0$, is (locally) constant and its graph is a horizontal hyperplane in the cylinder $T \times \mathbb{C}^n$. Such horizontal hyperplanes allow us to identify between themselves any two fibers $\{t = a\}$ and $\{t = b\}$, if the corresponding points belong to the same horizontal hyperplane.

The analogous notions for general bundles are defined using the covariant derivative ∇ instead of the exterior derivative d.

Definition 17.30. A *horizontal section* for a connexion ∇ on a holomorphic vector bundle π is a section satisfying the differential equation $\nabla s = 0$.

If ∇ is a connexion on the trivial bundle $U \times \mathbb{C}^n$ with a connexion form Ω, then horizontal sections $t \mapsto x(t)$ satisfy the Pfaffian linear equation $dx - \Omega x = 0$. Thus we see that connexions correspond to globally defined linear systems introduced in a geometric (coordinate-free) way.

Remark 17.31. Existence of horizontal *local* holomorphic sections over any simply connected chart free from singular points of a connexion, is automatic *only in the case where the base T is complex one-dimensional.* In all other cases even local existence of horizontal sections is guaranteed only under certain condition of *flatness* (absence of the *curvature*) of the connexion; see Problem 17.13.

Linear systems	**Meromorphic connexions**
Domain T (Riemann surface)	Base of the bundle T
Vector functions $\mathcal{M}(T) \otimes \mathbb{C}^n$	Sections of the bundle $\Gamma(\pi)$
Matrix 1-form $\Omega \in \operatorname{Mat}(n, \Lambda^1(T) \otimes \mathcal{M}(T))$	Meromorphic connexion $\nabla \colon \Gamma(\pi) \to \Gamma(\pi) \otimes \Lambda^1(T)$
Solutions of the linear system $dx = \Omega x$	Horizontal sections $\nabla s = 0$
Holonomy (monodromy), Cauchy operators	Parallel transport between fibers
Gauge transform	Bundle map

Table III.1. Glossary of terms: meromorphic connexions on holomorphic vector bundles vs. linear systems

In the same way as solutions of linear systems, horizontal sections are usually *multivalued*, i.e., exist only on the universal cover of $T \smallsetminus \Sigma$, where $\Sigma = \operatorname{Sing} \nabla$ is the singular locus of the connexion. On the other hand, if the base T is one-dimensional, Theorem 15.3 implies that horizontal sections can be constructed over any simply connected domain in the punctured base $T \smallsetminus \Sigma$. Moreover, partition of S on horizontal sections defines a *horizontal foliation* $\mathcal{F}_\nabla$ (with singularities) of the total space S, transversal to all fibers over nonsingular locus $T \smallsetminus \Sigma$.

Horizontal sections are "locally constant" with respect to the connexion ∇ and hence can be used to define the *parallel transport* between nearby fibers $\pi^{-1}(a)$ and $\pi^{-1}(a')$ over two sufficiently close points $a, a' \in T$. This transport is the precise equivalent of the holonomy map between two cross-sections τ_a and $\tau_{a'}$ to the null leaf of the foliation defined by an arbitrary linear system (15.3). In the same way as for the linear systems (connexions on the trivial bundles), the parallel transport along the leaves of horizontal foliation defines the *holonomy group* of the connexion. All these notions for connexions on trivial bundles coincide with their previously introduced homologues for linear systems. Table III.1 provides a glossary of parallel terms.

Theorem 17.32. *Let $\pi \colon S \to T$ be a holomorphic vector bundle of rank n and ∇ a meromorphic connexion on this bundle with the singular locus Σ.*

Then for any point a, any linearly independent vectors in the fiber $\pi^{-1}(a)$ and any simply connected domain $U \subseteq T \smallsetminus \Sigma$ there exist n holomorphic sections of π over U, linearly independent in each fiber.

The parallel transport along horizontal sections over closed paths γ from the fundamental group $\pi_1(S \smallsetminus \Sigma, a)$ defines a representation $\gamma \mapsto \Delta_\gamma$ of this group by linear holonomy operators $\Delta_\gamma \in \mathrm{GL}(\pi^{-1}(a))$.

If π, π' are two bundles over the same base and F is a holomorphic or meromorphic bundle map between them fibered over the identity, and ∇, ∇' are two F-related connexions on these two bundles, then the corresponding holonomy groups are also F-related, i.e., conjugated[3] *by the linear map $F(a)\colon \pi^{-1}(a) \to \pi'^{-1}(a)$.* □

17I. Connexions on line bundles. Trace of a meromorphic connexion. Connexions on line bundles (of rank 1) are determined by the scalar meromorphic 1-forms ω_j in each trivialization, i.e., each connexion ∇ is determined by its cochain of *scalar* 1-forms $\{\omega_\alpha\}$. Since 1×1-matrices commute, on the overlapping of domains U_i and U_j of two different trivializations, two forms ω_i, ω_j differ by an additive holomorphic term, the logarithmic derivative of the transition cocycle,

$$\omega_i = d \ln h_{ij} + \omega_j, \qquad d \ln h_{ij} = dh_{ij}/h_{ij}. \tag{17.19}$$

In particular, the residue $\operatorname{res}_a \nabla$ is well defined as the *scalar* residue of any of the two forms,

$$\operatorname{res}_a \nabla = \operatorname{res}_a \omega_i = \operatorname{res}_a \omega_j, \qquad a \in U_{ij}.$$

The total of residues of any meromorphic 1-form on a compact Riemann surface is zero: the sum makes sense since the residues are (complex) numbers that can be added between themselves. The following is a generalization of this fact for arbitrary line bundles.

Theorem 17.33. *The total of residues of any meromorphic connexion on a line bundle π over a compact Riemann surface T is the same for all connexions and equal to the degree of the bundle,*

$$\sum_{a \in T} \operatorname{res}_a \nabla = \deg \pi.$$

Proof. The difference between any two meromorphic connexions ∇, ∇' on the same line bundle is a globally well-defined meromorphic 1-form $\eta = \nabla - \nabla' \in \Lambda^1(T)$. Indeed, by (17.15) the difference is a well-defined operator-valued 1-form, but every linear self-map from $\mathrm{GL}(1, \mathbb{C})$ can be identified with

[3] In particular, if a point $a_j \in T$ is singular for one connexion and nonsingular for another, then the holonomy operators corresponding to a simple loop around this point are both trivial (identical).

its multiplicator which is a complex number (rather than an element of some fiber). From this observation it obviously follows that

$$\sum_a \operatorname{res}_a \nabla - \sum_a \operatorname{res}_a \nabla' = \sum_a \operatorname{res}_a \eta = 0,$$

since the total of residues of any meromorphic 1-form is zero (the total of integrals of η along all small loops around all singularities on T). Thus the sum of residues indeed does not depend on the connexion.

To show that it is equal to the degree, consider an arbitrary meromorphic section $s \in \Gamma(\pi)$ defined by a holomorphic cochain, $s \sim \{x_\alpha\}$, and let ∇ be the unique connexion for which s is horizontal (see Exercise 17.6). This connexion is defined by the cochain of logarithmic derivatives $\{\omega_\alpha\}$,

$$\nabla \cong \{\omega_\alpha\}, \qquad \text{where} \quad \omega_\alpha = dx_\alpha \cdot x_\alpha^{-1}.$$

The residue of the connexion ∇ at any point is the order of the section s at this point. Therefore

$$\sum_a \operatorname{res}_a \nabla = \sum_a \operatorname{res}_a \omega_\alpha = \sum_a \operatorname{ord}_a x_\alpha = \sum_a \operatorname{ord}_a s = \deg \pi$$

by (17.11). □

This result cannot be directly generalized to connexions on arbitrary bundles of rank greater than 1, since for such bundles the residues are linear self-maps of different fibers, hence cannot be simply added together. Thus the "total of all residues" for an arbitrary connexion is meaningless. The best one can get is a formula for the "total of *traces* of all residues" which is defined as follows.

Any meromorphic connexion ∇ on a holomorphic vector bundle π induces the *trace connexion*, denoted by $\operatorname{tr}\nabla$, on the determinant bundle $\det\pi$. If the connexion ∇ is trivialized by a cochain of meromorphic matrix 1-forms $\{\Omega_\alpha\}$, then $\operatorname{tr}\nabla$ is trivialized by the cochain $\{\omega_\alpha\}$,

$$\nabla \cong \{\Omega_\alpha\} \overset{\text{def}}{\Longleftrightarrow} \operatorname{tr}\nabla \cong \{\operatorname{tr}\Omega_\alpha\}. \tag{17.20}$$

Proposition 17.34. *The connexion* $\operatorname{tr}\nabla$ *is a well-defined meromorphic connexion on the bundle* $\det\pi$.

Two connexions ∇ *and* $\operatorname{tr}\nabla$ *are* det*-related: if* $s_1, \dots, s_n$ *are* n *linearly independent meromorphic sections of a rank* n *bundle* π, *horizontal for* ∇, *then their wedge product* $s_1 \wedge \cdots \wedge s_n$ *is a section of the line bundle* $\det\pi$, *horizontal for the connexion* $\operatorname{tr}\nabla$.

Both connexions have the same singular locus, and at every singular point

$$\operatorname{res}_a \operatorname{tr}\nabla = \operatorname{tr}\operatorname{res}_a \nabla. \tag{17.21}$$

Proof. To prove the first assertion, consider the cocycle $\mathcal{H} = \{H_{\alpha\beta}\}$ defining π and the respective cocycle $\det \mathcal{H} = \{h_{\alpha\beta}\}$, $h_{\alpha\beta} = \det H_{\alpha\beta}$. By the Liouville–Ostrogradskii formula (Problem 15.10),

$$\operatorname{tr} \Omega_\beta = \operatorname{tr}(dH_{\beta\alpha} \cdot H_{\alpha\beta}) + \operatorname{tr}(H_{\beta\alpha}\Omega_\alpha H_{\alpha\beta}) = dh_{\beta\alpha} \cdot h_{\alpha\beta} + \operatorname{tr} \Omega_\alpha,$$

that is, the cochain $\{\operatorname{tr} \Omega_\alpha\}$ representing $\operatorname{tr} \nabla$, is indeed a connexion on the bundle defined by the cocycle $\det \mathcal{H}$.

If $\{X_\alpha(t)\}$ is a fundamental (multivalued) matrix solution associated with the sections $s_1, \dots, s_n$, then $\{u_\alpha\} = \{\det_\alpha X(t)\}$ is a cochain defining the corresponding section of $\det \pi$. By the Liouville–Ostrogradskii formula,

$$\Omega_\alpha = \dot{X}_\alpha \cdot X_\alpha^{-1}, \qquad \operatorname{tr} \Omega_\alpha = \dot{u}_\alpha / u_\alpha,$$

which proves that the two connexions are det-related. □

By definition of degree of the arbitrary bundle, we have an immediate corollary from Theorem 17.33 and Proposition 17.34.

Corollary 17.35 (Index theorem for connexions on a vector bundle). *For any meromorphic connexion π on a holomorphic vector bundle π over a compact Riemann surface,*

$$\sum_a \operatorname{res}_a \operatorname{tr} \nabla = \sum_a \operatorname{tr} \operatorname{res}_a \nabla = \deg \pi. \quad \square \tag{17.22}$$

17J. Classification of holomorphic vector bundles over $\mathbb{P}$. We conclude this section by a complete description of all holomorphic vector bundles over the Riemann sphere.

Theorem 17.36. *Any holomorphic vector bundle over the open unit disk $\mathbb{D}$ or the affine line $\mathbb{C}$, is holomorphically trivial.*

Theorem 17.37. *Any holomorphic vector bundle π over the Riemann sphere $\mathbb{P}$ is holomorphically equivalent to the direct sum of standard line bundles of different degrees*

$$\xi_D \overset{\text{def}}{=} \xi_{d_1} \oplus \cdots \oplus \xi_{d_n}, \qquad D = \operatorname{diag}\{d_1, \dots, d_n\}, \ d_i \in \mathbb{Z}.$$

The collection of integer numbers $\{d_1, \dots, d_n\}$, called the splitting type, *is defined by the bundle uniquely modulo permutation.*

These results will be derived from assertions on solvability and equivalence of matrix cocycles.

Consider first the simplest cocycles inscribed in a covering by two charts $U_0, U_1 \subset \mathbb{P}$ (they may not cover the entire sphere $\mathbb{P}$). Assume that both U_i are topological disks with piecewise-smooth boundaries in $\mathbb{C}$ and their intersection U_{01} is connected.

There are then two topologically different possibilities: either the intersection U_{01} is also a topological disk (bounded by piecewise-smooth curve), or U_{01} is a topological annulus.

In the first case the holomorphic cocycle inscribed in such a covering will be referred to as a *Cartan cocycle*.

Lemma 17.38. *Any Cartan cocycle is holomorphically solvable.*

Matrix cocycles inscribed in the covering of the second type, in which the intersection U_{01} is a topological annulus, will be referred to as the Birkhoff–Grothendieck cocycle, cf. with §17**D**. Without loss of generality we will assume that the covering is standard (formed by two circular disks centered at $t = 0$ and $t = \infty$ respectively).

Lemma 17.39. *Any Birkhoff–Grothendieck matrix cocycle* $\mathcal{H} = \{H_{01}, H_{10}\}$ *is equivalent to a Birkhoff–Grothendieck cocycle defined by the diagonal matrix function* $\{t^D, t^{-D}\}$ *with an integer diagonal matrix* $D = \operatorname{diag}\{d_1, \dots, d_n\}$.

In other words, Lemma 17.39 asserts that any holomorphic function $H_{01}(t)$ in the annulus $U_{01} = A$ admits factorization

$$H_{01}(t) = F_0(t) \cdot \begin{pmatrix} t^{d_1} & & \\ & \ddots & \\ & & t^{d_n} \end{pmatrix} \cdot F_1(t) \tag{17.23}$$

with the matrix functions F_0, F_1 holomorphic and invertible in the disks U_0, U_1 around $t = 0$ and $t = \infty$ respectively.

This very deep result can be viewed from different angles. The treatment based on the operator theory and integral equations can be found in the article [**GK60**]. In this article the authors construct the factorization (17.23) of a matrix function H_{01} of very weak regularity (defined on the circle $|t| = 1$ and merely integrable on it), and obtain the factors $F_{0,1}$ holomorphic invertible inside and outside this circle, so that the identity (17.23) is understood on the circle in the sense of the limit values.

An alternative approach uses methods of analytic matrix functions. The first step is to show that any cocycle can be solved in meromorphic rather than in holomorphic functions. In other words, we show that there are no analytic (nonalgebraic) obstructions for solvability of matrix cocycles.

Theorem 17.40. *Any Cartan or Birkhoff–Grothendieck cocycle is meromorphically solvable: there exists a pair of meromorphic and meromorphically invertible matrix functions* F_i *defined in the domains* U_i, $i = 0, 1$, *such that*

$$F_0 = H_{01} F_1 \qquad \text{on the intersection} \quad U_{01}. \tag{17.24}$$

Idea of the proof. In the noncommutative (matrix) case one cannot reduce the "multiplicative" matrix equation (17.24) to the "additive" equation by simply taking logarithms as in the proof of Proposition 17.13. Yet if H_{01} is a near identical cocycle, $H_{01} = E + \varepsilon H$ for a small parameter ε, then one can use the ansatz $F_i = E+\varepsilon G_i$, $i = 0, 1$, and "linearize" the equation (17.24), rewritten as $\varepsilon G_0 = \varepsilon(H+G_1)+O(\varepsilon^2)$, by keeping only terms of first order in ε. This linearized equation $G_0 = H+G_1$ is additive and can always be solved with respect to G_0, G_1 in holomorphic matrix functions by literally reproducing the proof of Lemma 17.14. From this solvability after some additional efforts one can derive *holomorphic solvability of the matrix equation* (17.24) for all *near-identical* holomorphic matrix cocycles. This step resembles solving a nonlinear integral equation using the resolvent of a linearized equation. Somewhat unexpectedly, the resolvent operation for the Birkhoff–Grothendieck case is bounded and the corresponding nonlinear equation can be solved using contracting mapping principle. In the Cartan case the resolvent operator (given by the Cauchy integral) is unbounded and one has to use an appropriate modification of the Newton–Kolmogorov method of accelerated convergence to overcome this difficulty.

Once the problem is solved for any near-identical cocycles, any other (not necessarily near-identical) matrix cocycle can be approximated with any specified accuracy by a *rational matrix cocycle*. The rational cocycles are obviously meromorphically solvable (it is sufficient to collect factors with poles in the corresponding charts). From this observation one can easily derive meromorphic solvability of an arbitrary cocycle.

Accurate demonstrations can be found in [**GR65**, §VI.E], [**AB94**, §3.3] and in the recent book [**Bol00**, Lecture 9]. □

The second part of the proof transforms meromorphic solution of a cocycle into holomorphic solution of this cocycle or a into a holomorphic conjugacy of it with some standard cocycle. It is this step in which the difference between noncompact ($\mathbb{C}$ or $\mathbb{D}$) and compact ($\mathbb{P}$) base plays the key role. We will derive Lemmas 17.38 and 17.39 from Theorem 17.40 by elementary row and column operations with matrix functions.

Recall that an elementary operation on rows of a matrix is one of the following three:

(1) transposition of two rows of a matrix,

(2) adding to one of the rows a linear combination of other rows,

(3) multiplication of a row by a nonzero scalar.

Each elementary operation can be achieved by the left multiplication of the matrix by an appropriate *elementary matrix*. Except for the third type, the determinant of the corresponding elementary matrix is 1. Three parallel elementary operations on columns of a matrix can be achieved by an appropriate right multiplication.

In an obvious way, these elementary operations can be generalized for meromorphic matrix functions: transformations of the second type consist of adding to a row of a matrix function a linear combination of other rows with meromorphic coefficients. Transformations of the third type consist of multiplication of a row by a nonzero meromorphic function. Elementary operations on columns of meromorphic matrix functions are also self-explanatory.

Proof of Lemma 17.38. By Theorem 17.40, any Cartan cocycle can be resolved by a *meromorphic* cochain $\{F_0, F_1\}$. We will implement a series of modifications transforming this meromorphic cochain to a holomorphic cochain.

First, the meromorphic cochain can be modified so that all matrix functions $F_i(t)$ become holomorphic in the corresponding domains $U_i \subseteq \mathbb{C}$. To that end, all functions $F_i(t)$ should be multiplied by a suitable scalar power $(t - t_k)^{\nu_k}$, $\nu_k \in \mathbb{N}$, for each finite pole t_k of order ν_k. Clearly, the determinants of the holomorphic matrices $F_i(t)$ obtained by such multiplication, remain not identically vanishing, though they still may have isolated zeros of finite order.

In order to get rid of these zeros, we will further multiply F_i simultaneously by rational matrix functions from the right (this operation obviously will preserve the identity $H_{01}F_1 = F_0$). If t_* is an isolated root of, say, $\det F_1(t)$, then one of the columns of the matrix $F_1(t_*)$ is a linear combination of other columns, so that after the right multiplication by an appropriate constant matrix C one of the columns of $F_1(t_*)$ becomes zero. Then all entries from this column of the matrix function $F_1(t)C$ have the common factor $(t - t_*)$. After the right multiplication by the rational matrix function $R(t) = \operatorname{diag}\{1, \dots, (t - t_*)^{-1}, \dots, 1\}$, the modified matrix function $F_1(t)CR(t) = F_1'(t)$ remains holomorphic at t_*, and so apparently is $F_0'(t) = H_{01}(t)F_1'(t) = F_0(t)CR(t)$.

The total number of zeros of $\det F_i'(t)$, counted with multiplicities in $\mathbb{C}$, will decrease by 1 compared to that of $\det F_i(t)$. After a finite number of such steps we will get rid of all zeros of the determinant. The resulting cochain will resolve the cocycle, since by definition of the Cartan cocycle, both U_0 and U_1 belong to the finite part $\mathbb{C}$. The proof of Lemma 17.38 is complete. □

The proof of Lemma 17.39 requires the following result, known as the Sauvage lemma [**Har82**]. Let $(\mathbb{P}, \infty)$ be a small circular neighborhood of infinity. Any matrix function $H(t) = H_{01} \in \operatorname{Mat}(n, \mathfrak{M}(\mathbb{P}, \infty))$, meromorphic and not identically zero in this neighborhood, can be considered as a *cocycle* on the covering of the Riemann sphere by two charts, $U_0 = \mathbb{C}$ and $U_1 = (\mathbb{P}, \infty)$, which intersect by the punctured disk, itself a limit case of an annulus. We will refer to such cocycle as a *Sauvage cocycle*

Lemma 17.41 (Sauvage lemma). *Any Sauvage cocycle is holomorphically equivalent to a standard matrix cocycle $\{t^D\}$ with an appropriate diagonal integer matrix D, inscribed in the same covering.*

Proof. The proof is achieved by a series of suitable monopole gauge transforms which realize elementary matrix transformations bringing the Sauvage cocycle to a diagonal form.

1. If the germ $H(t)$ is holomorphic at $(\mathbb{P}, \infty)$ and degenerate at $t = \infty$, then there exists a *constant upper-triangular* matrix C and a holomorphic germ $H'(t)$ such that

$$CH(t) = t^{D'} H'(t), \qquad D' = \operatorname{diag}\{0, \dots, -1, \dots, 0\}. \tag{17.25}$$

Indeed, if $\det H(\infty) = 0$, then the rows of the *constant* matrix $H(\infty)$ must be linearly dependent, in particular, some row of it must be equal to a linear combination of the subsequent (relatively lower) rows. In other words, there exists an *upper*-triangular constant matrix C with determinant 1, such that the matrix $CH(\infty)$ has a zero row. But then this same row of the matrix function $CH(t)$ is divisible by t^{-1}, so that the matrix $H'(t) = t^{-D'} CH(t)$ is holomorphic at $t = \infty$.

Clearly, the order of zero of $\det H'(t)$ is strictly inferior (by one less) than the order of zero of $\det H(t)$:

$$\operatorname{ord}_\infty \det H'(t) = \operatorname{ord}_\infty \det H(t) - 1. \tag{17.26}$$

2. If D is an integer diagonal matrix $D = \operatorname{diag}\{d_1, \dots, d_n\}$ with non-increasing entries $d_1 \geqslant \cdots \geqslant d_n$, and $H(t)$ is holomorphic and degenerate at infinity, then the product $t^D H(t)$ is monopole equivalent to $t^{D+D'} H'(t)$ with D' and $H'(t)$ as above.

Indeed, by Step 1, there exists a constant upper-triangular matrix C such that $CH(t) = t^{D'} H'(t)$ with holomorphic $H'(t)$ satisfying (17.26). Consider the conjugacy of C by t^D, $\Pi(t) = t^D C t^{-D}$. Because of the upper-triangularity of C and monotonicity of the sequence $\{d_i\}$, the matrix function $\Pi(t)$ is an upper-triangular monopole. Since D and D' commute,

$$\Pi(t)\, t^D H(t) = t^D C t^{-D} \cdot t^D H = t^D CH = t^D t^{D'} H' = t^{D+D'} H'.$$

3. For an arbitrary diagonal matrix D one can find a *constant* permutation matrix $P \in \mathrm{GL}(n, \mathbb{C})$ (particular case of monopole) such that the diagonal entries of $D' = Pt^D P^{-1}$ will be monotonous as required in Step 2. This shows that the condition on the order of the diagonal entries d_i, imposed in Step 2, can be always achieved by a suitable monopole equivalence (left multiplication by P):

$$Pt^D H = Pt^D P^{-1} \cdot PH = t^{D'} H',$$

with a holomorphic H' degenerate at infinity together with H.

4. The proof of the Sauvage lemma can be achieved by simple induction. Any meromorphic germ $H(t)$ can be represented as $t^{D_1}H_1(t)$ with $H_1(t)$ holomorphic at infinity: it is sufficient to multiply $H(t)$ by a suitable (scalar) power of t. Since $\det H(t) \not\equiv 0$, the multiplicity of the root of $\det H_1(t)$ at $t = \infty$ is finite. The inductive application of the construction described above in Steps 1–3, allows us to construct a sequence of monopole transformations reducing $H_1(t)$ to the form of a product of two terms, $t^{D_k}H_k(t)$ as above (diagonal and holomorphic at infinity respectively), with strictly decreasing orders of the roots $\operatorname{ord}_\infty \det H_k(t)$. After finitely many steps the holomorphic term $H_m(t)$ becomes nondegenerate at infinity, and the Sauvage lemma is proved. □

Proof of Lemma 17.39. Proceeding as in the proof of Lemma 17.38, we may assume without loss of generality that the Birkhoff–Grothendieck cocycle H_{01} is already solved by the meromorphic cochain $\{F_0, F_1\}$ such that both F_0, F_1 are holomorphic and holomorphically invertible everywhere in their domains, possibly except for the point $t = \infty$, where F_1 has a finite order pole.

By the Sauvage lemma 17.41, the meromorphic matrix function germ $F_1^{-1}(t)$ can be represented as a composition $F_1^{-1} = \Pi(t)\,t^D G(t)$ with a polynomial and polynomially invertible (monopole) function $\Pi(t)$ and holomorphically invertible germ $G(t)$ at $t = \infty$. The matrix function $G_1 = t^{-D}\Pi^{-1}F_1^{-1}$ defined on the entire domain U_1, is holomorphic and holomorphically invertible in this domain. Indeed, since terms in the latter equality are holomorphically invertible in $U_1 \smallsetminus \{\infty\}$, while at the point $t = \infty$ the germ of this composition is G. Substituting the expression for $F_1 = G_1^{-1}t^{-D}\Pi^{-1}$ into the identity $H_{10}(t)F_0(t) = F_1(t)$, we get

$$H_{10}F_0 = G_1^{-1}t^{-D}\Pi, \qquad \text{i.e.,} \qquad H_{10}F_0\Pi^{-1} = G_1^{-1}t^{-D}.$$

In other words, the holomorphic cochain $\{F_0\Pi^{-1}, G_1^{-1}\}$, conjugates the initial Birkhoff–Grothendieck cocycle $\mathcal{H} = \{H_{01}\}$ with the standard cocycle $\{t^{-D}\}$. □

Proof of Theorems 17.36 and 17.37. The proof of both these theorems is achieved by literally the same arguments as the proof of Theorem 17.16. Namely, we consider a "triangulated" covering and consecutively resolve the Cartan cocycles using Lemma 17.38, until the disk $\mathbb{D}$ is exhausted. In the case of the Riemann sphere $\mathbb{P}$ we can replace the initial cocycle by an equivalent Birkhoff–Grothendieck cocycle. Then Lemma 17.39 proves that this cocycle is equivalent to one of the standard cocycles corresponding to the vector bundle ξ_D.

It remains only to prove the uniqueness of the splitting type D (clearly, bundles with permuted linear subbundles are equivalent). Assume that there exists a holomorphic bundle map between two bundles of different types D and D'. Then there exist a holomorphic matrix cochain $\{H_0, H_1\}$ inscribed in the Birkhoff–Grothendieck covering, such that

$$H_1 = t^D H_0 \, t^{-D'}, \qquad H_i \in \mathrm{GL}(n, \mathcal{O}(U_i)).$$

Consider an arbitrary matrix element of the form $h_{ij}^0(t)\, t^{d_i - d_j'}$ in the right hand side. If $d_i \geqslant d_j'$, then this element is holomorphic both in U_0, since h_{ij}^0 is holomorphic there, and in U_1, since it is equal to $h_{ij}^1(t) \in \mathcal{O}(U_1)$. This is possible only if h_{ij}^0 is a constant, necessarily zero if $d_i > d_j'$.

Assume that the two tuples of numbers $d_1, \dots, d_n$ and $d_1', \dots, d_n'$ are both arranged in the nonincreasing order. Consider the largest elements d_1 and d_1'. If $d_1 > d_1'$, then the matrices H_0, H_1 will have identically zero first row, contrary to their nondegeneracy. For reasons of symmetry, the strict inequality $d_1' < d_1$ is also impossible. This leaves only one possibility, $d_1 = d_1'$. Let k be the first place when the numbers d_k and d_k' are different.

If $d_k > d_k'$, then the matrix function $H_0(t)$ is block-upper-triangular with the upper $k \times k$-block having identically zero last row. Such a matrix is identically degenerate contrary to the assumption on the cochain $\{H_0, H_1\}$. Thus $d_k \leqslant d_k'$. For reasons of symmetry we also have $d_k' \leqslant d_k$, i.e., $d_k = d_k'$.

In other words, after arranging in the same nonincreasing order, both splitting types D and D' must coincide; but this means that they are permutations of each other. □

Exercises and Problems for §17.

Problem 17.1. Let $h_\alpha \colon U_\alpha \to \mathbb{C}^n$ be an atlas of charts for some open covering $\mathfrak{U}$ of a holomorphic manifold M. Write explicitly the trivializations for the tangent and cotangent bundles $\mathbf{T}M$ and $\mathbf{T}^*M$.

Exercise 17.2. Prove that two equivalent holomorphic cochains define two holomorphic equivalent vector bundles over the same base.

Problem 17.3. Let $\mathcal{H}, \mathcal{H}'$ be two cocycles (of different size matrices) corresponding to the vector bundles S, S' respectively. Construct explicitly the cocycles associated with the direct sum $S \oplus S'$ and the tensor product $S \otimes S'$.

Problem 17.4. Let $S' \subset S$ be a subbundle. Prove that the cocycle $\mathcal{H}$ associated with S, is equivalent to a cocycle of block upper-triangular matrices. Describe the cocycle associated with the quotient bundle $S'' = S/S'$.

Exercise 17.5. Prove that among all cocycles $\{t^d\}$ on the Riemann sphere, only the cocycle with $d = 0$ is solvable.

Exercise 17.6. Prove that for a given meromorphic section of a line bundle, there exists a unique connexion for which this section is horizontal.

Problem 17.7. Prove that the line bundle ξ_d over the projective line $\mathbb{P}$ admits nontrivial holomorphic sections if and only if its degree d is nonnegative.

Problem 17.8. Prove that the line bundle ξ_d over the projective line $\mathbb{P}$ admits nontrivial automorphisms different from multiplication by a constant factor, if and only if its degree d is negative.

Problem 17.9. Prove that the tangent bundle $\mathbf{T}\mathbb{P}$ and cotangent bundle $\mathbf{T}^*\mathbb{P}$ over the Riemann sphere have degrees 2 and -2 respectively.

Problem 17.10. Prove that a holomorphic bundle of rank n admits n holomorphic sections linearly independent in each fiber, if and only if the bundle is equivalent to the trivial one.

Exercise 17.11. Prove from the definition that the notion of connexion is local. More precisely, prove that for any two meromorphic sections s, s' of the same bundle, both holomorphic at a point $a \in T$ and with the same 1-jet, the respective values coincide, $\nabla s(a) = \nabla s'(a) \in \pi^{-1}(a)$.

Exercise 17.12. Prove that any connexion on a bundle of rank n is completely determined by n linearly independent horizontal sections: if two connexions have n common horizontal sections, then they coincide as differential operators.

Problem 17.13. Let $\pi_0 \colon T \times \mathbb{C}^n \to T$ be a trivial bundle over a simply connected holomorphic manifold T and ∇ a holomorphic connexion on it (holomorphic means meromorphic without singularities).

Prove that a collection of n *horizontal* holomorphic sections linearly independent in each fiber over a neighborhood U of a point a exists if and only if the connexion matrix form $\Omega = (\omega_{ij})_{i,j=1}^n$, $\omega_{ij} \in \Lambda^1(T) \otimes \mathcal{M}(T)$, satisfies the equation

$$d\Omega - \Omega \wedge \Omega = 0, \tag{17.27}$$

in a neighborhood of the point a, where

$$d\Omega = (d\omega_{ij})_{i,j=1}^n, \qquad \Omega \wedge \Omega = \Bigl(\sum\nolimits_k \omega_{ik} \wedge \omega_{kj}\Bigr)_{i,j=1}^n$$

are two matrix-valued 2-forms on T.

Problem 17.14 (**[Bol00]**). Find the splitting type (collection of the indices $d_1, \dots, d_n$) for the bundles defined by the Birkhoff–Grothendieck cocycles

$$\begin{pmatrix} t & \lambda \\ & t^{-1} \end{pmatrix}, \qquad \begin{pmatrix} t & \\ \lambda & t^{-1} \end{pmatrix} \tag{17.28}$$

Problem 17.15. Let $\mathcal{H}$ be a holomorphically solvable Birkhoff–Grothendieck cocycle (say, rational). Prove that any other rational cocycle $\mathcal{H}'$ sufficiently close to $\mathcal{H}$ in the annulus $A = U_0 \cap U_1$, is also solvable.

Problem 17.16 (Yu. L. Shmul'yan, 1954). Assume that the splitting type $d_1 \leqslant \dots \leqslant d_n$ of a Birkhoff–Grothendieck cocycle $\mathcal{H}$ has at most one gap, i.e., $d_n - d_1 \leqslant 1$. Prove that any close cocycle has the same splitting type. Give an example showing that this is not necessarily the case if $d_n - d_1 > 1$.

Exercise 17.17. Prove that the degree of the bundle ξ_D is equal to $|D| = d_1 + \cdots + d_n$.

Problem 17.18. Prove that any holomorphically invertible matrix function $F(t)$ in the annulus $A = U_0 \cap U_1$ can be factored out as $F(t) = H_0(t)H_1(t)\,t^D$ with the terms $H_i(t)$ holomorphically invertible in U_i, $i = 0, 1$, and an integer diagonal matrix D. It is this form that is sometimes called the *Birkhoff factorization.*

In particular, any nonzero meromorphic germ of a matrix function $F(t)$ at the infinity admits factorization $F(t) = \Pi(t)H(t)\,t^D$ with a monopole $\Pi(t)$ and a holomorphically invertible germ $H(t)$ at infinity.

Problem 17.19 ([**Bol00**]). Prove that a holomorphic vector bundle $\pi \colon S \to T$ is *topologically* trivial if and only if its degree is equal to zero. The topological triviality means that there exists a *homeomorphism* $F \colon S \to T \times \mathbb{C}^n$ fibered over the identity and linear on each fiber.

18. Riemann–Hilbert problem

> The problem is as follows: *To show that there always exists a linear differential equation of the Fuchsian class, with given singular points and monodromy group.* The problem requires the production of n functions of the variable z, regular throughout the complex z-plane except at the given singular points; at these points the functions may become infinite of only finite order, and when z describes circuits about these points the functions shall undergo the prescribed linear substitutions.
>
> *D. Hilbert*, 1901, reprinted from [**Hil00**]

The *Riemann–Hilbert problem*, also known as Hilbert's twenty-first problem, requires constructing a linear system with the prescribed monodromy group and positions of all singularities. The original formulation by Hilbert is somewhat confusing, since the clarification given in the text after it, describes only the regularity condition, while the main formulation explicitly mentions Fuchsian systems.

One can think of not one but rather *three* different accurate formulations, when a given monodromy group is required to be realized by:

(i) a Fuchsian linear nth order differential equation,

(ii) a linear system having only regular singularities, or

(iii) a Fuchsian system on the whole Riemann sphere $\mathbb{P}$.

In each case it is required that the equation (resp., the system) be nonsingular outside the preassigned points.

The negative answer in the first problem was known already by A. Poincaré: the reason is that the dimension of the space of all Fuchsian equations having m prescribed singular points on $\mathbb{P}$, is strictly smaller than the dimension of all admissible monodromy data, except for the case of second order equations with three singular points studied by Riemann. The corresponding problem is discussed in §19**F**.

J. Plemelj [**Ple64**] gave a solution of problem (ii) while claiming solution of the strongest problem (iii). The gap was discovered by Yu. Ilyashenko [**AI85**] and A. Treibich [**Tre83**] in the earlier eighties. The positive part of the Plemelj theorem is described in §18**B**.

Yet only recently it became clear that there is a substantial difference between the formulations (ii) and (iii). It was proved independently by A. Bolibruch [**Bol92**] and V. Kostov [**Kos92**] that an irreducible monodromy group can always be realized by a Fuchsian system. In this section we explain a remarkably simple proof of the Bolibruch–Kostov theorem which was communicated to us by the late A. Bolibruch.

However, for a reducible monodromy group the answer to problem (iii) may be negative. The counterexample, also due to Bolibruch, is described in §18**E**.

The way to understand reasons and obstructions for solvability of the Riemann–Hilbert problem passes through its generalization, the Riemann–Hilbert problem for meromorphic connexions on holomorphic vector bundles. The "elementary" (analytic) demonstration of these results, is given in 16.

18A. Riemann–Hilbert problem on abstract bundles. In invariant terms the Riemann–Hilbert problem can be formulated as follows: construct a meromorphic connexion on the *trivial* bundle over the Riemann sphere, having preassigned Fuchsian singular points at the specified points and the preassigned holonomy group.

In the category of abstract vector bundles the Riemann–Hilbert problem becomes in a sense trivial: *any collection of matrix connexion forms* can be realized by a meromorphic connexion on a suitable holomorphic vector bundle.

We start by choosing a special system of generators for the monodromy group. Consider m distinct points $a_1, \dots, a_m$ on the affine plane $\mathbb{C} \subset \mathbb{P}$. By choosing an appropriate affine chart one can always guarantee that $a_i/a_j \notin \mathbb{R}_+$, i.e., that the line segments $[0, a_j]$ connecting the origin with these points are pairwise disjoint except for the common origin.

Definition 18.1. The *canonical loops* generating the fundamental group of the Riemann sphere with finitely many deleted points $\mathbb{P} \smallsetminus \Sigma$, $\Sigma = \{a_1, \dots, a_m\}$ are the loops which consist of line segments connecting the origin with each singular point $a_i \in \mathbb{C}$, encircling the latter along a small counterclockwise circular arc and then returning along the same segment in the opposite direction; see Fig. III.2. All circular arcs are pairwise disjoint.

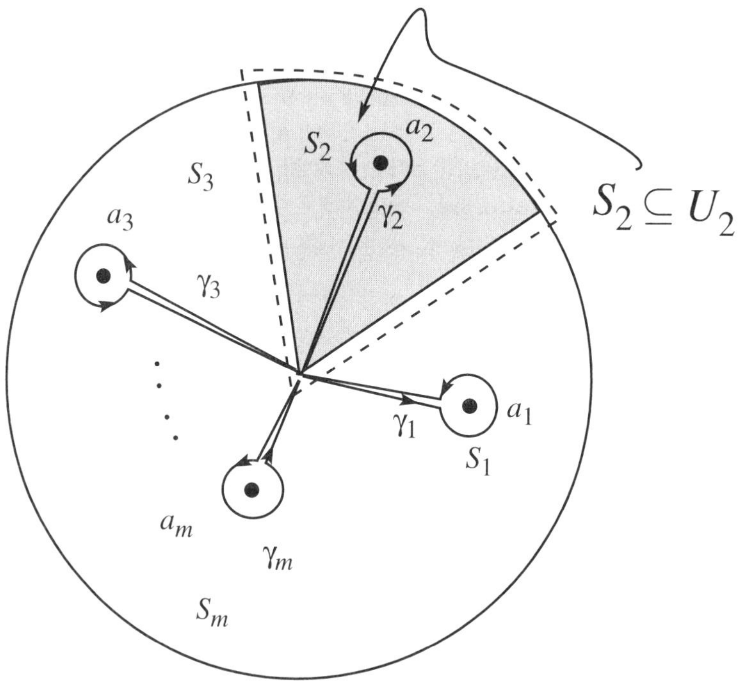

Figure III.2. Canonical loops and specification of the monodromy data

The fundamental group $\pi_1(\mathbb{P} \setminus \Sigma, 0)$ is generated by the canonical loops γ_i, $i = 1, \ldots, m$, related by a single identity $\gamma_1 \circ \cdots \circ \gamma_m = \mathrm{id}$. We will always assume that the points are numbered counterclockwise (see Fig. III.2) and cyclically, i.e., the point a_m follows after a_{m-1} and is in turn followed by a_1. Denote by $U_0 \subseteq \mathbb{C}$ the disk $\{|t| < R\}$ containing all points a_j.

Recall that the monodromy data is a collection of m points $a_1, \ldots, a_m$ as above and invertible linear operators $M_1, \ldots, M_m \in \mathrm{GL}(n, \mathbb{C})$ such that their product in the specified order is the identity; see (16.16).

Definition 18.2. The monodromy data is *realized* by a meromorphic connexion ∇ on a holomorphic vector bundle of rank n over $\mathbb{P}$, if the singular points of the connection are $a_j, j = 1, 2, ..., m$, and the holonomy Δ_j (the linear self-map of the fiber $\tau_0 = \pi^{-1}(0) \cong \mathbb{C}^n$ associated with each canonical loop γ_j) coincides with M_j for all $j = 1, \ldots, m$.

Example 18.3 (Realization of a single operator)**.** Every single holonomy operator can be immediately realized by the holonomy of a Fuchsian system. Indeed, let $U = U_j \subset \mathbb{C}$ be a simply connected domain containing both the origin and the point a_j. Then the holonomy operator for the Fuchsian matrix

1-form

$$\Omega_j = \frac{A_j\,dt}{t-a_j} \in \Lambda^1(U_j)\otimes \mathfrak{M}(U_j), \qquad \exp 2\pi i A_j = M_j, \tag{18.1}$$

associated with the fiber $\{0\}\times\mathbb{C}^n$, coincides with M_j. Recall that the equation $\exp 2\pi i A_j = M_j$ is solvable for any nondegenerate matrix M_j by Lemma 3.11.

Note that the realization is by no means unique: besides the freedom of choice for the matrix logarithm discussed in §3**D**, one can also construct (in the resonant case) a non-Euler system.

We show now how an arbitrary monodromy data for several singularities can be realized as a holonomy of a Fuchsian connexion on an abstract bundle. Consider a collection of meromorphic matrix 1-forms Ω_j, $j=1,\dots,m$, such that each form is meromorphic in U_0 and has a unique pole at a_j. The collection $\{\Omega_j\}_1^m$ of such meromorphic matrix 1-forms will be called *admissible*, if $\Delta_m\circ\cdots\circ\Delta_1=\mathrm{id}$. This happens automatically if each Ω_j realizes the holonomy operator M_j from the monodromy data $\{M_1,\dots,M_m\}$.

Theorem 18.4. *For any admissible collection of meromorphic 1-forms*

$$\begin{gathered}\Omega_j\in\Lambda^1(U_0)\otimes\mathfrak{M}(U_0), \qquad \operatorname{Sing}\Omega_j=\{a_j\}, \qquad j=1,\dots,m,\\ \Delta_j=\Delta_{\gamma_j}\in \mathrm{GL}(\tau_0), \qquad \Delta_m\circ\cdots\circ\Delta_1=\mathrm{id},\end{gathered} \tag{18.2}$$

there exists a holomorphic vector bundle $\pi\colon S\to\mathbb{P}$ of rank n over the Riemann sphere and a meromorphic connexion ∇ on this bundle such that the singular locus of ∇ coincides with $\Sigma=\{a_1,\dots,a_m\}$ and at each singular point a_k the connexion is locally biholomorphically equivalent to the connexion $d-\Omega_k$.

In other words, one can construct holomorphic bundles over $\mathbb{P}$ with any preassigned holonomy group, specifying in addition the types of singularities (regular, Fuchsian or even arbitrary irregular) as well as their position. Of course, there is no guarantee that the bundle obtained this way, will be trivial.

Proof. The assertion of the theorem is largely a tautology very similar to that asserted in Theorem 17.6. The accurate proof consists of two steps.

On the first step we construct a holomorphic bundle $\pi\colon S\to U_0$ over the large disk U_0 and a meromorphic connexion on it with the specified holonomy operators. Because of the admissibility, the holonomy associated with the boundary of the disk is identical. On the second step we "seal" the hole at infinity, constructing a holomorphic vector bundle over $\mathbb{P}$.

We construct explicitly the cocycle which defines the bundle π over the disk U_0 as follows. To define the covering, we slice the disk into sectors

$S_j = \{\alpha_j \leqslant \operatorname{Arg} t \leqslant \alpha_{j+1},\ |t| < R\}$ in such a way that each sector S_j contains only one singular point, and consider the covering of U_0 by the open domains U_j, $j = 1, \dots, m$, which are small ε-neighborhoods of these sectors. The number ε is chosen so small that the intersections $U_{j,j+1} = U_j \cap U_{j+1}$, the ε-neighborhoods of the rays $\operatorname{Arg} t = \alpha_j$, are all disjoint from the singular locus Σ. Note that the origin $t = 0$ belongs to all domains U_j.

If we slit each domain U_j along the radius connecting the corresponding point a_j with the boundary of the disk U_0, then none of these slits intersect the pairwise intersections U_{ij}. On the other hand, in the slit domains we may define holomorphic invertible matrix solutions $X_j(t)$ of the matrix differential equations $dX_j = \Omega_j X_j$ with the initial condition $X_j(0) = E$.

Define the holomorphic matrix cocycle

$$H_{ij} = X_i \cdot X_j^{-1} \qquad \text{on} \quad U_{ij}. \tag{18.3}$$

The cocycle identities are obviously satisfied, and differentiating (18.3), we conclude that

$$dH_{ij} = dX_i \cdot X_j^{-1} + X_i(-X_j^{-1} dX_j \cdot X_j^{-1}) = \Omega_i H_{ij} - H_{ij}\Omega_j.$$

Let $\pi\colon S \to U_0$ be the holomorphic vector bundle $\pi\colon S \to U_0$ over the disk U_0, described in Theorem 17.6, for which the cocycle $\mathcal{H} = \{H_{ij}\}$ is the collection of transition maps. Then the collection of the matrix forms Ω_i defines a meromorphic connexion ∇ on S with the polar locus Σ. Since $H_{ij}(0) = E$, the holonomy maps of this connexion, associated with the section $\pi^{-1}(0)$ and the loops γ_j, coincide with the prescribed linear operators Δ_j. In particular, the holonomy of the boundary circumference of the disk U_0 is trivial by (18.2).

To "seal the gap" and extend the bundle π just constructed over U_0 on the disk $\mathbb{P} \smallsetminus U_0$, we consider the trivial bundle of the same rank n over the disk $U_1 = \{|t| > R - \varepsilon\} \subset \mathbb{P}$ on the Riemann sphere, equipped with the trivial connexion $\nabla = d$.

Any linear invertible map of a fiber $\pi^{-1}(a) \to \mathbb{C}^n$, $a \in U_{01} = U_0 \cap U_1$, can be extended uniquely as a holomorphic gauge map $H_{01}\colon \pi^{-1}(U_{01}) \to U_{01} \times \mathbb{C}^n$ fibered over the identity map of the annulus U_{01}, which sends horizontal sections of ∇ to the horizontal (constant) sections of the trivial bundle. In a standard way we can now construct the holomorphic bundle over the union $U_0 \cup U_1 = \mathbb{P}$ with a holomorphic connexion on it, without singularities outside U_0 and the prescribed holonomy group in U_0. □

18B. Connexions on the trivial bundle. If the abstract bundle $\pi\colon S \to \mathbb{P}$ constructed in Theorem 18.4 is *holomorphically* equivalent to the trivial bundle $\pi_0\colon \mathbb{P} \times \mathbb{C}^n \to \mathbb{P}$, the globally defined connexion matrix would solve the Riemann–Hilbert problem in the classical sense. However, this holomorphic triviality may be only accidental, and in general the bundle will be nontrivial.

Nevertheless we can assume that the bundle is already in the standard Birkhoff–Grothendieck normal form, i.e., a pair of trivializations is chosen so that the transition cocycle between them is the standard matrix t^D.

For such a standard bundle we will construct an explicit *meromorphic trivialization*, a bundle map $F\colon S \to \mathbb{P} \times \mathbb{C}^n$ with a single pole at infinity. This bundle map is given by the cochain $\mathcal{F} = \{\mathrm{id}, t^{-D}\}$, where D is the splitting type.

The trivializing map F carries the connexion ∇ on S to a meromorphic connexion on the trivial bundle over $\mathbb{P}$. The resulting connexion has the same holonomy group, yet its singularity at infinity will in general only be regular non-Fuchsian.

The bundle π constructed in Theorem 18.4 is holomorphically equivalent to the standard Birkhoff–Grothendieck bundle, the equivalence being defined by a holomorphic matrix cochain $\mathcal{G} = (G_0, G_1)$. The meromorphic bundle map F which trivializes π, is the composition $\{F_0 = G_0, F_1 = t^{-D}G_1\}$, whose components are columns of the commutative diagram

$$
\begin{array}{ccc}
U_0 \supset A & \xrightarrow{H} & A \subset U_1 \\
{\scriptstyle G_0}\big\downarrow & & \big\downarrow{\scriptstyle G_1} \\
U_0 \supset A & \xrightarrow{t^D} & A \subset U_1, \\
{\scriptstyle \mathrm{id}}\big\downarrow & & \big\downarrow{\scriptstyle t^{-D}} \\
U_0 \supset A & \xrightarrow{\mathrm{id}} & A \subset U_1
\end{array}
\qquad
\begin{aligned}
A &= U_0 \cap U_1, \\
\mathbb{P} &= U_0 \cup U_1, \\
G_0 &\in \mathrm{GL}(n, \mathcal{O}(U_0)), \\
G_1 &\in \mathrm{GL}(n, \mathcal{O}(U_1)).
\end{aligned}
\tag{18.4}
$$

The upper square of this diagram is the holomorphic equivalence of the bundle π and the standard bundle ξ_D, the lower square is the meromorphic trivialization.

The Fuchsian connexion ∇ on the bundle π constructed in Theorem 18.4, is F-related with a connexion ∇_0 on the trivial bundle π_0. Yet this connexion is obviously regular, even if the point $t = \infty$ was singular (Fuchsian) for ∇. This immediately implies the affirmative solution of Problem (ii) (p. 312).

Theorem 18.5. *Any monodromy group can be realized by a regular linear system on the Riemann sphere.*

Moreover, the regular system can always be constructed with all singularities Fuchsian, except for at most one. □

Somewhat embarrassingly, the singularity of ∇_0 created at the point $t = \infty$, is non-Fuchsian even if this point was nonsingular for ∇. Yet the regular singular point at infinity for the connexion ∇_0 sometimes can be further simplified using monopole gauge transforms. Recall that the *monopole gauge transforms* are meromorphic gauge self-maps of the trivial bundle, which are nonsingular at all points of $\mathbb{P}$ except the point $t = \infty$ (cf. Definition 16.34).

The following result was first proved[4] by J. Plemelj in [**Ple64**].

Theorem 18.6. *If at least one monodromy operator M_j is diagonalizable, then the corresponding monodromy data can be realized by the holonomy of a Fuchsian system on $\mathbb{P}$.*

Proof. Consider the abstract bundle π realizing the specified holonomy group as in Theorem 18.4. Without loss of generality we may assume that the bundle is trivialized over two charts U_0, U_1 by a Birkhoff–Grothendieck cocycle H and the connexion ∇ is represented by two meromorphic matrix 1-forms Ω_0, Ω_1. Again without loss of generality we may assume that the diagonalizable monodromy operator corresponds to the singular point $t = \infty$ and the corresponding Fuchsian connexion form over U_1 is already diagonal and is an Euler system in the standard chart,

$$dX = \Omega_1 X, \qquad \Omega_1 = \Lambda \tfrac{dt}{t}, \quad \Lambda = \operatorname{diag}\{\lambda_1, \dots, \lambda_n\}. \tag{18.5}$$

Consider the meromorphic gauge transform (18.4) which trivializes the bundle π. This transform carries ∇ to the meromorphic connexion ∇_0 on the trivial bundle, defined by the single meromorphic matrix 1-form $\Omega \in \operatorname{Mat}(n, \Lambda^1(\mathbb{P}) \otimes \mathcal{M}(\mathbb{P}))$ with all singularities in the finite plane already Fuchsian.

The singularity at infinity is regular non-Fuchsian and has a fundamental (multivalued) matrix solution of the form $X(t) = t^{-D} G_1 t^{\Lambda}$, as follows from the explicit form (18.5) and the diagram (18.4), where G_1 is a holomorphic invertible matrix function near $t = \infty$.

Inverting the order of matrix terms by Lemma 16.36, we may rewrite the solution $X(t)$ under the form

$$X(t) = \Pi^{-1}(t) G'(t)\, t^{-D'} t^{\Lambda}, \qquad G' \in \operatorname{GL}(n, \mathcal{O}(\mathbb{P}, \infty)), \quad \Pi \in \operatorname{GL}(n, \mathbb{C}[t]).$$

After application of the monopole gauge transform Π we obtain a matrix form $\Omega' = d\Pi \cdot \Pi^{-1} + \Pi \Omega \Pi^{-1}$ with all finite singularities still Fuchsian (as the order of pole cannot be changed by a holomorphic local gauge equivalence) and the regular singular point at infinity, having a fundamental matrix solution $X'(t) = G'(t)\, t^{\Lambda - D'}$, since diagonal matrices commute and $t^{-D'} t^{\Lambda} = t^{\Lambda - D'}$.

[4] The assumption on diagonalizability was missing in [**Ple64**], as was noted by Ilyashenko and Treibich.

From this representation it follows immediately that Ω' has a first order pole at infinity with the principal term conjugate to $(\Lambda - D')\frac{dt}{t}$, that is, Ω' is Fuchsian also at infinity. □

18C. Invariant subbundles and irreducibility. Solvability of the Riemann–Hilbert problem for an arbitrary monodromy data is determined to a very large extent by existence and structure of *invariant subspaces* of the holonomy.

Let $\pi\colon S \to T$ be an arbitrary holomorphic vector bundle with a meromorphic connexion ∇ on it.

Definition 18.7. A subbundle $L \subset S$ is called ∇-invariant, if fibers of this subbundle are mapped into each other by all the horizontal transport operators.

In other words, L is invariant, if any parallel transport operator Δ_γ between two fibers τ_a, τ_b along any path γ connecting these points in $T \smallsetminus \Sigma$, maps $L_a = L \cap \tau_a$ into $L_b = L \cap \tau_b$.

A subspace $\ell \subset \mathbb{C}^n$ is invariant by a linear group $\mathfrak{G} \subset \mathrm{GL}(n, \mathbb{C})$, if it is invariant by all operators from the group. Obviously, if for a finitely generated group $\mathfrak{G} = \langle M_1, \dots, M_m \rangle$ it is sufficient to verify invariance only by the generators.

Since monodromy operators Δ_γ for all loops $\gamma \in \pi_1(T \smallsetminus \Sigma, a)$ are a special class of parallel transport maps, any subbundle $L \subset S$ invariant for a meromorphic connexion generates the invariant subspace $\ell_a = L \cap \tau_a$ for the monodromy group, regardless of the type of singular points. The inverse statement is true only for regular connexions (cf. with Problem 18.4).

Proposition 18.8. *Let ∇ be a* regular *meromorphic connexion on a holomorphic bundle $\pi\colon S \to T$.*

If $\ell \subset \tau_a = \pi^{-1}(a)$ is a linear subspace (sub-fiber) invariant by all holonomy operators Δ_γ, $\gamma \in \pi_1(T \smallsetminus \Sigma, a)$, then there exists a holomorphic subbundle $L \subset S$ invariant by ∇ and extending ℓ, so that $L \supset \ell$.

Proof. The only candidate for such a subbundle is the saturation of ℓ by horizontal sections. We show that this saturation is indeed a holomorphic subbundle of S, namely, it extends holomorphically at all regular singularities.

1. By parallel transport along a path connecting the base point a with any nonsingular point $t \notin \Sigma$, we can carry the subspace ℓ to a subspace $L(t)$ in the fiber $\pi^{-1}(t)$. The result of this transport does not depend on the choice of the path, since ℓ is invariant by all holonomy operators.

The subspaces $L(t)$, $t \notin \Sigma$, holomorphically depend on the base point: to see this locally near any point $b \in T \setminus \Sigma$, it is sufficient to choose a trivialization in which the connexion form is identically zero. In this trivialization $L(t)$ is independent of t.

2. It remains to prove that the subbundle L over $T \setminus \Sigma$, analytically extends to any singular point. This is a purely local problem that can be solved in a fixed trivialization $(\mathbb{C}, 0) \times \mathbb{C}^n$. Let $X(t)$ be a fundamental matrix solution of the corresponding linear system $dX = \Omega X$.

3. Consider first the case where the monodromy is trivial, i.e., $X(t)$ is a meromorphic matrix function. Without loss of generality we may assume that the subspace $L(t)$ is spanned by the first k columns (vector functions) of X. Our goal is to show that one can find some other k *holomorphic* vector functions, linearly independent for all $t \in (\mathbb{C}, 0)$.

If $k = 1$, then any meromorphic vector function $x_1(t)$ can obviously be uniquely represented as $x_1(t) = t^{\nu_1} y_1(t)$ with $y_1(\cdot)$ holomorphic and $y_1(0) \neq 0$. The function $y_1(t)$ spans the same subspace (line) and is holomorphic.

Assume that any k-dimensional meromorphic family of subspaces can be spanned by k holomorphic linearly independent vector functions. Making an additional holomorphic gauge transform, we may assume without loss of generality that these vector functions coincide with the first coordinate vector functions $y_1(t) = (1, 0, \dots, 0)^\top$, $y_2(t) = (0, 1, 0, \dots)^\top$, *etc.* Consider the meromorphic vector function $x_{k+1}(t)$. Without changing the subspace $L(t)$, we can replace it by another vector function $x'_k(t)$ whose first k coordinates are identically zero (subtracting a suitable linear combination of $x_1(t), \dots, x_k(t)$ with meromorphic coefficients). The vector function $x'_{k+1}(t)$ can again be uniquely represented as $x'_{k+1}(t) = t^{\nu_{k+1}} y_{k+1}(t)$ with $y_{k+1}(t)$ holomorphic and $y_{k+1}(0) \neq 0$. Since the first components of y_{k+1} are identically zeros, the vector functions $y_1, \dots, y_{k+1}$ are linearly independent.

4. Assume now that the monodromy of the singular point is nontrivial and the linear space generated by the first $k < n$ columns of the fundamental matrix solution $X(t)$ is invariant. If these columns are arranged in the form of a rectangular $n \times k$-matrix $Y(t)$, then the invariance means that for some constant $k \times k$-invertible matrix M the result of analytic continuation of Y around the origin is $Y(t)M$. Choosing any matrix logarithm $A \in \operatorname{Mat}(k, \mathbb{C})$ such that $\exp 2\pi i A = M$, we conclude that the matrix function $Z(t) = Y(t)t^{-A}$ is single-valued hence meromorphic at the origin. The columns of Z generate the same subspace as the columns of Y, thus by the previous arguments this subspace holomorphically depends on t at any regular singular point. $\square$

Definition 18.9. A meromorphic connexion on a holomorphic vector bundle is called *reducible*, if it admits a nontrivial invariant holomorphic subbundle. Otherwise the connexion is called *irreducible.*

From Proposition 18.8 it follows that a regular connexion is irreducible if and only if its holonomy group is irreducible as a linear representation of the fundamental group $\pi_1(T \smallsetminus \Sigma, a)$. In other words, (ir)reducibility is the property of the holonomy rather than of the connexion itself.

Example 18.10. Let Ω be a rational matrix 1-form on $\mathbb{P}$ defining a connexion on the trivial bundle over the Riemann sphere. If Ω has a block upper-triangular form, then the connexion $\nabla = d - \Omega$ is reducible. The corresponding invariant subbundle is the "constant" coordinate subbundle spanned by the first coordinate vectors.

Lemma 18.11. *Suppose that a rational $n \times n$-matrix 1-form Ω on the Riemann sphere $\mathbb{P}$ has $m \geqslant 1$ Fuchsian points and a regular non-Fuchsian point at the origin. Assume that locally near the origin the fundamental solution of the system admits representation*

$$X(t) = t^{\mathrm{N}}\, Y(t), \qquad \mathrm{N} = \operatorname{diag}\{\nu_1, \dots, \nu_n\}, \quad \nu_i \in \mathbb{Z},$$

where the multivalued matrix function $Y(t)$ is a fundamental solution for a Fuchsian singularity (so that $dY \cdot Y^{-1}$ has a first order pole at the origin) and ν_i are some integer numbers.

If the global monodromy group of the system is irreducible, then the difference between the numbers ν_i is explicitly bounded,

$$|\nu_i - \nu_j| \leqslant (m-2)(n-1), \qquad \forall i, j = 1, \dots, n. \tag{18.6}$$

Proof. The Pfaffian matrix of the system locally near the origin has the form

$$\Omega = \mathrm{N}\, t^{-1} dt + t^{\mathrm{N}} \Omega'\, t^{-\mathrm{N}},$$

where $\Omega' = dY \cdot Y^{-1}$ has a first order pole at the origin. Without loss of generality, we may assume that the entries of the integer diagonal matrix N are arranged in the nonincreasing order,

$$\nu_1 \geqslant \cdots \geqslant \nu_n$$

(one can always permute the rows by a global constant gauge transformation that preserves the irreducibility).

The idea of the proof is rather transparent: if two consecutive numbers ν_k, ν_{k+1} differ too much, then the matrix 1-form Ω will have a corner filled by rational forms of bounded degrees which are too flat to be nonzero. On the other hand, a zero corner implies reducibility which is forbidden by the assumptions of the lemma.

More accurate reasoning is as follows. If $\nu_k - \nu_{k+1} > m - 1$ for some k between 1 and $n-1$, then all entries in some upper right corner of the matrix Ω will have zero of order $> m-2$ at the origin. Indeed, if $i \leqslant k$ and $j \geqslant k+1$, then the (i,j)th matrix element of the Pfaffian matrix Ω is obtained by multiplying the corresponding element ω'_{ij} of Ω' by t^d, $d = \nu_i - \nu_j \geqslant \nu_k - \nu_{k+1} > m-1$. Since Ω' is Fuchsian, its entries have at most first order pole, thus the order of zero of all ω_{ij} with $i \leqslant k$ and $j \geqslant k+1$ will be greater than $m-2$.

On the other hand, since the form Ω is globally defined on the whole sphere, its entries are rational 1-forms. By assumptions, these forms have at most simple poles at no more than $m-1$ other points of $\mathbb{P}^1$. Thus the order of zero at the origin cannot be greater than $m-2$, unless the form is identically zero (the difference between the total number of poles and zeros for any rational form is always equal to 2). This necessarily implies that $\omega_{ij} \equiv 0$ for all combinations of i, j such that $i \leqslant k$ and $j \geqslant k+1$.

But the simultaneous occurrence of a corner of identical zeros as was described above, in the (rational, i.e., globally defined) Pfaffian matrix Ω means that the coordinate subspace $\{x_1 = \cdots = x_k = 0\}$ is invariant by the system, hence by all monodromy operators, contrary to the irreducibility assumption.

Thus for the case where the diagonal entries ν_i are arranged in the nonincreasing order, the difference between any two *consecutive* numbers cannot be greater than $m-2$. Hence the difference between *any* two ν_i is no greater than $(m-2)(n-1)$ in the absolute value, and this assertion is already independent on the order of these numbers. □

This lemma immediately implies an impossibility result of Riemann–Hilbert type. It provides for a wide class of holomorphic bundles on which the Riemann–Hilbert problem admits no solution.

Theorem 18.12. *An irreducible matrix group with m generators cannot be realized as a holonomy group of a meromorphic connexion with $m+1$ singular points on a holomorphic bundle with the splitting type $D = \{d_1, \dots, d_n\}$ over $\mathbb{P}$, unless the following inequalities hold,*

$$|d_i - d_j| \leqslant (m-2)(n-1) \qquad \forall i, j = 1, \dots, n. \tag{18.7}$$

Proof. Assume that such a connexion ∇ exists and the point at infinity is singular for it.

Consider the meromorphic trivialization of the bundle π by the cochain (18.4) described in §18**B**. This trivialization does not change the holonomy group, thus the connexion ∇_0 on the trivial bundle is also irreducible.

A fundamental matrix solution for horizontal sections $\nabla_0 X = 0$ near infinity has the form $X(t) = t^{-D}G(t)Y(t)$, where G is holomorphically invertible at infinity and $Y(t)$ is a fundamental solution of the equation $\nabla Y = 0$ near infinity. This follows from the explicit form of the trivialization (18.4).

By assumption, ∇ is Fuchsian, so the logarithmic derivative $dY \cdot Y^{-1}$ of the matrix function $Y(t)$ has a first order pole. Since G is holomorphic and invertible, the logarithmic derivative of the product GY also has a first order pole at infinity. If one of the equalities (18.7) is violated, after change of the independent variable $t \mapsto 1/t$ which sends infinity to the origin, it would contradict Lemma 18.11, since all other singularities of ∇_0 are Fuchsian. □

Remark 18.13. The assertion of Theorem 18.12 is remarkable for the following reason. In construction of the holomorphic bundle as in Theorem 18.4 each monodromy operator M_j can be realized by infinitely many different local connexion forms Ω_j. Even if only the Euler equations are used, still there is a freedom to choose matrix logarithms which can be used to produce infinitely many holomorphically nonequivalent types of singularities at each point $a_j \in \Sigma$. One could expect that combining these nonequivalent singularities and patching them together, one can produce infinitely many different splitting types of holomorphic bundles.

Theorem 18.12 claims that the *global condition of irreducibility of the monodromy group* imposes a global restriction that is compatible with *finitely many different splitting types* only. In the next subsection we will show that in fact one of these splitting types admits holomorphic trivialization.

18D. Bolibruch–Kostov theorem. The most remarkable positive result on solvability of the Riemann–Hilbert problem was discovered independently by A. Bolibruch [**Bol92**] and V. Kostov [**Kos92**].

Theorem 18.14. *Any irreducible matrix group can be realized as the holonomy group of a Fuchsian connexion on the trivial vector bundle over* $\mathbb{P}$.

In other words, any monodromy data $\{M_1, \dots, M_m\}$ such that the matrices M_j do not have a common nontrivial invariant subspace, can be realized by a linear system with rational matrix function Ω having only simple poles at the specified points and no other singularities. This is the strongest, the third form of the solvability of the Riemann–Hilbert problem on p. 312.

Proof. Unlike the previous demonstrations, when we started from an *arbitrary* Fuchsian connexion defined by a collection of connexion 1-forms $\{\Omega_\alpha\}$ on an abstract holomorphic vector bundle, realizing the specified holonomy group, this time we will use explicitly the freedom in the choice of the connexion forms Ω_j realizing each holonomy. It is sufficient to vary only one of the forms.

More precisely, we will assume that one of the preassigned singularities is at the point $t = \infty$ and the corresponding holonomy operator is *upper-triangular.* Such a singularity can be realized by the local connexion form $\Omega_m = A_m t^{-1}\,dt$ with an upper-triangular residue matrix $A = A_m$ with a fundamental matrix solution t^A. Yet without changing the holonomy Δ_m we can replace Ω_m by a meromorphically gauge equivalent 1-form which corresponds to replacing the matrix solution by another function $t^{\mathrm{N}}t^A$. More specifically, we consider the new connexion form of the structure

$$\Omega'_{\mathrm{N}} = \mathrm{N}t^{-1}\,dt + t^{\mathrm{N}}\Omega_m t^{-\mathrm{N}}. \tag{18.8}$$

The term $t^{\mathrm{N}}\Omega_m t^{-\mathrm{N}}$ has the first order at infinity by the usual arguments, if the matrix A is upper-triangular and the integer numbers ν_i follow in the ascending order, $\nu_1 < \cdots < \nu_n$.

Denote by π_{N} the holomorphic vector bundle obtained by gluing together the connexions $\Omega_1, \dots, \Omega_{m-1}, \Omega_{\mathrm{N}}$. This bundle carries the meromorphic connexion represented by the above cochain of 1-forms, which will be denoted ∇_{N}. The connexion ∇_{N} is irreducible by construction. Hence the splitting type $D = \operatorname{diag}\{d_1, \dots, d_n\} = D(\mathrm{N})$ of the bundle itself is constrained by the inequalities from Theorem 18.12.

Consider the meromorphic trivialization (18.4) of the bundle π_{N}. As usual, it has only Fuchsian singularities at all finite points, and a regular singularity at infinity with a fundamental matrix solution of the form

$$X(t) = t^{-D}G_1(t)\,t^{\mathrm{N}}t^A,$$

where the splitting diagonal matrix D and the holomorphic invertible matrix $G_1(t)$ depend on the diagonal matrix N. By the Permutation Lemma 16.36, there exists a monopole gauge transform that brings the fundamental solution X' into the form

$$X'(t) = G'(t)\,t^{-D'}t^{\mathrm{N}}t^A = G'(t)\,t^{-D'+\mathrm{N}}t^A, \qquad G' \in \mathrm{GL}(n, \mathcal{O}(\mathbb{P}, \infty)),$$

where the integer diagonal matrix D' has the same entries d_i but in a permuted order.

Yet (and this is the key step of the proof) if the sequence ν_i was ascending sufficiently fast and the sequence d'_i is constrained by the inequality $|d'_i - d'_{i+1}| \leqslant (m-2)(n-1)$ (cf. with (18.6)), then the sequence $\nu'_i = \nu_i - d'_i$ is also ascending (increasing). To ensure the monotonicity, it is sufficient to require that

$$\nu_{i+1} - \nu_i > (m-2)(n-1) \qquad \text{for all } i = 1, \dots, n-1.$$

The monotonicity of ν_i' is sufficient to guarantee that the singularity with the fundamental solution $t^{-D'+\mathrm{N}}t^A$ is Fuchsian (recall that A is upper-triangular). Left multiplication by the holomorphically invertible matrix G' does not change this fact: after meromorphic trivialization F and the subsequent monopole gauge transform we obtain the trivial bundle with a Fuchsian connexion on it. □

18E. Bolibruch counterexample. In this section we describe a reducible matrix group that cannot be realized as the holonomy of a Fuchsian connexion on the trivial bundle. More precisely, we describe an obstruction that prevents a given matrix group to be realized by a Fuchsian connexion on the trivial bundle. A similar obstruction is obtained for nontrivial bundles.

Recall that each linear operator $M \in \mathrm{GL}(n, \mathbb{C})$ over the field $\mathbb{C}$ always has at least one invariant subspace of each dimension $k = 1, \dots, n-1$ (Exercise 18.6). There are operators for which there are no other invariant subspaces.

Definition 18.15. A linear operator $M\colon \mathbb{C}^n \to \mathbb{C}^n$ will be called a *monoblock*, if its Jordan normal form consists of a single block of maximal size.

By definition, the spectrum of each monoblock is a *singleton*, i.e., the operator has a single eigenvalue ν and for any $k \leqslant n$ the power $(M - \nu E)^k$ has the rank *exactly* equal to $n - k$.

Lemma 18.16. *A monoblock operator on a complex n-space has exactly one invariant subspace of each intermediate dimension k between 1 and $n-1$. In a basis in which M has an upper-triangular matrix, this subspace is spanned by the first k vectors.*

Proof. Without loss of generality assume that the unique eigenvalue of M is zero, $\nu = 0$, that is, M is nilpotent.

If V is an invariant subspace of dimension $k \leqslant n$ for M, then the restriction of M on V must also be nilpotent, more precisely, $M^k|_V = 0$. But for a nilpotent operator of class B the rank of M^k is exactly $n - k$, which means that $\dim \operatorname{Ker} M^k = k$, and hence V must coincide with $\operatorname{Ker} M^k$, being thus uniquely defined.

It remains to notice that for an upper-triangular nilpotent matrix M, $\operatorname{Ker} M^k$ consists of the first k basic vectors. □

Monoblocks are rather rigid; for instance, any monoblock admits a unique matrix logarithm modulo a scalar matrix (cf. with Remark 18.13) which is also a monoblock (Problem 18.7). In other words, if a monoblock is realized as a holonomy of a Fuchsian singular point which is linearizable (i.e., equivalent to an Euler system), then the corresponding residue matrix is a monoblock as well.

In the class of non-Euler systems one may have really different (not locally holomorphically equivalent) Fuchsian realizations of a monoblock holonomy. In particular, a Fuchsian singular point with a monoblock holonomy can have different (though necessarily, resonant) eigenvalues. The following assertion may be considered as a true "nonlinear" analog of the fact that a monoblock matrix has a monoblock logarithm.

Lemma 18.17. *If a Fuchsian singular point of a connexion ∇ of rank n has a monoblock local monodromy, then for each intermediate dimension k, $1 \leqslant k \leqslant n-1$, there exists exactly one holomorphic subbundle $\pi_k \colon L_k \to (\mathbb{C}, 0)$, $\operatorname{rank} \pi_k = k$, invariant by ∇, and the residue of the restriction $\nabla_k = \nabla|_{L_k}$ of the connexion on the subbundle satisfies the inequalities*

$$\tfrac{1}{k} \operatorname{tr} \operatorname{res}_0 \nabla_k \geqslant \tfrac{1}{n} \operatorname{tr} \operatorname{res}_0 \nabla. \tag{18.9}$$

The equality is possible only for all values $k = 1, \dots, n-1$ simultaneously and only in the case where the residue matrix $\operatorname{res}_0 \nabla$ has a single eigenvalue.

Proof. The assertion is purely local, so it can be verified for a linear system in the Poincaré–Dulac–Levelt normal form (16.7).

Since the monodromy has a single eigenvalue, all eigenvalues $\lambda_1, \dots, \lambda_n$ of the residue $\operatorname{res}_0 \nabla$ fall in the same resonant group, i.e., differ only by integers, as follows from the explicit formula (16.11). Arranging the eigenvalues in the nonincreasing order $\lambda_1 \geqslant \cdots \geqslant \lambda_n$ (recall again that this means nonnegativity of all differences $\lambda_i - \lambda_j \geqslant 0$ for $i < j$). In these settings the connexion matrix $A(t)$ in (16.7) is upper-triangular (Remark 16.14).

For the system (16.7) in the upper-triangular form, each coordinate subspace $L_k = \{x_{k+1} = \cdots = x_n = 0\} \subset (\mathbb{C}, 0) \times \mathbb{C}^n$ generated by the first k coordinate vectors, is invariant and hence constitutes a "constant" invariant subbundle $\pi_k \colon L_k \to (\mathbb{C}, 0)$ of rank k. Moreover, the trace of the residue matrix restricted on L_k is the sum of the *first* k eigenvalues $\lambda_1, \dots, \lambda_k$ of the residue matrix. Yet since the *largest eigenvalues come first*, we instantly obtain the inequalities

$$\tfrac{1}{k} \operatorname{tr} \operatorname{res}_0 \nabla_k \geqslant \tfrac{1}{n} \operatorname{tr} \operatorname{res}_0 \nabla$$

for the restrictions ∇_k of ∇ on the subbundles L_k. The equality is possible if and only if the smallest and the largest eigenvalues are equal, i.e., if $\lambda_1 = \cdots = \lambda_n = \lambda$.

To prove the uniqueness, note that since the connexion is of Bolibruch type, each invariant subspace $\ell_k \in \mathbb{C}^n$ of rank k for the monodromy operator $M \in \mathrm{GL}(n, \mathbb{C})$ is unique and extends as a holomorphic invariant subbundle, necessarily coinciding with L_k. $\square$

Globalization of this construction leads to a very important notion which will play a central role in the construction of the counterexample.

Definition 18.18. A meromorphic connexion on a holomorphic vector bundle is called a *Bolibruch connexion*, if it has a nontrivial invariant subbundle, all the singular points of the connection are Fuchsian, and the local holonomy of each singular point is a monoblock operator.

The global analog of Lemma 18.17 then takes the following form. Note that, unlike the "inequality" between the complex numbers, understood in the "artificial" sense (11.3), the inequality (18.10) relates two rational numbers.

Theorem 18.19. *Suppose that a Bolibruch connexion ∇' on a holomorphic bundle π' over $\mathbb{P}$ has a nontrivial invariant subbundle π. Then the ratio of degree to rank for the subbundle is greater or equal to this ratio for the ambient bundle,*

$$\pi \subseteq \pi' \implies \frac{\deg \pi}{\operatorname{rank} \pi} \geqslant \frac{\deg \pi'}{\operatorname{rank} \pi'}. \tag{18.10}$$

The equality occurs if and only if the spectrum of each singularity of ∇' is a singleton.

Proof. Let $\nabla = \nabla'|_\pi$ be the restriction of ∇' on the subbundle $\pi\colon L \to \mathbb{P}$. Denote $k = \operatorname{rank} \pi$, $n = \operatorname{rank} \pi'$. By Corollary 17.35, the degree of both bundles is equal to the sum of traces of the residues of all singularities.

Adding together the local inequalities (18.9) over all singularities $a \in \Sigma$, we conclude that

$$\tfrac{1}{k} \deg \pi = \sum \tfrac{1}{k} \operatorname{res}_a \operatorname{tr} \nabla \geqslant \sum \tfrac{1}{n} \operatorname{res}_a \operatorname{tr} \nabla' = \tfrac{1}{n} \deg \pi'.$$

The equality occurs if and only if all spectra are singletons. □

Together with Corollary 17.25, Theorem 18.19 imposes rather strong restrictions on Bolibruch connexions on the trivial bundle.

Theorem 18.20. *For a Bolibruch connexion on a trivial bundle, the spectra of all singularities must necessarily be singletons, and the invariant subbundle itself must be trivial.*

Proof. If π_0 is a trivial bundle and π its subbundle invariant by a Bolibruch connexion ∇, then $\deg \pi_0 = 0$. By Theorem 18.19, we have $\deg \pi \geqslant 0$ and by Corollary 17.25, $\deg \pi \leqslant 0$. Together these inequalities leave only one possibility $\deg \pi = 0$, so that in both assertions the extreme cases occur. This implies both assertions. □

We have arrived at the main step of the impossibility proof. The assumptions of Theorem 18.20 (reducibility and Jordan block structure of the monodromy matrices) are imposed on the *holonomy group* of the connexion ∇ rather than on the *connexion itself*. However, the assertion concerns the connexion (more specifically, its residue matrices). In other words, Theorem 18.20 implicitly describes an obstruction to realizability of a reducible monodromy data of monoblock operators by a Fuchsian connexion on the trivial bundle. In particular, we arrive at the following result which is just a geometric reformulation of Theorem 16.33.

Theorem 18.21. *If a Fuchsian connexion ∇ on a holomorphic bundle π of rank 4 over $\mathbb{P}$ with three singular points has the monodromy matrices*

$$\begin{pmatrix} 1 & 1 & & \\ & 1 & 1 & \\ & & 1 & 1 \\ & & & 1 \end{pmatrix}, \quad \begin{pmatrix} 3 & 1 & 1 & -1 \\ -4 & -1 & 1 & 2 \\ & & 3 & 1 \\ & & -4 & -1 \end{pmatrix}, \quad \begin{pmatrix} -1 & & 2 & -1 \\ 4 & -1 & & 1 \\ & & -1 & \\ & & 4 & -1 \end{pmatrix}, \tag{18.11}$$

then the bundle π is necessarily nontrivial.

Proof. One can easily see that all three matrices (18.11) are monoblocks (with the respective eigenvalues $\mu_{1,2} = 1$, $\mu_3 = -1$) and have an invariant subspace spanned by the first two coordinates. Hence the connexion ∇ realizing the corresponding monodromy data, is necessarily a Bolibruch connexion. If the bundle π were trivial then by Theorem 18.20 each residue matrix $A_j = \operatorname{res}_{a_j} \nabla$ must have a singleton spectrum λ_j such that $\exp 2\pi i \lambda_j = \mu_j$ for all $j = 1, 2, 3$. Resolving the corresponding equations $\exp 2\pi i \lambda_{1,2} = 1$, $\exp 2\pi i \lambda_3 = -1$, we obtain the congruences

$$\lambda_1 \equiv \lambda_2 \equiv 0 \bmod \mathbb{Z}, \qquad \lambda_3 \equiv \tfrac{1}{2} \bmod \mathbb{Z}. \tag{18.12}$$

On the other hand, by the index theorem (Corollary 17.35) we would have for the trivial bundle the equality

$$\deg \pi = 0 = \operatorname{tr} A_1 + \operatorname{tr} A_2 + \operatorname{tr} A_3 = 4(\lambda_1 + \lambda_2 + \lambda_3).$$

The resulting impossible congruence $0 \equiv 2 \bmod 4\mathbb{Z}$ proves that the bundle π cannot be trivial. □

This argument gives an alternative proof of Theorem 16.33.

Exercises and Problems for §18.

Problem 18.1. Prove that the Riemann–Hilbert problem is solvable if all monodromy matrices commute, $[M_i, M_j] = 0$ for all i, j.

Exercise 18.2. Write a detailed proof of Theorem 18.5.

Exercise 18.3. Prove that the Riemann–Hilbert problem can be always solved by a Fuchsian linear system for any monodromy data if the meromorphic matrix form is allowed to have a single extra singular point with identical holonomy at any preassigned point off the singular locus Σ.

Problem 18.4. Construct an example of an irregular singularity and a subspace invariant by the (local) monodromy, which does not extend as an invariant holomorphic subbundle over a neighborhood of the singular point (cf. with Proposition 18.8).

Problem 18.5. Prove that any meromorphic rectangular matrix function $X(t)$ of size $n \times k$, $k < n$ can be locally near $t \in (\mathbb{C}, 0)$ represented under the form $X(t) = L(t)D(t)R(t)$, where $L(t)$ and $R(t)$ are holomorphic invertible square matrices of sizes $n \times n$ and $k \times k$ respectively, and $D(t)$ is the rectangular truncation (first k columns) of a diagonal matrix which has only integer powers t^{ν_i} or zeros on the diagonal.

Exercise 18.6. Prove that any operator $M \in \mathrm{GL}(n, \mathbb{C})$ has at least one invariant subspace $L_k \subset \mathbb{C}^n$ of each intermediate dimension $k = 1, \dots, n-1$.

Problem 18.7. Prove that any two matrix logarithms A, A' of the same monoblock operator differ by an integer multiple of the identity matrix modulo conjugacy:

$$\exp A = \exp A' \text{ is a monoblock } \implies A - CA'C^{-1} = 2\pi i k E$$

for a suitable integer number $k \in \mathbb{Z}$ and an invertible conjugacy matrix $C \in \mathrm{GL}(n, \mathbb{C})$. Prove that each logarithm is also a monoblock.

Problem 18.8. Prove that the Riemann–Hilbert problem is always solvable in the classical sense (i.e., on the trivial bundle) in dimension 2.

Problem 18.9. Prove that the monodromy data with one diagonal matrix can be realized by infinitely many nonequivalent Fuchsian systems.

Problem 18.10. Prove that any irreducible monodromy data can be realized by infinitely many nonequivalent Fuchsian systems.

Problem 18.11. Prove that the Riemann–Hilbert problem is nonsolvable in all dimensions greater than 4.

Problem 18.12. Prove the following generalization of Theorem 18.12. Let ∇ be a meromorphic *non-Fuchsian* connexion on a holomorphic vector bundle of rank n and the splitting type $D = \{d_1, \dots, d_n\}$ with at least one Fuchsian singularity. Denote by m the total order of poles of all singularities. Prove that if for some pair of indices $|d_i - d_j| \geqslant (m-2)(n-1)$, then the connexion ∇ is reducible, i.e., has an invariant subbundle.

19. Linear nth order differential equations

Linear high order scalar differential equation can be reduced to a rather special class of *companion* linear systems which are naturally defined connexions on the *jet bundle*. Because of the special form, regular singular points of such connexions can be easily identified and explicit meromorphic transformation bringing them to the Fuchsian form is well known since L. Fuchs himself. However, this meromorphic transformation is nontrivial and globally Fuchsian equations on the Riemann sphere $\mathbb{P}$ naturally "live" on nontrivial holomorphic vector bundles, whose type depends on the number of singular points.

An additional feature, an important tool of investigation, is the structure of (noncommutative) algebra on the set of linear differential operators, which implies the possibility of *factorization* of operators. The latter circumstance plays an important role when studying *roots of solutions* of linear ordinary differential equations.

At the end of the section we address several questions in the spirit of the Riemann–Hilbert problem for linear high order equations in the cases where these questions make sense.

19A. High order differential operators: algebraic theory. Let T be a Riemann surface (complex 1-dimensional manifold). Denote by $\mathfrak{M} = \mathfrak{M}(T)$ the field (commutative $\mathbb{C}$-algebra) of meromorphic functions on T. Any derivation $D \in \operatorname{Der}\mathfrak{M}$, a $\mathbb{C}$-linear self-map of $\mathfrak{M}$ into itself which satisfies the Leibnitz rule $D(fg) = f\,Dg + G\,Df$, is associated with a meromorphic vector field on T,

$$\operatorname{Der}\mathfrak{M} \cong \mathcal{D}(T) \otimes \mathfrak{M}.$$

Since T is one-dimensional, any two derivations differ by a meromorphic multiplier,

$$D, D' \in \operatorname{Der}\mathfrak{M} \iff D' = rD, \qquad \text{for some} \quad r \in \mathfrak{M}. \tag{19.1}$$

Definition 19.1. A *linear nth order differential operator* is any $\mathbb{C}$-linear operator $L\colon \mathfrak{M} \to \mathfrak{M}$, which admits a representation

$$\begin{gathered} L = a_0\,D^n + a_1\,D^{n-1} + \cdots + a_{n-1}D + a_n, \\ D \in \operatorname{Der}\mathfrak{M}, \quad a_0, a_1, \dots, a_n \in \mathfrak{M}, \quad a_0 \not\equiv 0. \end{gathered} \tag{19.2}$$

The operator a_0D^n is called the *leading term* of L. The operator L is called *monic* (more precisely, D-monic), if $a_0 = 1$. A linear nth order homogeneous differential equation is the equation of the form

$$Lf = 0. \tag{19.3}$$

This definition formally depends on the choice of the derivation D, yet one can immediately verify using (19.1) and the Leibnitz rule, that an expansion (19.2) can be re-expanded (with different coefficients, but of the same degree) in powers of any other derivation D'. We will denote

$$\mathfrak{D}(n,T) = \{L\colon \mathcal{M}(T) \to \mathcal{M}(T),\ \operatorname{ord} L = n\},$$
$$\mathfrak{D}(T) = \bigcup_{n\geqslant 0} \mathfrak{D}(n,T).$$

Differential operators of order 0 are multiplications by scalar functions and hence can be identified with the algebra $\mathcal{M} = \mathcal{M}(T)$ itself. The collection of differential operators of all orders is naturally filtered by the order.

The space of all differential operators $\mathfrak{D}(T)$ forms a noncommutative associative algebra by composition:

$$L, L' \in \mathfrak{D}(T) \implies LL', L'L \in \mathfrak{D}(T),$$
$$\operatorname{ord} LL' = \operatorname{ord} L'L = \operatorname{ord} L + \operatorname{ord} L'.$$

The only units of $\mathfrak{D}(T)$ are zero order operators corresponding to multiplication by a nonzero meromorphic function[5]. Though the algebra $\mathfrak{D}(T)$ is noncommutative, it has many features similar to that of the commutative algebra $\mathcal{M}[D]$ of polynomials in a single indeterminate D with coefficients in the ring $\mathcal{M} = \mathcal{M}(T)$ of meromorphic functions. Thus, the representation (19.2) can be considered now as a (noncommutative) polynomial expansion in $\mathfrak{D}(T)$ in powers of the derivation $D \in \operatorname{Der}\mathcal{M}(T)$ with all coefficients occurring *to the left* of all powers $D, D^2, \dots, D^n$. Another feature is the possibility of division with remainder similar to the division of univariate polynomials.

Lemma 19.2. *For any two operators $L \in \mathfrak{D}(n,T)$ and $Q \in \mathfrak{D}(k,T)$ of orders $n \geqslant k$, then there exist two operators P (the incomplete ratio) and R (the remainder), such that*

$$L = PQ + R, \qquad \operatorname{ord} P = \operatorname{ord} L - \operatorname{ord} Q, \quad \operatorname{ord} R < \operatorname{ord} Q. \tag{19.4}$$

Proof. The operators P, R can be constructed by the following algorithm which is a modification of the division algorithm for polynomials in one variable. If the operators L, Q are expanded in powers of any derivation $D \in \operatorname{Der}\mathcal{M}$ as follows:

$$\begin{aligned} L &= a_0 D^n + a_1 D^{n-1} + \cdots + a_n, \\ Q &= b_0 D^k + b_1 D^{k-1} + \cdots + b_k, \end{aligned} \qquad a_i, b_j \in \mathcal{M}, \tag{19.5}$$

then the leading term of the operator $D^{n-k}Q$ is $b_0 D^n$ and hence the operator $L_1 = L - P_0 Q$, where $P_0 = (a_0/b_0)D^{n-k}$, has the order $\leqslant n-1$. Repeating this step, we construct P_1 so that $L_2 = L_1 - P_1 Q$ is of the order strictly inferior to that of L_1, *etc.*

[5]The property of linear operators on the algebra $\mathcal{M}$ to be *differential operators* can be defined in purely algebraic terms of commutation with the units of the algebra of self-maps (Problem 19.1).

After at most $n-k$ steps we will be left with an operator of order strictly less than k, which is designated to be the residue R. The "partial incomplete ratios" $P_0, P_1, \dots$ add together to form the operator $P = P_0 + P_1 + \cdots$. □

Remark 19.3. Assume that all coefficients a_i, b_j of the operators L and Q in (19.5) are holomorphic at a given point $t_0 \in T$, and the leading coefficient b_0 of the divisor Q is nonvanishing at this point, $b_0(t_0) \neq 0$. Under these assumptions both the remainder and the incomplete ratio will be obtained as expansions in powers of D with coefficients holomorphic at t_0. This can be seen by direct inspection of the algorithm.

Definition 19.4. An operator $L \in \mathfrak{LO}(n,T)$ is *divisible* by $Q \in \mathfrak{LO}(k,T)$, if $L = PQ$ with $P \in \mathfrak{LO}(n-k,T)$. An operator L is *reducible*, if it is divisible by an operator $Q \in \mathfrak{LO}(k,T)$ with $0 < k < n$. Otherwise L is called *irreducible*.

19B. Linear ordinary differential equations: the naïve approach. Linear high order equations can be considered as a particular case of linear systems of first order differential equations of a special, so-called *companion form*.

In this section it will be convenient to enumerate coordinates of the complex space $\mathbb{C}^{n+1} = \{(x_0, \dots, x_n)\}$ starting from the zero index value, x_0. Let $D \in \operatorname{Der} \mathcal{M}$ be an arbitrary derivation; for instance, if $T = \mathbb{C}$ or $T = \mathbb{P}$, we may assume that $D = \frac{\partial}{\partial t}$. Then, denoting the unknown function by $y = x_0$ and its derivatives by $x_k = D^k y$, $k = 1, \dots, n$, we reduce the scalar equation (19.3) to the system

$$
\begin{aligned}
&Dx = A(t)x, \quad \text{where:}\\
&D \in \operatorname{Der} \mathcal{M},\\
&A \in \operatorname{Mat}(n, \mathcal{M}),
\end{aligned}
\qquad
A = \begin{pmatrix}
0 & 1 & & & \\
 & 0 & 1 & & \\
\cdots & \cdots & \cdots & \cdots & \cdots \\
 & & & 0 & 1 \\
-\frac{a_n}{a_0} & -\frac{a_{n-1}}{a_0} & \cdots & -\frac{a_2}{a_0} & -\frac{a_1}{a_0}
\end{pmatrix}, \tag{19.6}
$$

called the *companion system* for the linear equation (19.2)–(19.3). This reduction immediately allows us to reformulate for linear equations all results from §15.

Definition 19.5. A *regular* (*nonsingular*) *point* of a linear equation (19.2)–(19.3) is any point $t_0 \in T$ at which the vector field $D \in \mathcal{D}(T)$, associated with the derivation $D \in \operatorname{Der} \mathcal{M}(T)$, is nonsingular and all ratios $a_i(t)/a_0(t)$ are holomorphic (have no poles).

Nonregular points are naturally called *singularities* of the equation and denoted by $\operatorname{Sing} L$. A singular point is called *regular*, if it is regular for the companion system in the sense of Definition 16.2.

In other words, a singular point for the equation is regular, if all solutions of the equation together with their derivatives grow moderately (in the sense of Definition 16.1) as t tends to t_0.

Proposition 19.6. *Solutions of the linear equation* (19.2)–(19.3) *locally exist near any nonsingular point and admit unique analytic continuation along any path free from singularities of this equation.*

Dimension (over $\mathbb{C}$) of the space of solutions of this equation in any simply connected domain free from singularities of the equation, is equal to the order of the equation.

Proof. The first assertion is a reformulation of Theorem 15.3 for the companion system.

The second assertion immediately follow from the fact that the linear map which assigns to every holomorphic function $f(\cdot)$ the initial conditions,

$$f(\cdot) \longmapsto \big(f(t_0), \tfrac{d}{dt}f(t_0), \dots \tfrac{d^{n-1}}{dt^{n-1}}f(t_0)\big) \in \mathbb{C}^n,$$

becomes a linear isomorphism between *solutions* of the linear equation (19.2)–(19.3) and the space of initial conditions. Injectivity of this map is the uniqueness part, and surjectivity the uniqueness part of Theorem 15.3. $\square$

Proposition 19.6 implies that solutions of a linear equation $Lf = 0$ are holomorphic functions eventually ramified over the singular locus $\Sigma = \operatorname{Sing} L$. Since analytic continuation along paths preserves the space of solutions of this equation, the operator of analytic continuation Δ_γ along any loop $\gamma \in \pi_1(T \smallsetminus \Sigma, t_0)$ acts by a linear transformation on the row vector of functions,

$$\Delta_\gamma(f_1, \dots, f_n) = (f_1, \dots, f_n) \cdot M_\gamma, \qquad M_\gamma \in \mathrm{GL}(n, \mathbb{C}), \tag{19.7}$$

where M_γ are the *monodromy matrices*. In the future any tuple of holomorphic functions satisfying the monodromy property (19.7), will be called a *monodromic* tuple.

The monodromy property is almost sufficient for a collection of functions to satisfy a linear differential equation with meromorphic coefficients. The additional requirement is regularity of all singular points.

Theorem 19.7 (G. F. B. Riemann)**.** *A monodromic tuple of n functions regular at each ramification point of a finite set $\Sigma \subset T$, satisfies a linear ordinary differential equation $Lf = 0$ with meromorphic coefficients, $L \in \mathfrak{D}(k, T)$, $k \leqslant n$.*

This equation can be explicitly written using Wronskians; see Proposition 19.9.

Definition 19.8. The *Wronskian*, or Wronski determinant, of n functions is the determinant of the Wronski matrix,

$$W(f_1, \dots, f_n) = \det \begin{pmatrix} f_1 & f_2 & \dots & f_n \\ Df_1 & Df_2 & \dots & Df_n \\ \vdots & \vdots & \ddots & \vdots \\ D^{n-1}f_1 & D^{n-1}f_2 & \dots & D^{n-1}f_n \end{pmatrix}. \tag{19.8}$$

The Wronskian is a holomorphic (resp., meromorphic) function of $t \in U \subset T$ if all functions $f_1, \dots, f_n$ were holomorphic (resp., meromorphic) and D is holomorphic vector field in U.

The Wronskian depends multi-linearly (over $\mathbb{C}$) and anti-symmetrically on the functions f_j. In particular, it vanishes *identically* if the functions f_j are linearly dependent over $\mathbb{C}$. If $f_1, \dots, f_n$ are solutions of a linear equation (19.3), then $W(f_1, \dots, f_n)$ is the determinant of the matrix solution $X(t)$ of the associated companion system (19.6). By the Liouville–Ostrogradskii theorem (Problem 15.10),

$$Dw = -\frac{a_1(t)}{a_0(t)} w, \qquad w = W(f_1, \dots, f_n). \tag{19.9}$$

From this identity it follows that a Wronskian of n solutions of a linear equation is either nonvanishing everywhere outside the singular locus, or vanishes identically.

The Riemann theorem follows immediately from the following assertion.

Proposition 19.9 (gloss of Riemann Theorem 19.7)**.** *For any regular monodromic tuple $f_1, \dots, f_n$ such that the Wronskian $w(t) = W(f_1, \dots, f_n)(t)$ is not identically zero, the operator*

$$L = w^{-1} W(f_1, \dots, f_n, \cdot\,), \qquad Lf = w^{-1} W(f_1, \dots, f_n, f), \tag{19.10}$$

is a monic differential operator of order n with meromorphic coefficients,

$$L = D^n + a_1 D^{n-1} + \dots + a_{n-1} D + a_n \in \mathfrak{D}(n, T), \qquad a_i \in \mathcal{M}, \tag{19.11}$$

vanishing on all functions $f_1, \dots, f_n$.

Proof. To prove that L is a monic differential operator, we expand the "large" $(n+1) \times (n+1)$-determinant $W(f_1, \dots, f_n, f)$ in the elements of the last column containing the derivatives of f. The coefficients a_i of the expansion are $n \times n$-minors of the "large" matrix, formed by the first n columns. The leading coefficient (before the highest derivative) is exactly the minor $w = W(f_1, \dots, f_n)$. After division by w we conclude that L is a monic differential operator with the coefficients which are ratios of the minors.

All these minors have the same monodromy (the corresponding matrices are multiplied from the right by the same matrix factors M_γ), hence the ratios of their determinants are single-valued. Because of the regularity, the singularities of these ratios are finite order poles.

Since the Wronskian vanishes when any two columns coincide, each f_j belongs to the null space of L. □

Remark 19.10 (warning). The singular locus of the operator (19.11) can be larger than the ramification locus of the monodromic tuple $(f_1, \dots, f_n)$.

19C. Factorization of differential operators. Solutions of a linear differential equation in general do not belong to the field $\mathfrak{M} = \mathfrak{M}(T)$, but rather to some bigger field (extension) $\mathfrak{M}' \supseteq \mathfrak{M}$. This field can be obtained by formally adjoining these solutions and their derivatives of order $< n$. The extension field, denoted by

$$\mathfrak{M}' = \mathfrak{M}(f_1, \dots, f_n) = \mathfrak{M}(L),$$

is called the *Picard–Vessiot extension* of the initial field $\mathfrak{M} = \mathfrak{M}(T)$.

Picard–Vessiot extensions are *differential fields* (i.e., any derivation $D \in \operatorname{Der} \mathfrak{M}$ extends as a derivation to $\operatorname{Der} \mathfrak{M}'$) with the same subfield of constants (i.e., $Du = 0$, $u \in \mathfrak{M}'$, is possible if and only if $u = \text{const} \in \mathbb{C}$). Besides formally algebraic construction of such extensions, they can be identified with subfields of the field $\mathfrak{M}(T, t_0)$ of meromorphic germs at a nonsingular point $t_0 \notin T$.

In the same way as any polynomial admits factorization by linear terms over the field obtained by adjoining its roots to the field of the coefficients, every linear differential operator can be represented as a composition of first order operators with coefficients in $\mathfrak{M}' = \mathfrak{M}(L)$.

We start with an observation that divisibility of operators can be easily described in terms of common solutions.

Proposition 19.11. *An operator $L \in \mathfrak{L}\mathfrak{D}(T)$ is divisible by another operator $Q \in \mathfrak{L}\mathfrak{D}(T)$, if and only if any solution of $Qf = 0$ is also a solution of $Lf = 0$.*

Proof. The "if" part is obvious. To prove divisibility, consider a fundamental system $f_1, \dots, f_k$ of solutions of the equation $Qf = 0$ and divide L by Q with remainder R, $L = PQ + R$, as in Lemma 19.2. Being in the null space for L and Q by assumption, $f_1, \dots, f_k$ also belong to the null space of PQ and hence to the null space of R. Since $\operatorname{ord} R < k$, this is possible only when $R = 0$ by Proposition 19.6. □

For any meromorphic germ $0 \not\equiv f \in \mathfrak{M}(T, t_0)$ one can immediately construct a first order linear operator vanishing on this germ, e.g., in the form

$$Q = fD - f', \qquad f' = Df.$$

By Proposition 19.6, any operator L such that $Lf = 0$, can be divided by Q, $L = L'Q$. If another solution (germ) $g \in \mathcal{M}(T, t_0)$, is known, $Lg = 0$, then the germ $g' = Qg$ is a meromorphic solution of the equation $L'g' = 0$ and can be used to further factor the operator L'.

If all n solutions $f_1, \dots, f_n$ of the homogeneous nth order equation $Lf = 0$ are known, this procedure allows us to construct *complete factorization* of L as a composition of n first order operators with coefficients in $\mathcal{M}' = \mathcal{M}(f_1, \dots, f_n)$. The factorization involves Wronskians, or Wronski determinants of the functions.

Now we can describe the factorization of an arbitrary differential operator $L \in \mathfrak{D}(T)$ with a known system of n linearly independent solutions $f_1, \dots, f_n$, using the Wronskians of these functions. Assume that U is a simply connected domain without singularities of L, so that $f_1, \dots, f_n \in \mathcal{O}(U)$, and denote by

$$\begin{aligned} w_k &= W(f_1, \dots, f_k) \in \mathcal{O}(U), \qquad k = 1, \dots, n, \\ w_{-1} &= w_0 = 1, \qquad w_{n+1} = w_n, \end{aligned} \tag{19.12}$$

the Wronskians of the first k functions from the ordered tuple $f_1, \dots, f_n$ (the functions w_{-1}, w_0 and w_{n+1} are introduced for convenience).

Theorem 19.12. *If $f_1, \dots, f_n \in \mathcal{O}(U)$ are linearly independent solutions of the equation $Lf = 0$ with a monic operator $L = D^n + \cdots$, then L is a composition of n derivations D interspersed with $n+1$ multiplications $b_0, \dots, b_n \in \mathcal{M}(U) \cong \mathfrak{D}(0, U)$, as follows:*

$$\begin{aligned} L &= b_n\, D\, b_{n-1}\, D\, b_{n-2} \cdots b_2\, D\, b_1\, D\, b_0, \\ b_k &= \frac{w_k^2}{w_{k-1} w_{k+1}}, \qquad k = 0, 1, \dots, n. \end{aligned} \tag{19.13}$$

Proof. Consider the monic differential operators L_k of order $k = 0, 1, \dots, n$,

$$L_0 = \mathrm{id}, \quad L_k = w_k^{-1}(t) \cdot W(f_1, \dots, f_k, \cdot), \qquad k = 1, \dots, n.$$

We claim that these operators satisfy the operator identity

$$D\, \frac{w_{k-1}}{w_k}\, L_{k-1} = \frac{w_{k-1}}{w_k}\, L_k, \qquad k = 1, \dots, n. \tag{19.14}$$

Indeed, both parts are differential operators of the same order k with the same leading terms $(w_{k-1}/w_k)\, D^k$. The null spaces of both operators also coincide with the linear span of $f_1, \dots, f_k$ and hence with each other. Indeed, the functions $f_1, \dots, f_{k-1}$ obviously belong to the null space of both parts. On the last function f_k the operator L_k vanishes by definition, whereas $L_{k-1} f_k = w_k / w_{k-1}$, so the left hand side of (19.14) also vanishes. Being both monic and having the same null space, the operators occurring in the two sides of (19.14), must coincide.

The identity (19.14) can be rewritten as

$$L_k = \frac{w_k}{w_{k-1}}\, D\, \frac{w_{k-1}}{w_k} L_{k-1}, \qquad k = 1, \dots, n.$$

Applying it recursively to the monic operator $L = L_n$ which is what we are interested in by Proposition 19.9, we obtain its decomposition into n terms

$$L_n = \left(\frac{w_n}{w_{n-1}}\, D\, \frac{w_{n-1}}{w_n}\right) \cdots \left(\frac{w_2}{w_1}\, D\, \frac{w_1}{w_2}\right) \cdot \left(\frac{w_1}{w_0}\, D\, \frac{w_0}{w_1}\right) \cdot L_0,$$

which coincides with (19.13). □

The advantage of such "complete factorization" becomes clear when solving homogeneous or nonhomogeneous equations. Denote by D^{-1} *any* "primitive" operator, i.e., $D^{-1} f = \int f\, dt$ in the case $D = \frac{\partial}{\partial t}$ (defined modulo a constant). Then the general solution of the equation $Lf = g$ for L factored as in (19.13), is given by the symbolic formula

$$f = b_0^{-1}\, D^{-1}\, b_1^{-1}\, D^{-1} \cdots D^{-1}\, b_{n-1}^{-1}\, D^{-1}\, b_n^{-1} g. \tag{19.15}$$

In other words, *knowing a fundamental system of solutions of a homogeneous differential equation allows us to solve any nonhomogeneous equation by taking n quadratures.* This may be a convenient alternative to reducing the equation to the companion system and using the method of variation of constants.

In general, solutions of linear equations, are ramified at singular points hence the formal factorization (19.13) has in general multivalued coefficients. In other words, factorization (19.13) holds over the extension $\mathfrak{M}' \supsetneq \mathfrak{M}$ and *not* over the initial field $\mathfrak{M} = \mathfrak{M}(T)$. Reducibility of operators in over $\mathfrak{M}$ is closely related to reducibility of their monodromy group.

Theorem 19.13. *A linear operator $L \in \mathfrak{LO}(T)$ having only regular singularities in T, is reducible in the algebra $\mathfrak{LO}(T)$ if and only if its monodromy group is reducible (i.e., has a nontrivial invariant subspace).*

Proof. Assume that $L = PQ$ and $f_1, \dots, f_k$ is a fundamental system of solutions for $Qf = 0$. Then these functions also solve the equation $Lf = 0$ and span an invariant subspace of the monodromy group which is therefore reducible. Conversely, assume (without loss of generality) that an invariant subspace of the monodromy group for $Lf = 0$ is generated by the first k functions $f_1, \dots, f_n$ of some fundamental system of solutions. Then by the Riemann Theorem 19.7, there exists an operator $Q \in \mathfrak{LO}(T)$ of order k, annulled by these first functions. By Proposition 19.11, L is divisible by Q and hence reducible in $\mathfrak{LO}(T)$. □

Factorization of operators is compatible with regularity. For brevity we say that a differential operator $L \in \mathfrak{D}(T)$ is *regular* in $U \subset T$, if it has only regular singular points there.

Lemma 19.14. *Composition of two regular operators is regular. Conversely, if a regular operator is reducible in $\mathfrak{D}(T)$, then both factors are also regular.*

Proof. If $L = PQ$, then any solution of the equation $Lf = 0$ is a solution of the nonhomogeneous equation $Qf = g$, where g is some solution of the lower order equation $Pg = 0$. For any singular point $t_0 \in T$, the function g grows moderately at t_0 since P is regular. Since Q is also regular at this point, by Lemma 16.6 we conclude that f also grows moderately at t_0. This proves regularity of PQ.

Conversely, if $L = PQ$ is regular, then any function from the null space of Q grows moderately at any singular point t_0 regardless of regularity of P. To prove regularity of P, choose any solution g of the equation $Pg = 0$. As before, let f be any solution of $Qf = g$: by construction, f grows moderately as a solution of $Lf = 0$ and can be represented as

$$f(t) = (h_1, \dots, h_n)\,(t - t_0)^A (c_1, \dots, c_n)^\top,$$

where the row vector function $(h_1, \dots, h_n)$ is meromorphic at t_0, the column vector $(c_1, \dots, c_n)^\top$ has constant entries and A is any logarithm of the monodromy matrix around t_0. Any such function admits any number of derivations and multiplications by meromorphic functions while retaining the moderate growth at t_0. Therefore application of any operator $Q \in \mathfrak{D}(T)$ proves that $g = Qf$ grows moderately at t_0, so that P is regular. □

As an immediate application of this result, we have the local theorem on complete factorization.

Theorem 19.15. *Any differential operator $L \in \mathfrak{D}(T)$ having a regular singularity at a point $t_0 \in T$, admits complete factorization in a small neighborhood $U = (T, t_0)$ of this point,*

$$L = P_n P_{n-1} \cdots P_1, \qquad P_i \in \mathfrak{D}(U), \quad \operatorname{ord} P_i = 1, \tag{19.16}$$

with first order factors P_i having meromorphic coefficients in U and regular singularity at t_0.

Proof. The monodromy group of any operator in a punctured neighborhood U of an (isolated) singular point is cyclic and hence always admits a one-dimensional invariant subspace. By Theorem 19.13, $L = L_0$ is divisible from the right by a first order operator $P_1 \in \mathfrak{D}(U)$ whose leading term can be prescribed arbitrary. By Lemma 19.14, both P_1 and its left cofactor L_1 are regular at t_0. Thus the process can be continued by induction until the complete factorization is achieved. $\square$

Remark 19.16. Note that the leading terms of $P_1, \dots, P_{n-1}$ can be prescribed arbitrarily, as multiplication by a meromorphic germ is a unit of the algebra $\mathfrak{LO}(n, T)$.

19D. Fuchsian singularities of nth order equation. Similarly to the general case of linear systems, regular singularity is not necessarily a first order pole of the companion system if the derivation D itself is nonsingular at this point. However, unlike the general case, we can introduce the class of equations with "*first order pole*", which turns out to coincide with the class of regular equations.

The reason why the words above are enclosed by the quotation marks, is noninvariance of this notion. Indeed, the companion system (19.6) by definition has a singularity at a point $t_0 \in T$ if either the vector field D is singular at t_0, i.e., $D = r(t)\frac{\partial}{\partial t}$ in a local chart on T with $\operatorname{ord}_{t_0} r(t) > 0$, or D is nonsingular, $\operatorname{ord}_{t_0} r(t) = 0$, but some of the ratios a_i/a_0, $i = 1, \dots, n$ have a pole at t_0 (in such a case we denote by $\operatorname{ord} A$ the negative of the maximal order of the poles of all entries of a meromorphic matrix function $A(t)$). In both cases the order of the pole, understood as $\operatorname{ord} r - \operatorname{ord} A$, is positive. Yet this order *explicitly depends on the choice of the derivation D used to write the companion system.*

Definition 19.17. A differential operator $L \in \mathfrak{LO}(T)$ is *Fuchsian* at a singular point t_0, if in the companion form (19.6)

$$\operatorname{ord}_{t_0} D = 1, \qquad \operatorname{ord}_{t_0} A = 0.$$

This definition is equivalent to another, more transparent (though less invariant) description.

Proposition 19.18. *A differential operator L is Fuchsian at a finite point t_0, if after expansion in the powers of $D' = (t - t_0)\frac{\partial}{\partial t}$ and reduction to the monic form, it has holomorphic coefficients.* $\square$

Obviously, instead of the linear vector field D' one can use any other holomorphic germ with a simple singularity at t_0. Re-expanding an expression for the monic operator $D'^n + \cdots + a_{n-1}D' + a_n$ in powers of the "usual" differentiation $D = \frac{\partial}{\partial t}$, we obtain the property that is often used as the definition of finite Fuchsian singularity [**Inc44, Har82**].

Proposition 19.19. *A monic operator $L = D^n + \cdots + a_n \in \mathfrak{LO}(n, \mathbb{C})$, $D = \frac{\partial}{\partial t}$, has a Fuchsian singularity at a finite point $t = t_0 \in \mathbb{C}$, if and only if $\operatorname{ord}_{t_0} a_k(t) \geqslant -k$ for all $k = 0, \dots, n$.* $\square$

The advantage of the invariant Definition 19.17 is that it can automatically be reformulated for the case where the Fuchsian singularity is at infinity, $t_0 = \infty \in \mathbb{P}$ (Problem 19.6).

From the Sauvage Theorem 16.10 we immediately conclude that any Fuchsian singularity of an operator $L \in \mathfrak{LO}(T)$ is always regular. Somewhat unusual is the fact that for high order equations the inverse is also true.

Theorem 19.20 (L. Fuchs, 1868). *Any regular singularity of a linear ordinary differential equation with meromorphic coefficients, is Fuchsian.*

Proof. 1°. For equations of the first order the assertion of the theorem is verified by a straightforward computation. Assume that the regular singularity occurs at $t = 0$. Consider the equation $L'f = 0$, where $L = D' + a_1'(t)$ is expanded using the standard Euler derivation $D' = t\frac{\partial}{\partial t}$. If L has a regular singularity at $t = 0$, we can represent its solution as $f(t) = t^\lambda h(t)$ with an appropriate complex $\lambda \in \mathbb{C}$ and some meromorphic function $h(t)$. Changing λ by a suitable integer number, we can assume in addition that h is holomorphic and holomorphically invertible at $t = 0$. Substituting this representation for f into the equation $D'f + a_1'f = 0$, we obtain the formula $-a_1'(t) = D'f/f = \lambda + (D'h/h)$. Since h is holomorphically invertible and $D' = t\frac{d}{dt}$ holomorphic, we conclude that a_1' is holomorphic at t_0 and hence $L = D' + a_1'$ is Fuchsian.

2°. The case of an arbitrary order follows from the factorization Theorem 19.15. By this theorem, any regular operator L can be factored as $L = a_0'P_n \cdots P_1$ with each P_i being a first order operator regular at $t = 0$. Since the leading terms of P_i can be chosen arbitrarily (Remark 19.16), we assume that

$$P_i = tD + a_i' = D' + a_i', \qquad i = 1, \dots, n.$$

By Step 1°, each P_i is Fuchsian, that is, the free terms $a_1', \dots, a_n'$ are necessarily holomorphic at t_0. But then the composition $P_n \cdots P_1$ begins with the leading term D'^n and has all holomorphic coefficients after the complete expansion. In other words, L differs from a Fuchsian operator by a meromorphic factor a_0' and hence is also Fuchsian. □

The companion system can be rewritten in the Pfaffian form. Let $\omega \in \Lambda^1(T) \otimes \mathcal{M}(T)$ be the (scalar) meromorphic 1-form *dual* to the vector field D: by definition, this means that $\omega(D) \equiv 1$. By duality, D has a simple singularity at $t_0 \in T$ if and only if ω has a simple pole at this point. Using this form, the companion system can be written in the Pfaffian form as follows:

$$d\begin{pmatrix} x_0 \\ x_1 \\ \vdots \\ x_{n-2} \\ x_{n-1} \end{pmatrix} = \omega \begin{pmatrix} 0 & 1 & & & \\ & 0 & 1 & & \\ \cdots & \cdots & \cdots & \cdots & \cdots \\ & & & 0 & 1 \\ b_n & b_{n-1} & \cdots & b_2 & b_1 \end{pmatrix} \begin{pmatrix} x_0 \\ x_1 \\ \vdots \\ x_{n-2} \\ x_{n-1} \end{pmatrix}, \tag{19.17}$$

with holomorphic entries $b_1, \dots, b_n \in \mathcal{O}(T, t_0)$ and a form ω with the first order pole at t_0.

The matrix residue of the corresponding matrix 1-form $\Omega = \omega A$ is equal to $A(t_0) \cdot \operatorname{res}_{t_0} \omega$. Its eigenvalues are called characteristic exponents of the Fuchsian (regular) singularity.

Example 19.21. Any linear ordinary differential equation with a regular singularity at $t = 0$ can be written under the form $Lf = 0$, where

$$L = D'^n + a_1(t) D'^{n-1} + \cdots + a_{n-1}(t) D' + a_n(t), \qquad D' = t\frac{\partial}{\partial t}, \tag{19.18}$$

is the monic expansion in powers of the Euler derivation D' with the coefficients $a_j(t)$ holomorphic at the origin. The characteristic exponents of the corresponding singularity are roots of the polynomial

$$\lambda^n + a_1(0)\lambda^{-1} + \cdots + a_{n-1}(0)\lambda + a_n(0) = 0. \tag{19.19}$$

Obviously, instead of the Euler operator one can use any other operator D'' with a simple singularity and eigenvalue (linearization 1×1-matrix) equal to 1 (see also Problem 19.5).

Fuchsian singularities in the companion form (19.17) are considerably more rigid than general singularities of linear systems, for instance, analytic gauge transform to the Poincaré–Dulac–Levelt normal form destroys the "companion structure". Yet despite all that, one can apply Lemma 16.18 and obtain an ansatz for construction of analytic (ramified) solutions of linear equation near Fuchsian singularity under the form

$$\sum_1^n h_j(t)\, t^{\lambda_j} p_j(\ln t), \qquad h_j \in \mathcal{O}(\mathbb{C}, 0),$$

where $\lambda_1, \dots, \lambda_n$ are characteristic exponents and p_j are polynomials with constant coefficients. The degrees of the polynomials are determined by the resonance identities $\lambda_i \equiv \lambda_j \bmod \mathbb{Z}$ between the characteristic exponents, as encoded by the structure of the matrix I in (16.10).

19E. Jet bundles and invariant constructions. To describe the global structure of regular equations, we need geometric (invariant) description of the jet bundles. We recall briefly their construction; more details can be found in [**AVL91**].

Consider the n-jet space $J^n(T)$ which is the union of all jet spaces at all points of T. The space $J^n(T)$ is equipped with the natural projection $\tau_n \colon J^n(T) \to T$. This projection equips $J^n(T)$ with the structure of a holomorphic vector bundle as follows.

Let $U_\alpha \subset T$ be an open domain and $D_\alpha \in \mathcal{D}(U_\alpha)$ a holomorphic vector field (derivation) *nonsingular in* U_α, as usual identified with the derivation of the algebra $\mathcal{M}(U_\alpha)$. This derivation allows us to associate any jet of a function f at a point p with the (column) vector

$$(\text{jet of } f \text{ at } p \in U_\alpha) \overset{\Phi_\alpha}{\longmapsto} (f, Df, D^2 f, \dots, D^n f)^\top\big|_p, \qquad D = D_\alpha. \quad (19.20)$$

The map Φ_α defines a trivialization of $J^n(T)$ over the domain U_α.

If U_β is another domain and $D' = D_\beta$ another derivation holomorphic and nonsingular in U_β, then on the intersection $U_{\alpha\beta} = U_\alpha \cap U_\beta$ the two respective derivations $D = D_\alpha$ and $D' = D_\beta = r_{\beta\alpha} D_\alpha$ and their powers are related by the formulas

$$\begin{pmatrix} 1 \\ D' \\ D'^2 \\ \vdots \\ D'^n \end{pmatrix} = \begin{pmatrix} 1 & & & & \\ \vdots & r & & & \\ \vdots & \vdots & r^2 & & \\ \vdots & \vdots & \vdots & \ddots & \\ * & \cdots & \cdots & \cdots & r^n \end{pmatrix} \cdot \begin{pmatrix} 1 \\ D \\ D^2 \\ \vdots \\ D^n \end{pmatrix}, \qquad \begin{aligned} D &= D_\alpha \in \mathcal{D}(U_\alpha), \\ D' &= D_\beta \in \mathcal{D}(U_\beta), \\ r &= r_{\beta\alpha} \in \mathcal{O}(U_{\alpha\beta}). \end{aligned} \quad (19.21)$$

These formulas define the gauge transform

$$\Phi_\beta \circ \Phi_\alpha^{-1} \colon (t, x) \to (t, H_{\beta\alpha}(t)x), \qquad (19.22)$$

with the same matrix as in (19.21). The collection of matrices $H = H_{\beta\alpha} = H_{\alpha\beta}^{-1} \in \mathrm{GL}(n, \mathcal{O}(U_{\alpha\beta})$ form a matrix cocycle defining the bundle τ_n.

Definition 19.22. The bundle $\tau_n \colon J^n(T) \to T$, defined by the trivializations (19.20) (or, equivalently, by the matrix cocycle (19.21)) is called the *n-jet bundle* over the base T.

Example 19.23. The line bundle defined by the cocycle $r_{\alpha\beta}$ is equivalent to the cotangent bundle $\mathbf{T}^*T$ over the base T. Indeed, consider an arbitrary meromorphic cochain $\{f_\alpha\}$ associated with a section of this bundle. This means that $f_\beta = r_{\alpha\beta} f_\beta$ on any intersection $U_{\alpha\beta}$. We claim that this cochain consistently defines a meromorphic 1-form ω by the rules

$$\omega_\alpha(D_\alpha) = f_\alpha, \qquad \omega_\alpha \in \Lambda^1(U_\alpha) \otimes \mathcal{M}(U_\alpha).$$

Indeed, on the overlapping $U_{\alpha\beta}$ the forms coincide, $\omega_\alpha = \omega_\beta$ (can be verified either on D_α or D_β), thus the cochain $\{\omega_\alpha\}$ defines a global meromorphic 1-form $\omega \in \Lambda^1(T) \otimes \mathcal{M}(T)$.

Example 19.24. On the Riemann sphere $T = \mathbb{P}$ the two fields $D_0 = \frac{\partial}{\partial t} \in \mathcal{D}(\mathbb{C})$ and $D_1 = t^2 \frac{\partial}{\partial t} \in \mathcal{D}(\mathbb{P} \smallsetminus \{0\})$ define the Birkhoff–Grothendieck cocycle corresponding to the bundle $\tau_n(\mathbb{P})\colon J^n(\mathbb{P}) \to \mathbb{P}$ with the corresponding function $r_{10}(t) = t^2$. The determinant bundle $\det \tau_n$ is associated with the cocycle $\det H_{10} = t^{n(n+1)}$. Thus the degree of the bundle is nonzero,

$$\deg \tau_n = -n(n+1) \neq 0 \qquad \text{for } n \geqslant 1, \tag{19.23}$$

and hence the jet bundle is nontrivial for all $n \geqslant 1$. For $n = 0$ the bundle τ_0 is obviously trivial, $J^0(T) = T \times \mathbb{C}^{n+1}$ for any base T. The 1-jet bundle is described in Problem 19.8.

Every meromorphic function $u \in \mathcal{M}(T)$ defines a meromorphic section of the jet bundle $t \mapsto j^n_u(t)$, called the *jet extension* of u, which is holomorphic outside the polar locus of u. However, not every section of τ_n is the collection of jets of some function: there are integrability conditions that are necessary.

Let ω_α be holomorphic 1-forms dual to the vector fields D_α, $\omega_\alpha(D_\alpha) \equiv 1$. This is a holomorphic cochain of 1-forms. On any trivializing chart $\tau_n^{-1}(U_\alpha) \cong U_\alpha \times \mathbb{C}^{n+1}$, using the scalar form ω_α, we can construct a 2-dimensional distribution as the common null space of $n-1$ Pfaffian forms

$$dx_0 - x_1\omega_\alpha = 0,\ dx_1 - x_2\omega_\alpha = 0,\ \dots,\ dx_{n-1} - x_n\omega_\alpha = 0. \tag{19.24}$$

One can instantly verify that two such distributions defined over two different trivializations, are related by the same gauge transforms (19.21) (note that the formulas (19.24) "naively mean" that $D_\alpha x_k = x_{k+1}$).

Definition 19.25. The 2-dimensional distribution defined on the n-jet bundle $J^n(T)$ by the formulas (19.24) in the trivializing charts, is called the *Cartan distribution.*

The Cartan distribution singles out sections of the jet bundle, which are jet extensions of meromorphic functions. Namely, if $\mathcal{C}_q$ is the 2-dimensional subspace of the Cartan distribution at a point $q \in J^n(T)$ and $u \in \mathcal{O}(T,p)$ is a holomorphic germ at the point $p = \tau_n(q) \in T$ such that $j^n_u(p) = q$, then the graph of the section $t \mapsto j^n_f(t)$ is a holomorphic curve tangent to the plane $\mathcal{C}_q$. Moreover, one can easily verify that $\mathcal{C}$ can be "axiomatically" (invariantly) defined as the only 2-dimensional distribution on $J^n(T)$ which is tangent to graphs of all meromorphic sections of the form $t \mapsto j^n_u(t)$.

Conversely, any meromorphic section $s \in \Gamma(\tau_n)$ whose graph is tangent to the Cartan distribution at all points, is the graph of a jet extension of a meromorphic function $u \in \mathcal{M}(T)$, $s = j^n_u$. In the future we will refer to sections tangent to the Cartan distribution as the *integrable sections*.

Finally we make the following obvious observation: the bundles $J^n(T)$ are naturally "nested", more precisely, there exist bundle maps (all fibered over the identity) making the following diagram commutative:

$$\begin{array}{ccccccccc} J^0(T) & \xleftarrow{\tau^1_0} & J^1(T) & \xleftarrow{\tau^2_1} & \cdots & \xleftarrow{\tau^n_{n-1}} & J^n(T) & \xleftarrow{\tau^{n+1}_n} & \cdots \\ {\scriptstyle\tau_0}\downarrow & & {\scriptstyle\tau_1}\downarrow & & & & \downarrow{\scriptstyle\tau_n} & & \\ T & = & T & = & \cdots & = & T & = & \cdots \end{array} \tag{19.25}$$

The maps τ^k_{k-1} simply "forget" the last derivative. The kernel of each such map is one-dimensional. The corresponding one-dimensional subbundle $\mathcal{V}_k \subset J^k(T)$ will be referred to as *vertical* subbundle.

Now everything is ready to define in invariant terms linear ordinary differential equations.

Theorem 19.26. *For any holomorphic subbundle $\mathcal{L} \subset J^n(T)$ of codimension 1 in the n-jet bundle, transversal to the vertical subbundle $\mathcal{V} = \ker \tau^n_{n-1}$ almost everywhere, there exists a meromorphic connexion $\nabla = \nabla_{\mathcal{L}}$ on $J^n(T)$ with the following properties*:

1. *the subbundle $\mathcal{L}$ is invariant by ∇,*
2. *the singular locus of $\Sigma = \operatorname{Sing} \nabla$ consists of the points where $\mathcal{L}$ is nontransversal to the vertical bundle $\mathcal{V}$,*
3. *all ∇-horizontal sections of τ_n are integrable, i.e., are graphs of n-jet extensions of functions on T.*

The restriction of ∇ on $\mathcal{L}$ is uniquely defined.

Proof. The Cartan distribution restricted on the subbundle $\mathcal{L}$ (holomorphic submanifold of codimension 1) induces a 1-dimensional distribution (line field) on this bundle, eventually with singularities at the points of nontransversality between $\mathcal{L}$ and $\mathcal{C}$. The Cartan distribution always contains the vertical direction, hence transversality to $\mathcal{V}$ implies transversality to $\mathcal{C}$. Because of one-dimensionality, the constructed distribution is integrable. The integral curves (leaves of the integral foliation) by construction are tangent to the Cartan distribution $\mathcal{C}$. It remains to verify that the leaves of this foliation on $\mathcal{L}$ are horizontal sections for some meromorphic connexion ∇ on $J^n(T)$. We will explicitly construct the $(n+1) \times (n+1)$-matrix connexion 1-form Ω in any trivialization of $J^n(T)$, defined by a nonsingular vector field $D \in \mathcal{D}(U)$, $U \subseteq T$, or the dual form $\omega \in \Lambda^1(U) \otimes \mathcal{M}(U)$, as in (19.20).

The subbundle $\mathcal{L}$ in this trivializing chart is defined by a holomorphic equation $\sum_0^n a_i(t) x_{n-i} = 0$. Its differential (the tangent hyperplane to $\mathcal{L}$) modulo the Pfaffian equations (19.24) which define the Cartan distribution, is equal to

$$a_0\, dx_n + x_n(da_0 + a_1\omega) + x_{n-1}(da_1 + a_2\omega) + \cdots + x_1(da_{n-1} + a_n\omega) + x_0\, da_n.$$

If outside the singular locus $\{a_0 = 0\} \cap U$ this Pfaffian equation can be resolved with respect to dx_n. In conjunction with the Cartan equations this yields a meromorphic linear system over U, which by construction is tangent to the hypersurface $\{\sum_0^n a_i x_{n-i} = 0\}$ and the Cartan distribution. □

The connexion constructed in Theorem 19.26 is *not a companion connexion* on $J^n(T)$: its only advantage is the invariant construction. In practice the bundle $\mathfrak{L}$ satisfying the assumptions of the theorem, is projected along the vertical direction onto the $(n-1)$-jet bundle. The projection τ_{n-1}^n restricted on $\mathfrak{L}$, is a *meromorphic bundle map*, which carries the connexion $\nabla|_{\mathfrak{L}}$ to the meromorphic connexion defined by the Pfaffian companion system (19.17) with $b_i = -a_i/a_0$: the last equation is obtained by resolving the linear identity $\sum a_i x_{n-i} = 0$ with respect to x_n and substituting the result in the last Cartan equation $dx_{n-1} = \omega x_n$. Thus $\rho_{n-1}^n|_{\mathfrak{L}}$ carries $\nabla|_{\mathfrak{L}}$ into the companion connexion on $J^{n-1}(T)$.

For arbitrary (not regular) equations their interpretation as a connexion tangent to a subbundle $\mathfrak{L} \subset J^n(T)$ is as good (or as bad) as any other connexion meromorphically equivalent to it, in particular, as the companion connexion on the bundle $J^{n-1}(T)$ associated with an arbitrary meromorphic vector field $D \in \mathcal{D}(T)$ or the corresponding dual form $\omega \in \Lambda^1(T) \otimes \mathbf{M}(T)$. The "naive approach" described in §19**B**, corresponds to the choice of $D = \frac{\partial}{\partial t} \in \mathcal{D}(\mathbb{P})$ (note that the bundle $J^{n-1}(T)$ is also nontrivial, and this choice of D does not properly address the presence or absence of singularities at infinity).

However, if the connexion is *regular*, then it is natural to look for a bundle with *Fuchsian connexion* on it, *meromorphically* equivalent to the bundle $\mathfrak{L} \subset J^n(T)$ with the connexion $\nabla_{\mathfrak{L}}$.

Theorem 19.27. *If $L \in \mathfrak{LD}(\mathbb{P})$ is an arbitrary differential operator such that the linear equation $Lu = 0$ has $m \geqslant 0$ regular singularities, then the meromorphic connexion $\nabla|_{\mathfrak{L}}$ constructed in Theorem 19.26 is meromorphically gauge equivalent to a Fuchsian connexion on a holomorphic vector bundle π of rank n over $\mathbb{P}$. The degree of this bundle is equal to $(m-2)n(n-1)/2$.*

Proof. The existence of a Fuchsian connexion on an abstract bundle follows from the fact that any regular singularity at $t = t_j \in \mathbb{P}$ becomes Fuchsian after the local meromorphic gauge transform (re-expanding L in powers of $(t-t_j)\frac{\partial}{\partial t}$ rather than in powers of $\frac{\partial}{\partial t}$) by Theorem 19.20.

If $m = 2$, then there exists a holomorphic vector field D on $\mathbb{P}$ with exactly two simple (hyperbolic) singularities at two specified points. Expanding L in powers of D, we obtain expansion with holomorphic (hence constant) coefficients and nonvanishing leading term. Such an equation is necessarily an Euler equation (Problem 19.12) on the trivial bundle over $\mathbb{P}$.

If $m \neq 2$, such a vector field does not exist and the resulting bundle will be nontrivial. Assume that the point at infinity is nonsingular for the equation $Lu = 0$, and denote by $t_1, \dots, t_m \in \mathbb{C}$ distinct singular points of the equation, $\max_j |t_j| < R$. Consider two meromorphic vector fields on $\mathbb{P}$,

$$D_0 = \prod_{j=1}^{m}(t - t_j)\,\frac{\partial}{\partial t}, \qquad D_1 = t^{2-m} D_0.$$

They are holomorphic in the respective domains $U_0 = \mathbb{C}$, $U_1 = \mathbb{P} \smallsetminus \{|t| < R\}$ of the standard Birkhoff–Grothendieck covering and have singularities ("roots") only at the singular points of the equation.

By Theorem 19.20, after expansion in powers of D_0, D_1 and reduction to the corresponding companion form, we will obtain two meromorphic matrix functions Ω_0, Ω_1, with the following properties:

(1) Ω_0 has only Fuchsian singularities (simple poles) at the points $t_1, \dots, t_m$ and holomorphic at all other points of U_0,

(2) Ω_1 is holomorphic in U_1,

(3) in the annulus U_{01} the two forms are conjugated by the matrix function $H = H_{10}(t)$ as in (19.21) with the function $r = r_{10}(t) = t^{2-m}$.

The determinant $\det H_{10} = t^{(2-m)n(n-1)/2} = \det H_{01}^{-1}$ is the standard cocycle associated with the bundle ξ_d, $d = (m-2)n(n-1)/2$. Hence Ω_0, Ω_1 are two trivializations of a Fuchsian connexion on the holomorphic vector bundle associated with the cocycle $\{H_{01}, H_{10}\}$, which has degree d. □

From this result and Corollary 17.35 we immediately derive the assertion on the sum of all characteristic exponents.

Corollary 19.28. *The total of all characteristic exponents of a regular equation of order n with m singular points is equal to $(m-2)n(n-1)/2$.* □

19F. Riemann–Hilbert problem for higher order equations. The Riemann–Hilbert problem for scalar equations is to construct a Fuchsian equation of order n on $\mathbb{P}$ with the specified monodromy group. This problem is usually not solvable for one simple reason: the dimension of the variety of different monodromy data is larger than the dimension of the variety of Fuchsian equations.

Indeed, any equation with $m+2$ singular points $t_0 = 0, t_1, \dots, t_m \in \mathbb{C}$, $t_{m+1} = \infty \in \mathbb{P}$, is Fuchsian if and only if the corresponding linear operator can be written in the form

$$L = D^n + a_1 D^{n-1} + \cdots + a_n, \qquad D = \tfrac{\partial}{\partial t}, \quad a_k = \frac{p_k(t)}{\Delta^k(t)},$$
$$p_k \in \mathbb{C}[t],\ \deg p_k \leqslant mk, \quad k = 1, \dots, n, \quad \Delta(t) = \prod_1^m (t - t_j) \tag{19.26}$$

because of the restrictions on the order of the poles of coefficients at all singularities (note that D has a simple pole at both t_0 and t_{m+1}). The total number of parameters (assuming that the singular locus is fixed) is equal to

$$(m+1)+(2m+1)+\cdots+(nm+1)=\tfrac{1}{2}mn(n+1)+n.$$

The total number of entries in $m+1$ monodromy matrices is $(m+1)n^2$, (the last matrix is uniquely defined by the requirement that the product is equal to identity). In fact, one can assume that one of the matrices is reduced to the Jordan normal form which involves n diagonal terms (and the discrete choice 0 or 1 for the above-diagonal sequence). Thus the variety of all monodromy data has dimension equal to mn^2+n.

The second number is almost always greater than the first, thus the Riemann–Hilbert problem is not solvable for most monodromy data. The exceptional combinations when the equality occurs, are $m=0$ and $n=1$. The first case corresponds to Euler equations (Problem 19.12), the second to the scalar equation. In the second case the monodromy is commutative and clearly any collection of m multiplicators can be realized by a scalar first order equation with preassigned poles.

For the Euler equation the monodromy group is determined by a single matrix M.

Proposition 19.29. *Any invertible matrix $M\in \mathrm{GL}(n,\mathbb{C})$ can be realized (modulo conjugacy) as the monodromy matrix of an Euler operator*

$$D'^n+a_1D'^{n-1}+\cdots+a_{n-1}D'+a_n,\qquad D'=t\tfrac{\partial}{\partial t},\quad a_1,\dots,a_n\in\mathbb{C}. \tag{19.27}$$

Proof. We will show how a matrix in the Jordan normal form can be realized by the monodromy of an Euler equation.

One can immediately verify that the monodromy matrix of the operator D'^k, $k\geqslant 1$, is the (maximal) nilpotent Jordan $k\times k$-block in the basis $1,\ln t,\dots,\ln^{k-1}t$. The "conjugated" operator $(D'-\lambda)^k$ has the maximal Jordan block with the eigenvalue $\mu=\exp 2\pi i\lambda$ in the basis $t^\lambda\ln^j t$, $j=0,1,\dots,k-1$.

To build an arbitrary matrix with several Jordan blocks of various sizes, we use the composition of elementary factors of this form, which is again a monic Euler operator. Note that the Euler operators are always commuting between themselves, since their coefficients are constant.

If M consists of several Jordan blocks of sizes $\nu_1,\dots,\nu_s$ with *the same eigenvalue* $\mu\neq 0$, then this monodromy matrix is realized by the composition of *commuting* operators $L=\prod_{j=1}^s(D'-\lambda-j)^{\nu_j}$ for any fixed choice of the logarithm $\lambda=\frac{1}{2\pi i}\ln\mu$.

Finally, if $M = \operatorname{diag}\{M_1, \dots, M_r\}$ with the spectra M_j being singletons μ_j, then each block can be realized by an Euler operator L_{μ_j}, and the entire matrix is realized by the "product" (composition) of commuting operators $L = L_{\mu_1} \cdots L_{\mu_r}$. □

One can attempt to relax the Riemann–Hilbert problem for Fuchsian equations and demand less. For instance, the natural question would be whether one can realize a given collection of characteristic exponents by a suitable Fuchsian equation.

The "variety of exponents" of a Fuchsian system with m singularities has dimension $mn - 1$. This dimension is by one less than the product mn because the exponents are constrained by the equality from Corollary 19.28. Compared to the dimension of the variety of Fuchsian equations of the given order with the specified number of singularities, it is almost always less than the latter, which means that in general the solution should be nonunique.

There is only one case where the two dimensions coincide: $m = 3$, $n = 2$, i.e., for equations of second order with three singularities. The total sum of characteristic exponents in this case is equal to 1 by Corollary 19.28.

Theorem 19.30. *Any 6 numbers whose sum is equal to* 1*, can be realized as characteristic exponents of a Fuchsian equation of second order with three singular points.*

Proof. First we note that the characteristic exponents at each point can be shifted by an arbitrary constant, provided that these three constants added together give zero (Problem 19.16). Thus it is sufficient to realize the collection of exponents of the form

$$(0, \alpha), \quad (0, \beta), \quad \bigl(\gamma, 1 - (\alpha + \beta + \gamma)\bigr) \tag{19.28}$$

One can always use the method of indeterminate coefficients (19.26), expressing explicitly the characteristic exponents of this equation and show that the corresponding interpolation problem for polynomial coefficients indeed has a unique solution.

The freedom to choose the derivation allows us to reduce these computations very substantially. Assume (as is always done) that the three singularities are at the points $0, 1$ and ∞. Consider the vector field $D = t(t-1)\frac{\partial}{\partial t}$ which has simple singularities at $t = 0, 1$ with eigenvalues -1 and 1 respectively, and nonsingular point at infinity.

The operator

$$L = D^2 + p_1(t)\, D + q_2(t), \qquad D = t(t-1)\frac{\partial}{\partial t}, \tag{19.29}$$

is Fuchsian if p_1, q_2 are holomorphic functions in the entire finite part $\mathbb{C}$ with poles of respective orders at most 1 and 2 at infinity (Proposition 19.19).

This means that p_1 and q_2 are polynomials in t of the degrees 1 and 2 respectively.

The corresponding characteristic exponents at the points $t_0 = 0$ and $t_1 = 1$ are roots of the polynomials $(-\lambda)^2+p(t_0)(-\lambda)+q(t_0)$ and $\lambda^2+p(t_1)\lambda+q(t_1)$ respectively (changing λ to $-\lambda$ happens since the eigenvalue of D at t_0 is -1); see Example 19.21. Thus p is a linear polynomial taking values $-\alpha$ and β at the points $t_0 = 0$ and $t_1 = 1$ respectively, and q vanishes at both these points, $q = ct(t-1)$. To express the characteristic exponents at infinity, we re-expand the operator (19.29) in powers of the Euler operator $D' = (t-1)^{-1}D$ which has eigenvalue -1 at infinity. After division by $(t-1)^2$ we obtain a monic differential polynomial with the free term $ct/(t-1) \underset{t\to\infty}{\longrightarrow} c$, whose value at $t = \infty$ is equal to the product $\gamma_1\gamma_2$ of the characteristic exponents at the point $t_2 = \infty$.

Thus letting $c = \gamma(1-(\alpha+\beta+\gamma))$ we obtain the *hypergeometric equation* which solves the "relaxed Riemann–Hilbert problem" in the specific case of second order and three singularities,

$$L = D^2+\bigl(-\alpha+t\beta)\bigr)\,D+\gamma\bigl(1-(\alpha+\beta+\gamma)\bigr)\,t(t-1), \quad D = t(t-1)\tfrac{\partial}{\partial t}. \tag{19.30}$$

This expansion can be more easily memorized than the standard expansion [**Inc44**] of the hypergeometric equation

$$L = t(1-t)D'^2 + \bigl(\gamma' - (\alpha' + \beta' + 1)t\bigr)D' - \alpha'\beta', \qquad D' = \tfrac{\partial}{\partial t}, \tag{19.31}$$

which has characteristic exponents at the same three points $0, 1, \infty$ equal to

$$(0, 1-\gamma'), \quad (0, \gamma' - \alpha' - \beta'), \quad (\alpha', \beta').$$

The old-fashioned name for a general solution of this equation is the *Riemann P-function.* □

Remark 19.31. The term "hypergeometric *system*" is reserved for linear systems on $\mathbb{P}$ of a special form. Let $S \in \operatorname{Mat}(n, \mathbb{C})$ be a diagonalizable matrix with simple spectrum $\{s_, \dots, s_m\}$, and $A \in \operatorname{Mat}(n, \mathbb{C})$ an arbitrary matrix. Consider the linear system associated with the ordinary differential equation

$$(tE - S)\dot{x} = Bx, \qquad x \in \mathbb{C}^n,\ t \in \mathbb{C} \subset \mathbb{P}, \tag{19.32}$$

where E stands for the identical matrix. By a linear change of variables the matrix S can always be diagonalized. After inversion we have the meromorphic system

$$\dot{x} = \begin{pmatrix} (t-s_1)^{-1} & & \\ & \ddots & \\ & & (t-s_n)^{-1} \end{pmatrix} Bx. \tag{19.33}$$

This system has simple poles at the points $s_1, \dots, s_n$ and at the point $t = \infty$. The residue matrix A_j at each point has rank 1: the only nonzero row of this matrix is the jth row of the matrix B. Therefore the characteristic exponents at this point are all zeros, eventually except for the value $b_{jj} \in \mathbb{C}$.

The bridge between two notions, the hypergeometric systems and hypergeometric equations, is obvious. Each component of the hypergeometric 2×2-system

$$\begin{pmatrix} t & \\ & t-1 \end{pmatrix} \dot{x} = \begin{pmatrix} a & b \\ c & d \end{pmatrix} x. \tag{19.34}$$

satisfies a hypergeometric equation (19.30) (Problem 19.17).

Exercises and Problems for §19.

Problem 19.1 ([**VK75**], [**Kra97**]). Prove that a $\mathbb{C}$-linear self-map $L\colon \mathcal{M} \to \mathcal{M}$ is a linear differential operator of order $\leqslant n$, if and only if the iterated commutator $[g_0, [g_1, [\cdots, [g_n, L]\cdots]]]$ vanishes identically as a self-map of $\mathcal{M}$ for any $n+1$ multiplications $g_i\colon \mathcal{M} \to \mathcal{M}$, $f \mapsto g_i f$.

Exercise 19.2. Prove that the monodromy of a linear equation $Lf = 0$, $L \in \mathfrak{LO}(T)$, is reducible if and only if the holonomy of the respective companion system is reducible.

Problem 19.3. Let $f_1, \dots, f_n$ be functions holomorphic in a domain $U \subset T$. Prove that if $W(f_1, \dots, f_n) \equiv 0$, then these functions are linearly dependent over $\mathbb{C}$. Is this true for C^∞-smooth functions?

Exercise 19.4. Prove in detail Proposition 19.19.

Problem 19.5. Find characteristic exponents at the origin for a Fuchsian operator $L = D^n + a_1 D^{n-1} + \cdots + a_n$ with holomorphic coefficients $a_k \in \mathcal{O}(\mathbb{C}, 0)$ and a holomorphic vector field $D = (ct + \cdots)\frac{d}{dt}$ with $c \neq 0$.

Problem 19.6. Prove that the point $t_0 = \infty$ is Fuchsian for the monic linear operator (19.2) expanded in the powers of $D = \frac{\partial}{\partial t}$ with $a_0 \equiv 1$, if and only if $\operatorname{ord}_\infty a_k \geqslant k + 2 - n$.

Exercise 19.7. Let $s = \lambda_1 + \cdots + \lambda_n$ be the sum of characteristic exponents of a regular singularity of a linear equation $Lf = 0$. Prove that the Wronskian of a fundamental system of solutions $w(t) = W(f_1, \dots, f_n)$ can be represented as $w(t) = t^{s+n(n-1)/2} h(t)$, $h \in \mathcal{O}(\mathbb{C}, 0)$, $h(0) \neq 0$.

Problem 19.8. Prove that the 1-jet bundle $J^1(T)$ is equivalent to the direct sum of the trivial bundle of rank 1 and the cotangent bundle $\mathbf{T}^*T$ for any base T.

Problem 19.9. Let $\mathcal{C}'$ be a holomorphic 2-distribution on the jet bundle, which is tangent to graphs of all sections of the form $t \mapsto j^n_u(t)$ for all holomorphic germs $u \in \mathcal{O}(T, p)$, $p \in T$.

Prove that $\mathcal{C}'$ coincides with the Cartan distribution.

Problem 19.10. Prove that any integrable section $s \in \Gamma(\tau_n)$ of the jet bundle τ_n, is the jet extension of a meromorphic function $u \in \mathcal{M}(T)$, i.e., $s = j^n_u$.

Problem 19.11. Prove that the Cartan distribution itself is nonintegrable in the sense of Theorem 2.9.

Problem 19.12. Prove that a linear equation of order n with two regular singularities at $t = 0$ and $t = \infty$ is an Euler equation, i.e., it has the form

$$Lu = 0, \quad L = D^n + a_1 D^{n-1} + \cdots + a_{n-1} D + a_n, \quad a_j \in \mathbb{C}, \ D = t\tfrac{\partial}{\partial t}.$$

Find the complete factorization of the Euler equation into composition of first order Fuchsian operators.

Problem 19.13. Prove that for a regular linear equation with m singular points on a compact Riemann surface T, the sum of all characteristic exponents is equal to $(m - \chi)n(n-1)/2$, where $\chi = \deg \mathbf{T}^*T$ is the Euler characteristic (the degree of the cotangent bundle).

Exercise 19.14. Let $D = t\frac{\partial}{\partial t}$ be the Euler operator and u is the "operator of multiplication by t^λ", $\lambda \in \mathbb{C}$. Prove that the conjugated operator $u^{-1}Du$ is again a first order with meromorphic coefficients. Compute it.

Exercise 19.15. Let u be the operator of multiplication by a germ $c(t-t_0)^\lambda h(t)$, $h \in \mathcal{O}(\mathbb{C}, t_0)$, $h(t_0) \neq 0$. Prove that there exists a holomorphic vector field $D \in \mathcal{D}(\mathbb{C}, t_0)$ with a simple (hyperbolic) singular point, such that $u^{-1}Du = D + \lambda$ (cf. with the previous exercise).

Problem 19.16. Show that for an arbitrary Fuchsian operator L of order n with singularities at the points $t_1, \dots, t_m \in \mathbb{P}$ and arbitrary collection of the complex numbers $\lambda_1, \dots, \lambda_m$ such that $\sum \lambda_j = 0$, one can find another Fuchsian operator L' with the same singular points, such that the characteristic exponents $\alpha_{1,j}, \dots, \alpha_{n,j}$ at each singular point t_j are shifted by λ_j: $\alpha'_{i,j} = \alpha_{i,j} + \lambda_j$ for all i, j.

Problem 19.17. Find explicitly the hypergeometric equation (19.30) and the corresponding characteristic exponents for each component of the system (19.34).

20. Irregular singularities and the Stokes phenomenon

Unlike the Fuchsian singularities which can always be reduced to a simple formal normal form by means of a convergent gauge transform, irregular singularities have the formal classification considerably more involved and the normalizing transformations as a rule diverge.

20A. One-dimensional irregular singular points. Irregular singularities of scalar (one-dimensional) linear equations admit complete investigation. Consider the equation

$$t^m \dot{x} = a(t)x, \qquad m \geqslant 2, \quad a(t) = \lambda + a_1 t + a_2 t^2 + \cdots \in \mathcal{O}(\mathbb{C}, 0). \tag{20.1}$$

Its nontrivial solution is given by the explicit formula

$$x(t) = \exp \int \frac{a(t)}{t^m}\, dt = \exp[-t^{1-m}\lambda(1 + o(1))]. \tag{20.2}$$

The origin is an essential singularity of the function $x(t)$ holomorphic in the punctured neighborhood $(\mathbb{C}, 0) \smallsetminus \{0\}$.

Consider $2m-2$ rays from the origin on the complex plane $\mathbb{C}$, described by the condition

$$\operatorname{Re}(\lambda/t^{m-1}) = 0. \tag{20.3}$$

These rays subdivide the neighborhood $(\mathbb{C}, 0)$ into sectors of equal opening $\pi/(m-1)$.

Definition 20.1. An open sector bounded by two rays (20.3) is called the *sector of jump* (resp., *sector of fall*), if the real part of the ratio $\operatorname{Re}(\lambda/t^{m-1})$ is negative (resp., positive) in the interior of this sector.

In each proper subsector of these sectors the solution $x(t)$ of (20.2) grows exponentially fast (in the jump sectors) and is *flat* at $t=0$ (in the fall sectors). This explains the terminology, as follows from the formula (20.2).

Holomorphic classification of one-dimensional systems is very simple. Clearly, the order m is invariant; the following assertion shows that the $(m-1)$-jet of the coefficient $a(t)$ is a complete invariant of the classification, both formal and holomorphic.

Proposition 20.2. *Two meromorphic one-dimensional "linear systems" (equations) of the form* (20.1) *with the holomorphic coefficients* $a(t), a'(t) \in \mathcal{O}(\mathbb{C}, 0)$, *are holomorphically or formally gauge equivalent if and only the difference* $a(t)-a'(t)$ *is* m*-flat at the origin. In particular, any such equation is equivalent to a unique polynomial equation*

$$t^m \dot{x} = p(t), \qquad p \in \mathbb{C}[t], \quad \deg p \leqslant m-1, \quad p(0) = \lambda. \tag{20.4}$$

Proof. Any conjugacy $x \mapsto h(t)x$ between these equations must satisfy the condition $\dot{h}/h = (a-a')/t^m$ so h is holomorphic and invertible at the origin if and only if the right hand side is holomorphic at the origin. □

20B. Birkhoff standard form. A general (matrix) linear system of any dimension near a non-Fuchsian singular point can be reduced to a polynomial normal form, if the monodromy of the singular point is diagonalizable.

Consider a linear system of the form

$$t^m \dot{X} = A(t)X, \qquad A(t) \in \operatorname{Mat}(n, \mathcal{O}(\mathbb{C}, 0)), \quad A(0) = A_0, \tag{20.5}$$

with the *leading matrix* coefficient $A_0 \in \operatorname{Mat}(n, \mathbb{C})$. Recall that the integer number $m-1$ is the Poincaré rank of the singularity.

Theorem 20.3 (Birkhoff, 1913). *If the monodromy operator M of a system* (20.5) *is diagonal(izable), then this system is holomorphically gauge equivalent to a polynomial system*

$$t^m \dot{X} = A_0' + tA_1' + t^2 A_2 + \cdots + t^{m-1} A_{m-1}', \qquad A_i' \in \operatorname{Mat}(n, \mathbb{C}). \tag{20.6}$$

Proof. Let Λ be a diagonal matrix logarithm satisfying the condition $\exp 2\pi i\Lambda = M$. Then any fundamental matrix solution has the form $X(t) = F(t)\,t^{\Lambda}$, where F is a matrix function, single-valued and holomorphically invertible in the punctured neighborhood of the origin but eventually having an essential singularity at $t = 0$.

The function F considered as a Birkhoff–Grothendieck cocycle, is biholomorphically equivalent to a standard cocycle $t^{D'}$ inscribed in a covering

$$U_0 = \{|t| < r_0\},\ U_1 = \{|t| > r_1\}, \qquad U_i \subset \mathbb{P},$$

with sufficiently small values $0 < r_1 < r_0 \ll 1$. In other words, there exist a diagonal integer matrix D' and two holomorphic invertible matrix functions H'_0, H'_1 such that

$$F(t) = H'_0(t)\,t^{D'}H'_1(t), \quad H'_i \in \mathrm{GL}(n, U_i),\ i = 0, 1,\ D' = \operatorname{diag}\{d_1, \dots, d_n\},$$

Using the Permutation Lemma 16.36, we can find a monopole (matrix polynomial with constant nonzero determinant) Π such that $t^{D'}H'_1 = \Pi H_1 t^D$ with $H_1 \in \mathrm{GL}(n, U_1)$ still holomorphic at infinity and D a diagonal matrix obtained by permutation of entries from the diagonal matrix D'. The matrix $H'_0\Pi$ is holomorphic and invertible in U_0. Substituting, we obtain the representation[6]

$$F = H_0 H_1\,t^D, \qquad H_i \in \mathrm{GL}(n, U_i),\ i = 0, 1. \tag{20.7}$$

In fact, the function H_1 and its inverse are holomorphic in $\mathbb{P} \smallsetminus \{0\}$, i.e., both are entire functions of t^{-1}; its extension to the punctured neighborhood of the origin is given by rereading (20.7), $H_1 = H_0^{-1}F\,t^{-D}$.

Since the diagonal matrices Λ and D commute, the solution X of the irregular system can be represented as $X(t) = H_0 \cdot H_1\,t^{\Lambda'}$, $\Lambda' = D + \Lambda$.

After the gauge transform $X \mapsto X' = H_0^{-1}X$ holomorphic at the origin, the logarithmic derivative

$$\Omega' = dX' \cdot (X')^{-1} = dH_1 \cdot H_1^{-1} + t^{-1}\,H_1\Lambda' H_1^{-1}$$

can be extended on the whole Riemann sphere $\mathbb{P}$. This extension will have a simple pole at infinity and no other singularities except for $t = 0$.

The origin $t = 0$ is a pole of order m for Ω'. Indeed, it was a pole of order m for $\Omega = dX \cdot X^{-1}$; since Ω' and Ω are locally holomorphically conjugate at the origin by construction, this assertion is valid also for Ω'.

Thus the holomorphic gauge transform Ω' of the initial irregular system is a rational matrix 1-form on $\mathbb{P}$ with poles of order m at the origin and 1 at infinity. Thus the matrix coefficient $A'(t)$ of $\Omega' = A'\,dt$ must be a matrix

[6] Sometimes the factorization (20.7) itself is called the Birkhoff factorization of the matrix function F holomorphic in the annulus; see [**FM98**].

polynomial of degree m in t^{-1} without the free term (so that Ω' has at most a simple pole at infinity), exactly as was asserted. □

We wish to stress that Theorem 20.3 is a *global statement*, closely related to Theorem 18.6. If the monodromy is not diagonalizable, then the assertion is in general false [**Gan59**]. However, for *irreducible* irregular singularities the polynomial standard form still exists, as was shown in [**Bol94**]. In fact, this result is closely related to the Bolibruch–Kostov Theorem 18.14.

Recall that a meromorphic connexion (or a linear system) is reducible, if there exists an invariant holomorphic subbundle. Local reducibility means that the invariant subbundles exist locally near a singular point. After rectification of the corresponding subbundles by a suitable holomorphic gauge transform, a locally reducible system can always be brought into block upper-triangular form. A connexion (resp., linear system) is locally irreducible if it admits no nontrivial invariant holomorphic subbundles.

A regular (in particular, Fuchsian) singularity is always locally reducible: the monodromy operator M always has at least one invariant subspace in each dimension, and by Proposition 18.8, each such subspace spans an invariant subbundle. However, for *irregular* singularities Proposition 18.8 in general fails and there exist *locally irreducible singularities* (though this irreducibility is very difficult to check).

Theorem 20.4 (A. Bolibruch, [**Bol94**]). *A locally irreducible irregular singularity is holomorphically equivalent to a polynomial system* (20.6).

The proof of this assertion reproduces the proof of Theorem 18.14 with minimal modifications. The key argument is that a locally irreducible connexion on a holomorphic bundle over $\mathbb{P}$ is always globally irreducible.

Proof. We construct an abstract bundle π_{N} over $\mathbb{P}$ with a meromorphic connexion ∇_{N} on it, which has an irregular singular point at $t = 0$, biholomorphically equivalent to the given singularity $\Omega_0 = t^{-m}(A_0 + A_1 t + \cdots)\,dt$, and a Fuchsian singularity at $t = \infty$ with eigenvalues "well apart". Here $\mathrm{N} = \operatorname{diag}\{\nu_1, \cdots, \nu_n\}$ is a diagonal $n \times n$-matrix with sufficiently fast ascending sequence of integer numbers $\nu_1 \ll \nu_2 \ll \cdots \ll \nu_n$: for our purposes it is sufficient to guarantee that $\nu_{i+1} - \nu_i > (m-1)(n-1)$.

To construct this bundle, we assume that the holonomy operator M is upper-triangular and has an upper-triangular matrix logarithm $A = \frac{1}{2\pi i}\ln M$. Then for any choice of the matrix N the logarithmic derivative $\Omega_\infty = dY \cdot Y^{-1}$, where $Y(t) = t^{\mathrm{N}} t^{A}$, has a Fuchsian singularity at infinity (cf. with (18.8)).

Exactly as in the proof of Theorem 18.14, the two forms Ω_0 on $(\mathbb{C}^1, 0)$ and Ω_∞ on $\mathbb{P} \smallsetminus \{0\}$, considered as connexion forms, define a holomorphic bundle π_{N} and a meromorphic connexion ∇_{N} on it, with only two singularities, one of them Fuchsian. The total order of poles of ∇_{N} is equal to $m + 1$.

If the singularity at the origin is irreducible, then the connexion ∇_{N} is globally irreducible, hence the splitting type $D = \operatorname{diag}\{d_1, \dots, d_n\}$ of the bundle π_{N} is constrained by the inequality $|d_i - d_j| \leqslant (m-1)(n-1)$ (Problem 18.12, a slightly modified version of Theorem 18.12). Trivializing this bundle and making a suitable monopole transform

Π, we obtain (again exactly as in the proof of Theorem 18.14) a meromorphic connexion on the trivial bundle with an irregular singularity at $t = 0$ and a regular singularity with the fundamental solution $X(t) = G(t)\, t^{D'} t^{\mathrm{N}} t^{A} = G(t)\, t^{D'+\mathrm{N}} t^{A}$. In this expression the matrix function $G \in \mathrm{GL}(n, \mathcal{O}(\mathbb{P}, \infty))$ is holomorphically invertible at infinity, and D' is a diagonal matrix obtained from D by permutation of the diagonal entries. Because of the large gaps between the numbers ν_j, the entries of the diagonal matrix $D' + \mathrm{N}$ are still in the ascending order, hence the logarithmic derivative $dX \cdot X^{-1}$ is Fuchsian. Thus after the trivialization and the monopole gauge transform we obtain a rational matrix 1-form Ω' on $\mathbb{P}$ with a pole of order m at the origin and a simple pole at infinity. This gives the polynomial normal form (20.6). □

Remark 20.5. The "polynomial normal form" (20.6) is in general nonintegrable. Moreover, it is nonlocal: each matrix coefficient A'_k of the normal form depends on the entire series $\sum A_k t^k$ in (20.5). Thus its effectiveness in applications is rather limited.

20C. Resonances and formal diagonalization. The first step in the "genuine" classification of general irregular singularities is the formal classification similar to that described in §16**C** for Fuchsian systems with $m = 1$. Exactly as above, the linear system

$$t^m \dot{x} = A(t)x, \qquad A(t) \in \mathrm{Mat}(n, \mathcal{O}(\mathbb{C}, 0)), \tag{20.8}$$

associated with the matrix equation (20.5), can be reduced to a holomorphic vector field in $(\mathbb{C}^{n+1}, 0)$ corresponding to the "nonlinear" system of differential equations

$$\begin{cases} \dot{x} = A_0 x + tA_1 x + \cdots, & x \in (\mathbb{C}^n, 0), \\ \dot{t} = t^m, & t \in (\mathbb{C}, 0). \end{cases} \tag{20.9}$$

The spectrum of linearization of the system (20.9) at the singular point $(0,0)$ consists of the zero value $\lambda_0 = 0$ (since $m \geqslant 2$) and the eigenvalues $\lambda_1, \ldots, \lambda_n \in \mathbb{C}$ of the leading matrix coefficient $A_0 \in \mathrm{Mat}(n, \mathbb{C})$ (repetitions allowed).

Applying the Poincaré–Dulac technique to the nonlinear system (20.9), we can eliminate from its Taylor expansion all nonresonant terms. Exactly as in the case with Fuchsian systems in §16**C**, only occurrences of *cross-resonances* $\lambda_i = \lambda_j + k\lambda_0$ corresponding to the vector-monomials $t^k x_j \frac{\partial}{\partial x_i}$ will matter. As $\lambda_0 = 0$, this motivates the following definition.

Definition 20.6. The system (20.5) is said to be *nonresonant* at the origin, if all eigenvalues $\lambda_1, \ldots, \lambda_n$ of the leading matrix coefficient A_0 are pairwise different.

Theorem 20.7. *A non-Fuchsian system* (20.5) *at a nonresonant singular point* $t = 0$ *is formally gauge equivalent to a diagonal polynomial system of* *degree* m,

$$\begin{aligned} t^m\dot{x} &= \Lambda(t)x, \qquad \Lambda(t) = \operatorname{diag}\{p_1(t),\dots,p_n(t)\}, \\ p_i \in \mathbb{C}[t], \quad \deg p_i &= m, \qquad \Lambda(0) = \operatorname{diag}\{\lambda_1,\dots,\lambda_n\}. \end{aligned} \tag{20.10}$$

Proof. The same (literally) arguments that proved Theorem 16.15 in §16**C**, prove also that only resonant monomials of the form $c_{ijk}t^k x_j \frac{\partial}{\partial x_i}$ should be kept in the expansion (20.9), all others being removable. Elimination of the resonant monomials of degree $k \geqslant m$ can be achieved by Proposition 20.2 and the remark after it. □

As follows from the analysis of the scalar case in §20**A**, a system in the formal normal form (20.10) is integrable: there are diagonal matrix polynomial $B(t^{-1}) = B_0 t^{1-m} + B_1 t^{2-m} + \cdots + B_{m-2}t^{-1}$ and a constant diagonal matrix C, such that a fundamental matrix solution of (20.5) has the form $X(t) = t^C \exp B(t^{-1})$.

Remark 20.8. Note that the formal series that conjugate irregular singularities may diverge. Indeed, the nonresonant irregular system

$$t^2 \frac{d}{dt}\begin{pmatrix} y \\ z \end{pmatrix} = \begin{pmatrix} 1 & -1 \\ 0 & t \end{pmatrix}\begin{pmatrix} y \\ z \end{pmatrix}, \tag{20.11}$$

with a separating second variable can be reduced to the Euler equation (7.11) (Example 7.10). The Euler equation has a formal Taylor solution which diverges. Clearly, this would be impossible were the normalizing series convergent.

20D. Formal simplification in the resonant case. The direct proof of the formal diagonalization Theorem 20.7 looks as follows. The formal gauge transformation $X \mapsto X' = HX$ defined by a formal matrix series

$$H = E + \sum_{k>0} t^k H_k \in \mathrm{GL}(n, \mathbb{C}[[t]])$$

conjugates two systems (formal or convergent)

$$\begin{aligned} t^m\dot{X} &= A(t)X, & & & t^m\dot{X}' &= A'(t)X', \\ A(t) &= A_0 + \sum_{k>0} t^k A_k, & &\text{and} & A'(t) &= A_0 + \sum_{k>0} t^k A'_k, \end{aligned}$$

with the same principal part $A(0) = A'(0) = A_0$, if and only if H is a formal solution to the following matrix differential equation,

$$t^m\dot{H} = A'(t)H - HA(t). \tag{20.12}$$

Termwise, this equation is equivalent to the sequence of matrix equations involving the coefficients A_k, A'_k of the expansions for $A(t)$ and $A'(t)$ respectively,

$$\begin{aligned} 0 &= (A'_0 H_k - H_k A_0) + (A'_k - A_k) \\ &+ \sum_{i,j>0,\ i+j<k} (A'_i H_j - H_i A_j) - \begin{cases} kH_{k+1-m}, & k \geqslant m-1, \\ 0, & k < m-1. \end{cases} \end{aligned} \tag{20.13}$$

These equations can be rewritten in the form

$$[A_0, H_k] + A'_k = \text{matrix polynomial in } \{A'_j, H_j,\ 0 < j < k\}.$$

By Lemma 4.11, the image of the operator $\operatorname{ad}_{A_0}\colon B \mapsto [A_0, B]$ is a linear subspace in $\operatorname{Mat}(n, \mathbb{C})$ orthogonal (in the sense of some Hermitian structure) to the subspace of all matrices commuting with the conjugate matrix A_0^*. Thus the equations (20.13) are always solvable for suitable matrices H_k and A'_k such that $[A_0^*, A_k] = 0$.

If A_0 is nonresonant, it can be diagonalized, $A_0 = \Lambda = \operatorname{diag}\{\lambda_1, \dots, \lambda_n\}$, so that $\operatorname{Ker}\operatorname{ad}_{\Lambda^*}$ consists of diagonal matrices only. Thus a nonresonant irregular singularity is formally diagonalizable. Slightly more generally, if A_0 is block diagonal with each block having only one eigenvalue different for different blocks, then the complementary subspace can be chosen as matrices having the same block diagonal structure. This proves the following generalization of Theorem 20.7.

Theorem 20.9. *By a formal gauge transformation one can reduce an irregular system to the block-diagonal form with each block having the leading matrix with a single eigenvalue.*

Example 20.10. Assume that the leading matrix A_0 is a single Jordan block of size n with the eigenvalue λ_0, $A_0 = \lambda_0 E + J$. For an arbitrary matrix B commutation with J^* means that shifts of the columns of B to the left and shift of its rows downward produce the same result (in both cases the null column or row is added). Thus for any element B_{ij} the elements next to the right and one row above it coincide, the elements of the first row and the last column being all zeros. Thus $[B, J^*] = 0$, if and only if all elements on each secondary diagonal (parallel to the principal diagonal) are equal among themselves and equal to zero in the upper-right half (so that B is lower triangular).

Thus an irregular singularity with the leading matrix coefficient $A_0 = \lambda_0 + J$ can be brought to the form (20.8) in which

$$A(t) = \begin{pmatrix} \lambda_0 & 1 & & & \\ b_1(t) & \lambda_0 & 1 & & \\ \cdots & \cdots & \cdots & \cdots & \cdots \\ b_{n-2}(t) & \cdots & b_1(t) & \lambda_0 & 1 \\ b_{n-1}(t) & b_{n-2}(t) & \cdots & b_1(t) & \lambda_0 \end{pmatrix}.$$

In fact, one can further simplify the obtained normal form and get rid of all entries except those in the last row; see [**Arn83**, §30]. As a result, by a formal gauge transformation the system is reduced to the companion form modulo a scalar matrix,

$$A(t) = \lambda_0 E + \begin{pmatrix} 0 & 1 & & & \\ & 0 & 1 & & \\ \cdots & \cdots & \cdots & \cdots & \cdots \\ & & & 0 & 1 \\ a_n(t) & a_{n-1}(t) & \cdots & a_2(t) & a_1(t) \end{pmatrix} \tag{20.14}$$

with formal series $a_i \in \mathbb{C}[[t]]$. The eigenvalues of the matrix $A(t)$ are of the form $\lambda_0 + \lambda_i(t)$, where $\lambda_1(t), \dots, \lambda_n(t)$ are the roots of the characteristic equation

$$\lambda^n = a_1(t)\lambda^{n-1} + \cdots + a_{n-1}(t)\lambda + a_n(t).$$

Since $\lambda_1(0) = \cdots = \lambda_n(0) = 0$ by assumption, we see that the formal series $a_i \in \mathbb{C}[[t]]$ are all without the free terms.

Remark 20.11. If $f(t) = \exp(m\lambda_0/t^{m-1})$ is a solution of the equation $\dot f = -\lambda_0 t^{-m} f$, then the gauge transformation $X \mapsto f(t)X$ brings the system (20.14) to the true companion form (without the diagonal term $\lambda_0 E$). Being scalar, this transformation commutes with any other gauge equivalence, formal or convergent.

20E. Shearing transformations and ramified formal normal form. Further simplification of the system is possible only if we extend the class of formal gauge transformations, allowing for *ramified formal transformations* which are formal series in fractional powers of t. It was E. Fabry who realized (1885) the necessity of passing to fractional powers.

Example 20.12 (continuation of Example 20.10)**.** Consider again the case of a system whose leading matrix is a maximal size Jordan block. By Remark 20.11, without loss of generality we may assume that $\lambda_0 = 0$. Assume that $r \in \mathbb{Q}$ is a *positive* rational number, and consider the gauge transformation

$$H(t) = \operatorname{diag}\left\{1, t^{-r}, t^{-2r}, \dots, t^{(1-n)r}\right\}. \tag{20.15}$$

This transformation takes the system (20.5) with the matrix $A(t)$ as in (20.14), into that with the matrix

$$\begin{pmatrix} 0 & t^r & & & \\ & 0 & t^r & & \\ \cdots & \cdots & \cdots & \cdots & \cdots \\ & & & 0 & t^r \\ t^{(1-n)r}a_n & t^{(2-n)r}a_{n-1} & \cdots & t^{-r}a_2 & a_1 \end{pmatrix} - t^{m-1}R,$$

where $R = \operatorname{diag}\{0, r, 2r, \dots, (n-1)r\}$ is the diagonal matrix. The orders of zeros $\nu_k \in \mathbb{N}$ of the formal series $a_k(t)$ were all positive, since $a_k(0) = 0$. Choose r so that the orders of all terms $a'_k(t) = t^{-kr}a_k(t)$ are still nonnegative but the smallest of them is zero, $r = \min_k \nu_k/k$. The denominator of r is no greater than n.

After the conjugacy by H the matrix of the system will take the form

$$\dot{X} = [t^{-m+r}A'(t) + t^{-1}R]X, \qquad r > 0, \tag{20.16}$$

where $A'(t)$ is a companion matrix similar to (20.14) but with the entries $a'_k(t) \in \mathbb{C}[[t^{1/q}]]$, $k = 1, \dots, n$, now being formal series in *fractional* powers of t (and without the diagonal term λ_0). The leading (matrix) coefficient $A'(0)$ of $A'(t)$ is the companion matrix containing the complex numbers $a'_n(0), \dots, a'_1(0)$ as the last row. By the choice of r, *not all of them are simultaneously zero*, yet their sum is zero, since $\operatorname{tr} A'(0) = a'_1(0) = a_1(0) = 0$. Therefore if after the shearing transformation the system remains non-Fuchsian (i.e., if $r < m-1$), at least some of the leading eigenvalues must be nonzero.

Somewhat more elaborate computations allow us to prove similar statement also in the case where the leading matrix coefficient A_0 has several Jordan blocks with the common eigenvalue.

Notice now that the construction described in §20**D**, applies without any changes to the *ramified* formal series in fractional powers of t (i.e., when the indices i, j, k range over an arithmetic progression with rational noninteger difference). Applying Theorem 20.7 in these extended settings, we see that the system (20.16) can now be formally split into two subsystems.

By iteration of these two steps (splitting the system and subsequent shearing transformation) sufficiently many times, one can prove the following result.

Theorem 20.13 (Hukuhara (1942), Turritin (1955), Levelt (1975))**.** *By a suitable formal ramified gauge transformation an irregular singularity can be reduced to the diagonal form*

$$A(t) = t^{-r_1}P_1 + t^{-r_2}P_2 + \cdots + t^{-r_k}P_k + t^{-1}C,$$

where $r_1 > r_2 > \cdots > r_k > 1$ are rational numbers with the denominators not exceeding $n!$ and $P_1, \dots, P_k \in \operatorname{Mat}(n, \mathbb{C})$ are diagonal matrices commuting with C.

We will not give the proof in full details; see [**Var96**] and the references therein. Instead, we focus on the more transparent nonresonant case and study the problems of *holomorphic* rather than formal classification.

20F. Holomorphic sectorial normalization. Even in the nonresonant case there is a wide gap between formal and analytic classification. In this section we explain the geometric obstructions for convergence of formal normalizing transformations.

Definition 20.14. A *separation ray*[7] corresponding to a fixed value of m and a pair of complex numbers $\lambda \neq \lambda' \in \mathbb{C}$ is any of the $2(m-1)$ rays defined by the condition

$$\operatorname{Re}[(\lambda - \lambda')/t^{m-1}] = 0. \tag{20.17}$$

The following property is characteristic for separation rays, being an immediate consequence of the explicit formula (20.2). Consider two solutions $x(t), x'(t)$ of two scalar systems (20.1) with the same order m and the holomorphic coefficients $a(t), a'(t)$. Denote $\lambda = a(0)$, $\lambda' = a'(0)$. Recall that a function defined and holomorphic in a sector with the vertex at the origin is said to be *flat*, if it decreases faster than any power of the distance to this point, and the same is true for all its derivatives. A reciprocal $1/f$ of a flat nonvanishing function is called *vertical*.

Proposition 20.15. *If $R = \rho \cdot \mathbb{R}_+$, $|\rho| = 1$, is* not *a separation ray for the pair λ, λ', then out of the two reciprocal ratios $x(t)/x'(t)$ and $x'(t)/x(t)$ one after restriction on R is flat and the other is vertical, depending on whether $(\lambda - \lambda')/\rho^{m-1}$ is respectively negative or positive.* □

Everywhere here and below we always assume that any sector is bounded by two straight rays coming from the vertex (usually the origin); the angle between these rays is the *opening* of the sector. If $\widehat{H} \in \mathrm{GL}(n, \mathbb{C}[[t]])$ is a formal power series, we say that a holomorphic matrix function $H \in \mathrm{GL}(n, \mathcal{O}(S))$ *extends* this series, if $\widehat{H}$ is the *asymptotic series* for H in S, that is, the difference between $H(t)$ and any truncation $\widehat{H}_N(t) \in \mathrm{Mat}(n, \mathbb{C}[t])$ of $\widehat{H}$, the matrix polynomial of degree N, decreases faster than t^N,

$$\|H(t) - \widehat{H}_N(t)\| = o(|t|^N) \quad \text{as} \quad t \to 0,\ t \in S, \qquad \forall N \in \mathbb{N}.$$

Theorem 20.16 (sectorial normalization theorem, Y. Sibuya [**Sib62**]). *Assume that the leading matrix A_0 of the linear system* (20.5) *is nonresonant (i.e., has pairwise different eigenvalues $\lambda_1, \dots, \lambda_n$).*

If $S \subset (\mathbb{C}, 0)$ is an arbitrary sector not containing two separation rays for any pair of the eigenvalues λ_i, λ_j, then any formal gauge transformation $\widehat{H}(t) \in \mathrm{GL}(n, \mathbb{C}[[t]])$ conjugating (20.5) *with its polynomial diagonal normal form* (20.10), *can be extended to a holomorphic conjugacy $H_S(t) \in \mathrm{GL}(n, \mathcal{O}(S))$ between these systems in S.*

[7]The union of two separating rays in opposite directions is called a *Stokes line* in some sources.

The proof of this theorem is moved to the appendix; see §20**J** below. It differs both from the author's proof in [**Sib90**] and from that in [**Was87**].

20G. Sectorial automorphisms and Stokes matrices. If the sector is sufficiently wide, then the normalizing transform is necessarily unique. This can be seen by studying *automorphisms* of the system in the diagonal normal form. We will show that such systems admit no nontrivial automorphisms over such sectors.

More specifically, assume that $H'(t), H''(t)$ are two sectorial automorphisms conjugating an irregular singularity (20.5) with its diagonal formal normal form (20.10) in some sector $S \subset (\mathbb{C}, 0)$. Then the "superpositional ratio", the sectorial gauge transform with the matrix function $H(t) = H''(t) \cdot {H'}^{-1}(t)$, is an automorphism of the diagonal system (20.10).

Such automorphisms are most easily described by their action on a suitably chosen fundamental solution. In our case the diagonal system (20.10) admits a distinguished set of solutions which are themselves diagonal.

We fix a diagonal fundamental solution $W(t) = \operatorname{diag}\{w_1(t), \dots, w_n(t)\}$ for (20.10). Then any holomorphic sectorial automorphism $H(t)$ of the diagonal normal form, $H \in \mathrm{GL}(n, \mathcal{O}(S, 0))$, is uniquely determined by a constant matrix $C \in \mathrm{GL}(n, \mathbb{C})$ such that

$$H(t)W(t) = W(t)C. \tag{20.18}$$

This matrix will be referred to as the *Stokes matrix* of the sectorial automorphism $H(\cdot)$. This matrix depends on the choice of the diagonal fundamental solution W, yet because of the special growth pattern of solutions it can be rather accurately described.

Lemma 20.17. *Suppose that none of the two rays bounding a sector S is a separation ray for the system* (20.10) *in the formal normal form, and the eigenvalues of the leading matrix Λ_0 are ordered so that* $\operatorname{Re}\lambda_1 < \cdots < \operatorname{Re}\lambda_n$.

Then the Stokes matrix $C \in \mathrm{GL}(n, \mathbb{C})$ of any sectorial automorphism $H \in \mathrm{GL}(n, \mathcal{O}(S, 0))$ which is 0-tangent to the identity, $H(t) = E + o(1)$, possesses the following properties:

(1) *For any pair $i \neq j$ of indices, one of the matrix elements c_{ij}, c_{ji} must be zero, in particular,*
(2) *if $S \supset \mathbb{R}_+$, then $C - E$ is a nilpotent upper-triangular matrix.*
(3) *If S contains a separation ray for the pair $\lambda_i \neq \lambda_j$ then both $c_{ij} = c_{ji} = 0$, in particular,*
(4) *if S contains one separation ray for each pair of eigenvalues, then necessarily $C = E$.*

Proof. All assertions immediately follow by inspection of the asymptotic behavior of the sectorial automorphism written in terms of the Stokes matrix,

$$H(t) = W(t)CW^{-1}(t) = \|h_{ij}(t)\|, \qquad h_{ij}(t) = c_{ij}\, w_i(t)/w_j(t),$$

and the observation in Proposition 20.15.

Indeed, if the ratio $w_i(t)/w_j(t)$ along some ray in S is vertical, the corresponding coefficient c_{ij} must necessarily be zero. This proves the first two assertions.

To prove the remaining assertions, note that the two reciprocal ratios w_i/w_j and w_j/w_i have reciprocal asymptotical behavior along any two rays sufficiently close but separated by the separation ray for the eigenvalues λ_i and λ_j. By the preceding arguments, in this case both c_{ij} and c_{ji} must be absent. □

Proposition 20.18 (rigidity). *If a sector S has an opening bigger than $\pi/(m-1)$, then the sectorial normalization H_S described in Theorem 20.16, is unique.*

Proof. If there were two sectorial normalizations H', H'' with the same asymptotic series $\widehat{H}$, then their matrix ratio $H = H''{H'}^{-1}$ must be a sectorial automorphism of the formal normal form (20.10), tangent to the identity (i.e., of the form id + flat function). Since all separation rays for each pair of eigenvalues are separated by the angle $\pi/(m-1)$, the sector S of opening bigger than $\pi/(m-1)$ must contain at least one such ray for each pair. By the last assertion of Lemma 20.17, the corresponding Stokes matrix must be identity, which means that the ratio itself is identity. □

20H. Stokes phenomenon. Holomorphic classification of irregular singularities. Consider a linear system (20.5) of Poincaré rank $m-1$ at the nonresonant non-Fuchsian singular point $t = 0$, and let (20.10) be its formal normal form.

As before, we can assume without loss of generality that the leading matrix has eigenvalues ordered so that

$$\operatorname{Re}\lambda_1 < \cdots < \operatorname{Re}\lambda_n, \tag{20.19}$$

which means that neither the positive semiaxis $\mathbb{R}_+$ nor its rotated copies $\rho^k\,\mathbb{R}_+$, $k = 1, \dots, 2(m-1)$, where $\rho = \exp\frac{\pi i}{m-1}$, are separation rays for any two eigenvalues $\lambda_i \neq \lambda_j$.

The open sector S^* bounded by the rays $\mathbb{R}_+$ and $\rho\,\mathbb{R}_+$ of opening $\pi/(m-1)$ contains exactly one separation ray for each pair, none of them on the boundary. Thus one can enlarge slightly the opening of this sector to become $2\delta + \pi/(m-1)$ so that it still contains exactly one separation ray for each

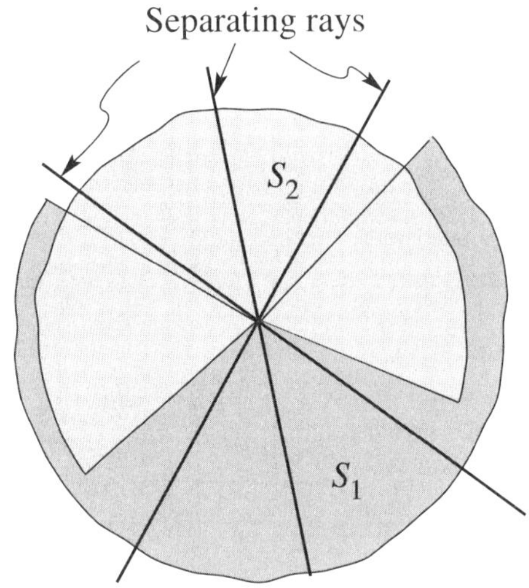

Figure III.3. Standard covering and separation rays in the simplest case $m = 2$

pair. Denote this enlarged sector by $S_1 = \{-\delta < \operatorname{Arg} t < \pi/(m-1)+\delta\}$, and let $S_2, \dots, S_{2(m-1)}$ be its rotated copies, $S_k = \rho^{k-1} S_1$. These sectors form a covering of the punctured neighborhood of the origin; the intersections are narrow flaps $S_{j,j+1} = \{|\operatorname{Arg} t - j\pi/(m-1)| < \delta\}$ of opening $2\delta > 0$ each. This collection of sectors will be referred to as the *standard covering* of the punctured neighborhood of the origin.

By Theorem 20.16, over each sector S_k there exists a holomorphic gauge conjugacy $H_k(t) \in \mathrm{GL}(n, \mathcal{O}(S_k))$ between the initial system (20.5) and its formal normal form (20.10). This conjugacy is unique by Proposition 20.18. The collection $\{H_k\}$ of these sectorial normalizing maps is called the *normalizing cochain* inscribed in the standard covering $\{S_k\}$.

Since all maps forming the normalizing cochain have the same common asymptotic series, the matrix ratios $F_{ij} = H_i H_j^{-1} = F_{ji}^{-1}$ defined on the nonempty intersections $S_i \cap S_j$, are sectorial automorphisms of the formal normal form (20.10). Clearly, the intersections $S_i \cap S_j$ are nonvoid if and only if $j = i + 1$ cyclically modulo $2(m-1)$; they are thin sectors around the rotated copies $\rho^j \mathbb{R}_+$ of the real axis.

Let $\{H_i\}$ be the (uniquely defined) normalizing cochain inscribed in the standard covering. Choose a diagonal fundamental matrix solution $W(t)$; since in general the normal form has a nontrivial monodromy, the solution $W(t)$ is multivalued. To avoid this, we slit the neighborhood along the ray $\{\operatorname{Arg} t = \pi/2(m-1)\}$ entirely belonging to the sector S_1 and disjoint with all overlapping sectors $S_{ij} = S_i \cap S_j$, $|i - j| = 1$, and consider a fundamental solution in the slit domain. Such a solution is defined uniquely modulo a diagonal transform

$$W(t) \mapsto DW(t) = W(t)D, \qquad D = \operatorname{diag}\{\alpha_1, \dots, \alpha_n\}, \tag{20.20}$$

and by construction it is holomorphic in all flaps S_{ij}.

Definition 20.19. The *Stokes collection* of a linear system at a nonresonant irregular singular point is the collection of Stokes matrices $\{C_j\}$, $j = 1, \dots, 2(m-1)$ of the sectorial automorphisms $F_{ij} = H_i H_j^{-1}$, $i + 1 = j$, corresponding to a diagonal solution $W(t)$ of the formal normal form.

Proposition 20.20. *The matrices C_j from the Stokes collection are unipotent.*

Proof. If S is a sector containing the positive semiaxis and the eigenvalues of A_0 are ordered as in (20.19), the assertion follows from the second assertion of Lemma 20.17. The general case can be brought to the former specific case by suitable rotation of the t-plane and re-enumeration of the eigenvalues. □

By Proposition 20.18, the Stokes collection is uniquely defined, as soon as the diagonal fundamental solution $W(t)$ is fixed. Replacing the diagonal solution $W(t)$ by another solution $DW(t) = W(t)D$ transforms the Stokes matrices by the simultaneous diagonal conjugacy

$$\begin{aligned} C_j \mapsto C_j' &= DC_jD^{-1}, \\ D &= \operatorname{diag}\{\alpha_1, \dots, \alpha_n\}, \end{aligned} \qquad \forall j = 1, \dots, 2(m-1). \tag{20.21}$$

The Stokes collections $\{C_1, \dots, C_{2m-2}\}$ and $\{C_1, \dots, C_{2m-2}\}$ related by the transformation (20.21), are called *equivalent* Stokes collections. Note that the *trivial* collection $C_1 = \cdots = C_{2m-2} = E$ is equivalent only to itself.

Theorem 20.21 (classification theorem for nonresonant irregular singularities). *Any two nonresonant irregular linear systems with a common formal normal form are locally holomorphically gauge equivalent if and only if their Stokes collections are equivalent in the sense* (20.21).

In particular, a linear system is holomorphically equivalent to its formal normal form, if and only if the Stokes collection is trivial.

Proof. Consider two systems with the same formal normal form. Without loss of generality we may assume that a common standard covering is chosen, and the uniquely defined normalizing cochains are denoted by $\{H_j\}$ and $\{H_j'\}$ respectively.

Let G be a holomorphic conjugacy between these systems. Together with the cochain $\{H_j'\}$, the cochain $\{H_jG\}$ clearly is also a normalizing cochain for the second system. By the uniqueness (Proposition 20.18), $H_j' = H_jG$ and hence $H_i'(H_j')^{-1} = DH_iH_j^{-1}D^{-1}$ for all $|i - j| = 1$. Coincidence of the transition cocycles means that the corresponding Stokes collections (apriori defined with respect to two different fundamental solutions W and $W' = DW$) are equivalent.

In the inverse direction this argument also works. If two Stokes collections are equivalent, then by choosing another diagonal fundamental solution we can guarantee that the corresponding Stokes operators simply coincide. Then the matrix quotients $G_j = H'_j H_j^{-1}$ and $G_i = H'_i H_i^{-1}$ coincide on the nonvoid intersections (when $|i-j| = 1$) and hence together define a matrix function G holomorphically invertible outside the origin. This function has an asymptotic series equal to the matrix ratio of two formal normalizing gauge transforms $\widehat{H}'\widehat{H}^{-1}$ for the two systems, hence extends at the origin. □

20I. Realization theorem. Proposition 20.20 describes the necessary property of Stokes operators associated with the given order m and a collection of eigenvalues $\lambda_1, \dots, \lambda_n$. It turns out that this is a unique requirement.

Theorem 20.22 (Birkhoff, 1909). *Any collection of unipotent upper-triangular matrices $\{C_i\}$ meeting the restrictions from Proposition 20.20, can be realized as the Stokes collection of a nonresonant irregular singularity with a preassigned formal normal form* (20.10).

Sketch of the proof. Consider the diagonal formal normal form (20.10), the standard covering S_j and the collection of holomorphic invertible matrix functions

$$F_{j,j+1}(t) = W(t)C_j W^{-1}(t) = F_{j+1,j}^{-1}(t), \qquad j = 1, \dots, 2(m-1),$$

defined in the corresponding nonempty intersections $S_{ij} = S_i \cap S_j$, $|i-j| = 1$. Here $W(t)$ is a diagonal fundamental solution of the formal normal form, holomorphic in the small neighborhood of the origin $(\mathbb{C}, 0)$ slit along the ray $\{\operatorname{Arg} t = \pi/2(m-1)\} \subset S_1$ as before. By our assumptions, the constant matrices C_j are related to the eigenvalues λ_j in such a way that the differences $F_{ij}(t) - E$ are flat in the thin sectors S_{ij}.

It can be shown that the cocycle $\mathcal{F} = \{F_{ij}\}$ is solvable by a holomorphic cochain $\mathcal{H} = \{H_j\}$ of holomorphic invertible matrix functions so that $F_{ij}H_j = H_i$ for $|i-j| = 1$. This means that the sectorial solutions $X_j(t) = H_j^{-1}(t)W(t) = X_i(t)C_j$ satisfy linear systems with the coefficient matrices

$$A_j(t) = t^m \tfrac{d}{dt}(H_j^{-1})H_j + H_j^{-1}(t)\Lambda(t)H_j(t)$$

coinciding on the intersections, $A_i(t) = A_j(t)$ for $t \in S_i \cap S_j$. The resulting matrix function $A(t)$, defined in the punctured neighborhood of the origin, is bounded hence holomorphic and by construction the system $t^m \dot{X} = A(t)X$ is holomorphically equivalent to the formal normal form $t^m \dot{W} = \Lambda(t)W$.

Clearly, the Stokes collection of the constructed system coincides with the prescribed data $\{C_j\}$.

Geometrically this construction consists of patching together linear systems defined over different sectors S_j, using the gauge maps F_{ij}, $|i-j| = 1$, for identification. The result will be a linear system defined on a holomorphic vector bundle over the punctured neighborhood $(\mathbb{C}, 0) \smallsetminus \{0\}$. Such a bundle is always holomorphically trivial, as any bundle over a noncompact Riemann surface [**For91**, §30]. The delicate circumstance is to verify that the linear system which appears after trivialization of this bundle, will have an irregular singularity of the prescribed formal type. The solvability of the "asymptotically trivial" cocycle $\{F_{ij}\}$ by a *holomorphic* cochain $\{H_i\}$ guarantees this automatically. Details can be found in [**BV89**]. □

As a corollary we conclude that there exist non-Fuchsian systems for which the formal diagonalizing series diverge. Moreover, in some sense this divergence is characteristic for the *majority* of non-Fuchsian singularities: Theorems 20.21 and 20.22 imply that classes of holomorphic gauge equivalence are parameterized by $(m-1)n(n-1)$ complex parameters (entries of the Stokes collections).

Appendix: Demonstration of Sibuya theorem

In this section we prove the Sectorial Normalization Theorem 20.16. This theorem can be reduced to an analytic claim asserting existence of flat solutions for a nonhomogeneous system of linear equations in a sector.

Throughout this appendix we fix a nonresonant linear system (20.5), its diagonal formal normal form (20.10) with $\Lambda(0) = \operatorname{diag}\{\lambda_1, \dots, \lambda_n\}$, $\lambda_i \neq \lambda_j$, and a formal transformation $\widehat{H} \in \mathrm{GL}(n, \mathbb{C}[[t]])$ conjugating the two. Given a sector S, we can then speak about sectorial conjugacy (or conjugacies) extending $\widehat{H}$ in this sector.

20J. Normalization in "acute" sectors. First we show that the problem of constructing holomorphic sectorial normalization conjugating an irregular singularity with its diagonal formal normal form, can be solved in any sufficiently "acute" sector, namely, if the opening of this sector is less than $\pi/(m-1)$. Enlarging this sector to wider sectors S_j of opening $\pi/(m-1)+2\delta$ forming the standard covering, is achieved relatively simply in §**20M**.

By the Borel–Ritt theorem [**Was87**, §9.2] (see also Problem 20.2), in any sector S there exists an analytic matrix function $F(t)$ whose asymptotic series in S is the prescribed normalizing series $\widehat{H}$. Conjugating the system (20.5) by F, we obtain a new system of the form $t^m \dot{X} = A'(t)X$ with the matrix $A'(t)$ holomorphic in S and having the same asymptotic series at the origin as the Taylor series $\Lambda(t)$ of the formal normal form $t^m \dot{X} = \Lambda(t)X$. Thus to construct the sectorial conjugacy between the system and its initial normal form, it is sufficient to remove by a suitable sectorial gauge transformation the *flat* nondiagonal part $B(t)$ from the system

$$
\begin{aligned}
t^m \dot{X} &= (\Lambda(t) + B(t))X, \qquad B(t) = \|b_{ij}(t)\|, \\
&b_{ij} \in \mathcal{O}(S), \quad b_{ii} \equiv 0, \quad b_{ij} \text{ flat in } S, \\
S &= \{\alpha < \operatorname{Arg} t < \beta,\ |t| < r\}, \qquad |\beta - \alpha| = \pi/(m-1) - 2\delta.
\end{aligned}
\tag{20.22}
$$

The diagonal entries of B can be assumed absent by Proposition 20.2. The positive parameters $1 \gg \delta > 0$ and $0 < r \ll 1$ characterizing the sector S, can be assumed as small as necessary.

A conjugacy $H(t)$ between (20.22) and (20.10), holomorphic in the sector S with the identical asymptotic series, satisfies the differential equation

$$t^m \dot{H} = \Lambda H - H(\Lambda + B) = [\Lambda, H] - HB. \tag{20.23}$$

The flat difference $Y(t) = H(t) - E$ satisfies the equation

$$\dot{Y} = [\Lambda, Y] - (E + Y)B, \qquad t \in S,\ B(\cdot) \text{ flat in } S. \tag{20.24}$$

Denote by $y = (y_1, \dots, y_k) \in \mathbb{C}^k$, $k = n(n-1)/2$, the collection of all off-diagonal entries of the matrix Y. The system (20.24) then takes the form

$$t^m \dot{y}(t) = [D + G(t)]y(t) + g(t), \qquad t \in S, \tag{20.25}$$

where D is a diagonal matrix corresponding to the commutator with the leading term $\Lambda_0 = \operatorname{diag}\{\lambda_1, \dots, \lambda_n\}$ of the formal normal form $\Lambda(t)$,

$$D\colon Y \mapsto DY = [\Lambda_0, Y].$$

Since the system was assumed nonresonant, all eigenvalues of D are nonzero,

$$D = \operatorname{diag}\{\mu_1, \dots, \mu_k\}, \qquad \mu_i \neq 0,\ i = 1, \dots, k,\ k = n(n-1)/2. \tag{20.26}$$

The term $G(t)$ corresponds to the commutator with the nonleading terms and the multiplication by the flat off-diagonal terms from the matrix B,

$$Y \mapsto GY = [\Lambda(t) - \Lambda_0, Y] + YB(t).$$

In our assumptions $G(t)$ tends to zero as $t \to 0$. The nonhomogeneity $g(t)$ consists of the off-diagonal terms of the matrix $B(t)$ and is also flat at the origin.

It is convenient to simplify the system further to reduce the Poincaré rank to the minimum and place the singular point at infinity so that the leading part would be a system with constant coefficients easy for explicit integration.

Changing the independent variable from $t \in S \subset (\mathbb{C}, 0)$ to $z = 1/t^{m-1} \in (\mathbb{P}, \infty)$ transforms the 1-form $t^{-m}\,dt$ to $(1-m)\,dz$. This transformation brings the system (20.25) to the form $dy/dz = (1-m)[D + G(z^{1/(1-m)})] + (1-m)g(z^{1/(1-m)})$ defined in a sector S' with the vertex at infinity and the opening strictly less than π. Rotating the z-plane if necessary, we can always assume that $S' = \{|z| > r,\ |\operatorname{Arg} z| < \frac{\pi}{2} - \delta\}$, where $\delta > 0$ is a small positive parameter.

Returning to the previous notations, we can rewrite the system (20.25) with respect to the new variable z as follows:

$$\begin{gathered}\tfrac{d}{dz}y = [D + G(z)]y + g(z), \qquad y \in \mathbb{C}^k, \\ z \in S' = \{|z| > r, \ |\operatorname{Arg} z| < \tfrac{\pi}{2} - \delta\}, \\ G(z) = o(1), \ g(z) = o(z^{-N}) \quad \text{for any } N \in \mathbb{N}, \quad \text{as } z \underset{S'}{\longrightarrow} \infty, \\ D = \operatorname{diag}\{\mu_1, \dots, \mu_k\}, \quad \mu_i \neq 0.\end{gathered} \tag{20.27}$$

Theorem 20.23. *The system* (20.27) *admits a flat solution holomorphic in the sector* S'.

Corollary 20.24. *The system* (20.24) *admits holomorphic flat solution* $Y \in \mathcal{O}(S)$ *in any "acute" sector* S *of opening less than* $\pi/(m-1)$.

The key idea of the proof of this theorem is to treat the system (20.27) as a perturbation of the linear diagonal equation

$$\frac{dy}{dz} = Dy, \qquad z \in S', \ D = \operatorname{diag}\{\mu_1, \dots, \mu_k\}.$$

Since the latter system is immediately integrable, we can explicitly describe the resolvent operator $\mathbf{S}$ for the corresponding nonhomogeneous equation,

$$\frac{dy}{dz} = Dy + h \iff y = \mathbf{S}h,$$

by the method of variation of constants. The resolvent $\mathbf{S}$ turns out to be a bounded linear integral operator for a suitable choice of the paths of integration, as explained in §20**K**. Using the resolvent $\mathbf{S}$, the initial equation (20.27) can be rewritten as a fixed point equation,

$$y = \mathbf{S}[Gy + g],$$

with the operator $y \mapsto \mathbf{G}y = Gy + g$ so strongly contracting that the composition $\mathbf{SG}$ is a contracting operator on a suitable Banach space.

Now we proceed with a detailed exposition.

20K. Core example. Consider first the particular one-dimensional case of the system (20.27),

$$\tfrac{d}{dz}y = \mu y + g(z), \qquad 0 \neq \mu \in \mathbb{C}, \quad y \in \mathbb{C}^1, \quad z \in S'. \tag{20.28}$$

with a flat nonhomogeneity $g(z) \in \mathcal{O}(S')$ and absent linear nonautonomous term, i.e., $G \equiv 0$. We are looking for a solution flat in the sector S'.

The solution of this system is given by the explicit formula obtained by variation of constants method (see Remark 15.6): for an arbitrary choice of the base point $b \in S'$,

$$y(z) = e^{\mu z}\left(y(b) + \int_b^z e^{-\mu\zeta} g(\zeta)\, d\zeta\right) = e^{\mu z} y(b) + \int_b^z e^{\mu(z-\zeta)} g(\zeta)\, d\zeta. \tag{20.29}$$

The upper limit of integration is the variable point z. The lower limit $b \in S'$ and the respective boundary condition $y(b)$ have to be chosen so that the solution (20.29) would be flat in S'.

Two cases have to be treated separately, depending on the relative position of $0 \neq \mu \in \mathbb{C}$ and S', namely,

(1) $\operatorname{Re} \mu a > 0$ for some $a \in S'$, that is, the solution of the homogeneous equation is unbounded in S'; this happens when S' overlaps with some sector of jump (in the sense of §20**A**), and

(2) $\operatorname{Re} \mu z < 0$ for all $z \in S'$, that is, the solution of the homogeneous equation decays exponentially fast in S' (i.e., when S' belongs to a fall sector).

The intermediate case where $\operatorname{Re} \mu z = 0$ along one of the boundary rays of S', will not be discussed, as we will not need it. We will refer to the sector of the first type as a *mixed sector with the growth direction* $a \in \mathbb{C}$, while calling the second case the sector of fall as before.

In the mixed sector we choose the base point at infinity in the growth direction, $b = +\infty \cdot a$. More precisely, we consider the ray $R_z = z + \mathbb{R}_+ a = \{\zeta = z + sa\colon s \in \mathbb{R}_+\}$ (with the orientation inherited from $\mathbb{R}_+$) and the integral operator $\mathbf{S}_+ \colon f \mapsto \mathbf{S}_+ f$,

$$\begin{aligned} \mathbf{S}_+ f(z) &= -\int_{R_z} e^{\mu(z-\zeta)} f(\zeta)\, d\zeta \\ &= -a \cdot \int_0^{+\infty} e^{-s\cdot\mu a} f(z+sa)\, ds, \qquad s \in \mathbb{R}_+. \end{aligned} \tag{20.30}$$

This integral converges since both the function $e^{-s\mu a}$ and $f(z+sa)$ decrease very fast as $s \to +\infty$. Note that since the sector S' was assumed acute, we can always delete a bounded subset so that the remaining infinite set is convex. For convex domains the construction is always well defined.

In the sector of fall we choose the base point $b = r$ on the "exterior circumference" of the sector S', and fix the initial condition $y(b) = 0$. Then the solution $y(\cdot)$ is given by the integral operator $\mathbf{S}_-$ along the segment $[r, z] = -[z, r] = \{z - sa\colon 0 \leqslant s \leqslant |z - r|\}$, where $a = a(z) = (z - r)/|z - r|$,

$$\begin{aligned} \mathbf{S}_- f(z) &= -\int_{[z,r]} e^{\mu(z-\zeta)} f(\zeta)\, d\zeta \\ &= -a \cdot \int_0^{|z-r|} e^{s\cdot\mu a} f(z-sa)\, ds, \quad a(z) = \frac{z-r}{|z-r|}. \end{aligned} \tag{20.31}$$

There is no question of convergence, since the segment is always finite.

Definition 20.25. Given the sector S' and a nonzero complex number μ such that $\operatorname{Re} \mu z \neq 0$ on the boundary of S', we denote by $\mathbf{S} = \mathbf{S}_{\mu,S'}$ the appropriate integral operator,

$$\mathbf{S}_{\mu,S'} = \begin{cases} \mathbf{S}_+, & \text{if } \operatorname{Re}\mu a > 0 \text{ for some } a \in S', \\ \mathbf{S}_-, & \text{if } \operatorname{Re}\mu z/|z| \leqslant \delta_0 < 0 \text{ for all } z \in S'. \end{cases} \tag{20.32}$$

Denote $\mathcal{O}(S'; N)$ the space of functions holomorphic in the sector S' and decreasing as fast as $O(|z|^{-N})$ for a nonnegative number $N \geqslant 0$. This space can be equipped with the weighted sup-like norm

$$\|f\|_N = \|f\|_{S';N} = \sup_{z\in S'} |z|^N \, |f(z)|. \tag{20.33}$$

Lemma 20.26. *The operator $\mathbf{S}_{\mu,S'}$ is bounded as a linear operator acting on the subspace $\mathcal{O}(S'; 0)$.*

Moreover, it remains bounded when considered as an operator on the space $\mathcal{O}(S'; N)$.

Proof. We fix the sector S' and treat separately the two possibilities of S' being mixed sector or fall sector, depending on the choice of μ. First we consider the case $N = 0$ corresponding to the usual sup-norm.

If S' is the mixed sector and $\|f\| = 1$, that is, $|f(z)| \leqslant 1$, then $|\mathbf{S}_+ f(z)| \leqslant |a| \int_0^\infty e^{-cs}\,ds = |a|/c$, $c = \operatorname{Re}\mu a > 0$.

If S' is the sector of fall, then $|\mathbf{S}_- f(z)| \leqslant |a| \int_0^{|z-r|} e^{cs}\,ds \leqslant 1/|c|$, where $c = c(z) = \operatorname{Re}\mu a(z)$. If z belongs to the translate $r + S'$ of the sector S', then $a(z) = (z - r)/|z - r|$ of modulus 1 belongs to S', hence by the second assumption (20.32) we have $|c(z)| \geqslant \delta_0 > 0$ bounded from below. This proves that $\mathbf{S}_- f$ is bounded in $r + S'$.

Moreover, one can replace S' by another sector $S'' \supset S'$ of slightly bigger opening but still a fall sector; the above arguments would prove then that $\mathbf{S}_- f$ is bounded in $r + S''$. It remains to notice that the difference $S' \smallsetminus (r + S'')$ is bounded, its diameter depending only on S', S'' and r, so the integral (20.31) is bounded there as well. Thus we have proved the boundedness of $\mathbf{S}_-$ with respect to the usual sup-norm $\|\cdot\|_0$ on S'.

To prove the boundedness with respect to the "weighted sup-norms" $\|\cdot\|_N$, assume that $\|f\|_N \leqslant 1$, i.e., $|f(z)| \leqslant |z|^{-N}$, and consider again both possibilities for S'.

Let S' be a mixed sector. Since S' is acute and $z, a \in S'$, we have $|z + sa| \geqslant c'\,|z|$ for some constant $c' > 0$ depending only on S' and all $s \in \mathbb{R}_+$, by obvious geometric considerations. Substituting this inequality into the integral (20.30), we majorize $\mathbf{S}_+ f$ in S' by $|c'z|^{-N} \cdot /|c|$. This proves the boundedness of $\mathbf{S}_+$.

To see why $\mathbf{S}_-$ is bounded in $r + S''$ with respect to this norm (where S'' is chosen as in the case $N = 0$), we split the segment of integration $[r, z]$

in (20.31) into two equal parts. On the initial part $\zeta \in [r, \frac{1}{2}(r+z)]$ the exponential factor $e^{\mu(z-\zeta)}$ is exponentially small, since $|z-\zeta| \geqslant \frac{1}{2}|z|$. On the distant part $\zeta \in [\frac{1}{2}(z+r), z]$ we have the inequality $|\zeta| \geqslant \frac{1}{2}|z|$ and hence by our assumption on f, $|f(\zeta)| \leqslant 2^{-N}|z|^{-N}$, so that the full integral $\mathbf{S}_- f(z)$ is bounded by $2^{-N}|z|^{-N}/|c(z)|$. Exactly as in the case $N = 0$, this implies that $\mathbf{S}_-$ is bounded in the $\|\cdot\|_N$-norm. □

Remark 20.27. In all these constructions the bound for the norm $\|\mathbf{S}_\pm\|_{S';N}$ may depend on N and the opening of the sector S' but does not depend on the "size" (the parameter r) of the sector. This can be verified independently by the rescaling arguments.

20L. Integral equation and demonstration of Theorem 20.23. If instead of the simple equation (20.28) we would have a slightly more general form

$$\tfrac{d}{dz}y = [\mu + G(z)]y + g(z), \tag{20.34}$$

then the method of variation of constants, instead of giving an explicit solution, would reduce (20.34) to an integral equation.

After the substitution $y(z) = e^{\mu z}y'(z)$ (20.34) is transformed to the equation $\frac{d}{dz}y'(z) = e^{-\mu z}[G(z)y(z) + g(z)]$, which after taking primitive and multiplication by $e^{\mu z}$ yields

$$y(z) = e^{\mu z}y(b) + \int_b^z e^{\mu(z-\zeta)}[G(z)y(z) + g(z)]\,dz.$$

Again the base point b can be chosen freely, and this freedom can be again used to ensure the flatness of solutions. As before, we conclude that

$$y = \mathbf{S}[Gy + g], \qquad \mathbf{S} = \mathbf{S}_{\mu,S'}, \tag{20.35}$$

if it exists, satisfies the differential equation (20.34).

A multidimensional generalization of this example for the k-dimensional system (20.27) is straightforward. Denote by $\mathbf{S}$ the diagonal integral operator defined on vector functions bounded in the sector S', as follows:

$$\mathbf{S}(y_1, \dots, y_k) = (\mathbf{S}_1 y_1, \dots, \mathbf{S}_k y_k), \qquad \mathbf{S}_i = \mathbf{S}_{\mu_i,S'},\ i = 1, \dots, k. \tag{20.36}$$

This operator, a Cartesian product of integral operators of the form (20.32), depends on the eigenvalues of the diagonal matrix $D = \operatorname{diag}\{\mu_1, \dots, \mu_k\}$, with the path of integration being in general different for each component.

In complete analogy with (20.35), solution of the system (20.27) can be constructed by solving the integral equation

$$y = \mathbf{S}[Gy + g], \qquad \mathbf{S} = \operatorname{diag}\{\mathbf{S}_1, \dots, \mathbf{S}_k\}. \tag{20.37}$$

The diagonal integral operator $\mathbf{S}$ is bounded by Lemma 20.26, if the boundary rays of S' are not exceptional for any μ_i, that is, not separation rays for the initial system (20.5). We show that the composition occurring in the right hand side of (20.37) is a contraction, if the sector $S' = \{|z| > r,\ |\operatorname{Arg} z| < \pi - \delta\}$ is sufficiently small, i.e., r is sufficiently large.

Proposition 20.28. *In the assumptions of Theorem 20.23 the operator*

$$y \mapsto \mathbf{G}y = Gy + g$$

is Lipschitz in the sense of any norm $\|\cdot\|_{S';N}$ on the space of vector functions holomorphic in $S'_r = S' \cap \{|z| > r\}$,

$$\|\mathbf{G}y - \mathbf{G}y'\|_{S'_r;N} < \rho\, \|y - y'\|_{S'_r;N}, \qquad \rho = \rho(r) > 0.$$

The Lipschitz constant $\rho(r)$ tends to zero as $r \to +\infty$.

Proof. The Lipschitz constant $\rho = \rho(r)$, actually independent of N, can be chosen as $\rho(r) = \sup_z\{|G(z)|\colon\ z \in S'_r\}$. Indeed,

$$\|\mathbf{G}y - \mathbf{G}y'\|_{S'_r;N} \leqslant \sup_{z\in S'_r} |z|^{-N}|G(z)| \cdot |y(z) - y'(z)| \leqslant \sup_{z\in S'_r} |G(z)| \cdot \|y - y'\|_{S'_r,N}.$$

By assumption, $G(z)$ tends to zero as $z \to \infty$ in S', hence $\rho(r) \to 0^+$ as $r \to +\infty$. □

Proof of Theorem 20.23. Our goal already has been reduced to showing that the integral equation (20.37) admits a solution flat in the sector S'. Without loss of generality we may assume that the rays bounding S' are not exceptional (otherwise one can increase slightly the opening while keeping the sector acute).

Let $N \geqslant 0$ be an arbitrary order of decay. As soon as r is sufficiently large, $r \geqslant r(N)$, the Lipschitz constant $\rho(r)$ of the operator $\mathbf{G}$ becomes smaller than the bound for the norm of the operator $\mathbf{S}$ with respect to any given N (recall that $\|\mathbf{S}\|_N$ does not depend on r; see Remark 20.27). In the corresponding $S'_r = S' \cap \{|z| > r(N)\}$ the composition $\mathbf{S} \cdot \mathbf{G}$ will be contracting in the $\|\cdot\|_N$-norm. Hence the fixed point-type integral equation (20.37) possesses a *unique* solution, a vector function with each component belonging to the space $\mathcal{O}(S'_N, N)$. Any such solution can in fact be extended to a function holomorphic in the entire sector S' by virtue of the differential equation (20.27) nonsingular in S'. By the uniqueness, any two such extensions necessarily coincide with each other on the intersection of their domains. Together they yield a vector function $y(z)$ holomorphic in S' and decreasing faster than $|z|^{-N}$ for any N as $|z| \to \infty$. In other words, the constructed solution $y(z)$ is flat as required. □

20M. Sector enlargement and the proof of Sibuya Theorem 20.16. Let S be an "acute" sector of opening $\pi/(m-1)-2\delta$ as in (20.22). Consider its rotations $S_{\pm} = e^{\pm 2i\delta}S$: the union of the three sectors $S \cup S_{+} \cup S_{-}$ is a sector of opening $\pi/(m-1)+2\delta$. By assumption, each sector $S_{\pm}$ may contain only those separation rays, that already were contained in S (and perhaps not all of them).

Since $S, S_{\pm}$ are all "acute", by Corollary 20.24 there exist normalizing cochains $H, H_{\pm}$ conjugating the initial system with its formal normal form. Therefore for suitable Stokes matrices $C_{\pm}$ (not to be confused with the Stokes collection of the initial system),

$$H(t) = H_{\pm}(t)WC_{\pm}W^{-1}(t) \quad \text{on the intersections } S_{\pm} \cap S, \tag{20.38}$$

where $W(t)$ is a fixed diagonal solution of the formal normal form. But since the flaps $S_{\pm} \smallsetminus S$ contain no separation rays, the difference $E - W(t)C_{\pm}W^{-1}(t)$ remains flat not only on the intersections $S_{\pm} \cap S$, but also on the sectors $S_{\pm}$. In other words, the right hand side of (20.38) extends the same series $\widehat{H}$ and provides an analytic continuation of H on the larger sector $S \cup S_{\pm}$.

Exercises and Problems for §20.

Problem 20.1. Let $J \in \operatorname{Mat}(n, \mathbb{C})$ be an upper-triangular standard nilpotent Jordan block of maximal size, and ad_J the linear operator of commutation with J. Prove that the linear subspace of matrices having zeros in all places except for the last row, is transversal (complementary) to the image of ad_J.

Problem 20.2 (Demonstration of the Borel–Ritt theorem after [**Was87**]). Let $\varphi(c, \beta; t) = 1 - \exp(-ct^{-\beta})$, $0 < \beta < 1$, $c > 0$, be a function holomorphic in a sector S of opening less than 2π. For an arbitrary formal series $\widehat{F} = \sum_{k=1}^{\infty} a_k t^k$ consider the series $F = \sum_{a_k \neq 0} a_k \varphi(|a_k|^{-1}, \beta; t)\, t^k$.

(a) Prove that $|1 - \exp z| < |z|$ if $\operatorname{Re} z < 0$. (b) Prove that for some $\beta \in (0,1)$ depending on S, the function $-t^{-\beta}$ has negative real part in S. (c) Prove that the series F is majorized by the series $\sum_{a_k \neq 0} |t|^{k-\beta}$ in the sector S. (d) Prove that the series F uniformly converges in S. (e) Prove that the asymptotic series for F coincides with $\widehat{F}$. (f) Prove the Borel–Ritt theorem.

Chapter IV

Functional moduli of analytic classification of resonant germs and their applications

21. Nonlinear Stokes phenomenon for parabolic and resonant germs of holomorphic self-maps

21A. Introduction and preliminaries. The relationship between formal and analytic classification of local holomorphic objects (germs of holomorphic vector fields at a singular point or germs of holomorphic self-maps at a fixed point) can be very delicate, as was already noted in Chapter I. While in some cases these two classifications coincide (either by the Poincaré Theorem 5.5 or by some of the more advanced results described in §5**E**), in the situation dangerously close to the resonance the normalizing series may diverge (cf. with Theorem 5.32 and Remark 5.33).

Conformal germs $\operatorname{Diff}(\mathbb{C}, 0)$ constitute probably the simplest class of objects for which these phenomena can be observed. Generic germs with multiplicator $\mu = \exp 2\pi i\lambda$ off the unit circle, $\lambda \notin \mathbb{R}$, are analytically linearizable, and the same is true if the number λ is real but irrational and does not admit too accurate applications by rational numbers. On the other hand, if λ is "almost rational" (violates the Brjuno condition given in Theorem 5.23), the germ is almost never linearizable.

The analytic classification of conformal germs violating the Brjuno condition is not known. Yet if instead of being "almost rational" the number λ is genuine rational, the corresponding Siegel resonant case is simpler in this respect. For resonant germs one can construct a complete analytic classification "on top" of the formal classification. This classification is especially simple in the "maximally resonant case" ($\lambda = 0$, i.e., $\mu = 1$) of *parabolic germs* $\mathrm{Diff}_1(\mathbb{C}, 0)$. The general resonant case $\lambda = m/n$ can be reduced to the parabolic case by studying a suitable iterational power $f^{\circ n} \in \mathrm{Diff}_1(\mathbb{C}, 0)$ ($\circ n$ means n times iterated composition).

The *Ecalle–Voronin modulus of analytic classification* $\mathcal{M}_f$ was discovered independently by J. Ecalle, B. Malgrange and S. Voronin in 1981. This modulus allows us to give answers to a number of natural questions about a parabolic germ $f \in \mathrm{Diff}_1(\mathbb{C}, 0)$, among them the following:

- **analytic classification:** when two resonant germs are analytically equivalent?
- **embedding in the flow:** when a given parabolic germ may be represented as the time one shift along a holomorphic vector field?
- **root extraction:** for what parabolic germs f and natural numbers $q \in \mathbb{N}$ the equation $g^{\circ q} = f$, involving the qth iteration, admits a convergent parabolic solution g?

The answers to these questions are given in this section. In particular, it will be shown that analytic equivalence of a parabolic germ to its formal normal form described in §4**I**, is an exceptionally rare event.

Recall that the set of parabolic germs $\mathrm{Diff}_1(\mathbb{C}, 0)$ is filtered by the order of tangency with the identity: we denote $\mathscr{A}_p = \{f \in \mathrm{Diff}_1(\mathbb{C}, 0) \colon f(z) = z + cz^{p+1} + \cdots,\ c \neq 0\}$; cf. with (6.3). The index p is a topological invariant of the germ and can be seen on the "phase portrait", the structure of orbits of action of the cyclic pseudogroup $\{f^{\circ Z}\} \subseteq \mathrm{Diff}_1(\mathbb{C}, 0)$.

Example 21.1. Consider the vector field $F(z) = z^2 \frac{\partial}{\partial z}$ and its time one map $f = \exp F \in \mathrm{Diff}_1(\mathbb{C}, 0)$. The field F is constant, $\frac{\partial}{\partial t}$ in the rectifying coordinate $t = -1/z$. In the initial coordinate z the real trajectories $\{\exp tF(a), t \in \mathbb{R}\}$, are circles.

Let U be a small circle $U = \{|z| < \varepsilon\}$. Its image in the chart t is a large circle $U' = \{|t| > 1/\varepsilon\}$. The orbits of the map f in the rectifying chart t are parts of arithmetic progressions of the form $a' + \mathbb{Z}$ that are disjoint with U'. If $|\operatorname{Im} a'|$ is sufficiently large, then these orbits are bi-infinite, otherwise the orbit through a' is infinite only in one direction (forward or backward).

Behavior of the orbits of the pseudogroup $\{f^{\circ \mathbb{Z}}\}$ in the initial chart z is shown on Fig. IV.1 and form two *Fatou petals*. In a similar way the orbits of the germ $f = \exp F$, $F(z) = z^{p+1} \frac{\partial}{\partial z}$, form $2p$ different petals.

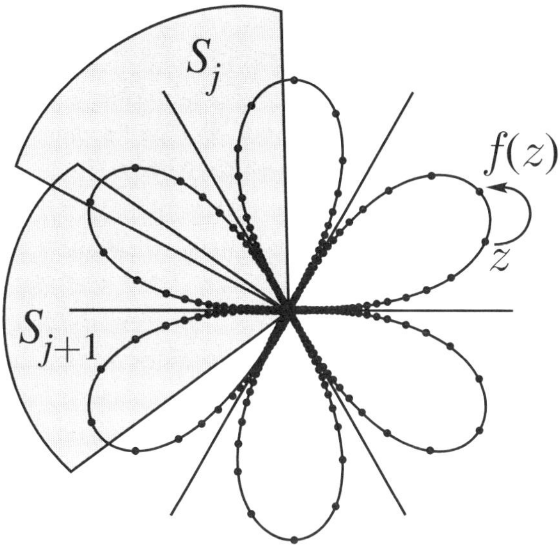

Figure IV.1. Fatou petals for the parabolic map with $p = 3$. Iterates of a map belong to the real flow curves only if the map is embeddable, otherwise the picture illustrates only the topology of iterations

Topological classification of parabolic germs is very simple.

Theorem 21.2 (C. Camacho, P. Sad, 1982; A. Shcherbakov, 1982). *Any parabolic germ from the class $\mathscr{A}_p$ is topologically equivalent to the time one map $f_{p,0} = \exp F_{p,0}$ of the standard vector field $F_{p,0} = z^{p+1}\frac{\partial}{\partial z}$.*

We will neither prove nor use this theorem.

21B. Sectorial normalization theorem. By Theorem 4.26, any germ from $\mathscr{A}_p$ is *formally* equivalent to the time one flow map

$$f_{p,\lambda} = \exp F_{p,\lambda}, \qquad F_{p,\lambda} = \frac{z^{p+1}}{1+\lambda z^p} \cdot \frac{\partial}{\partial z}, \tag{21.1}$$

of the *standard vector field* $F_{p,\lambda}$ for some complex value λ. The corresponding formal series $\widehat{H}$ conjugating the germ f with the embeddable model $f_{p,\lambda}$ is in general nonunique and may be divergent. However, with each such series $\widehat{H}$ one can associate a geometric object, *functional map-cochain*, similar to what was constructed in a different context by J.-P. Ramis and Y. Sibuya; see §20**F**, §20**G**. Functional map-cochains constitute a new class of local objects in complex analysis.

Construction of the normalizing map-cochains begins with the observation, due to L. Leau (1897), that parabolic germs can be holomorphically embedded into a flow albeit in domains smaller than the full neighborhood of a fixed point. The corresponding result is in many aspects parallel to the Sibuya Theorem 20.16.

Fix an arbitrary parabolic germ $f \in \mathscr{A}_{p,\lambda}$. Without loss of generality, making a linear transformation, we may and will always assume that the principal part of f is pre-normalized to the form

$$f_{p,\lambda}(z) = z + z^{p+1} + \cdots, \tag{21.2}$$

corresponding to the case $a = 1$ in (6.3).

Definition 21.3. Let $p \in \mathbb{N}$ be an integer number. A *nice p-covering* of a punctured neighborhood of the origin is the collection of $2p$ sectors of the form

$$S_j = \{z : |\operatorname{Arg} z - \pi j/p| < \alpha,\ |z| < r\}, \qquad j = 1, \dots, 2p, \tag{21.3}$$

where the angle α, $\pi/2p < \alpha < \pi/p$, and the radius $r > 0$ are two real parameters that are usually specified by the context.

Remark 21.4. The definition of nice p-covering is formally independent of any germ. In practice, it is a covering in which the *normalizing cochain* (see Definition 21.9 and Theorem 21.12) of a conformal germ $f \in \mathscr{A}_{p,\lambda}$ will be inscribed, under the tacit assumption that the germ is normalized as in (21.2). Without this condition the nice p-covering for an arbitrary germ $f(z) = z + az^{p+1} + \cdots$, $a \neq 0$, is obtained from the nice p-covering as it is described in Definition 21.3, by the appropriate rotation.

Any sector of a nice p-covering contains more than half of any of the two subsequent petals of the field $F_{p,\lambda}$; see Fig. IV.1 for $p = 3$. The characteristic property of these sectors is as follows: (a) every sector contains orbits of $f_{p,\lambda}$ infinite in exactly one direction (infinite forward orbits for even j, infinite backward orbits for odd j), and (b) none of the sectors contains bi-infinite orbits of f.

Theorem 21.5 (sectorial normalization theorem). *For any parabolic germ $f \in \mathscr{A}_{p,\lambda}$ normalized by the condition* (21.2)*, any formal series $\widehat{H}$ tangent to identity, which transforms f to the formal normal form $f_{p,\lambda}$, and any nice p-covering $\mathfrak{S} = \{S_1, \dots, S_{2p}\}$ there exists a holomorphic cochain $\mathcal{H} = \{H_1, \dots, H_{2p}\}$ subordinated to the covering $\mathfrak{S}$, such that:*

(1) *H_j conjugates f with its formal normal form $f_{p,\lambda}$ in S_j, and*

(2) *the formal series $\widehat{H}$ is a common asymptotic series for each function H_j in the respective sector S_j for all $j = 1, \dots, 2p$.*

The holomorphic cochain $\mathcal{H}$ satisfying both these conditions, is unique.

If the germ f analytically depends on auxiliary parameters $\varepsilon \in (\mathbb{C}^n, 0)$ while remaining in the same formal equivalence class, then the cochain $\mathcal{H}$ also analytically depends on these parameters.

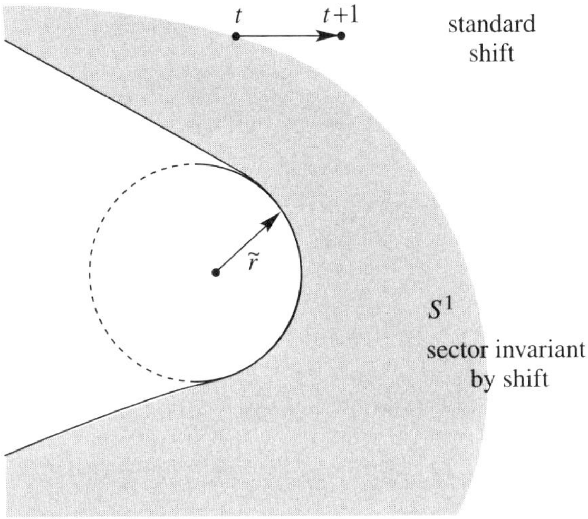

Figure IV.2. Sector S^1 in the t-plane. The sector is forward invariant by the standard shift. Its mirror image $S^0 = -S^1$ is backward invariant

It is convenient to prove the sectorial normalization theorem in the chart rectifying the standard vector field $F_{p,\lambda}$; see (21.1). This chart $t = t(z)$ can be found by integration of the differential equation

$$\frac{dt}{dz} = \frac{1 + \lambda z^p}{z^{p+1}}, \qquad t(z) = t_{p,\lambda}(z) = -\frac{1}{pz^p} + \lambda \ln z. \tag{21.4}$$

The field $F_{p,\lambda}$ in the chart t is constant, $\frac{\partial}{\partial t}$, hence the standard map $f_{p,\lambda}$ becomes the standard shift $t \mapsto t + 1$. The images of the sectors S_j of the nice covering can also be easily described: for j even the map $z \mapsto t_{p,\lambda}(z)$ transforms S_j to a domain which contains a sector with the vertex at infinity,

$$S^0 = \{t : |t| > \widetilde{r},\ |\operatorname{Arg} t - \pi| < \beta\} \tag{21.5}$$

for some β, $\frac{\pi}{2} < \beta < p\alpha$ and $\widetilde{r} = \widetilde{r}(\beta, r) \gg 1$. For j odd the image of S_j contains the sector $S^1 = -S^0$; see Fig. IV.2.

All the way around, the properly chosen branch of the inverse map $z = z_{p,\lambda}(t)$ transforms the sector S^0 (resp., S^1) into a domain on the z-plane, that contains a sector S_j' described by (21.3) with the parameters α, $\pi/2p < \alpha < \beta/p$, and $r > 0$ sufficiently small.

The "distortion" introduced by the rectifying chart t, is in some sense bounded. The following technical result provides some estimates that will be used in the proof.

Proposition 21.6. *If $u\colon S_j \to S_j$, is a map of the sector S_j into itself with the asymptotic behavior $u(z) = z + O(|z|^{N+1})$, then in the chart $t = t_{p,\lambda}(z)$*

the map $\widetilde{u} = t_{p,\lambda} \circ u \circ t_{p,\lambda}^{-1}$ *has the asymptotic behavior*

$$\widetilde{u}(t) = t + O(|t|^{-m+1}) \qquad \text{as } t \to \infty, \quad m = N/p, \tag{21.6}$$

as t *remains in* S^0 *or* S^1 *respectively. Conversely, a holomorphic map* $\widetilde{u}$ *defined in one of the sectors* $S^{0,1}$ *and satisfying there the asymptotical condition* (21.6), *in the* z*-chart differs from identity by an* $(N+1)$*-flat term as above.*

Proof of the proposition. The map $z \mapsto t = t_{p,\lambda}(z)$ from S_j to $S^{0,1}$ (which stands for S^0 or S^1 depending on the parity of j) as in (21.4) can be represented as the composition of three maps: pure fractional power, homothety and the map tangent to identity at infinity,

$$z \mapsto w = z^{-p} \mapsto v = -\tfrac{1}{p} w \mapsto t = v - \tfrac{\lambda}{p} \ln(-pv). \tag{21.7}$$

The fractional power $z \mapsto w = z^{-p}$ conjugates the automorphism $z \mapsto u(z) = z + O(|z|^{N+1}) = z(1 + O(|z|^N))$ of S_j with the automorphism of the form $w \mapsto w(1 + O(|w|^{-N/p}))^{-p} = w(1 + O(|w|^{-N/p}))$ of $S^{0,1}$. The homothetic conjugacy (linear rescaling) $w \mapsto v = -\frac{1}{p} w$ does not change the structure of the asymptotic behavior of any map u.

It remains to verify that conjugation by the ramified transformation $v \mapsto t = v + c\ln(-pv) = v\left(1 + c\frac{\ln v}{v} + \frac{c'}{v}\right)$, $c, c' \in \mathbb{C}$, preserves the order of tangency r between any automorphism $v \mapsto v + O(|v|^r)$ of $S^{0,1}$ with the identity, regardless of the choice of the branch of logarithm. This last remaining assertion follows from the fact that the terms $|\ln v|/|v|$ and $1/|v|$ tend to zero as $|v| \to \infty$ in the sector $S^{0,1}$. The details are left to the reader. □

Proof of the Theorem 21.5. Without loss of generality we may assume that f differs from its formal model $f_{p,\lambda}$ by terms of arbitrarily high order N. This can always be achieved by preliminary normalization of a finite jet of f by a suitable polynomial transformation.

0. In the rectifying chart t the problem reduces to constructing a holomorphic conjugacy H between the germ $\widetilde{f} \colon t \to t+1+R(t)$ and the standard shift $T \colon t \to t+1$, say, in the sector S^1. This conjugacy satisfies the identity

$$H \circ \widetilde{f} = T \circ H, \qquad T \colon t \mapsto t+1. \tag{21.8}$$

Substituting $H \colon t \to t + h(t)$, we obtain from (21.8) the following *Abel equation* for the holomorphic function h:

$$h = R + h \circ \widetilde{f}. \tag{21.9}$$

The series

$$h = \sum_{n=0}^{\infty} R \circ \widetilde{f}^{\circ n}, \tag{21.10}$$

if it converges uniformly, would obviously give a holomorphic solution to the Abel equation (21.9). Note that this series makes sense, since the sector S^1 is $\widetilde{f}$-invariant, so all iterates are well defined.

1. We will prove that the series (21.10) converges in a sector S^1 for a suitable choice of β and $\widetilde{r}$, if the flatness order m in (21.6) is sufficiently large. Indeed, for a sufficiently large $\widetilde{r}$ we have $\operatorname{Re}\widetilde{f}(t) > \operatorname{Re} t + \frac{1}{2}$, so that the iterates $\widetilde{f}^{\circ n}(a)$ of any point $a \in S^1$ with $|a| > \widetilde{r}$, remain in the sector $\{t : \operatorname{Arg}(t-a) < \pi/4\}$ and go to infinity fast enough: their absolute values are bounded below by an arithmetic progression with the difference $\frac{1}{2}$.

By Proposition 21.6 applied to the map $u = f \circ f_{p,\lambda}^{-1} \colon z \mapsto z + O(|z|^{N+1})$, we conclude that $\widetilde{f} \circ T^{-1}(t) = \widetilde{f}(t) - 1$ differs from the identity by the term $R(t) = O(t^{-m+1})$, $m = N/p$. If $1 - m < -2$ (by (21.6), this occurs if $N > 3p$), then the series (21.10) converges uniformly and its sum $h(t)$ is decreasing asymptotically as $h(t) = O(t^{-m+2})$ as $t \to \infty$. Thus an analytic solution to the Abel equation is constructed in the sector $S^1_{\widetilde{r}} = S^1 \cap \{|t| > \widetilde{r}\}$ for a sufficiently large $\widetilde{r}$.

Yet if a holomorphic solution of the equation (21.9) is defined in a "small" sector $S^1_{\widetilde{r}}$ with arbitrarily large $\widetilde{r}$, it can in fact be extended by iterations of $\widetilde{f}$ using the equation (21.9) in the entire domain obtained by saturation of $S^1_{\widetilde{r}}$ by orbits of the pseudogroup $\{\widetilde{f}^{\circ\mathbb{Z}}\}$ which is a much bigger sector (which corresponds to a smaller value $\widetilde{r}$).

Returning back to the initial chart z, we obtain a conjugacy $H_j = H_j^{(N)}$ between f and its normal form $f_{p,\lambda}$ defined in the sectors S_j of the given nice p-covering with the asymptotic behavior $H_j^{(N)}(z) - \widehat{H}(z) = O(z^{N/p})$. Still the entire asymptotic series for different $H_j^{(N)}$ may yet be different.

2. We prove that the normalization is *uniquely determined* by its asymptotic polynomial (truncation of the asymptotic series) of order $p+1$. Indeed, if there are two holomorphic normalizations H, H' whose difference is $(p+1)$-flat in the sector S, then the compositional ratio $G = H' \circ H^{-1}$ is an automorphism of the normal form $f_{p,\lambda}$. In the rectifying chart $t(z)$ the holomorphic map $\widetilde{G}$ is defined in the sector, say, S^1, commutes with the standard shift $t \mapsto t+1$. The difference between $\widetilde{G}$ and identity, the holomorphic function $g(t) = \widetilde{G}(t) - t$ is 1-periodic and bounded by Proposition 21.6. The sector S^1 contains a vertical strip of width 1 parallel to the imaginary axis: being bounded in this strip, g extends by 1-periodicity to a bounded function on $\mathbb{C}$. By the maximum principle, g must be a constant. The condition $\widetilde{G}(t) - t = o(1)$ in S^1 implies that $g \equiv 0$, hence $\widetilde{G} = \mathrm{id}$ and $G = \mathrm{id}$. The uniqueness is proved.

From this uniqueness it follows automatically that if the natural number N chosen at the beginning of the proof, is large enough, $N > 3p$, then the cochain $\mathcal{H}^{(N)}$ constructed on Step 1 of the proof, is in fact independent of N and hence asymptotic to the series $\widehat{H}$.

4. To prove analytic dependence on parameters, consider an analytic family $f_\varepsilon \in \mathscr{A}_{p,\lambda}$ depending holomorphically on the parameter $\varepsilon \in \mathbb{C}^k$. Then all terms of the series (21.10) are holomorphic both in t and ε. Since the series converges uniformly, its sum $h = h(t, \varepsilon)$ is holomorphic in ε. Therefore the cochain $\mathcal{H}_\varepsilon = \{H_{1,\varepsilon}, \dots, H_{2p,\varepsilon}\}$ depends analytically on the parameter ε. The proof of sectorial normalization theorem is complete. □

Remark 21.7. If two sectorial normalizations H, H' are not constrained by the common asymptotic polynomial, then they differ by the flow map of the standard field $F_{p,\lambda}$.

Indeed, in this case the compositional ratio $G = H' \circ H^{-1}$ commutes with $f_{p,\lambda}$. If the asymptotic series for $G - \mathrm{id}$ starts with terms of order q, then by Proposition 6.11 the commutator is nontrivial (differs from identity by nonzero terms of order $p+q$) unless $q = p+1$. In the latter case the difference $g(t) = \widetilde{G}(t) - t$ between the ratio and the identity in the rectifying chart is a 1-periodic *bounded* function, i.e., a constant: $\widetilde{G}(t) = t + s$, $s \in \mathbb{C}$. Thus $\widetilde{G}$ is the flow map of the standard constant field $\frac{\partial}{\partial t}$ in the rectifying coordinate, hence G itself is the flow map of the field $F_{p,\lambda}$.

On the overlapping of sectors of the nice p-covering the maps H_j in general differ from each other, though the difference

$$H_{j,j+1} = H_j(z) - H_{j+1}(z) \qquad \text{on } S_{j,j+1} = S_j \cap S_{j+1} \tag{21.11}$$

is flat (decreases faster than any finite power of z as $z \longrightarrow 0$), since all these maps have a common asymptotic Taylor series. In fact, the decay rate is exponential, as the following proposition shows. Besides, it is more natural for the reasons to be explained below, to estimate not the "additive coboundary" (21.11), but rather the "compositional coboundary", the difference between $H_\alpha \circ H_\beta^{-1}$ and the identity.

Proposition 21.8. *For any cochain $\mathcal{H}$ constructed in Theorem 21.5, the compositional ratios $R_{\alpha\beta} = H_\alpha \circ H_\beta^{-1}$ defined on slightly diminished nonempty intersection sectors $S_{\alpha\beta} = S_\alpha \cap S_\beta$, satisfy the inequalities*

$$|R_{\alpha\beta}(z) - z| < e^{-c|z|^{-p}}, \qquad z \in S_{\alpha\beta}. \tag{21.12}$$

for some positive constant $c > 0$.

Proof. Each map $R_{\alpha\beta}$ by construction is an automorphism of the standard germ $f_{p,\lambda}$. In the rectifying chart $t = t_{p,\lambda}$ the standard germ is the standard shift, thus the difference $r_{\alpha\beta} = R_{\alpha\beta} - \mathrm{id}$ between $R_{\alpha\beta}$ and the identity is a one-periodic function. The overlapping between two consecutive sectors of a nice covering $\mathfrak{S}_p$ in the rectifying chart is a sector with the vertex at infinity, bisected by the imaginary axis (if $p = 1$, then the intersection consists of two such sectors, each to be treated separately). By periodicity, $r_{\alpha\beta}$ can be extended on an upper (resp., lower) sufficiently remote half-plane $\{\pm \operatorname{Im} t \gg 1\}$ as a function that has zero limit as $\pm \operatorname{Im} t \to \infty$ (since $r_{\alpha\beta}$ is flat). All this guarantees that $r_{\alpha\beta}$ admits a *converging Fourier expansion*,

$$r_{\alpha\beta}(t) = \sum_k c_k e^{2\pi i k t}, \qquad \begin{cases} k \in \ \ \mathbb{N} & \text{in the upper half-plane,} \\ k \in -\mathbb{N} & \text{in the lower half-plane.} \end{cases} \tag{21.13}$$

The expansion (21.13) yields an exponential bound $|R_{\alpha\beta}(t) - t| < ce^{-2\pi|\operatorname{Im} t|}$ in the rectifying chart, which becomes the exponential bound (21.12) in the initial chart z. □

21C. Functional cochains, normalizing cochains. Motivated by the Sectorial Normalization Theorem 21.5, we introduce a new class of objects that are holomorphic "almost maps".

Definition 21.9. A *functional cochain* of type p is a holomorphic (scalar) cochain $\mathcal{F} = \{F_1, \dots, F_{2p}\}$ inscribed in a nice p-covering $\mathfrak{S} = \{S_1, \dots, S_{2p}\}$, such that all the functions forming this cochain have a common asymptotic Taylor series $\widehat{F} = \sum_1^\infty a_j z^j$. The differences of the consecutive components are exponentially small in their common domain of definition:

$$|F_j(z) - F_{j+1}(z)| < e^{-c|z|^{-p}} \quad \text{in } S_j \cap S_{j+1}.$$

The series $\widehat{F}$ will be referred to as the *Taylor series* of the functional cochain $\mathcal{F}$. The tuple $\{F_j - F_{j+1},\ j = 1, ..., 2p\}$ will be called the *additive coboundary of a functional cochain.*

Definition 21.10. A *map-cochain* of type p is a functional cochain of the same type with nonzero linear term of the Taylor decomposition: $a_1 \neq 0$. The *compositional coboundary* (or simply coboundary) of a map-cochain $\mathcal{F} = \{F_1, \dots, F_{2p}\}$ inscribed in a nice p-covering $\mathfrak{S}_p$, is the holomorphic cocycle

$$\delta\mathcal{F} = \mathcal{H}, \qquad \mathcal{H} = \{H_{\alpha\beta}\}, \quad H_{\alpha\beta} = F_\alpha \circ F_\beta^{-1} \quad \text{on } S_\alpha \cap S_\beta. \tag{21.14}$$

An arbitrary holomorphic self-map $f \in \mathscr{A}_{p,\lambda}$ defines a functional cochain $\mathcal{F} = \{f|_{S_j}, j = 1, \dots, 2p\}$ with trivial (identical) coboundary. Conversely, a functional cochain with trivial coboundary defines a holomorphic self-map of the punctured neighborhood of the origin, which extends holomorphically at the origin by the removable singularity theorem.

We wish to stress that the functional cochains are to be treated as single entities rather than collections of separate functions because of the very stringent conditions on the corresponding coboundary (which measures the difference between the cochain and an "ordinary" function).

In particular, map-cochains form a group by *sectorial composition*: if $\mathcal{F}$ and $\mathcal{G}$ are two such cochains inscribed in the same nice p-covering, then their composition is the functional cochain

$$\mathcal{F} \circ \mathcal{G} = \{F_1 \circ G_1, \dots, F_{2p} \circ G_{2p}\},$$

inscribed in a nice covering with eventually smaller opening α of the sectors and smaller radius r. The inverse cochain $\mathcal{F}^{-1}$ can be defined in a similar way.

Remark 21.11. To avoid reservations of this sort, it is very convenient to work with *germs* of cochains of the same type. The construction repeats the standard definition of germs with minor modifications.

Consider two different nice p-coverings $\mathfrak{S}'_p$ and $\mathfrak{S}''_p$ defined by the respective parameters (α', r') and (α'', r'') as in Definition 21.3. We will say that two functional cochains $\mathcal{H}'$ and $\mathcal{H}''$, inscribed in the respective coverings, are equivalent, if there exists a third covering $\mathfrak{S}_p$ with the parameters $r < \min(r', r'')$ and $\pi/2p < \alpha < \min(\alpha', \alpha'')$, such that the components H'_j and H''_j restricted on the sectors of this covering, coincide. The equivalence class of cochains is naturally called the *germ* of a functional cochain of the type p, and each of the cochains is naturally called the *representative* of this germ.

After the notion of germ of a cochain is introduced, we may immediately verify that for any two *germs* $\mathcal{G}, \mathcal{H}$ of map-cochains tangent to the identity, their composition $\mathcal{G} \circ \mathcal{H}$ and the inverse $\mathcal{H}^{-1}$ are well defined as germs of the composition of suitable representatives (Problem 21.2).

Such a localization transforms the map-cochains of the same type, tangent to the identity, into a group with the well-defined operation of composition. This group, which will be denoted $\widetilde{\operatorname{Diff}}^{p}_1(\mathbb{C}, 0)$, is an extension of the group $\operatorname{Diff}_1(\mathbb{C}, 0)$.

Yet in the future we will usually ignore (unless it may lead to confusion) both in notation and in argumentation the difference between cochains and their germs in the same way as it is often conveniently ignored when dealing with the usual maps, functions and their germs.

The coboundary of a map-cochain in some sense measures, the extent to which this "almost map" is different from a usual holomorphic map. We will show in §21**D** that the analytic classification modulus for parabolic germs is (modulo a minor modification) the coboundary of a normalizing cochain.

In the language of map-cochains, the sectorial normalization theorem together with the asymptotic estimate given in Proposition 21.8, can be reformulated as follows.

Theorem 21.12. *Any parabolic germ $f \in \mathscr{A}_{p,\lambda} \subset \operatorname{Diff}(\mathbb{C},0)$ pre-normalized by the condition* (21.2), *is conjugated in the group $\widetilde{\operatorname{Diff}}^p_1(\mathbb{C},0)$ of map-cochains to its formal normal form $f_{p,\lambda}$ by a suitable holomorphic map-cochain $\mathcal{H} \in \widetilde{\operatorname{Diff}}^p_1(\mathbb{C},0)$,*

$$\mathcal{H} \circ f = f_{p,\lambda} \circ \mathcal{H}, \qquad \mathcal{H} \in \widetilde{\operatorname{Diff}}^p_1(\mathbb{C},0). \tag{21.15}$$

The map-cochain $\mathcal{H}$ conjugating the germ with its normal form, is defined uniquely modulo the flow map of the standard vector field $F_{p,\lambda}$, i.e., any two cochains $\mathcal{H}, \mathcal{H}'$ satisfying (21.15), *satisfy also the identity*

$$\mathcal{H}' = g \circ \mathcal{H}, \qquad g = \exp sF_{p,\lambda} \in \operatorname{Diff}(\mathbb{C},0),\ s \in \mathbb{C}. \tag{21.16}$$

The coboundary $\mathcal{R} = \delta\mathcal{H}$ of the map-cochain $\mathcal{H}$ differs from identity by exponentially small terms as in (21.12).

If f depends analytically on some auxiliary parameters $\varepsilon \in (\mathbb{C},0)$ within the same formal class $\mathscr{A}_{p,\lambda}$, then the map-cochain $\mathcal{H}$ also can be chosen analytically depending on ε. □

Definition 21.13. A map-cochain $\mathcal{H} \in \widetilde{\operatorname{Diff}}^p_1(\mathbb{C},0)$ which conjugates a holomorphic germ $f \in \mathscr{A}_{p,\lambda}$ with its formal normal form as described in Theorem 21.12, is called a *normalizing cochain*[1].

The problem of analytic classification of parabolic germs can now be reformulated as the problem on the structure of the group $\widetilde{\operatorname{Diff}}^p_1(\mathbb{C},0)$ of map-cochains, in particular, in description of all *normalizing* map-cochains. A germ is analytically equivalent to its formal normal form if and only if its normalizing cochain belongs to the subgroup $\operatorname{Diff}_1(\mathbb{C},0) \subsetneq \widetilde{\operatorname{Diff}}^p_1(\mathbb{C},0)$. More generally, two germs $f, f' \in \mathscr{A}_{p,\lambda}$ are analytically equivalent if and only if the compositional ratio of the corresponding normalizing cochains $\mathcal{H}' \circ \mathcal{H}^{-1}$ is a "regular" (i.e., belonging to the subgroup $\operatorname{Diff}_1(\mathbb{C},0)$) holomorphism of the standard map $f_{p,\lambda}$. Development of these ideas naturally leads to the construction of Ecalle–Voronin modulus in the same way as the Stokes operators appear from the Sectorial Normalization Theorem 20.16.

21D. Ecalle–Voronin modulus and analytic classification of parabolic germs. Let $f \in \mathscr{A}_{p,\lambda}$ be a parabolic germ normalized by the condition (21.2), and $\mathcal{H} = \{H_1, \dots, H_{2p}\}$ a normalizing cochain transforming f to its

[1]Most assertions concerning normalizing cochains, remain valid for all map-cochains which conjugate two "regular" germs $f, f' \in \operatorname{Diff}(\mathbb{C},0)$ from the same formal equivalence class.

formal normal form. This cochain is defined uniquely modulo the composition with the flow map $g = \exp sF_{p,\lambda}$, $s \in \mathbb{C}$.

Consider the coboundary $\boldsymbol{\Phi} = \delta\mathcal{H}$ of the normalizing cochain: because of the special structure of the nice covering, the individual functions constituting this cocycle are more naturally numbered by the single index $j = 1, \dots, 2p$ cyclic modulo $2p$ rather than a pair of indices $\alpha, \beta \in \{1, \dots, 2p\}$ with $|\alpha - \beta| = 1$.

The components $\Phi_1, ..., \Phi_{2p}$ constituting the coboundary $\boldsymbol{\Phi} = \delta\mathcal{H}$, are defined in the sectors

$$S_{j,j+1} = \Sigma_j = \left\{ z : |z| < r, \ \left|\operatorname{Arg} z - \tfrac{2j-1}{2p}\pi\right| < \beta \right\}, \ j = 1, \dots, 2p, \quad (21.17)$$

for $r > 0$ and β sufficiently small. By Proposition 21.8, the components Φ_j are exponentially flat in the sectors Σ_j,

$$|\Phi_j(z) - z| \leqslant e^{-c|z|^{-p}}, \qquad z \in \Sigma_j, \quad (21.18)$$

and commute with the normal form $f_{p,\lambda} = \exp F_{p,\lambda}$ in these sectors,

$$\Phi_j \circ f_{p,\lambda} = f_{p,\lambda} \circ \Phi_j. \quad (21.19)$$

The action (21.16) of the complex flow $\exp \mathbb{C}F_{p,\lambda}$ on normalizing cochains induces the corresponding action on their respective coboundaries: if $\boldsymbol{\Phi} = \delta\mathcal{H}$ and $\boldsymbol{\Phi}' = \delta\mathcal{H}'$, then

$$\begin{aligned} \boldsymbol{\Phi} \circ g = g \circ \boldsymbol{\Phi}', \quad &\text{i.e.,} \quad \Phi_j \circ g = g \circ \Phi_j', \\ g = \exp sF_{p,\lambda}, \qquad &j = 1, \dots, 2p. \end{aligned} \quad (21.20)$$

Denote by $\mathscr{M}^\circ_{p,\lambda} \subseteq \delta\widetilde{\operatorname{Diff}}{}^p_1(\mathbb{C}, 0)$ the space of cocycles $\boldsymbol{\Phi} = (\Phi_1, \dots, \Phi_{2p})$ defined in the sectors (21.17) with some $r, \beta > 0$, and satisfying the conditions (21.18)–(21.19), and let $\mathscr{M}_{p,\lambda}$ be the quotient of this space by the action (21.20) of the complex flow of the standard field.

The space $\mathscr{M}^\circ_{p,\lambda}$ is a complex Banach space. This circumstance allows us to define continuous or analytic dependence of elements from the quotient on additional parameters if the latter are present: a parametric family of equivalence classes (elements from $\mathscr{M}_{p,\lambda}$) is said to depend analytically on the parameters, if it can be represented by an analytic parametric family of elements from $\mathscr{M}^\circ_{p,\lambda}$.

Definition 21.14. The *Ecalle–Voronin* modulus of a parabolic germ $f \in \mathscr{A}_{p,\lambda}$ is the equivalence class $\mathcal{M}_f \in \mathscr{M}_{p,\lambda}$ of the coboundary $\delta\mathcal{H}_f$ of any normalizing cochain $\mathcal{H}_f$ for f with respect to the equivalence (21.20).

This definition is parallel to the definition of the Stokes operators constructed in §20**G** as matrix cocycle consisting of automorphisms of the diagonal normal form, defined on the common domain of two sectorial normalization maps.

The principal result of this section is the following theorem which gives a complete description of classes of analytically equivalent parabolic germs.

Theorem 21.15 (Analytic classification theorem for parabolic germs).

1. (Invariant) *Every parabolic germ $f \in \mathscr{A}_{p,\lambda}$ is associated with a unique equivalence class $\mathcal{M}_f \in \mathscr{M}_{p,\lambda}$, the same for all analytically equivalent germs.*

2. (Equimodality vs. equivalence) *Conversely, two formally equivalent parabolic germs with the same invariant $\mathcal{M} \in \mathscr{M}_{p,\lambda}$, are analytically equivalent.*

3. (Realization) *Each equivalence class $\mathcal{M} \in \mathscr{M}_{p,\lambda}$, can be realized as the invariant of some parabolic germ $f \in \mathscr{A}_{p,\lambda}$.*

4. (Analytic dependence on parameters) *If the germ f analytically depends on finitely many complex parameters $\varepsilon \in (\mathbb{C}^k, 0)$ while remaining in the same class of formal equivalence, then the invariant $\mathcal{M}_f$ also depends analytically on ε.*

The easy three-quarters of the proof of Theorem 21.15.

1. *Invariance.* Let f and f' be analytically equivalent germs from the same class $\mathscr{A}_{p,\lambda}$, conjugated by an analytic conjugacy h, so that $f' = h^{-1} \circ f \circ h$. Let $\mathcal{H}$ be some normalizing cochain for f. Then $\mathcal{H}' = h^{-1} \circ \mathcal{H}$ is a normalizing cochain for f'. Coboundaries of these cochains coincide, therefore $\mathcal{M}_f = \mathcal{M}_{f'}$.

2. *Equimodality and equivalence.* Let $f, f' \in \mathscr{A}_{p,\lambda}$ be two germs with $\mathcal{M}_f = \mathcal{M}_{f'}$. Then there exist two normalizing cochains, $\mathcal{H}$ for f and $\mathcal{H}'$ for f', whose coboundaries are equivalent in the sense of (21.20): there exists $s \in \mathbb{C}$ such that

$$\delta\mathcal{H} = g \circ \delta\mathcal{H}' \circ g^{-1}, \qquad g = \exp sF_{p,\lambda}.$$

The cochain $\mathcal{F} = g \circ \mathcal{H}'$ is normalizing for f' together with $\mathcal{H}'$, and the coboundaries of $\mathcal{H}$ and $\mathcal{F}$ coincide.

The equality $\delta\mathcal{H} = \delta\mathcal{F}$ is equivalent to the equality $\delta(\mathcal{H} \circ \mathcal{F}^{-1}) = \mathrm{id}$. The latter means that the cochain $h = \mathcal{H} \circ \mathcal{F}^{-1}$ with the components $H_j \circ F_j^{-1}$ has trivial coboundary and hence is a well-defined map from $\mathrm{Diff}_1(\mathbb{C}, 0)$. By construction, h conjugates f and f' with each other in each sector S_j, hence in some full neighborhood of the origin.

3. *Realization.* The proof of this assertion is postponed until §21**F**; it is the only assertion of the theorem that does not follow immediately from the construction of the Ecalle–Voronin modulus.

4. *Analytic dependence.* Obviously, the coboundary of a map-cochain analytically depending on parameters as in Theorem 21.12, is also analytically depending on these parameters. □

21E. Almost complex structures and quasiconformal mappings. The last assertion on realization requires a new idea. Consider an arbitrary collection $\mathcal{M} \in \mathscr{M}_{p,\lambda}$, represented by a cocycle $\mathbf{\Phi}$. Starting from this collection, we will construct an abstract holomorphic curve S and an automorphism $F\colon (S,a) \to (S,a)$ in such a way that if S were a punctured neighborhood of the origin, the Ecalle–Voronin modulus for F would necessarily be $\mathcal{M}$. The most difficult part of this proof is to determine the conformal type of S; it is achieved below using the *quasiconformal mappings technique* in §21**F**. As a result, we construct a germ f and a normalizing cochain $\mathcal{H}$ which solves the equation $\delta\mathcal{H} = \mathbf{\Phi}$, i.e., prove the "nonlinear solvability" of any cocycle $\mathbf{\Phi}$, satisfying the conditions (21.18)–(21.19).

The problem of resolving a nonlinear δ-equation is rather similar to resolving the linear $\bar\partial$-equation of Poincaré in the class of C^∞-smooth functions of one complex variable. The central role in the solution of the δ-equation plays an analytic notion of almost complex structure and quasiconformal maps.

21**E**$_1$. *What remains on a complex manifold when the atlas on it is lost?* One of the possible answers may be the following. A complex manifold M^n becomes a real manifold $M := M^{2n} = {}^{\mathbb{R}}M^n$ of real dimension $2n$. What remains is the orientation and the *complex structure* on the tangent (or, equivalently, cotangent) bundle.

A *complex structure* on an $\mathbb{R}$-linear space L is an operator $I\colon L \to L$ such that $I^2 = -E$ (here E is the identity operator). Such an operator allows us to interpret L as a linear space over $\mathbb{C}$ with the action

$$(\lambda + i\mu)\cdot v = \lambda v + \mu\, Iv, \qquad \lambda,\mu \in \mathbb{R}, \quad v \in L. \tag{21.21}$$

One can easily verify that the dimension of the space L must be even. $\dim_{\mathbb{R}} L = 2n$, and the complex dimension of the space thus obtained, is $\dim_{\mathbb{C}} L = n$.

An *almost complex structure* on a smooth real even-dimensional manifold $M = M^{2n}$ is a smooth family of operators $I = \{I(p)\colon p \in M\}$,

$$I(p)\colon \mathbf{T}_pM \to \mathbf{T}_pM \qquad \text{such that} \quad I^2(p) = -E.$$

The operator $I = I(p)$ interpreted as multiplication by the imaginary unit i (root of -1), provides a linear complex structure on the tangent space $\mathbf{T}_pM$ at every point $p \in M$, making these spaces n-dimensional over $\mathbb{C}$.

Using the $\mathbb{C}$-action (21.21) on the tangent spaces $\mathbf{T}_pM$, one can split each respective complexified cotangent space ${}^{\mathbb{C}}\mathbf{T}_p^*M = \mathbf{T}_pM \otimes_{\mathbb{R}} \mathbb{C}$ (the space of $\mathbb{C}$-valued $\mathbb{R}$-linear functionals on $\mathbf{T}_pM$) into the direct sums of two complementary spaces of 1-forms, *"complex linear"* and *"complex anti-linear"* forms. We denote these subspaces by $L_p^{1,0}$ and $L_p^{0,1}$ respectively:

$$\omega_p(\lambda \cdot \xi) = \begin{cases} \lambda\omega_p(\xi), & \text{if } \omega_p \in L_p^{1,0}, \\ \bar{\lambda}\omega_p(\xi), & \text{if } \omega_p \in L_p^{0,1}, \end{cases} \qquad \forall \xi \in \mathbf{T}_pM,\ \lambda \in \mathbb{C}.$$

There are three natural requirements for these subspaces: first, $L_p^{1,0}$ should be "complex conjugate" to $L_p^{0,1}$, i.e., if the linear functional $\omega|_{\mathbf{T}_pM}$ belongs to $L_p^{1,0}$, then $\bar{\omega}$ should belong to $L_p^{0,1}$ and vice versa. Second, at every point these two subspaces should be complementary (transversal) to each other in ${}^{\mathbb{C}}\mathbf{T}_p^*M$. Finally, we need to retain the natural orientation: for any basis $\omega^{(1)}, \dots, \omega^{(n)}$ of the subspace $L_p^{1,0}$ over $\mathbb{C}$, the map

$$\mathbf{T}_pM \to \mathbb{C}^n, \qquad \xi \mapsto (\omega_p^{(1)}(\xi), \dots, \omega_p^{(n)}(\xi)),$$

should be orientation-preserving.

To summarize, an almost complex structure on M^{2n} is a subbundle $L = L^{1,0} \subset {}^{\mathbb{C}}\mathbf{T}^*M$ of the complexified cotangent bundle ${}^{\mathbb{C}}\mathbf{T}^*M$, such that the above three requirements are satisfied.

21$\mathbf{E}_2$. *Integrability of almost complex structure.*

Definition 21.16. A function $f\colon M \to \mathbb{C}$ on a manifold M^{2n} with an almost complex structure defined by a subbundle $L^{1,0}$ is called *holomorphic with respect to this structure*, if its differential df belongs to the subbundle at each point.

An almost complex structure is *integrable*, if there exists an atlas of charts $U_\alpha \to \mathbb{C}^n$, $\bigcup_\alpha U_\alpha = M$, such that every component of each chart is holomorphic with respect to the almost complex structure.

We discuss first the integrability conditions for the case $n = 1$. The higher dimension case $n > 1$ is treated in the next section.

For $n = 1$, in a complex chart $z \in \mathbb{C}$ any subbundle $L^{1,0}$ is spanned by a single form $\omega = a\,dz + b\,d\bar{z}$. The assumption on preserving the orientation implies that $|a| > |b|$, hence $a \neq 0$. Since ω makes sense only up to proportionality, we can without loss of assume that the 1-form defining an arbitrary almost complex structure on $\mathbb{C}$ or its subdomain, is

$$\omega = dz + \mu\,d\bar{z}, \qquad |\mu(z)| < 1. \tag{21.22}$$

It will be referred to as the *μ-complex structure.*

The sufficient condition for integrability of the μ-complex structure in dimension one is rather weak.

Theorem 21.17 (L. Ahlfors–L. Bers, [**AB60**]). *A μ-complex structure on the domain $\Omega \subset \mathbb{C}$ is integrable if $\mu = \mu(z)$ is an L^∞-measurable function with the norm*

$$\|\mu\|_{L^\infty(\Omega)} < 1. \tag{21.23}$$

In the most general case of measurable functions the differential of a function in Definition 21.16 should be understood in Sobolev sense. We will need only a smooth version of the Ahlfors–Bers integrability theorem.

Theorem 21.18 (A. Newlander–L. Nirenberg, [**NN57**]). *Any μ-complex structure with a C^∞-smooth function $\mu\colon \Omega \to \mathbb{C}$ satisfying the integrability condition* (21.23)*, is integrable: there exists an infinitely smooth chart $\Omega \to \mathbb{C}$ that is holomorphic in sense of this structure.*

By Definition 21.16, a nonzero smooth function $f\colon \Omega \to \mathbb{C}$ holomorphic in sense of the μ-complex structure, must have its differential proportional to $\omega = dz + \mu\, d\bar z$ and hence satisfy the partial differential equation

$$\frac{\partial f}{\partial \bar z} = \mu(z) \cdot \frac{\partial f}{\partial z}, \tag{21.24}$$

called the *Beltrami equation.* Any smooth solution f of the Beltrami equation (21.24) is a μ-holomorphic function and vice versa.

The analytic reformulation of the Newlander–Nirenberg theorem provides a sufficient condition for smooth solvability of the Beltrami equation (21.24).

Corollary 21.19. *The Beltrami equation* (21.24) *with a C^∞-smooth function μ satisfying the integrability condition $|\mu(z)| < 1$ everywhere in U, admits a C^∞-smooth solution.*

Remark 21.20. For future applications we will need the integrability conditions for almost complex structures in higher dimensions. Note that the differential of any form of type $(1,0)$ on a complex manifold is the sum of forms of the types $(2,0)$ and $(1,1)$. Denote the spaces of such forms by $L^{2,0}$ and $L^{1,1}$ respectively. Then we have the following identities:

$$dL^{1,0} \subset L^{2,0} \oplus L^{1,1} \tag{21.25}$$

$$L^{2,0} = L^{1,0} \wedge L^{1,0}, \tag{21.26}$$

$$L^{1,1} = L^{1,0} \wedge L^{0,1}, \tag{21.27}$$

$$L^{0,1} = \overline{L^{1,0}}. \tag{21.28}$$

The condition (21.23) is necessary for the integrability of an almost complex structure $L = L^{1,0}$ for $L^{2,0}$ and $L^{1,1}$ defined by (21.26)–(21.28). A sufficient condition for the integrability of finitely smooth almost complex structures is provided by the following theorem.

Theorem 21.21 (Newlander–Nirenberg theorem in the smooth category). *An almost complex structure that satisfies conditions (21.25)–(21.28) in $\mathbb{C}^2$ and is C^{2n+2}-smooth, is C^n-smoothly integrable: there exists a C^n-smooth chart $G_0 : (\mathbb{C}^2, 0) \to \mathbb{C}^2$ that is holomorphic with respect to this structure.* □

Now we turn back to the case of dimension one. We will need some simple properties of the Beltrami equation.

Proposition 21.22.

1. Let f be a solution to the Beltrami equation (21.24) and φ a holomorphic function defined on the range of f. Then $g = \varphi \circ f$ is a solution of the same Beltrami equation.

2. Let f and g be two solutions to the Beltrami equation (21.24), and $df(p) \neq 0$. Then there exists a holomorphic function φ such that $g = \phi \circ f$ near p.

Proof. The first assertion is obvious, since $dg = \psi\, df$, where ψ is the derivative of φ, and therefore dg is proportional to ω together with df.

To prove the second assumption, note that f is a local chart near p. Proportionality of df and dg means that the differential dg is $\mathbb{C}$-linear in this chart. Hence the composition $\varphi = g \circ f^{-1}$ has a complex linear differential and is holomorphic near $f(p)$. □

21F. Realization theorem for Ecalle–Voronin moduli. We will now prove the last remaining assertion of the Classification Theorem 21.15.

Theorem 21.23. *Every equivalence class $\mathcal{M} \in \mathscr{M}_{p,\lambda}$ may be realized as an Ecalle–Voronin modulus for some parabolic germ from the class $\mathscr{A}_{p,\lambda}$*

Proof. The proof follows the idea outlined at the beginning of §21**E**.

Consider a representative of the class $\mathcal{M}$, the cocycle $\mathbf{\Phi} = (\Phi_1, ..., \Phi_{2p})$ with the properties (21.18)–(21.19).

First, we will construct an abstract complex one-dimensional manifold (curve) using sectors of a nice covering as charts and the components $\Phi_1, \dots, \Phi_{2p}$ of the tuple $\mathbf{\Phi}$ as transition functions. The property (21.19) allows us to define a holomorphic map F of this curve into itself. Then we show that S is conformally equivalent to a punctured neighborhood of the origin $(\mathbb{C}, 0) \smallsetminus \{0\}$. This immediately implies that F can be holomorphically extended to the deleted point and is holomorphically equivalent to a germ $f\colon (\mathbb{C}, 0) \to (\mathbb{C}, 0)$. Finally, we verify that f is formally equivalent to the standard map $f_{p,\lambda}$ as in (21.1). The fact that the Ecalle–Voronin modulus of f (or F, what is the same) coincides with the class $\mathcal{M} \in \mathscr{M}_{p,\lambda}$ represented by the cochain $\mathbf{\Phi}$, is a tautology: it follows immediately from the construction of F.

21$\mathbf{F}_1$. *Construction of an abstract manifold with an automorphism.* Consider the disjoint union $S' = \bigsqcup_{j=1}^{2p} S_j$, where $S_j \subset \mathbb{C}$ are the sectors of a nice covering (21.3), and identify the points $z_j \in S_j$ with $z_{j+1} = \Phi_j(z_j) \in S_{j+1}$, where $z_j \colon S_j \to \mathbb{C}$ is the natural coordinate $z|_{S_j}$ inherited from $\mathbb{C}$. The quotient space is an abstract complex 1-dimensional manifold (complex curve) S which is diffeomorphic to a punctured disk.

The standard map $f_{p,\lambda}$ induces a map of S into itself. More precisely, consider somewhat smaller sectors $S'_j \subset S_j$ such that the standard map $f_{p,\lambda}$ maps S'_j into S_j, and such that their union still covers a small punctured neighborhood of the origin. Let $S' \subset S$ be the image of the disjoint union $\bigsqcup_{j=1}^{2p} S'_j$ after projection to the quotient space. Since all transition maps Φ_j used to construct the manifold S, commute with the standard map $f_{p,\lambda}$ by (21.19), the map $F\colon S' \to S$, defined in each "chart" z_j by the formula $F(z_j) = f_{p,\lambda}(z_j)$, is a well-defined map between the quotient spaces $S' \subseteq S$ and S itself. Slightly abusing the language, we will say that F is a conformal automorphism of S.

21$\mathbf{F}_2$. *Identification of the curve S.* A holomorphic curve S *diffeomorphic* to a punctured neighborhood of the origin, is not necessarily *conformally equivalent* to it: apriori, S is biholomorphically equivalent to one of the annuli $\{r < |z| < R\}$ with $0 \leqslant r < R \leqslant +\infty$. The realization theorem will be proved if we show that S is biholomorphic to a neighborhood $(\mathbb{C}, 0)$ with the deleted point 0 (which corresponds to the case $r = 0$, $R = 1$).

We construct first a C^∞-smooth (smooth, for simplicity) embedding of S into $(\mathbb{C}, 0)$. To do this, consider the covering of S by the sectors S'_j (more precisely, by their images in the quotient space by the action of $\mathbf{\Phi}$). As before, denote by $z_j \colon S'_j \to \mathbb{C}$ the local charts, and let $\{\psi_j\}_{j=0}^{2p}$ be a partition of unity subordinated to this covering: we assume that all derivatives of ψ_j grow no faster than some negative powers of $|z_j|$ as $|z_j| \to 0$ in the sectors. Define the map

$$H\colon S \to \mathbb{C} \smallsetminus \{0\}, \qquad H(z) = \sum_{j=0}^{2p} \psi_j z_j.$$

By construction, the map H is C^∞-smooth.

The inverse map $H^{-1}\colon (\mathbb{C}, 0) \smallsetminus \{0\} \to S$ is represented not by a single function, but rather by a tuple of coordinate functions $z_j \circ H^{-1}$. But since the transitions from a chart z_j to z_{j+1} are holomorphic, the Beltrami coefficient $\mu(z) = \partial_{\bar z} H^{-1}(z)/\partial_z H^{-1}(z)$ is well defined by Proposition 21.22.

We prove that this coefficient, which is a smooth function everywhere outside the origin, extends as a smooth function on the entire neighborhood $(\mathbb{C}, 0)$, flat at the origin. Indeed, since all functions z_j differ from each other by flat terms on the intersections of the consecutive sectors S'_j, the asymptotic Taylor series of H in powers of $z_j, \bar{z}_j$ coincides in fact with z_j (i.e., does not contain nonlinear terms, in particular, no powers involving $\bar{z}_j$).

Therefore all compositions $z_j \circ H^{-1}$ differ from each other by flat terms also, and by construction the asymptotic series at the origin for each of them is identity. Therefore the partial derivatives of $z_j \circ H^{-1}$ have the form

$$\frac{\partial(z_j \circ H^{-1})}{\partial z} = 1 + o(z^N), \qquad \frac{\partial(z_j \circ H^{-1})}{\partial \bar{z}} = o(z^N)$$

for any natural N.

Hence, the function $\mu(z)$ defined in the punctured neighborhood of the origin, extends smoothly at the origin as a flat function $\mu : \overline{U} \to \mathbb{C}$, where $U = H(S) \cup \{0\} \subset (\mathbb{C}, 0)$.

Consider an arbitrary solution $G \colon U \to \mathbb{C}$ of the Beltrami equation with the same Beltrami coefficient μ, normalized by the condition $G(0) = 0$: its existence is guaranteed by Corollary 21.19 from the Newlander–Nirenberg integrability theorem. By the second assertion of Proposition 21.22, the composition $h = G \circ H : S \to (\mathbb{C}, 0)$ is a holomorphic map between the two holomorphic curves, S and $(\mathbb{C}, 0) \smallsetminus \{0\}$. Moreover, it is a diffeomorphism, hence a biholomorphic equivalence. This completes identification of the surface S: it is biholomorphically equivalent to a punctured neighborhood of the origin.

The "abstract" map F is by construction biholomorphically equivalent to the self-map $f = h \circ F \circ h^{-1}$ of a punctured neighborhood of the origin. By the removable singularity theorem, the map f holomorphically extends to the origin. As a result, we conclude that after one-point completion of the curve S, the automorphism F is locally holomorphically equivalent to a holomorphic germ $f \in \operatorname{Diff}(\mathbb{C}, 0)$.

21**F**$_3$. *Formal and analytic type of the germ f.* All functions $z_j \circ H^{-1}$ differ from identity by flat functions. Besides, the map G is formally holomorphic (its Taylor series $\widehat{G}$ does not contain powers of $\bar{z}$), since G is a solution of the Beltrami equation with the flat function μ. The map h^{-1} conjugates f with $f_{p,\lambda}$ in sectors that contain no images of the intersections of sectors S'_j. Hence, the formal series $\widehat{h}^{-1}$ conjugates formal series for f with that for $f_{p,\lambda}$. This proves that f is formally equivalent to $f_{p,\lambda}$.

The maps $H_j = z_j \circ h^{-1}$, defined in the images $h(S_j')$ of the sectors S_j', form a normalizing cochain for f, as they conjugate f with $f_{p,\lambda}$ in these sectors. The proof of the Realization Theorem 21.23 is complete. $\square$

Corollary 21.24. *Any cocycle* $\mathbf{\Phi}$ *satisfying the conditions* (21.18)–(21.19), *is a coboundary of a normalizing cochain,* $\mathbf{\Phi} = \delta\mathcal{H} \in \widetilde{\operatorname{Diff}}_1^p(\mathbb{C}, 0)$. $\square$

21G. Fourier representation for the Ecalle–Voronin moduli. The sectorial gauge automorphisms $F_{j,j+1}(t) = H_{j+1}(t) \cdot H^{-1}(t)$ of of the diagonal normal form for an irregular linear singularity, as they are introduced in Definition 20.19, are in fact coboundaries of the corresponding sectorial normalizing cochains. These automorphisms are conveniently represent by constant Stokes matrices (20.18).

In the same way the coboundary of a normalizing cochain admits a concise description in the chart that rectifies the vector field $F_{p,\lambda}$. In general, this chart is not univalent. We discuss in detail the particular case $(p, \lambda) = (1, 0)$, where the rectifying chart has a simple form

$$t = t_{1,0}(z) = 1/z. \tag{21.29}$$

The general case will be treated later.

As explained in the proof of Proposition 21.8, the cocycle $\mathbf{\Phi} = \delta\mathcal{H}$ in the rectifying chart t consists of functions commuting with the standard shift $t \mapsto t+1$, that is, the differences $\varphi_j(t) = \Phi_j(t) - t$ are 1-periodic exponentially flat at infinity and defined initially in two sectors $S^{\pm}$.

Such functions can be expanded in the converging Fourier series, without the free terms, *converging* in the respective upper and lower half-planes $\operatorname{Im} t \gg 1$ and $\operatorname{Im} t \ll -1$:

$$\varphi_1(t) = \sum_{k>0} c_k e^{2\pi i k t}, \qquad \varphi_2(t) = \sum_{k<0} c_k e^{2\pi i k t}. \tag{21.30}$$

Replacing the cocycle $\mathbf{\Phi}$ by another cocycle equivalent to it in the sense (21.20), results in the argument shift of the functions φ_1, φ_2. More precisely, two tuples (φ_1, φ_2) and (φ_1', φ_2') correspond to equivalent cocycles, if there exists the shift $t \mapsto t + s$, $s \in \mathbb{C}$, which simultaneously conjugates the maps $\operatorname{id} + \varphi_j$ with $\operatorname{id} + \varphi_j'$ for $j = 1, 2$. This happens if and only if the respective Fourier coefficients $\{c_k\}$ and $\{c_k'\}$ from (21.30), are related by the identities

$$c_k = c_k' e^{2\pi i k s}, \qquad \forall k \in \mathbb{Z},\ k \neq 0. \tag{21.31}$$

The Analytic classification Theorem 21.15 in the Fourier representation implies the following corollary. Denote by $\mathscr{M}_{1,0}^{F\circ}$ the linear space of pairs of series of the form (21.13), converging respectively in the upper and lower half-plane $\operatorname{Im} t > C$, (resp., $\operatorname{Im} t < -C$), for a constant C depending on the series. Each pair from the space $\mathcal{M}_{1,0}^{F}$ can be aggregated into a bi-infinite string of the complex Fourier coefficients $\{c_k\}_{-\infty}^{+\infty}$, with $c_0 = 0$, and conversely, any bi-infinite string corresponding to a pair of *converging* series, represents an element from $\mathcal{M}_{1,0}^{F\circ}$. Denote the quotient space of $\mathcal{M}_{1,0}^{F\circ}$ by the equivalence relationship (21.31) by $\mathcal{M}_{1,0}^{F}$.

Corollary 21.25. *The modulus $\mathcal{M}_{1,0}$ of analytic equivalence of parabolic germs from the class $\mathcal{A}_{1,0}$ can be identified with the quotient space $\mathcal{M}_{1,0}^{F}$.* □

A similar description for arbitrary (p, λ) looks as follows: in the rectifying chart $t = t_{p,\lambda}(z)$ given by (21.4), the components Φ_j of the cocycle $\delta\mathcal{H}$ can be shown to take the form of converging Fourier series

$$\begin{aligned} \Phi_j(t) &= t + \sum_{\pm k=1}^{+\infty} c_{j,k} e^{2\pi i k t}, \qquad j = 1, \dots, 2p-1, \\ \Phi_{2p}(t) &= t + 2\pi i\lambda + \sum_{k=1}^{+\infty} c_{2p,k} e^{2\pi i k t}. \end{aligned} \tag{21.32}$$

The sign depends on the parity of j (plus for even j, minus for j odd), as well as the domains of convergence (upper or lower half-planes). On the collection $\mathcal{M}_{p,\lambda}^{F\circ}$ of all Fourier coefficients one has to introduce an equivalence relation similar to (21.31), and the corresponding quotient space $\mathcal{M}_{p,\lambda}^{F}$ could be identified with the space of the Ecalle–Voronin moduli $\mathcal{M}_{p,\lambda}$ for all parabolic germs from the class $\mathcal{A}_{p,\lambda}$.

21H. Directional derivative of the Ecalle–Voronin modulus. Like the Stokes operators constructed in §20**G**, the Ecalle–Voronin modulus cannot be computed in terms of any finite order jet of a parabolic germ. Indeed, any jet of parabolic germ of order greater than $2p+1$ admits two holomorphically nonequivalent representatives, yet all these representatives are formally equivalent. Thus the Ecalle–Voronin modulus depends on the entire "tail" of the Taylor series of a parabolic germ. In this section we will explicitly compute the first variation of the correspondence

$$\mathcal{A}_{p,\lambda} \to \mathcal{M}_{p,\lambda}^{F}, \quad f \mapsto \mathcal{M}_f, \qquad (p, \lambda) = (1, 0),$$

at the "point" corresponding to the standard (embeddable) formal normal form $f_{p,\lambda} = \exp F_{p,\lambda}$. The result will be given in terms of the Borel transform of the germ f; the germ of this transform depends on the entire Taylor series of f.

To simplify the exposition, we compute this first variation (the Gateaux derivative) only in the case $(p, \lambda) = (1, 0)$, where the rectifying map is single-valued. Consider an analytic family of parabolic germs $\widetilde{f}_\varepsilon(t)$, which from the very beginning is written in the rectifying chart t,

$$\widetilde{f}_\varepsilon(t) = t + 1 + \varepsilon R(t), \qquad R(t) = \sum_{k=0}^{\infty} a_k t^{-(k+1)}. \tag{21.33}$$

The Ecalle–Voronin modulus $\mathcal{M}(\varepsilon)$ of $\widetilde{f}_\varepsilon$ depends analytically on ε by the last assertion of the Analytic Classification Theorem 21.15. Consider the corresponding Fourier representation of this modulus, a pair of converging Fourier series

$$\varphi_j(t,\varepsilon) = \sum_{\pm k=1}^{\infty} c_k(\varepsilon)\, e^{2\pi i k t},$$

(the sign plus corresponds to φ_1, minus to φ_2); see (21.13). Since $\widetilde{f} = \widetilde{f}_0$ coincides with its formal normal form $f_{1,0} = \exp F_{1,0}$, by definition we have $\mathcal{M}(0) = 0$, and therefore

$$\mathcal{M}(\varepsilon) = \varepsilon\mathcal{M}_1 + O(\varepsilon^2), \qquad \mathcal{M}_1 = \frac{\partial\mathcal{M}(\varepsilon)}{\partial\varepsilon}\bigg|_{\varepsilon=0} \cong (\psi_1(t),\psi_2(t)) \in \mathscr{M}^{F\circ}_{1,0},$$
$$\psi_j = \sum_{\pm k=1}^{\infty} b_k e^{2\pi i k t}, \qquad j = 1,2. \tag{21.34}$$

The Fourier coefficients $b_k \in \mathbb{C}$, $k \in \mathbb{Z}$, $k \neq 0$, of the pair (ψ_1,ψ_2) are the derivatives at $\varepsilon = 0$ of the coefficients $c_k(\varepsilon)$. These derivatives can be explicitly computed from the Taylor coefficients of the series R in terms of the *Borel transform*.

Let $a(t) = \sum_{k=0}^{\infty} a_k\, t^{-(k+1)}$ be a converging Laurent series holomorphic in some neighborhood of $t = \infty$. Starting from this series, one can produce two functions of a new variable ζ, both analytic at $\zeta = 0$, as follows:

$$A_1(\zeta) = \sum_{k=0}^{\infty} \frac{a_k}{k!}\,\zeta^k, \qquad A_2(\zeta) = -\frac{1}{2\pi i}\oint_{\Gamma} a(t) e^{t\zeta}\,dt, \tag{21.35}$$

where Γ is a sufficiently large circle centered at the origin.

Proposition 21.26. *The germs of two functions $A_1(z)$ and $A_2(z)$ at the origin coincide.*

Proof. Consider the Laurent series for the function $a(t)\,e^{\zeta t}$ at $t = \infty$: the coefficient before t^{-1} (the residue of the 1-form $a(t)e^{\zeta t}dt$) is obtained by the termwise multiplication of the convergent Laurent series for $a(t) = \sum_{k=0}^{\infty} a_k/t^{k+1}$ and for $e^{\zeta t} = \sum_{k=0}^{\infty} \zeta^k t^k/k!$ respectively. One can instantly see that it is equal to $A_1(\zeta)$. The integral Cauchy formula gives the contour integral representation for the same residue. □

Definition 21.27. The Borel transform of a converging Laurent series $a(t) = \sum_{k=0}^{\infty} a_k\, t^{-(k+1)}$ defined near infinity, is the germ $\mathcal{B}a(\zeta)$ defined by any of the two equivalent representations (21.35).

Consider the analytic family of parabolic germs (21.33) from the class $\mathscr{A}_{1,0}$ and denote by $\mathcal{M}(\varepsilon) \in \mathscr{M}^F_{1,0}$ its Ecalle–Voronin modulus in the Fourier representation.

Theorem 21.28 (Tangential Ecalle–Voronin modulus). *The Gateaux derivative (21.34) of the Ecalle–Voronin modulus $\mathcal{M}(\varepsilon) \in \mathscr{M}^F_{1,0}$ has the Fourier coefficients*

$$b_k = -2\pi i(\mathcal{B}R)(-2\pi ik), \qquad k \in \mathbb{Z},\ k \neq 0. \tag{21.36}$$

Proof. We start with the explicit formula (21.10) for the normalizing cochain, as found in the proof of the Sectorial Normalization Theorem 21.5.

$$\widetilde{H}_1(t,\varepsilon) = t + \varepsilon \sum_{n=0}^{+\infty} R \circ \widetilde{f}_\varepsilon^{\circ n}(t), \qquad \widetilde{H}_2(t,\varepsilon) = t - \varepsilon \sum_{n=-1}^{-\infty} R \circ \widetilde{f}_\varepsilon^{\circ n}(t).$$

Computing the first variation of these functions in ε at $\varepsilon = 0$, when $\widetilde{f}^{\circ n}(t)$ becomes $t + n$, we conclude that

$$\left.\frac{\partial \widetilde{H}_1}{\partial \varepsilon}\right|_{\varepsilon=0} = \sum_{n=0}^{+\infty} R(t+n), \qquad \left.\frac{\partial \widetilde{H}_1}{\partial \varepsilon}\right|_{\varepsilon=0} = -\sum_{n=-1}^{-\infty} R(t+n),$$

the derivatives being well defined and holomorphic in S^0 and S^1 respectively. From these formulas we have for the transition functions $\widetilde{\Phi}_1 = \widetilde{H}_2 \circ \widetilde{H}_1^{-1}$ in S^+ and $\widetilde{\Phi}_2 = \widetilde{H}_1 \circ \widetilde{H}_2^{-2}$ in S^- respectively the formulas

$$\widetilde{\Phi}_1(t,\varepsilon) = t - \varepsilon \sum_{n=-\infty}^{+\infty} R(t+n) + O(\varepsilon^2), \quad \widetilde{\Phi}_2(t,\varepsilon) = t + \varepsilon \sum_{n=-\infty}^{+\infty} R(t+n) + O(\varepsilon^2),$$

and finally

$$\psi_1 = -\sum_{-\infty}^{+\infty} R(t+n), \qquad t \in S^+,$$

$$\psi_2 = \sum_{-\infty}^{+\infty} R(t+n), \qquad t \in S^-.$$

The assertion of the theorem now follows from a purely analytic statement expressing the above sums in terms of the Borel transform of $R(t)$.

Lemma 21.29. *Let $R(t)$ be a function holomorphic at infinity and having zero residue there, and $\psi(t) = \sum_{n=-\infty}^{+\infty} R(t+n)$. Then the kth Fourier coefficient b_k of $\psi(t) = \sum b_k\, e^{2\pi ikt}$ is $-2\pi i(\mathcal{B}R)(-2\pi ik)$.*

Proof. If $k > 0$, then for some large $\beta > 0$ we have

$$b_k = \int_{i\beta}^{i\beta+1} \psi(t)\, e^{-2\pi i k t}\, dt = \int_{i\beta-\infty}^{i\beta+\infty} R(t)\, e^{-2\pi i k t}\, dt$$
$$= \oint_\Gamma R(t)\, e^{-2\pi i k t}\, dt = -2\pi i (\mathcal{B}R)(-2\pi i k t).$$

The proof for $k < 0$ is completely analogous: one should take $\beta < 0$ with a sufficiently large absolute value. □

This computation completes the proof of Theorem 21.28. □

Using the "linear approximation" of the Ecalle–Voronin modulus, one can almost explicitly construct examples of formally equivalent but analytically nonequivalent parabolic germs.

Corollary 21.30. *Consider two analytic families of parabolic maps in the* t*-chart,* $\widetilde{f}_j(t, \varepsilon) = t + 1 + \varepsilon R_j(t)$, $j = 1, 2$ *with* R_j *being polynomials in* t^{-1} *of different degrees. Then for all* $\varepsilon \in \mathbb{C}$ *with the eventual exception of a discrete set,* $\widetilde{f}_{1,\varepsilon}$ *is not analytically equivalent to* $\widetilde{f}_{2,\varepsilon}$. *In particular,* $\widetilde{f}_{1,\varepsilon}$ *is not equivalent to* $\widetilde{f}_{2,\varepsilon}$ *for all sufficiently small values of* $\varepsilon \neq 0$.

Proof. In the opposite case the two *analytic* families should consist of analytically equivalent germs for *all* values of $\varepsilon \in \mathbb{C}$, hence the tangents of the corresponding derivatives should be equivalent in the sense that their Fourier coefficients must satisfy (21.13).

But the Borel transforms of the two polynomials of different degrees in t^{-1} are two polynomials of different degrees in ζ. This contradicts the assumption that they differ by a geometric progression, as should have been under the condition (21.13). □

21I. Applications: embedding, root extraction and computation of centralizer. The Ecalle–Voronin modulus constitutes a convenient tool for the solution of the problems listed at the beginning of this section.

21**I**$_1$. *Embedding in a flow.* When a parabolic germ may be embedded into a flow, i.e., be represented as the flow map of an analytic field? The complete answer is given by the following result.

Theorem 21.31. *A parabolic germ is embeddable if and only if its Ecalle–Voronin modulus is trivial, i.e., the coboundary of any normalizing cochain is identity,* $\delta\mathcal{H} = \{\mathrm{id}\}$.

Proof. The standard germ $f_{p,\lambda}$ is embeddable and triviality of the modulus $\mathcal{M}_f$ means analytic equivalence of f to the embeddable germ, i.e., embeddability. Conversely, any two embeddable germs from the same formal equivalence class are analytically equivalent to each other, since all holomorphic vector fields of the form $z^{p+1} + \lambda z^{2p+1} + \cdots$ on $(\mathbb{C}, 0)$ are holomorphically equivalent to each other (Theorem 5.25). □

21$\mathbf{I}_2$. *Centralizer and root.* The *centralizer* of a germ f is the (subgroup) $Z_f \subset \operatorname{Diff}(\mathbb{C}, 0)$ of all germs of conformal maps that commute with f. In general, the centralizer contains nonparabolic germs; see §6$\mathbf{B}_3$ below. We will refer by the name *parabolic centralizer* to the intersection $Z_f \cap \operatorname{Diff}_1(\mathbb{C}, 0)$, the collection of all *parabolic* germs in Z_f.

Obviously, the germ itself together with all its iterates $\{f^{\circ\mathbb{Z}}\}$ (both positive and negative), belongs to its parabolic centralizer. Moreover, if the equation

$$g^{\circ q} = f, \qquad g \in \mathscr{A}_{p,\lambda}, \tag{21.37}$$

admits a solution in the group $\operatorname{Diff}_1(\mathbb{C}, 0)$, then we say that g is a *root* of order $q \in \mathbb{N}$. The root is *maximal*, if $q > 1$ is the largest natural number for which the solution still exists. Note that the maximal root may not always exist, but if it exists, the entire group of *fractional iterates* $\{f^{\circ q^{-1}\mathbb{Z}}\} = \{g^{\circ\mathbb{Z}}\}$ also belongs to the parabolic centralizer of f.

It appears that the parabolic centralizer of f in fact *coincides* with the group of fractional iterates of f except for the case where the germ f is embeddable: in this case there is obviously no maximal root.

Theorem 21.32.

1. *For any nonembeddable parabolic germ its parabolic centralizer consists of its fractional iterates.*

2. *For all parabolic germs except for a set of infinite codimension, the maximal root is of order* 1, *i.e., the equation* (21.37) *has no parabolic solutions other than* $q = 1$, $g = f$.

3. *For an embeddable parabolic germ* $f = \exp F$, $F \in \mathcal{D}(\mathbb{C}, 0)$, *its parabolic centralizer consists of all flow maps* $\{f^{\circ\mathbb{C}}\} = \{\exp sF : s \in \mathbb{C}\}$.

Proof. On the formal level the structure of the (parabolic) centralizer of a parabolic germ is completely described in §6$\mathbf{B}_2$. In particular, the parabolic centralizer of a parabolic pre-normalized germ (21.2) also has the same form (21.2) with the same p.

Consider two parabolic commuting germs f and f', and let $\mathcal{H} = \{H_1, \dots, H_{2p}\}$ be a normalizing cochain which conjugates f with the formal normal form $f_{p,\lambda} = \exp F_{p,\lambda}$ in $\widetilde{\operatorname{Diff}}{}^p_1(\mathbb{C}, 0)$. Then $\mathcal{G} = \mathcal{H} \circ f'$ is another normalizing cochain for f. Indeed,

$$f_{p,\lambda} \circ \mathcal{G} = f_{p,\lambda} \circ \mathcal{H} \circ f' = \mathcal{H} \circ f \circ f' = \mathcal{H} \circ f' \circ f = \mathcal{G} \circ f.$$

By the uniqueness assertion of Theorem 21.12, the two normalizing cochains differ by a flow map of the vector field $F_{p,\lambda}$: there exists $g = \exp sF_{p,\lambda}$, $s \in \mathbb{C}$, such that $\mathcal{G} = g \circ \mathcal{H}$. By construction of $\mathcal{H}'$, this means that

$$\mathcal{H} \circ f' = \mathcal{G} = g \circ \mathcal{H},$$

that is, $\mathcal{H}$ also conjugates the second germ f' with a flow map $g = \exp sF_{p,\lambda}$. Therefore the compositional coboundary $\boldsymbol{\Phi} = \delta\mathcal{H}$ is an automorphism of the flow map g:

$$\Phi_j \circ g = g \circ \Phi_j, \qquad g = \exp sF_{p,\lambda}, \ s \in \mathbb{C}. \tag{21.38}$$

In the rectifying chart t the respective components $\Phi_j(t)$ commute with the shift $t \mapsto t+s$ *in addition* to commutation with the standard shift $t \mapsto t+1$. This means that the differences $\varphi_j(t) = \Phi_j(t) - t$ are holomorphic *double periodic* functions of the complex argument t. There are two possibilities.

Embeddable case. If the lattice $\mathbb{Z} + s\mathbb{Z} \subset \mathbb{C}$ has rank 2, then the only possibility for φ_j to be simultaneously holomorphic and "truly" double-periodic is to be constant. This means that $\boldsymbol{\Phi}$ is equivalent to the trivial cochain and the germ f is in fact analytically equivalent to an embeddable germ. The same is true if $s \in \mathbb{R} \smallsetminus \mathbb{Q}$: then the closure $\mathbb{Z} + s\mathbb{Z}$ is the line $\mathbb{R}$, and by the uniqueness theorem $\varphi_j = \text{const}$.

Nonembeddable case. If the germ f is nonembeddable, then φ_j should have a minimal period which divides simultaneously both 1 and s: this means that it should be of the form $1/q$ with $q \in \mathbb{N}$ and $s = r/q$ with $r \in \mathbb{Z}$. A $1/q$-periodic function φ_j must have all Fourier coefficients c_j vanishing unless q divides j. If $q > 1$, this would mean an infinite number of independent conditions imposed on φ_j, i.e., ultimately, on the Ecalle–Voronin modulus $\mathcal{M}_f$.

It remains to notice that if φ_j has period $1/q$, $q \in \mathbb{N}$, then the components Φ_j commute with the flow map $\exp {}^1\!/_q F_{p,\lambda}$ which is a root of order q from the normal form $f_{p,\lambda}$. This commutativity implies that the composition $h = \mathcal{H}^{-1} \circ \exp({}^1\!/_q F_{p,\lambda}) \circ \mathcal{H}$ has trivial coboundary and is, therefore, a holomorphic germ, $h \in \operatorname{Diff}(\mathbb{C}, 0)$. By construction, both f and f' are iterates of h: $f = h^{\circ q}$, $f' = h^{\circ r}$. Hence the parabolic centralizer consists of fractional iterates $f^{\circ\, r/q}$. $\square$

The proof of the above theorem gives in fact an explicit criterion of existence of the root of order $q > 1$ of a parabolic germ.

Corollary 21.33. *A parabolic germ $f \in \mathscr{A}_{p,\lambda}$ admits extraction of a root of order $q \in \mathbb{N}$, if and only if all components $\varphi_j(t) = \Phi_j(t) - t$ are $1/q$-periodic,* *or, equivalently, when the coboundary* $\mathbf{\Phi} = \delta\mathcal{H}$ *commutes with the flow map* $\exp(1/qF_{p,\lambda})$. *The root is given by the formula* $h = \mathcal{H}^{-1} \circ \exp(1/qF_{p,\lambda}) \circ \mathcal{H}$ *which is well defined under these assumptions.* □

In the Fourier representation, the corollary implies that a parabolic root of order $q > 1$ can be extracted from a parabolic germ f if and only if all Fourier coefficients of the modulus $\mathcal{M}_f \in \mathscr{M}^F_{p,\lambda}$ with numbers not divisible by q, vanish. Clearly, this condition is of codimension infinity for any $q > 1$.

21J. Resonant germs. Holomorphic classification of resonant germs with multiplicator $\exp 2\pi i\lambda$, $\lambda \in \mathbb{Q}$ a root of unity, can be reduced to the analytic classification of parabolic germs after passing to iterates. The classification can be regarded as an equivariant version of the Ecalle–Voronin modulus.

$21\mathbf{J}_1$. *Formal normal forms.* Let f be a resonant germ with the multiplier $\alpha = \alpha_{m,n} = e^{2\pi i m/n}$, with m and n mutually prime. Then its iterate $g = f^{\circ n}$ is a parabolic germ. The formal normal form for g is the series $\widehat{g} = \exp F_{p,\lambda}$ for some natural p and complex λ. We claim that p is necessarily divisible by n. Indeed, the resonant normal form for f involves only terms of powers divisible by n, $\widehat{f} = \alpha z(1 + \sum_1^\infty a_k z^{nk})$ by Theorem 4.21. The iterate also has the same structure, hence he number p, which determines the degree of the first nonlinear term in the series above, is divisible by n, i.e., $p = nk$ for some k.

Denote by $\mathscr{A}_{m,n,k,\lambda}$ the set of all resonant germs f with the multiplier $\alpha_{m,n} = e^{2\pi i m/n}$ such that $f^{\circ n} \in \mathscr{A}_{p,\lambda}$ with $p = kn$. Fix m, n, k, λ and consider the map

$$f^* = \alpha_{m,n} \cdot f_{p,\lambda} = e^{2\pi i m/n} \exp F_{kn,\lambda} \in \mathscr{A}_{m,n,k,\lambda}. \tag{21.39}$$

Note that multiplication by $\alpha_{m,n}$ commutes with the normal form $f_{p,\lambda}$, hence with f^*. Note also that all three commute with the flow map $\exp sF_{p,\lambda} = \exp sF_{p,\lambda}$ for any $s \in \mathbb{C}$.

$21\mathbf{J}_2$. *Normalizing cochain for* $f^{\circ n}$. We will show that f^* is the formal normal form for f: the proof will be derived as a consequence of a more important fact. Let $\mathcal{H}$ be an arbitrary normalizing cochain for the parabolic germ $g = f^{\circ n} \in \mathscr{A}_{kn,\lambda}$.

Lemma 21.34. *The cochain* $\mathcal{H}$ *conjugates the resonant germ* f *with the germ* f^*.

Proof. Consider the cochain $\mathcal{G} = \alpha^{-1}\mathcal{H} \circ f \in \widetilde{\mathrm{Diff}}{}_1^p(\mathbb{C}, 0)$, where $\alpha = \alpha_{m,n}$ and $p = nk$ are as above. It is another normalizing cochain for $g = f^{\circ n}$. Indeed, it is a cochain inscribed into the same nice cover, and its asymptotic series $\alpha^{-1} \circ \widehat{H} \circ \widehat{f}$ (here $\widehat{H}$ is a formal Taylor series for $\mathcal{H}$) has the identical linear term. Finally, $\mathcal{G}$ conjugates g with $f_{p,\lambda}$ in each sector. Indeed, f commutes with g, and the linear map α^{-1} commutes with $f_{p,\lambda}$. Hence,

$$\begin{aligned}\mathcal{G}\circ g\circ\mathcal{G}^{-1}&=\alpha^{-1}\circ\mathcal{H}\circ f\circ g\circ f^{-1}\circ\mathcal{H}^{-1}\circ\alpha\\&=\alpha^{-1}\circ\mathcal{H}\circ g\circ\mathcal{H}^{-1}\circ\alpha=\alpha^{-1}\circ f_{p,\lambda}\circ\alpha=f_{p,\lambda}.\end{aligned}$$

Therefore, $\mathcal{G}$ is another normalizing cochain for g. By the uniqueness of the normalizing cochain for parabolic germs, $\mathcal{G}=(\exp sF_{p,\lambda})\circ\mathcal{H}$ for some $s\in\mathbb{C}$. Let us prove that $s=1/n$.

Two previous equalities for $\mathcal{G}$ imply that

$$\alpha^{-1}\circ\mathcal{H}\circ f=(\exp sF_{p,\lambda})\circ\mathcal{H},\quad\text{hence}\quad\mathcal{H}\circ f^{\circ n}=(\exp nsF_{p,\lambda})\circ\mathcal{H},$$

since $\alpha^{-n}=1$. On the other hand, by definition $\mathcal{H}$ is known to conjugate $f^{\circ n}$ with $f_{p,\lambda}=\exp F_{p,\lambda}$. Therefore $sn=1$ and finally $\mathcal{H}\circ f\circ\mathcal{H}^{-1}=\alpha\circ(\exp{}^1\!/\!_n\,F_{p,\lambda})=f^*$, as was asserted. □

21J$_3$. *Functional moduli for resonant germs.* As in the case of parabolic germs, normalizing cochains for $f^{\circ n}$ form an equivalence class with the equivalence relation (21.20). Coboundaries of these cochains form an equivalence class with respect to relation (21.31) imposed on their Fourier coefficients. This class is the Ecalle–Voronin modulus of the germ $g=f^{\circ n}$. It appears that the same class is the functional modulus of f for the analytic classification of germs of the class $\mathscr{A}_{m,n,k,\lambda}$ (the class of formal equivalence of $m:n$-resonant germs whose nth iterate is in the formal class $\mathscr{A}_{nk,\lambda}$).

Yet not all coboundaries of normalizing cochains from the space $\mathscr{M}^{\circ}_{nk,\lambda}$ appear as moduli of analytic classification for the resonant germs from class $\mathscr{A}_{m,n,k,\lambda}$. To be a modulus of a germ of this class, the coboundary must satisfy additional very stringent restrictions.

Lemma 21.35. *If* $\mathbf{\Phi}$ *is the coboundary of a normalizing cochain* $\mathcal{H}$ *for the parabolic germ* $g=f^{\circ n}$, *and* f^* *is the formal normal form* (21.39), *then* $\mathbf{\Phi}$ *and* f^* *commute:*

$$f^*\circ\mathbf{\Phi}=\mathbf{\Phi}\circ f^*.\tag{21.40}$$

Proof. The proof is standard: the components of the cochain $\mathcal{H}$ conjugate f with f^*. Hence the component $\Phi_j=H_{j+1}\circ H_j^{-1}$ conjugates f^* with itself in the appropriate sectors. □

21J$_4$. *Analytic classification of resonant germs.* Holomorphic invariants of parabolic iterates are obviously holomorphic invariants of the resonant "roots". To construct a complete classification, it is necessary to verify that nonequivalent resonant germs cannot produce equivalent parabolic iterates. Besides, one has to describe all constraints imposed on the Ecalle–Voronin moduli from $\mathscr{M}_{p,\lambda}$ by the fact that the parabolic germs are iterates of resonant germs. Lemma 21.35 specifies one such constraint. The central result of this section claims that besides (21.40), there are no other constraints.

Theorem 21.36. *For every resonant germ $f \in \mathscr{A}_{m,n,k,\lambda}$ the Ecalle–Voronin modulus $\mathcal{M}_g = \mathcal{M}_{f^{\circ n}}$ of its iterate $g = f^{\circ n}$, a cochain $\mathbf{\Phi} = (\Phi_1, \dots, \Phi_{2p}) \in \mathscr{M}_{p,\lambda}$, $p = nk$, defined uniquely modulo the equivalence relationship* (21.20), *satisfies the following properties.*

1. (Invariance) *If two germs from the class $\mathscr{A}_{m,n,k,\lambda}$ are analytically equivalent, then their moduli coincide.*

2. (Equimodality and equivalence) *Conversely, two germs from $\mathscr{A}_{m,n,k,\lambda}$ with the same moduli, are analytically equivalent.*

3. (Realization) *Any tuple $\mathbf{\Phi} \in \mathscr{M}^\circ_{kn,\lambda}$ satisfying* (21.40) *may be realized as a modulus for some germ $f \in \mathscr{A}_{m,n,k,\lambda}$.*

4. (Analytic dependence on parameters) *If a family of germs $f_\varepsilon \in \mathscr{A}_{m,n,k,\lambda}$ depends analytically on a parameter ε, then the modulus of analytic equivalence also depends analytically on ε.*

Proof. This theorem can be derived from Theorem 21.15 which gives analytic classification of parabolic germs, by straightforward arguments.

1. If f and g are analytically equivalent, then so are $f^{\circ n}$ and $g^{\circ n}$. Statement 1 of Theorem 21.15 completes the proof of invariance.

2. If two coboundaries $\mathbf{\Phi}, \mathbf{\Psi} \in \mathscr{M}^\circ_{p,\lambda}$ are equivalent, then the respective normalizing cochains $\mathcal{H}$ for $f^{\circ n}$ and $\mathcal{G}$ for $g^{\circ n}$ differ by a flow map of the vector field $F_{p,\lambda}$, and after replacing G by another map-cochain $\mathcal{G}' = (\exp sF_{p,\lambda}) \circ \mathcal{G}$ they will have coinciding coboundaries and the composition $\mathcal{G}' \circ \mathcal{H}^{-1}$ has trivial coboundary and hence can be identified with a holomorphic self-map $h \in \operatorname{Diff}(\mathbb{C}, 0)$.

The cochain $\mathcal{H}$ conjugates f with f^*, the cochain $\mathcal{G}'$ conjugates g with f^*. Therefore the holomorphic map $h = \mathcal{G}' \circ \mathcal{H}^{-1}$ conjugates f and g.

3. Any cocycle $\mathbf{\Phi} \in \mathscr{M}^\circ_{p,\lambda}$ representing an arbitrary Ecalle–Voronin modulus $\mathcal{M} \in \mathscr{M}_{p,\lambda}$, can be realized as the coboundary $\delta\mathcal{H}$ of a cochain $\mathcal{H}$ normalizing the parabolic germ $g \in \mathscr{A}_{p,\lambda}$ (Corollary 21.24). Let f be the cochain *defined* by the composition

$$f = \mathcal{H}^{-1} \circ f^* \circ \mathcal{H}^{-1}, \qquad \text{i.e.,} \qquad f|_{S_j} = H^{-1}_{j+km} \circ f^* \circ H_j, \tag{21.41}$$

where the enumeration is cyclic modulo $2p$ as usual, $p = nk$. Apriori f is only a map-*cochain*, but the assumption (21.40) implies that in fact it is a well-defined conformal germ with the resonant multiplicator $\alpha = \alpha_{m,n}$. Indeed, componentwise the identity (21.40) has the form $\Phi_{j+km} = f^* \circ \Phi_j \circ (f^*)^{-1}$. On the intersection $S_j \cap S_{j+1}$ of two different sectors two expressions for f

coincide:

$$f|_{S_{j+1}} = H_{j+km+1}^{-1} \circ f^* \circ H_{j+1} = H_{j+km}^{-1} \circ \Phi_{j+km}^{-1} \circ f^* \circ \Phi_j \circ H_j$$
$$= H_{j+km}^{-1} \circ f^* \circ H_j = f|_{S_j}.$$

By construction, $\mathbf{\Phi}$ represents the Ecalle–Voronin modulus for $f^{\circ n}$, as required.

4. Analytic dependence on parameters follows immediately from the corresponding assertion of Theorem 21.15. □

From this theorem one can derive explicitly the description of Ecalle–Voronin moduli for resonant germs from the formal class $\mathscr{A}_{m,n,k,\lambda}$: the modulus consists of holomorphic cochains $\mathbf{\Phi} = (\Phi_1, \dots, \Phi_{2k})$ commuting with f^* and the linear map α simultaneously.

Exercises and Problems for §21.

Problem 21.1. Prove that sums and products of sectorial cochains of the same type are sectorial cochains again (the operations here are taken componentwise).

Problem 21.2. Prove that *germs* of functional map-cochains as they are introduced in Remark 21.11, indeed form the group denoted by $\widetilde{\operatorname{Diff}}_1^p(\mathbb{C}, 0)$.

Problem 21.3. Let $\mathcal{H} \in \widetilde{\operatorname{Diff}}_1^p(\mathbb{C}, 0)$ be a *normalizing* map-cochain (i.e., $\mathcal{H}$ conjugates two holomorphic germs from $\operatorname{Diff}_1(\mathbb{C}, 0)$). Assume that one of the sectorial components of $\mathcal{H}$ is identity. Prove that $\mathcal{H}$ is identity itself.

Hint. Two germs conjugated by $\mathcal{H}$, must coincide.

In the Problems 21.4–21.9 it is required to describe topologically the *space of orbits* U/f of a cyclic pseudogroup of holomorphic maps $\{f^{\mathbb{Z}}\}$, defined on a domain U. Two points $x, y \in U$ are called f-related, if $f(x) = y$, and this partial relation is maximally extended by symmetry and transitivity. The quotient space may be very pathological, for instance, non-Hausdorff, or even not a topological space (e.g., the quotient of the unit disk $\mathbb{D}$ by an irrational rotation $z \mapsto e^{2\pi i\lambda}z$, $\lambda \notin \mathbb{Q}$).

Problem 21.4. Prove that the space of orbits $\mathbb{C}/f$ of the shift $f(t) = t + 1$ on the complex line $\mathbb{C}$ is a holomorphic curve equivalent to the punctured plane $\mathbb{C}^* = \mathbb{C} \smallsetminus \{0\}$.

Describe the spaces of orbits $\mathbb{C}_+/f$ and $i\mathbb{C}_+/f$, where $\mathbb{C}_+ = \{\operatorname{Re} t > 0\}$ (the map f remains the same).

Problem 21.5. Describe the quotient spaces of $\mathbb{C} \smallsetminus 100\mathbb{D} = \{|t| > 100\}$ by the shift. Describe the quotient of the unit disk $\mathbb{D}$ by the flow map $f = \exp F_{1,0}$; cf. (21.1).

Problem 21.6. Describe the space of orbits $(\mathbb{C}_+, 0)/f$, where $f(z) = z + z^2 + \cdots$ is a parabolic germ. Can one replace $\mathbb{C}_+$ by the sector $\{|\operatorname{Arg} z| < \frac{3}{4}\pi\}$ without changing the answer?

Problem 21.7. Prove that the space of orbits in Problem 21.6 is quasiconformally equivalent to the punctured plane $\mathbb{C}^*$.

Problem 21.8. Using Theorem 21.18, prove that a parabolic germ $f \in \mathscr{A}_{p,\lambda}$ is holomorphically equivalent to its formal normal form $\exp F_{p,\lambda}$ in sectors around the positive and the negative real semiaxes.

This is an alternative proof of the Sectorial Normalization Theorem 21.5.

Problem 21.9. Prove that the space of orbits $(\mathbb{C}^*, 0)/f$ of a parabolic germ in a punctured neighborhood of the origin is a topological quotient of the disjoint union $(\mathbb{C}_1^* \sqcup \mathbb{C}_2^*)/(\varphi_+ \sqcup \varphi_-)$ by two conformal germs $\varphi_+ \colon (\mathbb{C}_1^*, 0) \to (\mathbb{C}_2^*, 0)$, $\varphi_- \colon (\mathbb{C}_1^*, \infty) \to (\mathbb{C}_2^*, \infty)$ (the points of the first punctured plane are identified with their images in the second plane).

Prove that this is a non-Hausdorff topological space equipped nevertheless with an atlas of holomorphic charts. Express the Ecalle–Voronin modulus of f in terms of the germs $\varphi_\pm$.

Problem 21.10. Describe the Ecalle–Voronin modulus of a germ from the class $\mathscr{A}_2$ which commutes with the involution $\sigma \colon z \mapsto -z$.

The same question about a germ f of class $\mathscr{A}_1$ which is conjugated by σ to its inverse f^{-1}.

Problem 21.11. Let $f, g \in \operatorname{Diff}_1(\mathbb{C}, 0)$ be two parabolic germs related by the square map $z \mapsto z^2$, i.e., $g(z) = \sqrt{f(z^2)}$. How are the formal classes of f, g related? Their Ecalle–Voronin moduli?

Problem 21.12. Assume that the Cartesian map $(z, w) \mapsto \big(f(z), g(w)\big)$ with two parabolic components $f, g \in \operatorname{Diff}_1(\mathbb{C}, 0)$, admits a holomorphic nonsingular invariant curve through the origin on $(\mathbb{C}^2, 0)$. What can be said about the formal types of f, g and their Ecalle–Voronin moduli?

Problem 21.13. Prove that for any combination of the natural parameters $p, q \in \mathbb{N}$ a generic Cartesian map (f, g) as in Problem 21.12 with $f \in \mathscr{A}_p$, $g \in \mathscr{A}_q$, is not embeddable in a holomorphic flow.

Problem 21.14. Let γ be the germ of a smooth (nonsingular) real analytic curve on the complex plane $(\mathbb{C}, 0)$. Prove that there is a conformal germ h rectifying γ (mapping it to the real axis).

Reflection in γ is an anti-holomorphic map which is conjugated by h with the symmetry $z \mapsto \bar z$ in the real axis. Prove that this reflection does not depend on the choice of h.

Problem 21.15. Let $\gamma_\pm \subset (\mathbb{C}, 0)$ be germs at zero of two smooth real analytic curves which have a simple (quadratic) tangency between themselves. Find necessary and sufficient conditions for existence of a third germ of real analytic curve at zero ("mirror") γ_0, also tangent to the same direction, such that the reflection in γ_0 permutes γ_+ and γ_-.

Hint. Reduce this question to a root extraction problem.

Problem 21.16. Prove that any antiholomorphic germ $(\mathbb{C}, 0) \to (\mathbb{C}, 0)$ is a symmetry with respect to some real analytic curve.

Problem 21.17. Let g and h be two symmetries from Problem 21.15, and $f = gh$. Prove that any of these symmetries conjugates f with its inverse. Conversely, if f is a holomorphic germ, and g is a symmetry with respect to an analytic curve passing through zero that conjugates f with its inverse, then f is a product of two symmetries.

Problem 21.18. Consider two pairs of germs of real analytic curves at zero from Problem 21.15. Prove that these two pairs are holomorphically equivalent (that is, one pair may be transformed into another by a holomorphic coordinate change) if and only if the product of the symmetries corresponding to the first pair is holomorphically equivalent to the product corresponding to the second pair.

Problem 21.19. A *cuspidal curve* on the complex 2-plane $(\mathbb{C}^2, 0)$ is the image of a holomorphic map $t \mapsto \big(z(t), w(t)\big)$ with two generic components $z(t) = t^2 + \cdots$, $w(t) = t^2 + \cdots$. Cartesian maps introduced in Problem 21.12, naturally act on cuspidal curves.

Describe analytic classification of generic cuspidal curves by the Cartesian action.

Hint. Reduce the problem to Problem 21.10.

22. Complex saddles and saddle-nodes

In this section we describe orbital analytic classification of resonant complex saddles and saddle-nodes. Together with the analytic normal forms from Chapter I, this almost completes analytic classification of all elementary planar singularities. The only type of elementary singularities, for which the classification is not known, is that of *Cremer saddles*. Thus the term "elementary" receives the second justification: besides being "elementary atoms" into which all isolated singularities can be blown up, as explained in §8, the elementary singularities indeed have "simple" nature.

22A. Complex saddles revisited. A singular point of a complex planar vector field is a *complex saddle* provided that the ratio of its eigenvalues is real negative[2], $\lambda_1/\lambda_2 = -\lambda \in -\mathbb{R}_+$. The main problem that we deal with in this section is *orbital analytic classification of complex saddles* or, what is the same, analytic classification of the corresponding singular foliations. The results of this section later will be applied to nonlocal problems. The Realization Theorem 22.9 is the core in the solution of the nonlinear Riemann-Hilbert problem in §23. Some technical results developed in this section are crucial for the proof of the Nonaccumulation theorem in §24.

Any complex saddle has two smooth holomorphic separatrices by the Hadamard–Perron Theorem 7.1. The holonomy (monodromy) map associated with a loop on a separatrix making one turn around the singular point

[2]From this point of view centers with two eigenvalues $\pm i\omega$, $\omega > 0$, are complex saddles. This understanding was clear already to H. Dulac.

is *elliptic*, i.e., tangent to the linear rotation $w \mapsto \nu w$, with the multiplicator $\nu = \exp 2\pi i\lambda$ on the unit circle[3].

Somewhat unexpectedly, the inverse statement is also true: analytic equivalence of monodromy maps of two saddles *with the same linear parts* implies their orbital analytic equivalence (Theorem 22.7). Moreover, *any elliptic germ of a conformal mapping may be realized as the monodromy map of a complex saddle.* This reduces orbital analytic classification of complex saddles to the analytic classification of germs of conformal maps in dimension one. In the resonant case this classification was constructed in §21. Nonresonant elliptic germs automatically belong to the Siegel domain; known results on their linearizability are briefly listed in §5**E**.

Remark 22.1. To avoid trivial disclaimers when passing from saddles to their monodromy, we consider complex saddles with *marked separatrices.* This means that we always work in local complex coordinates (z, w) chosen in such a way that the separatrices belong to the coordinate axes, and the monodromy map of a saddle always corresponds to the small loop on the z-axis. By this convention the monodromy map is obviously an invariant of the orbital analytic classification: two holomorphically orbitally equivalent marked saddles have analytically conjugate monodromy maps. Otherwise one could consider a holomorphism swapping the role of coordinate axes.

22B. Saddles and their monodromy: formal normal forms.

Formal normal forms for saddles were described in Proposition 4.29; see Table I.1. Recall that for a nonresonant saddle the formal orbital normal form is linear:

$$F_0 = z\frac{\partial}{\partial z} - \lambda w\frac{\partial}{\partial w}. \tag{22.1}$$

The monodromy transformation of this linear field is also linear,

$$f_0(w) = \nu w, \qquad \nu = e^{-2\pi i\lambda}. \tag{22.2}$$

For a resonant saddle, the orbital formal normal form is either linear as in (22.2), or rational:

$$F_0 = z\frac{\partial}{\partial z} + w(-\lambda + q(u))\frac{\partial}{\partial w}, \qquad q = \frac{u^{p+1}}{1+\alpha u^p}, \qquad \begin{matrix} u = z^m w^n, \\ \lambda = \frac{m}{n}. \end{matrix} \tag{22.3}$$

where m, n, p are positive integers, $\alpha \in \mathbb{C}$. Denote by $\mathscr{B}_{m,n,p,\alpha}$ the class of all complex saddles with the same formal normal form (22.3). Denote by $\mathfrak{F}_0$ the singular holomorphic foliation defined by the vector field F_0 in the normal form (22.1) or (22.3). Recall that in §21$\mathbf{J}_1$ we introduced the notation $\mathscr{A}_{n,m,p,\lambda}$ for the class of conformal germs with the multiplicator $\exp 2\pi i m/n$ whose nth iteration is formally equivalent to the time one of the flow (21.1) with some $p \in \mathbb{N}$ and $\lambda \in \mathbb{C}$.

[3]Here parabolic germs with $\nu = 1$ are considered as a particular case of elliptic germs.

Lemma 22.2.

1. *Monodromy transformation of a nonresonant or formally orbitally linearizable resonant germ of a foliation* (22.1), *is formally equivalent to the rotation* (22.2).

2. *Monodromy transformation for a foliation from the class* $\mathscr{B}_{m,n,p,\alpha}$ *is a conformal germ of the formal type* $\mathscr{A}_{-m,n,p,\beta}$ *with* $\beta = \alpha/2\pi i$.

The proof of this lemma goes beyond a mere integration of the normal form which would be quite elementary (Remark 4.30). Indeed, the difference between two vector fields with the common N-jet at the origin, may still be not small on the loop along which the holonomy operator is considered. Thus the two monodromy operators might differ in all terms, no matter how large N is. To compute the jet of a high order of the monodromy map, one has to ensure that the vector field differs from its normal form by a field sufficiently flat *on the separatrix.*

22C. Normalization on the separatrix cross. We prove in this section that a saddle germ of holomorphic vector field can be brought to its normal form on the level of any finite order jets on both separatrices. The corresponding result was already known to H. Dulac [**Dul23**].

Lemma 22.3. *Any germ of a saddle vector field can be analytically transformed to a form that differs from the formal normal form* (22.1) *or* (22.3) *respectively, by the field that vanishes on the coordinate cross together with any preassigned number of derivatives.*

In other words, for any $N \in \mathbb{N}$ a saddle resonant germ $F \in \mathcal{D}(\mathbb{C}^2, 0)$ is orbitally analytically equivalent to the germ

$$F = F_0 + z^N w^N R \frac{\partial}{\partial w}, \qquad R \in \mathcal{O}(\mathbb{C}^2, 0), \tag{22.4}$$

where R is the germ of a function holomorphic at the origin (depending on the order N) and $F_0 \in \mathcal{D}(\mathbb{C}^2, 0)$ is the formal normal form.

Proof. Note that rectification of the two separatrices of the germ of a (marked) saddle brings this germ to the form

$$z(\lambda_1 + g_1(z,w))\frac{\partial}{\partial z} + w(\lambda_2 + g_2(z,w))\frac{\partial}{\partial w}, \qquad g_1(0,0) = g_2(0,0) = 0,$$

where $\lambda_1, \lambda_2 \in \mathbb{C}$ are nonzero eigenvalues. Replacing this field by an orbitally equivalent one,

$$\begin{aligned} F &= z\tfrac{\partial}{\partial z} + w\big(-\lambda + h(z,w)\big)\tfrac{\partial}{\partial w} \\ &= F_0 + R_0\tfrac{\partial}{\partial w}, \qquad R_0(z,w) = wf(z) + O(w^2), \end{aligned} \tag{22.5}$$

which corresponds to $N = 0$ in (22.4). We will only consider vector fields whose z-component is $z\frac{\partial}{\partial z}$.

We will prove by induction that for any N by an analytic coordinate change, R_0 may be replaced by $z^N w^N R_N$ with R_N holomorphic at the origin in $\mathbb{C}^2$. Only the resonant case with $\lambda = \frac{m}{n}$ will be considered; the nonresonant case $\lambda \notin \mathbb{Q}_+$ is simpler and treated in exactly the same way.

Note that the germ R can be assumed as flat *at the origin*, as necessary, since all nonresonant terms of order $\leqslant M$ can be removed from F by polynomial transformations of the Poincaré–Dulac algorithm.

Assume by induction that in (22.5) the term R is already divisible by w^l and is M-flat at the origin for $M = N(m+n+2)$:

$$R = w^l f(z) + O(w^{l+1}), \quad R(z,w) = o(|z|+|w|)^M, \quad M = N(m+n+2).$$

For $l = 1$ this coincides with (22.5). We want to achieve divisibility by w^{l+1} after a suitable transformation

$$\operatorname{id}+h : (z,w) \mapsto (z, w + w^l g(z)), \qquad g \in \mathcal{O}(\mathbb{C}^1, 0),\ h(z,w) = (0, w^l g(z)).$$

Denote

$$\widetilde{F} = \left(E + \frac{\partial h(z,w)}{\partial(z,w)}\right) F \circ (\operatorname{id}+h)^{-1}$$

the transformed vector field. To achieve the normalization of jets of order $l+1$ so that $\widetilde{F} = F_0 + O(w^{l+1})$, we have to meet the condition

$$\frac{\partial h}{\partial \mathbf{z}} \begin{pmatrix} z \\ -\lambda w \end{pmatrix} + \begin{pmatrix} 0 \\ \lambda h + R \end{pmatrix} = O(w^{l+1}).$$

Given the explicit form of $\operatorname{id}+h$, this translates into the following functional equation on g

$$z\frac{dg}{dz} - \lambda(l-1)g + f = 0.$$

This linear ordinary differential equation may be solved explicitly. Substituting a Taylor series for $f = \sum_k f_k z^k$, we immediately determine the Taylor series for $g = \sum_k g_k z^k$,

$$g_k = \frac{f_k}{\lambda(l-1) - k}. \tag{22.6}$$

Some of the denominators in (22.6) may vanish for a rational $\lambda = \frac{m}{n}$. But all such cases correspond to small $k = \frac{m}{n}(l-1) < mN$, while the flatness assumption implies that f is flat of order at least $M - N$ at the origin. If $M = N(n+m+2)$, all the coefficients f_k, for $k < mN$ are zeros. Thus zero denominators never occur in (22.5) for nonzero numerators f_k, and the formula makes sense. The series $\sum_{k=mN}^{\infty} g_k z^k$ converges together with $\sum_{k=mN}^{\infty} f_k z^k$ as the denominators in (22.6) tend to infinity.

In a completely similar way, iterating the coordinate changes of the form $(z,w) \mapsto ((z,w)+z^l g(w))$, one may transform the field $F_0 + R\frac{\partial}{\partial w}$ with $R = R(z,w)$ divisible by $w^N z^l$, to a field of the same form with l replaced by $l+1$. Note that at every step the germ g is N-flat at the origin. Hence normalization along the w-axis does not affect previously achieved normal form structure along the z-axis.

At the end the difference between the field F and its formal normal form is divisible by $z^N w^N$ as required. □

22D. Proximity of leaves of complex saddles and their normal forms. In this subsection we prove estimates on divergence of solutions that are a key tool in the study of complex saddles.

The Gronwall inequality (Lemma 15.5) can be easily modified for nonlinear differential equations. Consider two nonautonomous systems of differential equations of the form

$$\frac{dw_i}{dt} = F_i(t, w_i), \qquad i = 1, 2, \quad w_i \in \mathbb{C},\ t \in [0, T], \tag{22.7}$$

on the finite segment of the real time $[0,T]$, and let $w_1(t), w_2(t)$ be two solutions with the common initial condition $w_1(0) = w_2(0)$, which belong to some domain $D \subset [0,T] \times \mathbb{C}$ of the Cartesian product, such that for each $t \in [0,T]$ the intersection $D \cap \{t\} \times \mathbb{C}$ is a convex set. We want to establish a quantitative degree of proximity between these solutions, assuming that the right hand sides are close to each other.

Lemma 22.4 (Nonlinear Gronwall inequality). *Assume that in the domain $D \subset [0,T] \times \mathbb{C}$ the complex functions $F_{1,2}(t,w)$ satisfy the inequalities*

$$|F_1(t,w) - F_2(t,w)| \leqslant C, \quad |F_i(t,w) - F_i(t,w')| \leqslant L\,|w - w'|. \tag{22.8}$$

Then

$$|w_1(t) - w_2(t)| \leqslant CT\,e^{LT}, \qquad \forall t \in [0,T]. \tag{22.9}$$

Proof. Differentiating the difference $\delta(t) = |w_1(t) - w_2(t)|$, we obtain

$$\begin{aligned}
\tfrac{d}{dt}\delta(t) &\leqslant |\tfrac{d}{dt}w_1(t) - \tfrac{d}{dt}w_2(t)| \\
&= |F_1(t, w_1(t)) - F_2(t, w_2(t))| \\
&\leqslant |F_1(t, w_1(t)) - F_1(t, w_2(t))| + |F_1(t, w_2(t)) - F_2(t, w_2(t)| \\
&\leqslant L\,|w_1(t) - w_2(t)| + C = L\delta(t) + C.
\end{aligned}$$

Comparing this means that the function $\delta(t)$ does not exceed the solution of the nonhomogeneous linear differential equation

$$\tfrac{d}{dt}y(t) = Ly(t) + C, \qquad y(0) = 0.$$

The last equation can be immediately integrated, yielding the bound $\delta(t) \leqslant Ct\,e^{Lt}$ for all $t \in [0,t]$. □

Corollary 22.5. *Any finite order N-jet of a holonomy map Δ along a leaf L of a holomorphic foliation $\mathcal{F}$, defined by a holomorphic vector field F, is completely determined by the N-jet of the field F on the leaf L.*

Proof. The assertion of the corollary means that if two foliations $\mathcal{F}_1, \mathcal{F}_2$ defined by holomorphic vector fields F_1, F_2, have a common leaf L and their difference is N-flat on L, then the difference between the respective holonomy operators Δ_1, Δ_2 (associated with the same path $\gamma \subset L$ and the same cross-sections at the endpoints of the path) is N-flat.

Restricting the vector fields on a tubular neighborhood of the path γ and choosing convenient local coordinates, we obtain two complex differential equations on the product $[0,1] \times (\mathbb{C}, 0)$ with $F_i(t,0) \equiv 0$ and $|F_1(t,w) - F_2(t,w)| < |w|^N$. Two trajectories of the two fields starting at the same point $(0,w)$ with $|w| = r$, will both remain in the cylinder $D = \{|w| < re^{Lt}\}$, where L is the common Lipschitz constant of the two fields.

Thus we can apply the inequality (22.9) with $T = 1$ $C = e^{LN} r^N$ and conclude that $|\Delta_1(w) - \Delta_2(w)| < C'|w|^N$, where C' is a constant depending on L, N. Thus the difference between the holonomy maps is N-flat. $\square$

Now the proof of Lemma 22.2 becomes an easy exercise.

Proof of Lemma 22.2. By direct computation we can verify that for the field F_0 in the rational formal normal form (22.3) the monodromy operator is a germ from the corresponding resonant class. Indeed, consider the multivalued function $v = z^\lambda w$, $\lambda = m/n$ (the root of the resonant monomial). Evolution of this function by virtue of the system is governed by the quotient equation $\dot v = q(v^n)$ with the rational function q the same as in (22.3). Thus after continuation over the loop $z = e^{2\pi i t}$, $t \in [0,1]$, the initial value of v will change to $\exp 2\pi i F_{pn,\alpha}$, where $F_{pn,\alpha}$ is the vector field in the rational normal form (21.1) on the complex line and $\exp 2\pi i F$ denotes the flow map for the complex time $2\pi i$. On the other hand, because of the multivaluedness, the function v itself after continuation along the loop will be multiplied by a constant ν (the corresponding root of unity $\nu = e^{2\pi i m/n}$) which must be factored out. Thus the monodromy operator in the chart $w = v|_{z=1}$ takes the form

$$f(w) = \nu^{-1} \exp 2\pi i F_{pn,\alpha}, \qquad F_{pn,\alpha} = w^{pn+1}(1 + \alpha w^{pn})^{-1} \tfrac{\partial}{\partial w}. \tag{22.10}$$

After the linear rescaling $w \mapsto Cw$ with $C = (2\pi i)^{-1/n}$ the monodromy takes the form $e^{2\pi i n/m} \exp F_{pn,\beta}$ with $\beta = \alpha/2\pi i$ (we distinguish the exponent of a number from the flow of the vector field in this combined notation). This coincides with the definition of the class $\mathscr{A}_{-m,n,p,\beta}$.

To show that formally orbitally equivalent saddle vector fields have formally equivalent monodromies, we fix an arbitrary N and reduce the field to its formal normal form modulo the difference which is N-flat on the separatrix $\{w = 0\}$, using Lemma 22.3.

By Corollary 22.5, the N-jet of a holonomy map along a leaf L of a holomorphic foliation is completely determined by the N-jet of the vector field generating this foliation. Therefore the N-jet of the monodromy coincides with the N-jet of the formal normal form, which was computed above. Since N is arbitrary, the coincidence holds on the level of formal series also. □

For trajectories defined on the infinite interval, the bound (22.9) is usually meaningless because of the exponential growth of the right hand side. In such a situation rather than measuring the divergence between any two particular trajectories, we will measure the difference from identity for a homeomorphism which sends solutions of one equation to those of the other. The subtle difference in the construction translates into a different type of boundary conditions. The accurate construction goes as follows.

Consider the saddle vector field $F_0 \in \mathcal{D}(\mathbb{C}^2, 0)$ in the rational normal form (22.3) and another field F_1 which differs from F_0 by the difference N-flat on the coordinate cross as in (22.4). The corresponding ordinary differential equations written in the logarithmic chart $t = -\ln z$, take the form

$$\begin{aligned}&\frac{dw_i}{dt} = F_i(t, w_i), \qquad F_i(t, w) = \lambda w(1 + e^{-pt}w^p + \cdots),\\ &|F_0(t, w) - F_1(t, w)| \leqslant e^{-Nt}w^N \qquad i = 0, 1, \quad p \in \mathbb{N}.\end{aligned} \tag{22.11}$$

We will consider these equations on the cylinder $D = \{t \in \mathbb{R}_+,\ |w| < 1\}$.

We expect that the conjugacy $\mathbf{H}\colon D \to D$, $(s, r) \mapsto (s, H(s, r))$ between these equations, which preserves the t-coordinate and is identical on the slice $s = 0$, is small, more precisely, that the function H remains bounded on D and the difference $H(s, r) - r$ tends to zero sufficiently fast as N is large enough.

This is an assertion on proximity between two trajectories, $w_0(t)$ and $w_1(t)$, which both depend on two parameters (r, s), $|r| < 1$ and $s \geqslant 0$ through the boundary conditions

$$w_0(0) = w_1(0), \quad w_0(s) = r, \quad \text{so that} \quad H(r, s) = w_1(s). \tag{22.12}$$

Lemma 22.6 (Proximity lemma). *The component $H(s, r)$ of the map conjugating the two equations* (22.11), *satisfies the estimates*

$$|H(s, r) - r| \leqslant e^{-Ns/2}, \qquad \text{uniformly over} \quad s \geqslant 1,\ |r| < 1/2. \tag{22.13}$$

Proof. We will change the dependent variables in the equations (22.11), substituting

$$u_i = w_i e^{-\lambda t}, \qquad i = 0, 1,$$

which essentially corresponds to replacing the initial vector fields in $(\mathbb{C}^2, 0)$ by the respective quotient equations with the "skew resonance monomial" $u = z^\lambda w$. The advantage is that solutions of the quotient equations are "almost constant" and can be easily controlled.

More accurately, we obtain from (22.11) two complex differential equations on the real line,

$$\begin{gathered}\frac{du_i}{dt} = G_i(t, u_i), \quad |G_i(t,u)| \leqslant u^2, \qquad \left|\frac{\partial G_i}{\partial u}(t,u)\right| \leqslant |u|, \\ |G_0(t,u) - G_1(t,u)| \leqslant u^N e^{-tN}, \qquad i = 0, 1,\end{gathered} \tag{22.14}$$

(to simplify computations, we normalized all constants to 1 which is always possible by rescaling, and consider the worst case $p = 2$).

Consider the point (s, r) with $|r| = \frac{1}{2}$ (again we consider only one "worst" point). The corresponding trajectory $u_0(t)$ of the first equation in (22.14) is determined by the boundary condition $|u_0(s)| < \frac{1}{2}e^{-s}$. We first observe, integrating this equation in the reverse time, that $|u_0(0)| < \frac{3}{4}e^{-s}$. Indeed, the speed of evolution in the chart $1/u$ is bounded by 1, thus

$$|u_0(0)| < (2e^s - s)^{-1} \leqslant \tfrac{3}{4}e^{-s}.$$

For the same reason the trajectory $u_1(t)$ with the same initial condition $u_1(0) = u_0(t)$, $|u_1(0)| < \frac{3}{4}e^{-s}$, will satisfy the inequality $|u_1(s)| < e^{-s}$.

Thus both trajectories u_0, u_1 belong to the exponentially thin cylinder $D_s = \{|u| < e^{-s},\ 0 \leqslant t \leqslant s\}$. The Lipschitz constant of the right hand sides in this equation do not exceed $L_s = e^{-s} < 1$ in D_s, and the difference between the right hand sides does not exceed $C_s = e^{-Ns}$. Substituting this into the Gronwall inequality (22.9), we conclude that the solutions $u_0(t), u_1(t)$ at the last moment $t = s$ are very close,

$$|u_0(s) - u_1(s)| \leqslant se^{-Ns}e^s, \text{ hence } |w_0(s) - w_1(s)| \leqslant se^{(1+\lambda)s - Ns} < e^{Ns/2}.$$

Thus for all sufficiently large N, s the uniform bound (22.13) is proved. □

22E. Monodromy as the modulus of analytic classification. Now we have all the necessary tools to prove that the monodromy is a modulus of analytic classification of complex saddles in the following precise sense.

Theorem 22.7. *Suppose that two germs of complex saddle vector fields have the same linear part* (22.1) *and their monodromy maps corresponding to the z-axis are analytically equivalent.*

Then the germs of these vector fields are orbitally analytically equivalent.

Proof. First we prove that the vector fields are formally orbitally equivalent.

By Lemma 22.2, the formal class of the monodromy determines almost uniquely the formal class of the saddle. The only uncertainty is in restoring the characteristic ratio m/n from the elliptic multiplicator $\nu = \exp 2\pi i^m/n$. In general this is impossible, and the ratio m/n can be restored only modulo an integer term, yet in the assumptions of the theorem the linear part is explicitly specified. The other formal invariants of the map (the order of the first nonlinear terms p or their absence, and the formal invariant β) determine uniquely the corresponding formal invariants of the resonant saddle.

Second, we show that two saddles with analytically conjugated monodromies are themselves holomorphically orbitally equivalent by a map that keeps the marked separatrices into each other. The construction of the conjugacy is rather straightforward: we extend the holomorphism between the cross-sections, which conjugates the holonomy maps, to a biholomorphism between the foliations, defined on the complement to the second (unmarked) separatrices. This extension is possible, since by Lemma 11.14 saturation of the cross-section by leaves of the saddle foliation fills the entire complement. This extension is well defined since the two holonomy maps are conjugated.

Finally we show that the constructed biholomorphic conjugacy between complements to the unmarked separatrices, extends holomorphically on these separatrices as well by the removable singularity theorem. The estimates required to apply this theorem, are derived from the Proximity lemma 22.6.

To construct the conjugacy, we assume that the two vector fields are already brought by a biholomorphic transformation to the form (22.4) which differs from the formal normal form (22.3) by terms N-flat on the separatrix cross, and the monodromy of the separatrices simply *coincide* in the respective charts. The conjugacy will preserve the first coordinate.

The corresponding differential equation can be rewritten with respect to the variable $t = -\ln z$, but now we consider the complex value of t in the right half-plane $\mathbb{C}_+ = \{\operatorname{Re} t \geqslant 0\}$. The corresponding differential equations have the form (22.11) with the only difference that the bounds take a slightly different form,

$$|F_0(t) - F_1(t)| \leqslant e^{-N \operatorname{Re} t} w^N, \qquad t \in \mathbb{C}_+. \tag{22.15}$$

The conjugacy $\mathbf{H}\colon \mathbb{C}_+ \times \mathbb{D} \to \mathbb{C}_+ \times \mathbb{D}$, $(s,r) \mapsto (s, H(s,r))$ will be constructed exactly as in the Proximity lemma, namely, we consider the solution $w_0(t)$ of the first equation, defined by the initial (more precisely, *terminal*[4]) condition $w(s) = r$, and the solution $w_1(t)$ of the second equation with the same initial condition $w_1(0) = w_0(0)$, and let $H(s,r) = w_1(s)$ as in (22.12). This conjugacy by construction sends solutions of the first equation to solutions of the second equation and is holomorphic.

[4]The difference is more of a psychological than mathematical nature.

The function $H(s,r)$ is $2\pi i$-periodic in s, $H(s+2\pi i, r) \equiv H(s,r)$. Indeed, both differential equations are $2\pi i$-periodic in t. Thus if one replaces the condition $w_0(0) = w_1(0)$ by the condition

$$w_0(2\pi i) = w_1(2\pi i) \tag{22.16}$$

in the choice of the two solutions w_0, w_1, the result will certainly be the same. Yet the conditions $w_0(0) = w_1(0)$ and (22.16) are equivalent, since the monodromy of the two equations coincide.

Because of this periodicity, the function H descends as a well-defined function $H(z,w)$, holomorphic in the punctured disk $\{0 < |z| < 1\}$. From the Proximity Lemma 22.6, we have $|H(z,w) - w| \leqslant |z|^{N/2}$ as $z \to 0$, thus if $N \geqslant 3$, the corresponding conjugacy extends holomorphically by the identity map on the deleted separatrix $\{z = 0\} \times \mathbb{D}$. This completes the proof of the theorem. □

22F. Orbital analytic classification of resonant saddles. By a *modulus of orbital analytic classification of a marked saddle resonant germ of planar vector field*, we mean the Ecalle–Voronin modulus of analytic classification of the (resonant conformal) monodromy map associated with the marked separatrix. This modulus is described by the Classification Theorem 21.36: the corresponding classification space is a subspace of $\mathscr{M}^{\circ}_{np,\lambda}$ satisfying the additional relation (21.40).

As an immediate consequence of Theorem 22.7, we obtain the statements 1, 2 and 4 of the following result that gives complete classification of resonant saddles.

Theorem 22.8 (Analytic classification theorem for parabolic germs).

1. (Invariant) *If two germs of saddle resonant vector fields F and F' with the same linear part* (22.1) *are orbitally analytically conjugate by a transformation that preserves the coordinate axes, then their moduli coincide.*

2. (Equimodality vs. equivalence) *Conversely, two saddle resonant germs F and F' from $\mathscr{B}_{-m,n,p,\beta}$ with the same modulus are orbitally analytically equivalent.*

3. (Realization) *Any tuple $\Phi \in \mathscr{M}^{0}_{p,\lambda}$ that satisfies* (21.40) *may be realized as the modulus for some saddle resonant germ $v \in \mathscr{B}_{-m,n,p,\beta}$.*

4. (Analytic dependence on parameters) *If a family of germs F_ε from the same formal $\mathscr{B}_{-m,n,k,\beta}$ class depends analytically on a parameter ε, then the modulus of orbital analytic equivalence also depends analytically on ε.*

To prove Theorem 22.8 completely, we need to show that any resonant conformal germ can be realized as the monodromy map of a resonant saddle with a preassigned linear part compatible with the monodromy.

Theorem 22.9. *For any conformal elliptic germ* $f\colon z \mapsto e^{2\pi i\varphi}z + O(z^2)$, $\varphi \in \mathbb{R}$ *and any* $\lambda < 0$ *such that* $\lambda = \varphi \bmod \mathbb{Z}$, *there exist a saddle germ of a planar vector field with the linear part* (22.1) *whose monodromy map coincides with* f.

Theorem 22.9 in the resonant case was proved by Martinet-Ramis [**MR83**], and in the general case by J.-C. Yoccoz and R. Perez-Marco [**PMY94**]. The proof presented below goes back to [**EISV93**].

22G. Realization of monodromy: proof of Theorem 22.9.

22G$_1$. *Main idea and preparations.* The proof is based on the idea which is crucial for the study of nonlinear Stokes phenomena. The foliation with the assigned monodromy is constructed as an abstract complex manifold M not embedded in any complex linear space. The construction is similar to the construction of a suspension of a self-map (Theorem 2.31; see §2**F**) in which the principal features of the construction are already present.

The manifold M is topologically equivalent to a product of a punctured disc and another disc, yet the foliation on it is given not by one vector field in $(\mathbb{C}^2, 0)$ but rather by several analytic vector fields defined in different charts on M. The main part of the proof is to identify M as a neighborhood of the origin in $\mathbb{C}^2$ with a w-axis deleted, and the foliation as a phase portrait of some germ (22.17).

As the first step of this construction, we need some preparation.

Let f be the a conformal elliptic germ. Denote by f_0 its formal normal form (22.2) or (22.10). Without loss of generality we may assume that f has the form

$$f = (\mathrm{id} + h) \circ f_0, \qquad h(w) = o(w^N), \tag{22.17}$$

for as large N as necessary.

As follows from Lemma 22.2, the formal normal form of the monodromy map and the linear part of the complex saddle determine uniquely the formal normal form of this saddle. Let F_0 be the corresponding formal normal form (22.1) or (22.3) of the vector field F that we are attempting to construct. We will construct F by a surgery on the corresponding foliation $\mathcal{F}_0$ defined by F_0: the phase space will be slit along the set $(\mathbb{R}_+, 0) \times (\mathbb{C}, 0)$ and sealed back in such a way that the monodromy will coincide with the preassigned germ f instead of f_0.

22G$_2$. *Construction of an abstract holomorphic foliation with the preassigned monodromy.* Let us introduce the following notations:

(1) $\mathbb{D}_z = \{|z| < 1\}$, $\mathbb{D}_w = \{|w| < 1\}$ the open unit disks on the corresponding axes,

(2) $K_0 = \mathbb{D}_z \smallsetminus \{0\}$ the punctured disk, $\widetilde{K}_0$ the universal cover of K_0 with the coordinates $\widetilde{z} = (r, \varphi) \in \mathbb{R}_+ \times \mathbb{R}$;

(3) $\widetilde{K} \subset \widetilde{K}_0$ the domain on the universal cover,

$$\widetilde{K} = \{\widetilde{z} \in \widetilde{K}_0 : \widetilde{z} = re^{i\varphi},\ r \leqslant 1,\ -\tfrac{\pi}{4} < \varphi < 2\pi + \tfrac{\pi}{4}\},$$

(4) $M_0 = K_0 \times \mathbb{D}_w$ the unit bidisk without the w-axis,

(5) $\widetilde{M} = \widetilde{K} \times \mathbb{D}_w$ the corresponding domain in the covering space,

(6) $\Pi \colon \widetilde{K} \to K_0$ the natural projection onto the z-axis; we also use Π to denote the projection $\Pi \colon \widetilde{M} \to M_0$,

(7) $S_0 = \{z \in K_0 : |\operatorname{Arg} z| < \frac{\pi}{4}\}$. The preimage $\Pi^{-1}(S_0) \subset \widetilde{K}$ consists of two connected components

$$S = \{\widetilde{z} \in \widetilde{K} : -\tfrac{\pi}{4} < \varphi < +\tfrac{\pi}{4}\},$$
$$S' = \{\widetilde{z} \in \widetilde{K} : 2\pi - \tfrac{\pi}{4} < \varphi < 2\pi + \tfrac{\pi}{4}\}.$$

Let $\mathcal{F}_0$ be the foliation on M_0 determined by the vector field F_0 (in the form (22.1) or (22.3) respectively; the latter form is determined by the linear part and the normal form (22.3) of the monodromy transformation f as explained in Lemma 22.2).

Let $\widetilde{F}$ and $\widetilde{\mathcal{F}}$ be the pullback of F_0, and $\mathcal{F}_0$ on $\widetilde{M}$ respectively. For $\widetilde{z} \in S$ denote $\widetilde{z}' \in S'$ the point with the same projection on K_0: $\Pi(\widetilde{z}) = \Pi(\widetilde{z}')$.

We will now construct the *sealing map*

$$\Phi : S' \times \mathbb{D}_w \to S \times \mathbb{C}$$

with the following properties:

(1) Φ preserves the first coordinate, i.e., $\Phi(\widetilde{z}, w) = \big(\widetilde{z}', \Phi_z(w)\big)$ (the notation is consistent since $z = \Pi(\widetilde{z}) = \Pi(\widetilde{z}')$);

(2) Φ respects the vector field $\widetilde{F}$ and the foliation $\widetilde{\mathcal{F}}$, bringing leaves to leaves.

The first property of the sealing map Φ allows us to define the quotient space $M = \widetilde{M}/\Phi$ by identifying points of $S' \times \mathbb{D}_w$ with their images ("sealing the two flaps") so that the quotient space is naturally equipped with the projection on the punctured disk K_0. The second property means that the field $\widetilde{F}$ and the foliation $\widetilde{\mathcal{F}}$ defined by it, correctly define a vector field F and the respective foliation $\mathcal{F}$ on M. The leaves of this foliation project without critical points on the base K_0 (i.e., are transversal to all lines $\{z = \text{const}\}$), and hence the loop γ generating the fundamental group of K_0 defines the holonomy map for the quotient foliation $\mathcal{F}$ on M (for the cross-section $\{z = 1\}$), referred to as the *monodromy* map.

Our immediate goal is to construct the sealing map Φ so that the monodromy of the foliation $\mathcal{F}$ coincides with the preassigned germ f.

In order to achieve the Property (2), we extend this map along the leaves of the foliation $\widetilde{\mathcal{F}}$. More precisely, for an arbitrary point $\widetilde{z}' \in S'$ choose a simple arc γ_z connecting $z = \Pi(\widetilde{z}')$ with 1 in the sector $S_0 = \{|\varphi| < \frac{\pi}{4}\}$. The holonomy map $P_z \colon \{z\} \times \mathbb{D}_w \to \{1\} \times \mathbb{D}_w$ along the leaves of the foliation $\widetilde{\mathcal{F}}_0$ over the curve γ_z is covered by two holonomy maps $\widetilde{P}_z \colon \{\widetilde{z}\} \times \mathbb{D}_w \to \{1\} \times \mathbb{D}_w$ and $\widetilde{P}'_z \colon \{\widetilde{z}'\} \times \mathbb{D}_w \to \{1'\} \times \mathbb{D}_w$ for the pullback foliation $\widetilde{\mathcal{F}}$. Since the sectors S, S' are simply connected, this map is well defined (independent of the choice of the arc γ_z with the same endpoints) for $|w|$ sufficiently small.

Define the extension of Φ on $S' \times \mathbb{D}_w$ by the formula

$$\Phi(\widetilde{z}', w) = (\widetilde{z}, \Phi_z(w)), \qquad \Phi_z(w) = \widetilde{P}_z^{-1} \circ (\mathrm{id} + h) \circ \widetilde{P}'_z(w). \tag{22.18}$$

We will prove later that this map is indeed well defined in the domain $S' \times \{|w| \leqslant r\}$ and biholomorphic on its image for $r > 0$ small enough.

Remark 22.10. In fact by the above arguments the sealing map Φ may be extended to a larger domain

$$\Omega = S_1 \times \{|w| \leqslant r\}, \qquad S_1 = \{\widetilde{z} = re^{i\varphi} \in \widetilde{K} : \pi < \varphi < 2\pi + \tfrac{\pi}{4}\}.$$

By construction, the sealing map Φ respects the foliation $\widetilde{\mathcal{F}}$. Denote by M the quotient space $\widetilde{M}/\Phi$ (the points of $\widetilde{M}$ are identified if and only if one is the Φ-image of the other). Since $\Phi_* \widetilde{F} = \widetilde{F}$, the vector field $\widetilde{F}$ defines a quotient vector on M denoted by F. The corresponding foliation will be denoted $\mathcal{F}$. Note that $\Phi(\widetilde{z}', 0) = (\widetilde{z}, 0)$, hence the leaf $\{w = 0\} \subseteq \widetilde{M}$ projects into a separatrix of the foliation $\mathcal{F}$.

It is easy to see that the monodromy of $\mathcal{F}$ along the loop coincides with f; cf. with Theorem 2.31. Indeed, the monodromy of the foliation $\widetilde{\mathcal{F}}$ on $\widetilde{M}$ is f_0, hence the lift of the curve $z = e^{2\pi i t}$, $t \in [0, 1]$, on the leaf of $\widetilde{\mathcal{F}}$ in $\widetilde{M}$ passing through $(1, w)$, ends at $(1', f_0(w))$. The identification map Φ brings the latter point to $\bigl(1, (\mathrm{id} + h) \circ f_0(w)\bigr) = (1, f(w))$ by (22.17).

The construction of M and F is over, and it remains to identify them. In fact, we will identify not the manifold M itself, but its smaller open subset. Let $\widetilde{M}_\rho \subset \widetilde{M}$ be the preimage of the bidisk $\{|z| \leqslant \rho,\ |w| < \rho\} \subseteq \mathbb{C}^2$ on $\widetilde{M}$ and denote M_ρ its natural projection onto the quotient space M. We will prove that for $\rho > 0$ sufficiently small, M_ρ is biholomorphically equivalent to a neighborhood of the origin without the axis $(\mathbb{C}^2, 0) \smallsetminus \{w = 0\}$, while the vector field F in this biholomorphic chart extends on the deleted axis to a holomorphic saddle vector field on $(\mathbb{C}^2, 0)$ from the preassigned formal class. This requires some technical estimates.

22**G**$_3$. *Asymptotic properties of the sealing map Φ.* In this subsection we prove that for sufficiently large N in (22.17) the map Φ tends to identity together with some derivatives as $z \to 0$.

Proposition 22.11. *If the function h in* (22.17) *is N-flat at $w = 0$, then the sealing map* (22.18) *admits the asymptotic estimate*

$$|\Phi_z(w) - w| = O(|z|^{\lambda(N-2)}) \qquad \text{as } z \to 0. \tag{22.19}$$

uniformly over $|w| < \frac{1}{2}$.

Corollary 22.12. *For any natural k the number N in* (22.17) *may be so chosen that $\Phi_z(w) - w$ would tend to zero as $z \to 0$ together with its derivatives of order $\leqslant k$ uniformly in the disk $|w| < \frac{1}{2}$.*

Proof of the corollary. It follows from Proposition 22.11 and the Cauchy estimates. □

Proof of Proposition 22.11. The estimates are essentially the same in both the resonant and nonresonant cases, the latter being more transparent. In this case the germ F_0 is linear, and the holonomy map is linear as well,

$$P_z(w) = \alpha(z)w, \qquad \alpha(z) = z^{-\lambda}.$$

Denote by α the map $w \mapsto \alpha w$. Then, by (22.18),

$$\Phi_z = \alpha(z) \circ (\mathrm{id} + h) \circ \alpha^{-1}(z),$$

and therefore by (22.17)

$$\Phi_z = \mathrm{id} + O(z^{\lambda(N-1)})$$

In the resonant case,

$$\tfrac{1}{2}|z^{-\lambda}w| < |P_z(w)| < 2|z^{-\lambda}w|$$

for all sufficiently small z, w. Hence by (22.17)–(22.18),

$$P_z^{-1} \circ (\mathrm{id} + h) \circ P_z = P_z^{-1} \circ (P_z + h \circ P_z) = \mathrm{id} + O(z^{\lambda(N-2)})$$

(we simplified the notation in (22.18) ignoring the difference between the points of the z-plane and its universal covering). □

22**G**$_4$. *Identification of the manifold M.* We will now construct a diffeomorphism between G and a neighborhood of the origin in $\mathbb{C}^2$ without the w-axis. This nonanalytic diffeomorphism nevertheless carries the complex structure on M into an almost complex structure on its image. The asymptotic estimates of the previous subsection allow us to extend this almost complex structure on the removed axis, obtaining an almost complex structure *full neighborhood* of the origin. By the Newlander–Nirenberg theorem, this complex structure is integrable and M turns out biholomorphically equivalent to a neighborhood of the origin in $\mathbb{C}^2$ without the axis. This will complete identification of M.

The diffeomorphism G is constructed via a smooth map $\widetilde{G} : \widetilde{M} \to \mathbb{C}^2$ respecting the sealing map Φ,

$$\widetilde{G}|_{S'\times\mathbb{D}_w} = \widetilde{G} \circ \Phi|_{S\times\mathbb{D}_w}. \tag{22.20}$$

Then the quotient map $G\colon \widetilde{M}/\Phi \to \mathbb{C}^2$ will be well defined.

The property (22.20) is achieved by the standard smooth interpolation. Let χ be a smooth real nonnegative *cutoff function* of one variable $\varphi \in (-\frac{\pi}{4}, 2\pi + \frac{\pi}{4})$, equal to zero on $(-\frac{\pi}{4}, +\pi]$ and one on $[2\pi - \frac{\pi}{4}, 2\pi + \frac{\pi}{4})$. Denote $\widetilde{\chi}(\widetilde{z}) = \chi(\varphi)$ for $\widetilde{z} = re^{i\varphi} \in \widetilde{K}$ and define

$$\widetilde{G}(\widetilde{z}, w) = \bigl(\widetilde{z}, w + \widetilde{\chi}(\widetilde{z})\Phi_z(w)\bigr). \tag{22.21}$$

This definition is correct, since in the part of $\widetilde{M}$ where Φ_z is undefined, the cutoff function $\widetilde{\chi}$ is identically zero (cf. with Remark 22.10).

Consider the pullback of the complex structure on M by the map $H = G^{-1}$. This is an almost complex structure defined by the pullback of the "$(1,0)$-subbundle" (forms of type $(1,0)$ of the complexified cotangent bundle on M), as described in §21**E**.

The $(1,0)$-subbundle on M is spanned by two Φ-invariant $(1,0)$-forms on $\widetilde{M}$,

$$\zeta_1 = d\widetilde{z}, \qquad \zeta_2 = dw + \widetilde{\chi}(\widetilde{z})\, d(\Phi_z(w) - w).$$

The form $d\Phi_z(w)$ is holomorphic on its domain; the factor $\widetilde{\chi}(\widetilde{z})$ is zero outside this domain. Hence, both forms are of type $(1,0)$ on the whole of $\widetilde{M}$. The form ζ_2 has two representations over $\varPi^{-1}(S_0) \times \mathbb{D}_w$,

$$\zeta_2 = \begin{cases} \zeta_2^0 = dw, & \text{on } S \times \mathbb{D}_w, \\ \zeta_2^1 = d\Phi_z(w) & \text{on } S' \times \mathbb{D}_w. \end{cases}$$

This implies the required Φ-invariance, since $\zeta_2^1 = \Phi^*\zeta_2^0$. Denote by ζ_2^* the form induced by ζ_2 on the quotient space M.

The almost complex structure on M_0 induced by the map $G : M \to M_0$ is defined by the two forms,

$$\omega_1 = dz, \quad \omega_2 = H^*\zeta_2^*, \qquad H = G^{-1}. \tag{22.22}$$

Proposition 22.13. *If the number N in (22.17) is sufficiently large, then ω_2 and dw have the same 4-jet: $\omega_2 - dw \to 0$ as $z \to 0$ together with its derivatives of orders $\leqslant 4$.*

Proof. By (22.21),

$$\widetilde{G}(\widetilde{z}, w) - (\widetilde{z}, w) = \widetilde{\chi}(\widetilde{z})(\Phi_z(w) - w).$$

The function $\widetilde{\chi}$ depends on $\operatorname{Arg} \widetilde{z}$ only. Hence, for any multi-index k with $|k| < 4$ the kth derivative of $\widetilde{\chi}$ grows no faster than r^{-4} as $r = |\widetilde{z}| \to 0$. On the other hand, for N in (22.17) large enough, all the derivatives of $\Phi_z(w) - w$ of order less than 5 tend to zero faster than r^5 as $r \to 0$. Hence, all the derivatives of the product $\widetilde{\chi}(\widetilde{z})(\Phi_z(w) - w)$ of order less than 5 tend to zero as $r \to 0$. □

Proposition 22.14. *For any sufficiently small ρ the domain M_ρ is biholomorphically equivalent to a neighborhood of the origin in $\mathbb{C}^2$ without the w-axis.*

Proof. Note that the closure of M_0 is a closure of an open domain $U \subset \mathbb{C}^2$; topologically, M_0 is diffeomorphic to U without the axis. We can continue the almost complex structure generated by ω_1, ω_2 to U by postulating that $\omega_2 = dw$ on the w-axis. This extended almost complex structure is integrable. Indeed, the almost complex structure (22.22) is integrable on M_0 because it is induced from a true complex structure on M. Let $L^{1,0}$ be the span of the forms ω_1, ω_2 Then the integrability condition (21.23) holds on M_0. By continuity, it remains valid after the extension on U. Hence by the Newlander–Nirenberg Theorem 21.18, there exists a C^1 smooth chart $G_0 \colon M_{0,\rho} \to \mathbb{C}^2$ for a sufficiently small ρ, which is holomorphic in the sense of the almost complex structure (22.22). Without loss of generality we may assume that this chart preserves the z-coordinate and is tangent to identity at the origin.

The composition map $G_1 = G_0 \circ G : M_\rho \to \mathbb{C}^2$ between complex analytic manifolds is (truly) biholomorphic. This map identifies manifold M_ρ with an open subset of $\mathbb{C}^2$ without the w-axis. □

22**G**$_5$. *Identification of the singular foliation $\mathcal{F}$.* Let us prove that the vector field F_* obtained from the field F on M_ρ by the holomorphic transformation G_1, may be extended as a saddle vector field with the orbital normal form F_0. To that end, it is sufficient only to prove that F_* and F_0 have the same linear parts is appropriate coordinates: since F and F_0 have the same (i.e., analytically conjugate) monodromy, by Lemma 22.2 this will be sufficient to guarantee that the fields are from the same formal class.

Since the first component of G_0 is identically z, the first component of F_* is $z\frac{\partial}{\partial z}$. The second component is bounded and holomorphic outside the w-axis. By the theorem on removable singularity, it may be holomorphically extended to this axis.

The field F_* is obtained by transferring the field F_0 from $\widetilde{M}$ first on M_ρ by passing to the quotient and then on $M_{\rho,0}$ by the map G_0. Since both maps are tangent to identity by construction, the linear parts of F_0 and F_* coincide. This proves that F_* is formally equivalent to F_0 as requested.

The proof of the Realization Theorem 22.9 is therefore complete. □

22H. Complex saddle-nodes: preparation. The only two remaining classes of elementary singularities of complex holomorphic foliations, not yet considered from the point of view of analytic classification, are *Cremer saddles* (formally but not analytically linearizable saddles) and degenerate elementary singularities with one zero and one nonzero eigenvalue of the linearization at the singular point. While not much can be said about the former type, the latter admits complete analytic classification very similar to that of resonant saddles. This classification was achieved by J. Martinet and J.-P. Ramis. In order to shorten the terminology, we will abbreviate the term *degenerate elementary singularity* to (complex) *saddle-node*; cf. with §9.

By the Poincaré–Dulac theorem, a saddle-node is formally orbitally equivalent to the integrable normal form

$$\frac{z^{r+1}}{1+az^r}\cdot\frac{\partial}{\partial z}-w\,\frac{\partial}{\partial w};\tag{22.23}$$

see Table I.1. We will show that the normalizing series transforming saddle-nodes to their formal normal forms, usually diverge. In order to make the subsequent exposition maximally transparent, we will consider only the formal class corresponding to $r=1$ and $a=0$. The class of saddle-nodes with such formal invariants will be denoted $\mathscr{E}_{0,1}$.

The normal form (22.23) has two separatrices (the coordinate axes). However, by Theorem 7.8, only one separatrix is always a smooth holomorphic curve. The Euler Example 7.10 shows that the formal series for the invariant curve tangent to the eigenvector with zero eigenvalue (the *center manifold*), may indeed be diverging.

Yet, as in the case of parabolic germs, there always exists a biholomorphism conjugating the saddle node with its formal normal form in the sectors of special form.

Lemma 22.15. *A complex saddle-node from the formal equivalence class $\mathscr{E}_{1,0}$ is analytically equivalent to the foliation generated by the holomorphic vector field*

$$z^2\tfrac{\partial}{\partial z}-\bigl(w-z^N f(z,w)\bigr)\tfrac{\partial}{\partial w}.\tag{22.24}$$

Instead of the proof. This statement can be considered as a normalization along the holomorphic invariant manifold (curve) in complete analogy with Lemma 22.3. The proof is achieved by minor modification of the proof of that lemma, details are left to the reader. □

22I. Sectorial normalization. Let $S_\pm$ be two sectors of the complex plane,

$$S_+=\{|\operatorname{Arg} z-\tfrac{\pi}{2}|<\alpha\},\quad \tfrac{\pi}{2}<\alpha<\pi,\qquad S_-=-S_+.\tag{22.25}$$

Each of these sectors is symmetric by the imaginary axis and has opening greater than π. These two sectors overlap by the two sectors symmetric by the real axis. They will play different roles and hence have different names. The *fall sector* $S_\downarrow$ is the component of the intersection containing the negative semiaxis, while the *sector of jump* $S_\uparrow$ is the component containing the positive semiaxis:

$$S_\uparrow = \{|\operatorname{Arg} z| < \alpha - \tfrac{\pi}{2}\}, \qquad S_\downarrow = \{|\operatorname{Arg} z - \pi| < \alpha - \tfrac{\pi}{2}\}. \tag{22.26}$$

The names correspond to the behavior of solutions $w = ce^{-1/x}$, $c \in \mathbb{C}$, of the formal normal form $\frac{dw}{dz} = -\frac{w}{z^2}$.

The following result is due to M. Hukuhara, T. Kimura and T. Matuda (1961).

Theorem 22.16 (sectorial normalization theorem [**HKM61**]). *The saddle-node foliation generated by the vector field* (22.24)*, is biholomorphically equivalent to the standard foliation generated by its formal normal form* (22.23) *in each of the two cylinders $S_\pm \times \mathbb{D}$, where $\mathbb{D} = \{|w| < 1\}$.*

The conjugating biholomorphisms can be chosen preserving the z-coordinate and continuously extendable by the identity on the w-axis $\{z = 0\}$.

As before, on the overlapping of the two cylindrical domains $\widetilde{S_\downarrow} = S_\downarrow \times \mathbb{D}$ and $\widetilde{S_\uparrow} = S_\uparrow \times \mathbb{D}$ two sectorial normalizations need not coincide. The "disagreement", an automorphism of the standard foliation, is the modulus of analytic equivalence, called the *Martinet–Ramis modulus*.

The construction is based on the (easy) investigation of properties of the normal form. This normal form for saddle-nodes for the class $\mathscr{E}_{1,0}$ becomes even more transparent in the chart $t = 1/z$ (in the general case one has to use the rectifying chart for the z-component of the formal normal form). The corresponding differential equation takes the form

$$\frac{dw}{dt} = w, \qquad t^{-1} \in S_\pm. \tag{22.27}$$

Note that the sectors $S_\pm$ after the inversion simply exchange their (indistinguishable) roles. We will keep the notation $S_\uparrow, S_\downarrow$ for the sectors of jump and fall also in the t-plane.

Definition 22.17. We shall say that a biholomorphic map $\mathbf{H}$ is a *distinguished automorphism* of the standard saddle-node (22.27) over a connected simply connected domain $U \subseteq \mathbb{C}$ containing infinity in its closure, $\infty \in \overline{U}$, if $\mathbf{H}$ maps the cylindric domain $U \times \frac{1}{2}\mathbb{D}$ into $U \times \mathbb{D}$, preserves the t-coordinate and extends continuously at infinity by the identical map of $\frac{1}{2}\mathbb{D}$ into itself,

$$\mathbf{H}(t, w) = \big(t, H(t, w)\big), \qquad t \in U,\ |w| < \tfrac{1}{2}, \qquad \lim_{t\to\infty} H(t, \cdot) = \mathrm{id}\,. \tag{22.28}$$

Each distinguished automorphism H of the equation (22.27), preserving the t-coordinate is completely determined by the self-map $\varphi_a = H(a,\cdot)$ of any fixed cross-section $\tau_a = \{t = a\}$ into itself, obtained by the restriction of $\mathbf{H}$. Indeed, the restriction of $\mathbf{H}$ on any other cross-section τ_b is obtained by the linear conjugacy,

$$\varphi_b = \lambda \circ \varphi_a \circ \lambda^{-1}, \qquad \lambda\colon \mathbb{C} \to \mathbb{C},\ w \mapsto \lambda w, \qquad \lambda = e^{b-a}. \tag{22.29}$$

The following proposition describes the distinguished automorphisms of the standard saddle-node (22.27) over three types of sectors.

Proposition 22.18.

1. *The only distinguished automorphism over the sector of fall $S_\downarrow$ is generated by the shift, $\varphi_a(w) = w + c_a$, $c \in \mathbb{C}$.*

2. *All distinguished automorphisms over the sector of jump $S_\uparrow$ are parabolic maps fixing the origin, $H(t,w) = w + O(w^2)$.*

3. *The only distinguished automorphism over any of the sectors $S_\pm$ is the identity $H(t,\cdot) = \mathrm{id}$.*

Proof. If $\mathbf{H}$ is a distinguished automorphism over a given sector S, then for any choice of the point $b \in S$ the linear conjugacy φ_b from (22.29) must be a holomorphic germ *with a certain limit* as b tends to infinity in the sector. Expanding φ_a in the Taylor series, we see that after the linear conjugacy (22.29) the kth Taylor coefficient of φ_a is multiplied by λ^{1-k}:

$$\varphi_a(w) = c_0 + c_1 w + c_2 w^2 + \cdots \longmapsto \lambda c_0 + c_1 w + \lambda^{-1} c_2 w^2 + \cdots = \varphi_b(w).$$

1. If $S = S_\downarrow$ is a sector of fall, then the value $\lambda = \lambda_b$ takes arbitrarily small values, thus multiplication by λ^{1-k} forces all nonaffine (e.g., with $k \geqslant 2$) coefficients of φ_b to grow to infinity as $|\lambda_b| \to 0$, unless they are identically zero. This is compatible with existence of the limit as $b \to \infty$ in S only if the germ φ_a is affine (i.e., all higher coefficients vanish), $\varphi_a = c_{0,a} + c_{1,a} w$.

Since all maps φ_a are linear conjugate, the constant $c_{1,a}$ is the same and does not depend on a, i.e., $c_{1,a} = c_1$. The condition that φ_a tends to the identity as $a \to \infty$ means that $c_1 = 1$, i.e., that φ_a is a shift: $f_a(w) = w + c_a$.

2. If $S = S_\uparrow$ is the sector of jump, then the multiplier $\lambda_b = e^{b-a}$ takes arbitrarily large values as b tends to infinity in S. Thus the free term of φ_b, corresponding to $k = 0$, will grow to infinity unless it is identically zero. Thus $\varphi_a(w) = c_{1,a} w + O(w^2)$. The condition that φ_a tends to the identity as $a \to \infty$ means that $c_{1,a} = 1$, i.e., that $\varphi_a(w) = w + O(w^2)$ is a parabolic germ.

3. The last assertion follows from the two previous claims, since each sector $S_\pm$ contains both a sector of jump and a sector of fall. □

Remark 22.19. If in the definition of the distinguished automorphism the condition that $\lim \varphi_a = \mathrm{id}$ as $a \to \infty$ is relaxed to the assumption that the limit exists, then the distinguished automorphisms over the sector of fall will have affine restrictions $\varphi_a(w) = Cw + c_a$, while over the sectors $S_\pm$ they will necessarily be linear automorphisms $\varphi_a(w) = Cw$.

Note also that replacing a point a by any other point a' results in a linear conjugacy of the self-map φ_a by a linear transformation.

22J. Martinet–Ramis modulus and holomorphic classification of the saddle-nodes. Now everything is ready to describe the holomorphic classification of complex saddle-nodes. Consider the space $\mathscr{MR}^\circ_{1,0} \cong \mathbb{C} \times \mathrm{Diff}_1(\mathbb{C}, 0)$ of pairs (φ, ψ): the first element of each pair is a shift, $\varphi(w) = w + c$, the second a parabolic germ $\psi(w) = w + O(w^2)$. Two such pairs will be called *equivalent*, if there exists a linear transform $w \mapsto Cw$, simultaneously conjugating both components of these pairs. The *Martinet–Ramis modulus* is the equivalence class of pairs from $\mathscr{MR}^\circ_{1,0}$. The space of equivalence classes will be denoted by $\mathscr{MR}_{1,0}$.

The sectorial normalization theorem allows us to assign to each saddle-node from the formal class $\mathscr{E}_{1,0}$ an element of the space $\mathscr{MR}_{1,0}$ as the quotient of the two sectorial normalizations. Let $\mathcal{H} = (H_+, H_-)$ be the *normalizing cochain*, the pair of maps realizing the sectorial normalization in the cylinders $\widetilde{S_\pm} = S_\pm \times \mathbb{D}$. The coboundary $\delta\mathcal{H} = H_+ \circ H_-^{-1}$ is an automorphism of the normal form defined in the union $\widetilde{S_\uparrow} \cup \widetilde{S_\downarrow}$ of cylinders over the sectors $S_\uparrow$ and $S_\downarrow$. Without loss of generality we may assume that this automorphism is distinguished. By Proposition 22.18, the cocycle $\Phi = \delta\mathcal{H} = (\varphi, \psi)$ can be identified with an element from the space $\mathscr{MR}^\circ_{1,0}$. Using the different cross-sections and/or a different normalizing chart results in replacing the pair (φ, ψ) by an equivalent pair (φ', ψ'), that is, by the same modulus from $\mathscr{MR}_{1,0}$.

Thus we constructed the *Martinet–Ramis correspondence*

$$\mathscr{E}_{1,0} \to \mathscr{MR}_{1,0}, \qquad \mathcal{F} \mapsto \delta\mathcal{H}_{\mathcal{F}}, \tag{22.30}$$

which associates with each saddle-node from the formal class $\mathscr{E}_{1,0}$, the equivalence class of coboundary of its normalizing cochain $\mathcal{H}_{\mathcal{F}}$, i.e., the Martinet–Ramis modulus. The components of this modulus play different roles and will be separately referred to as the *shift component* and the *parabolic component* of the Martinet–Ramis modulus.

Theorem 22.20 (Analytic classification theorem for saddle-nodes)**.**

1. (Invariant). *The Martinet–Ramis moduli of holomorphically equivalent saddle-node foliations coincide.*

2. (Equimodality and equivalence). *Conversely, two formally equivalent saddle-nodes with the same Martinet–Ramis modulus, are analytically equivalent.*

3. (Realization). *Any element from the space* $\mathscr{MR}_{1,0}$ *is the Martinet–Ramis modulus of a suitable saddle-node from the formal class* $\mathscr{E}_{1,0}$.

Martinet–Ramis modulus depends analytically on parameters under reasonable assumptions, but we will not discuss the proof here.

Idea of the proof. The first two assertions of the theorem are fairly standard and their proof coincides almost literally with that of the parallel statements from Theorems 21.15 and 22.8.

The third statement is proved using the surgery, i.e., patching together two pieces of the same standard foliation (22.23) defined on two cylinders $\widetilde{S_\pm}$, using the given pair (φ, ψ) for constructing the transition maps over the sectors $S_\uparrow, S_\downarrow$. These maps are constructed to preserve the standard foliation. The result of this surgery (the quotient space of the disjoint union by the transition maps) is an abstract complex 2-dimensional manifold with a foliation on it. To identify this manifold as the bidisk and the formal type of the foliation as the class $\mathscr{E}_{1,0}$, the technique of quasiconformal maps is used in a way similar to that in the proof of Theorem 22.8. □

22K. Application of the Martinet–Ramis moduli: existence of center manifold and topological classification. Topological classification of saddle-nodes is much more subtle than the topological classification of parabolic self-maps and saddles. The reason is the (non)existence of the center manifold that cannot be determined by any finite order jet of the foliation at the singular point.

Proposition 22.21. *A saddle-node has a holomorphic center manifold if and only if the shift component of the Martinet–Ramis modulus is trivial (the identity).*

In this case the holonomy of the center manifold coincides with the parabolic component of the modulus.

Proof. The center manifold in both sectorial normalizing charts has the form $w = 0$ (all other leaves are escaping the neighborhood over the sector of jump). The leaf $w = 0$ is always automatically preserved by the parabolic component of the modulus, thus two local representations correspond to a single leaf if and only if $\varphi(0) = 0$. The second assertion is trivial, since solutions of the normal form are single-valued over the sectors $S_\pm$. □

If a saddle-node has an analytic center manifold, then the topological invariant of the corresponding holonomy (the natural number p; see Theorem 21.2) is obviously a topological invariant of the singular foliation. It turns out that there are no other topological invariants.

Theorem 22.22 (P. Elizarov [**Eli88**]). *All saddle-nodes without analytic center manifold from the formal class $\mathscr{E}_{1,0}$, are topologically equivalent.*

Two saddle-nodes from this formal class, both having analytic central manifolds, are topologically equivalent if and only if the corresponding holonomy germs are topologically equivalent.

General saddle-nodes from the formal class $\mathscr{E}_{p,\lambda}$ can also be studied using this technique; see [**Eli88, Sad87**]. In this case the theory is more involved, yet at the end a complete classification (both analytic and, based on it, topological) is available.

22L. Concerning the sectorial normalization theorem. The sectorial normalization Theorem 22.16 is very similar to the Sibuya Theorem 20.16. We will explain only the basic ideas of the proof of this theorem, which can be found in the original book [**HKM61**] and in [**Żoł06**, §9.59].

The sectorial transformation which preserves the z-coordinate and transforms the equation in the preliminary normal form (22.24) to the normal form (22.23) can be obtained as a solution to certain integral equation. The integral operator associated with this equation is the composition of a resolvent for the linear differential equation, and an operator of argument shift.

This equation is easier to write with respect to the new independent variable $t = 1/z$. In this case we have the following two differential equations:

$$\frac{dw}{dt} = w + t^{-N} f(t, w), \qquad \frac{dy}{dt} = y; \tag{22.31}$$

and look for a transformation conjugating them, under the form

$$(t, y) \mapsto (t, w), \qquad w = y + h(t, y), \tag{22.32}$$

where h is a function holomorphic over the sector $t \in S_+$, $|t| > R \gg 1$, $|y| < 1$.

Construction of the conjugacy consists of two steps: first, we show that there exists a piece of the center manifold over the sector S_+ which can be rectified to become the z-axis. This allows us to assume that the function f vanishes identically on $w = 0$. In the second step we construct the conjugacy between the two systems (22.31) assuming that $f(t, 0) \equiv 0$.

1. In the first step we look for a solution of the first (nonlinear) equation (22.31). Denote by $\mathbf{S}'$ the resolvent (linear integral operator) which maps

a function $g(t)$, $t \in S_+$, into the unique solution $r = r(t)$ of the differential equation

$$\frac{dr}{dt} = r + g(t), \qquad r(t) \to 0, \quad \text{as } \operatorname{Re} t \to +\infty.$$

The explicit expression for the resolvent $\mathbf{S}'$ follows from the method of variation of constants,

$$\mathbf{S}' g(t) = \int_{\gamma'(t)} e^{t-\tau} g(\tau)\, d\tau,$$

and it is defined for all functions g bounded in S_+, if the path of integration $\gamma'(t)$ is chosen as a horizontal ray $\operatorname{Im} \tau = \text{const}$ connecting $t + \infty$ with t for all points t with $\operatorname{Im} t \gg 1$. For other points a vertical segment is added, as shown on Fig. IV.3.

Let $\mathbf{G}'$ be the operator that takes the function $r(t)$ into the function $t^{-N} f\big(t, r(t)\big)$. Using this operator, we can describe the solution of the equation (22.31), corresponding to the central manifold (i.e., which tends to zero as $|t| \to \infty$, $t \in S_+$, as the fixed point of the composition,

$$r = (\mathbf{S}' \circ \mathbf{G}')\, r.$$

One can easily verify, that the operator $\mathbf{G}'$ is strongly contracting in the same sense as in Lemma 5.14, while $\mathbf{S}'$ is bounded on the space of holomorphic functions decreasing at infinity, if N is sufficiently large. Thus the composition is contracting and has a unique fixed point, corresponding to a solution. Without loss of generality, we may assume that this solution coincides with the axis $w = 0$.

2. To conjugate the two differential equations (22.31), the function $h(t, y)$ from (22.32) must satisfy the first order partial differential equation,

$$\mathbf{L}h = h + t^{-N} f\big(t, y + h(t, y)\big), \qquad \mathbf{L}h = \left(\frac{\partial h}{\partial t} + y \frac{\partial h}{\partial y}\right). \tag{22.33}$$

The left hand side of this equation is the Lie derivative $\mathbf{L}h$ of the unknown function along the vector field corresponding to the formal normal form $\frac{dy}{dt} = y$.

Again we consider this nonlinear equation as the perturbation of the linear nonhomogeneous equation. More precisely, we will define the resolvent operator $\mathbf{S}$ such that for any function $g(t, y)$, analytic in $S_+ \times \mathbb{D}$ and decreasing as $|t| \to \infty$, the function $\mathbf{S}g$ is the solution $u(t, y)$ of the equation

$$\mathbf{L}u = u + g, \qquad g = g(t, y),\ t \in S_+,\ |y| < 1. \tag{22.34}$$

To define the operator $\mathbf{S}$, we use the method of characteristics. The homogeneous equation $\mathbf{L}u = u$ immediately admits the solution $u(t, y) = v(t, y)\, e^t$. The function $v(t, y)$ satisfies then the simplest equation $\mathbf{L}v = e^{-t} g$, which

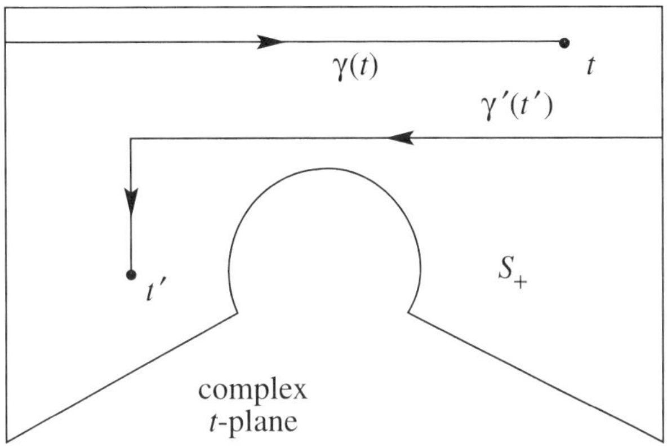

Figure IV.3. Paths of integration $\gamma'(t')$, $t' \in S_+$, for construction of the center manifold and $\gamma(t)$ for the sectorial normalization

can be solved by the taking the primitive of the function g along the trajectories $\tau \mapsto (\tau, e^\tau)$ of the standard field.

We define the solution to (22.34) by the integral

$$u = \mathbf{S}g, \qquad u(t,y) = \int_{\gamma(t)} e^{t-\tau}\, g(\tau, ye^\tau)\, d\tau. \tag{22.35}$$

However, the path of integration $\gamma(t)$ cannot be chosen as before, since ye^τ should remain bounded along this path (otherwise the value of the function g will not be defined). On the other hand, if $\operatorname{Re}\tau$ tends to $-\infty$ along the path, the exponential factor in the integral is growing to infinity, thus we should restrict the class of functions g. Assume that the function g vanishes identically on $y = 0$ and satisfies the inequality $|g(t,y)| < C|y|\,|t|^{-2}$ in $S_+ \times \mathbb{D}$. Then the integral (22.35) converges absolutely, provided that the path of integration is chosen as $t - \mathbb{R}_+$ for $|t| \gg 1$, and a vertical segment added for the remaining values of t as shown on Fig. IV.3.

Now the rest of the proof is rather standard. We define the operator of argument shift

$$\mathbf{G}\colon u(t,y) \mapsto t^{-N} f\big(t, y + u(t,y)\big), \tag{22.36}$$

and look for the solution of the partial differential equation (22.33) in the form of a fixed point for the integral operator,

$$h = (\mathbf{S} \circ \mathbf{G})\, h, \qquad h = h(t,y),\ t \in S_+, |y| < 1. \tag{22.37}$$

The composition in the right hand side is a well-defined operator on all functions h such that $h(t,0) \equiv 0$, which decrease faster than $|t|^{-2}$ as $|t| \to \infty$, $t \in S_+$. Easy estimates similar to those from the proof of Sibuya Theorem 20.16, show that this operator is contracting and hence admits a holomorphic solution $h = h(t,y)$ over the sector S_+. This solution gives sectorial normalization in Theorem 22.16. □

Exercises and Problems for §22.

Exercise 22.1. Describe the Ecalle–Voronin modulus for the monodromy of a resonant saddle *analytically* equivalent to its formal normal form (22.3).

Exercise 22.2. Prove that two resonant complex saddle foliations are topologically equivalent, if they have the same linear parts and topologically conjugate monodromies.

Exercise 22.3. Let $F \in \mathcal{D}(\mathbb{C}^{n+1}, 0)$ be the germ of a vector field whose linearization matrix A has n eigenvalues in the left half-plane $\operatorname{Re}\lambda < 0$ and one eigenvalue $\lambda = 1$ in the right half-plane ("saddle of index 1"). Prove that the corresponding singular foliation has a one-dimensional "positive" separatrix S tangent to the eigenvector with the positive eigenvalue, and compute the linearized holonomy associated with a small loop on S in terms of the block structure of A. Prove that the corresponding self-map belongs to the Poincaré domain.

Problem 22.4. For any holomorphic map $f \in \operatorname{Diff}(\mathbb{C}^n, 0)$ of Poincaré type, construct a saddle of index 1 as in the previous exercise, which has the holonomy coinciding with f.

Problem 22.5. Assume that two saddles of index 1 have the same linear parts and analytically equivalent holonomy maps along the "positive" separatrices, are analytically equivalent as singular foliations.

Problem 22.6. Prove that a topologically linearizable *resonant* saddle (singular foliation on $(\mathbb{C}^2, 0)$ topologically equivalent to the foliation defined by its linear part), is holomorphically linearizable.

Problem 22.7. Prove that the holonomy map associated with a holomorphic invariant curve of a saddle-node (tangent to the eigenvector with nonzero eigenvalue), is a parabolic germ.

Problem 22.8. Describe the formal type of the holonomy in Problem 22.7 via the formal normal form of the saddle-node.

Problem 22.9. Write a detailed proof of Lemma 22.15.

Problem 22.10. Prove that a saddle-node of the formal type $\mathscr{E}_{1,0}$ is analytically equivalent to its formal normal form if and only if the Martinet–Ramis modulus is trivial, $(\varphi, \psi) = (\mathrm{id}, \mathrm{id})$.

Problem 22.11. Describe the Ecalle–Voronin modulus of the holonomy of a saddle-node, as it was introduced in Problem 22.7, in terms of the Martinet–Ramis modulus of the saddle-node.

23. Nonlinear Riemann–Hilbert problem

The nonlinear Riemann–Hilbert problem is a natural analog of its linear counterpart considered in detail in §18. Analytically the problem consists of reconstruction of a *nonlinear* differential equation from its monodromy data. We restate it in a geometric language.

23A. Statement of the problem. Consider a holomorphic one-dimensional singular foliation $\mathcal{F}$ on a complex analytic manifold M^{n+1} of any dimension $n+1$ greater than or equal to 2, near a separatrix $S \subset M$ (a compact invariant holomorphic curve carrying one or more singular points of $\mathcal{F}$). The holonomy group of this foliation is a finitely generated subgroup $H = H_{\mathcal{F}}$ in the group of germs of automorphisms $\operatorname{Diff}(\mathbb{C}^n, 0)$ of a generic cross-section to S. The general form of the Riemann–Hilbert problem is the inverse problem of constructing the foliation $\mathcal{F}$ starting from the separatrix S, the manifold M and the finitely generated subgroup $H \subset \operatorname{Diff}(\mathbb{C}^n, 0)$ subject to certain restrictions on the types of singular points.

In our considerations the separatrix will always be the Riemann sphere $\mathbb{P}$, though the ambient manifold can vary; cf. with §23**D**.

The (linear) Riemann–Hilbert problem considered in §18, corresponds to the case where $M = \mathbb{P} \times \mathbb{C}^n$ is the trivial holomorphic vector bundle over $S = \mathbb{P}$ and H is a linear subgroup of $\mathrm{GL}(n, \mathbb{C})$. The singular points of the corresponding differential equation were required to be of Fuchsian type. We introduce now a nonlinear analog of this type. Recall that (by definition) the foliation $\mathcal{F}$ near each singularity is locally defined by the germ of a holomorphic vector field v having an isolated singular point. The vanishing order $\varkappa_a(\mathcal{F}, S)$ of $\mathcal{F}$ along S was introduced in Definition 14.25.

Definition 23.1. A singular point $a \in S$ is called *nonlinear Fuchsian*, if the vanishing order of $\mathcal{F}$ along S is equal to 1, i.e., if the restriction $v|_S$ has a simple singular point at a.

Since S is smooth, this condition means that the linearization of the vector field at this point is a linear operator $A = v_{*,a}$ having a *nonzero* eigenvalue associated with the eigenvector tangent to S at a. Without loss of generality we can assume that the eigenvalue in question is equal to 1. Then in the local coordinates (t, x), in which the separatrix coincides with the t-axis, the vector field v generating $\mathcal{F}$, takes the form

$$\dot{t} = t + \cdots, \qquad \dot{x} = Ax + \cdots,$$

where the dots denote nonlinear terms of order 2 and higher; cf. with the formula (16.4) describing *linear* Fuchsian systems.

Definition 23.2. A foliation is said to be of the class NF ("nonlinear Fuchsian") along a separatrix S, if all its singular points on S are nonlinear Fuchsian.

In the most simple settings, the manifold M is just a Cartesian product $\mathbb{P} \times (\mathbb{C}^n, 0)$, a thin cylinder over $\mathbb{P}$. Denote by $\gamma_0, \dots, \gamma_m$ the loops generating the fundamental group of the leaf $\mathbb{P} \smallsetminus \{a_0, \dots, a_m\}$, where a_j are preassigned singular points (the product $\gamma_0 \circ \cdots \circ \gamma_m$ is trivial). If $\mathcal{F}$ is a foliation on M

with the leaf $\mathbb{P} \smallsetminus \{a_0, \dots, a_m\}$, then it defines the collection of the holonomy operators $\Delta_{\gamma_j, \mathcal{F}} \in \operatorname{Diff}(\mathbb{C}, 0)$ associated with the loops γ_j.

Nonlinear Riemann–Hilbert problem (Cartesian version). *Given a collection of several holomorphic germs $g_0, \dots, g_m \in \operatorname{Diff}(\mathbb{C}^n, 0)$ such that their composition is identity, $g_0 \circ g_1 \circ \cdots \circ g_m = \mathrm{id}$, construct a holomorphic foliation $\mathcal{F}$ of the class NF on $M = \mathbb{P} \times (\mathbb{C}^n, 0)$ with nonlinear Fuchsian singularities only at the points a_j and the preassigned holonomy operators $\Delta_{\gamma_j, \mathcal{F}} = g_j$, $j = 0, \dots, m$.*

If $\mathcal{F}$ is the foliation solving the Riemann–Hilbert problem, its linearization along the separatrix is a meromorphic connexion on the trivial n-dimensional vector bundle over $\mathbb{P}$, whose monodromy operators associated with the loops γ_j will be linearizations M_j of the nonlinear germs g_j. Therefore, the corresponding linear Riemann–Hilbert problem would be solvable. This observation yields an obvious *necessary condition* for the solvability of the Nonlinear Riemann–Hilbert problem, solvability of the corresponding linear problem.

The natural question arises: *whether this necessary condition is also sufficient*? in other words, *are there essentially nonlinear obstructions for solvability of the nonlinear Riemann–Hilbert problem*? We show below that even in the most simple case $n = 1$, when any linear problem is trivially solvable, there exist nonlinear obstructions.

23B. One-dimensional case: the example. In the linear one-dimensional case the holonomy group is commutative generated by $m + 1$ linear maps $x \mapsto \nu_j x$, $j = 0, \dots, m$, such that

$$\nu_0 \nu_1 \cdots \nu_m = 1. \tag{23.1}$$

A meromorphic connexion defined by the differential equation

$$dx = \omega x, \qquad \omega = \sum_{j=0}^{m} \lambda_j \frac{dt}{t - a_j}, \tag{23.2}$$

with the meromorphic form ω having simple poles at the points $a_0, \dots, a_m$ has the preassigned monodromy group, if

$$\exp 2\pi i \lambda_j = \nu_j, \qquad j = 0, 1, \dots, m. \tag{23.3}$$

The point at infinity must be nonsingular, which translates into the condition

$$\lambda_0 + \cdots + \lambda_m = 0. \tag{23.4}$$

The linear Riemann–Hilbert problem would be solved, if a collection of residues $\{\lambda_0, \dots, \lambda_m\}$ meeting the conditions (23.3)–(23.4) can be constructed for the given collection of multipliers $\{\nu_0, \dots, \nu_m\}$. Clearly, (23.1) is a necessary condition for solvability of the system (23.3)–(23.4). Yet it is also sufficient. Indeed, if $\nu_0 \cdots \nu_m = 1$, then $\lambda_0 + \cdots + \lambda_m = k \in \mathbb{Z}$ for any choice of the logarithms (solutions of (23.3)). But one can always replace λ_0 by $\lambda_0 - k$ to ensure that (23.4) holds.

However, when passing from a linear context to the nonlinear one, we may loose the freedom of choosing the additive integer term k of the residue arbitrarily. If g is a conformal germ with a multiplicator ν, $g(x) = \nu x + \cdots$, then it can be realized as a holonomy operator of a nonlinear Fuchsian singular point with the ratio of eigenvalues λ only if $\exp 2\pi i\lambda = \nu$. If $|\nu| = 1$ and ν is nonresonant (not a root of unity), then the residue λ must be real irrational. The case $\lambda > 0$ corresponds to a singular point of the Poincaré type (cf. §5**A**), hence the corresponding monodromy germ g should necessarily be analytically linearizable by Poincaré Linearization Theorem 5.5. Thus a germ g *which is not analytically linearizable*, can be realized as a monodromy map of only a nonlinear saddle corresponding to $\lambda < 0$. This inequality, if it holds for *every singularity*, is an obstruction to the condition (23.4).

In more details, the above argument shows that the group generated by nonresonant *analytically nonlinearizable* germs $g_0, \ldots, g_m$ cannot be realized as the holonomy group of a holomorphic foliation; if this were possible, then the sum of all residues of the linearization of the vector field would be *strictly negative*, contradicting (23.4). Such examples can occur already for $m = 2$ (i.e., with three singular points). In the following section a general necessary and sufficient condition for solvability of the one-dimensional Nonlinear Riemann–Hilbert problem is given.

23C. Local Riemann–Hilbert problem. Since a hyperbolic conformal germ is always analytically linearizable, we consider only nonhyperbolic germs with the multiplicators on the unit circle. Such multiplicators have the form $\nu = \exp 2\pi i\lambda$ with λ real; the germ is resonant if and only if $\lambda \in \mathbb{Q}$. It is convenient to introduce the *normalized logarithm* of numbers on the unit circle, choosing the branch as follows:

$$|\nu| = 1, \quad \ln^{-} \nu = \lambda \iff \exp 2\pi i\lambda = \nu, \quad -1 \leqslant \lambda < 0. \tag{23.5}$$

A nonresonant conformal germ is always formally linearizable, but may be not analytically linearizable. Such "pathological" germs will be referred to as *Cremer germs* (the term "Cremer point" being common in holomorphic dynamics [**Mil99**]). Their existence (and even abundance for certain irrational values of λ) can be proved by the methods described in §5**G**.

A resonant germ may be formally linearizable, but in this case an appropriate iterational power of the germ is a formally linearizable map tangent to identity, i.e., the identity map itself. In such a case the formal conjugacy is in fact analytic and the initial germ (a root of identity) is analytically linearizable. Resonant nonlinearizable germs were discussed in detail in §22. By Theorem 22.9, any such germ can be realized as the monodromy map of a holomorphic separatrix for a nonlinear resonant saddle with a negative ratio of eigenvalues $\lambda \in -\mathbb{Q}_+$. Yet *some* resonant germs can *also* be realized as monodromies of a *nonlinear resonant node*; see Table I.1,

$$\begin{cases} \dot{x} = nx + ay^n, \\ \dot{y} = y. \end{cases} \tag{23.6}$$

with a positive ratio of eigenvalues $1 : n$. Such a node, always analytically equivalent to its formal normal form (23.6), has a *unique* holomorphic smooth separatrix through the origin. We will refer to germs that can be realized as monodromies of resonant nodes as *Dulac germs*. Clearly, a necessary condition for being a Dulac germ is $\nu = \exp 2\pi i/n$ for some $n \geqslant 2$.

Remark 23.3. The property of being a Dulac germ cannot be determined by any finite order jet, yet their existence is obvious.

This classification is designed to make the following inequalities true.

Lemma 23.4. 1. *If a nonhyperbolic analytically nonlinearizable germ with multiplicator ν on the unit circle is realized as the monodromy map of a nonlinear Fuchsian singular foliation with the ratio of eigenvalues $\lambda \in \mathbb{R}$, then*

$$\lambda \leqslant \begin{cases} \ln^- \nu + 1, & \textit{for Dulac germs,} \\ \ln^- \nu & \textit{otherwise.} \end{cases} \tag{23.7}$$

2. *Conversely, any nonhyperbolic analytically nonlinearizable conformal germ can be realized as the monodromy of a nonlinear Fuchsian singular foliation with the ratio of eigenvalues satisfying the inequality* (23.7).

Proof. If a Dulac germ is realized as the monodromy map of a resonant node (23.6), then the corresponding ratio of eigenvalues is $\frac{1}{n}$ and the multiplicator $\nu = \exp 2\pi i/n$. By definition of $\ln^-$, the branch of the normalized logarithm should be chosen so that $\ln \nu = -1 + \frac{1}{n}$ and we have the equality $\lambda = \ln^- \nu + 1$. If the germ (Dulac or Cremer) is realized as the monodromy map of a nonlinear resonant saddle, then $\ln^- \nu$ is the maximal value for the ratio of eigenvalues that is still negative; choosing any bigger value would mean that the singularity is a node rather than a saddle.

The second assertion of the lemma follows immediately from Theorem 22.9. □

Lemma 23.4 immediately implies the necessity assertion of the following theorem giving a complete solution of the Nonlinear Riemann–Hilbert problem in the one-dimensional case.

Theorem 23.5. *A collection of conformal germs* $g_0, \dots, g_m \in \operatorname{Diff}(\mathbb{C}^1, 0)$ *satisfying the condition* $g_0 \circ \cdots \circ g_m = \mathrm{id}$ *can be realized as generators of the holonomy group of a foliation of the class NF on the trivial bundle* $\mathbb{P} \times (\mathbb{C}^1, 0)$ *if and only if one of the following two conditions hold*:

(1) *at least one germ* g_j *is linearizable, or*

(2) *the collection contains* k *Dulac germs, and*

$$k + \sum_0^m \ln^- \nu_j \geqslant 0, \qquad \nu_j = \frac{dg_j}{dx}(0). \tag{23.8}$$

Indeed, if the linearization of the foliation $\mathcal{F}$ realizing the prescribed holonomy group is described by the linear equation (23.2), and all singularities are nonlinearizable, then Lemma 23.4 applies to all of them. Combining the equality (23.4) with the inequalities (23.7), we obtain the inequality

$$0 = \sum_0^m \lambda_j \leqslant \sum_{\text{Dulac}} (1 + \ln^- \nu_j) + \sum_{\text{other}} \ln^- \nu_j = k + \sum_{\text{all}} \ln^- \nu_j.$$

In the next subsection we derive the global sufficiency assertion of Theorem 23.5 from local sufficiency assertions of Lemma 23.4.

23D. Sufficiency of the solvability conditions. The proof of the sufficiency part of Theorem 23.5 is organized along the same lines as in §22: we construct a singular holomorphic foliation on an abstract holomorphic 2-manifold M, which realizes the specified holonomy group, and then identify M as a neighborhood of the Riemann sphere $\mathbb{P} \times \{0\}$ in the Cartesian product $\mathbb{P} \times (\mathbb{C}^1, 0)$.

In the first step we construct a nonsingular foliation on the open neighborhood M' of the holed sphere U obtained by deleting from $\mathbb{P}$ small disjoint disks $D'_0, \dots, D'_m$ around the singular points $a_0, \dots, a_m$ of the singular locus $\Sigma \subset \mathbb{P}$. The holed sphere U itself will be the leaf L_0 of this foliation, and the holonomy group of it will coincide with H (note that the fundamental groups of U and $\mathbb{P} \smallsetminus \Sigma$ coincide). The construction, the "simultaneous suspension" of several holomorphic self-maps, is organized along the same lines as the suspension of a single self-map in Theorem 2.31.

In the second step we seal the holes in M' with the cylinders $D_j \times (\mathbb{C}^1, 0)$ carrying singular foliations $\mathcal{F}_j$, in such a way that their separatrices $D_j \times \{0\}$ will be sealing the holes in the leaf L_0; here $D_j \Supset D'_j$ are slightly bigger disks sealing the holes on U. The singular foliations $\mathcal{F}_j$ of the class NF, constructed in the second assertion of Lemma 23.4, have the preassigned monodromy maps associated with these separatrices. The freedom of choice of the ratios λ_j of the corresponding eigenvalues is constrained by the inequalities (23.7). In the assumptions of the theorem one can use the remaining freedom to guarantee that $\sum_j \lambda_j$ is zero. As a result, we obtain a singular foliation $\mathcal{F}$ defined on an abstract 2-dimensional manifold M, which is of class NF with respect to the leaf $L \cong \mathbb{P}$ carrying only singularities of the class NF at the specified points.

The sealing step can be implemented preserving the "horizontal coordinate in the direction of L", so that leaves of the singular patches $\mathcal{F}_j$ are graphs of solutions of suitable ordinary differential equations, except for the vertical separatrices through the singular points.

Thus together we obtain an abstract manifold M with a singular foliation $\mathcal{F}$ on it, a "horizontal" separatrix $L \cong \mathbb{P}$ of this foliation, carrying all nonlinear Fuchsian singularities of $\mathcal{F}$, and a holomorphic projection $\pi\colon M \to L$ of constant rank 1, which is transversal to the foliation over all nonsingular points.

Consider the *normal bundle* of the embedded curve L, i.e., by definition, the quotient bundle $\mathbf{T}M/\mathbf{T}L$ over $\mathbb{P}$. The linearization of the 1-form determining $\mathcal{F}$ yields a meromorphic connexion on the normal bundle with Fuchsian singularities at the points a_j only; the residues of this connexion are the ratios λ_j. The degree of this bundle is equal to the sum of all ratios $\sum_j \lambda_j$ by Theorem 17.33. In our construction this sum is equal to zero, that is, the normal bundle of L in M is trivial, $\mathbf{T}M/\mathbf{T}L \cong \mathbb{P} \times \mathbb{C}^1$, as explained in §17**D**.

At the final stage of the proof we use the Savel′ev–Grauert Theorem 23.6 to show that if the normal bundle is trivial, then the manifold M itself is biholomorphically equivalent to the cylinder $\mathbb{P} \times (\mathbb{C}^1, 0)$ as requested.

We pass on to the detailed exposition. Let $\widehat{U}$ be the universal covering over U. This is a Riemann surface whose points are pairs (t, γ), where t is a point in U and ρ is the homotopy class of a path connecting t with a fixed base point $a_* \in U$. The fundamental group $\pi_1(U, a_*)$ naturally acts on the universal covering: a loop $\gamma \in \pi_1(U, a_*)$ sends (t, ρ) to $(t, \rho \circ \gamma)$. The automorphisms of $\widehat{U}$ defined in such a way are called *covering transformations* or *deck transformations*; see [**For91**].

Any representation $H\colon \pi_1(U, a_*) \to \operatorname{Diff}(\mathbb{C}^1, 0)$ of the fundamental group of U by conformal germs defines the action of the fundamental group $\pi_1(U, a_*)$ on the Cartesian product $\widehat{M} = \widehat{U} \times (\mathbb{C}^1, 0)$: a loop γ acts by the transformation G_γ as follows:

$$G_\gamma\colon (t, \rho, z) \mapsto (t, \rho \circ \gamma, g^{-1}(z)), \qquad g = H(\gamma) \in \operatorname{Diff}(\mathbb{C}^1, 0). \tag{23.9}$$

The quotient space $\widehat{M} = \widehat{U} \times (\mathbb{C}^1, 0)/G$ (the space of orbits of this action) is a holomorphic 2-manifold equipped with the natural projection $\widehat{\pi}$ on U.

The Cartesian product $\widehat{U} \times (\mathbb{C}^1, 0)$ carries the trivial holomorphic (nonsingular) foliation by the curves $\{z = \text{const}\}$. These curves are locally preserved by the action (23.9) and hence the quotient space M' gets equipped with a well-defined foliation $\mathcal{F}'$. Since all germs $H(\gamma)$ fix the origin, $U \times \{0\}$ is a well-defined embedded curve in M' which is a leaf L of the foliation $\mathcal{F}_M$. By construction, the holonomy of $\mathcal{F}'$ associated with the leaf L, coincides with the group H. The projection $\widehat{\pi}\colon \widehat{M} \to U$ factors through the natural projection $\pi'\colon M' \to U$ which makes M' into a one-dimensional (nonlinear) bundle over the holed sphere U.

In the next step we seal the holes by bidisks $D_j \times (\mathbb{C}^1, 0)$, $j = 0, 1, \dots, m$, where $D_j \supset D'_j$ are slightly bigger (but still disjoint) disks around the deleted singularities. On each such bidisk we consider a holomorphic foliation $\mathcal{F}_j$ with a unique nonlinear Fuchsian singular point, whose holonomy map realizes the preassigned conformal germ g_j.

It is important that under the assumption (23.8) the foliations $\mathcal{F}_j$ can be chosen so that the corresponding ratios of eigenvalues λ_j satisfy the equality (23.4). Indeed, under this condition the natural number $l = -\sum_0^m \ln^- \nu_j$ does not exceed the number k of Dulac germs in the given collection $\{g_0, \dots, g_m\}$. We choose any l Dulac germs and realize them as holonomy maps of resonant nodes (this is possible by the definition of Dulac germs), while all other germs will be realized as holonomy maps of saddles (resonant or nonresonant) with the ratios of eigenvalues λ_j exactly equal to the respective normalized logarithms $\ln^- \nu_j$. Finally, if one of the germs, say, g_0 is holomorphically linearizable, then one can always realize it as the holonomy group of a linear singularity $\mathcal{F}_0$ with the ratio of eigenvalues λ_0 such that (23.4) holds no matter what the other ratios were.

Formally, the sealing of the holes in M' is organized in a way resembling the construction of suspension (Theorem 2.31). Denote by $\pi_j\colon D_j \times (\mathbb{C}^1, 0) \to D_m$ the projections parallel to the second Cartesian components. Consider the disjoint union

$$M' \sqcup D_0 \times (\mathbb{C}^1, 0) \sqcup \cdots \sqcup D_m \times (\mathbb{C}^1, 0) \tag{23.10}$$

with the following identification of points. The intersection $\big(D_j \times (\mathbb{C}^1, 0)\big) \cap M$ is biholomorphically equivalent to the cylinder over the annulus $K_j = D_j \cap (\mathbb{P} \smallsetminus D'_j)$. Take any two cross-sections ${\pi'}^{-1}(t, 0)$ and $\pi_j^{-1}(t, 0)$ to $\mathcal{F}_j$ and $\mathcal{F}'$ respectively at the same point $(t, 0)$, $t \in K_j$, and identify these cross-sections in an arbitrary (holomorphically invertible) way. This identification can be uniquely extended along the leaves of the foliations by analytic continuation, conjugating at the same time π' with π_j. Since the germs of the holonomy maps associated with the middle circle loop of K_j for both $\mathcal{F}'$ and $\mathcal{F}_j$ are equal to the same germ g_j (by construction), the identification of points on the transversals extends to identification (biholomorphic maps) of cylinders $\pi'^{-1}(K_j)$ with $\pi_j^{-1}(K_j)$ sending leaves to leaves.

As a result of sealing the holes, we obtain the quotient foliation $\mathcal{F}$ on the quotient space M of the disjoint union (23.10) by the equivalence relation obtained via the above identification of points. The space M inherits the natural structure of a holomorphic (nonlinear) bundle over $L \cong \mathbb{P}$: the projections π' and $\pi_1, \dots, \pi_m$ together define a well-defined holomorphic projection $\pi\colon M \to L$. The holonomy group of this foliation by construction coincides with the group generated by the specified germs g_j. What remains is to show that the surface M itself is biholomorphically equivalent to the trivial cylinder $\mathbb{P} \times (\mathbb{C}^1, 0)$.

The holomorphic curve $L \cong \mathbb{P}$ is regularly embedded into the surface M. Consider its *normal bundle*, a linear holomorphic vector bundle over L whose fibers are the quotient spaces $\mathbf{T}_a M/\mathbf{T}_a L$ of complex dimension 1. The holomorphic type of any line bundle is completely determined by its degree; see §17**D**. In particular, a line bundle of degree 0 is trivial, i.e., biholomorphically equivalent to the cylinder $L \times \mathbb{C}$. By Theorem 17.33, this degree is equal to the sum of residues of any meromorphic connexion on this bundle.

Linearization of the foliation $\mathcal{F}$ along the curve L yields, as explained in §14**B**, such a meromorphic connexion. In any local chart this connexion is defined by a meromorphic differential 1-form with poles at the singular points $a_j \in \Sigma$; cf. with (23.2). The corresponding residues are the ratios λ_j of eigenvalues of nonlinear Fuchsian singularities of $\mathcal{F}$. By construction of the foliation $\mathcal{F}$, the sum of residues is equal to zero, hence *the normal bundle of L in M has degree* 0.

The assertion of the theorem now follows from the following theorem due to H. Grauert (1962, for negative degree) and V. Savel′ev (1982, for zero degree).

Theorem 23.6 (H. Grauert [**Gra62**], V. I. Savel′ev[**Sav82**]). *If the normal bundle of an embedded Riemann sphere $\mathbb{P} \cong L \subset M$ has a nonpositive degree, then a small neighborhood of L in M is biholomorphically equivalent to the neighborhood of the null section in the normal bundle.*

Indeed, a zero degree line bundle is trivial, hence M near L is locally biholomorphically equivalent to the cylinder, as required. The proof of Theorem 23.5 is complete. □

Appendix I. Nonlinear Riemann–Hilbert problem on the exceptional divisor. In the formulation of the nonlinear Riemann–Hilbert problem as stated in §23**A**, choosing the manifold M to be a Cartesian product of the separatrix and the (poly)disk is not the only natural one.

Recall (cf. with §11**C**) that any germ of a vector field of order m which has a nondicritical blow-up, has a naturally defined *vanishing holonomy group* (the holonomy of the exceptional divisor after a simple blow-up), which for a generic germ of order m is generated by exactly $m+1$ conformal germs; see p. 122. In this appendix we give necessary and sufficient conditions for a subgroup of $\operatorname{Diff}(\mathbb{C}^1,0)$ of conformal germs to be realizable as the vanishing holonomy.

Denote by $\mathscr{N}_m$ the class of (germs of singular holomorphic) foliations generated by holomorphic vector fields $F = (p_m + \cdots)\frac{\partial}{\partial x} + (q_m + \dots)\frac{\partial}{\partial y}$ of order m such that the homogeneous polynomial $h_{m+1} = yp_m - xq_m$ is square-free.

Theorem 23.7. *A collection of conformal germs $g_0,\dots,g_m \in \operatorname{Diff}(\mathbb{C}^1,0)$ satisfying the condition $g_0 \circ \cdots \circ g_m = \mathrm{id}$ can be realized as generators of the vanishing holonomy group of a foliation of the class $\mathscr{N}_m$, if and only if one of the two conditions hold*:

(1) *at least one germ g_j is linearizable, or*

(2) *the collection contains k Dulac germs, and*

$$k + \sum_0^m \ln^- \nu_j \geqslant -1, \qquad \nu_j = \frac{dg_j}{dx}(0). \tag{23.11}$$

Proof. After a simple blow-up $\sigma\colon (\mathbb{M},\mathbb{E}) \to (\mathbb{C}^2,0)$ a foliation from the class $\mathscr{N}_m$ on $(\mathbb{C}^2,0)$ becomes a holomorphic singular foliation on the complex Möbius band $\mathbb{M}$ near the exceptional divisor $\mathbb{E}$ (cf. with Definition 8.11). The exceptional divisor is a separatrix of this foliation, and all singularities are nonlinear Fuchsian by Proposition 8.18. The sum of residues of the connexion linearizing any foliation having $\mathbb{E}$ as the separatrix, is equal to -1, as explained in Theorem 14.7.

Exactly the same arguments that prove Theorem 23.5, show that the assumptions of the theorem are necessary and sufficient for existence of a singular holomorphic foliation on a neighborhood of zero section of the line bundle of degree -1 over $\mathbb{P}$ with the specified holonomy. By the Grauert Theorem 23.6, any such bundle is locally biholomorphically equivalent to the bundle $(\mathbb{M},\mathbb{E}) \to \mathbb{E}$.

The blow-up projection $\sigma\colon (\mathbb{M},\mathbb{E}) \to (\mathbb{C}^2,0)$ carries the constructed holomorphic foliation to a holomorphic foliation on the punctured neighborhood of the origin in $\mathbb{C}^2$. By the removable singularity theorem, such a foliation holomorphically extends to the origin and necessarily is of the class $\mathscr{N}_m$. □

Appendix II. Demonstration of the Savel′ev theorem. We give here the proof of the Savel′ev theorem in the particular form we need. Let

$\pi\colon M \to \mathbb{P}$ be a holomorphic one-dimensional bundle (holomorphic projection of constant rank one) over the embedded Riemann sphere $\mathbb{P} \hookrightarrow M$. This nonline bundle can be linearized: the linear fiber over a point $t \in \mathbb{P}$ is the tangent space to the fiber $\pi^{-1}(t)$, i.e., the kernel $\operatorname{Ker} d\pi \subset \mathbf{T}_t M$. Because of the condition on the rank of π, this kernel is always transversal to the tangent subspace to $\mathbb{P}$; this allows us to identify the above bundle with the normal bundle $N = \mathbf{T}M/\mathbf{T}\mathbb{P}$ of the embedded curve $\mathbb{P} \hookrightarrow M$.

Theorem 23.8. *Assume that the normal bundle N of an embedded projective line $\mathbb{P} \hookrightarrow M$ has degree 0 and hence is trivial. Then the bundle $\pi\colon M \to \mathbb{P}$ itself is locally holomorphically trivial, i.e., there exist a biholomorphism between a neighborhood of $\mathbb{P}$ in M and a cylinder $\mathbb{P}\times(\mathbb{C}^1, 0)$ which conjugates π with the Cartesian projection on $\mathbb{P}$.*

Proof. Consider the covering of the Riemann sphere $\mathbb{P}$ by two open circular disks $U_\pm$ intersecting by an annulus $K \subset \mathbb{P}$; we will work in the affine chart such that $K = \{\frac{1}{2} < |t| < \frac{3}{2}\}$.

By the Y.-T. Siu theorem [**Siu77**, Corollary 2], we may assume that the bundle π is trivialized over these disks[5]. In other words, each of the open sets $\pi^{-1}(U_\pm)$ can be equipped with the local coordinates $(t, x_\pm) \in U_\pm \times (\mathbb{C}^1_\pm, 0)$ such that the π is the projection parallel to the respective $x_\pm$-coordinate on the $t_\pm$-axis.

The transition function between the two charts respects the map π defined globally, hence must have the form

$$(t, x_-) \mapsto (t, x_+), \qquad x_+ = F(t, x_-) = x_- + f(t, x_-). \tag{23.12}$$

The linearization $\varphi(t) = \frac{\partial F}{\partial x}(t, 0) = 1 + \frac{\partial f}{\partial x}(t, 0)$ defines a scalar Birkhoff-Grothendieck cocycle $\varphi\colon U_+ \cap U_- \to \mathbb{C} \smallsetminus \{0\}$ and determines the linear (1-dimensional) normal bundle N, as explained above. By assumption of the theorem, this bundle is trivial, $\varphi = \varphi_+/\varphi_-$ for appropriate holomorphic functions $\varphi_\pm$ nonvanishing in $U_\pm$ respectively. Replacing the coordinate functions x_+, x_- by $\varphi_+(t)\,x_+$ and $\varphi_-(t)\,x_-$ respectively, one may guarantee that the function $f(t, x)$ has no linear terms in its Taylor expansion in x,

$$f(t, x) = q(t)x^2 + \cdots, \qquad t \in K = U_+ \cap U_-. \tag{23.13}$$

The problem of trivialization of the bundle $\pi\colon M \to \mathbb{P}$ globally over the union of the charts $U_+ \cup U_-$ reduces to finding two new holomorphic charts which would agree over the intersection $\pi^{-1}(U_-) \cap \pi^{-1}(U_+)$. Denoting this (common) chart by x, we are hereby looking for the holomorphic functions $x_\pm = x + h_\pm(t, x)$ satisfying (23.12). This latter condition is a functional equation

$$x + h_-(t, x) + f\big(t, x + h_-(x)\big) = x + h_+(t, x), \tag{23.14}$$

[5] In fact, the Siu theorem holds in any dimension for any embedded Stein manifold.

which has to be solved with respect to the pair of functions $h_\pm(t,x)$, holomorphic in $U_\pm \times (\mathbb{C}^1, 0)$ respectively.

Note the similarity between the equation (23.14) and (5.6) that arises in the proof of the Poincaré theorem on analytic linearization. Not surprisingly, the method of the proof is similar.

Consider first the homological equation obtained by "linearization" (ignoring the argument shift) of (23.14). Omitting the (common after such "linearization") arguments, we obtain the linear functional equation

$$h_- - h_+ = f. \tag{23.15}$$

This equation can be instantly solved by expanding f in the (convergent) Laurent series and taking h_- as the sum of its Taylor part and H_+ as the sum of all negative powers of t. The operator

$$L\colon f \mapsto h = (h_-, h_+) \tag{23.16}$$

is bounded (this can also be seen from the Cauchy representation; see Problem 23.1).

Consider now the operator of argument shift

$$S = S_f\colon h \mapsto f \circ (\mathrm{id} + h_-), \tag{23.17}$$

defined on pairs of holomorphic functions and taking values in functions holomorphic in $K \times (\mathbb{C}^1, 0)$. More specifically, we introduce the scale of Banach spaces $\mathcal{B}^\pm_\rho$, $\mathcal{B}^0_\rho$ of functions holomorphic on $U^\pm \times \{|x| \leqslant \rho\}$ and $K \times \{|x| \leqslant \rho\}$ respectively, equipped with the maximum modulus norm. Then the operator of argument shift S_f is strongly contracting in the sense of Definition 5.13. Indeed, since the function $f = f(t,x)$ has no constant and linear terms in x for all t, the arguments of Lemma 5.14 apply almost *verbatim*: $\|S_f(0)\|_\rho = \|f\|_\rho = O(\rho^2)$, and similarly $\|S_f(h) - S_f(h')\|_\rho \leqslant O(\rho)\|h - h'\|_\rho$ if $\|h\|_\rho, \|h'\|_\rho \leqslant \rho$.

The functional equation (23.14) can be rewritten as the equation for the fixed point $h = (h_+, h_-)$ of the composition operator $L \circ S_f$,

$$h = L \circ S_f(h), \qquad h \in \mathcal{B}^+_\rho \times \mathcal{B}^-_\rho. \tag{23.18}$$

The composition of a bounded operator L with a strongly contracting operator S_f is contracting for all sufficiently small $\rho > 0$. By the fixed point theorem, the equation (23.18) (and together with it (23.14)) has a unique holomorphic solution. This completes the proof of Savel′ev's theorem. □

Exercises and Problems for §23.

Problem 23.1. Let $K = \{1/2 < |z| < 3/2\}$ be the annulus, and $P_\pm$ the integral Cauchy operators,

$$(P_+ f)(z) = \frac{1}{2\pi i} \oint_{|z|=3/2} \frac{f(\zeta)\, d\zeta}{\zeta - z}, \qquad (P_- f)(z) = -\frac{1}{2\pi i} \oint_{|z|=1/2} \frac{f(\zeta)\, d\zeta}{\zeta - z},$$

representing a function $f \in \mathcal{A}(K)$ as a difference of two functions $h_\pm = P_\pm f$, holomorphic and bounded in the disks $\{|z| < 3/2\}$ and $\{|z| > 1/2\}$ on the Riemann sphere $\mathbb{P}$. Denote by $\|f\| = \max_{\zeta \in K} |f(\zeta)|$ the supremum-norm.

Prove that $|P_+ f(z)| \leqslant 3\|f\|$ if $|z| \leqslant 1$ and $|P_- f(z)| \leqslant 3\|f\|$ if $|z| \geqslant 1$.

Prove that $|P_+ f(z)| \leqslant \|f\| + 3\|f\| = 4\|f\|$ for $|z| > 1$. Prove that the norm of the operator P_+ is no greater than 4.

Problem 23.2. Assume that a foliation of the class NF on the Cartesian product $\mathbb{P} \times (\mathbb{C}, 0)$ with the coordinates (z, w) has all vertical separatrices $\{z = a_j\}$, $a_1, \dots, a_n \in \mathbb{P}$, and the point at infinity is nonsingular.

Prove that this foliation can be defined by an ordinary differential equation

$$\frac{dw}{dz} = \sum_{j=1}^{n} \frac{f_j(w)}{z - a_j}, \qquad \sum_{j=1}^{n} f_j(w) \equiv 0, \tag{23.19}$$

with holomorphic germs $f_1, \dots, f_n \in \mathcal{O}(\mathbb{C}, 0)$.

Exercise 23.3. Prove that for $n \geqslant 4$ *not all foliations* from the class NF on the Cartesian cylinder $\mathbb{P} \times (\mathbb{C}, 0)$ can be brought into the form (23.19). What happens for $n \leqslant 3$?

Problem 23.4. Prove that in the assumptions of Theorem 23.5 the holonomy group can be constructed in the class of foliations described in Problem 23.2.

Problem 23.5. Prove that Dulac germs are holomorphically embeddable.

The following series (Problems 23.6–23.13) distilled from the beautiful paper [**Lor06**], proves existence of the converging normal form for cuspidal singularities; see p. 72.

Problem 23.6. Prove that a cuspidal vector field with the linearization $w\frac{\partial}{\partial z}$ can be analytically transformed to the preliminary normal form $F_0 = f(z, w)\frac{\partial}{\partial z} + g(z, w)\frac{\partial}{\partial w}$ with $f(0, 0) = g(0, 0) = 0$ and $f(0, w) = w$ in a sufficiently small bidisk $(\mathbb{C}, 0) \times D_0$, $D_0 = \{|w| < r\} \subset \mathbb{P}$.

Problem 23.7. Consider the vector field $F_1 = w\frac{\partial}{\partial z}$ on the bidisk $(\mathbb{C}, 0) \times D_1$, $D_1 = \{|w| > r/2\} \subset \mathbb{P}$. Prove that the restriction of the two fields F_0, F_1 on the cylinder over the annulus $(\mathbb{C}, 0) \times K$, $K = D_0 \cap D_1$, are biholomorphically equivalent by a conjugacy $H(z, w)$ that fixes the points of the annulus $\{z = 0\}$, $H(0, w) = (0, w)$.

Problem 23.8. Let $\widetilde{M} = \big((\mathbb{C}, 0) \times D_0\big) \sqcup \big((\mathbb{C}, 0) \times D_1\big)$ be the disjoint union of the two bidisks and $M = \widetilde{M}/H$ is the quotient space by the map H constructed

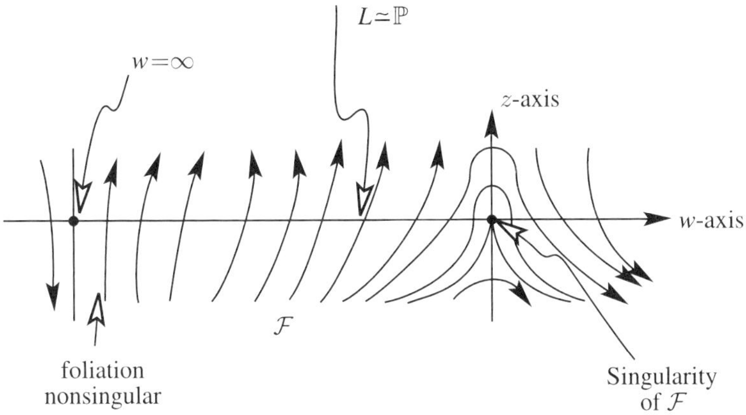

Figure IV.4. Foliation with a cuspidal singularity and a pole, obtained by "globalization"

in Problem 23.7. Prove that the manifold M carries a meromorphic vector field F (coming from the fields F_0, F_1 in the respective charts $(\mathbb{C}, 0) \times D_i$) and hence a singular holomorphic foliation $\mathcal{F}$. Prove that the foliation $\mathcal{F}$ has a unique singular point on the embedded projective line $L \cong \mathbb{P}$ which appears from the union of the two disks $\big(\{0\} \times D_0\big) \cup \big(\{0\} \times D_1\big)$; cf. Fig. IV.4.

Problem 23.9. Using the field F, construct a meromorphic section of the normal bundle $N_L = \mathbf{T}M/\mathbf{T}L$ of the embedded curve L. Prove that this section has one simple zero at $w = 0$ and one simple pole at $w = \infty$.

Problem 23.10. Prove that the normal bundle N of the embedding L in M is trivial. Prove, using Savel′ev's Theorem 23.6, that there exists a neighborhood of L in M, biholomorphically equivalent to the Cartesian product $(\mathbb{C}, 0) \times \mathbb{P}$.

Problem 23.11. Show that the tangency curve S between F and the foliation $\{z = \text{const}\}$ is a smooth holomorphic curve transversal to L.

Find a biholomorphism bringing S to the line $\{w = 0\}$ and the polar locus of F to $\{w = \infty\}$.

Problem 23.12. In the coordinates $(z, w) \in (\mathbb{C}, 0) \times \mathbb{P}$ the vector field F has all coefficients rational in w. Prove that, in fact,

$$F = f(z)w\tfrac{\partial}{\partial z} + \big(g_0(z) + wg_1(z)\big)\tfrac{\partial}{\partial w} \tag{23.20}$$

with three holomorphic germs $f, g_0, g_1 \in \mathcal{O}(\mathbb{C}, 0)$, $f(0) \neq 0$.

Problem 23.13. Find the transformation $(z, w) \mapsto \big(h_1(z), w + h_2(z)\big)$ which brings the field (23.20) into the normal form which appears in Table I.1.

Hint. First bring the function f to $f \equiv 1$ by changing only the z-coordinate.

24. Nonaccumulation theorem for hyperbolic polycycles

> [*What can be said about*] the maximal number of and position of Poincaré's boundary cycles (*cycles limites*) for a differential equation of the first order and degree of the form $\frac{dy}{dx} = \frac{Y}{X}$, where X and Y are rational integral functions of nth degree in x and y?
>
> *D. Hilbert*, 1901, reprinted from [**Hil00**]

This second part of Hilbert's sixteenth problem appears to be one of the most elusive in his famous list [**Hil00**], second only to the Riemann ζ-function Conjecture. In the introductory subsection §24**A** based on [**Ily02**], we briefly describe the current status of this problem.

The body of the section is devoted to investigation of limit cycles of analytic vector fields[6]. The central result of this section, Theorem 24.24 on finiteness of limit cycles of analytic vector fields having only nondegenerate singular points, was proved by Yu. Ilyashenko in [**Ily84**].

24A. Legends and truth on the limit cycles. As most problems from the Hilbert's list, the sixteenth problem is formulated very broadly and can be made precise in a variety of ways.

24**A**$_1$. *Various flavors of Hilbert's sixteenth.* By different placement of quantifiers the Hilbert's question can be transformed into three problems in increasing order of strength as follows.

Problem I. *Is it true that a planar polynomial vector field may have only finitely many limit cycles?*

Problem II. *Can the number of limit cycles be bounded by a constant depending only on the degree n of the vector field?*

Assuming the affirmative answer to Problem II, denote by $H(n)$ the *Hilbert number*, the conjectural bound for the number of limit cycles that a polynomial vector field of degree n may exhibit. Linear vector fields have no limit cycles, hence $H(1) = 0$. Finiteness of $H(2)$ is already an open problem.

Problem III. *Give an upper bound for $H(n)$.*

Only Problem I is solved now. The affirmative answer was proved independently in [**Ily91**] and [**Eca92**].

To separate analytic and algebraic aspects of the Hilbert problem, we will consider the following two questions concerning *real analytic* rather than polynomial vector fields.

---

[6]Recall that the limit cycle of a vector field is an isolated compact leaf of the real foliation defined by the vector on the real plane, real 2-sphere or the projective plane $\mathbb{R}P^2$; cf. with Definition 9.11

Problem IV. *Is it true that a real analytic vector field on the 2-sphere* $\mathbb{S}^2$ *has only a finite number of limit cycles?*

The "purely analytic" counterpart of Problem II has the following form.

Problem V. *Given a parametric family of real analytic vector fields on the 2-sphere, analytically depending on finitely many parameters varying over a compact subset in the parameter space, is it true that the number of limit cycles in this family is uniformly bounded?*

An affirmative answer in Problem V implies solutions of the Problems I, II and IV, since polynomial vector fields can be extended as real analytic foliations of the 2-sphere, and constitute a finite-parametric family parameterized by the coefficients of the vector fields varying over the projective space (see §25**A** for detailed explanations). In fact, it is the solution of Problem IV that is achieved in [**Eca92**] and [**Ily91**]. In other words, the known *individual finiteness* of limit cycles for polynomial vector fields has analytic rather than algebraic nature.

Clearly, all these questions reformulated literally for C^∞-smooth rather than real analytic vector fields, have negative answers; see §9**F**. Yet somewhat surprisingly there are meaningful questions which are reasonable "smooth analogs" of the above analytic problems. The following formulation is an implicit conjecture that the exotic smooth vector fields with infinitely many limit cycles constitute a subset of *infinite codimension* in the total space of C^∞-smooth vector fields on the sphere.

Problem VI (Hilbert–Arnold problem). *Given a* generic *n-parametric family of C^∞-smooth vector fields on the 2-sphere, smoothly depending on parameters varying over a compact subset in the parameter space, is it true that the number of limit cycles in this family is uniformly bounded?*

A restricted version of this problem (under the additional assumption on the types of singular points that are allowed to occur in the family) is solved in [**IY95**]. We wish to stress that this formulation is unrelated (neither implies nor is implied by) to any of the algebraic/analytic Problems I to V.

24$\mathbf{A}_2$. *Historical sketch.* As is typical for most of the problems from Hilbert's list, the sixteenth problem lies on the crossroads of many different directions and served as a motivation for many developments. Yet its own history is rather dramatic: several times it was believed to be proved only to later discover gaps.

Before Hilbert, Henri Poincaré considered polynomial vector fields in the plane, in the framework of his geometric theory of differential equations. He introduced the notion of limit cycle and proved that a planar polynomial vector field *without saddle connexions* has only a finite number of limit cycles.

In 1923, Dulac [**Dul23**] claimed a solution of Problem I in full generality. In the mid-fifties of the twentieth century, Petrovskii and Landis published a solution to Problem III [**PL55, PL57**]. They claimed that $H(n)$ is bounded by a certain polynomial of degree 3 in n, and $H(2) = 3$. In the early sixties a severe error in the arguments by Petrovskii and Landis was revealed by S. Novikov and Yu. Ilyashenko. Later quadratic vector fields with 4 limit cycles were explicitly constructed in [**CW79, Shi80b**].

In 1981, a ruinous gap was found in Dulac's solution of Problem I (cf. [**Ily85**]): Dulac was operating with asymptotic series as if they were convergent. Thus after eighty years of intense efforts our knowledge on Hilbert's sixteenth problem was still almost the same as at the time when the problem was formulated.

24**A**$_3$. *Some recent progress on the Hilbert's sixteenth problem.* The principal achievement is the general theorem solving Problems I and IV.

Theorem 24.1 (Individual finiteness theorem, [**Ily91, Eca92**]). *A polynomial vector field in the plane has only a finite number of limit cycles. The same is true for analytic vector fields on the 2-sphere.*

After some preliminary work described in §24**B**–§24**D**, the Finiteness Theorem 24.1 follows from the Nonaccumulation Theorem 24.23 formulated below. It is the Nonaccumulation theorem that is the most difficult result, whose proof occupies hundreds of pages. We will not discuss it, though the analytic normal forms for parabolic singularities and saddle resonant vector fields obtained in §21–§22 play the key role in this analysis. Ecalle's theory of resurgent functions is presented in [**Eca85, Eca92**].

The infinitesimal Hilbert's sixteenth problem deals with limit cycles that appear by perturbation of Hamiltonian vector fields that do not have limit cycles at all. Its main tool is investigation of Abelian integrals considered as analytic multivalued functions of complex parameters. These questions are discussed in detail in §26 below.

Bifurcation theory is intimately related to Hilbert's sixteenth. Indeed, the function "number of limit cycles of the equation" has points of discontinuity corresponding to equations whose perturbations generate limit cycles via bifurcations. Limit cycles may bifurcate from separatrix polygons, also known as *polycycles* (defined in §24**C**). The *cyclicity* of a polycycle in a family of equations is the maximal number of limit cycles that may bifurcate from the polycycle in this family, very much like cyclicity of singular point introduced in §12**A**, p. 201. Using the notion of cyclicity, one can formulate the Hilbert-type problems in the language of bifurcations theory.

Problem VII. *Is it true that a polycycle occurring in a finite parameter family of planar analytic vector fields has only finite cyclicity?*

Problem VIII. *Is it true that a polycycle occurring in a generic k-parameter family of smooth planar vector fields may generate only a finite number of limit cycles, with an upper bound depending on k only? (This latter quantity is denoted by $B(k)$.)*

The affirmative answer in Problem VII would imply a solution of Problem II and existence of the Hilbert number $H(n)$ for any finite n (without giving the slightest idea of how this number can be computed). The affirmative answer in Problem VIII would lead to an instant solution of the Hilbert-Arnold Problem VI. These implications are proved by using simple compactness arguments due to R. Roussarie [**Rou98**]. Both Problems VII and VIII remain unsolved, yet the latter seems to be easier than the former, in light of the recent achievements.

More precisely, denote by $E(k)$ the maximal cyclicity of a polycycle that can occur in a *generic* k-parameter family of smooth vector fields, under the additional assumption that all singular points on this polycycle are *elementary*.

Theorem 24.2 (Ilyashenko and Yakovenko [**IY95**]). *For any k, the number $E(k)$ is finite and bounded from above by an elementary function of k.*

As a corollary, one can immediately conclude that the Hilbert–Arnold problem has the affirmative answer if restricted on the smooth vector fields having only elementary singularities on the 2-sphere.

The proof of Theorem 24.2 is constructive and yields an algorithmic expression for the upper bound. Further elaborating this construction, V. Kaloshin in [**Kal03**] obtained a simple explicit upper bound,

$$E(k) \leqslant 2^{25k^2}. \tag{24.1}$$

The Kaloshin bound is apparently very much excessive, yet it is one of the first *Hilbert-type* numbers (bounds pertinent to the number of limit cycles) obtained during the hundred years of quest.

In the rest of this section we illustrate the power of the analytic normal forms theory and prove the Individual Finiteness Theorem 24.1 under the additional assumption that all singular points of the vector field and nondegenerate saddles. To present the complete proof, we have to go back to the early times of the geometric theory of differential equations.

24B. Poincaré–Bendixson theory revisited. One of the highlights of the geometric theory of real planar vector fields is the Poincaré–Bendixson theorem. It describes the limit behavior of phase trajectories of vector fields without singular points in domains on the 2-sphere using purely topological arguments. In the next three subsections we apply similar methods to describe limit sets of aperiodic trajectories for spherical vector fields with singularities and the accumulation sets for their periodic trajectories.

Here and below we consider smooth real vector fields on the sphere and their trajectories parameterized by real values of the time. Since the sphere is compact, any such trajectory can be extended for all values of the time $t \in \mathbb{R}$. Let $v \in \mathcal{D}(\mathbb{S}^2)$ be a vector field and $\varphi\colon \mathbb{R} \to \mathbb{S}^2$ its trajectory.

Definition 24.3. An ω-limit set of a trajectory φ is the set of all points $y \in \mathbb{R}^2$ which are limits of sequences of points $\varphi(t_n)$ corresponding to sequences of time $t_n \to \infty$. An α-limit set of a trajectory $\varphi(t)$ is the ω-limit set of the trajectory $\varphi(-t)$, i.e., after the time reversal.

We will denote these limits by $\omega(\varphi)$ and $\alpha(\varphi)$ respectively.

Remark 24.4. The definition of an ω- (resp., α-) limit set can be modified for noncomplete vector fields or for fields defined in noninvariant domains. It is sufficient to require that φ be defined for all sufficiently large *positive* (resp., *negative*) values of time.

One can give an alternative description for $\omega(\varphi)$. For any $T > 0$ denote by φ_T the restriction of the phase curve on the semi-interval $[T, +\infty)$. This is a forward invariant set whose closure $\overline{\varphi_T} \subset \mathbb{S}^2$ is also forward invariant (forward invariance is invariance by the real flow maps $\varPhi_t = \exp tv$ of the field $v \in \mathcal{D}(\mathbb{S}^2)$ for nonnegative times $t \in \mathbb{R}_+$). These sets form a family of nested connected compacts on the sphere, whose intersection, as one can easily see, coincides with $\omega(\varphi)$:

$$\varnothing \neq \omega(\varphi) = \bigcap_{T>0} \overline{\varphi_T} \Subset \mathbb{S}^2. \tag{24.2}$$

From the description (24.2) one can easily derive the following properties of limit sets on the sphere.

Proposition 24.5. *The ω-limit set of a trajectory of the spherical vector field is a closed connected set invariant by both positive and negative flow of the field.* □

Remark 24.6. The same definitions can be given for a vector field on the plane $\mathbb{R}^2$, but in this case the sets φ_T can be unbounded, $\overline{\varphi_T}$ noncompact and, as a result, the ω-limit set can be empty or nonconnected.

Example 24.7. The phase portraits sketched on Fig. IV.5 show that ω-limits of trajectories on the sphere can be singular points, cycles (periodic orbits) or more complicated objects which consist of several singular points

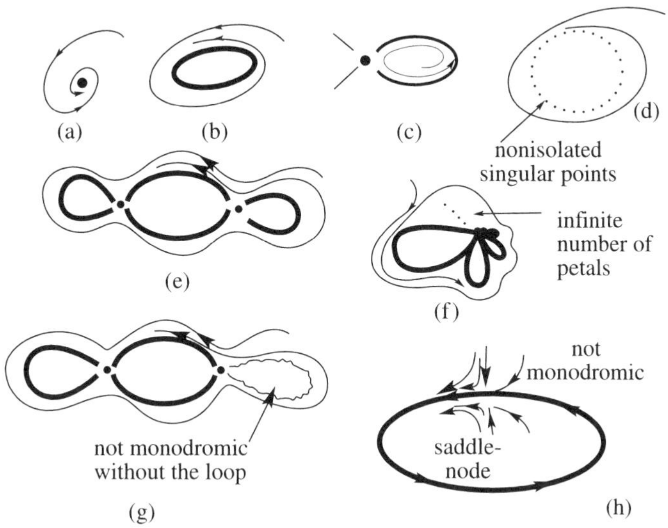

Figure IV.5. Zoo of limit periodic sets. (a) Isolated singular point. (b) Periodic orbit. (c) Separatrix loop. (d) Curve of nonisolated singular points. (e) Monodromic polycycle. (f) Singular point with infinitely many homoclinic trajectories. (g) Part of a polycycle is a polycycle but not monodromic. (h) Oriented but not monodromic saddle-node loop

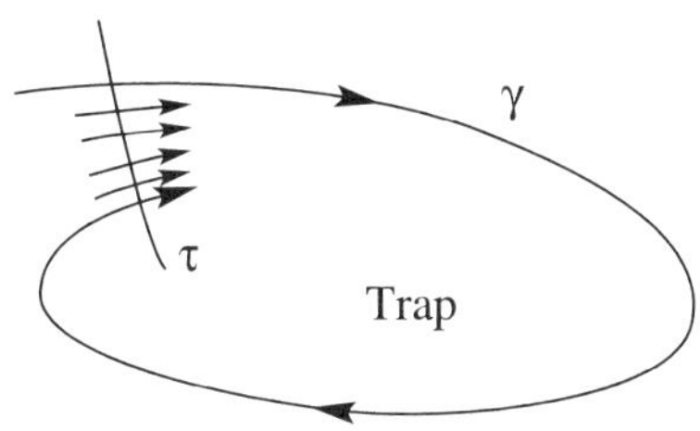

Figure IV.6. Bendixson trap

together with several orbits which are bi-asymptotic to these singular points as $t \to \pm\infty$.

In order to describe ω-limit sets, we introduce a simple but powerful construction designed by Bendixson.

Definition 24.8. A *Bendixson trap* for a vector field v on the sphere is a closed oriented piecewise-smooth curve which consists of two smooth parts:

(1) a piece of *nonperiodic* phase trajectory γ oriented by the field and thus defining the orientation of the trap, and
(2) a smooth arc τ transversal to the field at all its points.

By the Jordan theorem, any Bendixson trap divides the sphere into two connected domains, one of them invariant by the flow of v in the forward time (it will be referred to as *interior* to justify the term "trap"), the other ("*exterior*") invariant in the reverse time. Note that the orientation of the trap can be opposite to the orientation of the boundary of the interior part.

Lemma 24.9. *No point on the transversal arc of a Bendixson trap can belong to an ω-limit set of any trajectory.*

In particular, the invariant arc of the trap cannot be an ω-limit set.

Proof of the lemma. Any orbit starting on the transversal arc enters the interior domain either immediately, or at worst after traversing the invariant arc of the trap, and never leaves it since that moment. In particular, it can never return to a sufficiently small neighborhood of the arc τ. □

As an immediate consequence, we can prove that a trajectory accumulates to its ω-limit set from one side only.

Proposition 24.10. *If $\gamma = \omega(f)$ contains a nonsingular point a and $\tau\colon (\mathbb{R}^1, 0) \to \mathbb{S}^2$ is a cross-section to γ at a, then all intersections of φ with τ occur only on one side of the cross-section.*

Proof. If φ intersects τ at two points p and q on two different sides of τ, then the closed line formed by the arc $\varphi|_p^q$ of φ from p to q and the arc $\tau|_q^p$ of τ from q to p is a trap. The point $a \in \tau|_q^p$ is hence a point of a limit set which lies on the transversal arc of a trap, in contradiction with Lemma 24.9. □

The following result constitutes the most familiar part of the Poincaré–Bendixson theory.

Theorem 24.11 (H. Poincaré, 1886, I. Bendixson, 1901). *An ω-limit set which does not contain singular points of the field, is necessarily a periodic orbit.*

Proof. Let $\gamma = \omega(\varphi)$ be the limit set and $a \in \gamma$ a nonsingular point on it. Consider a cross-section τ to γ at a as in Proposition 24.10. The trajectory φ crosses τ infinitely many times. Consider the positive orbit $\psi \subseteq \gamma$ starting at a. It must intersect τ some time in the future. Indeed, otherwise the closure $\overline{\psi(t)|_{[1,+\infty)}}$ would be a compact subset of the sphere disjoint from τ, and since the orbit φ must remain in a neighborhood of this compact, it would be unable to cross τ infinitely many times.

Hence ψ crosses τ again. If this intersection occurs at a point b different from a, then the closed curve formed by $\psi|_a^b$ and $\tau|_b^a$ would be a trap in contradiction with Lemma 24.9.

The only remaining possibility is that ψ crosses τ at the same point $a \in \tau \cap \psi$. Then ψ and hence γ is a periodic orbit of v. □

In the presence of singular points the limit sets can be more complicated, as mentioned in Example 24.7. Still these limit sets admit a rather simple description.

A trajectory φ of a vector field is called bi-asymptotic to two points a, b if $\{a\} = \alpha(\varphi)$, $\{b\} = \omega(\varphi)$. Clearly, in such a case both a and b must be singular points; the case $a = b$ is *not* excluded.

Theorem 24.12. *Any limit set of a vector field on the sphere consists of singular points and entire trajectories of the field, bi-asymptotic to some of these singular points.*

To prove this theorem, we reformulate it in the language of *iterated* limit sets. Being invariant, an ω-limit set of any orbit φ consists of entire trajectories of the field. This allows us to iterate the construction of limit sets.

Definition 24.13. The iterated limit set $\omega^2(\varphi)$ is the union of ω-limit sets of *all* trajectories forming $\omega(\varphi)$.

If γ is a singular or periodic orbit, then $\omega(\gamma) = \omega^2(\gamma) = \gamma$. The set $\omega^2(\varphi)$ is also closed and invariant by the flow, but may well be nonconnected.

In the same way higher iterated ω-limit sets can be defined inductively as unions of limit sets of all trajectories forming a previous iteration. By construction, they constitute a sequence of nested compacts. Yet it turns out that on the plane this generalization does not lead to anything new. The core statement of the Poincaré–Bendixson theory asserts that the iterated ω-limit sets on the sphere in fact stabilize from the second step. The following statement has no analogs for vector fields on higher-dimensional manifolds.

Lemma 24.14. *For any vector field with isolated singular points on the sphere, the ω^2-limit of any trajectory is either a periodic orbit, or a collection of singular points.*

Proof. Suppose that $\Gamma = \omega^2(\varphi)$ contains a nonsingular point a of the field, and let τ be a cross-section to Γ at a. This means that some invariant trajectory γ from $\omega(\varphi)$ must cross τ infinitely many times. But the contour formed by an arc of γ between two *subsequent* crossings and a segment of the cross-section will be a Bendixson trap unless γ is periodic. This would contradict Lemma 24.9. □

Proof of Theorem 24.12. By Lemma 24.14, both α- and ω-limit sets of any nonconstant trajectory $\gamma \subseteq \omega(\varphi)$ are singular points. □

We conclude this section by an example showing that on surfaces other than the sphere (with its simple topological properties), Theorem 24.12 fails completely.

Example 24.15. The constant vector field $dy/dx = \alpha$, $\alpha \in \mathbb{R}$ on the 2-torus $\mathbb{T}^2 = (\mathbb{R} \bmod \mathbb{Z})^2$ has all trajectories periodic for $\alpha \in \mathbb{Q}$. However, if $\alpha \notin \mathbb{Q}$ is irrational, the ω-limit of any orbit coincides with the entire torus $\mathbb{T}^2$.

24C. Polycycles, monodromy, correspondence maps. Without further assumptions on the vector field it is difficult to describe more precisely possible limit sets of trajectories on the sphere.

From this moment on we will assume that all vector fields satisfy the following two finiteness assumptions:

(1) the field has only isolated singular points on the sphere, and

(2) each singular point has only finitely many hyperbolic sectors (cf. Definition 9.2).

These assumptions are automatically satisfied for *real analytic* vector fields.

By Theorem 24.12, in these assumptions the ω-limit set Γ of any trajectory φ is a planar (more accurately, spherical) finite graph consisting of finitely many vertices (singular points) connected by edges (trajectories bi-asymptotic to these vertices).

This graph is co-oriented: by Proposition 24.10, every edge $\gamma \subset \Gamma$ has a "positive" side, from which the trajectory φ accumulates to Γ, and the "negative" side. Therefore among the connected components of $\mathbb{S}^2 \setminus \Gamma$ (faces of the spherical graph) there is a distinguished component Ω containing φ; see Fig. IV.7. Each connected component C of the boundary $\partial\Omega \subseteq \Gamma$ is an "almost circle", i.e., the image of the circle $\mathbb{S}^1 = \mathbb{R}/\mathbb{Z}$ by a continuous map $\iota \colon \mathbb{S}^1 \to C$ bijective except finitely many points that are mapped into singular points of v. Since the curve φ cannot (again by Jordan theorem) approach any point from $\partial\Omega \setminus \Gamma$, we conclude that $\partial\Omega = \Gamma$ and hence $\partial\Omega$ must be connected, $\partial\Omega = C$. In other words, $\Gamma = \omega(\varphi)$ which is not a singular point or a cycle, is a closed continuous curve bounding a spherical domain, whose self-intersections can occur only at singularities. Such an object is called a *polycycle*.

Definition 24.16. A *polycycle* of a vector field is a finite oriented spherical graph Γ such that:

(1) topologically Γ is a continuous image of the circle $\mathbb{S}^1$,

(2) vertices of Γ are at the singular points of the field,

(3) edges of Γ are infinite trajectories of the field.

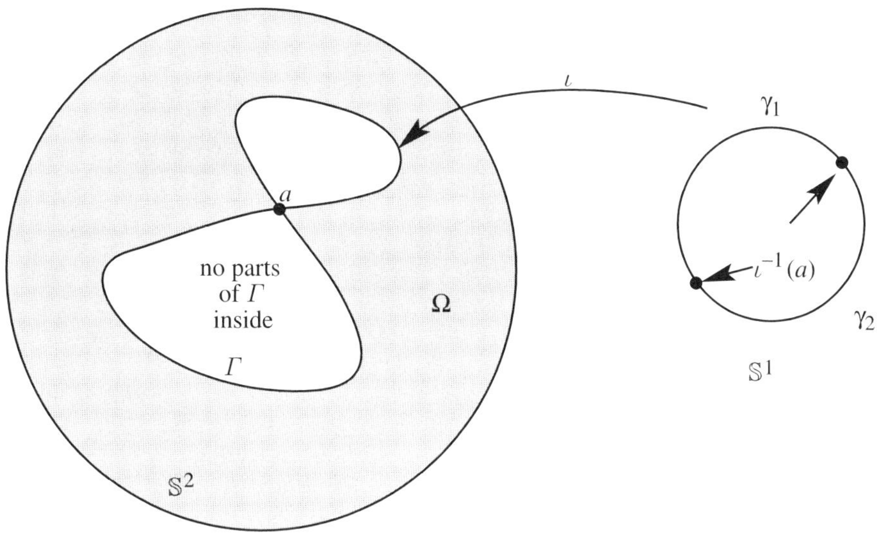

Figure IV.7. The continuous "almost one-to-one" image of the circle Γ bounding the connected domain Ω

Note that among the singular points (also cyclically enumerated) repetitions are allowed whereas the edges are all distinct.

Let $\tau_+ \colon (\mathbb{R}^1_+, 0) \smallsetminus \{0\} \to (\mathbb{S}^2, a)$ be a *semi-section*, the restriction of a cross-section τ at a nonsingular point $a \in \Gamma$, on the "positive" open semi-interval (i.e., such that $\varphi \cap \tau_+$ is nonvoid).

Proposition 24.17. *There is a well-defined first return map (also called* monodromy map*)* $\Delta_\Gamma \colon \tau_+ \to \tau_+$ *such that for any point* $p \in \tau_+$ *the orbit of* v *starting at* p*, intersects* τ_+ *for the first time again at* $\Delta_\Gamma(p)$.

Proof. Consider the infinite sequence of points $x_1, x_2, \dots,$ which are *subsequent* intersections of the trajectory φ with the semi-section τ_+; this sequence converges to the base point a of the semi-section.

Consider the trap T formed by the arc of φ from x_1 to x_2 and a piece of τ_+ between these points. The trajectory φ starting from the point x_2 entirely belongs to the annulus $T \smallsetminus \Omega$, where Ω is the spherical domain containing φ. Without loss of generality we may assume that this annulus contains no singular points of the field other than belonging to the polycycle (recall that singularities of v are isolated).

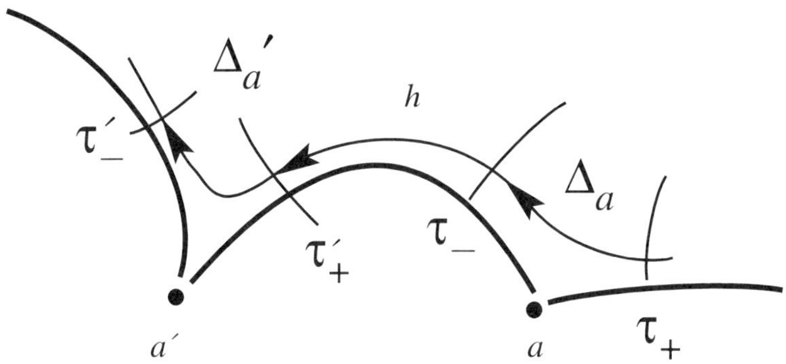

Figure IV.8. Correspondence maps

Consider the strip Π formed by two arcs $\varphi' = \varphi|_{x_1}^{x_2}$ and $\varphi'' = \varphi|_{x_2}^{x_3}$ of the trajectory φ and two segments $\tau' = \tau_+|_{x_1}^{x_2}$ and $\tau'' = \tau_+|_{x_2}^{x_3}$ on the cross-section. We claim that any other trajectory ψ starting on τ', crosses τ'' at some time in the future.

Indeed, ψ cannot cross the arcs φ', φ'' as they are phase curves of the field. If ψ does not cross τ'', then its ω-limit must be nonvoid. Since Π does not contain singular points, the ω-limit set must be a cycle by the Poincaré–Bendixson Theorem 24.11. But by the Poincaré–Hopf index theorem, each cycle must contain a singular point in its interior, leading again to the contradiction.

Therefore the first return map Δ_Γ is well defined on τ' and takes values on τ''. For the same reasons Δ_Γ is well defined on any segment $\tau_+|_{x_n}^{x_{n+1}} \subset \tau_+$. Since these segments together cover the entire semi-section τ_+, the proposition is proved. $\square$

Remark 24.18 (terminological)**.** Note that the first return map Δ_Γ constructed in the proof of Proposition 24.17, possesses the following property: for all points $p \in \tau_+$ sufficiently close to a, the orbit connecting p with $\Delta_\Gamma(p)$ remains in an arbitrarily small neighborhood of the polycycle. This condition excludes some polycycles, e.g., those sketched on Fig. IV.5 (g), (h), from being limit sets of trajectories. In the future we will call a polycycle Γ *monodromic*, if it admits the first return map along orbits that remain in an arbitrarily small neighborhood of Γ.

Consider a singular point $a \in \Gamma$ on a monodromic polycycle Γ, and let $\gamma_+, \gamma_- \subseteq \Gamma$ be two trajectories such that $\omega(\gamma_+) = a = \alpha(\gamma_-)$ (the loop case where $\gamma_+ = \gamma_-$ is not excluded). Let $\tau_\pm$ be two semi-sections to the curves $\gamma_\pm$ at two points $a_\pm$ respectively, from the "positive" side of each of them. The same arguments as in the proof of Proposition 24.17 show that each trajectory starting on τ_+ sufficiently close to a_+, crosses also τ_- somewhere near a_-. This allows us to define the *correspondence map* $\Delta_a\colon \tau_+ \to \tau_-$ associated with the singular point $a \in \Gamma$. This map, in general, not analytically extendable to the point a_+, remains continuous after setting $\Delta_a(a_+) = a_-$. By this construction, Δ_a is defined modulo the freedom in choosing the cross-sections $\tau_\pm$, i.e., modulo a conjugacy by analytic germs $h_\pm \in \operatorname{Diff}(\mathbb{R}^1, 0)$ from left and right, $h_- \circ \Delta_a \circ h_+$.

We will summarize the results of this section as follows.

Theorem 24.19. *Assume that a smooth vector field on the sphere has only isolated singular points, each of them having at most finitely many hyperbolic sectors.*

Then an ω-limit set of any orbit of this field is either a singular point, or a cycle (periodic orbit) or a finite monodromic polycycle Γ.

In the latter case the first return map of this polycycle Δ_Γ is well defined on any semi-section τ_+ to Γ at a nonsingular point of the latter, and expands as a finite composition of the form

$$\Delta_\Gamma = h_n \circ \Delta_{a_n} \circ h_{n-1} \circ \Delta_{a_{n-1}} \circ \cdots \circ h_1 \circ \Delta_{a_1} \circ h_0. \tag{24.3}$$

Here Δ_{a_i} are correspondence maps associated with the singular points $a_i \in \Gamma$, and h_i are some real analytic maps. □

24D. Accumulation of limit cycles. Recall (see Definition 9.11) that a limit cycle is an *isolated* periodic trajectory of a vector field.

As the first step towards the solution of Problem I (finiteness problem for limit cycles) we will describe possible accumulation sets for limit cycles of smooth vector fields. Such fields may indeed have an infinite number of limit cycles, but these cycles must accumulate to a monodromic polycycle. To make this statement precise, we need the notion of the Hausdorff distance.

Definition 24.20. Let A, B be two subsets of a metric space M. The Hausdorff distance between them is the nonnegative number

$$\operatorname{dist}(A, B) = \max[\sup_{a\in A} \operatorname{dist}(a, B),\ \sup_{b\in B} \operatorname{dist}(b, A)], \tag{24.4}$$

where $\operatorname{dist}(x, Y) = \inf_{y\in Y} \operatorname{dist}(x, y)$ is the distance between a point x and any subset $Y \subset M$.

One can easily verify (see [**BBI01**, Chapter 7]) that the Hausdorff distance satisfies the triangle inequality and defines a *metric* on the space of *closed* subsets: if A, B are closed and $\operatorname{dist}(A, B) = 0$, then $A = B$.

A sequence of subsets $A_1, A_2, \dots, A_n, \dots \subseteq M$ converges in the sense of Hausdorff distance to a limit A, if every point of $a \in A$ is the limit of a sequence of points $a_1, a_2, \dots$ such that $a_i \in A_i$. An alternative description of the limit is similar to (24.2)

$$A = \bigcap_{n=1}^{\infty} \overline{\bigcup_{i=n}^{\infty} A_i};$$

see [**BBI01**, Exercise 7.3.4]. The following result is elementary but very useful.

Theorem 24.21 (W. Blaschke). *If the metric space M is compact, then the space of compact subsets of M equipped with the Hausdorff distance is also compact.*

Proof. See [**BBI01**], Theorem 7.3.8. □

Theorem 24.22. *Assume that a smooth vector field on the sphere $\mathbb{S}^2$ has only isolated singular points, each of them having at most finitely many hyperbolic sectors.*

If this field has infinitely many limit cycles, then there exists an infinite sequence of these cycles $\{\gamma_i\}_{i=1}^{\infty} \subset \mathbb{S}^2$ converging in the sense of the Hausdorff distance to a singular point, a cycle (periodic orbit) or a monodromic polycycle Γ.

In the latter case if $\Delta_\Gamma \colon \tau_+ \to \tau_+$ is the monodromy map of the polycycle, then the intersection points $p_i = \gamma_i \cap \tau_+$ are isolated fixed points for Δ_Γ accumulating to the base point of the semi-section τ_+.

Proof. By Blaschke Theorem 24.21, an infinite number of limit cycles on the compact 2-sphere must contain an infinite sequence of cycles that accumulates in the sense of the Hausdorff distance to a compact subset $\Gamma \subseteq \mathbb{S}^2$. We show that if Γ contains a nonsingular point of v, then Γ is either a cycle or a monodromic polycycle.

To do this, one can modify slightly the arguments leading to the proof of Theorem 24.19. Yet we can reduce Theorem 24.22 to Theorem 24.19 directly, using a *plug* as on Fig. IV.9.

Let $a \in \Gamma$ be a nonsingular point. Consider *two* close semi-sections τ_+, τ_+' at the points $a \neq a'$ to the trajectory γ passing through a, and denote by p_i, p_i' the corresponding intersection points between the cycles γ_i with these cross-sections.

Consider the narrow strip Π ("plug") bounded by $\gamma|_a^{a'}$ and the two semi-sections τ_+, τ_+' (the outer bound can be chosen rather arbitrarily). Let w be a C^∞-smooth vector field which coincides with v everywhere outside of Π and on the boundary $\tau_+ \cup \tau_+' \cup \gamma|_a^{a'}$ of the latter, such that its orbits which begin at p_i pass through Π and end at p_{i+1}'.

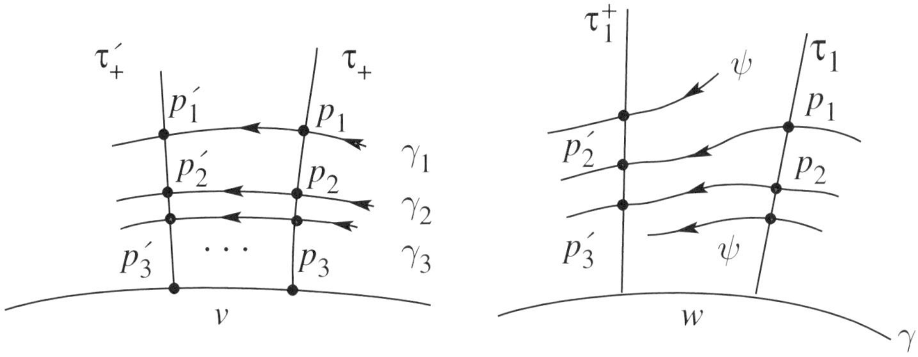

Figure IV.9. The plug: modification of a vector field in a small strip Π between two semi-sections

Then all cycles γ_i of the initial field v belong to the single trajectory ψ of the field w. Obviously, the ω-limit set of ψ coincides with the Hausdorff limit set Γ for the sequence of the limit cycles γ_i. By Theorem 24.19, the former is a cycle or a monodromic polycycle. □

Finiteness Theorem 24.1 for limit cycles of planar and spherical analytic vector fields follows now from a purely analytic local property of the monodromy map of polycycles of such fields.

Theorem 24.23 (General finiteness theorem, Yu. Ilyashenko [**Ily91**], J. Écalle [**Eca92**]). *The monodromy map of a polycycle of an analytic vector field in the plane cannot have an infinite number of isolated fixed points.*

We will prove here this theorem under a simplifying assumption that the polycycle is *hyperbolic*, i.e., it carries only nondegenerate saddles at the vertexes. This implies the following theorem which is the main result of this section.

Theorem 24.24 (Easy finiteness theorem). *A real analytic vector field on the 2-sphere, having only nondegenerate singular points, may have only finitely many limit cycles.*

The proof is based on investigation of the individual correspondence maps for analytic hyperbolic saddles and their compositions with holomorphic germs.

24E. Almost regular germs and monodromy of hyperbolic polycycles. Developing the ideas of Dulac [**Dul23**], we introduce a class of germs with two competing properties. On one hand, this class is large enough as to include monodromy transformations of hyperbolic polycycles $\Delta_\Gamma\colon (\mathbb{R}^1_+,0) \to (\mathbb{R}^1_+)$, $z \mapsto \Delta_\Gamma(z)$, which are in general not analytic at $z=0$.

On the other hand, this class is so close to the class of analytic functions that germs of this class are uniquely determined by their asymptotic expansions.

In this section we will mostly work in the *logarithmic chart* $\zeta = -\ln z$: in this chart the interval $z \in (0, \varepsilon)$ becomes a neighborhood of infinity, $\zeta \in (\frac{1}{\varepsilon}, +\infty)$.

Definition 24.25. A *standard* (*quadratic*) *domain* Ω_C is the image of the right half-plane $\mathbb{C}_+ = \{\operatorname{Re}\zeta > 0\}$ by the map

$$\varphi_C\colon \zeta \mapsto \zeta + C\sqrt{1+\zeta}, \qquad C > 0. \tag{24.5}$$

The constant C is a parameter determining the "size" of the standard domain Ω_C.

Definition 24.26. An *exponential series*, or *Dulac series*, is the formal series

$$S = \alpha\zeta + \beta + \sum_{j=1}^{\infty} p_j(\zeta)\exp(-\nu_j\zeta), \qquad \alpha,\beta \in \mathbb{R}, \quad p_j \in \mathbb{R}[\zeta], \tag{24.6}$$

in which

$$\alpha \geqslant 0, \qquad 0 < \nu_1 < \nu_2 < \cdots < \nu_n < \cdots, \qquad \lim \nu_j = +\infty.$$

No assumptions on convergence of the series (24.6) is made.

A function f defined in some standard domain Ω_C is said *to admit an expansion in the Dulac series* (24.6) (to be *expandable*, for short), if for any order $\nu > 0$ there exists a partial sum S_ν of this series, such that

$$|f(\zeta) - S_\nu(\zeta)| = o\big(\exp(-\nu\zeta)\big) \qquad \text{as } |\zeta| \to \infty \text{ in } \Omega_C. \tag{24.7}$$

Definition 24.27. The germ of a real analytic map $f\colon (\mathbb{R}^1_+, 0) \to (\mathbb{R}^1_+, 0)$ is called *almost regular*, if in the logarithmic chart the germ $-\ln f\big(\exp(-\zeta)\big)$ has a representative that can be extended as a biholomorphic map between two standard domains and expanded in a Dulac exponential series there.

Remark 24.28. Apriori in the Definition 24.27 one can allow dependence of the Dulac series on the order ν to which it approximates the almost regular germ f. Yet the asymptotic series (24.6), if it exists, is unique, and this is proved exactly like the uniqueness of the Taylor asymptotic series. Assume that for any ν there exists a Dulac polynomial $S_\nu(\zeta)$ (a finite sum of the form (24.6) with positive exponents ν_j not exceeding ν) such that the difference $f - S_\nu$ is decreasing as $o\big(\exp(-\nu\zeta)\big)$. Then all polynomials S_ν are necessarily truncations of a single Dulac series S as in (24.6) which is an asymptotic series for the function f. Indeed, if $\nu' > \nu$, then S_ν is a truncation of $S_{\nu'}$. Otherwise their difference cannot be decreasing as $o\big(\exp(-\nu'\zeta)\big)$ as $\zeta \to \infty$ in Ω_C.

The condition of almost regularity is weaker than analyticity at the point $z = 0$. Indeed, any converging Taylor series $f(z) = a_1 z + a_2 z^2 + \cdots$ in the logarithmic chart becomes a uniformly *convergent* Dulac series

$$\begin{aligned} -\ln f(\zeta) &= \ln a_1 + \zeta + \ln\bigl(1 + \tfrac{a_2}{a_1}\exp(-\zeta) + \tfrac{a_3}{a_1}\exp(-2\zeta) + \cdots\bigr) \\ &= \zeta + \beta + (\text{Dulac series without affine part}). \end{aligned}$$

Yet the following property means that in some respects almost regular germs are similar to analytic germs which were called *regular* in the old-fashioned language of the nineteenth century (this explains the choice of the term "almost regular").

Theorem 24.29. *An almost regular germ is uniquely determined by its asymptotic Dulac series: two almost regular germs with the same series coincide identically in their common domain.*

In other words, not only the Dulac asymptotic series is uniquely defined by an almost regular germ as Remark 24.28 notes, but the germ itself is completely determined by its series.

It turns out that the class of almost regular germs is large enough for our purposes.

Theorem 24.30. *The germ of the monodromy map of a hyperbolic polycycle is almost regular.*

The Nonaccumulation Theorem 24.24 is an almost direct consequence of these two theorems, as the following argument shows.

Proof of Theorem 24.24. Suppose that limit cycles accumulate to a hyperbolic polycycle Γ. Then the monodromy map $\Delta = \Delta_\Gamma \colon (\mathbb{R}^1_+, 0) \to (\mathbb{R}^1_+)$ has an infinite number of isolated fixed points accumulating to $z = 0$, as explained in §24**D**.

By Theorem 24.30, in the logarithmic chart $\zeta = -\ln z$ the monodromy map $f(\zeta) = -\ln \Delta(\exp -\zeta)$ admits an exponential asymptotic series S of the form (24.6) and has infinitely many real fixed points accumulating to $\zeta = +\infty$. We claim, following Dulac [**Dul23**], that this series is in fact an identity, $S = \zeta$.

Indeed, consider the difference $S - \zeta$ which also admits the exponential series (24.6). If this difference is nonzero, then its leading term is either affine $(\alpha - 1)\zeta + \beta$, or exponential $p_1(\zeta)\exp(-\nu_1\zeta)$. In both cases the difference between the monodromy map $f(\zeta)$ itself and the identity ζ has the form $g(\zeta)(1 + o(1))$, where $g(\zeta)$ is a real analytic function on $\mathbb{R}_+$ *with only finitely many* (real) *zeros*, which contradicts the assumption that these zeros are accumulating to infinity. Hence the series S must be identical, $S = \zeta$.

Thus the asymptotic series S of the map f is identity. On the other hand, Δ is almost regular by Theorem 24.30. Theorem 24.29 implies that in this case the map f itself is identity, $f(\zeta) \equiv \zeta$, and hence $\Delta(z) \equiv z$. Thus Δ cannot have isolated fixed points at all. The contradiction proves the Nonaccumulation Theorem 24.24. □

Remark 24.31. In [**Dul23**] Dulac tacitly assumed that the monodromy map with the identical Dulac series, is itself identity, circumventing Theorem 24.29. However, this assertion is wrong in absence of hyperbolicity of the polycycle. In [**Ily84**] one can find an example of a (nonhyperbolic) polycycle whose monodromy differs from identity by a flat (decaying faster than any exponential of ζ) nonzero function.

The rest of this section is devoted to the proof of the two key facts: Theorem 24.29 is proved in §24**G**, while the proof of Theorem 24.30 is postponed until subsection §24**H**. In order to carry out the proofs, we need some elementary properties of almost regular maps.

24F. Elementary properties of almost regular maps. The class of almost regular germs is rather natural. As was already noted, it contains all germs regular at $z = 0$.

Example 24.32. The power map $z \mapsto cz^\lambda$ for $\lambda > 0$ is almost regular. Indeed, in the logarithmic chart this map becomes *affine*, $\zeta \mapsto \lambda\zeta + \beta$, $\beta = -\ln c$. The corresponding Dulac series is finite, and it remains only to verify that it maps any standard domain into another standard domain. One can easily verify that the image of the standard domain Ω_C belongs to the standard domain $\Omega_{C'}$ if $C' = \alpha^{1/2}C + C_0$ for C_0 sufficiently large.

Rather expectedly, the class of almost regular germs is closed by composition.

Lemma 24.33. *Composition of two almost-regular germs is again an almost regular germ.*

Proof. It is convenient to treat separately the *affine germs* of the form $\zeta \mapsto \alpha\zeta + \beta$, $\alpha \geqslant 0$, $\beta \in \mathbb{C}$, and the *parabolic almost regular germs* whose Dulac series starts with the identical term,

$$S = \zeta + \sum_{\nu > 0} p_\nu(\zeta) \exp(-\nu\zeta). \tag{24.8}$$

Let us check that if $f(\zeta)$ is a function holomorphic in a standard domain Ω_C and admits there an estimate $|f(\zeta) - \zeta| < \exp(-\varepsilon\zeta)$ for some $\varepsilon > 0$, then the image of Ω_C by f contains a standard domain $\Omega_{C'}$ for C' sufficiently large. Indeed, the exponential small "perturbation" cannot change the asymptotic behavior of the curve

$$\operatorname{Re}\zeta = C\,|\operatorname{Im}\zeta|^2 + O(1), \qquad \operatorname{Im}\zeta \to \pm\infty \tag{24.9}$$

which is the boundary $\partial\Omega_C$. Preservation of the class of standard domains under action of affine maps is discussed in Example 24.32.

Thus composition of almost regular germs is defined (after analytic continuation) in some standard domain and takes it into another standard domain. It remains to verify the existence of an asymptotic Dulac expansion for a composition of two almost regular maps.

Note that if $R = \sum_{\nu>0} p_\nu(\zeta)\exp(-\nu\zeta)$ is a Dulac series *without the affine part* (with only positive exponents), then all its powers $R^2, R^3, \dots$ and any product $\exp(-\mu\zeta)R$, $\mu > 0$, are also of the same form. Therefore the formal exponent

$$\exp(-\mu R) = 1 + \sum_{k>0}(-\mu R)^k/k!$$

is also a well-defined Dulac series. The direct substitution now shows immediately that the composition of two parabolic series

$$(\zeta + R') \circ (\zeta + R) = (\zeta + R) + \sum_{\mu>0} p_\mu(\zeta + R)\exp(-\mu\zeta)\exp(-\mu R) = \zeta + R''$$

is a parabolic Dulac series.

It remains only to notice that composition of a parabolic Dulac series with an affine map $a\colon \zeta \mapsto \alpha\zeta + \beta$ (in any order) is obviously a Dulac series, and moreover, parabolic germs constitute a normal subgroup: if $f(\zeta) = \zeta + R$ is a parabolic germ, then $a^{-1} \circ f \circ a$ is again a parabolic germ. □

Remark 24.34. Since the maps holomorphic at infinity are automatically almost regular, the definition of the almost regular maps does not depend on the coordinate chart: by Lemma 24.33, the composition $g^{-1} \circ f \circ g$ is again a map defined in a standard domain and asymptotic to a Dulac series there.

24G. Phragmén–Lindelöf principle for almost regular germs. In this subsection we prove Theorem 24.29. It is a purely analytic fact closely related to the enhanced version of the maximum modulus principle known as the *Phragmen–Lindelöf principle.*

Recall that the maximum modulus principle asserts that a function $f = f(z)$ holomorphic in a (bounded) domain $z \in D$ and continuous on the boundary achieves the maximal value of its modulus $|f(z)|$ somewhere on the boundary ∂D. If the continuity assumption fails at a single point of the boundary, the function may well be unbounded.

Example 24.35. The function $f(z) = \exp(1/z)$ is holomorphic in the disk $|z - 1| < 1$ and continuous on its boundary except the single point $\{z = 0\}$. Yet this function is unbounded in D, despite the fact that its modulus is constant on the boundary $\partial D \smallsetminus \{1\}$. The latter fact becomes obvious in the conformal chart $\zeta = 1/z$ which transforms the function f into the exponent $\exp\zeta$ and the domain into the half-plane $\operatorname{Re}\zeta > 1/2$. The restriction of f on the boundary has constant modulus $m = \exp\frac{1}{2}$.

This example illustrates the phenomenon that lies at the core of the Phragmén–Lindelöf principle: the maximum modulus principle may fail if the boundary of the domain contains a point a near which f is unbounded, but only if the growth of f when approaching such a point is sufficiently fast; the "critical threshold" for the growth rate depends on the geometry of the boundary ∂D near a.

For our applications it is sufficient to consider only domains on the Riemann sphere, bounded by two circular arcs. In a suitable chart they become sectors with the vertex at the origin with an opening angle $2\pi/\alpha$, symmetric with respect to the real ray $\mathbb{R}^1_+ \subset \mathbb{C}$.

Theorem 24.36 (Phragmén–Lindelöf, 1908). *Assume that a function $f(z)$ is holomorphic in the sector $S_\alpha = \{z : |\operatorname{Arg} z| < \frac{\pi}{2\alpha}\}$ for some $\alpha \geqslant 1$ and is continuous and bounded on the boundary of this sector,*

$$|f(z)| \leqslant M \qquad \text{for all } z \text{ such that } \operatorname{Arg} z = \pm\tfrac{\pi}{2\alpha}. \tag{24.10}$$

If the growth of f admits a uniform apriori bound

$$|f(z)| = O(\exp|z|^\beta), \qquad |z| \to \infty, \quad z \in S_\alpha, \tag{24.11}$$

for some $\beta < \alpha$, then in fact f is bounded in S_α by the same constant, $|f(z)| \leqslant M$ for all $z \in S_\alpha$.

Proof. Consider, following [**Tit39**, §5.6], the auxiliary function $g(z) = \exp(-\varepsilon z^\gamma) \cdot f(z)$ with an arbitrary small positive $\varepsilon > 0$ and some γ between α and β. We have

$$|g(z)| = \exp\bigl(-\varepsilon|z|^\gamma \cdot \cos(\gamma \operatorname{Arg} z)\bigr)\,|f(z)|.$$

Since $\gamma < \alpha$, we have $\cos(\frac{\pi\gamma}{2\alpha}) > 0$ and hence

$$|g(z)| \leqslant M \qquad \forall z \in \partial S_\alpha = \{\operatorname{Arg} z = \pm\tfrac{\pi}{2\alpha}\}.$$

On the circular arcs $\{|z| = r\} \cap S_\alpha$ by the growth assumption on f we have the estimates

$$|g(z)| \leqslant \exp\bigl(-\varepsilon r^\gamma \cos\tfrac{\gamma\pi}{2\alpha}\bigr) \cdot |f(z)| \leqslant C\exp\bigl(r^\beta - \varepsilon r^\gamma \cos\tfrac{\gamma\pi}{2\alpha}\bigr).$$

As $\gamma > \beta$ and $\varepsilon > 0$, the latter expression tends to zero as $r \to \infty$, hence the maximum modulus principle applied to the bounded sector $S_\alpha \cap \{|z| < r\}$ for all sufficiently large r yields the inequality $|g(z)| \leqslant M$ there. Since r can be arbitrarily large, $|g(z)| \leqslant M$ everywhere in S_α.

The last inequality, transformed to the form $|f(z)| \leqslant M \exp(\varepsilon|z|^\gamma)$, for any finite $z \in S_\alpha$ admits passing to the limit as $\varepsilon \to 0^+$, yielding the inequality $|f(z)| \leqslant M$ in S_α. □

To apply this result to the half-plane $\mathbb{C}_+$ corresponding to $\alpha = 1$, we would have to require that f grows subexponentially as $|z| \to \infty$. Yet this growth condition can be relaxed if f is controlled along the real axis.

Lemma 24.37. *Let f be a function holomorphic in the half-plane $\mathbb{C}_+$ and continuous and bounded on the imaginary axis $i\mathbb{R} = \partial\mathbb{C}_+$. Assume that f grows at most exponentially in $\mathbb{C}_+$, i.e., $|f(z)| \leqslant C\exp(\mu|z|)$ for some $\mu > 0$.*

Then under this apriori growth assumption:

1. *if f is bounded on the real axis $\mathbb{R}_+ \subset \mathbb{C}_+$, then f is bounded everywhere in $\mathbb{C}_+$ and the maximum of its absolute value is achieved somewhere on the boundary;*
2. *if f decreases faster than any exponent along the real axis $\{z > 0\}$, $|f(z)| \leqslant C_\rho \exp(-\rho z)$ for any large $\rho > 0$, then f is identically zero, $f \equiv 0$.*

Moreover, these assumptions hold if the half-plane $\mathbb{C}_+$ is replaced by the standard domain Ω_C.

Proof. By Theorem 24.36 applied with $\alpha = 2$, $\beta = 1$ to each of the quarter-planes $\mathbb{C}_+ \cap \{\pm \operatorname{Im} z > 0\}$, we conclude that f is bounded in each of them, proving thus the first assertion of the lemma.

To prove the second assertion, consider the family of functions $f_\varepsilon(z) = f(z)\exp(z/\varepsilon)$ for arbitrarily small $\varepsilon > 0$. Any such function still has exponential growth in $\mathbb{C}_+$. Since the exponent has modulus equal to 1 on $i\mathbb{R}$ for any $\varepsilon > 0$, the maximum absolute value M achieved by f_ε on the boundary, does not depend on ε. Finally, if f decreases faster than any exponent along $\mathbb{R}_+$, so does each f_ε. Applying the first assertion of the lemma to f_ε, we arrive at the inequality $|f_\varepsilon(z)| \leqslant M$ for all $z \in \mathbb{C}_+$ and all $\varepsilon > 0$. Rewriting this inequality in the form $|f(z)| \leqslant M\,|\exp(-z/\varepsilon)|$ and passing to the limit as $\varepsilon \to 0^+$, we conclude that $f(z)$ must vanish identically in $\mathbb{C}_+$.

Finally, if f satisfies the assumptions of the lemma in a standard domain Ω_C, then $f \circ \varphi_C$ obviously satisfies the same assumptions in $\mathbb{C}_+$, where $\varphi_C \colon \mathbb{C}_+ \to \Omega_C$ is the map (24.5) occurring in the definition of the standard domain. □

Proof of Theorem 24.29. Theorem 24.29 is an immediate corollary to Lemma 24.37.

If two almost regular germs g and h have the same asymptotic expansions (24.6), then their difference germ $g-h$ has zero asymptotic expansion. Let f be a representative of this difference. By Definition 24.27, it can be holomorphically extended to some standard domain Ω_C, and grows no faster than a linear function there. On the other hand, f decays at infinity faster than any exponential, since its asymptotic series is identically zero. By Lemma 24.37, $f \equiv 0$, hence, $g \equiv h$. □

24H. Correspondence map of a hyperbolic saddle. The proof of Theorem 24.30 rests upon the following result.

Theorem 24.38. *The correspondence map of a hyperbolic saddle is almost regular.*

To prove this theorem, we first note that the correspondence map of a hyperbolic saddle *in the formal normal form* (22.3) is almost regular; moreover, in this case the corresponding Dulac series is convergent.

If the normal form is linear, then the correspondence map is a pure power, $w = cz^\lambda$, which becomes affine $\zeta \mapsto \lambda\zeta + \ln c$, in the logarithmic charts. Thus only a nonlinear normal form should be studied.

Consider the saddle vector field in the formal normal form, defined by the ordinary differential equations

$$\begin{cases} \dot w = -\lambda w(1+q(u)), & u(z,w) = z^m w^n, \\ \dot z = z, & \lambda = \frac{m}{n}, \end{cases} \qquad q(u) = \frac{u^{p+1}}{1+\alpha u^p}. \tag{24.12}$$

Let τ_+ and τ_- be the cross-sections $\{w=1\}$ and $\{z=1\}$ to the vector field (24.12) with the charts z and w on them respectively. The correspondence map $\Delta\colon \tau_+ \to \tau_-$ is well defined for $z>0$ and takes positive values.

Proposition 24.39. *The correspondence map Δ between the cross-sections τ_+ and τ_- for the vector field* (24.12) in the formal normal form, *is almost regular.*

Moreover, the corresponding Dulac series in this case is convergent: there exists a real analytic function $G \in \mathcal{O}(\mathbb{R}^2, 0)$ such that

$$-\ln \Delta(\zeta) = \lambda\zeta + G\bigl(\exp(-m\zeta), \zeta\exp(-m\zeta)\bigr), \qquad \zeta = -\ln z. \tag{24.13}$$

Proof. The assertion follows from integrability of the vector field (24.12) which allows us to compare the value of the resonant monomial on the intersections $(z,1)$ and $(1,w)$ of an arbitrary integral trajectory of (24.12) with the cross-sections $\tau_\pm$. One has to prove that the solution of the initial value problem for the quotient differential equation

$$\frac{du}{dt} = -n\lambda\frac{u^{p+2}}{1+\alpha u^p}, \qquad u(0) = z^m,$$

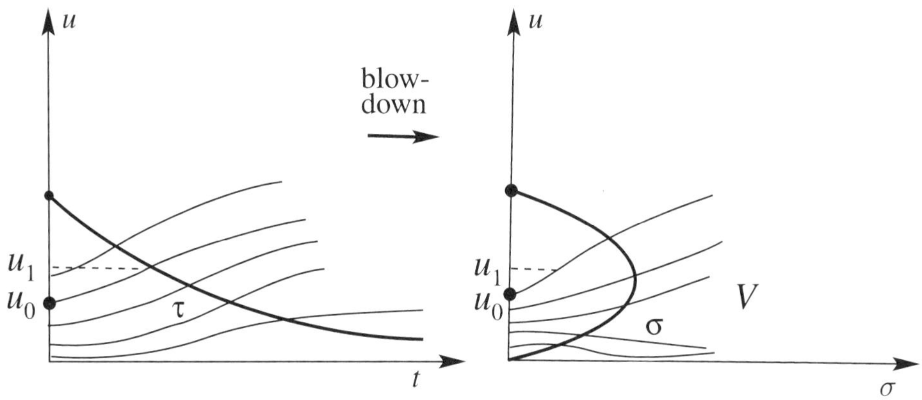

Figure IV.10. Integration of the quotient equation via blow-down

evaluated at the moment $t = -\ln z$, is (after extracting of the mth order root) an almost regular function $w(z)$ of z. The proof can be achieved by explicit integration and investigation of the resulting algebraic relation between z, $\ln z$ and w.

Yet one can avoid intermediate calculations applying the following geometric construction (a particular nonparametric case of [**IY91**, Lemma 11]). The quotient equation can be coupled with the trivial equation $\dot t = 1$, resulting in a vector field in the positive quadrant of the (t, u)-plane,

$$\begin{cases} \dot u = u[-n\lambda\, q(u)], \\ \dot t = 1, \end{cases} \qquad t, u \geqslant 0. \tag{24.14}$$

Suppose that a trajectory γ of the initial field (24.12) crosses τ_+ at the point $(1, z)$ corresponding some value $u_0 = z^m$. Then the travel time necessary to reach τ_- is equal to $-\ln z = -\frac{1}{m}\ln u_0$.

Consider on the (t, u)-plane the curve $\tau = \{t = -\frac{1}{m}\ln u\}$; the value u at the moment of intersection between γ and τ_- is the u-coordinate of the intersection of the respective trajectory of (24.14) with τ (Fig. IV.10).

The system (24.14) *admits a simple blow-down of the t-axis*: after passing to the coordinates u and $v = tu$, we obtain

$$\dot u = u\bigl(-n\lambda\, q(u)\bigr), \qquad \dot v = u + v\bigl(-n\lambda\, q(u)\bigr) = u\left(1 - n\lambda\frac{vu^p}{1+\alpha u^p}\right) \tag{24.15}$$

(we use the fact that $q(u)$ is divisible by u). After division by u we obtain a *nonsingular vector field* V in a neighborhood of the origin on the (u, v)-plane, tangent to the v-axis. The curve τ blows down to the curve σ defined by the equation $v = -\frac{1}{m}u\ln u$ which tends to the origin as $u \to 0^+$.

The vector field V, being transversal to the u-axis and tangent to the v-axis, admits a real analytic first integral $\Phi(u,v) = uF(u,v)$, $F(0,0) = 1$, uniquely defined by the Cauchy boundary data $\Phi(u,0) \equiv u$. From the above description of the correspondence map, we conclude that the u-value u_1 at the moment when the trajectory crosses the exit section τ_-, is equal to the value of Φ restricted on σ, i.e., $u_1 = u_0 F(u_0, -\frac{1}{m} u_0 \ln u_0)$.

Returning to the coordinates $z = u_0^{1/m}$ and $w = u_1^{1/n}$, we conclude that the correspondence map for the saddle in the formal normal form can be expressed as

$$\begin{aligned} w &= [z^m F(z^m, -z^m \ln z)]^{1/n} && G \in \mathcal{O}(\mathbb{R}^2, 0), \\ &= z^\lambda G(z^m, z^m \ln z), && G(0,0) = 1, \end{aligned} \tag{24.16}$$

where the function $G = F^{1/n}$ is real analytic in its two variables since $F(0,0) = 1$.

In the logarithmic chart $\zeta = -\ln z$ the correspondence map $-\ln w$ defined by the expressions (24.16) becomes a convergent Dulac series. □

For a saddle *not* in the formal normal form, we can no longer claim that the correspondence map is represented by a *convergent* Dulac series; for instance, this is impossible for formally linearizable but analytically nonlinearizable saddles (for more examples see [**Tri90**]). Nevertheless, we will show that this map extends analytically into sufficiently large domain in the logarithmic chart and admits an *asymptotic* Dulac series there.

Lemma 24.40. *The correspondence map of a saddle in the logarithmic chart extends to a standard domain Ω_C for a sufficiently large $C > 0$.*

Proof. The meromorphic nonlinear differential equation

$$\frac{dw}{dz} = -\lambda \frac{w}{z} \cdot (1 + \Psi(z,w)), \qquad z, w \in \mathbb{C}, \tag{24.17}$$

in the logarithmic chart takes the form

$$\frac{dw}{d\zeta} = -\lambda w (1 + \psi(w, \zeta)). \tag{24.18}$$

The function ψ holomorphic in the product $\mathbb{C}_+ \times \{|w| < 1\}$ can without loss of generality be assumed uniformly arbitrarily small there, in particular, it is sufficient if $|\psi| < \lambda/2$.

Consider the function $W(\zeta, \eta)$ of two complex variables, which is initially only locally defined near the diagonal $\{\zeta = \eta\}$ as the solution of the equation (24.18) with the initial condition $W(\eta, \eta) = 1$.

An oriented path $\gamma = \gamma(\eta)$ in the ζ-half-plane $\mathbb{C}_+$, connecting the point $\zeta = \eta$ with the point $\zeta = 0$ will be called *admissible*, if $W(\cdot, \eta)$ can be

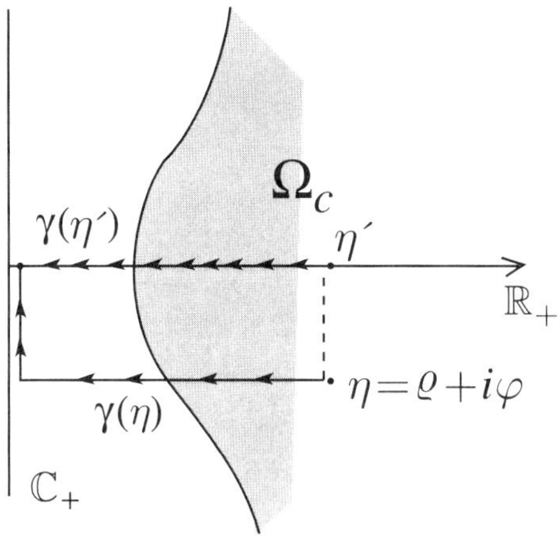

Figure IV.11. Analytic continuation of saddle correspondence map

analytically continued along this path from its initial value $W(\eta, \eta) = 1$, and this continuation satisfies the restriction $\left|W(\,\cdot\,, \eta)|_\gamma\right| \leqslant 1$ along this path.

If γ is an admissible path, then it defines the germ at $\zeta = \eta$ of *some* branch $\Delta(\eta) = W(0, \eta)$ of the complexified correspondence map; here the right hand side is obtained by the above continuation along γ.

If $\eta_+ \in \mathbb{R}_+$ is a point on the real axis, then the path $\gamma(\eta_+) = [\eta_+, 0]$ (the real segment) is admissible, since $W(\cdot, \eta_+)$ is increasing on the real axis. The function $\Delta(\eta_+) \colon \mathbb{R}_+ \to \mathbb{R}_+$ obtained by continuation along these paths defines the *real branch* of the correspondence map.

In order to obtain the analytic continuation of this real branch to a point $\eta \in \mathbb{C}_+$, one should find an admissible path $\gamma(\eta) = \gamma_0$ which can be continuously deformed *within a family of admissible paths* γ_s, $s \in [0, 1]$, into a real segment $[0, \eta'] = \gamma_1$.

Let $\eta = \varrho + i\varphi$ be a point in the half-plane $\mathbb{C}_+$. We claim that the path $\gamma(\eta)$ which consists of the segment of length ϱ from η to the point $i\varphi \in i\mathbb{R} = \partial\mathbb{C}_+$, and the segment of length $|\varphi|$ on imaginary axis, continuing the path to the origin, is admissible provided that η belongs to some standard domain Ω_C.

Indeed, for points $\varrho + i\varphi$ inside the standard domain Ω_C, we have the asymptotic representation $|\varphi| = (\varrho/C)^{1/2} + O(1)$ as $\varrho \to +\infty$. Along the first segment of the corresponding path $\gamma = \gamma(\varrho + i\varphi)$ the modulus of W decreases exponentially from 1 to a small value not exceeding $\exp(-\lambda\varrho/2)$ if $|\psi| < \frac{1}{2}$, since $\operatorname{Re}(\lambda + \psi(z, w)) \geqslant \lambda/2$ along this path.

The Cauchy operator $F^{i\varphi}$ of analytic continuation (flow) along the vertical segment can be represented in the form

$$F^{i\varphi} = F^{i\theta} \circ \Delta_z^{nk}, \qquad 0 \leqslant \theta < 2\pi n, \ k \in \mathbf{N},$$

where $\Delta_z(w) = (\exp 2\pi i \frac{m}{n})w + O(w^2) \in \operatorname{Diff}(\mathbb{C}, 0)$ is the holonomy (monodromy) operator associated with the standard loop $z = \exp 2\pi i t$, $t \in [0,1]$, on the z-axis. The linear part of Δ_z is a rational rotation, so that the nth iterate $\Delta_z^n(w) = w + O(w^2) \in \operatorname{Diff}_1(\mathbb{C}, 0)$ is parabolic (tangent to the identity). Because of the inequality between $|\varphi|$ and ϱ implied by the condition $\varrho + i\varphi \in \Omega_C$, we have an upper bound $k = O\big((\varrho/C)^{1/2}\big)$.

Let L be the maximal Lipschitz constant of the flow map $F^{i\theta}$ over $0 \leqslant \theta \leqslant 2\pi n$ on the disk $\{|w| \leqslant 1\}$. Clearly, $L < +\infty$.

The growth of iterates of $\Delta_z^{nk}(w)$ of the parabolic germ Δ_z^n as $k \to \infty$ can be estimated comparing the growth of the cascade of iterates $\sigma_a \colon r \mapsto r + ar^2$ with the growth of solutions of the auxiliary differential equation $\dot r = br^2$ on the time interval $t \in [0, k]$ (here a, b are real parameters). Indeed, since Δ_z^n is parabolic, $|\Delta^n(z) - z| < a|z|^2$, thus the absolute value of the iterates $\Delta_z^{kn}(z)$ does not exceed $\sigma_a^k(r)$, $r = |z|$. The flow $\sigma_b' = \exp br^2 \frac{\partial}{\partial r}$ of the auxiliary equation can also be immediately computed: $\sigma_b'(r) = r + br^2/(1 - br)$. Thus for each $a > 0$ one can find $b > 0$ such that the flow majorizes the cascade, $\sigma_a^k < \sigma_b'^k$ for all k on a sufficiently small interval $r < \varepsilon$.

The flow of this equation with the initial condition $r(0) = |W(i\varphi, \eta)| \leqslant \exp(-\lambda\varrho/2)$ at the moment $k = O\big((\varrho/C)^{1/2}\big)$ does not exceed the reciprocal $\big|\exp(\lambda\varrho/2) - O(\varrho/C)^{1/2}\big|^{-1}$ (the auxiliary flow is constant in the chart $1/w$). Thus we conclude that along the path $\gamma(\eta)$ the function $W(\,\cdot\,, \eta)$ is bounded in the absolute value by $L\big|\exp(\lambda\varrho/2) - O(\varrho/C)^{1/2}\big|^{-1}$ which is less than 1 if $\varrho > C$ (as is the case if $\eta \in \Omega_C$) and C is sufficiently large.

The path $\gamma(\eta)$, $\eta = \varrho + i\varphi$, can be deformed to a segment of the real axis as follows: its endpoint $\eta_s = \varrho + i\varphi(1 - s)$ moves parallel to the imaginary axis towards $\varrho \in \mathbb{R}$, and $\gamma(\eta_s)$ as before consists of a horizontal segment of the same length ρ and contracting vertical segments of length $(1 - s)|\varphi|$. All estimates remain the same during this deformation, hence the paths $\gamma(\eta_s)$ are admissible for all $s \in [0, 1]$. □

Proof of Theorem 24.38. After the existence of analytic continuation of the saddle correspondence map into a standard domain is proved, the Proximity lemma 22.6 together with Proposition 24.39 allow us to prove that this map admits an asymptotic expansion in the Dulac series. This will complete the proof of Theorem 24.38.

Consider an arbitrary saddle vector field F. By Proposition 24.39, without loss of generality we may assume that the coordinates are chosen that F differs from its formal normal form F_0 by N-flat terms as in (22.4).

We compare the correspondence maps Δ and Δ_0 for the two saddle fields, F and F_0 respectively. Both maps are defined in some standard domain Ω_C, and the correspondence map Δ_0 for F_0 is represented as a convergent Dulac series there.

By the Proximity Lemma 22.6, the correspondence map Δ for F differs from Δ_0 in Ω_C by the term that decays sufficiently fast to infinity,

$$\Delta(\zeta) - \Delta_0(\zeta) = O(\exp(-N\zeta/2)) \qquad \text{as } |\zeta| \to \infty, \quad \zeta \in \Omega_C.$$

This means that the Dulac series Δ_0 approximates Δ with an accuracy corresponding to $\nu = N/2$ in (24.7). Since N can be arbitrary, this (together with Remark 24.28) proves that the correspondence map for any saddle vector field is almost regular.

The assertion of Theorem 24.30 follows immediately from Lemma 24.33, as a composition of almost regular germs is almost regular. $\square$

Exercises and Problems for §24.

Exercise 24.1. What step of the proof of the Poincaré–Bendixson fails when attempting to literally reproduce it for the 2-torus?

Problem 24.2. Let $H(x+iy) = y^2 - x^2 + y^4$ be a polynomial on the plane $\mathbb{R}^2 \cong \mathbb{C}^1$. Plot the phase portrait of the vector field $\dot z = ie^{iH(z)}\,(\frac{\partial H}{\partial x} + i\frac{\partial H}{\partial y})$ (in the complex notation).

Exercise 24.3. Modify the previous example to construct explicitly a polynomial vector field with an ω-limit set which carries any number of singular points on it.

Problem 24.4. Prove that sums and products of almost regular germs are almost regular.

Problem 24.5. Give an example of a saddle, for which the correspondence map has a divergent Dulac series.

Problem 24.6. Prove, using the Phragmen–Lindelöf theorem, that if one of the sectorial components of a *normalizing* map-cochain (Definition 21.9) is identical, then all other components are also identical (cf. with Problem 21.3).

Chapter V

Global properties of complex polynomial foliations

The last chapter focuses on the global properties of singular holomorphic foliations on the complex projective plane $\mathbb{P}^2$. Any such foliation is necessarily algebraic. This algebraicity plays the central role when discussing properties of algebraic leaves of the foliation (existence, apriori bounds on their degree, integrability) in §25.

Integrable, in particular, Hamiltonian foliations form a very special class, yet this class is very important for applications. Besides, though real foliations from this class cannot have limit cycles (all periodic trajectories occur in continuous families and are nonisolated), these cycles can be created by small perturbation of integrable systems. Location of these newborn cycles is determined by zeros of certain Abelian integrals, which are transcendental functions of the parameters. Investigation of properties of these functions is the central subject of §26, in which algebraic and transcendental methods work hand in hand.

The last two sections of the book deal with topology of linear foliations in $\mathbb{C}^n$ and generic properties of planar polynomial foliations in $\mathbb{P}^2$. The tools used in these two sections, are completely transcendental.

In §27 we study *linear* foliations in higher dimensions and show that their topology is radically different in the Poincaré and in the Siegel cases. While the Poincaré case corresponds to structurally stable foliations (their topology is not affected by small perturbations), in the Siegel case foliations have continuous moduli of topological classification. This can be seen as yet another manifestation of *rigidity*: a foliation topologically equivalent to the given one and sufficiently close to it, is in fact conjugated to the given foliation by a very special, almost linear homeomorphism.

Unlike the real foliations that are severely constrained by the topology of the real 2-sphere, planar *complex* foliations generically exhibit rather complicated behavior, in particular, they have countably many "complex limit cycles", their leaves are everywhere dense. Besides, generic foliations do not admit nontrivial deformations preserving topology, again displaying the same rigidity phenomenon similar to the one discussed in the linear case. All these phenomena are studied in §28.

25. Algebraic leaves of polynomial foliations on the complex projective plane $\mathbb{P}^2$

Any singular foliation defined by a *polynomial* 1-form on the complex plane $\mathbb{C}^2$, can be extended as a singular foliation on the complex *projective* plane $\mathbb{P}^2$. Conversely, any singular foliation on the complex projective plane with a finite number of singular points may be obtained in this way.

In this section we discuss the notion of degree of a polynomial foliation on $\mathbb{P}^2$ and two natural classes of "foliations of the given degree". The *generic* foliations from these two classes have similar though not coinciding properties.

Then we switch to the study of algebraic leaves of polynomial foliations, focusing on determination of their degrees: the problem goes back to Poincaré. Its recent solution is the central result of the section. After that we show that *generically* polynomial foliations have no algebraic leaves. On the other hand, abundance of algebraic leaves implies integrability.

25A. Extension of polynomial foliations on $\mathbb{P}^2$ and the projective degree. The affine plane $\mathbb{C}^2$ with coordinates (x, y) can be associated with points of the projective plane $\mathbb{P}^2$ with homogeneous coordinates $[x : y : 1]$. The complement $\mathbb{P}^2 \smallsetminus \mathbb{C}^2 \cong \mathbb{P}$ consisting of points with homogeneous coordinates $[x : y : 0]$, is the *projective line at infinity* denoted by $\mathbb{I}$. A foliation defined by the Pfaffian equation $\{\omega = 0\}$ with a polynomial 1-form ω, can be naturally extended as a holomorphic foliation with isolated singularities on the whole $\mathbb{P}^2$.

25$\mathbf{A}_1$. *Extension on the infinite line.* Consider a polynomial 1-form

$$\omega = p(x,y)\,dx + q(x,y)\,dy, \quad p,q \in \mathbb{C}[x,y], \ \max(\deg p, \deg q) = r, \quad (25.1)$$

and let $\mathcal{F}$ be the foliation of $\mathbb{C}^2$ defined by the Pfaffian equation $\omega = 0$. As usual, we assume that $\gcd(p, q) = 1$, i.e., that all singularities of ω are isolated.

To study the foliation $\mathcal{F}$ in a neighborhood of $\mathbb{I}$, we pass to the coordinates $u = 1/x$, $v = y/x$, and consider a neighborhood of the line $\{u = 0\} = \mathbb{I}$. In these coordinates ω becomes meromorphic,

$$\begin{aligned}\omega &= -p\left(\frac{1}{u}, \frac{v}{u}\right)\frac{du}{u^2} + q\left(\frac{1}{u}, \frac{v}{u}\right)\frac{u\,dv - v\,du}{u^2} \\ &= -\frac{1}{u^{r+2}}\big(p_r(1,v) + vq_r(1,v)\big)\,du \\ &\quad + \frac{1}{u^{r+1}}\big[-\big(p_{r-1}(1,v) + vq_{r-1}(1,v)\big)\,du + q_r(1,v)\,dv\big] \\ &\quad + \frac{1}{u^r}[\cdots] + \cdots\end{aligned} \tag{25.2}$$

(grouped are the Laurent terms of different degrees, while p_k, q_k, $k = 0, \dots, r$ denote the homogeneous components of the polynomials p, q respectively). Depending on various possible relationships between different homogeneous components of the form ω, one can have a pole of orders $r + 2$ or less on $\{u = 0\}$. More precisely, we have the following alternative, depending on the homogeneous polynomial

$$h_{r+1}(x, y) = x\,p_r(x, y) + y\,q_r(x, y) \in \mathbb{C}[x : y] \tag{25.3}$$

depending on the principal homogeneous component of the coefficients of ω.

Nondicritical case. If the homogeneous polynomial h_{r+1} does not vanish identically, then the order of pole of ω on the infinite line $\mathbb{I}$ is *exactly* $r + 2$. Multiplying ω by u^{r+2} we obtain a polynomial 1-form ω' of degree $r + 1$ defining the same foliation $\mathcal{F}$ in the coordinates (u, v),

$$\omega' = \big(h_{r+1}(1, v) + O(u)\big)\,du + u\big(q_r(1, v) + O(u)\big)\,dv. \tag{25.4}$$

The following facts can be immediately verified by direct inspection.

(i) The polynomial Pfaffian equation $\{\omega' = 0\}$ of degree $r + 1$ has isolated singularities on $\mathbb{I}$ at the points $a_i = [x_i : y_i : 0]$ corresponding to the roots of the homogeneous polynomial h_{r+1}.

(ii) The infinite line $\mathbb{I} = \{u = 0\}$ is a separatrix of the foliation $\mathcal{F}$ extended on $\mathbb{P}^2$.

(iii) The linearization of the equation $\{\omega' = 0\}$ along the infinite line (as described in §14**C**) yields the linear equation

$$du = \theta u, \qquad \theta = -\frac{q_r(1, v)}{h_{r+1}(1, v)}\,dv. \tag{25.5}$$

This equation defines a meromorphic connexion on a line bundle over $\mathbb{I}$ with singularities at the points a_i. The residues of the connexion form θ at these points are the *characteristic numbers*

$$\lambda_i = -\frac{q_r(1, v_i)}{s_r(v_i)}, \qquad s_r(v) = \frac{\partial h_{r+1}(1, v)}{\partial v}, \quad v_i = y_i/x_i. \tag{25.6}$$

(iv) The sum of all residues $\sum_{i=1}^{r+1} \lambda_i$ does not depend on the foliation $\mathcal{F}$ (Theorem 17.33). One can easily check that

$$\lambda_1 + \cdots + \lambda_{r+1} = 1, \qquad \lambda_i = \operatorname{res}_{a_i} \theta. \tag{25.7}$$

This shows that the embedding $\mathbb{I} \hookrightarrow \mathbb{P}^2$ is different from the embedding $\mathbb{E} \hookrightarrow \mathbb{M}$ (cf. with (10.11)): the normal bundle in the former case has degree $+1$.

(v) Any meromorphic 1-form θ with $r + 1$ simple poles and arbitrary residues λ_i constrained by the single condition (25.7) can be obtained as the connection form induced on the infinite line $\mathbb{I}$ by an appropriate polynomial 1-form ω of degree r.

Dicritical case. If $h_{r+1} \equiv 0$, then the order of the pole of ω on $\mathbb{I}$ is no more than $r + 1$, and the foliation $\mathcal{F}$ in the coordinates (u, v) is defined by a polynomial 1-form ω' of degree $\leqslant r$,

$$\begin{gathered} \omega' = -\bigl(p_{r-1}(1, v) + vq_{r-1}(1, v)\bigr)\, du + q_r(1, v)\, dv + u\omega'', \\ \omega'' \text{ holomorphic on } \{u = 0\}. \end{gathered} \tag{25.8}$$

In fact, the degree of ω' must be *exactly equal* to r; otherwise the univariate polynomial $q_r(1, v)$ in (25.8) must vanish identically. Together with the condition $h_{r+1}(1, v) = p_r(1, v) + vq_r(1, v) \equiv 0$ this would imply that $p_r(1, v) \equiv 0$ as well, which is impossible. As a result, we have the following characterization of the dicritical case:

(vi) the polynomial Pfaffian equation $\{\omega' = 0\}$ of degree exactly r may have isolated singularities on the infinite line $\mathbb{I}$, yet

(vii) the infinite line itself is never a separatrix of the foliation $\mathcal{F}$ extended on $\mathbb{P}^2$.

The same conclusions obviously hold for the third affine chart on $\mathbb{P}^2$ corresponding to the variables $w = x/y$, $z = 1/y$.

25$\mathbf{A}_2$. *Projective degree. Classes $\mathcal{A}_r$ and $\mathcal{B}_r$.* The above computations show that passage from one affine chart on $\mathbb{C}^2$ to another may change the degree of a polynomial field (resp., form) defining the foliation. This fact prompts for several definitions of the degree of a polynomial foliation on $\mathbb{P}^2$. Note that the projective plane without *any line* $\mathbb{P}^1 \cong \ell \subset \mathbb{P}^2$ admits an affine chart for which ℓ plays the role of the infinite line $\mathbb{I}$.

Definition 25.1. The class $\mathcal{A}_r$ consists of all foliations of $\mathbb{P}^2$ which in a *fixed affine neighborhood* $\mathbb{C}^2 \subset \mathbb{P}^2 \smallsetminus \ell$, are defined by polynomial forms $\omega \in \Lambda^1[x,y]$ of degree $\leqslant r$ with isolated singularities.

Clearly, the class $\mathcal{A}_r = \mathcal{A}_r(\ell)$ is defined independently of the choice of the affine chart on the affine neighborhood $\mathbb{C}^2$, but is not invariant by projective transformations of $\mathbb{P}^2$. However, since any two lines in $\mathbb{P}^2$ can be superposed by a projective transformation, for any other choice of the "infinite line" $\ell' \subset \mathbb{P}^2$ the corresponding class $\mathcal{A}_r(\ell')$ will be naturally isomorphic to $\mathcal{A}_r(\ell)$.

For the fixed affine chart, the class $\mathcal{A}_r$ can be identified with the *complex projective space* of all polynomial vector fields (resp., polynomial 1-forms) of degree $\leqslant r$: two fields (forms) which differ by a constant multiplier, define the same foliation. This observation allows us to discuss *generic* properties of foliations from the class $\mathcal{A}_r$. For instance, a *generic foliation from the class $\mathcal{A}_r$ is nondicritical and hence has an invariant leaf at infinity.*

On the other hand, one can attempt to give an invariant definition of *projective degree* via homogeneous coordinates in $\mathbb{C}^3$.

Consider the space $\mathbb{C}^3 \smallsetminus \{0\}$ equipped with the homogeneous coordinates $[X : Y : Z]$, and the *Euler vector field*

$$V = X\frac{\partial}{\partial X} + Y\frac{\partial}{\partial Y} + Z\frac{\partial}{\partial Z} \tag{25.9}$$

on it. For any homogeneous 1-form Ω of degree r

$$\begin{aligned} &\Omega = A(X,Y,Z)\,dX + B(X,Y,Z)\,dY + C(X,Y,Z)\,dZ, \\ &A, B, C \text{ homogeneous}, \quad \deg A = \deg B = \deg C = r, \end{aligned} \tag{25.10}$$

consider the distribution of 2-planes (Pfaffian equation) $\{\Omega = 0\}$.

The canonical projection $\pi\colon \mathbb{C}^3 \smallsetminus \{0\} \to \mathbb{P}^2$ along the lines tangent to trajectories of V, correctly defines a 1-dimensional distribution (the line field) on $\mathbb{P}^2$ if and only if Ω vanishes on V identically, i.e., when

$$\Omega(V) = XA + YB + ZC \equiv 0. \tag{25.11}$$

Under this condition one can define the quotient distribution on $\mathbb{P}^2 = \mathbb{C}^3 \smallsetminus \{0\}/\mathbb{C}\smallsetminus\{0\}$ which in any affine chart will be defined by a suitable polynomial 1-form.

Definition 25.2. A polynomial foliation $\mathcal{F}$ on $\mathbb{P}^2$ has the *projective degree* r, if in the homogeneous coordinates $[X : Y : Z]$ it is defined by a homogeneous 1-form Ω of degree r as in (25.10) satisfying the identity (25.11), and the coefficients A, B, C of the form have *no common factor.*

Remark 25.3. If the coefficients A, B, C have a common polynomial factor (necessarily homogeneous) f, i.e., $A = fA'$, $B = fB'$, $C = fC'$, then the foliation $\mathcal{F}$ has projective degree $\leqslant r - \deg f$: obviously, the homogeneous form $\Omega' = f^{-1}\Omega$ also satisfies the identity (25.11). Thus the equality $\deg \Omega = r$ without excluding the reducible cases defines a polynomial foliation *of the projective degree* $\leqslant r$.

Example 25.4. The affine chart $x = X/Z$, $y = Y/Z$ on $\mathbb{P}^2$ can be identified with the affine subspace $\mathbb{C}^2 \cong \Pi = \{Z = 1\} \subset \mathbb{C}^3 \smallsetminus \{0\}$. In this chart the distribution obtained by the projection π can be described by the Pfaffian equation $\omega = 0$, where the form ω is the restriction of Ω on the plane Π,

$$\omega = \Omega|_{\{Z=1\}} = A(x, y, 1)\,dx + B(x, y, 1)\,dy. \tag{25.12}$$

It is a polynomial 1-form of degree $\leqslant r$.

Conversely, any distribution of lines defined by a polynomial form $\omega = p(x, y)\,dx + q(x, y)\,dy \in \Lambda^1[x, y]$ of degree r can be "lifted" to a (singular) distribution of 2-planes on $\mathbb{C}^3$, containing the Euler field, if the coefficients of the polynomial 1-form $\Omega = A\,dX + B\,dY + C\,dZ$ of degree $\leqslant r + 1$ are chosen as follows:

$$\begin{aligned} A(X, Y, Z) &= \quad Z^{r+1}\,p(X/Z, Y/Z), \\ B(X, Y, Z) &= \quad Z^{r+1}\,q(X/Z, Y/Z), \\ C(X, Y, Z) &= -Z^{-1}\big(X\,A(X, Y, Z) + Y\,B(X, Y, Z)\big). \end{aligned} \tag{25.13}$$

Note that in general Ω cannot be constructed in the class of homogeneous forms of degree $\leqslant r$, since the coefficient C may not be polynomial in the latter case.

However, if the homogeneous polynomial $xp_r(x, y) + yq_r(x, y) \in \mathbb{C}[x : y]$ vanishes identically (i.e., in the dicritical case), the coefficients of the form Ω of degree $r + 1$ restored as in (25.13), will all be divisible by Z and hence the foliation can be defined by a homogeneous 1-form $\Omega' = Z^{-1}\Omega$ of degree r vanishing on the Euler field. The restriction of Ω' on Π still coincides with ω.

Definition 25.5. The class $\mathcal{B}_r$ is the collection of foliations of the projective degree $\leqslant r$ on $\mathbb{P}^2$.

Similarly to the class $\mathcal{A}_r$, the class $\mathcal{B}_r$ can be identified with a suitable projective space of homogeneous 1-forms Ω as in (25.10), constrained by the linear equalities (25.11) and considered modulo a constant multiplier. This circumstance allows us to introduce on $\mathcal{B}_r$ a Lebesgue measure and study *generic* properties of polynomial foliations, that hold for all $\mathcal{F}$ except a subset of $\mathcal{B}_r$ of zero measure. Besides, one can describe properties valid for Zariski open subsets of $\mathcal{B}_r$.

The class $\mathcal{B}_r$ is invariant by projective transformations. As follows from Example 25.4, it can be described as the class of foliations defined by 1-forms of degree $\leqslant r$ in any affine chart.

Proposition 25.6.

$$\mathcal{B}_r = \bigcap_{\ell\subset\mathbb{P}^2} \mathcal{A}_r(\ell). \quad \square$$

Example 25.7. There are no foliations of projective degree 0: a form $\alpha\,dX+\beta\,dY+\gamma\,dZ$ vanishes on the Euler field V if and only if $\alpha=\beta=\gamma=0$.

Any foliation of projective degree 1 in an affine chart containing a singularity at (x_0, y_0), is given by the 1-form

$$(x-x_0)\,dy-(y-y_0)\,dx=0. \tag{25.14}$$

If the singularity lies on the infinite line $\mathbb{I}$, then the form is constant, $\omega = \alpha\,dx+\beta\,dy$.

All other linear vector fields define (after extension on $\mathbb{P}^2$) polynomial foliations of projective degree 2[1].

From the definitions of the two classes of polynomial foliations it follows that for each fixed affine chart we have the following *proper* inclusions,

$$\varnothing=\mathcal{B}_0\subset\mathcal{A}_0\subset\cdots\subset\mathcal{B}_r\subset\mathcal{A}_r\subset\mathcal{B}_{r+1}\subset\cdots. \tag{25.15}$$

The difference between the classes $\mathcal{A}_r$ and $\mathcal{B}_r$ is largely about existence of the invariant line at infinity.

Proposition 25.8.

1. *The difference $\mathcal{A}_r\smallsetminus\mathcal{B}_r$ consists of all foliations from $\mathcal{A}_r$ tangent to the infinite line $\ell=\mathbb{I}$ (i.e., nondicritical at infinity).*

2. *The difference $\mathcal{B}_{r+1}\smallsetminus\mathcal{A}_r$ consists of all foliations from $\mathcal{B}_{r+1}$ transversal to the infinite line $\ell=\mathbb{I}$ almost everywhere.* $\square$

Thus a generic foliation from $\mathcal{A}_r$ has an invariant line at infinity, while a generic foliation from $\mathcal{B}_r$ has no invariant lines at all. Later we will prove a stronger statement: a generic foliation form $\mathcal{A}_r$ has no algebraic leaves *besides* the infinite line, whereas a generic foliation from $\mathcal{B}_r$ has no invariant algebraic leaves *at all.* The former claim was proved by I. G. Petrovskiĭ and E. M. Landis in 1955 (see [**PL55**] and the appendix to this section), and can be modified to prove the latter claim as well. Yet we give a different, more transparent demonstration for the class $\mathcal{B}_r$; see Theorem 25.18.

[1]This creates some awkwardness, since linearity is firmly associated with the first degree polynomials. To avoid it, in some sources, e.g., in [**CLN91**], the degree of a polynomial foliation on $\mathbb{P}^2$ is defined as $r-1$, where r is the projective degree introduced above.

25$\mathbf{A}_3$. *Degree and tangency between foliations and lines.* Recall that the degree of a projective curve can be defined as the number of intersections with a generic line. In a similar way the projective degree of a polynomial foliation $\mathcal{F}$ can be described by the total order of contacts between this foliation and a noninvariant line $\ell \subset \mathbb{P}^2$.

The tangency order between $\mathcal{F}$ and ℓ at a nonsingular point $a \notin \Sigma = \operatorname{Sing}(\mathcal{F})$ is the tangency order between the leaf L_a of $\mathcal{F}$ passing through a, and the line ℓ. The formula (8.38) generalizes this definition for the case where $a \in \Sigma$ but ℓ is not a separatrix: one has to take the 2-form $\omega \wedge dl = f(x, y)\, dx \wedge dy$, where $\{l = 0\}$ is a linear local equation of the line ℓ, and compute the order of zero at the point a of the coefficient $f(x, y)$ of this 2-form after restriction on the line ℓ. (An equivalent definition can be given in terms of the Lie derivative of l along a vector field defining the foliation.)

This order will be denoted by $\tau_a(\ell, \mathcal{F})$. The *total tangency order* between ℓ and $\mathcal{F}$ is by definition the sum

$$\tau(\ell, \mathcal{F}) = \sum_{a \in \ell} \tau_a(\ell, \mathcal{F}). \tag{25.16}$$

Proposition 25.9. *The total tangency order between a foliation $\mathcal{F} \in \mathcal{B}_r$ and a noninvariant line $\ell \subset \mathbb{P}^2$ is equal to $r - 1$.*

Proof. Choose an affine coordinate system in which ℓ is the axis $\{y = 0\}$ and the infinite line is also not invariant. In the corresponding coordinates the form $\omega = p\, dx + q\, dy$ defining $\mathcal{F}$ has degree r, and the number of contacts between $\mathcal{F}$ and ℓ is the number of roots (counted with multiplicities) of the univariate polynomial $p(x, 0) \in \mathbb{C}[x]$.

We claim that this polynomial has degree $r - 1$ and not r. Indeed ω is dicritical at infinity since $\mathbb{I}$ is not invariant by the choice of the affine coordinates. Hence $xp_r(x, y) + yq_r(x, y) \equiv 0$, where p_r, q_r as usual denote the homogeneous terms of p, q respectively. Restricting this identity on $\{y = 0\}$, we conclude that $xp_r(x, 0) \equiv 0$, i.e., $p_r(x, 0) \equiv 0$. Thus the polynomial $p(x, 0)$ has no terms of order r.

One can easily verify by direct inspection of the formulas (25.2) that if the point $\ell \cap \mathbb{I}$ is not a point of contact between $\mathcal{F}$ and ℓ, then $\deg p(x, 0)$ is exactly $r - 1$. □

Since the total order of contact does not depend on the choice of the line ℓ, it can be chosen for the geometric definition of the projective degree of polynomial foliations on $\mathbb{P}^2$; cf. with [**CLN91**].

25$\mathbf{A}_4$. *Ubiquity of polynomial foliations.* Computations of §25**A** show that any singular foliation of $\mathbb{C}^2$ generated by a polynomial vector field, can be extended as a singular foliation of the projective plane. The inverse is also true, as the following theorem shows.

Theorem 25.10. *Any singular foliation on $\mathbb{P}^2$ in any affine chart is generated by a suitable polynomial vector field or* 1*-form.*

Recall that by Definition 2.24, the singular locus of a foliation must be an analytic set of codimension $\geqslant 2$, i.e., a finite point set of $\mathbb{P}^2$.

Proof of the theorem. The proof is a straightforward application of the Chow theorem [**Mum76**] asserting that any analytic subset of a projective variety is algebraic.

Consider the tangent bundle $\mathbf{T}\mathbb{P}^2$ and its projectivization $\mathbf{P}^3 = P\mathbf{T}\mathbb{P}^2$: by definition, it is the quotient space of all pairs (a, v), $0 \neq v \in \mathbf{T}_a\mathbb{P}^2$, by the equivalence relation $(a, v) \sim (a', v')$ if and only if $a = a'$ and $v' = \lambda v$ for some $\lambda \neq 0$.

The singular foliation $\mathcal{F}$ on $\mathbb{P}^2$ with the singular locus Σ defines a map $s\colon \mathbb{P}^2 \smallsetminus \Sigma \to \mathbf{P}^3$ associating with each nonsingular point the direction of the line through it, tangent to $\mathcal{F}$. The image $s(\mathbb{P}^2 \smallsetminus \Sigma)$ belongs to a closed analytic subset of $\mathbf{P}^3$. Indeed, near a singular point $a \in \Sigma$ the foliation is spanned by a vector field $F(x)$ by Theorem 2.22. The graph of s is locally defined then by a single analytic equation. In the local chart $x = (x_1, x_2)$ on $\mathbb{P}^2$ and the homogeneous coordinates $[v_1 : v_2]$ on the fiber, this equation takes the form

$$v_1 F_2(x_1, x_2) - v_2 F_1(x_1, x_2) = 0,$$

where $F_{1,2}(x_1, x_2)$ are the coordinates of the vector field F in the local chart.

Thus the closure of the graph $S = \overline{s(\mathbb{P}^2 \smallsetminus \Sigma)}$ is an analytic subset of the projective manifold $\mathbf{P}^3$. By the Chow theorem, the submanifold S is itself algebraic. The map s and the projection $\pi : \mathbf{P}^3 \to \mathbb{P}^2$ restricted on the graph of s, are mutually inverse, defined on Zariski open subsets and hence are birational isomorphisms.

The assertion of the theorem is in fact valid for any foliation on a projective algebraic variety [**Ily72b**]. □

Because of Theorem 25.10, the classes of singular holomorphic foliations on $\mathbb{P}^2$ and foliations defined by polynomial forms/fields, coincide. For brevity we will speak about *polynomial foliations* (on $\mathbb{P}^2$).

25B. Algebraic leaves and Poincaré problem: a synopsis. The global analog of a complex separatrix of a holomorphic foliation $\mathcal{F}$ (as it was introduced in Definition 2.27) is a compact analytic (hence algebraic) curve $C \subset \mathbb{P}^2$ which is tangent to $\mathcal{F}$ at all nonsingular points of C and $\mathcal{F}$. Any such curve can be defined in the homogeneous coordinates $[X : Y : Z]$

on $\mathbb{C}^3$ by a homogeneous polynomial $f(X, Y, Z)$ of some degree m. We will always assume (unless explicitly stated otherwise) that f is *square-free* (*reduced*), i.e., is the product of *pairwise different* irreducible factors.

Assume that Ω is a homogeneous 1-form on $\mathbb{C}^3$ of degree r defining the foliation in the homogeneous coordinates $[X : Y : Z]$. Then the algebraic curve $C \subset \mathbb{P}^2$ defined by the square-free equation $\{f = 0\}$ of degree m is invariant by $\mathcal{F}$, if Ω and df are collinear on C, i.e., if

$$\Omega \wedge df = f \cdot \Phi, \tag{25.17}$$

where Φ is a homogeneous 2-form in $\mathbb{C}^3$ of degree $r-1$, called a *cofactor form* associated with the polynomial f which is sometimes called "*invariant factor*". Conversely, any nonzero homogeneous solution to this equation, *even if not square-free*, corresponds to an invariant algebraic curve C (cf. with Lemma 25.28 below).

The central theme of this section is two-fold:

- *scarcity* of polynomial foliations having algebraic leaves, and
- explicit upper bounds for the degree of algebraic leaves in terms of the (projective) degree of the foliation.

The second question was first raised by H. Poincaré in 1891 and since then is usually referred to as the *Poincaré problem*. The important results in this direction, which we formulate and prove in this section, were obtained recently by D. Cerveau, A. Lins Neto, C. Camacho, P. Sad and M. Carnicer.

The global Poincaré problem has a local analog that was discussed in §14**G**. Rather surprisingly, the solution of the global problem heavily uses the solution of the local one.

Example 25.11. A foliation $\mathcal{F} \in \mathcal{B}_2$ defined by the 1-form $x\,dy - \lambda y\,dx$ for λ irrational has only three algebraic leaves of degree 1: two coordinate axes and the infinite line $\mathbb{I}$.

On the contrary, if $\lambda = p/q$ is a nonzero rational number, then the foliation has all other leaves also algebraic, $y^q - cx^p = 0$, $c \neq 0$. If λ is positive rational different from an integer or inverse integer, these leaves have a singularity at the origin. If λ or $1/\lambda$ is an integer number, all leaves are smooth in $\mathbb{C}^2$. However, the singularity of the leaves in this case reappears at infinity.

Example 25.12. Let $f(X, Y, Z)$ be a homogeneous polynomial of degree r in three variables. The differential df is a homogeneous form in $\mathbb{C}^3$ but it does not vanish on the Euler field V as is required to define a foliation of the projective plane; cf. with (25.11). The analog of a "Hamiltonian" foliation on $\mathbb{P}^2$ is the foliation of projective degree r defined by the rational 1-form $\Omega = \frac{df}{f} - r\frac{dl}{l}$, where $l = l(X, Y, Z)$ is an arbitrary *linear* form. Choosing an affine chart in which $\{l = 0\}$ is the infinite line $\mathbb{I}$, we see that in this affine chart the foliation is defined by the polynomial form df. All its leaves $L_\alpha = \{f - \alpha l^r = 0\}$ are algebraic of degree r, except for the "infinite" line $L_\infty = \{l = 0\}$.

Thus we see that for the Hamiltonian foliation of projective degree r, any finite union of the algebraic curves L_α and/or L_∞ is a (reducible) algebraic curve. However, if we demand that the invariant curve have *only transversal self-intersections*, then any such curve must necessarily be of the form $L_\alpha \cup L_\infty$ and hence its degree cannot exceed $r + 1$. Degree of an *irreducible* algebraic curve of a Hamiltonian foliation is no greater than r.

Example 25.13. A generalization of the previous example is the *Darbouxian* foliation defined by the form $\Omega = \sum_{i=1}^k \lambda_i \frac{df_i}{f_i}$, where $f_i(X, Y, Z)$ are homogeneous mutually prime polynomials of degrees r_i and λ_i are complex numbers such that $\sum \lambda_i r_i = 0$. Such a foliation always has a reducible algebraic separatrix $C = \bigcup_i \{f_i = 0\}$; existence of other algebraic leaves depends on the arithmetical properties of the tuple $[\lambda_1 : \cdots : \lambda_k] \in \mathbb{P}^{k-1}$.

Example 25.11 suggests that without some restrictions either on the foliation or on the properties of the leaf one cannot expect any bound on the degree of the leaf in terms of the degree of the foliation. The additional conditions may have a rather simple form.

Theorem 25.14 (D. Cerveau and A. Lins Neto, 1991, [**CLN91**])**.** *Let $\mathcal{F} \in \mathcal{B}_r$ be a polynomial foliation of projective degree r on $\mathbb{P}^2$ and $C \subset \mathbb{P}^2$ an algebraic separatrix of degree m for $\mathcal{F}$.*

If the curve C is smooth or has at worst transversal self-intersection points, then $m \leqslant r + 1$.

In fact, in [**CLN91**] a stronger result is proved. If a foliation of the projective degree r has an algebraic separatrix C of degree $r+1$ with normal crossings, then C is necessarily reducible and the foliation must be of a very special type: in homogeneous coordinates it is defined by a *logarithmic* form $\Omega = \sum_i \lambda_i \frac{df_i}{f_i}$, where $f_i \in \mathbb{C}[X, Y, Z]$ are homogeneous polynomials of degree $r_i \in \mathbb{N}$, $\lambda_i \in \mathbb{C}$ and $\sum \lambda_i r_i = 0$. In particular, if C is irreducible then $m = \deg(C) \leqslant r$. We will prove later in §25**E** a slightly weaker result.

Theorem 25.15. *A* smooth *projective curve $C = \{f(X, Y, Z) = 0\} \subset \mathbb{P}^2$ can be invariant for a foliation $\mathcal{F} \in \mathcal{B}_r$ only if* $\deg C \leqslant r$.

If C is smooth and the equality $\deg C = r$ *is achieved, then the foliation $\mathcal{F}$ is Hamiltonian (defined by a rational* 1*-form $\frac{df}{f} - r\frac{dl}{l}$, where $l = l(X, Y, Z)$ is a linear form) and C is a part of the reducible separatrix $C \cup \{l = 0\}$ of degree $r + 1$.*

The smoothness assumption can be replaced by the irreducibility assumption (Problem 25.9).

To avoid apriori assumptions on the curve, one may impose certain assumptions on the foliation itself. For instance, if all singularities are irrational saddles, then *any* invariant algebraic curve must have only normal crossings, since there are only two smooth separatrices through each singularity, which are transversal to each other (at nonsingular points all invariant curves are smooth). However, this assumption is way too restrictive and can be considerably relaxed.

The strongest known result in this direction is the following theorem.

Theorem 25.16 (M. Carnicer [**Car94**]). *If a foliation $\mathcal{F} \in \mathcal{B}_r$ of projective degree r has no generalized dicritical singularities, then any algebraic separatrix of this foliation has degree $m \leqslant r+1$.*

Remark 25.17. In fact, for the inequality $\deg C \leqslant r+1$ to hold it is sufficient to require that there are no generalized dicritical singularities only on the separatrix C itself. Yet without knowing C this relaxed condition makes no sense.

Assumptions of Theorem 25.16 hold for a generic foliation: it sufficient to require, e.g., that the foliation has no singularities with the ratio of eigenvalues equal to 1. Yet in fact a *generic* foliation from the class $\mathcal{B}_r$ for $r \geqslant 2$ has no algebraic leaves *at all.*

Theorem 25.18. *A generic foliation from the class $\mathcal{B}_r$ has no algebraic leaves.*

More precisely, there is an open dense semialgebraic *subset in $\mathcal{B}_r$, which corresponds to holomorphic foliations without algebraic leaves.*

Remark 25.19. A generic *real* foliation (defined by real polynomial forms) also does not have algebraic leaves (neither real nor complex). Yet in this case the notion of genericity is weaker: we can assert only that the exceptional foliations have Lebesgue measure zero.

The proof of scarcity of foliations with algebraic leaves is *implicit*: it is rather difficult to construct explicitly examples of foliations without algebraic leaves. One such example is given by the following theorem.

Theorem 25.20 (J.-P. Jouanolou, 1979; see also [**CLN91**]). *For any $n \geqslant 2$, the foliation on $\mathbb{P}^2$ defined by the 1-form*

$$(x^n - y^{n+1})\,dx - (1 - xy^n)\,dy \tag{25.18}$$

has no algebraic invariant curves.

In this section we arbitrarily switch between the terms "algebraic leaves", "algebraic separatrices" and "algebraic invariant curves". This is justified by the following result.

Theorem 25.21. *Any projective curve invariant by a polynomial foliation on $\mathbb{P}^2$, carries at least one singularity of this foliation.*

Demonstration of all these theorems occupies the sections §25**C**–§25**E**. The two Theorems 25.14 and 25.16 share essentially the same proof exposed in §25**C**. The supplementary ingredient for the stronger Theorem 25.16 is the local inequality established in Theorem 14.28 which was derived in §14**G** from Theorem 14.20. The latter theorem in turn can be considered as a solution of some local version of the Poincaré problem.

25C. Global analysis and upper bounds for degrees of algebraic invariant curves. In this subsection we prove Theorems 25.14 and 25.16.

The proof is based on the following observation. Consider a compact Riemann surface and a meromorphic vector field on it. If the surface is nonsingular, then the number of zeros of the field minus the number of its poles, both counted with multiplicities, is equal to the Euler characteristic of the surface, and does not depend on the field. This follows from the Poincaré–Hopf theorem applied to Riemann surfaces. The idea is to apply this theorem to two vector fields defined on an algebraic leaf C of a polynomial foliation, and compare the results.

25**C**$_1$. *Outline of the proof.* To construct these vector fields we need to choose a special affine chart on $\mathbb{P}^2$. This chart is characterized by the condition that the corresponding infinite line $\mathbb{I} \subset \mathbb{P}^2$ should intersect the algebraic leaf C transversally. Then the number of intersection points is equal to the degree of the leaf $m = \deg C$. Besides, we assume that the infinite line itself is not a separatrix of $\mathcal{F}$. In this chart, the foliation is defined by a polynomial vector field denoted by $\mathbf{F}$ of degree r equal to the projective degree of the foliation $\mathcal{F}$. The field $\mathbf{F}$ is tangent to C. Denote by F the restriction of $\mathbf{F}$ on C, a meromorphic vector field with poles only at infinity $\mathbb{I} \cap C$.

Let $f \in \mathbb{C}[x, y]$ be the *minimal polynomial* of the curve C in the chosen affine chart: by definition, f is square-free and $C \smallsetminus \mathbb{I} = \{f = 0\}$. Let $\mathbf{H} = \left(\frac{\partial f}{\partial y}, -\frac{\partial f}{\partial x}\right)$ be the corresponding Hamiltonian vector field. This field obviously is tangent to C. Denote by H be the restriction of $\mathbf{H}$ on C.

Let P_F, Z_F, P_H, Z_H be the numbers of poles and zeros of F and H respectively, all counted with multiplicities. Then

$$P_F - Z_F = P_H - Z_H = -\chi(C), \tag{25.19}$$

where $\chi(C)$ is the Euler characteristic (the equality makes sense when C is a smooth curve).

The order of poles is easy to calculate exactly for H and estimate from above for F: below we prove that

$$P_H = m(m-3), \qquad P_F \leqslant m(r-2). \tag{25.20}$$

The parallel result for $\widetilde{F}$ and $\widetilde{H}$ is the same, since the poles of both F and H occur at the smooth points of C which are not desingularized by φ.

The main statement to prove is the inequality

$$Z_H \leqslant Z_F, \tag{25.21}$$

and its generalization for the case of nonsmooth curves. The proof is elementary when C is smooth or has only simple self-intersections, and requires an involved proof using desingularization in the general case of arbitrary singularities on C.

In all cases (25.19) together with (25.21) implies $P_H \leqslant P_F$. Substituting the values (25.20) we obtain the inequality $m \leqslant r+1$ asserted in the two theorems.

$25\mathbf{C}_2$. *Multiplicities of the poles.* We now pass to the detailed proofs, starting with the bounds on the number of poles (25.20).

Recall that we consider a projective curve C of degree m intersecting a line $\ell \subset \mathbb{P}^2$ transversally, and choose once and for all an affine chart for which ℓ serves as the infinite line $\mathbb{I}$.

Proposition 25.22. 1. *For any polynomial vector field* $\mathbf{F} = a\frac{\partial}{\partial x} + b\frac{\partial}{\partial y}$ *of degree* r *on* $\mathbb{C}^2$, *tangent to the curve* C, *its restriction on* C *is a meromorphic vector field having poles of order* not greater than $r-2$ *at each infinite point* $a \in C \cap \mathbb{I}$.

2. *If* H *is the restriction on* C *of the Hamiltonian vector field* $\mathbf{H} = \frac{\partial f}{\partial x}\frac{\partial}{\partial y} - \frac{\partial f}{\partial y}\frac{\partial}{\partial x}$ *of degree* $m-1$, *where* f *is the minimal polynomial equation of* C, *then all these poles have order* exactly equal to $m-1-2 = m-3$.

Proposition 25.22 immediately implies the (in)equalities (25.20) for the case where C is a smooth or nodal curve nonsingular at infinity, since each such curve has exactly m smooth branches at infinity.

Proof of the Proposition 25.22. 1. Denote the principal homogeneous components of the field $\mathbf{F}$ by a_r and b_r respectively. Since C crosses transversally the infinite line, the reciprocal $u = 1/x$ can be used as a local coordinate on each branch of C (after a linear change of coordinates x, y if necessary). Direct computation yields $\dot{u} = -u^2 a(1/u, v/u) = u^{2-r}[a_r(1,v) + O(u)]$, which means that the order of the pole of F on each of the m branches of C near ℓ does not exceed $r-2$.

2. By our choice of the affine chart, the principal homogeneous part f_m of the polynomial f is square-free (all linear factors are distinct). We claim that in this case the order of pole on each branch of C is exactly equal to $m-3$.

Indeed, if $f = f_m$ were homogeneous, this can be established by direct calculation. Denote $f_m = \prod_{j=1}^{m}(y-\alpha_j x)$, $\alpha_i \neq \alpha_j$, then for the x-component of $\mathbf{H}$ we have $\dot{x} = -\sum_{j=1}^{m}\prod_{i\neq j}(y-\alpha_i x)$, and the restriction of the right hand side on every line $y = \alpha_j x$ is equal to $c_j x^{m-1}$, $c_j = \prod_{i\neq j}(\alpha_j - \alpha_i) \neq 0$. Therefore $\dot{u} = -c_j\, u^{3-m}$ and the order of the pole is exactly $m-3$. The presence of lower degree components in the expansion $f = f_m + f_{m-1} + \cdots$ cannot change this order. □

Proposition 25.23. *Smooth points of an affine curve are noncritical for the minimal polynomial of this curve.*

Corollary 25.24. *The Hamiltonian vector field* $\mathbf{H} = \big(\frac{\partial f}{\partial y}, -\frac{\partial f}{\partial x}\big)$ *restricted on the curve* $C = \{f = 0\}$ *is nonvanishing at all smooth points of* C. □

Proof of the proposition. If C is smooth at a and defined by a reduced polynomial equation $\{f = 0\}$, then df does not vanish on C. Indeed, by the smoothness assumption, the germ of C at each its point a can be defined by a holomorphic equation $\{\varphi_a = 0\}$ with $d\varphi_a$ nonvanishing at a. The *germ* of f at a is divisible by φ_a, $f = \psi_a\varphi_a$. The germ of the curve $\{\psi_a = 0\}$ can be neither different from the germ of C (this would mean that C has at least two different local branches, which would imply nonsmoothness of C), nor coincide with it. The latter assumption implies that $df \equiv 0$ near a on C, hence, all over C. This contradicts to the minimality of f. The only remaining possibility is $\psi_a(a) \neq 0$, hence $df(a) = \psi_a(a)\, d\varphi_a(a) \neq 0$ as asserted. □

25**C**$_3$. *Demonstration of Theorem 25.14, smooth case.* Let the curve C with the two meromorphic fields F and H on it be as above. If C is smooth, then by Proposition 25.23 the field H has no zeros on C. Hence, $0 = Z_H \leqslant Z_F$ (the latter number is a nonnegative by definition). On the other hand, Proposition 25.22 implies the relations (25.20). Substituting them into (25.19), we prove the theorem in the smooth case.

25**C**$_4$. *Demonstration of Theorem 25.14 for curves with normal crossings.* The same arguments can be applied to nodal curves (curves with nodal singularities, i.e., normal self-intersections), if zeros on each smooth branch through the nodal point are counted separately. Formally this can be achieved by *normalization* of the curve (blowing up all its singular points): by Problem 25.1, the result of such a blow-up will be a nonsingular (eventually reducible) smooth analytic curve $\widetilde{C}$ on a 2-dimensional surface.

As before, we consider the vector fields F and H on the curve. The total number of poles of each field remains the same as it was in the smooth case, the difference concerns zeros.

At each normal crossing $a \in C$ the Hamiltonian function f has a Morse (nondegenerate) critical point by Problem 25.2, hence the Hamiltonian vector field has a simple zero after restriction on each smooth branch of C. Thus we have the total number of zeros $Z_F = 2|\Sigma|$, where Σ is the nodal (singular) locus of the curve C.

The field F also must vanish at a with multiplicity $\geqslant 1$ after restriction on each branch of C, so that $Z_F \geqslant 2|\Sigma| = Z_H$. The proof of Theorem 25.14 is achieved exactly as in the smooth case, if we notice that for both fields the difference $Z_H - P_H = Z_F - P_F = \chi(\widetilde{C})$ is the Euler characteristic of the normalized curve $\widetilde{C}$. □

25**C**$_5$. *Demonstration of Theorem 25.16.* To prove the most general result, we need a suitable generalization of the inequality (25.21) for singular curves. The treatment of the nodal case in §25**C**$_4$ suggests that one has to deal separately with each locally irreducible branch γ of the curve C. The corresponding local inequality (14.29) was already prepared in §14**J**

Definition 25.25. A *global desingularization* of a projective curve C with the singular locus Σ is a holomorphic map $\varphi\colon \widetilde{C} \to C$ such that:

- $\widetilde{C}$ is a smooth compact holomorphic curve,
- φ is one-to-one over the smooth part $C \smallsetminus \Sigma$, and
- considered as an embedding into $\mathbb{P}^2$, the map φ is holomorphic.

Theorem 25.26. *The global desingularization exists.*

Proof. We will construct $\widetilde{C}$ in an abstract way using the local uniformization Theorem 2.26; see [**Chi89**, §6]. Let $\gamma_i\colon U_i \to \mathbb{P}^2$ be finitely many maps as in (2.8) defined on open disks $U_i \subset \mathbb{C}$ such that their union covers the entire curve C. Without loss of generality one may assume that the disks are so small that the differentials $d\gamma_i$ vanish only at the centers of some disks that are mapped to singular points of C.

Consider the disjoint union $\bigsqcup U_i$ and the obvious equivalence relationship, such that points in different disks are identified if and only if their images represent the same point on C. The quotient space is an abstract (smooth) holomorphic curve, denoted by $\widetilde{C}$, and the maps γ_i together define a well-defined map $\varphi\colon \widetilde{C} \to C$ which is biholomorphic and invertible outside the singular locus $\Sigma \subset C$. □

Since the global desingularization is one-to-one outside a discrete set, the vector field $\mathbf{F}$ which is tangent to C can be pulled back as a meromorphic vector field $\widetilde{F} = \varphi_*^{-1}(\mathbf{F}|_C)$ on $\widetilde{C}$. As before, the poles of $\widetilde{F}$ and $F = \mathbf{F}|_C$ are "the same": the multiplicity of $\widetilde{F}$ at a pole $a \in \widetilde{C}$ is equal to the multiplicity of F at $\varphi(a) \in \mathbb{C} \cup \mathbb{I}$. The same obviously holds for $H = \mathbf{H}|_C$ and its pullback $\widetilde{H}$ on $\widetilde{C}$.

As for the zeros, for any locally irreducible branch $\gamma \subseteq C$ through a singular point $a \in \Sigma$, the orders of zero of $\widetilde{F}$ and $\widetilde{H}$ at $\widetilde{a} \in \varphi^{-1}(a)$ by construction coincide with the vanishing orders of the foliations $\mathcal{F}$ and $\mathcal{H}$ along γ at a (see Definition 14.25).

In the assumptions of Theorem 25.16 we can apply Theorem 14.28 and conclude that the vanishing order of $\widetilde{H}$ at any point from $\widetilde{\Sigma} = \varphi^{-1}(\Sigma)$ is less or equal to that of $\widetilde{F}$. Adding these inequalities together, we obtain the inequality $Z_{\widetilde{F}} \geqslant Z_{\widetilde{H}}$.

It remains to notice that though the curve $\widetilde{C}$ may be reducible and not connected, the equality $Z_{\widetilde{F}} - P_{\widetilde{F}} = Z_{\widetilde{H}} - P_{\widetilde{H}} = \chi(\widetilde{C})$ remains true if the right hand side is understood as the sum of the Euler characteristics of all smooth components. The end of the proof of Theorem 25.16 is standard, exactly as in §25$\mathbf{C}_3$–§25$\mathbf{C}_4$. □

25D. Scarcity of algebraic leaves for foliations of the class $\mathcal{B}_r$. As a corollary to Theorem 25.14, one can obtain the following result.

Theorem 25.27. *If all singular points of a foliation $\mathcal{F}$ from the class $\mathcal{B}_r$ on $\mathbb{P}^2$ are hyperbolic and the ratios of the two eigenvalues at each point are nonreal, then such a foliation has no algebraic separatrices of degree greater than $r+1$.*

Proof. An invariant curve of a foliation is smooth as long as it does not pass through singularities. Every hyperbolic singularity with the nonreal ratio of eigenvalues is analytically linearizable by the Poincaré Theorem 5.5 and hence admits *exactly two* analytic invariant curves (local separatrices) which intersect transversally.

Thus any algebraic invariant curve of a foliation satisfying the assumptions of the theorem, must be smooth or have at worst normal crossings. By Theorem 25.14, such a curve may have degree at most $r+1$. □

In fact, *generic* foliations from the class $\mathcal{B}_r$ do not have algebraic invariant curves *at all*. The proof is based on the following observation.

Lemma 25.28. *For any combination of natural numbers $r \geqslant 2$ and $m \geqslant 1$ the foliations $\mathcal{F} \in \mathcal{B}_r$ having invariant algebraic separatrices of degree $\leqslant m$, constitute an algebraic (projective) subvariety in the projective space $\mathcal{B}_r$.*

Proof. Consider the complex linear space of homogeneous 1-forms Ω of degree r in the homogeneous coordinates $[X:Y:Z] \in \mathbb{C}^3 \smallsetminus \{0\}$ constrained by the condition $\Omega(V) = 0$ (see (25.11)): the space $\mathcal{B}_r$ is the projectivization of this linear space. If the algebraic curve given by the reduced (square-free) homogeneous equation $\{f = 0\}$, $f \in \mathbb{C}[X:Y:Z]$, $\deg f = m$, is a separatrix of the foliation $\mathcal{F} \in \mathcal{B}_r$ defined by the Pfaffian equation $\Omega = 0$, then for some homogeneous 2-form Φ on $\mathbb{C}^3$ we have

$$\Omega \wedge df = f\,\Phi. \tag{25.22}$$

Conversely[2], *any* homogeneous polynomial solution $(f, \Phi) \in \mathfrak{L}_f \times \mathfrak{L}_\Phi$ of (25.22) corresponds to an invariant algebraic curve C of $\mathcal{F}$, though the degree of this curve may be *smaller* than m, if f is not square-free; if $f = \prod_j f_j^{\nu_j}$ with $\nu_j \geqslant 1$, $\sum \nu_j = m$, then C is defined by the equation $\prod f_j = 0$.

We claim that the subspace of homogeneous forms Ω of degree r (vanishing on the Euler field) for which the equation (25.22) is solvable, constitutes an algebraic subvariety of the corresponding projective space $\mathcal{B}_r$. The proof is achieved using proper projections of algebraic sets. We will need two well-known facts on projections of projective algebraic varieties.

Recall that for any linear subspace L^k in the complex space $\mathbb{C}^p$ the *projection* $\pi_L \colon \mathbb{C}^p \to \mathbb{C}^{p-k}$ *parallel to* L can be defined as follows: choose any complementary subspace $M \cong \mathbb{C}^{p-k}$ transversal to L and let π_L be the Cartesian projection $L \oplus M \to M$. The projection π_L is defined modulo a composition: if M is replaced by another complementary subspace with the same properties, then π_L is replaced by a map $h \circ \pi_L$, $h \in \mathrm{GL}(n-k, \mathbb{C})$.

For a projective space $\mathbb{P}^p$ its projection along a projective subspace $L = L^k \subset \mathbb{P}^p$ is an algebraic map $\pi_L \colon \mathbb{P}^p \smallsetminus L \to \mathbb{P}^{p-k-1}$ defined as follows.

Consider the linear space $\mathbb{C}^{p+1}$ which is the homogeneous model for $\mathbb{P}^p$, and the linear subspace L' which is a homogeneous model for L. Choose a complementary subspace $M' \cong \mathbb{C}^{p-k}$ and the Cartesian projection $\rho \colon \mathbb{C}^{p+1} \to \mathbb{C}^{p-k}$. For any point $a \in \mathbb{P}^p \smallsetminus L$ the line $\mathbb{C}a \subset \mathbb{C}^{p+1}$ is disjoint with L' (except for the origin) and its image $\rho(\mathbb{C}a)$ is a nontrivial (i.e., not reducible to one point) line in $\mathbb{C}^{p-k}$. This allows us to pass to projectivizations and construct the map $\pi_L \colon \mathbb{P}^p \smallsetminus L \to \mathbb{P}^{p-k-1}$ which sends the point a into the equivalence class represented by the line $\rho(\mathbb{C}a)$. This map is called *projection with center* L. It is defined modulo a projective automorphism of the target space $\mathbb{C}^{p-k-1}$.

After recalling this basic construction, we can formulate the following fundamental properties of projections of algebraic sets.

Proposition 25.29. *1. If $X \subset \mathbb{C}^n \times \mathbb{P}^p$ is an algebraic variety, then its projection on $\mathbb{C}^n$ parallel to $\mathbb{P}^p$ is an algebraic variety in $\mathbb{C}^n$.*

2. If $X \subset \mathbb{P}^p$ is an algebraic variety and $L^k \subset \mathbb{P}^p$ is a projective k-dimensional subspace disjoint with X, *then the projection $\pi_L(X)$ of X on $\mathbb{P}^{p-k-1}$ with the center on L, is an algebraic variety.*

References for the proposition. Both results follow from the classical elimination theory and algebraically reflect the compactness of the projective space $\mathbb{P}^m$. The first assertion appears under the name of the Principal Theorem of Elimination Theory in [**Mum76**, §2B, (2.24)].

The assertion that the projective image of X is closed if X is disjoint with L, is known as M. Noether's lemma on normalization [**Mum76**, Corollary 2.29].

---

[2] This converse assertion fails in the nonhomogeneous settings: if ω is a polynomial Pfaffian 1-form on $\mathbb{C}^2$ and $f \in \mathbb{C}[x, y]$ a nonzero polynomial such that $\omega \wedge df = f\Phi$ for some 2-form Φ, then the solution $f = \text{const}$, $\Phi = 0$ does not correspond to a foliation having an invariant curve.

The analytic counterpart of these results is the Remmert theorem on proper projections of analytic sets [**GR65**, Ch. V, § C, Theorem 5]. □

Consider the linear spaces $\mathfrak{L}_1 = \{\Omega\colon \deg\Omega = r,\ \Omega(V) = 0\}$ of homogeneous 1-forms on $\mathbb{C}^3$, $\mathfrak{L}_2 = \{f \in \mathbb{C}[X:Y:Z]\colon \deg f = m\}$ of homogeneous polynomials and $\mathfrak{L}_3 = \{\Phi \in \Lambda^2[X:Y:Z]\}$ of the corresponding homogeneous cofactors (the degree of the cofactor is completely determined by r, m). Denote by $P(\mathfrak{L}_i)$ the corresponding projectivizations, the quotient spaces by the complex multiplicative action,

$$P(\mathfrak{L}_j) = \mathfrak{L}_j \smallsetminus \{0\} \big/ \mathbb{C} \smallsetminus \{0\}, \qquad j = 1, 2, 3.$$

The equation (25.22) determines an algebraic variety in $\mathfrak{L}_1 \times \mathfrak{L}_2 \times \mathfrak{L}_3$. Together with f it is also satisfied by λf, $\lambda \neq 0$, thus the variety in fact sits in $\mathfrak{L}_1 \times P(\mathfrak{L}_2) \times \mathfrak{L}_3$. Denote it by Q.

By Proposition 25.29, the projection of Q parallel to the middle term in $\mathfrak{L}_1 \times P(\mathfrak{L}_2) \times \mathfrak{L}_3$ produces an algebraic variety Q' in the space $\mathfrak{L}_1 \times \mathfrak{L}_3$. The projection Q' is a cone; together with (Ω, Φ) the equation (25.22) is also satisfied by the pair $(\lambda\Omega, \lambda\Phi)$ for any $\lambda \neq 0$. The projectivization Q'' of Q' is an algebraic subset of the *projective* space $P(\mathfrak{L}_1 \times \mathfrak{L}_3)$.

The linear subspace $\mathfrak{L}_3 \cong \{\Omega = 0\} \times \mathfrak{L}_3$ in $\mathfrak{L}_1 \times \mathfrak{L}_3$ corresponds to a projective subspace $P(\mathfrak{L}_3) \subset P(\mathfrak{L}_1 \times \mathfrak{L}_3)$ disjoint with Q''; indeed, if $\Omega \equiv 0$, then the equation (25.22) cannot be satisfied unless either Φ or f is identically zero.

The projection of $P(\mathfrak{L}_1 \times \mathfrak{L}_3)$ with the center $P(\mathfrak{L}_3)$ takes Q'' into an algebraic set $Q''' \subset P(\mathfrak{L}_1)$ by the second assertion of Proposition 25.29. This variety is precisely the projectivization of the cone of homogeneous 1-forms, for which the equation (25.22) admits a nontrivial solution. □

Proof of Theorem 25.18. An algebraic subset of a projective space has either measure zero or coincides with the whole space. To exclude the latter possibility, it suffices to construct a single example of a polynomial foliation without algebraic separatrices. Thus the Jouanolou example (Theorem 25.20) implies that for any finite m, the foliations from the class $\mathcal{B}_r$ which have algebraic invariant curve of degree $\leqslant m$, constitute a proper algebraic subvariety X^r_m in the projective space $\mathcal{B}_r$. The countable union $\bigcup_{m\geqslant 0} X^r_m$ has zero measure, hence its complement, corresponding to foliations without algebraic leaves, has full measure. The argument works for both complex and real algebraic foliations alike.

One can refine the assertion on genericity for complex foliations. Indeed, the *finite* union $\bigcup_{m=0}^{r+1} X^r_m$ is a closed proper algebraic subset, whose complement $\mathcal{B}'_r$ is open in $\mathcal{B}_r$; foliations from the class $\mathcal{B}'_r$ may have only algebraic solutions of degree $\geqslant r+2$.

Consider the open dense subset $\mathfrak{B}''_r \subset \mathcal{B}'_r$ of foliations which have only nondegenerate singular points with nonreal characteristic ratios. These extra conditions define a semialgebraic, open and dense subset U_r in $\mathcal{B}_r$, so that $\mathcal{B}''_r$ is also open dense and semialgebraic. By Theorem 25.27, foliations from the class $\mathcal{B}''_r$ cannot have algebraic leaves of degree $\geqslant r+2$. Hence the entire open dense subset $\mathcal{B}''_r$ consists of polynomial foliations that cannot have algebraic leaves at all.

Note that the subset U_r excludes the real foliations, thus the assertion on generic absence of algebraic leaves in this case is weaker than in the complex case (Remark 25.19). □

25E. Smooth invariant curves. Unlike Theorem 25.14, the stronger result claimed by Theorem 25.15 is proved using more analytic tools. We follow the exposition in [**CLN91**] with some modification.

Lemma 25.30 (Division lemma). *If a smooth projective curve C defined by the square-free homogeneous equation $\{f(X,Y,Z)=0\}$ of degree m is a separatrix of a polynomial foliation of the projective degree r defined by a homogeneous 1-form Ω on $\mathbb{C}^3$, then there exist a homogeneous polynomial $g(X,Y,Z)\in\mathbb{C}[X,Y,Z]$ and a homogeneous 1-form $\mu\in\Lambda^1[\mathbb{C}^3]$ such that*

$$\Omega = g\,df + f\mu, \qquad \deg g = r-m+1, \quad \deg\mu = r-m. \tag{25.23}$$

Proof. The equation (25.23) in any dimension is *locally* solvable near any smooth point of an analytic hypersurface $\{f=0\}$. Indeed, one can always choose a holomorphic coordinate system so that the hypersurface takes the form $C=\{x_1=0\}\subset(\mathbb{C}^n,0)$. A 1-form tangent to C admits a local representation $\sum_1^n a_i(x)\,dx_i$ with the analytic coefficients $a_2(x),\dots,a_n(x)$ vanishing on C and hence divisible by x_1.

Consider the cone $K=\{f=0\}\smallsetminus\{0\}$ in $\mathbb{C}^3\smallsetminus\{0\}$ which is a smooth hypersurface (the origin is deleted). Because of this smoothness, near each point K one may choose a covering of a punctured neighborhood of the origin in $\mathbb{C}^3$ by, say, small polydisks U_α so that in each polydisk

$$\Omega = g_\alpha\,df + f\,\mu_\alpha \qquad \text{on } U_\alpha.$$

On the intersections $U_{\alpha\cap\beta}=U_\alpha\cap U_\beta$ we have $(g_\alpha-g_\beta)\,df+f(\mu_\alpha-\mu_\beta)=0$, that is, the analytic functions $g_\alpha-g_\beta$ are divisible by f:

$$g_\alpha - g_\beta = f\,h_{\alpha\beta}, \qquad h_{\alpha\beta}\in\mathcal{O}(U_{\alpha\beta}).$$

The holomorphic cochain $h_{\alpha\beta}$ is a cocycle: $h_{\alpha\beta}+h_{\beta\gamma}+h_{\gamma\alpha}=0$ on all triple intersections $U_{\alpha\beta\gamma}$.

Solvability of this cocycle constitutes the assertion of H. Cartan's theorem on triviality of the cohomology $H^1(\mathbb{C}^3 \smallsetminus \{0\}, \mathcal{O})$ [**Car38**]. Applying this theorem, we conclude that there exists a holomorphic cocycle $\{h_\alpha\}$ such that $h_{\alpha\beta} = h_\alpha - h_\beta$. Substituting this into the definition of $h_{\alpha\beta}$, we conclude that

$$g_\alpha + h_\alpha f = g_\beta + h_\beta f \qquad \text{on } U_{\alpha\beta},$$

i.e., the functions $g_\alpha + h_\alpha f$ together define a global function g holomorphic on $\mathbb{C}^3 \smallsetminus \{0\}$. Similarly,

$$\mu_\alpha - h_\alpha\, df = \mu_\beta - h_\beta\, df \qquad \text{on } U_{\alpha\beta},$$

which allows us to construct a 1-form μ holomorphic on $\mathbb{C}^3 \smallsetminus \{0\}$. By the removable singularity theorem, both g and μ extend holomorphically at the origin. Together g and μ solve the equation (25.23).

Apriori, g and μ can be nonhomogeneous, since the decomposition (25.23) is generally nonunique. However, since f and df are homogeneous of degrees m and $m-1$ respectively, one can choose the homogeneous components of the constructed g and μ of degrees $r-m+1$ and $r-m$ respectively: they would constitute a *homogeneous* solution for (25.23). □

Remark 25.31. The Division Lemma 25.30 is projective (i.e., deals with homogeneous forms and polynomials in three variables). It admits an affine analog concerning nonhomogeneous forms and polynomials in two variables. The proof of this affine division lemma has a similar structure, globalization of local representations, but unlike its projective counterpart, the globalization is achieved using the Max Noether "$AF + BG$ theorem" rather than the Cartan theorem. At the end we explain how the affine result can be used to prove the projective one, this providing an alternative demonstration of Lemma 25.30.

Lemma 25.32 (Affine division lemma). *If a smooth affine curve $C = \{f = 0\} \subset \mathbb{C}^2$, $f \in \mathbb{C}[x,y]$ of degree m, is transversal to infinity $\mathbb{I}$ and invariant for a foliation defined by a polynomial 1-form ω of degree r, dicritical at infinity, then*

$$\omega = g\, df + f\mu, \tag{25.24}$$

where $g \in \mathbb{C}[x,y]$ is a polynomial of degree $r-m+1$ and μ a polynomial 1-form of degree $r-m$.

Sketch of demonstration. Since C is invariant, $\omega \wedge df$ is a 2-form that vanishes on C, hence one can write $\omega \wedge df = fg\, dx \wedge dy$, where $h \in \mathbb{C}[x,y]$ is a polynomial coefficient. Since df does not vanish on C, the polynomial h must vanish at all critical points of f, defined by the algebraic equations $\{a \in \mathbb{C}^2 \colon df(a) = 0\}$. Moreover, the germ of h at each critical point a belongs to the ideal $\langle \frac{\partial f}{\partial x}, \frac{\partial f}{\partial y} \rangle$ generated by the partial derivatives of f in the corresponding local ring $\mathcal{O}(\mathbb{C}^2, a)$ of holomorphic germs. After some technical work one can derive from the Max Noether theorem [**GH78**, Chapter 5,§3] that h globally belongs to the ideal generated in $\mathbb{C}[x,y]$ by the partial derivatives of f, in other words, that the 2-form $h\, dx \wedge dy$ is divisible by df, $h\, dx \wedge dy = df \wedge \mu$, where μ is a polynomial 1-form. This identity implies that $(\omega - f\mu) \wedge df \equiv 0$. Since df has only isolated singularities, the last condition means that $\omega - f\mu$ is divisible by df, $\omega - f\mu = g\, df$ for some polynomial $g \in \mathbb{C}[x,y]$. An accurate analysis shows that the degrees of g and μ are indeed as asserted. □

Lemma 25.30 can in turn be derived from its affine counterpart, Lemma 25.32, as follows. Consider the affine hyperplane $\Pi = \{Z = 1\} \subset \mathbb{C}^3$ and restrict on it the homogeneous 1-form Ω and the homogeneous polynomial f, denoting these restrictions by ω and φ respectively. Without loss of generality we may assume that φ is transversal to infinity and ω is dicritical at infinity. By Lemma 25.32, the form ω can be represented as follows, $\omega = \psi\, d\varphi + \varphi\sigma$, where $\psi \in \mathbb{C}[x, y]$ and $\sigma = \alpha\, dx + \beta\, dy \in \Lambda^1[x, y]$ are a polynomial of degree $r - m + 1$ and polynomial 1-form of degree $r - m$ respectively.

Any polynomial of degree k in two variables considered as a function on $\Pi \subset \mathbb{C}^3$ can be extended as a homogeneous polynomial of three variables of the same degree. Extending this way the polynomial $\psi(x, y)$ and the coefficients $\alpha(x, y), \beta(x, y)$, we obtain the polynomial g and two of the three coefficients of the form $\mu = a(X, Y, Z)\, dX + b(X, Y, Z)\, dY + c(X, Y, Z)\, dZ$. The remaining coefficient $c \in \mathbb{C}[X, Y, Z]$ must be chosen so that the Euler identity in $\mathbb{C}^3$ holds: evaluating both parts of (25.23) on the Euler vector field V transversal to Π, we obtain the equation

$$0 = mg + \mu(V), \qquad g = g(X, Y, Z), \quad \mu(E) = Xa + Yb + Zc.$$

This equation allows us to restore $c(X, Y, Z)$ only as a *rational* homogeneous function. Yet an accurate analysis shows that in fact c is a homogeneous *polynomial* of degree $r - m$ if (a) the form ω is dicritical at infinity and (b) the polynomial $\varphi = f|_\Pi$ is transversal to infinity. Both conditions can be achieved by a suitable choice of homogeneous coordinates in $\mathbb{C}^3$, as was already mentioned. An interested reader will easily restore the omitted computations.

Proof of Theorems 25.15 and 25.21. Theorem 25.15 is an immediate corollary of the Division Lemma 25.30. Indeed, assume that a smooth algebraic curve of degree m given by its homogeneous equation $C = \{f = 0\}$ is a separatrix of the foliation defined by a homogeneous form Ω of degree r. Then by Lemma 25.30 we have a representation (25.23).

By definition, Ω vanishes on the Euler field V and f is homogeneous of degree m. Evaluating (25.23) on the Euler field and using the Euler identity, we conclude with the identity

$$mg + \mu(V) \equiv 0. \tag{25.25}$$

Thus the form μ cannot vanish identically; indeed, in such a case we would have $g \equiv 0$ and hence by (25.23) $\Omega \equiv 0$ in contradiction with our assumptions.

Therefore we have an inequality between the degrees $r = \deg \Omega = \deg(f\mu) = m + \deg \mu \geqslant m$.

If $r = m$, then $\deg \mu = 0$, i.e., $\mu = \text{const}$, and $\deg g = 1$, i.e., dg is also a constant 1-form, $dg(V) \equiv g$. From this and (25.25) we conclude that $\mu(V) = -m\, dg(V)$, and since both forms μ and $-m\, dg$ are constant and take the same values of V, they coincide identically: $\mu = -m\, dg$. Substituting this to (25.23), we obtain the representation $\Omega = g\, df - mf\, dg = g^{m-1}\, d(f/g^m)$ meaning that Ω is a differential of a rational function.

In other words, a foliation having a smooth algebraic separatrix of maximal possible degree, *must* be Hamiltonian; cf. with Example 25.12. This completes the proof of Theorem 25.15.

To prove Theorem 25.21, note that the polynomial g must be of degree at least 1, i.e., nonconstant. The points of C where g vanishes, are singular for $\mathcal{F}$. Hence a *smooth* projective curve necessarily carries a singularity of $\mathcal{F}$. If C is itself singular, then the singularity of C must be a singularity of $\mathcal{F}$ by the very definition of foliation. The proof of Theorem 25.21 is also complete. □

25F. Jouanolou example. Now we have all the necessary tools to prove Theorem 25.20 and show that the foliation (25.18) has no algebraic leaves on the projective plane. One of the reasons is a very high degree of symmetry of this foliation.

The main idea behind the proof is rather simple. Suppose a foliation of degree n has a maximal number of singular points, $\approx n^2$ by the Bézout theorem, all of them complex saddles, and symmetries of the field act transitively (cyclically) on these points. An invariant curve, if it exists, must pass through one of these points, hence by symmetry through all of them. This means that the curve is either smooth, or nodal (has only the normal crossings) and its degree is explicitly bounded by Theorem 25.14. Yet a nodal curve of a relatively low degree $m \approx n$ cannot have so many ($\approx n^2$) self-intersections by the Plücker formula.

This leaves the only possibility that an algebraic leaf $C = \{f = 0\}$ of degree m is necessarily smooth, then $m \leqslant n$ and by the Division Lemma 25.30, these singular points occur only at the intersections of C with the auxiliary polynomial g from (25.23) of degree $n - m$. The number of roots of a system of two algebraic equations $f = 0$ and $g = 0$ of degrees m and $n - m$ respectively is no greater than $\approx n^2/2$, which is again less than the initial estimate $\approx n^2$. The contradiction shows that there are no algebraic leaves.

The accurate argument goes as follows.

Proof of Jouanolou Theorem 25.20. Consider the foliation $\mathcal{F}$ of the projective degree $n + 1$ on $\mathbb{P}^2$ which in the affine chart (x, y) is defined by the form

$$\omega = (x^n - y^{n+1})\,dx - (1 - xy^n)\,dy. \tag{25.26}$$

This foliation is very symmetric: for any ε which is a root of unity of degree $\nu = n^2 + n + 1$ the transformation

$$\sigma\colon (x, y) \mapsto (\varepsilon x, \varepsilon^{n+1} y)$$

preserves $\mathcal{F}$, since $\sigma^*\omega = \varepsilon^{n+1}\omega$.

The foliation (25.26) has ν singular points $a_1, \dots, a_\nu$ belonging to the σ-orbit of the obvious singularity $a_1 = \{x = y = 1\}$. One can immediately verify that all these singularities are hyperbolic and there are no other singularities of $\mathcal{F}$ on $\mathbb{P}^2$.

Because of σ-equivariance, any algebraic separatrix of σ, if it exists, is part of a larger σ-invariant separatrix C defined by some square-free polynomial $f \in \mathbb{C}[x, y]$. We claim that C must be smooth and carry all ν singular points of $\mathcal{F}$.

Indeed, if C is nonsmooth, then it must have only normal crossings (also called nodal points), since all singularities of $\mathcal{F}$ are hyperbolic. If $\deg C = m$, then the number s of such points is related to the genus g of the curve by the Plücker formula [**Mir95**, Proposition 2.6],

$$g = \tfrac{1}{2}m(m-1) - s. \tag{25.27}$$

Since the genus g is always nonnegative, we obtain the inequality $s \leqslant \frac{1}{2}m(m-1)$.

On the other hand, since C has only normal crossings, $m \leqslant n + 2$ by Theorem 25.14. Combining this with the Plücker inequality, we conclude that $s \leqslant \frac{1}{2}(n+2)(n+1)$. This number is *strictly less* than $\nu = n^2 + n + 1$ for $n > 1$, which means that the self-intersections are impossible and C is smooth.

For the *smooth* projective curve $C = \{f(X, Y, Z) = 0\}$, a stronger assertion concerning the degree holds. By Theorem 25.15,

$$m = \deg C \leqslant n + 1. \tag{25.28}$$

On the other hand, by Lemma 25.30, the homogeneous form Ω which represents the foliation (25.26) in the homogeneous coordinates in $\mathbb{C}^3$, can be divided by df, i.e., $\Omega = g\, df + f\, \mu$, where g, μ are the homogeneous function and the homogeneous form as in (25.23) with $\deg g = n + 2 - m$. Since C is smooth, df does not vanish on $\{f = 0\}$ (cf. Proposition 25.23). Hence the singularities of $\mathcal{F}$ on C may occur only at the points where the coefficient g vanishes.

By Theorem 25.21, there is at least one singularity of $\mathcal{F}$ on C. But because of the σ-invariance, *all* ν singularities of $\mathcal{F}$ also lie on C. Note that they do not belong to one line for $n > 0$, therefore $m = \deg C$ should be greater or equal to 2.

But this contradicts to the Bézout theorem. Indeed, the singularities are given by solutions of the system $\{f = 0,\ g = 0\}$ of two algebraic equations of degree m and $n + 2 - m$ respectively; their number is therefore no greater than $m(n + 2 - m) \leqslant (n + 1)(n + 2 - m) \leqslant (n + 1)n < \nu$ by (25.28). The resulting contradiction shows that the Jouanolou foliation (25.26) has no algebraic separatrices for $n \geqslant 2$. $\square$

25G. Darboux integrability. So far we discussed the question of *existence* and the *maximal degree* of algebraic separatrices of a polynomial foliation. The natural question would be to ask about their *number*. Of course, there are trivial situations when *all leaves* of the foliation are algebraic, e.g., in the Hamiltonian case; see Example 25.12. To exclude such situations, one may ask about the number of *isolated* algebraic leaves. Note that in the *complex* projective space the notions of *compactness* and algebraicity coincide, therefore the question may be formulated as follows: *how many isolated compact complex invariant curves may have a holomorphic singular foliation of degree r on $\mathbb{P}^2$?* Despite the apparent similarity between this question and Hilbert's sixteenth problem (about limit cycles which are isolated compact leaves of the *real* polynomial foliation on $\mathbb{R}P^2$; cf. with §24**A**)), the "complex" version is by far more simple. The answer is given by the Darboux integrability theory. This theory implies that a polynomial foliation having too many algebraic leaves, is necessarily integrable.

25**G**$_1$. *Classical Darboux approach.* We begin the exposition in the simplest settings. Consider a polynomial vector field $F \in \mathcal{D}[x,y]$ of degree r on the affine plane $\mathbb{C}^2$, and its invariant algebraic curve $C = \{f = 0\} \subset \mathbb{C}^2$ of degree m, as usual, defined by a square-free polynomial $f \in \mathbb{C}[x,y]$. The invariance condition (25.17) written in terms of the Lie derivative Ff, takes the form

$$Ff = fg, \qquad f, g \in \mathbb{C}[x,y], \quad \deg f = m, \ \deg g \leqslant r-1, \tag{25.29}$$

where g is the polynomial *cofactor* associated with the polynomial "invariant factor" f. Note that the degree of the cofactor does not exceed $r-1$ no matter what the degree of the invariant factor was. This observation lies at the heart of the Darboux theory.

Theorem 25.33. *If a planar polynomial vector field $F \in \mathcal{D}[\mathbb{C}^2]$ of degree r has $n \geqslant \frac{1}{2}r(r+1)+1$ different irreducible invariant curves $C_1, \dots, C_n$, then it admits a (multivalued) first integral of the form $\Phi = f_1^{\lambda_1} \cdots f_n^{\lambda_n}$, where $f_j \in \mathbb{C}[x,y]$, $j = 1, \dots, n$, are irreducible polynomials determining the respective curves C_j and $\lambda_j \in \mathbb{C}$ the complex exponents, not all equal to zero.*

Proof. The dimension of the linear space of all polynomials in two variables x, y of degree $\leqslant r-1$ is $\frac{1}{2}r(r+1)$. Thus if the field F has as many invariant factors $f_1, \dots, f_n$ as is assumed in the theorem, $Ff_j = f_j g_j$, then the corresponding cofactors $g_1, \dots, g_n$ must necessarily be linearly dependent: there exist complex numbers $\lambda_1, \dots, \lambda_n$, not all equal to zero, such that $\lambda_1 g_1 + \cdots + \lambda_n g_n = 0$. Direct computation shows that the nonconstant multivalued function Φ is the first integral of F for any choice of the branches:

$$F\Phi = \Phi \cdot \sum_{j=1}^{n} \lambda_j \frac{Ff_j}{f_j} = \Phi \cdot \sum_{j=1}^{n} \lambda_j g_j \equiv 0.$$

The proof is complete. □

This theorem is the first in a chain of results linking integrability with the presence of many invariant algebraic curves. For instance, one extra algebraic invariant curve implies that the first integral can be chosen *rational.*

Theorem 25.34 (J.-P. Jouanolou, 1979). *If a polynomial vector field F of degree r has $\frac{1}{2}r(r+1)+2$ algebraic irreducible invariant curves, then it has a rational first integral.*

Proof. By Theorem 25.33, the field F admits a number of multivalued integrals in the form of products of complex powers of the polynomials $f_1, \dots, f_{n+1}$. Choose two such integrals Φ, Φ' which are different in the sense that, say, the first does not involve the power of f_{n+1} while the second does not involve the power of f_n.

The two closed 1-forms $\omega = d\Phi/\Phi$ and $\omega' = d\Phi'/\Phi'$ are *rational*:each of them is a linear combination of the logarithmic derivatives $df_1/f_1, \dots, df_{n+1}/f_{n+1}$. Since both Φ and Φ' are first integrals of the same foliation generated by the field F, the forms ω and ω' are *proportional* at each point of $\mathbb{P}^2$, i.e., differ by a *rational* factor $h \in \mathfrak{M}(x,y)$. The ratio h is obviously nonconstant (otherwise the integrals would involve the powers of the same terms).

We claim that h is the first integral of the field F. Indeed, differentiating the identity $\omega' = h\omega$ and using the fact that both ω, ω' are exact, we conclude that $0 = dh \wedge \omega$, i.e., all three forms dh, ω and ω' are proportional.

Thus h is a rational first integral, $Fh = 0$, as required. □

Corollary 25.35. *A foliation defined by a polynomial vector field of degree r, may have at most $\frac{1}{2}r(r+1)+1$* isolated *compact invariant curves.*

Proof. Otherwise the foliation would admit a rational first integral, hence all leaves would be algebraic and none of them can be isolated. □

Another application of Theorem 25.34 is the following finiteness result.

Corollary 25.36. *For any polynomial foliation $\mathcal{F}$ on $\mathbb{P}^2$, the degree of its irreducible algebraic separatrices is uniformly bounded by a constant depending only on $\mathcal{F}$.*

Proof. If this degree is unbounded, then the number of different irreducible separatrices is infinite. By Theorem 25.34, the foliation has a rational first integral. The degree of this integral is an upper bound for the degrees of *all* algebraic leaves of $\mathcal{F}$, contrary to the assumption. $\square$

Remark 25.37. Sometimes it is more natural to describe polynomial foliations by their Pfaffian equations $\{\omega = 0\}$, $\omega \in \Lambda^1[\mathbb{C}^2]$ and treat the cofactors of an invariant curve $C = \{f = 0\}$, $f \in \mathbb{C}[x, y]$, as a 2-form Θ such that

$$\omega \wedge df = f\,\Theta, \qquad \Theta \in \Lambda^2[\mathbb{C}^2]. \tag{25.30}$$

25$\mathbf{G}_2$. *Generalized Darboux integrability.* The above exposition relies on some very explicit and particular form of integrability. We will present a more general approach, partially based on [**CL00**].

Definition 25.38. A polynomial foliation $\mathcal{F}$ of the projective plane $\mathbb{P}^2$ is *Darboux integrable*, if it is generated by a *closed* meromorphic (rational) 1-form ω on $\mathbb{P}^2$.

By this definition, a Darboux integrable foliation admits an analytic first integral which is a multivalued function on $\mathbb{P}^2$ ramified over an algebraic curve $\Sigma \subset \mathbb{P}^2$ (algebraic subvariety of positive codimension).

The definition of integrability established in Theorem 25.33 is indeed a particular case of the general Definition 25.38. This follows from Lemma 11.27 giving the explicit description of exact rational 1-forms on $\mathbb{P}^2$. In fact, the same arguments that prove the local Lemma 11.27, prove its global counterpart.

Lemma 25.39. *If $\Sigma \subseteq \mathbb{P}^2$ is an algebraic curve whose irreducible components are given in an affine chart by irreducible polynomial equations $\{f_i = 0\} \subset \mathbb{C}^2$, $f_i \in \mathbb{C}[x, y]$, then any rational closed 1-form with the polar locus on Σ is cohomologous to a linear combination of logarithmic derivatives,*

$$\omega = \sum_{j=1}^{n} \lambda_j \frac{df_j}{f_j} + d\left(\frac{g}{f_0}\right), \qquad f_0, f_1, \dots, f_n, g \in \mathbb{C}[x, y],\ \lambda_j \in \mathbb{C}, \tag{25.31}$$

where f_0 is a polynomial nonvanishing off Σ. $\square$

Corollary 25.40. *Any Darboux integrable foliation admits a multivalued "first integral" of the form $\Phi = \exp(g/f_0) \cdot \prod_1^n f_j^{\lambda_j}$.* $\square$

Now we can give an invariant definition of the *invariant differentials*, generalizing the notion of invariant curves. Consider a polynomial vector field F of degree r on $\mathbb{C}^2$ and the foliation $\mathcal{F}$ on $\mathbb{P}^2$ generated by this field.

Definition 25.41. An *invariant differential* for the vector field F is a *closed* rational 1-form α on $\mathbb{P}^2$, with the pole of order $\leqslant 1$ on the infinite line, such that the rational function $h = \alpha(F)$ has no singularities in the affine plane $\mathbb{C}^2$, i.e., is a *polynomial*. This polynomial is called the *cofactor* associated with the invariant differential α:

$$\alpha(F) = h \in \mathbb{C}[x, y], \qquad d\alpha = 0. \tag{25.32}$$

The invariant differential is *simple*, if it has only first order poles in the affine part $\mathbb{C}^2 \subset \mathbb{P}^2$, otherwise it is called *multiple*[3].

The invariant differentials for a given field F obviously form a *complex linear space* $\mathfrak{D}_F \subseteq \Lambda^1[\mathbb{C}^2]$. The corresponding cofactors form a subspace in the space of all polynomials $\mathfrak{C}_F \subseteq \mathbb{C}[x, y]$.

Example 25.42. 1. A nonzero *polynomial* closed 1-form α is exact and cannot have a pole of order $\leqslant 1$ on the infinite line unless being identically zero. Therefore each invariant differential α for a polynomial vector field F should be a *rational* form with the nonvoid polar locus $C = C_\alpha \subset \mathbb{C}^2$ which is an algebraic curve.

2. As follows from Lemma 11.27, any simple invariant differential is a linear combination of logarithmic differentials,

$$\alpha = \sum_{j=1}^{n} \lambda_j \frac{df_j}{f_j}, \tag{25.33}$$

for some irreducible polynomials f_j and complex numbers λ_j. Conversely, if (25.33) is an invariant differential with a cofactor h, then each logarithmic derivative df_j/f_j also is a invariant differential with some cofactor h_j, and each algebraic curve $C_j = \{f_j = 0\}$ is an invariant algebraic curve for F.

This observation gives a complete description of all simple invariant differentials which are in one-to-one correspondence with algebraic invariant curves of the field F.

3. The multiple invariant factors correspond to *divisors* with nontrivial multiplicities (greater than 1). Indeed, by the same Lemma 11.27, α is the sum of a simple Darboux part and the exact rational form $d(g/f_0)$ which has poles of order $\geqslant 2$ on the polar locus $\{f_0 = 0\} \subseteq \bigcup_{j=1}^{n}\{f_j = 0\}$. More precisely, if f_0 has a pole of some order $k \geqslant 1$ on an irreducible curve C, then the form α has a pole of order $k + 1$ there. If the exact term $d(g/f_0)$ is present, then at least on one of the irreducible curves C_j the invariant differential α has a pole of order $\geqslant 2$.

[3]We do not discuss here the question of *multiplicity* which is to be assigned to multiple invariant differentials. This question is addressed in [**CLP07**], we only mention here that this multiplicity does not coincide with the order of pole of the corresponding invariant differential.

Remark 25.43. As with divisors, the multiple invariant differentials can be understood as limits of one or several confluent simple invariant differentials. Indeed, if $\alpha = df/f$ and $\beta = dg/g$ are two simple invariant differentials corresponding to two close polynomials, $g/f = 1 + \varepsilon w$, where $w \in \mathbb{C}(x, y)$ is a rational function and ε a small parameter, then the linear space spanned by these two simple invariant differentials coincides with the linear spaces spanned, say, by df/f and $dw/(1 + \varepsilon w)$; the limit position of this space is spanned by the simple invariant differential df/f and the exact 1-form dw (which has a pole of order $\geqslant 2$).

The cornerstone of the Darboux method remains the same as in the classical context.

Theorem 25.44. *If the linear map*

$$i_F \colon \mathfrak{D}_F \to \mathfrak{C}_F, \qquad \alpha \overset{i_F}{\longmapsto} \alpha(F) \tag{25.34}$$

from the space of invariant differentials $\mathfrak{D}_F$ for the field F to the space $\mathfrak{C}_F$ of polynomial cofactors has a nontrivial kernel, then F is Darboux integrable.

Proof. Any nonzero closed rational form α such that $\alpha(F) = 0$, generates the same foliation as the field F itself. □

To apply the (obvious) Theorem 25.44, one has to produce an upper bound for the dimension of the space of the cofactors and construct sufficiently many linearly independent invariant differentials. It turns out that the first task can be implemented without any explicit knowledge of the field.

Apriori, the definition of a cofactor $h \in \mathbb{C}[x, y]$ does not impose any restriction on its degree. Yet it turns out that this degree is *automatically* no greater than $r - 1$, where $r = \deg F$ is the affine degree of the polynomial field F.

Proposition 25.45. *If $F \in \mathcal{D}[\mathbb{C}^2]$ is a polynomial vector field on $\mathbb{C}^2$ and α is an invariant differential for F with the cofactor $h = \alpha(F)$, then*

$$\deg \alpha(F) \leqslant \deg F - 1. \tag{25.35}$$

Proof. The inequality is obvious for the simple invariant differentials: if $\alpha = df/f$ with $f \in \mathbb{C}[x, y]$ and $\alpha(F) = h$ is a polynomial, then $Ff = fh$.

For multiple invariant differentials one can assume without loss of generality that α is exact, $\alpha = d(g/f_0)$. For the form α to have a pole of order $\leqslant 1$ on the infinite line (as required by the definition of the invariant differential), the rational primitive g/f_0 should have no pole on the infinite line, which is possible only if $\deg g \leqslant \deg f_0$. The assumption that $d(g/f_0) = h$ has a polynomial cofactor, means that

$$f_0\,(Fg) - g\,(Ff_0) = hf_0^2,$$

which is possible only if $\deg h \leqslant \deg f_0 + \deg F + \deg g - 1 - 2\deg f_0 = \deg F - 1 + \deg g - \deg f_0 \leqslant \deg F - 1$. □

This proposition implies that

$$\dim_{\mathbb{C}} \mathfrak{D}_F \leqslant \tfrac{1}{2}r(r+1), \qquad \text{where } r = \deg F. \tag{25.36}$$

Corollary 25.46 (Generalized Darboux theorem). *If a polynomial vector field of degree r has $\frac{1}{2}r(r+1)+1$ linearly independent invariant differentials, then it is Darboux integrable.*

Corollary 25.47 (Generalized Jouanolou theorem). *If a polynomial vector field of degree r has $\frac{1}{2}r(r+1)+2$ linearly independent invariant differentials, then it has a rational first integral.*

Proof of both Corollaries. The dimension of polynomials of degree $\leqslant r-1$ in two variables is $\frac{1}{2}r(r+1)$, so any given $\frac{1}{2}r(r+1)+1$ cofactors are linearly dependent and Theorem 25.44 applies. If there is an extra invariant differential independent from the first one, then there can be constructed two nonproportional closed rational 1-forms ω, ω' tangent to the same foliation $\mathcal{F}$. Their ratio is a nonconstant rational first integral in the same way as in Theorem 25.34. □

These results generalize the results by J. Llibre and C. Christopher [**CL00**] for the case where the field admits invariant differentials[4] of multiplicity higher than 2.

Moreover, one can further improve the two Corollaries, if the polynomial vector field possesses invariant differentials *"not passing"* through singular points of F, i.e., not containing points of $\operatorname{Sing}\mathcal{F}$ in their singular loci. Indeed, in this case *any* cofactor must vanish at these points: this vanishing condition is a linear constraint that further reduces the dimension of the target space $\mathfrak{C}_F$ of the map (25.34).

In particular, assume that a polynomial vector field F of degree r has n invariant differentials, none of which contains some given l singular points of F in the polar locus. If these l points are in general position (so that the subspace of polynomials of degree $\leqslant r-1$ vanishing at all these points has codimension l), and $n+l \leqslant \frac{1}{2}r(r+1)$, then the field F is Darboux integrable; occurrence of yet another independent invariant differential with the same properties implies that F admits a rational first integral; cf. with [**CL00**].

[4] In [**CL00**] the authors explicitly require that the exponential factor has the cofactor of degree $\leqslant r-1$, where r is the degree of the vector field F.

Appendix: Foliations with invariant lines and algebraic leaves of foliations from the class $\mathcal{A}_r$

One of the principal results of this section is that generic polynomial foliations from the class $\mathcal{B}_r$ on $\mathbb{P}^2$ have no compact separatrices (such separatrices would automatically be algebraic) for $r \geqslant 2$. This makes application of the tools related to holonomy groups, very problematic. However, if we change the point of view and consider all foliations given in a *fixed* affine chart by polynomial 1-forms of a given degree r, then generically such foliations possess the invariant line at infinity which is a unique algebraic separatrix with (generically) rather reach fundamental group. This paves the way to rigidity theorems of §28. On the other hand, many properties of the class $\mathcal{A}_r$ are parallel to those of the class $\mathcal{B}_r$.

From now on we fix a line ℓ in $\mathbb{P}^2$ and any affine chart (x, y) on $\mathbb{C}^2 = \mathbb{P}^2 \setminus \ell$ for which this line is the infinite line denoted by $\mathbb{I}$.

Recall (cf. Definition 25.1) that the class $\mathcal{A}_r$ has the natural structure of a (complex) projective space $\mathbb{P}^N$ of dimension $N = (r+1)(r+2) - 1$ with the homogeneous coordinates being coefficients of the polynomial 1-form ω. This again allows us to speak about generic properties of foliations from this class.

Example 25.48. A generic foliation $\mathcal{F} \in \mathcal{A}_r$ has an invariant line $\mathbb{I}$ carrying exactly $r+1$ hyperbolic singular points.

Indeed, the sufficient condition for having an invariant line at infinity is described by Proposition 25.8. If $\omega = p\,dx + q\,dy$ is the Pfaffian form defining the foliation, this condition takes the form $xp_r(x, y) + yq_r(x, y) \not\equiv 0$, where $p_r\,dx + q_r\,dy$ is the principal homogeneous part of degree r of p, q respectively. The singularities on $\mathbb{I}$ correspond to roots of the homogeneous polynomial $h_{r+1} = xp_r + yq_r$ which are generically all distinct.

The ratios of eigenvalues of linearization (characteristic numbers) at each such point are given by the expressions (25.6), and are all nonzero if the homogeneous polynomials p_r, q_r have no common roots.

Definition 25.49. Denote by $\mathcal{A}'_r \subset \mathcal{A}_r$ the class of all foliations having invariant line $\mathbb{I}$ at infinity and exactly $r+1$ distinct hyperbolic singularities on it. This class constitutes a Zariski open subset in the complex (linear or projective) space $\mathcal{A}_r$.

We can prove now another assertion illustrating scarcity of algebraic leaves of polynomial foliations. The following theorem is a direct counterpart of Theorem 25.27.

Theorem 25.50. *If all $r+1$ exponents at infinity of a foliation $\mathcal{F} \in \mathcal{A}'_r$ are nonreal, then this foliation has no algebraic leaves of degree greater than $r+1$.*

Proof. Near each singular point on $\mathbb{I}$, the foliation is linearizable by the Poincaré Theorem 5.5 and hence there exists a local biholomorphism between $\mathcal{F}$ and a foliation $(v - v_j)du - \lambda_j u\,dv = 0$. The only local leaves of the latter which can belong to an algebraic leaf of the initial foliation, are two invariant curves (separatrices), one of which is a part of the line $\mathbb{I}$ and the other is transversal to it. All other local leaves have logarithmic ramification.

Thus an algebraic invariant curve of $\mathcal{F}$, if it exists, must intersect the infinite line $\mathbb{I}$ transversally at some of the $r+1$ singular points at infinity. Yet an algebraic curve of degree d in $\mathbb{P}^2$ intersects any line, in particular, $\mathbb{I}$, at exactly d points counted with their multiplicity. Thus $d \leqslant r+1$. □

Remark 25.51. The arguments proving Theorem 25.50, also show that for the foliations satisfying the assumptions of this theorem, the principal homogeneous part of the polynomial equation defining an algebraic leaf of degree $\leqslant r+1$, if such a leaf exists, must be a product of linear factors corresponding to lines passing through singular points on $\mathbb{I}$. The multiplicity of any such factor must not exceed 1 so that the intersection of the leaf with $\mathbb{I}$ remains transversal.

The assumptions of Theorem 25.50 on the exponents at infinity can be relaxed to cover generic real foliations from the class $\mathcal{A}'^{\mathbb{R}}_r$, if we require that exponents at infinity are not rational, $\lambda_j \in \mathbb{C} \smallsetminus \mathbb{Q}$. In these relaxed assumptions each singularity on the infinite line still has a unique smooth separatrix transversal to $\mathbb{I}$.

Theorem 25.50 places an apriori upper bound for the degree of algebraic leaves of a *generic* foliation from the class $\mathcal{A}'_r$. We explain now an algorithm allowing to determine *all* algebraic leaves of degrees $\leqslant s$ for an arbitrary foliation from the class $\mathcal{A}'_r$ and any given s.

It will be convenient to assume that the foliation $\mathcal{F} \in \mathcal{A}'_r$ is defined by a polynomial vector field $F \in \mathcal{D}[\mathbb{C}^2]$, represented as the sum of homogeneous terms of degrees from 0 to r, $F = F_r + F_{r-1} + \cdots + F_1 + F_0$. Assume that $f = f_s + f_{s-1} + \cdots + f_0$ is the polynomial equation of an algebraic leaf (separatrix), also represented as the sum of homogeneous components of degree $\leqslant s$ (we do not assume that $s \leqslant r+1$). Then there exists a polynomial *cofactor* $g = g_{r-1} + g_{r-2} + \cdots + g_1 + g_0 \in \mathbb{C}[x, y]$, such that

$$Ff = fg \tag{25.37}$$

(the left hand side is the derivation of f along the field F; cf. with §1**G**). Collecting the homogeneous terms from two sides, we arrive at the system of equations

$$F_r f_s = f_s g_{r-1}, \tag{25.38}$$

$$F_r f_{s-1} = f_{s-1} g_{r-1} + f_s g_{r-2} - F_{r-1} f_s, \tag{25.39}$$

$$F_r f_{s-2} = f_{s-2} g_{r-1} + f_{s-1} g_{r-2} + f_s g_{r-3} - F_{r-2} f_s - F_{r-1} f_{s-1}, \tag{25.40}$$

....

$$F_r f_1 = f_1 g_{r-1} + f_2 g_{r-2} + \cdots - F_{r-1} f_2 - F_{r-2} f_3 - \cdots \tag{25.41}$$

This system can be explicitly solved.

The identity (25.38) is *bilinear* with respect to the unknown homogeneous polynomials f_s, g_{r-1}. It admits solutions of any degree s.

Lemma 25.52. *If the foliation $\mathcal{F}$ belongs to the class $\mathcal{A}'_r$, then every solution f_s of* (25.38) *has the form*

$$f_s = \prod_{j=1}^{r+1} l_j^{\nu_j}, \qquad \nu_j \in \mathbb{Z}_+, \quad \sum_j \nu_j = s, \tag{25.42}$$

where $l_j \in \mathbb{C}[x,y]$ are linear homogeneous polynomials defining the lines $\ell_j \subset \mathbb{C}^2$ passing through the origin and the infinite singular points S_j of $\mathcal{F}$. The corresponding cofactor g_{r-1} is uniquely defined by the choice of f_s.

Remark 25.53. For foliations of the class $\mathcal{B}_r$, $F_r = a_{r-1}(x,y)V$, where V is the Euler field and $a_{r-1} \in \mathbb{C}[x,y]$, hence *any* homogeneous polynomial f_s satisfies the equation (25.38) $g_{r-1} = s a_{r-1}$.

The remaining equations have a "triangular" structure which allows us to solve them inductively starting from any solution f_s, g_{r-1} of the equation (25.38). Solvability of these equations can be described as follows.

Lemma 25.54. *If f_s is a square-free solution of* (25.38), *i.e., if $\nu_j \leqslant 1$ for all $j = 1, \ldots, r+1$, then in the assumptions of Lemma 25.52 the system of equations* (25.39)–(25.40) *is generically not solvable.*

More precisely, in the space of all polynomial vector fields F of the given degree $r \geqslant 2$ with the fixed principal part F_r, the fields which admit polynomial integrals starting with f_s, constitute a proper algebraic subvariety.

Proof of both lemmas. All assertions are verified by the direct computations in the homogeneous coordinates x and $v = y/x$; because of the homogeneity, the variables separate. In doing this it is convenient to view the left hand sides of the equations as the ratios of the appropriate 2-forms

$\frac{\omega_r \wedge df_j}{dx \wedge dy}$, $j = s, s-1, \dots, 1$, which in turn are equal to the ratios

$$\frac{x^{r+j-1}[(p_r(1,v) + vq_r(1,v))\,dx + xq_r(1,v)\,dv] \wedge (j\,f_j(1,v)\,dx + x\frac{df_j(1,v)}{dv}dv)}{x\,dx \wedge dv}$$
$$= x^{r+j-1}\left(h_{r+1}(1,v)\frac{df_j(1,v)}{dv} - jf_j(1,v)q_r(1,v)\right), \qquad h_{r+1} = p_r + vq_r.$$

After passing to the new coordinates the system of the equations (25.38)–(25.40) takes the form

$$h_{r+1}\tfrac{d}{dv}f_s - sq_rf_s = f_sg_{r-1}, \tag{25.43}$$
$$h_{r+1}\tfrac{d}{dv}f_{s-1} - (s-1)q_rf_{s-1} = f_{s-1}g_{r-1} + f_sg_{r-2} + w_{r+s}, \tag{25.44}$$
$$h_{r+1}\tfrac{d}{dv}f_{s-2} - (s-2)q_rf_{s-2} = f_{s-2}g_{r-1} + f_sg_{r-3} + w_{r+s-1}, \tag{25.45}$$

....

$$h_{r+1}\tfrac{d}{dv}f_1 - q_rf_1 = f_1g_{r-1} + f_sg_{r-s-2} + w_r \tag{25.46}$$

where all polynomials depend only on the single variable v and we abbreviated $h_{r+1}(1,v)$, $q_r(1,v)$, $f_j(1,v)$ and $g_j(1,v)$ to $h_{j+1}(v), q_r(v), f_j(v)$ and $g_j(v)$ respectively. The terms denoted by w_j stand for polynomials of degree $\leqslant j$ in v, which are linear combinations of the polynomials f_i and g_k and their derivatives occurring in the preceding lines of the system. This triangular structure allows us to solve the system with respect to the homogeneous components f_j, g_j, starting from the first equation (25.43).

The equation (25.43) implies that the roots $v_1, \dots, v_{r+1}$ of polynomial $h_{r+1}(v)$ (corresponding to the singularities $S_1, \dots, S_{r+1}$ on the infinite line) should cancel the poles of the logarithmic derivative $\frac{d}{dv}f_s/f_s$ of the principal term f_s. Since the latter is the sum of simple fractions, this means that all roots of f_s should be among the set $\{v_1, \dots, v_{r+1}\}$, i.e., $f_s(v) = \sum_{j=1}^{r+1} \nu_j/(v - v_j)$ with $\nu_j \geqslant 0$, $\sum_j \nu_j = s$. Conversely, any polynomial of the form (25.42) yields a solution to (25.43).

The remaining equations are linear with respect to the polynomials f_{s-j} and g_{r-j-1} respectively, assuming that all higher order homogeneous components of both the integral f and the cofactor g are already known. We show that the solution of those equations reduces to solving interpolation problems for univariate polynomials. In the *square-free case* when $s \leqslant r+1$ and f_s is a product of pairwise different linear factors, we show that the second equation (25.44) is generically solvable whereas the third equation (25.45) is not solvable.

Indeed, evaluating (25.44) at any root v_k of f_s which must also be the root of h_{r+1}, we conclude with the equations

$$c_kf_{s-1}(v_k) + w_{r+s}(v_k) = 0, \text{ where } c_k = (s-1)q_r(v_k) + g_{r-1}(v_k),\ k = 1, \dots, s.$$

These equations uniquely prescribe the values f_{s-1} at v_k provided that $c_k \neq 0$ which generically holds true. The problem of recovering the polynomial f_{s-1} of degree $s-1$ is hereby reduced to the s-point interpolation. The latter problem is always solvable, and the initial equation (25.44) can be used now to determine uniquely the polynomial g_{r-2}.

The same arguments literally apply also to the subsequent equations starting from (25.45), yet the interpolation problem to be solved would require restoring a polynomial f_j of degree $j < s-1$, by its arbitrarily assigned values at s distinct points. Generically this problem is not solvable unless a certain polynomial relation between coefficients of the system holds. This condition can be explicitly stated as the requirement that the rank of the extended matrix of the nonhomogeneous system is equal to the rank of the matrix of the homogeneous system. □

Remark 25.55. The case where f_s is not square-free, is treated by similar arguments involving interpolation with derivatives.

Combining Theorem 25.50 with Lemma 25.54, we arrive at the direct analog of Theorem 25.18 for foliations of the class $\mathcal{A}_r$.

Theorem 25.56 ([**PL55**]). *A generic polynomial foliation from the class $\mathcal{A}_r$ has no algebraic leaves besides the infinite line $\mathbb{I}$.*

Proof. First, it is sufficient to consider only polynomials from the Zariski open subset $\mathcal{A}'_r$. The assumptions of Theorem 25.50 select a full-measure subset in the space of foliations $\mathcal{A}'_r$ for which the algebraic leaf, if it exists, must have degree $s \leqslant r+1$ and the principal homogeneous part f_s of the corresponding polynomial must be square-free by Remark 25.51. By Lemma 25.54, outside a proper algebraic subset the corresponding system (25.37) is not solvable except for the trivial solution $f = c$, $g = 0$ and hence has no algebraic leaves of any degree $s \leqslant r+1$. □

Remark 25.57. The proof of Theorem 25.56 is based on the exposition in [**Pet96**], where a number of gaps from the first publication [**PL55**] was sealed.

In fact, one can compute directly the dimension of the space of polynomial foliations having only invariant curves of degree $\leqslant r+1$ with normal self-intersections, as explained in [**BL88**]. This allows us to avoid explicit computations proving Lemmas 25.52 and 25.54.

Exercises and Problems for §25.

Problem 25.1. Let $\sigma\colon (\mathbb{M}, \mathbb{E}) \to (\mathbb{C}^2, 0)$ be the standard blow-up and $C \subset (\mathbb{C}^2, 0)$ the germ of an analytic curve having normal crossing at the origin.

Prove that the blow-up $\widetilde{C}$ consists of two smooth analytic connected components, and σ is a biholomorphic equivalence between each component and a smooth branch of C.

Problem 25.2. Simple self-intersections of an affine planar algebraic curve are nondegenerate (Morse) critical points for the minimal polynomial of this curve.

Prove that the corresponding Hamiltonian vector field H restricted on the curve, has a simple zero on each smooth component of a normal self-intersection.

Exercise 25.3. Give a direct proof of the equality (25.7) for the sum of exponents at infinity for foliations of the class $\mathcal{A}_r$.

Exercise 25.4. Give a direct proof of the assertion (v) from §25$\mathbf{A}_1$, p. 472, i.e., prove that any collection of complex numbers constrained by the equality (25.7), can be realized as exponents at infinity of a suitable foliation from the class $\mathcal{A}_r$.

Exercise 25.5. Compute explicitly the effective dimensions of the classes $\mathcal{A}_r$ and $\mathcal{B}_r$ (dimensions of the projective spaces in which these classes naturally reside as open dense subsets).

Exercise 25.6. Give a complete proof of Proposition 25.8.

Problem 25.7. Prove that any singular foliation by analytic curves in $\mathbb{C}P^n$ in any affine chart may be represented by a suitable polynomial vector field.

Problem 25.8. Compute the total tangency order between a foliation of the class $\mathfrak{M}_r$ and a smooth projective curve of degree m.

Problem 25.9. Prove that degree of an *irreducible* separatrix of a foliation of the class $\mathcal{B}_r$ without generalized dicritical singularities does not exceed r.

Hint. Use Theorem 14.20.

Problem 25.10. A function R is said to be an *integrating factor* for a polynomial Pfaffian equation $\{\omega = 0\}$, $\omega \in \Lambda^1[\mathbb{C}^2]$, if $R\omega$ is closed, $d(R\omega) \equiv 0$.

Assume that a polynomial foliation $\{\omega = 0\}$ admits several invariant curves $C_i = \{f_i = 0\}$ such that the corresponding cofactor forms Θ_i (see Remark 25.37) generate a subspace containing the differential $d\omega$, so that $d\omega = \sum_i \lambda_i \Theta_i$ for some $\lambda_i \in \mathbb{C}$. Prove that the equation admits a *Darbouxian integrating factor* of the form $R = \prod_i f_i^{\lambda_i}$.

Exercise 25.11. Assume that the cofactor 2-forms $\Theta_1, \dots, \Theta_m$ of a Pfaffian equation $\{\omega = 0\}$ together with the differential $d\omega$ are linearly dependent in $\Lambda^2[\mathbb{C}^2]$. Prove that the foliation either admits a Darbouxian integral or a Darbouxian integrating factor.

Problem 25.12. Prove that polynomial integrable foliations of codimension 1 on $\mathbb{P}^n$ admit a uniform upper bound for the number of algebraic leaves, unless they have a Darbouxian integral. More precisely, show that for each combination of $n, r \in N$ there exists a bound $N \in \mathbb{N}$ such that any integrable polynomial foliation $\{\omega = 0\}$ of degree r on $\mathbb{P}^n$ either has at most N different irreducible algebraic leaves, or admits a Darbouxian first integral, i.e., $\omega \wedge \omega' = 0$, where $\omega' = \sum \lambda_i \frac{df_i}{f_i}$, f_i homogeneous polynomials in $n+1$ variables.

Exercise 25.13. Prove that a quadratic planar vector field with three invariant lines in general position admits a Darbouxian first integral.

Exercise 25.14. Prove that the bounds in Theorems 25.16 and 25.15 are sharp: there exist foliations from the class $\mathcal{B}_r$ which have algebraic invariant curve of degree $r+1$ and smooth algebraic curve of degree r.

26. Perturbations of Hamiltonian vector fields and zeros of Abelian integrals

Limit cycles are very difficult to track in general. The problem can be considerably simplified by *localization* in the phase space and/or parameters. For instance, restricting the domain in the phase plane to a neighborhood of an elliptic singular point allows us to track small amplitude limit cycles, as explained in §12. Another possibility implicitly explored in §24, is the study of limit cycles near separatrix polygons (*polycycles*).

One of the most powerful methods of analysis in general is localization in the parameter space: starting from an object with known simple properties, investigate what happens after small perturbation. In application to the study of vector fields, appearance and disappearance of limit cycles goes by the name of *bifurcation*.

In this section we consider *bifurcations of limit cycles from nonisolated periodic orbits*. The number and location of these cycles in the most important cases is determined by zeros of a special class of functions, Abelian integrals. Recall that Abelian integrals are integrals of rational 1-forms over cycles on algebraic curves, and if considered as functions of the parameters (coefficients of polynomials defining the curves), they are transcendental functions of several complex variables.

We study the algebraic and topological structure of Abelian integrals, proving several fundamental results that are widely used but not yet available in a complete and elementary exposition. The central algebraic result is description of the *module* of Abelian integrals over the ring of polynomials and explicit computation of the basis of this module. The topological study allows us to compute the monodromy group of continuous branches of Abelian integrals. Finally we bring together the two theories and derive a *Picard–Fuchs system* of linear ordinary differential equations for Abelian integrals, establish the type of its singularities and almost irreducibility of its monodromy. This opens the way to apply the "linear" tools developed in Chapter III to investigation of bifurcations of nonlinear systems (though this way remains unexplored in the book[5]).

[5]For further developments in this direction see the publications [**IY96, NY99, NY01, NY03, NY04, Yak05, Yak06**]. An alternative approach can be found in [**GI06, GI07**].

26A. Poincaré–Pontryagin criterion and generalizations. If γ is a closed (periodic) *nonisolated* orbit of a real analytic vector field F, then it must be an identical cycle by Theorem 9.12: some sufficiently narrow annulus-like neighborhood U of γ is entirely filled by closed orbits of F. In this case F is *analytically* integrable in U: there exists a real analytic function $f\colon U \to \mathbb{R}$ without critical points, such that $Ff = 0$ (Problem 26.1). The Pfaffian equation for the foliation takes the form $\{df = 0\}$.

Let $\varepsilon \in (\mathbb{R}^1, 0)$ be a small real parameter and f a real analytic function without critical points as above. Consider a real analytic perturbation of the initial integrable foliation $\{df = 0\}$, written in the Pfaffian form as

$$df + \varepsilon\omega = 0, \qquad \omega \in \Lambda^1(U), \quad \varepsilon \in (\mathbb{R}^1, 0), \tag{26.1}$$

where $\omega \in \Lambda^1(U)$ is a real analytic 1-form. Denote by

$$\Delta = \Delta_\gamma\colon (\mathbb{R}^1, 0) \times (\mathbb{R}^1, 0) \to (\mathbb{R}^1, 0), \qquad (z, \varepsilon) \mapsto \Delta(z, \varepsilon),$$

the holonomy map of the cycle γ considered as a function of the parameter ε. Since all elements of the construction are real analytic, Δ can be expanded in the converging series,

$$\Delta(z, \varepsilon) = z + \varepsilon I_1(z) + \cdots + \varepsilon^k I_k(z) + \cdots, \tag{26.2}$$

where $I_k(z)$ are real analytic functions defined in some common neighborhood of the origin $z = 0$. Since the case $\varepsilon = 0$ corresponds to the integrable system, the term I_0 is absent in (26.2).

The first not identically zero function in the sequence $I_1, I_2, \ldots,$ plays a special role.

Proposition 26.1. *Assume that the first nonzero function $I_k(z)$ for some $k \geqslant 1$ has n isolated zeros (counted with their multiplicities) in the closed interval $\{|z| \leqslant \rho\}$. Then there exists a small positive value $r > 0$ such that the foliation* (26.1) *in $\{|z| < \rho\} \subset U$ has no more than n limit cycles for all $|\varepsilon| < r$.*

Proof. Limit cycles correspond to the roots of the equation $\Delta(z, \varepsilon) - z = 0$. In the assumptions of the proposition the left hand side is divisible by ε^k: $\Delta(z, \varepsilon) - z = \varepsilon^k I'(z, \varepsilon)$. The number of isolated roots of the real analytic function $I'(z, \varepsilon) = I_k(z) + \varepsilon I_{k+1}(z) + \cdots$ for all sufficiently small ε does not exceed the total number of roots of its limit $I_k(z) = \lim_{\varepsilon\to 0} I'(z)$. □

Remark 26.2. The number of geometrically distinct zeros of the first nonzero function I_k can provide also a *lower* bound for the number of limit cycles, if the former are all of an odd multiplicity (e.g., all simple). If the first nonzero function I_k has a real root of an *odd* order, then the Poincaré function has *at least one* real root which is a limit cycle (this obviously follows from the intermediate value theorem). Other roots in general may be complex and do not correspond to limit cycles. For roots of I_k of an even order *all* roots of the displacement function may escape to the nonreal domain and do not manifest themselves as limit cycles.

The analytic expression of the first variation for the perturbation (26.1) is very simple.

Theorem 26.3 (Poincaré [**Poi90**], Pontryagin [**Pon34**]).

$$I_1(z) = -\oint_{\{f=z\}} \omega. \tag{26.3}$$

If the integral (26.3) is not identically zero, it may have only finitely many zeros on the cross-section τ. By Proposition 26.1, this number is an upper bound for the number of limit cycles of the perturbed foliation (26.1) for all sufficiently small values of the parameter ε.

Proof. Denote by $\gamma_{z,\varepsilon}$ the arc of an integral curve of the perturbed foliation between the point with the coordinate z on the cross-section τ and the next intersection with τ. By the choice of the chart $z = f|_\tau$ and the definition of the displacement,

$$\Delta(z,\varepsilon) - z = \int_{\gamma_{z,\varepsilon}} df = -\varepsilon \int_{\gamma_{z,\varepsilon}} \omega.$$

The last equality holds since $df + \varepsilon\omega$ vanishes identically on $\gamma_{z,\varepsilon}$ for any z. As $\varepsilon \to 0$, the arc $\gamma_{z,\varepsilon}$ tends uniformly in the C^1-sense to the closed curve $\gamma_{z,0} = \{f = z\}$. Hence the integral $\int_{\gamma_{z,\varepsilon}} \omega$ converges to the integral in (26.3). □

26B. Higher variations of the holonomy. If the Poincaré integral (26.3) vanishes identically, the higher variations I_k, $k = 2, 3, \dots$ should be computed until either a not identically vanishing variation is found, or for some reason it becomes clear that the family (26.1) entirely consists of integrable foliations for all small values of ε.

We describe an analytic procedure expressing the first nonzero function $I_k(z)$ as an integral of a certain analytic 1-form ω_k along the level ovals $\{f = z\} \subset U$. To describe this procedure, we need the following simple analytic observation.

Consider a domain $U \subset \mathbb{R}^2$ and a real analytic function $f\colon U \to \mathbb{R}$ without critical points in it.

Definition 26.4. A real analytic 1-form $\alpha \in \Lambda^1(U)$ is relatively exact with respect to the integrable foliation $\mathcal{F} = \{df = 0\}$ in a domain U, if

$$\alpha = h\,df + dg, \qquad h, g \in \mathcal{O}(U), \tag{26.4}$$

with two functions g, h real analytic in U.

The integral of a relatively exact form α along any closed oval on any level curve $\{f = z\} \subset U$ is obviously zero:

$$\forall \text{ oval } \delta \subseteq \{f = z\} \qquad \oint_\delta \alpha = 0. \tag{26.5}$$

The inverse assertion (and especially the complexification thereof) is considerably more delicate. It holds true only under some additional topological assumptions; see §26**D** below. The simplest case, however, is rather easy.

Lemma 26.5. *If U is the topological annulus formed by ovals of the level curves $\{f = z\}$ transversal to a global cross-section τ and a form $\alpha \in \Lambda^1(U)$ satisfies the condition* (26.5), *then α is relatively exact in U.*

Proof. For any $x \in U$ denote by $\gamma(x)$ an oriented arc of the level curve passing through x between x and the point of its intersection with τ. This arc is defined modulo an integer multiple of the loop (oval) $\delta = \{f = z\}$, $z = f(x)$, yet because of the condition (26.5), the integral $g(x) = \int_{\gamma(x)} \alpha$ is a well-defined analytic function in U. By construction, the forms α and dg take the same values on each vector tangent to any level curve $\{f = z\} \subset U$, i.e., the difference $\alpha - dg$ at each point is proportional to df. Since df never vanishes in U, the proportionality coefficient is a real analytic function: $\alpha - dg = h\, df$ for some $h \in \mathcal{O}(U)$. □

Remark 26.6. The representation (26.4) is not unique. One can replace $g(x)$ by $g(x) + u(f(x))$ with an arbitrary function u.

To compute the first function I_i which does not vanish identically, we construct inductively, using the representation (26.4), the sequence of real analytic 1-forms $\omega_1, \omega_2, \dots \in \Lambda^1(U)$ as follows.

1°. (*Base of induction*). $\omega_1 = \omega$ is the perturbation form from (26.1).

2°. (*Induction step*). If the forms $\omega_1, \dots, \omega_j$ are already constructed and turned out to be relatively exact, then by Lemma 26.5, $\omega_j = h_j\, df + dg_j$. In this case we define

$$\omega_{j+1} = -h_j\, \omega. \tag{26.6}$$

Theorem 26.7. *If ω_k, $k \geqslant 2$, is the first not relatively exact 1-form in the sequence $\omega_1, \dots, \omega_{k-1}, \omega_k$ constructed inductively by* (26.6), *then*

$$I_k(z) = -\oint_{\{f=z\}} \omega_k. \tag{26.7}$$

This theorem generalizes the Poincaré–Pontryagin Theorem 26.3. The algorithm of inductive construction of the forms ω_k, sometimes referred to as the *Françoise algorithm*, was independently suggested in [**Yak95**] and [**Fra96**], but probably was known much earlier.

Proof. Denote by U' the annulus U slit along the cross-section τ. This is a simply connected domain (curvilinear rectangle) foliated by the level curves of the function f, transversal to the two sides (denoted by τ_- and τ_+). Denote $\mathcal{F}_\varepsilon$ the foliation defined by the Pfaffian equation (26.1) in U'.

1. Let $u = u(x, \varepsilon) = u_\varepsilon(x)$ be the first integral of the foliation $\mathcal{F}_\varepsilon$ in U',

$$(df + \varepsilon\omega) \wedge du_\varepsilon \equiv 0, \tag{26.8}$$

which for $\varepsilon = 0$ coincides with f and analytically depends on the parameter ε; such an integral exists because the topology of the foliation $\mathcal{F}_\varepsilon$ in the slit annulus U' is trivial. (This integral is not uniquely defined for $\varepsilon \neq 0$.)

Denote by $z_\varepsilon = u_\varepsilon|_{\tau_+}$ the restriction of u_ε on the "terminal" side τ_+ of the cross-section τ. Being a small analytic perturbation of the chart $z = u_0|_\tau$, z_ε is also an analytic chart on τ. Then the restriction of the same solution u_ε on the "initial" side $u_\varepsilon|_{\tau_-}$ is the numeric value of the holonomy map $\Delta(\cdot, \varepsilon)$ related to the chart z_ε. Indeed, since u_ε is constant along integral curves, for points on τ_- it yields the value of the chart z_ε at the moment of the next hit.

In other words, the displacement function $\Delta(z_\varepsilon, \varepsilon) - z_\varepsilon$ related to the chart z_ε, is given by the difference $u_\varepsilon|_{\tau_-} - u_\varepsilon|_{\tau_+}$ of the first integral u_ε. Since level curves of f in U' are oriented in the direction from τ_- to τ_+, it is more natural to compute the *negative* of this expression, the difference

$$u_\varepsilon\Big|_{\tau_-}^{\tau_+} = -\big(\Delta(z_\varepsilon, \varepsilon) - z_\varepsilon\big). \tag{26.9}$$

2. The convenience of expressing the displacement function in terms of solutions of the partial differential equation (26.8) stems from the *linearity* of the latter. In particular, one can look for its solution in terms of the converging series,

$$u_\varepsilon = f + \varepsilon u_1 + \varepsilon^2 u_2 + \cdots, \tag{26.10}$$

where u_k are real analytic functions in the slit annulus U'. Substituting this series into the equation (26.8), we obtain the following system of equations on the respective components.

$$\begin{aligned} \omega \wedge df + df \wedge du_1 &= 0, \\ \omega \wedge du_1 + df \wedge du_2 &= 0, \\ \cdots\cdots\cdots\cdots\cdots\cdots \\ \omega \wedge du_{k-1} + df \wedge du_k &= 0. \end{aligned} \tag{26.11}$$

By assumption, the forms $\omega_1, \dots, \omega_{k-1}$ are all relatively closed, hence they admit representations

$$\omega_j = h_j\, df + dg_j, \qquad j = 1, \dots, k-1.$$

We claim that the functions $u_j = g_j$ satisfy the first $k-1$ equations of system (26.11). Indeed, direct substitution yields for all $j = 1, \dots, k-2$

$$\begin{aligned} \omega \wedge du_j + df \wedge du_{j+1} &= \omega \wedge (\omega_j - h_j\, df) + df \wedge (\omega_{j+1} - h_{j+1}\, df) \\ &= -h_j \omega \wedge df + df \wedge \omega_{j+1} = df \wedge (\omega_{j+1} + h_j \omega) = 0. \end{aligned}$$

The fact that the first $k-1$ components of the solution (26.10) are well-defined functions in the *nonslit* annulus U means that their contribution to the difference (26.9) is zero, $u_k\big|_{\tau_-}^{\tau_+} \equiv 0$, and all Melnikov functions $I_1, \dots, I_{k-1}$ are vanishing identically.

3. The kth equation of the system (26.11) can be used to determine the component u_k. The same computation as above reduces this equation to the form

$$\begin{aligned} 0 &= \omega \wedge (-h_{k-1}\, df) + df \wedge du_k = df \wedge (du_k + h_{k-1}\omega) \\ &= df \wedge (du_k - \omega_k). \end{aligned}$$

This means that the 1-form $du_k - \omega_k$ vanishes on all level curves $f = \text{const}$, i.e., that u_k can be restored as the primitive along these curves,

$$u_k(x) = \int_{\tau_-}^{x} \omega_k,$$

where the path of integration is the arc $\gamma(x)$ of the level curve, connecting an appropriate point on the slit τ_- with the variable point $x \in U'$. The difference (increment) of u_k from τ_- to τ_+ is then equal to the integral along the entire oval,

$$u_k(z_\varepsilon)\bigg|_{\tau_-}^{\tau_+} = \oint_{\{f = z_\varepsilon\}} \omega_k.$$

From (26.9) we conclude that

$$\Delta(z_\varepsilon, \varepsilon) - z_\varepsilon = -\varepsilon^k u_k(z_\varepsilon)\bigg|_{\tau_-}^{\tau_+} + O(\varepsilon^{k+1}) = -\varepsilon^k \oint_{\{f = z_\varepsilon\}} \omega_k + O(\varepsilon^{k+1}).$$

As $\varepsilon \to 0$, the chart z_ε converges uniformly to the chart $z = f|_\tau$, and we obtain the assertion of the theorem. □

Remark 26.8. The functions h_k playing the key role in the inductive construction, can also be restored as integrals of appropriate forms. Indeed, they satisfy the equations $d\omega_k = dh_k \wedge df$ and hence can be restored as the primitives

$$h_k(\cdot) = \int^{\bullet} \frac{d\omega_k}{df} \tag{26.12}$$

along the level curves of f (the form $\frac{d\omega_k}{df}$ is the Gelfand–Leray derivative of ω; see §26**G** below). If h_k is any solution of the equation $d\omega_k = dh_k \wedge df$, then the form $\omega_k - h_k\,df$ is closed and, under the condition that periods of ω_k are all zero, is exact in U, since its integral over the oval $f = \text{const}$ generating the homology group of U, is zero.

The representation (26.12) allows us to write down all forms ω_k as *iterated integrals* along the level curves of f. The details can be found in [**Gav05**].

Note also that the above approach can be almost literally applied to a perturbation more general than (26.1),

$$df + \varepsilon\theta_1 + \varepsilon^2\theta_2 + \cdots = 0, \qquad \theta_1, \theta_2, \dots \in \Lambda^1(U). \tag{26.13}$$

The system (26.11) becomes nonhomogeneous but still retains the triangular form allowing for an explicit solution.

26C. Infinitesimal Hilbert's sixteenth problem. Proposition 26.1 and Theorems 26.3 and 26.7 indicate that in order to bound the number of limit cycles which appear by polynomial perturbation $\mathcal{F}_\varepsilon = \{\theta + \varepsilon\omega = 0\}$ of a polynomial *integrable* foliation $\mathcal{F}_0 = \{\theta = 0\}$ on $\mathbb{R}P^2$, it is necessary to estimate the number of zeros of integrals of the rational 1-form ω over the nonisolated ovals of the unperturbed foliation $\mathcal{F}_0$. The problem of finding an explicit upper bound for this number in terms of the degrees of θ and ω is referred to by numerous names: infinitesimal Hilbert problem, relaxed Hilbert problem, Hilbert–Arnold problem, tangential Hilbert problem, *etc.* In the above formulation the problem appeared between the lines in [**Ily69**] and since then repeatedly mentioned by Arnold in his seminar; see [**Arn04**].

However, when the initial foliation is defined by a *closed rational* 1-form θ, the first integral f can be *nonalgebraic* (cf. with §25**G**$_1$). Limit cycles can also be born from *separatrix polygons* of $\mathcal{F}_0$ rather than from ovals, in which case an additional analysis is required (Proposition 26.1 does not apply in this case). Finally, the problem which turns out to be transcendentally difficult, is to determine *how many identically zero Melnikov functions should be computed before one can guarantee that the perturbation in fact preserves the integrability.* Even in the most simple case where the foliation is by circles, $f(x, y) = x^2 + y^2$, so that all integrals I_k are in fact polynomial functions of z, this question is open (the so-called *Poincaré problem*). All these difficulties force us to concentrate on the first really nontrivial case of Abelian integrals which appear as follows.

Consider the important class of *Hamiltonian foliations* defined by *exact polynomial* form df with a real polynomial $f \in \mathbb{R}[x, y]$ of some degree $\deg f = n+1$. If the perturbation form ω in (26.1) is also polynomial, then we arrive at the following restricted formulation of a problem continuing the series of Hilbert-type problems from §24**A**.

Problem IX (infinitesimal version of Hilbert's sixteenth problem). *Find an upper bound for the number of* isolated *zeros of the integral* $I(z) = \oint_{\{f=z\}} \omega$ *of a polynomial* 1*-form* ω *over the algebraic ovals* $\{f = z\}$ *in terms of the degrees* $\deg df$ *and* $\deg \omega$.

Since all other settings are practically unexplored, we will refer to this restricted formulation as *the* infinitesimal Hilbert problem.

Definition 26.9. A (complete) Abelian integral is the integral $\oint_\delta \omega$ of a rational 1-form ω over an oval of an algebraic curve $\delta \subseteq \{f = 0\}$. This integral depends on the coefficients of the form $\omega \in \Lambda^1[x, y]$ and of the polynomial $f \in \mathbb{R}[x, y]$ as the parameters.

In most cases we will fix the form ω and all coefficients of f except for the free term, and consider the corresponding Abelian integral as the function of only one parameter,

$$I_{f,\omega}(z) = \oint_{\{f=z\}} \omega. \tag{26.14}$$

The infinitesimal Hilbert problem as it appears above, is the *problem on the maximal possible number of real isolated zeros of the Abelian integral* (26.14).

Note that the function $I_{f,\omega}(z)$ is in general *multivalued*, since the real level curve of f may consist of several ovals (besides noncompact components). However, any compact real oval $\delta \subseteq \{f = z_0\}$ can be continuously deformed to a uniquely defined compact oval on all sufficiently close level curves $\{f = z\}$, $z \in (\mathbb{R}, z_0)$. This allows us to define unambiguously *continuous real branches* of the Abelian integral (26.14). Simple arguments show that each continuous branch is real analytic in the interior of its domain (Problem 26.3).

26D. Relative cohomology and integrals: algebraic vs. analytic.

The global algebraic nature of the infinitesimal Hilbert problem justifies introduction of a special algebraic language of *relative* cohomology. This language is parallel to the *de Rham cohomology* which describes the difference between closed and exact differential forms; see [**War83**].

26**D**$_1$. *Relative de Rham complex and its cohomology.* The condition (26.5) can be interpreted in cohomological terms as follows. Consider the de Rham complex[6]

$$0 \longrightarrow \Lambda^0(U) \xrightarrow{d} \Lambda^1(U) \xrightarrow{d} \Lambda^2(U) \xrightarrow{d} 0 \tag{26.15}$$

[6] In general, an algebraic complex is a chain of modules $A_0 \xrightarrow{d} A_1 \xrightarrow{d} A_2 \xrightarrow{d} A_3 \cdots$ with a derivation d whose square $d \circ d$ is zero.

formed by the modules $\Lambda^k = \Lambda^k(U)$ of real analytic k-forms in the domain U and the exterior derivative d (we deal only with the 2-dimensional domain U, but all constructions can be instantly generalized for an arbitrary dimension). The exterior derivative d takes the submodule $df \wedge \Lambda^{k-1} \subseteq \Lambda^k$ into $df \wedge \Lambda^k \subseteq \Lambda^{k+1}$, which means that d descends to an operator (also denoted by d) between the *modules of relative k-forms*, the quotient modules $\Lambda^k_f(U) = \Lambda^k(U)/df \wedge \Lambda^{k-1}(U)$. Passing to the quotients transforms the de Rham complex (26.15) into the *relative de Rham complex*

$$0 \longrightarrow \Lambda^0(U) \xrightarrow{d} \Lambda^1_f(U) \xrightarrow{d} \Lambda^2_f(U) \xrightarrow{d} 0. \tag{26.16}$$

The cohomology of this complex, the quotients $\operatorname{Ker} d/\operatorname{Im} d$, is called the *relative cohomology* $H^k_f(U)$, $k = 0, 1, \dots, n$ (to make the term precise, one has to specify the ring of functions—polynomial, real analytic, smooth, *etc.*; see below).

The zero relative cohomology module $H^0_f(U) = \{g \in \mathcal{O}(U) : dg = h\,df\}$ can be identified with functions constant along the level curves of f. If f has no critical points in U, *any* 2-form is divisible by df. To prove that, it is sufficient to construct just one area form (nonvanishing 2-form) divisible by df; any other 2-form will then be proportional to it hence also divisible by df. If $\theta \colon U \to \mathbb{R} \bmod 2\pi$ is the cyclic variable ("polar angle") along the ovals $f = \text{const}$, then the required area form is $df \wedge d\theta$. Thus in the absence of critical points of f, $\Lambda^2_f(U) = 0$ and hence $H^2_f(U) = 0$.

The only dimension when the relative cohomology is nontrivial, is 1. The definition of relative exactness was given earlier (Definition 26.4). On the other hand, since $\Lambda^2_f(U) = 0$, any 1-form is relatively closed. In terms of the relative cohomology, Lemma 26.5 asserts that the *period map*

$$H^1_f(U) \to H^0_f(U), \qquad \alpha \mapsto g(x) = \oint_{\gamma \ni x} \alpha$$

(the integral is taken over the oval $f = \text{const}$ passing through the point x), is an isomorphism.

Remark 26.10. The notion of relative cohomology can be defined for any *closed* (not necessarily exact) form $\theta \in \Lambda^1$ as the cohomology of the complex $\Lambda^k_\theta(U) = \Lambda^k(U)/\theta \wedge \Lambda^{k-1}(U)$. However, in this case the analysis is considerably more subtle; see [**BC93**].

The construction of relative cohomology depends on the base ring, from which the function f and the coefficients of the form α are taken. Thus far we were dealing with real analytic forms and functions in an annulus. However, when dealing with Abelian integrals, it is natural to assume that the base ring is $\mathbb{R}[x, y]$ or $\mathbb{C}[x, y]$, and all forms are also polynomial.

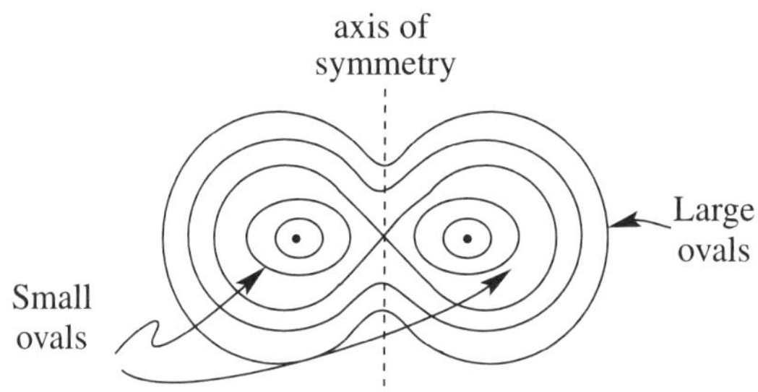

Figure V.1. Three continuous families of ovals

The straightforward generalization of Lemma 26.5 for *polynomial* rather than real analytic 1-forms fails. The integral of a polynomial 1-form ω over a continuous family of ovals on the level curves of a polynomial f may vanish identically, yet the form ω may not admit the representation (26.4) with *polynomial* g and h.

Example 26.11. Consider the symmetric polynomial $f(x,y) = y^2 - x^2 + x^4$. The real level curves $\{f = z\}$ are empty for $z < -\frac{1}{2}$, carry two "small" ovals for $z \in (-\frac{1}{2}, 0)$ and only one "large" oval for $z > 0$; see Fig. V.1. The large ovals are all symmetric with respect to both axes, while the "small" ovals are symmetric only in the y-axis.

Integral of the 1-form $\omega = x^2\,dy$ over the family of "large" ovals vanishes identically, since the integrals over the two parts in the half-planes $\{x > 0\}$ and $\{x < 0\}$ mutually cancel each other.

Yet the form ω cannot be represented under the form (26.4) with *polynomial* g and h; if this were the case, then the integral of ω over any of the two families of "small" cycles would also vanish identically. Yet $d\omega = -2x\,dx \wedge dy$ keeps constant sign in each half-plane $\{\pm x > 0\}$, hence the integral over each of these two families is nonzero.

The explanation of this phenomenon lies in the fact that after analytic continuation Abelian integrals become multivalued functions ramified over a certain finite set. One branch can be identical zero, while others not. However, if *all complex* branches vanish identically, the validity of the assertion is restored, as asserts Theorem 26.13 below.

26$\mathbf{D}_2$. *Analytic relative cohomology.* To achieve a proper complexification of Abelian integrals, consider the real polynomial $f \in \mathbb{C}[x,y]$ as a complex function $f\colon \mathbb{C}^2 \to \mathbb{C}$, and denote by $L_z = f^{-1}(z) \subset \mathbb{C}^2$ the *complex affine* level curves. These curves taken together form the (singular holomorphic) Hamiltonian foliation $\mathcal{F} = \{df = 0\}$ on $\mathbb{P}^2$.

The projective compactification of the leaves $\overline{L_z} \in \mathbb{P}^2$ may be singular or not; singular curves correspond to isolated values of z from some *critical locus* $\Sigma \subset \mathbb{C}$.

Any polynomial 1-form ω restricted on a nonsingular curve L_z is a closed form (for reasons of dimension) with the poles at infinity. The restriction $\omega|_{L_z} \in \Lambda^1(L_z)$ on L_z is exact, if and only if the integral of ω along any cycle δ on L_z is zero.

Definition 26.12. A rational 1-form is *analytically relatively exact* with respect to an integrable (Hamiltonian) foliation $\mathcal{F} = \{df = 0\}$ on $\mathbb{P}^2$ with $f \in \mathbb{C}[x, y]$, if the integral of ω along any cycle $\delta \in H_1(L, \mathbb{Z})$ on any leaf $L \in \mathcal{F}$ is zero,

$$\forall L \in \mathcal{F},\ \forall \delta \in H_1(L, \mathbb{Z}), \quad \oint_\delta \alpha = 0. \tag{26.17}$$

Clearly, any form that is *algebraically relatively exact*, i.e., representable under the form

$$\omega = h\,df + dg, \qquad h, g \in \mathbb{C}[x, y], \tag{26.18}$$

(cf. with Definition 26.4), is also analytically relatively exact with respect to the Hamiltonian foliation $\mathcal{F} = \{df = 0\}$. The inverse statement, a genuine algebraic counterpart of Lemma 26.5, is the following theorem.

Theorem 26.13 (Yu. Ilyashenko [**Ily69**], L. Gavrilov [**Gav98**]). *Assume that the polynomial $f \in \mathbb{C}[x, y]$ satisfies the following two conditions:*

(1) *all affine level curves $L_z = f^{-1}(z) \subset \mathbb{C}^2$ are connected, and*
(2) *all critical points of f in $\mathbb{C}^2$ are isolated.*

Then a polynomial 1-form ω is algebraically relatively exact with respect to the Hamiltonian foliation $\mathcal{F} = \{f = 0\}$ if and only if it is analytically relatively exact with respect to it.

In other words, for Hamiltonian foliations satisfying the assumptions of the theorem, any 1-form with zero periods on all leaves is representable under the form (26.18).

Proof. In one direction the theorem is trivial. To prove the other direction, denote by n the degree of f. Without loss of generality we may assume that the polynomial f restricted to the y-axis $Y = \{x = 0\}$ has the same degree n (this can always be achieved by a suitable affine change of variables x, y).

Each level curve L_z intersects Y by the same number of points $p_1(z), \dots, p_n(z)$ (every point is counted as many times as the multiplicity of the root of $f(0, y) - z$). For a point $(x, y) \in \mathbb{C}^2$ let $\gamma_j(x, y)$ be a path connecting this point with the jth point $p_j(z)$ on the intersection of the level curve L_z, $z = f(x, y)$, passing through it, with Y. Existence of such paths follows from the first assumption of the theorem.

The paths $\gamma_1(x, y), \dots, \gamma_n(x, y)$ are defined only modulo elements from $H_1(L_z, \mathbb{Z})$, but if the restriction $\omega|_{L_z}$ is exact, then the integrals $\int_{\gamma_j(x,y)} \omega$ are uniquely defined. The function

$$g(x, y) = \frac{1}{n} \sum_{j=1}^{n} \int_{\gamma_j(x,y)} \omega \tag{26.19}$$

is correctly defined on the complement to the union of critical level curves $S = f^{-1}(\Sigma) \subset \mathbb{C}^2$, since it does not depend on the (noninvariant) way of enumeration of the points $p_j(z)$ and the freedom in the choice of the paths $\gamma_j(x, y)$. Moreover, it is holomorphic on $\mathbb{C}^2 \smallsetminus S'$, where S' is the union of singular level curves S and the level curves tangent to the axis Y. Indeed, in this case one can choose the paths $\gamma_j(x, y)$ analytically depending on (x, y), say, as lifts on the level curves of some paths on the x-plane, connecting the origin $x = 0$ with the variable point x while avoiding the critical points of the projection $(x, y) \mapsto x$ restricted on L_z.

The function $g(x, y)$ remains bounded near the algebraic set $S' \subset \mathbb{C}^2$ of codimension 1; therefore, it extends analytically on S' as an entire function. Moreover, as (x, y) tends to infinity, both the length of the paths and the integrand in (26.19) grow at most polynomially in $|x| + |y|$, therefore the averaged primitive given by this expression, is a *polynomial*: $g \in \mathbb{C}[x, y]$.

The polynomial 1-form $\omega - dg$ by construction vanishes on all vectors tangent to the level curves. Since the form df does the same, we conclude that $\omega - dg = h\,df$, where h is a meromorphic function on $\mathbb{C}^2$ defined on the complement to the set of critical points where df vanishes. If this set is zero-dimensional (of codimension 2 in $\mathbb{C}^2$), as follows from the second assumption of the theorem, then h necessarily extends analytically to these points and hence is a *polynomial*, $h \in \mathbb{C}[x, y]$. The required representation is constructed. □

26**D**$_3$. *Bonnet theory.* Assumptions of Theorem 26.13 are rather nonrestrictive: the first is guaranteed automatically, if the Hamiltonian foliation $\mathcal{F}$ has only simple (in some sense) singularities on the infinite line, while the second assumption holds true for any square-free polynomial. If one of these assumptions is violated, the analytic relative exactness may not imply the algebraic one. Theorem 26.13 is in fact the particular case of a more general assertion concerning the relative cohomology.

Definition 26.14. The *Bonnet set* $\operatorname{Bs}(f)$ of a polynomial $f \in \mathbb{C}[x, y]$ is the set of values z such that the affine level curve $L_z = f^{-1}(z) \subset \mathbb{C}^2$ is either nonconnected or carries a nonisolated critical point of f.

In the assumptions of Theorem 26.13, the Bonnet set is empty.

Theorem 26.15 (P. Bonnet, 1999 [**Bon99, BD00**]). *Assume that the Bonnet set of a polynomial f is finite.*

Then for any polynomial 1-form which is exact on each level curve L_z, there exist a pair of polynomials $g, h \in \mathbb{C}[x, y]$ and a polynomial $b \in \mathbb{C}[z]$, nonvanishing outside the Bonnet set $\mathrm{Bs}(f)$, *such that*

$$(b \circ f)\,\omega = h\,df + dg. \tag{26.20}$$

Proof. For any $z \in \mathbb{C} \smallsetminus \mathrm{Bs}(f)$ the form $\omega|_{L_z}$ is exact. Since L_z is assumed connected, there exist a polynomial $g_z(x, y)$ such that $\omega - dg_z$ vanishes on all vectors tangent to L_z. This means that

$$\omega - dg_z = (f - z)\xi_z + u\,d(f - z) = (f - z)\theta_z, \qquad \theta_z \in \Lambda^1[x, y]$$

(we integrated by parts). The polynomial forms θ_z are defined for all values $z \in \mathbb{C} \smallsetminus \Sigma$ (uncountably many of them), whereas the number of different monomial 1-forms is countable. Therefore there must exist a linear dependence between the forms θ_z, involving only finitely many of them:

$$\sum_{j=1}^{m} \lambda_j \theta_{z_j} = 0, \qquad \text{for some } z_1, \dots, z_m \in \mathbb{C} \smallsetminus \Sigma. \tag{26.21}$$

The identity (26.21) can be rewritten as follows:

$$\omega \cdot \sum_{j=1}^{m} \frac{\lambda_j}{f - z_j} = \sum_{j=1}^{m} \frac{\lambda_j\, dg_j}{f - z_j},$$

or, after getting rid of the denominators,

$$B_0(f)\,\omega = \sum_j B_j(f)\,dg_j, \qquad B_0, B_1, \dots, B_m \in \mathbb{C}[z],$$

with the appropriate polynomials $B_0, \dots, B_m \in \mathbb{C}[z]$. Integrating by parts the right hand side, we obtain the required identity,

$$B_0(f)\,\omega = dg + h\,df, \qquad g = \sum_{j=1}^{m} B_j(f) g_j, \quad h = -\sum_{j=1}^{m} g_j \frac{dB_j}{dz}(f).$$

Apriori, the roots of B_0 may be arbitrary. We will show now that all of them except for those from the Bonnet set, can be eliminated by an appropriate division. Indeed, assume that $z \notin \mathrm{Bs}(f)$ and consider the representation

$$(f - z)\alpha = h\,df + dg, \qquad h, g \subset \mathbb{C}[x, y] \tag{26.22}$$

for an arbitrary polynomial form $\alpha \in \Lambda^1[x, y]$. This representation implies that the polynomial g is locally constant along L_z, as its differential vanishes on the tangent vector to L_z at any smooth point of the latter. Since $z \notin \mathrm{Bs}(f)$, L_z is connected and therefore g is globally constant on L_z; without loss of generality we may assume that $g|_{L_z} = 0$. But the primary decomposition of $f - z$ in $\mathbb{C}[x, y]$ contains no multiple factors (otherwise L_z would carry nonisolated critical points of f). Therefore vanishing of g on L_z implies that g is divisible by $f - z$, $g = (f - z) \cdot g'$. Substituting this into (26.22), we conclude that

$$(f - z)\alpha = (f - z)dg' + [h + g']\,df. \tag{26.23}$$

Since df is nonvanishing at all noncritical points of L_z, the coefficient $h + g'$ must vanish on L_z and, hence, as before, it must be divisible by $f - z$: $h + g' = (f - z)h'$. Substituting it into (26.23), we see that the factor $f - z$ can be cancelled out from all three terms of it, yielding a new representation $\alpha = dg' + h'\,df$ with $g', h' \in \mathbb{C}[x, y]$.

Applying inductively this division procedure to the form $\alpha = B_0(f)\,\omega$, we eliminate all roots of the polynomial B_0 except for those that belong to the Bonnet set. At the end we arrive at the representation (26.20) with the product $b(x, y) = \prod_{\zeta_k \in \mathrm{Bs}(f)} (f(x, y) - \zeta_k)^{m_k}$ in the left hand side. □

26D$_4$. *Polynomials transversal to infinity.* The condition of connectedness of affine level curves $L_z = f^{-1}(z)$ is intimately related to the behavior of the polynomial f "at infinity". There is a simple sufficient condition on the principal homogeneous terms of f, guaranteeing certain regularity of f at infinity, in particular, entailing that all curves L_z are connected. This condition will repeatedly appear in the future.

Definition 26.16. A polynomial $f \in \mathbb{C}[x, y]$ of degree $n + 1 \geqslant 2$ is called *transversal to infinity*, if one of the two equivalent conditions holds:

(1) its principal homogeneous part factors out as the product of $n + 1$ pairwise different linear forms;

(2) its principal homogeneous part has an isolated critical point of multiplicity n^2 at the origin.

The term "transversality" is explained by the following proposition, which is proved by an elementary computation in the affine chart covering the infinite line.

Proposition 26.17. *If f satisfies any of the two above conditions, then the projective compactification $\overline{L_z}$ of any level curve $L_z = f^{-1}(z)$ intersects transversally the infinite line $\mathbb{I} \subset \mathbb{P}^2$.* □

Corollary 26.18. *If f is transversal to infinity, then all affine level curves $L_z = f^{-1}(z) \subset \mathbb{C}^2$ are connected.*

Proof. Consider the irreducible decomposition of $\overline{L_z} = \bigsqcup_j C_j$. Any irreducible component $C_j \in \mathbb{P}^2$ is always connected, and any two irreducible components in $\mathbb{P}^2$ necessarily intersect (by the number of points equal to the product of their degrees, if counted with multiplicities; see [**Mum76**, §5B]). The intersection points of different components are necessarily singular and hence cannot lie on the infinite line by Proposition 26.17. Thus any two components intersect somewhere at the finite (affine) part $\mathbb{C}^2 \subset \mathbb{P}^2$, which means that the affine level curves curves are all connected. □

26E. Brieskorn lattice and Petrov modules. Theorem 26.13 allows us to describe algebraically the space of Abelian integrals as multivalued functions. They constitute a module over the ring $\mathbb{C}[z]$. The basis of this module will be computed in this section.

Let $f \in \mathbb{C}[x, y]$ be a polynomial.

Definition 26.19. The *Brieskorn lattice* is the quotient space

$$\mathbf{B}_f = \frac{\Lambda^2}{df \wedge d\Lambda^0}, \qquad \Lambda^{1,2} = \Lambda^{1,2}[x, y]. \tag{26.24}$$

The *Petrov module* is the quotient space

$$\mathbf{P}_f = \frac{\Lambda^1}{df\cdot\Lambda^0 + d\Lambda^0}, \qquad \Lambda^0 \cong \mathbb{C}[x,y] \tag{26.25}$$

of all polynomial 1-forms modulo the subspace of algebraically relatively exact forms. In the assumptions of Theorem 26.13, the Petrov module can be identified with the space of Abelian integrals.

Both $\mathbf{B}_f$ and $\mathbf{P}_f$ can be considered as $\mathbb{C}[f]$-modules: the generator f of the ring $\mathbb{C}[f]$ acts on equivalence classes of forms as the multiplication by the polynomial $f(x,y)\in\mathbb{C}[x,y]$. Definition of this action is correct, since

$$f\,df\wedge dg = df\wedge d(fg), \qquad f\cdot(h\,df+dg) = (fh-g)\,df + d(fg).$$

The exterior derivative d is a linear *bijection* $d\colon \mathbf{P}_f\to\mathbf{B}_f$ but *not* a $\mathbb{C}[f]$-module homomorphism.

In this subsection we explicitly construct the bases for these modules (actually, we will be mostly interested in $\mathbf{P}_f$)for polynomials f transversal to infinity.

Let f be a polynomial of degree $n+1$ transversal to infinity, with the principal homogeneous part f_{n+1}. The quotient space

$$Q_{df_{n+1}} = \frac{\Lambda^2}{df_{n+1}\wedge\Lambda^1} \cong \frac{\mathbb{C}[x,y]}{\left\langle\frac{\partial f_{n+1}}{\partial x},\frac{\partial f}{\partial y}\right\rangle} \tag{26.26}$$

is a finite-dimensional complex algebra (cf. with Definition 8.22).

The (complex) dimension of the quotient (26.26) is equal to n^2. Indeed, both partial derivatives $\partial f_{n+1}/\partial x$ and $\partial f_{n+1}/\partial y$ are homogeneous polynomials and factor as products of exactly n *linear* forms each. No linear factor of $\partial f/\partial x$ can occur in $\partial f/\partial y$, otherwise the singularity of f_{n+1} will be non-isolated. By Proposition 8.25 the dimension of the quotient algebra is n^2.

Let $\omega_1,\dots,\omega_m$, $m=n^2$, be any homogeneous monomial 1-forms whose differentials $d\omega_1,\dots,d\omega_m$ generate the basis of $Q_{df_{n+1}}$. We will show that these differentials also generate the full quotient algebra $Q_{df}=\Lambda^2/df\wedge\Lambda^1$ over $\mathbb{C}$ and the Brieskorn lattice $\mathbf{B}_f$ over $\mathbb{C}[f]$, while the forms themselves generate the Petrov module $\mathbf{P}_f$. This result has several different demonstrations (see, e.g., [**KP06**] which treats the multidimensional case as well). We follow the exposition in [**Yak02**] which has an advantage of being purely algebraic and effective.

Proposition 26.20. *Assume that the polynomial $f=f_{n+1}+\cdots$ is transversal to infinity.*

Then any collection of 2-forms $d\omega_j$, $j=1,\dots,n^2$, which generates the quotient $Q_{df_{n+1}}=\Lambda^2/df_{n+1}\wedge\Lambda^1$ over $\mathbb{C}$, generates also the quotient $Q_{df}=\Lambda^2/df\wedge\Lambda^1$.

Proof. By the choice of the forms $d\omega_j$, any 2-form $\mu \in \Lambda^2$ can be "divided with remainder" by df_{n+1} as follows:

$$\mu = \sum_1^m c_j \, d\omega_j + df_{n+1} \wedge \eta, \qquad \eta \in \Lambda^1$$

with the "incomplete ratio" η. Substitute in this equality $f_{n+1} = f - \varphi$, where φ is the collection of all nonprincipal terms of f, $\deg \varphi \leqslant n$. Then $\mu = \sum_1^m c_j \, d\omega_j + df \wedge \eta - \mu'$, where $\mu' = \varphi \wedge \eta$ is a 2-form of degree *strictly smaller* than $\deg \mu$. The division process can therefore be continued inductively until the "incomplete ratio" disappears. □

Theorem 26.21. *Let $f = f_{n+1} + \cdots \in \mathbb{C}[x, y]$ be a polynomial of degree $n+1$ transversal to infinity and $d\omega_1, \dots, d\omega_m$, a collection of $m = n^2$ monomial 2-forms generating the quotient algebra $Q_{df_{n+1}} = \Lambda^2 / df_{n+1} \wedge \Lambda^1$.*

Then any 1-form $\omega \in \Lambda^1$ can be represented as follows:

$$\omega = \sum_{j=1}^m (p_j \circ f) \, \omega_j + h \, df + dg \tag{26.27}$$

with some polynomials $g, h \in \mathbb{C}[x, y]$ and the univariate polynomials $p_1, \dots, p_m \in \mathbb{C}[z]$. The degrees of the coefficients p_j satisfy the equalities

$$(n+1) \deg_z p_j + \deg \omega_j = \deg \omega, \qquad j = 1, \dots, m. \tag{26.28}$$

Corollary 26.22. *In the assumptions of the Theorem 26.21, the forms $\omega_1, \dots, \omega_m$ generate the Petrov module $\mathbf{P}_f$ over $\mathbb{C}[f]$.* □

We begin the proof with the following lemma which may be considered as an analog of the Euler identity in the Brieskorn lattice. Let f_{n+1} be an arbitrary *homogeneous* polynomial of degree $n + 1$.

Lemma 26.23. *Any polynomial 2-form divisible by df_{n+1}, has a primitive divisible by f_{n+1} in the Brieskorn lattice $\mathbf{B}_{f_{n+1}}$.*

In other words, for any 1-form $\eta \in \Lambda^1$ there exists $\omega \in \Lambda^1$ and such that

$$df_{n+1} \wedge \eta = d(f_{n+1} \, \omega) \mod df_{n+1} \wedge d\Lambda^0. \tag{26.29}$$

Proof of the lemma. By the Euler identity, we have $f_{n+1} = \frac{1}{n+1} i_V \, df_{n+1}$, where $V = x \frac{\partial}{\partial x} + y \frac{\partial}{\partial y}$ is the Euler field and $i_V \colon \Lambda^k \to \Lambda^{k-1}$ the antiderivation substituting V as the first argument of a differential k-form.

Since any 3-form on the 2-plane vanishes, $df_{n+1} \wedge \mu = 0$ for any 2-form μ. Applying the antiderivation i_V, we conclude from this and the Euler formula above, that for any 2-form μ,

$$\begin{aligned} 0 = i_V(df_{n+1} \wedge \mu) &= (i_V \, df_{n+1}) \wedge \mu - df_{n+1} \wedge (i_V \mu) \\ &= (n+1) f_{n+1} \mu - df_{n+1} \wedge (i_V \mu). \end{aligned} \tag{26.30}$$

Using (26.30), the equation (26.29) with respect to the unknown 1-form ω can be transformed as follows,

$$df_{n+1} \wedge \eta = df_{n+1} \wedge \omega + \tfrac{1}{n+1}\, df_{n+1} \wedge (i_V\, d\omega) + df_{n+1} \wedge d\xi.$$

The latter is an equation to be solved now with respect to ω and ξ. It will be obviously satisfied if

$$\eta = \tfrac{1}{n+1}\, i_V\, d\omega + \omega + d\xi,$$

or, after applying the derivation d to both sides,

$$d\eta = \tfrac{1}{n+1}\, d(i_V \mu) + \mu, \qquad \mu = d\omega,$$

(the derivation results in an equivalent condition since any closed polynomial form on $\mathbb{C}^2$ is exact).

To show that the last equation is always solvable with respect to μ for any 2-form $d\eta$, we transform it for the last time using the homotopy formula $L_V = di_V + i_V d$ [**Arn97**] and the fact that $d\mu = 0$ (as a 3-form on the 2-plane). Finally the equation (26.29) is reduced to the equation

$$d\eta = \left(\tfrac{1}{n+1}\, L_V + \mathrm{id}\right)\mu, \qquad \mu = d\omega, \tag{26.31}$$

where L_V is the Lie derivative acting on 2-forms. Since V is the Euler field, each monomial 2-form $x^p y^q\, dx \wedge dy$ is an eigenvector for L_V with the positive eigenvalue $p + q + 2$. Thus $\frac{1}{n+1} L_V + \mathrm{id}$ is a diagonalizable operator with positive eigenvalues on the space of polynomial forms of any degree. Such an operator is invertible, which yields a solution to (26.31) and ultimately to (26.29). □

Remark 26.24. A similar argument shows that a 2-form divisible by df_{n+1} is also divisible by f_{n+1} in the Brieskorn lattice (i.e., modulo $df_{n+1} \wedge d\Lambda^0$); see [**Yak02**].

Proof of Theorem 26.21. The proof imitates demonstration of Proposition 26.20. By assumption, the forms $d\omega_1, \dots, d\omega_m$ form a basis of the quotient algebra $Q_{df_{n+1}}$ associated with the principal homogeneous part f_{n+1} of the polynomial f. This means that the 2-form $d\omega \in \Lambda^2[\mathbb{C}^2]$ can be uniquely represented as

$$d\omega = \sum_{j=1}^{m} c_j\, d\omega_j + df_{n+1} \wedge \eta, \tag{26.32}$$

where $\eta \in \Lambda^1[\mathbb{C}^2]$ is a polynomial 1-form.

By Lemma 26.23, the "incomplete ratio" $df_{n+1} \wedge \eta$ can be rewritten as $d(f_{n+1}\omega') + df_{n+1} \wedge dg$ for some polynomial $g \in \Lambda^0$. Passing to the primitives, we conclude that

$$\omega - \sum_{j=1}^{m} c_j \omega_j = f_{n+1}\omega' - g\, df_{n+1} + dh.$$

Substitute f_{n+1} by $f - \varphi$, where $\varphi \in \mathbb{C}[x, y]$ is the collection of all nonprincipal terms of f, $\deg \varphi \leqslant n$. After collecting terms, we obtain the equality $\omega - \sum_{j=1}^{m} c_j\omega_j = f\omega' - g\, df + dh - \omega''$, $\omega'' = \varphi\omega' - g\, d\varphi$. In other words, we have

$$\omega - \sum c_j\omega_j = f\omega' - \omega''$$

in the $\mathbb{C}[f]$-module $\mathbf{P}_f$; cf. with (26.25). The degrees of both 1-forms ω', ω'' are strictly less than that of ω, therefore the process can be continued by induction, resulting at the end in the representation (26.28).

The assertion on the degrees follows directly from inspection of this division algorithm. □

Remark 26.25. A similar argument shows that any 2-form $\mu \in \Lambda^2[\mathbb{C}^2]$ can be represented as the sum $\sum_{j=1}^{m} (q_j \circ f)\, d\omega_j \bmod df \wedge d\Lambda^0$. Moreover, both assertions admit natural generalizations for polynomials $f \in \mathbb{C}[x_1, \dots, x_k]$, $k \geqslant 2$, whose principal *quasi*homogeneous part has an isolated critical point at the origin. Details can be found in [**Yak02**].

26F. Polynomials as topological bundles. In this section we study the analytic continuation of Abelian integrals as multivalued functions of one complex variable. The ramification locus of any Abelian integral and its monodromy are completely determined by the Hamiltonian f. In what follows we denote by Σ the set of *critical values* of f,

$$\Sigma = \bigl\{z \in \mathbb{C}\colon \exists (x, y) \in \mathbb{C}^2,\ \tfrac{\partial f}{\partial x} = \tfrac{\partial f}{\partial y} = f - z = 0 \text{ at } (x, y)\bigr\}. \tag{26.33}$$

Theorem 26.26. *If a polynomial $f \in \mathbb{C}[x, y]$ is transversal to infinity, then the map $f\colon \mathbb{C}^2 \to \mathbb{C}$ defines a topological bundle over the set of all noncritical values $\mathbb{C} \smallsetminus \Sigma$.*

In other words, for any $a \in \mathbb{C} \smallsetminus \Sigma$ there exists a small neighborhood $U \ni a$ in $\mathbb{C} \smallsetminus \Sigma$ such that the full preimage $f^{-1}(U)$ is homeomorphic to the Cartesian product $f^{-1}(a) \times U$ and f restricted on this preimage is topologically conjugate by this homeomorphism to the projection of $f^{-1}(a) \times U$ on the second term.

This result obviously follows from the general Theorem 26.27 below. Consider the complex *affine* space P of all bivariate polynomials of degree $n + 1$ with the fixed principal square-free homogeneous part f_{n+1} (all such polynomials are by definition transversal to infinity). The dimension of this space is $r = (n+2)(n+3)/2$, $P \cong \mathbb{C}^r$, and it can be identified with the space of nonprincipal coefficients λ_{ij} in the expansion

$$\Phi(\lambda; x, y) = f_{n+1}(x, y) + \sum_{0 \leqslant i+j \leqslant n} \lambda_{ij} x^i y^j. \tag{26.34}$$

In the product space $P \times \mathbb{C}^2$ fibered over P consider the algebraic hypersurface $X = \{\Phi(\lambda, x, y) = 0\}$ and its fiberwise compactification $\overline{X} \subset P \times \mathbb{P}^2$. Denote by $\pi \colon \overline{X} \to P$ the natural projection $P \times \mathbb{P}^2 \to P$ restricted on the surface $\overline{X}$. The preimages $\pi^{-1}(\lambda) \subset \{\lambda\} \times \mathbb{P}^2$ are projectively compactified zero level curves of the polynomial $\Phi_\lambda(x, y) = \Phi(\lambda, x, y) \in \mathbb{C}[x, y]$. The preimages by the projection $\pi \colon \overline{X} \to P$ run over all level curves of all polynomials with the fixed principal homogeneous part.

Let $\boldsymbol{\Sigma} \subset P$ be the set of all parameters λ for which the affine curve $\{\Phi_\lambda(x, y) = 0\} \subset \mathbb{C}^2$ is singular (nonsmooth). By definition (compare with (26.33)),

$$\boldsymbol{\Sigma} = \bigl\{\lambda \in P \colon \exists (x, y) \in \mathbb{C}^2 \colon \tfrac{\partial \Phi}{\partial x} = \tfrac{\partial \Phi}{\partial y} = \Phi = 0 \text{ at } (\lambda, x, y)\bigr\}. \tag{26.35}$$

Theorem 26.27. *If the principal homogeneous part f_{n+1} is square-free, then the projection $\pi \colon \overline{X} \to P$ and its restriction on the affine part $X \subset \overline{X}$ are topologically locally trivial bundles over the complement $P \smallsetminus \boldsymbol{\Sigma}$.*

Proof. 1. For any point $a = (\lambda, x, y) \in \overline{X}$ over $\lambda \notin \boldsymbol{\Sigma}$ the complex tangent space $\mathbf{T}_a X$ at this point projects surjectively onto the tangent space $\mathbf{T}_\lambda P \cong P \cong \mathbb{C}^r$ at the point $\lambda = \pi(a)$.

For points in the affine part $X \subset \overline{X}$ this follows from the fact that one of the partial derivatives $\partial\Phi/\partial x$ or $\partial\Phi/\partial y$ is nonvanishing at $a \in X$ by the assumption $\lambda \notin \boldsymbol{\Sigma}$.

For points at infinity the above surjectivity assertion holds regardless of the choice of λ, since in the suitable coordinates $v = 1/x$, $u = y/x$ the equation of $\overline{X}$ takes the form $\{\Psi = 0\}$, where $\Psi(\lambda; u, v) = f_{n+1}(1, u) + v g_1(\lambda; u, v)$. Since $f_{n+1}(1, u)$ has $n+1$ distinct *simple* roots, the derivative $\partial\Psi/\partial u$ does not vanish on the infinite line $v = 0$ for all λ.

2. Let $F_1, \dots, F_{2r}$ be commuting vector fields on the base P, spanning the tangent bundle $\mathbf{T}P$ (e.g., the fields $\partial/\partial\lambda_{ij}$ and $\sqrt{-1}\,\partial/\partial\lambda_{ij}$ for all i, j). The above surjectivity means that the preimage $\pi^{-1}(U) \subset U \times \mathbb{P}^2$, where $U \subset P \smallsetminus \boldsymbol{\Sigma}$ is a sufficiently small open set disjoint with $\boldsymbol{\Sigma}$, can be covered by a union of open sets $U_\alpha \subset \overline{X}$ such that in each neighborhood there exist $2r$ real analytic vector fields $\overline{F}_{k,\alpha}$ tangent to $\overline{X}$ and π-related with F_k for all k; see Fig. V.2. Moreover, the fields $\overline{F}_{k,\alpha}$ can be assumed tangent to the intersection of $\overline{X}$ with the infinite line in each fiber. Indeed, since $\partial\Psi/\partial u \neq 0$, the v-component of such a lift (in the chart $v = 1/x$, $u = y/x$ as above) can be chosen arbitrarily, in particular, zero.

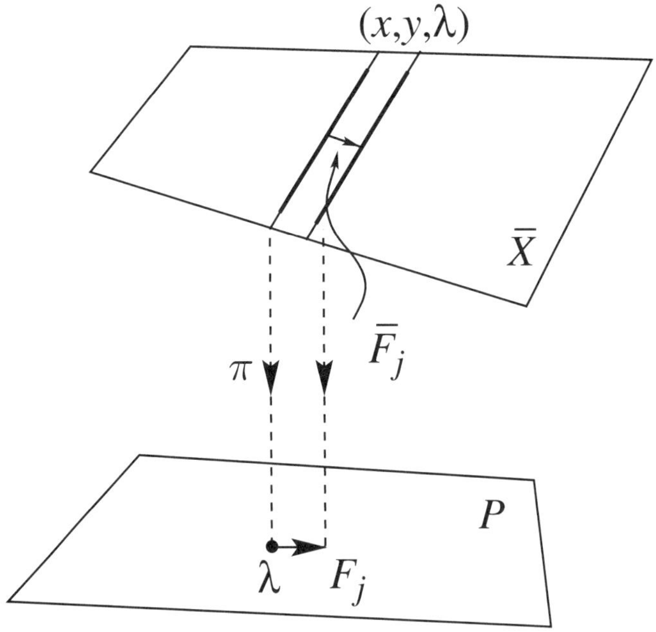

Figure V.2. Topological trivialization of the map

3. Since π is proper (all preimages of points are compact projective curves), one may assume that the covering is finite. Let $\{\psi_\alpha \geqslant 0\}$ be a partition of unity subordinated to the covering U_α. Then the vector fields $\overline{F}_k = \sum_\alpha \psi_\alpha \overline{F}_{k,\alpha}$ are also π-related with F_k and tangent to $\overline{X}$. Since F_k commute, the commutators $[\overline{F}_k, \overline{F}_j]$ are tangent to the fibers $\pi^{-1}(\lambda)$. By a standard modification, one can make the fields $\overline{F}_k$ also commuting in $\mathbf{T}\overline{X}$ by adding appropriate vector fields tangent to the fibers; see [**War83**].

4. Shifts along the commuting vector fields $\overline{F}_k$ realize homeomorphisms between all fibers $\pi^{-1}(\lambda)$, $\lambda \in U$, and trivialize the map $\pi\colon \overline{X} \to P$. $\square$

Corollary 26.28. *In the assumptions of Theorem 26.26, any cycle $\delta(z_*) \in H_1(L_{z_*}, \mathbb{Z})$ can be continued along any path $\gamma\colon [0,1] \to \mathbb{C} \smallsetminus \Sigma$, $\gamma(0) = z_*$, avoiding critical values of f. The result is defined uniquely as a cycle in the homology group and does not change when γ is replaced by a homotopy equivalent path.* $\square$

This corollary allows us to consider the effect of continuous deformations of affine level curves L_z as z goes along closed loops in the z-plane avoiding the critical set Σ. With any such closed loop γ (beginning and ending at a certain fixed regular base point $z_* \notin \Sigma$), the *topological monodromy operator*

$$\Delta_\gamma\colon H_1(L_{z_*}, \mathbb{Z}) \to H_1(L_{z_*}, \mathbb{Z}), \tag{26.36}$$

can be associated: together these operators constitute the representation of the fundamental group $\pi_1(\mathbb{C} \smallsetminus \Sigma, t_*)$ (as usual, the choice of the base point is not important). We will show that in general the operators Δ_γ are *nontrivial*: after continuation along closed paths the cycles on the level curves in general do not return to their initial positions.

We conclude this analysis with the following basic result.

Theorem 26.29 (analytic continuation of Abelian integrals). *If f is a polynomial transversal to infinity, then for any polynomial 1-form ω and any cycle $\delta \in H_1(L_{z_*}, \mathbb{Z})$, the Abelian integral $\oint_\delta \omega$ can be extended as an analytic multivalued function along any path avoiding the critical values of f.*

Proof. The restriction of any polynomial 1-form on L_z is closed (holomorphic), hence its integral is the same for all homotopic loops. By Theorem 26.26, for any fixed $z_* \notin \Sigma$ one can choose a representative of the cycle $\delta(z)$ continuously depending on z in a sufficiently small neighborhood of z_* in such a way that its projection, say, on the x-plane parallel to the y-axis is the same curve denoted by D. Then one can choose an analytic branch $y = y(x, z)$ of solution of the equation $f(x, y) = z$ over D. The integral of any form along $\delta(z)$ can be reduced to an integral over D of an analytic 1-form depending analytically on z. The rest follows from the standard theorem on (complex) differentiability of integrals depending on parameters. □

Remark 26.30. Without some assumptions "on infinity" the assertion of Theorem 26.26 and all its corollaries fails. In general there may exist regular values of f such that the preimages $f^{-1}(z)$ change their topological type at these points, exhibiting singularities on the infinite line $\mathbb{I}$. Such values are called *atypical values* of the polynomial f, but they also always form a finite subset of $\mathbb{C}$ (empty when f is transversal to infinity).

Theorem 26.31. *If f is a polynomial of degree $n+1$ transversal to infinity, then the rank of the first homology of any nonsingular fiber is equal to n^2.*

Proof. Consider first the case where the polynomial f is homogeneous and coincides with its principal part f_{n+1}. For such a polynomial the affine level curves are L_z are all affine equivalent to any one of them, say, to L_1. The surface L_1 is a compact Riemann surface of some genus g with $n+2$ deleted points. The genus g can be computed by the Riemann–Hurwitz formula [**For91**]. The projection $(x, y) \mapsto x$ restricted on L_1 defines the ramified covering $\overline{L_1} \to \mathbb{P}^1$ of multiplicity $m = n + 1$. The ramification points of the covering are defined by the equation $\frac{\partial f}{\partial y} = 0$ which is a homogeneous polynomial equation of degree n. The system of equations $f(x, y) = 1$, $\frac{\partial f}{\partial y}(x, y) = 0$ has therefore $n(n + 1)$ solutions, none of them at infinity and all simple. Indeed, the polynomial $\frac{\partial f}{\partial y}$ factors as the product of n linear terms corresponding to distinct lines on the (x, y)-plane. Restriction of the homogeneous polynomial f on each line is a homogeneous polynomial of degree $n+1$ in one variable (the local parameter along the line), which must be nonzero since in the opposite case f would have a multiple linear factor. All roots of the equation $ct^{n+1} = 1$, $c \neq 0$, are simple and distinct, hence each of n lines contributes exactly $n+1$ simple solutions to the system $f = 1$, $\frac{\partial f}{\partial y} = 0$.

Each simple solution of the system $f = 1$, $\frac{\partial f}{\partial y} = 0$ corresponds to a ramification point of index $k_j = 2$. By the Riemann–Hurwitz formula the total genus is

$$g = 1 - m + \sum_j \frac{k_j - 1}{2} = -n + \frac{1}{2}n(n+1) = \tfrac{1}{2}n(n-1). \tag{26.37}$$

The first homology group of each fiber L_z is generated by $2g$ *canonical loops* forming the basis of the homology of the projective compactification $\overline{L_z}$ and any n out of $n+1$ small loops around the deleted points at infinity (the sum of all $n+1$ small loops is homologous to zero). Any closed loop on L_z is homologous to a linear combination of these cycles with integral coefficients (since $H_1(\overline{L_z}, \mathbb{Z})$ is the free group generated by the canonical loops). Ultimately we have

$$\operatorname{rank} H_1(L_z, \mathbb{Z}) = 2g + n = n(n-1) + n = n^2.$$

This proves the assertion on the genus of curves when f is a homogeneous polynomial.

If f is not homogeneous and $f - f_{n+1} \neq 0$, the level curves $L_z = \{f = z\}$ for large values of z are perturbations of the level curves $\widetilde{L}_z = \{f_{n+1} = z\}$. Since the latter are nonsingular for $z \neq 0$, by the implicit function theorem L_z and $\widetilde{L}_z$ are diffeomorphic for sufficiently large values of z. But *all* curves L_z with $z \notin \Sigma$ are diffeomorphic to each other, therefore the rank of any homology group is the same *everywhere*, $\operatorname{rank} H_1(L_z, \mathbb{Z}) = \operatorname{rank} H_1(\widetilde{L}_1, \mathbb{Z}) = n^2$. □

26G. Gelfand–Leray derivative. The derivative of an Abelian integral is again an Abelian integral. More precisely, we have the following rule of derivation of Abelian integrals.

Theorem 26.32. *Let ω and η be two rational 1-forms such that*

$$d\omega = df \wedge \eta, \tag{26.38}$$

and $\delta(z) \in H_1(L_z, \mathbb{Z})$, $z \notin \Sigma$, a continuous family of cycles on noncritical level curves of f, not passing through poles of neither ω nor η. Then

$$\frac{d}{dz} \oint_{\delta(z)} \omega = \oint_{\delta(z)} \eta. \tag{26.39}$$

Proof. Consider the real 2-dimensional cylindrical surface M^2 in $\mathbb{C}^2$, parameterized by real parameters $s \in \mathbb{R} \bmod \mathbb{Z}$ and $t \in [0, \varepsilon]$ so that each circle $\{t = \text{const}\}$ parameterizes the cycle $\delta(z+t)$ on the level surface L_{z+t}. Such a surface exists for sufficiently small $\varepsilon > 0$, since the foliation $\{df = 0\}$ has trivial holonomy along the loop $\delta(z)$.

The boundary of M consists of two cycles, $-\delta(z)$ and $\delta(z+\varepsilon)$, hence by the Stokes theorem,

$$\oint_{\delta(z+\varepsilon)} \omega - \oint_{\delta(z)} \omega = \iint_M d\omega = \iint_M df \wedge \eta. \tag{26.40}$$

The form df vanishes on all cycles $\delta(z+t)$, so that the double integral in (26.40) reduces to the iterated integral

$$\int_0^\varepsilon dt \cdot \oint_{\delta(z+t)} \eta.$$

Dividing both parts of the equality by ε and passing to the limit as $\varepsilon \to 0$, we conclude with the formula (26.39): the convergence follows from the assumptions on the cycle $\delta(z)$. □

Example 26.33. Let $f(x,y) = x^2 + y^2$ and $\delta(z)$ for $z > 0$ be the real circle oriented counterclockwise. Then the Abelian integral $\oint_{\delta(z)} \omega$ of the 1-form $\omega = y\,dx$ is equal to $-\pi z$ (the area of the circle). The equation (26.38) is satisfied by the meromorphic 1-form $\eta = \frac{1}{2}\frac{dx}{y}$. This form has poles on the real cycles $\delta(z)$, but the restriction of η on all level curves L_z is holomorphic ($\eta|_{L_z}$ has removable singularity at the points with $y = 0$). Hence the cycles $\delta(z)$ can be moved off the polar locus of η without changing the integrals, while permitting application of Theorem 26.32. The integral of η along the cycles is identically equal to $-\pi$. This example allows us to recall the order of terms in the wedge product (26.38).

The Gelfand–Leray derivative of a polynomial 1-form is only *rational* on $\mathbb{C}^2$ and *nonunique*. However, its *restriction* on the nonsingular affine level curves $L_z = \{f = z\}$ is a uniquely defined holomorphic 1-form from $\Lambda^1(L_z)$. Indeed, the derivative is defined uniquely modulo a rational 1-form on $\mathbb{C}^2$ having zero restriction on the fibers (the only solutions of the equation $\eta \wedge df = 0$). On the other hand, locally near a point $a \in L_z$ at which $\frac{\partial f}{\partial y} \neq 0$, the Gelfand–Leray derivative of a form ω with $d\omega = A(x,y)\,dx \wedge dy$ can be obtained by restriction on L_z of the holomorphic form $\eta = -\frac{A(x,y)}{\frac{\partial f}{\partial y}(x,y)}\,dx$.

The Gelfand–Leray derivative is often denoted by $\frac{d\omega}{df}$: while this notation is ambiguous if used for a rational form in $\mathbb{C}^2$, the ambiguity disappears after restriction on the level curves.

Remark 26.34. If $d\omega = A(x,y)\,dx \wedge dy$ has a polynomial coefficient $A \in \mathbb{C}[x,y]$ of degree m and f is of degree $n+1$ transversal to infinity, then the Gelfand–Leray derivative $\frac{d\omega}{df}$ restricted on the level curves L_z has a pole of order $\leqslant m-n+2$ at each of the $n+1$ points at infinity. Indeed, the partial derivative $\frac{\partial f}{\partial y}$ restricted on each analytic branch of the curve $\{f=z\}$, has a pole of order exactly n and no smaller. This follows from computations with principal homogeneous terms: arguments similar to those proving Theorem 26.31, show that the leading coefficient of the partial derivative cannot vanish. Thus the derivative η has the form $O(x^{m-n})\,dx$, at infinity, i.e., the pole of order $\leqslant m-n+2$ in the local chart $u=1/x$.

In particular, if $\omega = P\,dx + Q\,dy$ with $\deg P, Q \leqslant n$, and f is transversal to infinity of degree $n+1$, then the order of the pole of $\frac{d\omega}{df}$ is at most 1, i.e., all poles at infinity are simple.

26H. Picard–Fuchs system and its properties. Consider a polynomial $f \in \mathbb{C}[x,y]$ of degree $n+1$, transversal to infinity, and let $\boldsymbol{\omega} = (\omega_1,\dots,\omega_m)$, $m=n^2$, be the tuple of monomial 1-forms generating the Petrov module $\mathbf{P}_f$ as in Theorem 26.21.

Theorem 26.35. *For any continuous family of cycles $\delta(z)$ on the level curves, the column vector $X=X(z)$ of the periods $\oint_{\delta(z)}\omega_j$, $j=1,\dots,m$, satisfies the following system of linear ordinary differential equations,*

$$(zE - A)\cdot\frac{dX}{dz} = (B_0 + zB_1)\cdot X. \tag{26.41}$$

Here A, B_0, B_1 are constant $m\times m$-matrices.

Proof. Consider the 2-forms $f\,d\omega_j \in \Lambda^2$, for all $j=1,\dots,m$. Each of them can be "divided with remainder" by df: by Proposition 26.20 there exist polynomial 1-forms η_j of degrees $\deg\eta_j \leqslant \deg\omega_j$ such that

$$f\,d\omega_j = df\wedge\eta_j + \sum_{k=1}^{m} a_{jk}\,d\omega_k \tag{26.42}$$

with some complex numbers $a_{jk}\in\mathbb{C}$ forming an $m\times m$-matrix A. These identities can be rewritten under the form

$$d\Big(f\omega_j - \sum_k a_{jk}\omega_k\Big) = df\wedge(\eta_j+\omega_j). \tag{26.43}$$

The 1-forms η_j can be in turn expanded using Theorem 26.21, as

$$\eta_j = \sum_{k=1}^{m} (b_{jk} \circ f) \cdot \omega_k \mod d\Lambda^0 + \Lambda^0 \, df, \tag{26.44}$$

with some univariate polynomials $b_{jk}(z)$ which have to be composed with f. Since $\deg \eta_j \leqslant 2n < 2(n+1)$, the degrees of these polynomials by (26.28) cannot exceed 1. Together these polynomials can be arranged into a linear matrix polynomial $B(z) = \|b_{jk}(z)\|_{j,k=1}^{m} = B_0 + zB_1$.

Note that the f restricted on $\delta(z)$ is identically equal to z. Therefore applying Theorem 26.32, we conclude that

$$\frac{d}{dz}(zX(z) - A \cdot X(z)) = (E + B_0 + zB_1) \cdot X(z). \tag{26.45}$$

Since A is constant, the relationship (26.45) is equivalent to (26.41). □

Many properties of the matrices A, B_0, B_1 can be seen from their explicit construction in the proof of Theorem 26.35.

Corollary 26.36. *If f has only nondegenerate critical points, then A is a diagonalizable matrix whose spectrum consists of the corresponding critical values of f.*

Proof. By construction, A is the matrix of multiplication by f in the quotient algebra $\Lambda^2/df \wedge \Lambda^1 \cong Q_{df}$. If f has only nondegenerate critical points (as usual, being transversal to infinity), then the quotient algebra is isomorphic to the algebra of functions on m distinct points forming the critical locus of f in $\mathbb{C}^2$ and the spectrum of A consists of the critical values of f, counted with their multiplicities if some critical values at distinct critical points coincide. □

This in turn implies the following conclusion.

Corollary 26.37. *If f is a Morse polynomial transversal to infinity, then the linear system* (26.41) *has only Fuchsian singularities at the critical values of the polynomial f.*

Proof. In the assumptions of the corollary, the determinant $\det(zE - A)$ has only simple roots at the critical values of f, hence the inverse matrix $(zE - A)^{-1}$ has simple poles there. □

Remark 26.38. In a similar way the matrix coefficients of the decomposition B_0, B_1 can be described. In particular, their norms can be, if necessary, estimated from above in terms of the relative magnitude of principal and nonprincipal homogeneous coefficients of f.

Remark 26.39. The singular point of the system (26.41) at infinity is in general non-Fuchsian (though obviously always regular); see [**Nov02**]. However, if instead of $m = n^2$ forms $d\omega_j$ generating the algebra $\mathbf{Q}_f$ we take all $\nu = n(2n-1)$ monomial 2-forms of degree $\leqslant 2n$, then in the first division (26.42) one can achieve $\deg \eta_j \leqslant \deg \omega_j \leqslant 2n$ and hence one can always represent $d\eta_j$ as a linear combination of the forms $d\omega_k$ with *constant complex* coefficients. This will produce a *redundant* system of linear differential equations satisfied by the vector of periods $X'(z)$ of all ν 1-forms in the *hypergeometric form*,

$$(zE - A')\frac{d}{dz}X'(z) = B'X'(z), \qquad A', B' \in \operatorname{Mat}(\nu, \mathbb{C}). \tag{26.46}$$

Such a system always has a Fuchsian singular point at infinity. For details see [**NY01**].

26I. Vanishing cycles and Picard–Lefschetz formulas. In this section we compute the monodromy operators (26.36) associated with a particular case where the loop γ is a small path encircling just one Morse critical value of the polynomial f.

As z tends to a nondegenerate critical value $a \in \Sigma$ of f, among all cycles on the curve L_z one can distinguish a certain cycle $\delta_a(z) \in H_1(L_z, \mathbb{Z})$ called the *vanishing cycle*. This cycle is defined uniquely modulo orientation (i.e., up to multiplication by -1 in the group $H_1(L_z, \mathbb{Z})$).

To describe the vanishing cycle accurately, assume that the critical value of the polynomial f is $a = 0$ and the corresponding nondegenerate critical point is at the origin. Without loss of generality we may further assume that $f(x, y) = x^2 - y^2 + \cdots$. All this can be achieved by affine changes of the variables x, y and z.

Consider the parallel projection on the x-plane parallel to the y-axis, restricted on different level curves L_z. These restrictions have critical points depending on z, when the derivative $\frac{\partial f}{\partial y}$ vanishes on L_z: near all other points the curve L_z locally biholomorphically covers the x-plane.

Out of those critical points, there are exactly two points near the origin, defined by the equations

$$x^2 - y^2 + \cdots = z, \qquad 2y + \cdots = 0$$

(as usual, the dots denote higher order terms). Resolving these equations, we conclude that L_z can be locally described as the Riemann surface covering the x-plane $(\mathbb{C}, 0)$ with ramification at the two near the origin, $x_\pm(z) = \pm\sqrt{z + \cdots}$, the ramification having order 2. The loop on the x-plane, which encircles these points, can be lifted to a cycle $\delta_0(z)$ on the level curve L_z for all $z \neq 0$ sufficiently close to zero; see Fig. V.3.

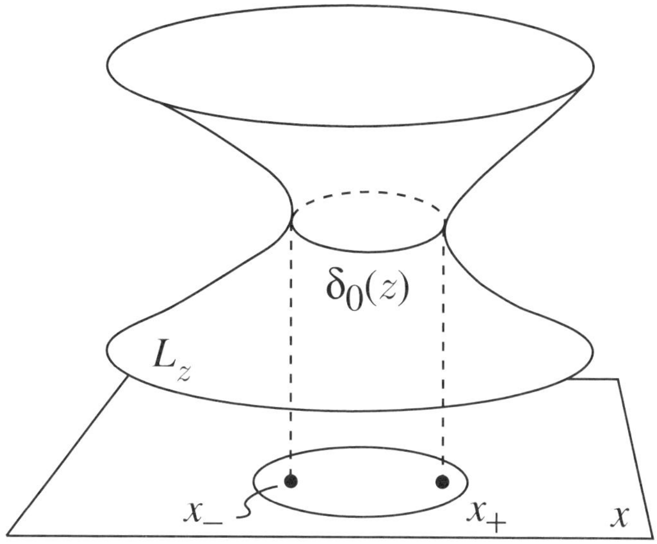

Figure V.3. Vanishing cycle at a Morse singularity

Definition 26.40. The cycle $\delta_0(z) \in H_1(L_z, \mathbb{Z})$ defined by this construction for all $z \neq 0$ sufficiently close to the critical value $z_0 = 0$, is called the *vanishing cycle* (more precisely, the cycle vanishing at the critical value z_0).

Remark 26.41. The vanishing cycle (modulo orientation and the free homotopy deformation on L_z) can be characterized by the following purely topological property: *as $z \to 0$, the cycle $\delta_0(z)$ can be represented by a continuous family of loops on L_z of length that tends to zero.* This description explains the terminology.

Now we can describe the monodromy operator for a small loop encircling a Morse critical value. Suppose that the regular value z varies along the small circular loop γ around the origin, $z(t) = \rho e^{2\pi i t}$, $0 < \rho \ll 1$, parameterized by the real variable $t \in [0, 1]$. Then the two points $x_\pm(z(t))$ also rotate along two curves approximating two half-circles $x_\pm(t) = \pm\sqrt{\rho}e^{\pi i t}(1 + o(1))$ and at the end exchange their places; see Fig. V.4.

Looking at this figure, one can construct a continuous isotopy of the plane, which is identical outside the disk of radius, say, $3\sqrt{\rho}$ and a rotation by π on the disk of radius $2\sqrt{\rho}$. This isotopy of the plane lifts as an isotopy of the fiber L_ρ on itself, identical outside a small disk centered at the critical point, called the *Dehn twist*.

The action of the Dehn twist on the vanishing cycle $\delta_0(\rho)$ itself is trivial: the cycle "rotates" along itself. However, if $\delta'(\rho)$ is another cycle which intersects the vanishing cycle, then on the level of homology the Dehn twist

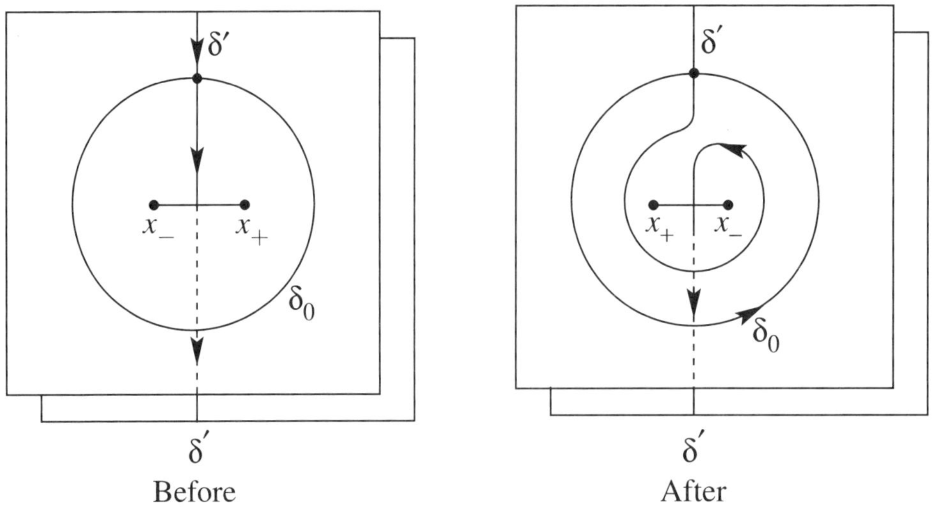

Figure V.4. Vanishing cycles and monodromy around a Morse singularity

acts by adding to $\delta'(\rho)$ the vanishing cycle $\delta_0(\rho)$ with the sign $\pm$ depending on the intersection index $\delta_0 \cdot \delta'$ between δ_0 and δ'. If δ' is a simple curve, this is instantly clear from Fig. V.4; for cycles having multiple intersections with δ_0 one has to use the additivity of the intersection index to prove the following result, called the *Picard–Lefschetz formula*:

$$\Delta_\gamma \delta = \delta + (\delta \cdot \delta_0)\, \delta_0. \tag{26.47}$$

Remark 26.42. The formal construction of the Dehn twist and the proof of Picard–Lefschetz formulas can be found in numerous sources, among them [**AGV88, Pha67, DFN85**]. In the local context, where only the intersection of the level curves with a small ball around the critical point are considered, one should exercise a certain care distinguishing between the *absolute* and *relative* (modulo the boundary) homology; see [**AGV88**, §1].

As an immediate corollary from the Picard–Lefschetz formulas, we obtain the following. Choose any basis $\boldsymbol{\delta} = \{\delta_1, \dots, \delta_m\}$ in the homology $H_1(L_a, \mathbb{Z})$ for some regular value $a \in \mathbb{C} \smallsetminus \Sigma$, considered as a row vector. Then analytic continuation Δ_γ along any loop $\gamma \in \pi_1(\mathbb{C} \smallsetminus \Sigma, a)$ is represented by a matrix M_γ as follows:

$$\Delta_\gamma \boldsymbol{\delta} = \boldsymbol{\delta} \cdot M_\gamma, \qquad M_\gamma \in \operatorname{Mat}(m, \mathbb{Z}). \tag{26.48}$$

Proposition 26.43. *All monodromy matrices M_γ defined above, are unimodular,* $\det M_\gamma = 1$.

Proof. If γ is a small loop around a Morse critical value, then the equality $\det M_\gamma = 1$ follows immediately from (26.47) (the vanishing cycle δ_0 can always be chosen as the first element in the row $\boldsymbol{\delta}$).

An arbitrary loop is a product of several small loops as above. $\square$

26J. Period matrices. Let f be a polynomial of degree $n+1$ transversal to infinity, and $\boldsymbol{\omega} = (\omega_1, \dots, \omega_m)$, $m = n^2$, an arbitrary tuple of polynomial 1-forms. With this tuple, considered as a column vector, and any choice of a locally constant basis $\boldsymbol{\delta}(z) = (\delta_1(z), \dots, \delta_m(z))$ in the homology of the level curves L_z, $z \in \mathbb{C} \smallsetminus \Sigma$, one can associate the *period matrix*

$$X(z) = \boldsymbol{\omega} \otimes \boldsymbol{\delta}(z) = \begin{pmatrix} \oint\limits_{\delta_1(z)} \omega_1 & \cdots & \oint\limits_{\delta_m(z)} \omega_1 \\ \vdots & \ddots & \vdots \\ \oint\limits_{\delta_1(z)} \omega_m & \cdots & \oint\limits_{\delta_m(z)} \omega_m \end{pmatrix} \tag{26.49}$$

which is an analytic multivalued matrix-function ramified over the locus Σ.

As follows from (26.48), the analytic continuation of the period matrix results in the right multiplication by the monodromy matrices M_γ,

$$\Delta_\gamma X(z) = X(z) \cdot M_\gamma. \tag{26.50}$$

Proposition 26.44. *If f is a polynomial transversal to infinity, then the determinant* $\det X(z)$ *of any period matrix, regardless of the choice of the forms $\boldsymbol{\omega}$, is a polynomial in z with zeros at the points of Σ.*

Proof. Together with Theorem 26.26, Proposition 26.43 implies that the determinant $\det X(z)$ is a single-valued function on $\mathbb{C} \smallsetminus \Sigma$. As z tends to infinity, the integrals occurring as the entries of $X(z)$ grow no faster than polynomially in $|z|$ in any sector. This implies that $\det X(z)$ is a polynomial. As z tends to a point $a \in \Sigma$, at least one cycle (a linear combination of the basis $\boldsymbol{\delta}(z)$) vanishes, hence the determinant tends to zero. $\square$

From Proposition 26.44 it follows that (in the standing assumption that f is a Morse polynomial transversal to infinity) the determinant of *any* period matrix is divisible by the discriminant polynomial $D_f(z) = \prod_{z_j \in \Sigma}(z - z_j)$. The natural question is whether this description is precise, i.e., whether there are additional points at which the determinant of periods vanish. The answer is given by the following result.

Theorem 26.45. *Assume that the polynomial $f = f_{n+1} + \cdots$ of degree $n+1$ is transversal to infinity and the tuple of 1-forms $\boldsymbol{\omega} = (\omega_1, \dots, \omega_m)$ satisfies the assumptions of Theorem 26.21 (their differentials generate the quotient* *algebra* $Q_{df_{n+1}} = \Lambda^2/df_{n+1} \wedge \Lambda^1$). *Then*

$$\det X(z) = c(\boldsymbol{\omega}) D_f(z), \qquad D_f(z) = \prod_{z_j \in \Sigma} (z - z_j), \quad c(\boldsymbol{\omega}) \neq 0. \tag{26.51}$$

Proof. By the de Rham theorem [**War83**], for any basis of the homology of an arbitrary level curve L_a, $a \notin \Sigma$, there exist m forms on L_a such that the respective period matrix is nondegenerate. These m forms on L_a are restrictions of some tuple of polynomial 1-forms $\Omega_1, \dots, \Omega_m \in \Lambda^1[\mathbb{C}^2]$.

By Theorem 26.21, there exists a polynomial $m \times m$-matrix $P(z)$ of coefficients of expansion of Ω_j in the basis $\boldsymbol{\omega}$ of the module $\mathbf{P}_f$, such that

$$Y(z) = P(z)X(z), \qquad Y(z) = \begin{pmatrix} \oint_{\delta_1(z)} \Omega_1 & \cdots & \oint_{\delta_m(z)} \Omega_1 \\ \vdots & \ddots & \vdots \\ \oint_{\delta_1(z)} \Omega_m & \cdots & \oint_{\delta_m(z)} \Omega_m \end{pmatrix}. \tag{26.52}$$

The matrix $Y(a)$ is nondegenerate by construction, therefore $X(a)$ must also be nondegenerate. We conclude that the determinant $\det X(z)$ is a polynomial *without roots* outside Σ. Such a polynomial can be only of the form (26.51). □

Remark 26.46. In most expositions, Theorem 26.45 together with the Proposition 26.44, established by analytical and topological arguments, is the starting point of the construction of a basis for the module of Abelian integrals.

The usual strategy of proving the formula (26.51) is to compute the sectorial growth rate of all entries of the matrix $X(z)$ and show that $\det X(z) = O(|z|^n)$ as $|z| \to \infty$. This shows that $\det X(z)$ is a polynomial with roots at all n points of the critical locus Σ, which leaves the only possibility $\det X(z) = c\, D_f(z)$, where $c = c(\boldsymbol{\omega}) \in \mathbb{C}$ is a constant. However, the accurate proof that $c \neq 0$ requires some effort; see [**Nov02**] where all details are explicitly supplied. The explicit value of $c(\boldsymbol{\omega})$ was recently obtained by A. Glutsyuk [**Glu06**]. For earlier results; see [**Var89**].

Staring from Theorem 26.45, one can derive (using the Cramer rule for finding the indeterminate coefficients), that the integral of any polynomial form Ω over any cycle $\delta(z)$ is a polynomial combination of integrals of the basic forms $\omega_1, \dots, \omega_m$ over this cycle, the coefficients being independent of the choice of the cycle,

$$\oint_{\delta(z)} \Omega = \sum_{j=1}^{m} p_j(z) \oint_{\delta(z)} \omega_j.$$

By Theorem 26.13, the 1-form $\Omega - \sum (p_j \circ f)\, \omega_j \in \Lambda^1$ is algebraically relative cohomologous to zero. This gives an alternative (analytic) proof of Theorem 26.21. For details see [**Gav98, Nov02**].

We choose an alternative strategy based on Theorem 26.21, since it is algorithmic and provides explicit bounds for the operator of decomposition of any polynomial 1-form in the elements of the basis for $\mathbf{P}_f$.

26K. Monodromy of Abelian integrals. Monodromy of the Picard–Fuchs system (26.41) is a linear representation of the fundamental group $\pi_1(\mathbb{C} \smallsetminus \Sigma, a)$ by automorphisms of solutions of the system. It turns out that the easiest way to study this representation is via topology of the map $f\colon \mathbb{C}^2 \to \mathbb{C}$.

26$\mathbf{K}_1$. *Completeness of the system of vanishing cycles.* A generic (Morse) polynomial of degree $n+1$ transversal to infinity, has n^2 critical points and hence every nonsingular level curve carries exactly n^2 (topological continuations) of vanishing cycles. On the other hand, the homology group of a generic leaf L_z also is of rank n^2 over $\mathbb{Z}$, as shown in Theorem 26.31. The two numbers are equal, and this equality is not accidental.

Theorem 26.47. *Vanishing cycles generate the first homology group of any fiber $L_z = \{f = z\} \subset \mathbb{C}^2$ of a Morse polynomial transversal to infinity.*

Clarification and references. The precise meaning of this theorem is as follows. For a fixed regular value $a \in \mathbb{C} \smallsetminus \Sigma$ of f consider simple paths α_j, $j = 1, \dots, m$, connecting a with each of the critical values $a_1, \dots, a_m$ of f, $m = n^2$. Each vanishing cycle $\delta_j \in H_1(L_z, \mathbb{Z})$ is well defined for all $z \in (\mathbb{C}, a_j)$ and can be uniquely continued along α_j^{-1} to a cycle $\delta_j \in H_1(L_a, \mathbb{Z})$. Besides, each path α_j defines a loop $\gamma_j \in \pi_1(\mathbb{C} \smallsetminus \Sigma, a)$ which corresponds to going along α_j, encircling a_j by a sufficiently small positive circular loop and returning back along the same path α_j^{-1} (i.e., inverting the direction). The collection of paths $\{\alpha_1, \dots, \alpha_m\}$ is called *proper*, if the simple loops $\gamma_1, \dots, \gamma_m$ generate the fundamental group $\pi_1(\mathbb{C} \smallsetminus \Sigma, a)$.

Theorem 26.47 asserts that for any polynomial f and any regular value a one can always construct a proper system of paths $\alpha_1, \dots, \alpha_m$ such that the corresponding continuations of vanishing cycles $\delta_1(a), \dots, \delta_m(a)$ generate the entire homology $H_1(L_a, \mathbb{Z})$.

This assertion can be derived from the corresponding local result, Theorem 1 from [**AGV88**, Chapter I, §2]. Alternatively, one can use the results by A. B. Zhizhchenko [**Žiž61**]. A very good exposition of these things is given in the recent paper [**MMJR97**, Sect. 4]: Theorem 26.47 is an immediate corollary to the Theorem 4.4 of the latter paper. □

26$\mathbf{K}_2$. *Transitivity of the monodromy on vanishing cycles.* The global structure of the topological monodromy group is characterized by the following property.

Theorem 26.48. *The monodromy group acts transitively on the collection of all vanishing cycles: for any two such cycles $\delta_1, \delta_2 \in L_a$, $a \notin \Sigma$, there exists a loop $\gamma \in \pi_1(\mathbb{C} \smallsetminus \Sigma, a)$ such that $\Delta_\gamma \delta_1 = \pm\delta_2$.*

Proof. Assertion of this theorem follows from the fact that the discriminant variety $\boldsymbol{\Sigma}$, introduced in (26.35), is irreducible and hence its smooth part, parameterizing in a certain sense all vanishing cycles of all polynomials with a fixed principal homogeneous part, is connected. We follow the exposition in [**Pus97**]; see also [**AGV88**, Theorem 4, Chapter I, §3].

Let $f \in \mathbb{C}[x,y]$ be a Morse polynomial transversal to infinity, and $a_1, a_2 \in \Sigma$ the critical values, $a_i = f(x_i, y_i)$, corresponding to the two cycles δ_i vanishing at the two critical points $C_i = (x_i, y_i) \in \mathbb{C}^2$, $i = 1,2$. Theorem 26.48 will be proved, if we find a path γ_{12} connecting a_1 with a_2 and avoiding Σ everywhere else, such that the parallel transport (continuation) of $\delta_1(z)$, $z \in (\mathbb{C}, a_1)\smallsetminus\Sigma$, along this path coincides with $\delta_2(z)$, $z \in (\mathbb{C}, a_2)\smallsetminus\Sigma$, modulo orientation of the latter.

We will show first that there exists a continuous deformation of the (singular) zero level curve $L_{a_1} = \{f - a_1 = 0\}$ onto the other singular level curve $L_{a_2} = \{f - a_2 = 0\}$, which sends the respective (uniquely defined) critical points into each other, if one is allowed to change continuously *all nonprincipal coefficients* of the polynomial f rather than only its free term.

To that end, consider the universal deformation $\Phi(\lambda; x, y) = f_{n+1} + \sum_{0\leqslant i+j\leqslant n} \lambda_{ij} x^i y^j$ as in (26.34) and the discriminant variety $\boldsymbol{\Sigma}$ introduced in (26.35). Let $\boldsymbol{\Sigma}^\circ \subset \boldsymbol{\Sigma}$ be the set of parameters $\lambda \in P \cong \mathbb{C}^r$ such that the zero level curve of the polynomial $f_\lambda = \Phi(\lambda; \cdot, \cdot)$ carries only one nondegenerate critical point. This is the principal stratum of the algebraic variety $\boldsymbol{\Sigma}$, a relatively open subset with a complement which is an algebraic variety of lower dimension.

Lemma 26.49.

1. *All points of $\boldsymbol{\Sigma}^\circ$ are smooth on $\boldsymbol{\Sigma}$.*
2. *The set $\boldsymbol{\Sigma}^\circ$ is connected.*

Proof of the lemma. To prove the first assertion, consider the polynomial $\Phi' = \Phi - \lambda_{00}$ which in fact depends only on the parameters λ_{ij} with $i+j > 0$. By the implicit function theorem, Morse critical points are *stable*: if $\lambda \in \boldsymbol{\Sigma}^\circ$ and $C_* \in \mathbb{C}^2$ is the corresponding Morse critical point of $\Phi'|_\lambda$, then for any sufficiently close combination of the parameters $\lambda' = \{\lambda_{ij},\ i+j > 0\}$, the polynomial $f'_{\lambda'} = \Phi'(\lambda', \cdot, \cdot)$ has a nearby Morse critical point $C(\lambda')$ analytically depending on λ'. The critical value $s(\lambda')$ of $f'_{\lambda'}$ at this point will also depend analytically on λ', which means that $\boldsymbol{\Sigma}^\circ$ is the graph of an analytic function, $(\lambda', -s(\lambda')) \in \boldsymbol{\Sigma}^\circ$.

To prove connectedness of $\boldsymbol{\Sigma}^\circ$, note that $\boldsymbol{\Sigma}$ is the image of the surface

$$S = \{(\lambda, x, y) \in \mathbb{C}^r \times \mathbb{C}^2 \colon \partial\Phi/\partial x = \partial\Phi/\partial y = \Phi = 0\} \qquad (26.53)$$

by the projection $(\lambda, x, y) \mapsto \lambda$. The part S' of S given by the inequality $\{\det\big(\partial^2\Phi/\partial(x,y)\big) \neq 0\}$, by the first assertion of the lemma, parameterizes smooth points of $\boldsymbol{\Sigma}$, including self-intersections of several smooth components.

The projection $\pi\colon (\lambda, x, y) \mapsto (x, y)$ restricted on S is a holomorphic *affine* subbundle of the trivial bundle $\pi\colon (\lambda, x, y) \mapsto (x, y)$. Indeed, the equations (26.53) for any fixed (x, y) are affine with respect to λ and define an affine subspace in P. The local triviality follows from the fact that any translation in the (x, y) plane corresponds to an affine transformation of the parameters λ: the nonprincipal coefficients of the polynomial $\Phi(\lambda; x{+}a, y{+}b)$ for any $(a, b) \in \mathbb{C}^2$ are affine functions of λ.

The degeneracy condition $\det\big(\partial^2 f/\partial(x,y)^2\big) = 0$ as well as the occurrence of another critical point determine a *proper* complex affine subbundle in S. This properness guarantees that the complementary set

$$S^\circ = \big\{(\lambda, x, y) \in S\colon \det\big(\partial^2 f/\partial(x,y)^2\big) \neq 0,\ \forall (x', y') \in \mathbb{C}^2,\ (\lambda, x', y') \notin S\big\}$$

which parameterizes $\boldsymbol{\Sigma}^\circ$, is connected. The lemma is proved. □

We return to the proof of Theorem 26.48. Let $\mathbf{a}_1, \mathbf{a}_2 \in \boldsymbol{\Sigma} \subset \mathbb{C}^r$ be the points corresponding to the nonprincipal coefficients of the polynomials $f - a_1$ and $f - a_2$ respectively, where $f = f_{n+1} + \cdots$ is the initial polynomial. By Lemma 26.49, these points appear at the intersection of the smooth part $\boldsymbol{\Sigma}^\circ$ with the complex line $\ell = \{\lambda' = \text{const}\}$.

Since $\boldsymbol{\Sigma}^\circ$ is connected, the points $\mathbf{a}_1$ and $\mathbf{a}_2$ can be connected by a path $\boldsymbol{\gamma}_{12}$ parameterized by $t \in [1, 2]$. Since $\boldsymbol{\Sigma}^\circ$ is a smooth hypersurface, this path can be deformed to a path that avoids $\boldsymbol{\Sigma}$ everywhere except for the endpoints, but remains sufficiently close to $\boldsymbol{\Sigma}$ so that the vanishing cycle on all the level curves $L_t = L_{\boldsymbol{\gamma}_{12}(t)} \subseteq \mathbb{C}^2$ is uniquely determined. The corresponding deformation of zero level curves of the polynomial $\Phi(\boldsymbol{\gamma}_{12}(t), \cdot, \cdot)$ carries the unique vanishing cycle δ_1 on $\{\Phi(\mathbf{a}_1, \cdot, \cdot) = 0\}$ onto the unique vanishing cycle δ_2 on $\{\Phi(\mathbf{a}_1, \cdot, \cdot) = 0\}$.

During this deformation all nonprincipal coefficients of the polynomial are changed, not just the free term. However, by the global Zariski theorem, the path $\boldsymbol{\gamma}_{12} \subset \mathbb{C}^r \smallsetminus \boldsymbol{\Sigma}$ with the endpoints on the line ℓ can be deformed (by a homotopy with the fixed endpoints) to a path γ_{12} entirely belonging to $\ell \smallsetminus \boldsymbol{\Sigma} = \ell \smallsetminus \Sigma$. In this deformation only the free term of f is changed, hence the corresponding deformation coincides with the monodromy action described in Corollary 26.28. The proof of Theorem 26.48 is complete. □

26$\mathbf{K}_3$. *Almost irreducibility of the monodromy.* The transitivity established in Theorem 26.48, allows us to prove that the topological monodromy group is almost irreducible in the following precise sense. Consider the first homology $G = H_1(L_a, \mathbb{Z})$ of a generic level curve L_a, $a \notin \Sigma$. This is a $\mathbb{Z}$-module equipped with an antisymmetric intersection form $G^2 \ni (\delta, \delta') \mapsto \delta \cdot \delta' \in \mathbb{Z}$. Since L_a is $n+1$ times punctured Riemann surface of some genus g, the module G is generated by $2g$ canonical cycles ℓ_j, ℓ'_j, $j = 1, \dots, g$ and any n "small loops" $s_1, \dots, s_n$ around $n+1$ punctures at infinity. The intersection form in this basis looks very simple:

$$\ell_i \cdot \ell'_j = \begin{cases} 1, & i = j, \\ 0, & i \neq j, \end{cases} \qquad s_k \cdot G = 0. \tag{26.54}$$

Any loop $\gamma \in \pi_1(\mathbb{C} \smallsetminus \Sigma, a)$ in the base defines a topological monodromy operator $\Delta_\gamma \colon G \to G$ which is $\mathbb{Z}$-linear and preserves the intersection form. Together the operators determine $\mathbb{Z}$-linear representation $\gamma \mapsto \Delta_\gamma$ of the fundamental group by automorphisms of the module G. If $S \subseteq G$ is a submodule *invariant by all monodromy operators*, the quotient representation by automorphisms of the quotient module G/S is well defined by the action

$$\widetilde{\Delta}_\gamma(\delta \bmod S) = (\Delta_\gamma \delta) \bmod S, \qquad \forall (\delta \bmod S) \in G/S.$$

Theorem 26.50. *For a polynomial f of degree $n+1$ transversal to infinity, the submodule $S \subset G = H_1(L_a, \mathbb{Z})$ generated by the "small loops" $s_1, \dots, s_n$ is invariant and each monodromy operator Δ_γ acts identically on it.*

The action of the topological monodromy $\gamma \mapsto \widetilde{\Delta}_\gamma \in \operatorname{Aut}(G/S)$ on the quotient $\mathbb{Z}$-module G/S is irreducible, i.e., has no nontrivial invariant submodules.

Proof. The first assertion follows from the Picard–Lefschetz formulas (26.47) and the structure of the intersection form (26.54): the "small loops" s_k have zero intersection index with any vanishing cycle.

To prove the second assertion, consider an arbitrary invariant submodule $G' \neq 0 \bmod S$ in G/S. By factoring out all "small loops" the quotient G/S inherits the intersection form which is *nondegenerate*: if a cycle δ is orthogonal to all G (i.e., has zero intersection index with all loops ℓ_j, ℓ'_j for all $j = 1, \dots, g$), then δ itself is zero modulo S.

We first claim that G' contains one of the vanishing cycles mod$S$. Indeed, an element $\delta \notin S$ cannot be orthogonal to all vanishing cycles: since the latter generate the entire G, this would contradict to the nondegeneracy of the intersection form on G/S. If the intersection index $\delta \cdot \delta_i \neq 0$, then together with δ the submodule G' also contains the element $\delta + c\delta_i$ with $c \neq 0$ because of the invariance and the Picard–Lefschetz formula (26.47). Then $\delta_i \in G' \bmod S$. But because of the transitivity of the action of the topological monodromy (Theorem 26.48), G' contains all other vanishing cycles.

This means that $G' = G \bmod S$, that is, a nonzero invariant submodule necessarily coincides with the entire module G. This proves the irreducibility of the quotient action. □

Since the topological monodromy acts on the period matrix $X(z)$ by right multiplications as in (26.50), it coincides with the monodromy of the corresponding Picard–Fuchs system (26.41), and we obtain a rather important property of the latter system.

Corollary 26.51. *The monodromy group of the Picard–Fuchs system* (26.41) *is almost irreducible: it has an n-dimensional subspace on which all monodromy operators are identical, while the quotient representation $\gamma \mapsto \widetilde{\Delta}_\gamma$ is irreducible.* □

26**K**$_4$. *Gauss–Manin connexion.* The first homology $H_1(L_a, \mathbb{Z})$ and cohomology $H^1(L_a, \mathbb{C})$ of a generic fiber $L_a = f^{-1}(a) \in \mathfrak{F}$ are a lattice ($\mathbb{Z}$-module) and a complex space of the rank (resp., complex dimension) equal to n^2. To achieve uniformity, we consider the linear spaces $H_1(L_a, \mathbb{C}) = H_1(L_a, \mathbb{Z}) \otimes \mathbb{C}$ of formal combinations of cycles with complex coefficients. Then we obtain two families of complex spaces of the same dimension, indexed by nonsingular values $a \notin \Sigma$.

Each of these families is a holomorphic vector bundle over the set of regular values $X = \mathbb{C} \smallsetminus \Sigma$, equipped with meromorphic connexions. To see this, consider first the *homology bundle*: because of the local topological triviality (Theorem 26.26), we can choose any basis in the homology of a nonsingular fiber and then carry it continuously to all nearby regular fibers. This gives local trivialization of the homology bundle $H_1(\cdot, \mathbb{C}) \to X$ and a locally flat connexion $\nabla_\circ$ on it: sections horizontal in the sense of this connexion are continuous sections of the projection $H_1(\cdot, \mathbb{Z}) \to X$.

To define local trivializations of the *cohomology bundle* $H^1(\cdot, \mathbb{C}) \to X$, note that any polynomial 1-form $\omega \in \Lambda^1[\mathbb{C}^2]$ defines a section $[\omega]\colon X \to H^1(\,\cdot\,, \mathbb{C})$: for any point $z \in X$, the value $[\omega](z)$ is the cohomology class of the form ω restricted on the leaf $L_z \subset \mathbb{C}^2$. For any $n^2 = m$ 1-forms $\omega_1, \dots, \omega_m$, whose restrictions on a given nonsingular fiber L_a are cohomologically independent, the independence also persists on all nearby fibers L_z for all $z \in (X, a)$. The corresponding sections $[\omega_1], \dots, [\omega_m]$, clearly locally holomorphic, furnish a local trivialization of the cohomology bundle $H^1(\cdot, \mathbb{C}) \to X$.

Integration $(\alpha, \delta) \mapsto \oint_\delta \alpha$ is a natural pairing (duality) between the two line bundles over $X = \mathbb{C} \smallsetminus \Sigma$. This duality allows us to carry different structures from one bundle to the other. In particular, there is a natural connexion ∇° on the cohomology bundle, dual to the $\nabla_\circ$ on the homology bundle. This connexion, called the *Gauss–Manin connexion*, is uniquely determined the identity

$$d(\alpha, \delta) = (\alpha, \nabla_\circ \delta) + (\nabla^\circ \alpha, \delta) \qquad \forall \alpha\colon X \to H^1(\cdot, \mathbb{C}),\ \delta\colon X \to H_1(\cdot, \mathbb{C})$$

valid for any two sections α and δ of the homology and cohomology bundles respectively. Choosing a basis of *horizontal* sections $\delta_1(z), \dots, \delta_m(z)$ with $\nabla_\circ \delta_j = 0$, we see that for the sections $[\omega_1], \cdots, [\omega_m]$ their covariant derivatives are expansions of the *differential* of the period matrix $(\omega_i, \delta_j(z))$ in the periods of the forms ω_j themselves. In other words, the Picard–Fuchs system of linear equations (26.41) in this geometric language is nothing more than the matrix form of the Gauss–Manin connexion with respect to the chosen trivialization of the cohomology bundle.

Note that though the form ω is "constant" (does not explicitly depend on z), its restriction on L_z is not constant in the sense of the Gauss–Manin connexion, i.e., the section $[\omega]$ is not horizontal.

26L. Real branches of Abelian integrals and lower bounds for the number of limit cycles. Now we can return to the Example 26.11 and show that this type of behavior of Abelian integrals is impossible if f is a Morse polynomial transversal to infinity.

Theorem 26.52 (Yu. Ilyashenko [**Ily69**], I. Khovanskaya (Pushkar′) [**Pus97**]). *Let $f \in \mathbb{R}[x,y]$ be a real polynomial transversal to infinity whose complexification is a Morse function, and $\gamma(t)$ a continuous family of real ovals on the level curves of f.*

If ω is an arbitrary polynomial 1-form with identically zero integral over $\gamma(t)$, then ω is relatively exact (can be represented as in (26.18)). If $\deg\omega < \deg df$, then ω is exact, $\omega = dg$, $g \in \mathbb{R}[x,y]$.

To prove this result, we need the following topological lemma.

Lemma 26.53. *Any nonsingular real oval on the level curve of a real Morse polynomial, either is itself the continuation of a vanishing cycle, or has a nonzero intersection index with at least one vanishing cycle.*

Proof. Any real oval $\gamma \subset \mathbb{R}^2$ belongs to a topological annulus on the plane filled by real ovals of level curves. The inner boundary of this annulus cannot be empty. If the inner boundary is a critical point of f, then γ itself is a vanishing cycle. Otherwise the inner boundary is a singular oval carrying a (Morse) critical point of f which is a saddle. The cycle vanishing at this saddle is purely imaginary and intersects γ with the index ± 1. □

Proof of the Theorem 26.52. Suppose that $I(t) = \oint_{\gamma(t)} \omega \equiv 0$. By the Picard–Lefschetz formula (26.47) and Lemma 26.53, for at least one vanishing cycle $\delta_1(z)$ the integral $I_1(z) = \oint_{\delta_1(z)} \omega$ is also vanishing identically. But then by Theorem 26.48, the integral of ω over *any* vanishing cycle of f is identically zero. Since vanishing cycles generate the homology group of any fiber (Theorem 26.47), this means that the 1-form ω is *relatively closed.*

Application of Theorem 26.13, whose assumptions are automatically satisfied if f is a Morse polynomial transversal to infinity, shows then that ω is (algebraically) relatively exact: $\omega = h\,df + dg$ with *some* polynomials h, g. Symmetrizing this identity (adding it with its complex conjugate), we can assume without loss of generality that both h, g are real.

To prove the second assertion of the theorem, note that the polynomial "primitive" g obtained by the integral (26.19) of a form of degree $\deg\omega < \deg df$ grows no faster than $o(|x|+|y|)^{n+1}$ (as follows from direct estimates) and hence is a polynomial form of degree not exceeding n. The difference $\omega - dg$ is a polynomial 1-form divisible by the form df of a *higher* degree. This means that $h = 0$ and ω is exact, $\omega = dg$. □

Remark 26.54. The second assertion of Theorem 26.52 can be proved using a more direct argument as in [**Pus97**]. Consider the Gelfand–Leray derivative $\frac{d\omega}{df}$ of the form ω. By Remark 26.34, this derivative has poles of order at most 1 at infinity. If the integral of ω along any cycle on L_z is zero, then the residues at these points are all zeros. Since the poles are simple, absence of the residues means that the derivative $\frac{d\omega}{df}$ is in fact holomorphic at these points and its primitive on each compactified fiber $\overline{L_z}$, is a holomorphic function, which is necessarily a constant. Hence the derivative $\frac{d\omega}{df}$ is itself zero restricted on each fiber, and thus $d\omega = 0$.

As an application of Theorem 26.52, we construct a polynomial foliation of the plane from the class $\mathcal{A}_n$ having $\frac{1}{2}(n+1)(n-2)$ limit cycles.

Theorem 26.55 (Yu. Ilyashenko, [**Ily69**], I. Khovanskaya (Pushkar′), [**Pus97**]). *If $f \in \mathbb{R}[x,y]$ is a Morse polynomial of degree $n+1$ transversal to infinity, then for any $N = \frac{1}{2}(n+1)(n-2)$ real ovals of the integrable foliation $\{df = 0\}$ on $\mathbb{R}^2$ one can construct a form*

$$\omega = P(x,y)\,dx + Q(x,y)\,dy, \qquad P,Q \in \mathbb{R}[x,y], \quad \deg P,Q \leqslant n, \tag{26.55}$$

such that the perturbation $\{df + \varepsilon\omega = 0\}$ (cf. with (26.1)) produces at least N limit cycles which converge to the specified ovals as $\varepsilon \to 0$.

Proof. Consider the linear space of all polynomial 1-forms ω of the specified degree: the dimension of this space is $n(n+1)$. The exact forms constitute a subspace of dimension $\frac{1}{2}(n+2)(n+1) - 1$ in it. The quotient space has dimension $n(n+1) - \frac{1}{2}(n+2)(n+1) + 1$ which is exactly equal to $N+1$.

For any choice of N real ovals $\delta_i \subseteq \{f = c_i\}$, $i = 1, \dots, N$, the condition $\oint_{\delta_i} \omega = 0$ constitutes a linear restriction on the form ω. As soon as the number of restrictions is less than the dimension of the (quotient) space, there exists at least one form, by construction *not exact*, whose integral is zero along all the specified ovals.

By Theorem 26.52, *all these zeros are isolated.* Indeed, otherwise the integral must have a real branch which is identically zero, which is possible only if the form is exact.

If all these zeros are simple, then by Remark 26.2, the corresponding perturbation will produce at least N limit cycles.

If some of the zeros are of even orders, then the corresponding limit cycles can "escape" into the nonreal domain. In this case the perturbation form should be produced in the following way. Assume without loss of generality that all ovals $\delta_1, \dots, \delta_N$ are oriented positively so that the form $\omega_0 = y\,dx$ has *negative* integral over each such oval. Let k_+ denote the number of ovals such that the corresponding real branch of the integral $\oint \omega$ has a local

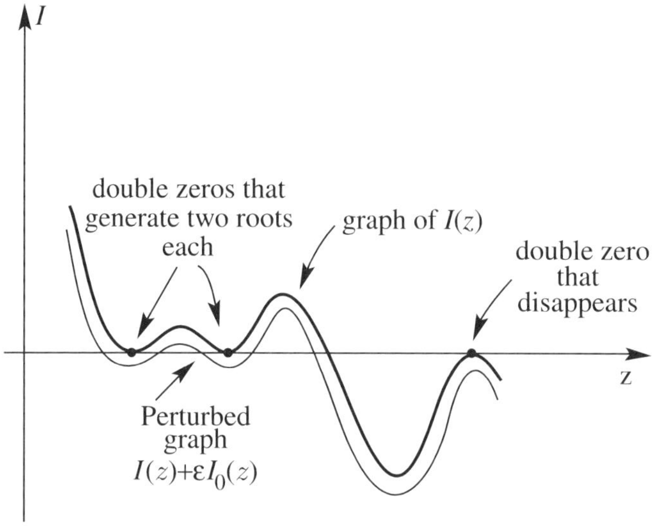

Figure V.5. Construction of the perturbation with the specified number of simple roots

minimum there, and by k_- the number of ovals yielding the local maximum (recall that in any case the roots are isolated). The total number $k_+ + k_-$ is equal to the number of different ovals where the integral $I(z) = \oint_{\delta(z)} \omega$ has a local extremum. The remaining $N - (k_+ + k_-)$ ovals correspond to the roots where I changes its sign. If $k_+ \geqslant k_-$, consider the form $\omega + \epsilon\omega_0$, where $1 \gg \epsilon > 0$ is an auxiliary parameter: the corresponding integral is obtained by subtracting a small everywhere positive quantity $I_0(z) = \oint_{\delta(z)} \omega_0$ from the function $I(z)$.

In particular, each of the k_+ local minima of I will produce at least two odd order roots, roots at the local maxima will disappear, and every odd order root from the remaining $N - (k_+ + k_-)$ will produce at least one odd order root again if ϵ is sufficiently small; see Fig. V.5. Since $k_+ \geqslant k_-$, the total order of odd order roots that appear after this small variation *of the form* $\omega_\epsilon = \omega + \epsilon y\,dx$ of the corresponding integral $I(z) + \epsilon I_0(z)$ will have at least $N - (k_+ + k_-) + 2k_+ \geqslant N$ odd order roots. The same Remark 26.2 shows now that the number of limit cycles in this degenerate case will be again no less than N. □

Remark 26.56. The lower bound for the number of zeros of Abelian integrals (and the respective limit cycles) is not sharp. N. F. Otrokov constructed in [**Otr54**] examples with a larger number of limit cycles. However all of them encircle a unique singular point of the foliation, whereas Theorem 26.55 allows to place them much more freely.

The principal term of both the Otrokov's lower bound and the bound achieved in Theorem 26.55 is of the same form $\frac{1}{2}n^2 + O(n)$. However, for special *rather symmetric* polynomials f of degree $n+1$ one can construct Abelian integrals of forms of degree n having more zeros. For instance, there are known examples with as many as $n^2 - 1$ isolated real zeros for $n = 2, \dots, 10$.

Exercises and Problems for §26.

Problem 26.1. Prove that a real foliation is really analytically integrable near an identical cycle.

Exercise 26.2. Why Proposition 26.1 does not apply to a neighborhood of a *critical* level curve $\{f = 0\}$, carrying a critical point?

Problem 26.3. Prove that a continuous branch of any real Abelian integral (over a continuous family of compact ovals of f) is real analytic on any interval free from real critical values of the polynomial f.

Problem 26.4. Let γ be a real oval of a cubic ultra-Morse polynomial. Prove that for any ε there exists a quadratic vector field with a limit cycle of multiplicity 2 whose Hausdorff distance from γ is smaller than ε.

Problem 26.5. Prove that for any r there exists a real polynomial vector field from the class $\mathcal{A}_r$ with a limit cycle of multiplicity $\frac{1}{2}(r+1)(r-2)$.

Problem 26.6. Prove that the Bonnet set of any polynomial is an algebraic subset in $\mathbb{C}$. Give an example of a polynomial $f \in \mathbb{C}[x, y]$ with an infinite Bonnet set $\operatorname{Bs}(f) = \mathbb{C}$.

Problem 26.7. Prove Proposition 26.17.

Exercise 26.8. Find atypical values of the polynomial $f(x, y) = xy(xy - 1)$.

Problem 26.9. Prove that for polynomials not transversal to infinity, the period matrix remains single-valued and has at worst poles at the atypical values.

Problem 26.10. Prove that for a polynomial $f = f_{n+1} + \cdots$ transversal to infinity, the level curves $f = z$ and $f_{n+1} = z$ are homeomorphic for all sufficiently large $|z|$ (cf. with the end of the proof of Theorem 26.31).

Problem 26.11. Consider a compact Riemann surface C and a noncontractible simple loop γ on it. Because of the noncontractibility, the difference $C \smallsetminus \gamma$ is connected and has a boundary which is homeomorphic to two disjoint circles. Sealing the two holes by topological disks results in a new surface $\widetilde{C} = C/\gamma$ called *pinching* of C along γ.

Compare the Euler characteristic of C and C/γ.

Problem 26.12. If f is a complex polynomial transversal to infinity with only nondegenerate critical points (some of the critical values can coincide), then any critical level curve can be obtained from a nearby nonsingular level curve by pinching along the corresponding vanishing cycles. Prove this statement.

Problem 26.13. Prove the Plücker formula (25.27), using Problems 26.11 and 26.12.

Problem 26.14. For any collection of cycles $c_1, \dots, c_m$ on a Riemann surface their *intersection graph* is a graph with m vertices which are connected by an edge if and only if the corresponding intersection index $c_i \cdot c_j$ is nonzero.

Prove that for any polynomial transversal to infinity, one can construct a basis of vanishing cycles (as in the Clarification to Theorem 26.47) such that the corresponding intersection graph is connected.

Hint. Consider a small perturbation of the degenerate polynomial $f = \prod_{j=1}^{n+1} l_j$ with generic affine polynomials of degree 1.

Problem 26.15. Prove that for *any* polynomial and *any* proper system of paths the intersection graph constructed in Problem 26.14 is always connected.

Problem 26.16. Derive from the Problems 26.14 and 26.15 the assertion on transitivity of action (Theorem 26.48).

The Problems 26.17–26.22 together give an upper bound for the *multiplicity* of an isolated zero of an Abelian integral. This result was obtained by P. Mardešić [**Mar91**].

Problem 26.17. Let f be a Morse polynomial of degree $n+1$ transversal to infinity, and $\omega \in \Lambda^1[\mathbb{C}^2]$ a polynomial form of degree n. Consider the integrals $J_k(z) = \oint_{\delta_k(z)} \omega$, where $\delta_1(z), \dots, \delta_m(z)$ are vanishing cycles constructed as in Theorem 26.47. Let $W(z)$ be a Wronski determinant of the functions $J_1, \dots, J_m$.

Prove that $W(z)$ is a rational function of z and describe its polar locus.

Problem 26.18. Prove that $W \equiv 0$ if and only if ω is closed.

Hint.: Use the Problem 26.15 and exactness theorem.

Problem 26.19. Estimate from above the order of a pole of W at any critical point of the ultra-Morse polynomial.

Problem 26.20. Prove that the integral J_k in Problem 26.17 has an algebraic singular point at infinity: a branch point of order equal either to $r+1$, or to a divisor of $r+1$.

Problem 26.21. Give an upper bound for the order of the pole of W at infinity.

Problem 26.22. Give an upper bound for the *multiplicity* of an isolated zero of any of the integrals $J_k(z)$.

Problem 26.23. Consider a real ultra-Morse polynomial H with a compact component Γ of a critical level that contains a critical point A and is not a singleton. Prove that this component is an eight shaped figure. Let the corresponding critical value be zero, and a level curve $\{H = \varepsilon\}$ has a smooth component Γ_ε such that Γ lies inside Γ_ε for any small positive ε. Then the level curve $\{H = -\varepsilon\}$ has two components for the same ε, denoted by $\Gamma^1_{-\varepsilon}$ and $\Gamma^2_{-\varepsilon}$, one contained in one loop of Γ, another in another one. Let $\delta_\varepsilon \subset \{H = \varepsilon\}$ be a vanishing cycle close to A. Consider a loop $\Gamma_{-\varepsilon} \subseteq L_{-\varepsilon} = \{H = -\varepsilon\}$ obtained from Γ_ε by continuation over a half-circle of radius ε centered at zero in the set of noncritical values of

H. Find the expression of the corresponding element $[\Gamma_{-\varepsilon}] \in H_1(L_{-\varepsilon}, \mathbb{Z})$ through $[\Gamma^1_{-\varepsilon}], [\Gamma^2_{-\varepsilon}], [\delta_{-\varepsilon}]$.

Problem 26.24. Prove that for any r there exists a real polynomial vector field from the class $\mathcal{A}_r$ with a limit cycle of multiplicity $\frac{1}{2}(r+1)(r-2)$.

Problem 26.25. Modifying the proof of Theorem 26.35, write explicitly the Picard–Fuchs system for the cubic polynomial $f(x,y) = \frac{1}{2}(x^2+y^2) - \frac{x^3}{3}$ and the two forms $\omega_1 = y\,dx$ and $\omega_2 = xy\,dx$.

Problem 26.26. Let $f(x,y) = y^2 + x^{n+1} + p_n(x)$ be a *hyperelliptic* polynomial. Prove that the forms $y\,dx, xy\,dx, \dots, x^{n-1}y\,dx$ form a basis of the corresponding Petrov module $\mathbf{P}_f$. Modifying the proof of Theorem 26.35, write explicitly the Picard–Fuchs system for the corresponding *hyperelliptic integrals*.

27. Topological classification of complex linear foliations

The famous Grobman–Hartman theorem [**Gro62, Har82**] asserts that any real vector field whose linearization matrix is hyperbolic (i.e., has no eigenvalues with zero real part), is topologically orbitally equivalent to its linearization. An elementary analysis shows that two hyperbolic linear real vector fields are orbitally topologically conjugated if and only if they have the same number of eigenvalues to both sides of the imaginary axis.

This section describes the complex counterparts of these results. From the real point of view a holomorphic 1-dimensional singular foliation on $(\mathbb{C}^n, 0)$ by phase curves of a holomorphic vector field is a 2-dimensional real analytic foliation on $(\mathbb{R}^{2n}, 0)$. If the singularity at the origin is in the Poincaré domain, this foliation induces a *nonsingular real* 1-*dimensional* foliation (*trace*) on all small $(2n-1)$-dimensional spheres $\mathbb{S}^{2n-1}_\varepsilon = \{|x_1|^2 + \cdots + |x_n|^2 = \varepsilon > 0\}$. Under the *complex* hyperbolicity-type conditions excluding resonances, the trace is generically structurally stable. Poincaré resonances manifest themselves via *bifurcations* of this trace foliation.

On the contrary, if the singularity is in the Siegel domain, the corresponding foliations exhibit *rigidity*: two foliations are topologically equivalent if and only if there is a rather special conjugacy between them which is completely determined by n complex numbers. This rigidity implies that there are *continuous invariants* (moduli) of topological classification.

27A. Trace of the foliation on the small sphere. Consider the real sphere of radius $\varepsilon > 0$,

$$\mathbb{S}_r = \{r^2(x) = \varepsilon\} \subseteq \mathbb{C}^n, \qquad r^2(x) = |x|^2 = \sum_1^n x_i \bar{x}_i. \tag{27.1}$$

The differential of the (nonholomorphic) function $r^2\colon \mathbb{C}^n \to \mathbb{R}$ is a complex-valued 1-form, $dr^2 = x\,d\bar{x} + \bar{x}\,dx$, which on the complex vector field $F(x) = (v_1(x), \dots, v_n(x))$ takes the value

$$dr^2 \cdot v(x) = \sum_{i=1}^{n} x_i\,\bar{v}_i + \bar{x}_i\,v_i = 2\,\mathrm{Re}\big(\sum x_i \bar{v}_i\big) \in \mathbb{R}.$$

If $F(x) = \Lambda x$ is a linear diagonal vector field with the eigenvalues $\lambda_1, \dots, \lambda_n \in \mathbb{C}$, then

$$dr^2 \cdot F = 2\,\mathrm{Re} \sum \lambda_i |x_i|^2.$$

The following observation due to V. Arnold [**Arn69**] gives a topological characterization of the Poincaré type holomorphic foliations.

Proposition 27.1. *All complex phase curves of the diagonal linear vector field Λx of Poincaré type in $\mathbb{C}^n$ are transversal as 2-dimensional embedded surfaces, to all spheres $\mathbb{S}_\varepsilon$, $\varepsilon > 0$.*

Proof. The tangent space to any trajectory considered as a real 2-dimensional surface in $\mathbb{R}^{2n} = \mathbb{C}^n$, is spanned over $\mathbb{R}$ by the vectors $v(x) = \Lambda x$ and $v'(x) = i\Lambda x$. To prove the transversality, it is sufficient to verify that the 1-form dr^2 cannot vanish on both vectors simultaneously for $x \neq 0$.

If the spectrum belongs to the Poincaré domain, then without loss of generality we may assume that

$$\mathrm{Re}\,\lambda_i < 0, \qquad i = 1, \dots, n. \tag{27.2}$$

Indeed, replacing the field Λx by the orbitally equivalent field $\alpha\Lambda x$, $|\alpha| = 1$, preserves all holomorphic phase curves but rotates the spectrum of Λ as a whole.

Under the assumption (27.2) the expression

$$dr^2 \cdot F = s(x) = \sum \lambda_i |x_i|^2 \in \mathbb{C} \tag{27.3}$$

is in the left half-plane, moreover,

$$\mathrm{Re}\,s(x) \leqslant \delta\,|x|^2 < 0, \qquad \delta > 0. \tag{27.4}$$

This implies the required transversality. □

Remark 27.2. Transversality is an open condition: sufficiently small perturbations of the vector field leave it transversal to the compact sphere.

In particular, if $F(x) = \Lambda x + w(x)$ is a nonlinear vector field, then the rescaling $x \mapsto \varepsilon x$ conjugates its restriction on the ε-sphere $\mathbb{S}^{2n-1}_\varepsilon$ with the restriction of the field $F_\varepsilon(x) = \Lambda x + \varepsilon^{-1} w(\varepsilon x)$ on the unit sphere $\mathbb{S}^{2n-1}_1$. But since the nonlinear part $w(x)$ is at least of second order, the field F_ε is ε-uniformly close on the unit sphere to the linear field $F_0(x) = \Lambda x$. Thus we conclude that the *nonlinear* vector field F is transversal to all sufficiently small spheres $\mathbb{S}_\varepsilon^{2n-1}$.

Definition 27.3. Let $\mathcal{F} = \{L_\alpha\}$ be a foliation on a manifold M. The *trace of the foliation* on a submanifold $N \subset M$ is the partition of N into connected components of intersection of the leaves L_α with N, $\mathcal{F}|_N = \{L_\alpha \cap N\}$.

In general, the trace of a foliation need not itself be a foliation; the intersections $L_\alpha \cap N$ can be nonsmooth in general. Even in the analytic context one cannot exclude the appearance of singularities.

Corollary 27.4. *The trace of the holomorphic foliation $\mathcal{F}$ induced by a linear vector field of Poincaré type on any sphere $\mathbb{S}_\varepsilon^{2n-1}$ is a smooth (actually, real analytic) nonsingular real 1-dimensional foliation $\mathcal{F}' = \mathcal{F}|_{\mathbb{S}_\varepsilon}$.*

Proof. By the implicit function theorem, intersection of each leaf with the sphere is a smooth curve. □

Moreover, for singularities of Poincaré type the trace of the foliation on a (sufficiently small) sphere determines completely the foliation up to the topological equivalence, even if the vector field spanning the foliation is nonlinear.

Definition 27.5. A (topological) *cone* over a set $K \subset \mathbb{C}^n \setminus \{0\}$ is the set $\mathsf{C}K = \{rx \colon 0 \leqslant r \leqslant 1,\ x \in K\} \subseteq \mathbb{C}^n$. If $\mathcal{F}'$ is a foliation on the sphere $\mathbb{S}_1^{2n-1} \subset \mathbb{C}^n$, then the *cone over the foliation* $\mathsf{C}\mathcal{F}'$ is the foliation of $\mathbb{C}^n \setminus \{0\}$ whose leaves are the cones over the leaves of $\mathcal{F}'$.

Theorem 27.6. *A singular foliation $\mathcal{F}$ on $(\mathbb{C}^n, 0)$, generated by a vector field of Poincaré type, is topologically equivalent to the cone over its trace $\mathcal{F}'_\varepsilon = \mathcal{F}|_{\mathbb{S}_\varepsilon^{2n-1}}$ on any sufficiently small sphere.*

Proof. Under the normalizing assumption (27.2) the real flow of the vector field Λx, the one-parametric subgroup of linear maps $\{\Phi^t = \exp t\Lambda \colon t \in \mathbb{R}\}$ is locally contracting: orbits $\Phi^t(x)$, $x \in \mathbb{S}_1^{2n-1}$ of all points uniformly converge to the origin as $t \to +\infty$. This follows again from (27.4): if ε is so small that $|w(x)| < \frac{\delta}{2}|x|$ for $|x| < \varepsilon$, we have $|\Phi^t(x)| < \exp(-\delta t/4)\,|x|$ for all $t > 0$.

The real flow Φ^t is tangent to the foliation $\mathcal{F}$. Thus the map h of the small ε-ball $\{|x| \leqslant \varepsilon\}$ into itself, defined by the formulas

$$h(rx) = \Phi^{-\ln r}(x), \qquad 0 < r \leqslant 1,\ x \in \mathbb{S}_\varepsilon^{2n-1}, \qquad h(0) = 0,$$

is a homeomorphism conjugating $\mathsf{C}(\mathcal{F}|_{\mathbb{S}_\varepsilon})$ with $\mathcal{F}$. □

In particular, Theorem 27.6 implies that all foliations $\mathcal{F}'_\varepsilon$ are topologically equivalent to each other. Yet without the additional assumptions they may be nonequivalent to the foliation $\mathcal{F}'_0$ which is the trace of the linear foliation $\mathcal{F}_0$ on (any) sphere. This additional assumption is called *complex hyperbolicity.*

27B. Structural stability of the trace of hyperbolic foliation.

Definition 27.7. A holomorphic germ of a vector field $\dot{x} = Ax + \cdots$ in $(\mathbb{C}^n, 0)$ is *complex hyperbolic* (or just *hyperbolic*[7] if this does not lead to confusion), if no two eigenvalues λ_i, λ_j of the linearization matrix A differ by a real factor,

$$\lambda_i/\lambda_j \notin \mathbb{R} \qquad \text{for all} \quad i \neq j. \tag{27.5}$$

In particular, A must be nondegenerate and diagonalizable.

Under the additional assumption of complex hyperbolicity we can completely describe the trace of the linear diagonal foliation and show that it is structurally stable: any C^1-small perturbation produces a foliation that is topologically equivalent to the initial one.

Everywhere below in this section $\mathcal{F}$ is a singular foliation of $\mathbb{C}^n$ by phase curves of the complex hyperbolic vector field Λx with the eigenvalues $\lambda_1, \ldots, \lambda_n$ of the diagonal matrix Λ in the Poincaré domain and, if necessary, normalized by the condition (27.2). We denote by $\mathcal{F}'$ its restriction on $\mathbb{S}_1^{2n-1}$.

The first immediate consequence of complex hyperbolicity is the fact that the only multiply-connected leaves of the foliation $\mathcal{F}$ by complex phase curves of a diagonal linear system, are its separatrices.

Proposition 27.8. *The only multiply-connected leaves of a foliation generated by complex hyperbolic linear system $\dot{x} = Ax$ in $\mathbb{C}^n$ are its separatrices which are lines spanned by the eigenvectors of A. All other leaves of $\mathcal{F}$ are simply connected.*

Proof. Without loss of generality we may assume that A is diagonal, $A = \Lambda = \operatorname{diag}\{\lambda_1, \ldots, \lambda_n\}$. The application

$$t \mapsto x(t) = \bigl(c_1 \exp(\lambda_1 t), \ldots, c_n \exp(\lambda_n t)\bigr) = \Phi^t(c), \qquad c \in \mathbb{C}^n, \tag{27.6}$$

parameterizes the phase curve passing through a point $c \in \mathbb{C}^n$. This parametrization is not injective, if $\exp t\lambda_j = 1$ for some t and all j corresponding to nonzero coordinates of the point c. If there is only one such coordinate, then the noninjectivity is indeed possible if $t = 0 \bmod T_j$, where T_j is the corresponding period. If a has at least two nonzero coordinates j and k, then the simultaneous occurrence $t = 0 \bmod T_j$ and $t = 0 \bmod T_k$ is impossible: it would mean that the ratio T_j/T_k is rational hence real. □

[7] In order to distinguish this from the *real hyperbolicity* of self-maps, introduced in Definition 7.2. The reasons why two seemingly different notions are called by similar names, are clarified by Proposition 27.10 below.

Assume that in addition to the normalizing condition (27.2), the enumeration of the eigenvalues $\lambda_1, \dots, \lambda_n$ is chosen in the increasing order of their arguments in the interval $[0, 2\pi)$,

$$\operatorname{Arg}\lambda_1 < \operatorname{Arg}\lambda_2 < \cdots < \operatorname{Arg}\lambda_{n-1} < \operatorname{Arg}\lambda_n \tag{27.7}$$

(this is possible since by the hyperbolicity assumption $\lambda_j/\lambda_k \notin \mathbb{R}$, so all values of the arguments are distinct).

Since the coordinate axes are leaves of $\mathcal{F}$, the big circles $C_i = \{x_j = 0,\ j \neq i,\ |x_i| = 1\}$ are leaves of $\mathcal{F}'$. We show that all other leaves are bi-asymptotic to these circles.

Proposition 27.9. *If A is hyperbolic, then the limit set $\overline{\gamma} \smallsetminus \gamma$ of any leaf $\gamma \in \mathcal{F}'$ different from C_j, is the union of two big circles $C_j \cup C_k$, $j \neq k$.*

Proof. Any leaf L_c of the "large" foliation $\mathcal{F}$ passing through a point $c \in \mathbb{C}^n$ is parameterized by the map (27.6). The intersection $\gamma_c = L_c \cap \mathbb{S}_1^{2n-1}$ is defined by the equation

$$|c_1|^2 \exp 2\operatorname{Re}(\lambda_1 t) + \cdots + |c_n|^2 \exp 2\operatorname{Re}(\lambda_n t) = 1. \tag{27.8}$$

As follows from the transversality property, this is a smooth curve parameterized by a smooth curve $\widetilde{\gamma}_c$ on the t-plane, defined by the equation (27.8).

The curve $\widetilde{\gamma}_c$ *apriori* may have compact and noncompact components. But any compact component must bound a compact set in $\mathbb{C} \cong L_c$ so that the function $|x(t)|$ has critical points inside. Such critical points correspond to nontransversal intersections that are forbidden by Proposition 27.1.

Thus γ_c may consist of only noncompact components (eventually, several) along which $|t|$ tends to infinity. But as $|t| \to \infty$, the growth rate of each exponential term $\exp\bigl(2\operatorname{Re}(\lambda_j t)\bigr)$ is determined by the angular behavior of t. In particular, since all exponentials in (27.8) should be bounded (unless the corresponding coefficients c_j vanish), we have the necessary condition that all limit directions $\lim\{t/|t| : t \in \widetilde{\gamma}_c,\ |t| \to +\infty\}$ must be within the sector $S_c = \bigcap_{j:\ c_j \neq 0} \{\operatorname{Re}\lambda_j t \leqslant 0\}$. However, if t tends to infinity (asymptotically) *in the interior* of this sector, then all exponents will tend to zero in violation of (27.8).

Thus if L_c is *not* a separatrix (i.e., more than one coefficient c_j is nonzero), the curve $\widetilde{\gamma}_c$ must be bi-asymptotic to the two boundary rays of the sector S_c. This in turn means that the corresponding trajectory γ_c is bi-asymptotic to the two cycles $C_j \neq C_k$. □

Behavior of leaves near each cycle C_j is determined by the iterations of the corresponding holonomy map of the foliation $\mathcal{F}'$ which can be easily expressed in terms of the holonomy of the corresponding complex separatrix $\mathbb{C}e_j$, $e_j = (0, \dots, \underset{j}{1}, \dots, 0) \in \mathbb{C}^n$, of the initial holomorphic foliation $\mathcal{F}$.

Consider the circular leaf $C_j \subset \mathbb{S}_1^{2n-1}$ of the foliation $\mathcal{F}'$ with the orientation induced by the counterclockwise (positive) direction of going around the origin in the jth coordinate axis. Then for any (smooth) $(2n-2)$-dimensional cross-section $\tau_j' \colon (\mathbb{R}^{2n-2}, 0) \to (\mathbb{S}_1^{2n-1}, e_j)$ transversal to the trace foliation $\mathcal{F}'$ at the point $e_j \in C_j$, one can define the first return map (holonomy) $h_j = \Delta_{C_j} \colon (\tau_j', 0) \to (\tau_j', 0)$.

Proposition 27.10. *The holonomy $h_j \in \operatorname{Diff}(\mathbb{R}^{2n-2}, 0)$ of each cycle C_j is differentiably conjugate to the diagonal linear map $\Lambda_j \in \operatorname{Diff}(\mathbb{C}^{n-1}, 0)$ hyperbolic in the sense of Definition 7.2: its eigenvalues $\{2\pi i\lambda_k/\lambda_j\}$, $k \neq j$ are all off the unit circle.*

Proof. Since the sphere $\mathbb{S}_1^{2n-1}$ is transversal to the foliation $\mathcal{F}$, any smooth (nonholomorphic) cross-section $\tau_j' \colon (\mathbb{R}^{2n-2}, 0) \to (\mathbb{S}_1^{n-1}, e_j)$ transversal to the trace foliation $\mathcal{F}'$ at the point $e_j \in C_j$ *inside* $\mathbb{S}_1^{2n-1}$, will also be transversal to the complex separatrix of $\mathcal{F}$ lying on the jth coordinate axis.

The holonomy maps for the foliation $\mathcal{F}$ associated with the two cross-sections, τ_j' and the "standard" cross-section $\tau_j \colon (\mathbb{C}^{n-1}, 0) \to (\mathbb{C}^n, e_j)$, are smoothly conjugate, in fact, the conjugacy is real analytic as a germ between $(\mathbb{R}^{2n-2}, 0)$ and $(\mathbb{C}^{n-1}, 0)$. The holonomy for the "standard" cross-section was already computed; see Example 2.28. □

Proposition 27.11. *The unstable (resp., stable) manifold of the cycle C_j is the sphere $\mathbb{S}_1^{j-1} = \{x_{j+1} = \dots = x_n = 0\} \cap \mathbb{S}_1^{2n-1}$ (resp., the sphere $\mathbb{S}_1^{n-j-1} = \{x_1 = \dots = x_{j-1} = 0\} \cap \mathbb{S}_1^{2n-1}$).*

Proof. The corresponding complex coordinate planes $\mathbb{C}^{j-1}$ and $\mathbb{C}^{n-j-1}$ in $\mathbb{C}^n$ are invariant by the foliation $\mathcal{F}$ and the computations of the preceding proof show that the restriction of the first return map on the corresponding spheres (in intersection with the cross-section τ_j') has only eigenvalues $\exp 2\pi i\lambda_k/\lambda_j$. All these numbers are of modulus less than one (resp., greater than one). Since the stable (unstable) manifolds are uniquely defined, this proves the proposition. □

The properties of the foliation $\mathcal{F}'$ established by these three propositions, imply its *structural stability*: any sufficiently close foliation is topologically equivalent to $\mathcal{F}'$.

Theorem 27.12 (J. Guckenheimer, 1972 [**Guc72**]). *Assume that the diagonal matrix Λ is complex hyperbolic and in the Poincaré domain.*

Then the holomorphic vector field $F(x) = Ax + w(x)$ is topologically orbitally linearizable, i.e., the holomorphic singular foliation of $(\mathbb{C}^n, 0)$ by complex phase curves of the holomorphic vector field is locally topologically equivalent to the foliation defined by the linear vector field $F_0(x) = Ax$.

Moreover, any sufficiently close vector field is locally topologically orbitally equivalent to F.

Proof. Consider the rescaling $F_\varepsilon(x) = \varepsilon^{-1}F(\varepsilon x)$, the corresponding foliation $\mathcal{F}_\varepsilon$ in the ball $\{|x| < 1\}$ and its trace $\mathcal{F}'_\varepsilon$ on the unit sphere $\mathbb{S}_1^{2n-1} = \varepsilon^{-1}\mathbb{S}_1^{2n-1}$.

By Theorem 27.6, both foliations $\mathcal{F}_\varepsilon$ and $\mathcal{F}_0$ are topological cones over their traces $\mathcal{F}'_\varepsilon$ and $\mathcal{F}'_0$. The assertion of the theorem will follow from the topological equivalence of the latter two foliations on $\mathbb{S}_1^{2n-1}$.

By the Palis–Smale theorem [**PS70a**], a vector field on the compact manifold is structurally stable (i.e., its phase portrait is topologically orbitally equivalent to that of any sufficiently C^k-close vector field) if it meets the following *Morse–Smale conditions*:

(1) its singular points and limit cycles are hyperbolic (i.e., all eigenvalues of the linearization at any singular point have nonzero real parts, and all multiplicators of any limit cycle have modulus different from 1);

(2) its orbits can accumulate only to singular points or limit cycles;

(3) all stable and unstable invariant manifolds of singular points and limit cycles (which exist by the hyperbolicity assumption) intersect transversally.

All these conditions for the foliation $\mathcal{F}'_0$ are verified in Propositions 27.10, 27.9 and 27.11 respectively. Therefore the foliation $\mathcal{F}'_0$ is structurally stable and hence topologically equivalent to $\mathcal{F}'_\varepsilon$ for all small ε.

Returning to the initial nonlinear vector field $F = F_1$, we conclude that it is topologically orbitally equivalent to its linearization in all sufficiently small balls $\{|x| < \varepsilon\}$. □

Corollary 27.13. *Any two* linear *complex hyperbolic vector fields of Poincaré type in $\mathbb{C}^n$ generate globally topologically equivalent singular foliations.*

Any two nonlinear *holomorphic vector fields in $(\mathbb{C}^n, 0)$, whose linearizations are complex hyperbolic vector fields of Poincaré type, generate* locally *topologically equivalent singular holomorphic foliations.*

Proof. Since topological equivalence is transitive, by Theorem 27.12 the second assertion of the corollary follows from the first one.

To prove the assertion on linear systems, note that any two complex hyperbolic matrices in the Poincaré domain can be continuously deformed into each other within this class. Indeed, any such matrix can be first diagonalized and all its eigenvalues brought into the open left half-plane. Then all absolute values of these eigenvalues can be made equal to 1 without changing their arguments; this will affect neither hyperbolicity nor the Poincaré property. Finally, the arguments of the eigenvalues can be assigned any positions, say, at equal angles between $\pi/2$ and $-\pi/2$. In this normal form the two diagonal matrices of the same size differ only by reordering of the coordinate axes. □

27C. Resonances in the Poincaré domain. Without complex hyperbolicity the foliation traced by a *linear* system on the unit sphere is still nonsingular, but may have nontrivial recurrence. Indeed, in this case the first return map for one of the cycles will have a multiplicator $\exp 2\pi i\lambda_1/\lambda_2$ which has modulus 1. The corresponding foliation $\mathcal{F}'$ on the sphere will then have a family of invariant 2-tori foliated by periodic or quasiperiodic orbits, depending on whether the ratio $\lambda_1/\lambda_2 \neq 1$ is rational or not. Since both rational and irrational numbers are dense, two nonhyperbolic linear systems in the Poincaré domain can be arbitrarily close to each other but topologically nonequivalent.

Generically, occurrence of multiple eigenvalues leads to the linearization matrix with a nontrivial Jordan normal form. Consider for simplicity the case $n = 2$, where such a form is necessarily a block of size 2. Then the corresponding foliation has only one complex separatrix. The same arguments as were used in the proof of Proposition 27.10 show that this separatrix leaves the trace in the form of a cycle on the sphere $\mathbb{S}^3$ whose first return map is conjugate to the complex holonomy of the separatrix.

Somewhat surprisingly and in contrast with the previously discussed diagonal cases, the holonomy map of this separatrix is *essentially nonlinear*: it cannot be linearized by a suitable choice of the cross-section (or, what is the same, a chart on it), as explained in Example 2.30. The computation below for the case where $n = 1$ shows that the holonomy has a fixed point of multiplicity exactly equal to 2 and thus a small perturbation will produce two close fixed points corresponding to two cycles of the trace foliation.

Occurrence of nonlinearities affects the situation in a similar way when (Poincaré) resonances occur, as was observed in [**Arn69**]. Consider the simplest Poincaré resonance in $\mathbb{C}^2$ and compute the holonomy map.

Proposition 27.14. *Consider a planar resonant singularity of the Poincaré type in the formal normal form*

$$\dot{x} = nx + a\,y^n, \quad \dot{y} = y, \qquad a \in \mathbb{C},\ n \geqslant 1. \tag{27.9}$$

Then the holonomy Δ of the unique separatrix $y = 0$, computed for the standard cross-section $\tau = \{x = 1\}$, is tangent to a rotation by the rational angle $2\pi/n$ and its nth iteration has an isolated fixed point of multiplicity $n + 1$ at the origin.

Proof. The system (27.9) is integrable: its general solution is $y(t) = C \exp t$, $x = (C' + aC^n t) \exp nt$, with arbitrary constants $C, C' \in \mathbb{C}$. The initial condition $(x(0), y(0)) = (1, s) \in \tau$ yields for the corresponding solution the formula

$$x(t) = (1 + as^n\, t) \exp nt, \qquad y(t) = s \exp t.$$

For $s = 0$ the x-component of the solution (separatrix) is $2\pi/n$-periodic. For small $s \in (\mathbb{C}, 0)$, the solution with this initial condition crosses again the section τ at the moments $t_k(s) = 2\pi ik/n + \delta_k(s)$, $\delta_k(s) = o(1)$, $k = 1, 2, \dots$, where $\delta_k(s)$ is the complex root of the equation

$$1 + as^n(2\pi ik/n + \delta_k(s)) = \exp\bigl(-n\delta_k(s)\bigr) = 1 - n\delta_k(s) + \cdots, \quad \lim_{s\to 0} \delta_k(s) = 0.$$

This equation can be resolved with respect to $\delta_k(s)$ defining the latter as an analytic function of s by the implicit function theorem. Computing the Taylor terms, we see immediately that

$$\delta_k(s) = -\frac{2\pi ika}{n^2}\, s^n + \cdots, \qquad t_k(s) = \frac{2\pi ik}{n} + \delta(s).$$

The iterated power of the holonomy map Δ^k is therefore

$$\Delta^k(s) = s \exp t_k(s) = \lambda^k\, s \exp \delta(s) = \lambda^k s(1 - kA\, s^n + \cdots),$$

$$\lambda = \exp \tfrac{2\pi i}{n}, \qquad A = \tfrac{2\pi ia}{n^2} \neq 0.$$

The nth iterated power of Δ is tangent to the identity and has an isolated fixed point of multiplicity exactly $n + 1$. □

Corollary 27.15. *The resonant node corresponding to the resonance $(1 : n)$, $n \geqslant 2$, can be analytically linearized if and only if it can be topologically linearized in the complex domain.*

Proof. Consider the trace of the foliation on the unit sphere. The first return map is a topological invariant of the foliation. For the nonlinear Jordan node (27.9) with $a \neq 0$ the holonomy map is nontrivial (its nth power has an isolated fixed point), whereas the holonomy map for the linear node is linear and its nth power identical. □

Note that in the real domain all nodes are topologically equivalent to each other.

Remark 27.16. The resonant conformal germ $\Delta \in \operatorname{Diff}_1(\mathbb{C}, 0)$ has a fixed point at the origin and its nth iteration Δ^n is tangent to identity with order $n+1$.

Hence the iterated power $\widetilde{\Delta}^n$ of any sufficiently close conformal germ $\widetilde{\Delta}$ will have $n+1$ fixed points near the origin. One of these points is a fixed point for $\widetilde{\Delta}$ by the implicit function theorem. The remaining n points form a tuple of *n-periodic* points that are positioned approximately at vertices of a regular n-gon and permuted by $\widetilde{\Delta}$ cyclically.

In terms of the traces of the foliations, this means that a vector field obtained by a sufficiently small perturbation of the nonlinearizable resonant node, produces a foliation on $\mathbb{S}^3_1$ which has two cycles close to each other and linked with the index $n \geqslant 2$. All other leaves of the foliation are bi-asymptotic to these cycles. This gives the complete topological description for the *bifurcation of complex topological type* for passing through a Poincaré resonance. The assertion remains true also for the Jordan node (linear or not) with the ratio of eigenvalues equal to 1.

27D. Topological classification of linear complex flows in the Siegel domain. As opposite to the Poincaré case, the topological classification of holomorphic foliations generated by Siegel-type linear flows involves a number of continuous invariants. This means that in general an arbitrary small variation of the linear system results in a topologically different holomorphic foliation. This is a manifestation of the phenomenon known as *rigidity*.

Consider a hyperbolic linear vector field $\dot{x} = Ax$ of Siegel type in $\mathbb{C}^n$, i.e., assume that the origin belongs to the convex hull of the eigenvalues $\lambda_1, \dots, \lambda_n$ of A; see §5**A**. The complex hyperbolicity in the sense of Definition 27.7 implies that the matrix A can be assumed diagonal, $A = \Lambda = \operatorname{diag}\{\lambda_1, \dots, \lambda_n\}$, and the origin is necessarily *in the interior* of the convex hull $\operatorname{conv}\{\lambda_1, \dots, \lambda_n\} \subseteq \mathbb{C}$. In particular, hyperbolic Siegel systems exist only when $n \geqslant 3$.

Hyperbolicity means that the invariant axes (diagonalizing coordinates) of the linear vector field can be ordered to meet the following condition:

$$\begin{aligned} &\dot{x} = \Lambda x, \qquad x \in \mathbb{C}^n, \ \Lambda = \operatorname{diag}\{\lambda_1, \dots, \lambda_n\}, \quad n \geqslant 3, \\ &\operatorname{Im} \lambda_{j+1}/\lambda_j < 0, \quad j = 1, \dots, n, \qquad 0 \in \operatorname{conv}\{\lambda_1, \dots, \lambda_n\}. \end{aligned} \tag{27.10}$$

Here and everywhere below the enumeration of coordinates is cyclical modulo n, so that the assumption (27.10) includes the condition $\operatorname{Im} \lambda_n/\lambda_1 < 0$ as well. Denote by $\varPhi^t = \exp t\Lambda \colon \mathbb{C}^n \to \mathbb{C}^n$ the *complex* flow of the linear system $\dot{x} = \Lambda x$ and $\mathcal{F}$ the (singular) holomorphic foliation by phase curves of this flow:

$$\mathcal{F} = \{L_x\}_{x \neq 0}, \qquad L_x = \{\varPhi^t(x) \colon t \in \mathbb{C}\}.$$

Definition 27.17. The (complex) *periods* of the linear system (27.10) are the complex numbers $T_j = 2\pi i/\lambda_j \in \mathbb{C}$, $j = 1, \dots, n$.

For a hyperbolic system, the ratios of periods are never real.

Definition 27.18. Two tuples of complex numbers $\boldsymbol{T} = (T_1, \dots, T_n)$, and $\boldsymbol{T}' = (T'_1, \dots, T'_n)$ are called *affine equivalent*, if after an eventual rearrangement, one of the following two equivalent conditions holds:

(1) The exists an $\mathbb{R}$-linear map $M: \mathbb{C} \to \mathbb{C}$ such that $MT_j = T'_j$ for all $j = 1, \dots, n$,

(2) The rank of the $(4 \times n)$-matrix V whose columns are real 4-tuples $v_j = (\operatorname{Re} T_j, \operatorname{Im} T_j, \operatorname{Re} T'_j, \operatorname{Im} T'_j) \in \mathbb{R}^4$, is equal to 2.

The equivalence of the two conditions is immediate. If the rank of the matrix V is equal to 2 and the nonzero 2×2-minor occurs in the first two columns v_1, v_2, then any other column v_j, $j > 2$, can be represented as a real combination $\alpha v_1 + \beta v_2$, so that $T_j = \alpha T_1 + \beta T_2$ and $T'_j = \alpha T'_1 + \beta T'_2$ with the same $\alpha, \beta \in \mathbb{R}$. If M is an $\mathbb{R}$-linear map taking T_1 and T_2 to T'_1 and T'_2, then it will automatically map all other complex numbers (planar vectors) $T_3, \dots, T_n$ into $T'_3, \dots, T'_n$ respectively: $MT_j = M(\alpha T_1 + \beta T_2) = \alpha T'_1 + \beta T'_2 = T'_j$.

Conversely, if the there exists a map M mapping T_j into T'_j, then the last two rows of V are linear combinations of the first two rows, so that the rank of V is $\leqslant 2$. The equality occurs under the hyperbolicity assumption.

Theorem 27.19 (N. Ladis [**Lad77**], C. Camacho–N. H. Kuiper–J. Palis [**CKP76, CKP78**], Yu. Ilyashenko [**Ily77**]). *Assume that the singular holomorphic foliations $\mathcal{F}, \mathcal{F}'$ generated by two hyperbolic linear systems of Siegel type are topologically equivalent.*

Then the collections of the complex periods $\boldsymbol{T} = (T_1, \dots, T_n)$ and $\boldsymbol{T}' = (T'_1, \dots, T'_n)$ of the corresponding linear systems are affine equivalent: there exists an affine map $M: \mathbb{C} \to \mathbb{C}$ such that $MT_j = T'_j$ for all $j = 1, \dots, n$.

The proof of this theorem begins in §27**E** and occupies the rest of §27. The inverse statement is straightforward.

Theorem 27.20. *If two collections of periods for two diagonal linear systems are affine equivalent, the corresponding holomorphic singular foliations on $\mathbb{C}^n$ are topologically equivalent.*

Proof. Without loss of generality we may assume that the $\mathbb{R}$-linear map $M: \mathbb{C} \to \mathbb{C}$ taking the collection $\{\lambda_1^{-1}, \dots, \lambda_n^{-1}\}$ into $\{\lambda_1'^{-1}, \dots, \lambda_n'^{-1}\}$, is orientation-preserving. Otherwise replace one of the foliations by its image by the total conjugacy $(x_1, \dots, x_n) \mapsto (\bar x_1, \dots, \bar x_n)$: the latter is generated by the linear system with the eigenvalues $\{\bar\lambda_1, \dots, \bar\lambda_n\}$ (note that the map $\lambda \mapsto \bar\lambda$ reverts the orientation).

Consider, following Proposition 6.46, the map $h_\gamma\colon \mathbb{C} \to \mathbb{C}$, $x \mapsto x\,|x|^\gamma$, $\gamma \in \mathbb{C}$, extended as $h_\gamma(0) = 0$ at the origin. If $\operatorname{Re}\gamma > -1$, it is a homeomorphism of the complex plane into itself, since $\big||x|^\gamma\big| = |x|^{\operatorname{Re}\gamma}$ and therefore $|h_\gamma(x)| = |x|^{1+\operatorname{Re}\gamma}$.

We are looking for a diagonal homeomorphism H of the form $H(x) = \big(h_{\gamma_1}(x_1), \dots, h_{\gamma_n}(x_n)\big)$ which would conjugate two linear holomorphic foliations with the diagonal matrices $\Lambda = \operatorname{diag}\{\lambda_1, \dots, \lambda_n\}$ and $\Lambda' = \operatorname{diag}\{\lambda'_1, \dots, \lambda'_n\}$ as follows:

$$H \circ \exp t\Lambda = \exp t'\Lambda' \circ H, \qquad t' = t'(t), \tag{27.11}$$

where $t \mapsto t'(t)$ is a suitable $\mathbb{R}$-affine map, and $\gamma_1, \dots, \gamma_n$ are appropriate complex parameters with $\operatorname{Re}\gamma_j > -1$.

Since all maps H, Λ, Λ' are diagonal, the condition (27.11) consists of n independent "scalar" conditions,

$$h_{\gamma_j}(z \exp t\lambda_j) = h_{\gamma_j}(z) \exp t'\lambda'_j, \qquad j = 1, \dots, n, \tag{27.12}$$

which must hold identically for all $z \in \mathbb{C}$ and t; the affine map $t \mapsto t'$ must be the same for all j. Substituting the explicit formula for h_{γ_j}, we obtain after cancellation of $z\,|z|^{\gamma_j}$ the conditions

$$\exp[\lambda_j t + \gamma_j \operatorname{Re}(\lambda_j t)] = \exp t'\lambda'_j,$$

which will all be satisfied once we solve the system of *linear* equations

$$\lambda_j'^{-1}[\lambda_j t + \gamma_j \operatorname{Re}(\lambda_j t)] = t', \qquad j = 1, \dots, n. \tag{27.13}$$

Notice that any $\mathbb{R}$-affine map has the form $t \mapsto t' = at + b\bar t$ with uniquely determined complex numbers $a, b \in \mathbb{C}$. This map is orientation-preserving if and only if $|a| > |b|$. Substituting these formulas into the equations (27.13), we obtain the system of equations

$$\tfrac12 \lambda_j'^{-1}\lambda_j(2+\gamma_j) = a, \quad \tfrac12 \lambda_j'^{-1}\bar\lambda_j\gamma_j = b, \qquad j = 1, \dots, n.$$

The necessary and sufficient condition of solvability of these equations is obtained by elimination of the variables γ_j as follows. First we transform the equations to the form

$$2 + \gamma_j = 2a\lambda'_j\lambda_j^{-1}, \qquad \gamma_j = 2b\lambda'_j\bar\lambda_j^{-1}. \tag{27.14}$$

Subtracting one equation from the other yields the identities $1 = \lambda'_j(a\lambda_j^{-1} - b\bar\lambda_j^{-1})$ which in turn can be rewritten under the form

$$\lambda_j'^{-1} = a\lambda_j^{-1} - b\bar\lambda_j^{-1}, \qquad j = 1, \dots, n. \tag{27.15}$$

This solvability condition means affine equivalence of periods of the two systems in the sense of Definition 27.18: the inverse eigenvalues are simultaneously conjugated by the $\mathbb{R}$-affine map $M\colon w \mapsto aw - b\bar w$.

Conversely, if there exist complex a, b satisfying all identities (27.15), then one can resolve simultaneously all equations (27.14):

$$\gamma_j = 2b\frac{\lambda'_j}{\bar\lambda_j} = 2a\frac{\lambda'_j}{\lambda_j} - 2 = \lambda'_j(a\lambda_j^{-1} + b\bar\lambda_j^{-1}) - 1 = \frac{a\lambda_j^{-1} + b\bar\lambda_j^{-1}}{a\lambda_j^{-1} - b\bar\lambda_j^{-1}} - 1, \tag{27.16}$$

(the last transformation uses (27.15)). The corresponding conjugacy $H_\gamma = \big(h_{\gamma_1}(x_1), \dots, h_{\gamma_n}(x_n)\big)$ satisfies (27.11). It remains to verify that H is a homeomorphism, i.e., $\operatorname{Re}\gamma_j > -1$.

The direct computation yields

$$\operatorname{Re}\gamma_j + 1 = \operatorname{Re}\frac{a\lambda_j^{-1} + b\bar\lambda_j^{-1}}{a\lambda_j^{-1} - b\bar\lambda_j^{-1}} = \frac{|\lambda_j|^{-1}|^2(|a|^2 - |b|^2)}{|a\lambda_j^{-1} - b\bar\lambda_j^{-1}|^2}$$

(it is sufficient and easier to compute for $|\lambda_j| = 1$). This expression is positive if the $\mathbb{R}$-linear map M is orientation-preserving; indeed, in this case $|a| > |b|$, and hence $\operatorname{Re}\gamma_j > -1$ as required. □

Note that the sufficiency of affine equivalence of periods for topological equivalence of foliations is independent of whether the system is in the Siegel domain or not.

27E. Complex transition time and topology of linear hyperbolic maps in $\mathbb{C}^2$. In this section we begin the proof of topological rigidity of linear systems in the Siegel domain (Theorem 27.19)

As follows from Proposition 27.8, all nontrivial (other than separatrices) solutions of the system (27.10) are simply connected. Therefore for each leaf $L \in \mathcal{F}$ of the foliation, other than one of the separatrices, the complex function

$$t(x, y) = t \iff \varPhi^t(x) = y, \qquad x, y \in L \tag{27.17}$$

is correctly defined on *pairs* of points of that leaf. We will refer to $t(x, y)$ as the (complex) *transition time* from x to y. This function is holomorphic: indeed, $|\partial\varPhi^t(x)/\partial t| \neq 0$ on the leaf, so the implicit function theorem applies.

The transition time satisfies the obvious *cocycle identity*: for any n points on *the same* leaf,

$$t(x_1, x_2) + \cdots + t(x_{n-1}, x_n) + t(x_n, x_1) = 0, \qquad x_1, \dots, x_n \in L. \tag{27.18}$$

The transition time depends continuously on the leaf unless it grows to infinity. More accurately, if x_m, y_m are two sequences of points on simply connected leaves L_m that converge to the limits $x = \lim x_m$, $y = \lim y_m$, then the transition times $t(x_m, y_m)$ converge to a finite limit *provided that x and y belong to the same simply connected leaf L*:

$$x, y \in L \neq S_j \implies \lim_{m\to\infty} t(x_m, y_m) - t(x, y).$$

Indeed, in this case there exists a curve $\gamma \subset L$ connecting x with y. Trivializing the foliation near this curve, we see that x_m can be connected by a close curve γ_m with y_m on L_m.

On the contrary, each separatrix S_j is a multiply-connected domain and the flow $\varPhi^t$ restricted on S_j, is T_j-periodic, $\varPhi^{t+T_j} \equiv \varPhi^t$ for any $t \in \mathbb{C}$ (whence the term "period").

Consider the case $n = 3$ and denote by τ_j the *standard cross-section* $\{x_j = 1\} \cong \mathbb{C}^2$ to the separatrix $S_j = \mathbb{C}e_j$, $j = 1, 2, 3$, equipped with the coordinates (x_{j-1}, x_{j+1}) (recall that the enumeration of coordinates is cyclical). Denote by Δ_j the corresponding holonomy map *along the separatrix*: because of the periodicity and the choice of the cross-sections,

$$\Delta_j = \varPhi^{T_j}\big|_{\tau_j}, \qquad j = 1, 2, 3.$$

The operators Δ_j are linear diagonal with the eigenvalues $\exp 2\pi i \frac{\lambda_{j\pm1}}{\lambda_j}$. Given the assumption (27.10), we have

$$|\exp(2\pi i\lambda_{j-1}/\lambda_j)| < 1 < |\exp(2\pi i\lambda_{j+1}/\lambda_j)|.$$

This is the (real) hyperbolicity condition from Definition 7.2

Denote by $W_j^{\pm}$ the corresponding stable and unstable subspaces in τ_j: Δ_j is contracting on W_j^- and expanding on W_j^+ for all $j = 1,2,3$; see Fig. V.6.

This hyperbolic structure immediately implies the following lemma.

Lemma 27.21. *If $P = (1,0,p) \in W_1^-$, $P' = (1,p',0) \in W_1^+$ are two points, $pp' \neq 0$, then one can find two converging sequences of points $P_m \to P$ and $P'_m \to P'$ in the cross-section τ_1 such that $\Delta_1^m(P_m) = P'_m$. The number m of iterates grows to infinity.*

Proof. If μ and ν are the contracting and expanding eigenvalues of Δ_1, $|\mu| < 1 < |\nu|$, then the points

$$P_m = (1, \nu^{-m}p', p), \quad P'_m = (1, p', \mu^m p),$$

obviously meet all requirements. □

Before proceeding with the formal proof of this theorem, we briefly discuss the differences which occur between the Poincaré and Siegel hyperbolic cases as seen on the trace left by a linear foliation $\mathcal{F}$ on the unit sphere $\mathbb{S}^{2n-1} \subset \mathbb{C}^n$. We will deal with the simplest case $n = 3$.

First, the transversality of $\mathcal{F}$ to $\mathbb{S}^5 = \{|x_1|^2 + |x_2|^2 + |x_3|^2 = 1\}$ no longer holds: if $(\boldsymbol{\rho}, \boldsymbol{T}) = 0$ and $\boldsymbol{\rho} \in \mathbb{R}^3_+$, then on the 3-torus $\mathbb{T}^3 = \{|x_j| = \rho_j\}$ the leaves are tangent to the sphere. However, the coordinate axes (separatrices) are transversal to $\mathbb{S}^5$ and leave their traces on this sphere as the cycles $C_1, C_2, C_3 \subset \mathbb{S}^5$. These cycles are hyperbolic, and their corresponding invariant manifolds are 3-spheres $\mathbb{S}_j^{\pm} \cong \mathbb{S}^3$ for each $j = 1,2,3$: $\mathbb{S}_j^+ = \mathbb{S}^5 \cap \{x_{j-1} = 0\}$, $\mathbb{S}_j^- = \mathbb{S}^5 \cap \{x_{j+1} = 0\}$.

Here the similarity ends. First, the invariant manifolds do not intersect transversally. Quite contrary, $\mathbb{S}_j^+$ *coincides* with $\mathbb{S}_{j+1}^-$ and all trajectories inside this 3-sphere are bi-asymptotic to C_j and C_{j+1}. Behavior of the trace foliation $\mathcal{F}$ on this sphere is of Poincaré type.

All other trajectories on $\mathbb{S}^5 \smallsetminus \{x_1x_2x_3 = 0\}$, i.e., outside of the union of all invariant manifolds, are *closed*. Indeed, if $\lambda_1, \lambda_2, \lambda_3$ form a triangle, then at least one of the absolute values $|\exp \lambda_j t|$ tends to infinity as $|t| \to \infty$ along any ray. By (27.8), the trace of any leaf $L \in \mathcal{F}$ on $\mathbb{S}^5$ is compact (periodic). In particular, there are singular points of $\mathcal{F}|_{\mathbb{S}^5}$.

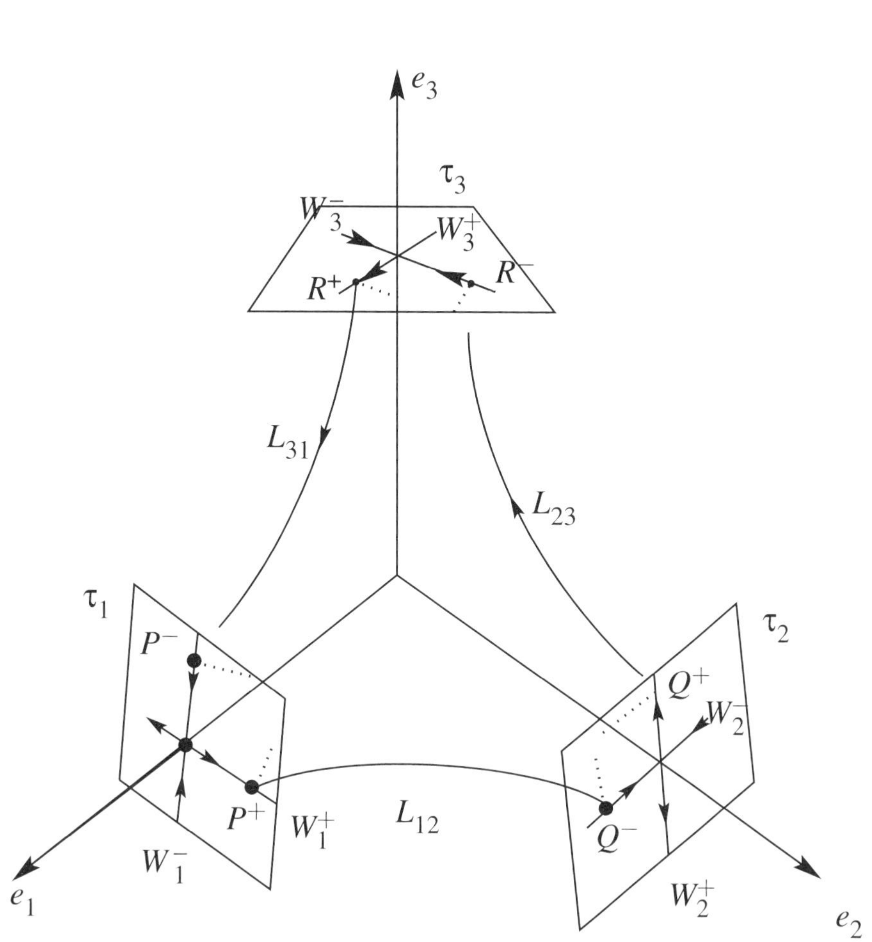

Figure V.6. Demonstration of Theorem 27.19: construction of the sequences $P_m^\pm, Q_m^\pm, R_m^\pm$

Thus we see that the trace of the foliation has singularities and nontrivial recurrence on the sphere $\mathbb{S}^5$.

27F. Main construction. The proof of Theorem 27.19 for $n = 3$ is based on construction of a sequence of leaves L_m of the foliation $\mathcal{F}$ that accumulate to *all three complex separatrices* simultaneously as $m \to \infty$. It is the *relative portions of time* spent near each separatrix, which constitute the continuous invariant underlying Theorem 27.19. The traces of these leaves on the unit sphere $\mathbb{S}^5$ will be very long but closed curves, that "spend most of their length" near the separatrix cycles C_j.

Assume that $n = 3$ and the three eigenvalues $\lambda_1, \lambda_2, \lambda_3$ form a triangle on the complex plane, containing the origin in the interior. Then their respective periods $\boldsymbol{T} = (T_1, T_2, T_3)$ also form the triangle with the same property.

There exists a unique positive vector $\boldsymbol{\rho} = (\rho_1, \rho_2, \rho_3) \in \mathbb{R}^3_+$, such that

$$0 = (\boldsymbol{\rho}, \boldsymbol{T}) = \rho_1 T_1 + \rho_2 T_2 + \rho_3 T_3, \qquad |\boldsymbol{\rho}| = 1. \tag{27.19}$$

Approximating the positive numbers $\rho_i > 0$ (27.19) by rational numbers as in the proof of Proposition 5.2, we can construct a sequence of *natural* vectors $\mathbf{k}_m = (k_{1,m}, k_{2,m}, k_{3,m}) \in \mathbb{N}^3$ such that

$$(\mathbf{k}_m, \boldsymbol{T}) \to 0, \quad \frac{\mathbf{k}_m}{|\mathbf{k}_m|} \to \boldsymbol{\rho} \qquad \text{as } m \to \infty. \tag{27.20}$$

In the hyperbolic Siegel case $|\mathbf{k}_m| \to +\infty$ implies $k_{m,j} \to +\infty$ for all $j = 1, 2, 3$.

Choose two arbitrary points $P^\pm \in W_1^\pm$ on the invariant subspaces in the cross-section τ_1 and let $P_m^\pm$, $m = 1, 2, \dots$, be two sequences of points satisfying the condition

$$t(P_m^-, P_m^+) = k_{m,1}, \qquad \lim_{m\to\infty} P_m^\pm = P^\pm.$$

Existence of such a sequence is asserted by Lemma 27.21.

The leaf $L_{12} \in \mathcal{F}$ passing through P^+ belongs to the invariant plane $x_3 = 0$ and intersects (transversally) the cross-section τ_2 at some point Q^- belonging to the Δ_2-invariant subspace W_2^-. By transversality arguments and continuity of the transition time along the leaf L_{12}, all nearby leaves L_m passing through P_m^+, cross τ_2 at some points Q_m^- that converge to Q^- so that the transition time between P_m^+ and Q_m^- has a limit as $m \to +\infty$, denoted by T_{12}:

$$\lim_{m\to\infty} t(P_m^+, Q_m^-) = t(P^+, Q^-) = T_{12}.$$

In the same way we can construct a sequence of points $R_m^+ \in \tau_3$ converging to $R^+ \in W_3^+$ such that P^-, R^+ belong to the same leaf of $\mathcal{F}$ denoted by L_{31}, and $t(R_m^+, P_m^-)$ has a limit,

$$\lim_{m\to\infty} t(R_m^+, P_m^-) = t(R^+, P^-) = T_{31}.$$

Now we construct two remaining sequences, $R_m^- \in \tau_3$ and $Q_m^+ \in \tau_2$, as follows:

$$R_m^+ = \Delta_3^{k_{m,3}}(R_m^-), \qquad Q_m^+ = \Delta_2^{k_{m,2}}(Q_m^-)$$

(more accurately, R_m^- should be defined starting from R_m^+ that were already constructed, iterating the inverse of the holonomy map, $R_m^- = \Delta_3^{-k_{m,3}}(R_m^+)$).

In contrast with the previous steps of the construction, convergence of the sequences R_m^-, Q_m^+ to some limits R^-, Q^+ that belong to the respective subspaces W_3^-, W_2^+ requires verification. Computation of the following lemma is a central step of the entire construction.

Lemma 27.22. *In the above settings, the sequences of points R_m^- and Q_m^+ converge,*

$$\lim_{m\to\infty} R_m^- = R^- \in W_3^-, \qquad \lim_{m\to\infty} Q_m^+ = Q^+ \in W_2^+.$$

The limit points Q^+ and R^- belong to the same leaf $L_{23} \in \mathcal{F}$, and the transition time $t(Q^+, R_-) = T_{23}$ satisfies the cocyclic identity

$$T_{12} + T_{23} + T_{31} = 0. \tag{27.21}$$

Proof. The proof of convergence is nearly identical for the two sequences. By construction, $Q_m^+ \in \tau_2$, so the second coordinate is identically 1 along this sequence. Next, since the first coordinate is contracting by iterations of Δ_2 and $k_{m,2} \to \infty$, from the definition $Q_m^+ = \Delta^{k_{m,2}}(Q_m^-)$ it follows that the first coordinate of the points Q_m^+ tends to zero. It remains to show only that the third coordinate has nonzero limit.

By construction of the points and taking into account the condition (27.20), we have

$$t(P_m^-, Q_m^+) = k_{m,1}T_1 + t(P_m^+, Q_m^-) + k_{m,2}T_2 = -k_{m,3}T_3 + T_{12} + o(1).$$

Since the third coordinate $x_3(t) = x_3(0)\exp\lambda_3 t$ is T_3-periodic along any solution $x(t) = (x_1(t), x_2(t), x_3(t))$, we conclude that the third coordinate tends to the nonzero limit equal to $[\exp\lambda_3 T_{12}]p$, where p is the third coordinate of the point $P^- = (1, 0, p)$.

The proof of the second limit is completely similar. For exactly the same reasons, the only coordinate whose convergence requires a proof, is the second coordinate x_2 that is T_2-periodic on leaves of $\mathcal{F}$. By construction, we have

$$t(P_m^+, R_m^-) = -k_{m,1}T_1 - T_{31} - k_{m,3}T_3 + o(1) = k_{m,2}T_2 - T_{31} + o(1),$$

and the limit exists: $x_2(R_m^-) \to [\exp(-\lambda_2 T_{31})]\,p'$, where p' is the second coordinate of the point $P^+ = (1, p', 0)$.

It remains to show that the points R^- and Q^+ belong to the same leaf of $\mathcal{F}$. This again follows from the same computation:

$$t(Q_m^+, R_m^-) = (\mathbf{k}_m, \boldsymbol{T}) - (T_{12} + T_{31}) + o(1).$$

By uniform continuity of the flow $\Phi^t(x)$ in x for all bounded values of t, the points R^- and Q^+ belong to the same leaf of $\mathcal{F}$. The identity (27.21) follows from (27.20). □

Remark 27.23. The construction depends on the initial choice of the two points $P^{\pm}$ as the parameters. A simple inspection shows that if these points are chosen sufficiently close to e_1, then the points $Q^{\pm}$ and $R^{\pm}$ will be arbitrarily close to e_2 and e_3 respectively.

27G. Topological functoriality of the main construction and the proof of Theorem 27.19. Consider two complex hyperbolic linear flows of Siegel type in $\mathbb{C}^3$ and denote the corresponding holomorphic singular foliations by $\mathcal{F}$ and $\mathcal{F}'$ respectively. Let $\boldsymbol{T} = (T_1, T_2, T_3)$ and $\boldsymbol{T}' = (T_1', T_2', T_3')$ be the corresponding periods.

Assume that $H\colon \mathbb{C}^3 \to \mathbb{C}^3$ is a homeomorphism conjugating the foliations. By Proposition 27.8 the complex separatrices are uniquely characterized by being multiply-connected, hence H must map coordinate axes into coordinate axes. Without loss of generality we may assume that $H(e_j) = e_j$, where e_j, $j = 1, 2, 3$, are the three unit vectors in $\mathbb{C}^3$.

The construction described in §27**F** associates with the three positive real numbers $\boldsymbol{\rho} = (\rho_1, \rho_2, \rho_3)$ satisfying the condition (27.19), a sequence of leaves $L_m \in \mathcal{F}_m$ that accumulate to the union of three separatrices S_1, S_2, S_3 and the three "heteroclinic" leaves L_{12}, L_{23}, L_{31}. More precisely, each leaf L_m carries six points $P_m^{\pm}, Q_m^{\pm}, R_m^{\pm}$ each converging as $m \to \infty$ to the respective limits $P^{\pm}, Q^{\pm}, R^{\pm}$, in such a way that the transition times are as follows (see Fig. V.7),

$$\begin{aligned} t(P_m^-, P_m^+) &= k_{m,1} T_1, & t(P_m^+, Q_m^-) &= t(P^+, Q^-) + o(1), \\ t(Q_m^-, Q_m^+) &= k_{m,2} T_2, & t(Q_m^+, R_m^-) &= t(Q^+, R^-) + o(1), \\ t(R_m^-, R_m^+) &= k_{m,3} T_3, & t(R_m^+, P_m^-) &= t(R^+, P^-) + o(1). \end{aligned} \tag{27.22}$$

Denote by L_m' the images of the leaves $L_m' = H(L_m)$. Let τ_j', $j = 1, 2, 3$ be three standard cross-sections to the separatrices S_j' of the second foliation $\mathcal{F}'$. (Note that τ_j' coincide with τ_j if we identify the phase spaces of the two foliations $\mathcal{F}, \mathcal{F}'$). The homeomorphism H in general does not map the cross-sections τ_j to τ_j', but in any case the images $H(\tau_j')$ are "topologically transversal" to the separatrices S_j': each nearby local leaf of $\mathcal{F}'$ in a small neighborhood of e_j intersects $H(\tau_j)$ only once. This allows us to define the local holonomy correspondences $h_j\colon (H(\tau_j), e_j) \to (\tau_j', e_j)$ between the two cross-sections, at least in sufficiently small neighborhoods of the points e_j. They are local homeomorphisms.

Consider the following six points on the leaves L_m',

$$\begin{aligned} \widetilde{P}_m^{\pm} &= h_1 \circ H(P_m^{\pm}) \in \tau_1', \\ \widetilde{Q}_m^{\pm} &= h_2 \circ H(Q_m^{\pm}) \in \tau_2', \\ \widetilde{R}_m^{\pm} &= h_3 \circ H(R_m^{\pm}) \in \tau_3'. \end{aligned} \tag{27.23}$$

All these sequences are converging, since $h_j \circ H\colon \tau_j \to \tau_j'$ are homeomorphisms and the preimages were converging by construction. Denote by $\widetilde{P}^{\pm}, \widetilde{Q}^{\pm}, \widetilde{R}^{\pm}$ their respective limits.

Let $t'(\cdot,\cdot)$ be the transition time function defined on pairs of points on the same leaf of the second foliation $\mathcal{F}'$ via the flow of the vector field $\dot{x} = \Lambda' x$ generating $\mathcal{F}'$.

Lemma 27.24.

$$
\begin{aligned}
t'(\widetilde{P}_m^-, \widetilde{P}_m^+) &= k_{m,1}T_1', &\qquad t'(\widetilde{P}_m^+, \widetilde{Q}_m^-) &= t'(\widetilde{P}^+, \widetilde{Q}^-) + o(1),\\
t'(\widetilde{Q}_m^-, \widetilde{Q}_m^+) &= k_{m,2}T_2', &\qquad t'(\widetilde{Q}_m^+, \widetilde{R}_m^-) &= t'(\widetilde{Q}^+, \widetilde{R}^-) + o(1),\\
t'(\widetilde{R}_m^-, \widetilde{R}_m^+) &= k_{m,3}T_3', &\qquad t'(\widetilde{R}_m^+, \widetilde{P}_m^-) &= t'(\widetilde{R}^+, \widetilde{P}^-) + o(1).
\end{aligned}
\tag{27.24}
$$

Proof. The three left equalities follow from the fact that $h_j \circ H$ conjugates the holonomy Δ_j of the foliation $\mathcal{F}$ on the cross-section τ_j, with the holonomy Δ_j' of the foliation $\mathcal{F}'$ on the cross-section τ_j'. To obtain P_m^+ from P_m^-, one has to iterate $k_{m,1}$ times the map Δ_1, therefore $\widetilde{P}_m^+ = (\Delta_j')^{k_{m,2}}(\widetilde{P}_m^-)$. Since $t'(x, \Delta_j'(x)) = T_1'$, we conclude that $t'(\widetilde{P}_m^-, \widetilde{P}_m^+) = k_{m,1}T_1'$. The other three equalities are completely similar.

To prove the remaining three limits, we note that the limit points, say, $\widetilde{P}^+$ and $\widetilde{Q}^-$ belong to the same leaf $L_{12}' = H(L_{12})$, again by continuity of H. Therefore $t'(\widetilde{P}^+, \widetilde{Q}^-)$ is the finite limit of $t'(\widetilde{P}_m^+, \widetilde{Q}_m^-)$ as $m \to \infty$. The other two transition times $t'(\widetilde{Q}_m^+, \widetilde{R}_m^-)$, $t'(\widetilde{R}_m^+, \widetilde{P}_m^-)$ have finite limits in exactly the same way. □

Proof of Theorem 27.19 for $n = 3$. The cocycle identity

$$
\begin{aligned}
t'(\widetilde{P}_m^-, \widetilde{P}_m^+) + t'(\widetilde{P}_m^+, \widetilde{Q}_m^-) + t'(\widetilde{Q}_m^-, \widetilde{Q}_m^+) + t'(\widetilde{Q}_m^+, \widetilde{R}_m^-)&\\
+ t'(\widetilde{R}_m^-, \widetilde{R}_m^+) + t'(\widetilde{R}_m^+, \widetilde{P}_m^-) &= 0
\end{aligned}
$$

together with (27.24) implies that

$$
(\mathbf{k}_m, \boldsymbol{T}') = O(1), \qquad \text{as } m \to \infty.
$$

Dividing this identity by $|\mathbf{k}_m| \to \infty$ yields in the limit the equality

$$
(\boldsymbol{\rho}, \boldsymbol{T}') = 0, \qquad \boldsymbol{\rho} = (\rho_1, \rho_2, \rho_3) \in \mathbb{R}_3^+.
$$

In other words, the positive vector $\boldsymbol{\rho} \in \mathbb{R}_+^3$ satisfying the condition $(\boldsymbol{\rho}, \boldsymbol{T}) = 0$, satisfies also the condition $(\boldsymbol{\rho}, \boldsymbol{T}') = 0$.

Thus the system of four linear equations (over $\mathbb{R}$), equivalent to the two complex equalities,

$$
(\boldsymbol{\rho}, \boldsymbol{T}) = 0, \quad (\boldsymbol{\rho}, \boldsymbol{T}') = 0, \tag{27.25}
$$

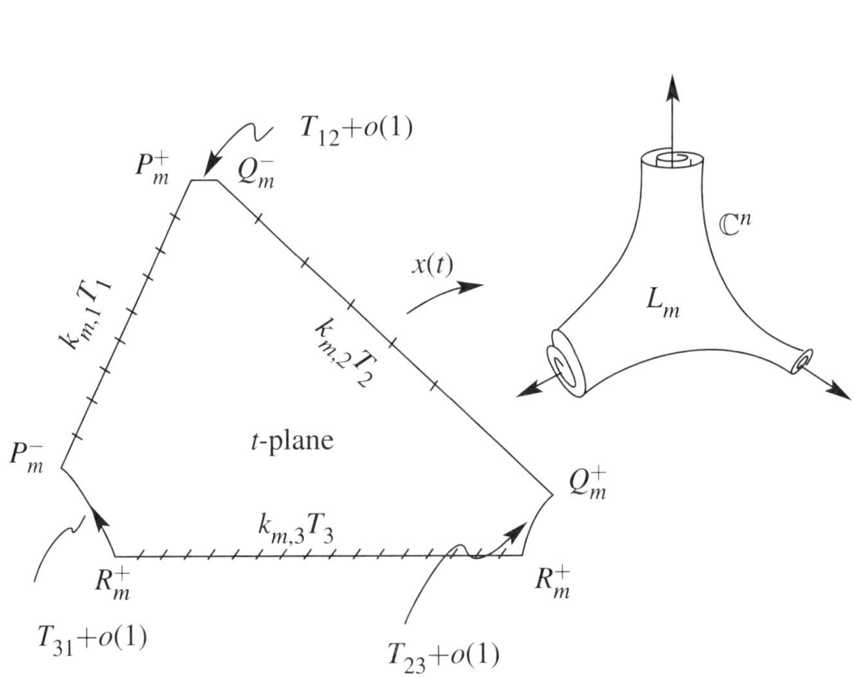

Figure V.7. Demonstration of Theorem 27.19: topological functoriality

has a nontrivial solution. This means that the rank of its coefficient matrix is 2. By Definition 27.18 (2), the two collections of periods $\boldsymbol{T}$ and $\boldsymbol{T}'$ are affine equivalent. □

Remark 27.25. The three-dimensional construction used in the above proof, in fact implies some multidimensional corollaries. Consider two linear hyperbolic Siegel-type systems in $\mathbb{C}^n$, $n > 3$, with the complex periods $\boldsymbol{T}$ and $\boldsymbol{T}'$ respectively, which are topologically orbitally equivalent (i.e., the corresponding foliations $\mathcal{F}$ and $\mathcal{F}'$ are topologically equivalent). By Proposition 27.8, without loss of generality (changing the enumeration of coordinates if necessary) we may assume that the conjugating homeomorphism H sends the complex separatrices S_j (the coordinate axes) to the separatrices S_j' for all $j = 1, \dots, n$.

Assume that the first three eigenvalues $\lambda_1, \lambda_2, \lambda_3 \in \mathbb{C}$ of the first system already form a triangle containing the origin strictly inside. Then the respective triplets of periods (T_1, T_2, T_3) and (T_1', T_2', T_3') are affine equivalent in the sense of Definition 27.18.

Indeed, the coordinate plane $\mathbb{C}^3$ spanned by the first three coordinates in $\mathbb{C}^n$, is invariant by the complex flow of the first system hence, the construction of the leaves $L_m \subset \mathbb{C}^3$ can be carried out without any changes.

On the other hand, the three-dimensional proof of Theorem 27.19 does not use the fact that the images $L'_m = H(L_m)$ belong to any coordinate subspace invariant for the second system: the only fact required for the proof is accumulation of the leaves L'_m to the three complex separatrices S'_1, S'_2, S'_3 of the second system. The conclusion on affine equivalence of the respective periods obviously holds in this case.

One may be tempted to prove Theorem 27.19 for $n > 3$ by studying all 3-dimensional (invariant) coordinate planes the restriction on which is of Siegel type, based on the above Remark. However, the accurate proof goes along slightly different lines.

First we make some simple topological observations. It was already noted that the coordinate axes of a diagonal hyperbolic linear system are topologically distinguished from other leaves. On the other hand, *not every* (invariant) coordinate subspace is topologically distinguished: homeomorphisms conjugating two Siegel foliations may not map them into the corresponding subspaces of other foliations. Yet some coordinate subspaces are topologically distinguished.

If $\lambda_1, \dots, \lambda_n \in \mathbb{C}$ is a point set, its element is called a *corner point* if it can be separated from the rest of the set by a real line.

Lemma 27.26. *Assume that two diagonal hyperbolic linear systems in the same space $\mathbb{C}^n$ are topologically conjugated by a homeomorphism H preserving the coordinate axes (separatrices).*

If λ_n is a corner point of the spectrum of the first system, then H maps the coordinate hyperplane $\mathbb{C}^{n-1} = \{x_n = 0\} \subset \mathbb{C}^n$ into itself.

Proof. The coordinate hyperplane $\{x_n = 0\}$ is distinguished by the following topological description: all leaves *not belonging* to this plane, accumulate to nonsingular points on the complex separatrix $S_n = \mathbb{C}e_n$. By our assumption on the enumeration of the coordinates, the separatrices S_n and S'_n are H-related, hence their "complementary" hyperplanes are also H-related. □

Proof of Theorem 27.19 for any $n > 3$. The proof goes by induction in n. The basis at $n = 3$ is already established.

Consider a hyperbolic Siegel-type linear system in $\mathbb{C}^{n+1}$ with the spectrum $\lambda_1, \dots, \lambda_{n+1}$ containing the origin strictly inside its convex hull. As before, we can assume without loss of generality that the system is diagonal, so any coordinate subspace of any (complex) dimension between 1 and n is invariant.

Assume that the enumeration of the axes is so chosen that 0 is inside the convex hull $\operatorname{conv}(\lambda_1, \dots, \lambda_n)$, while the last remaining eigenvalue λ_{n+1} is a corner point. Elementary geometric considerations show that this is always possible.

By Lemma 27.26, the invariant hyperplane $\{x_{n+1} = 0\} \subset \mathbb{C}^{n+1}$ is topologically invariant: any homeomorphism H between $\mathcal{F}$ and another such foliation $\mathcal{F}'$ defined by a diagonal hyperbolic linear system, necessarily conjugates the restrictions of these foliations on the respective hyperplanes $\{x_{n+1} = 0\}$ and $\{x'_{n+1} = 0\}$.

By the inductive assumption, the truncated collections of the periods $(T_1, \dots, T_n)$ and $(T'_1, \dots, T'_n)$ are affine equivalent: there exists an $\mathbb{R}$-linear map M of $\mathbb{C}$ into itself, taking one collection into the other.

To show that this map takes the last period T_{n+1} into T'_{n+1}, notice that for elementary reasons at least one of the triangles $\operatorname{conv}(\lambda_{n+1}, \lambda_j, \lambda_k)$, $1 \leqslant j \neq k \leqslant n$, also contains the origin in its interior (the union of these triangles contains the convex hull of all $n+1$ eigenvalues). By Remark 27.25, the triplets (T_{n+1}, T_j, T_k) and (T'_{n+1}, T'_j, T'_k) are affine equivalent by an $\mathbb{R}$-linear map $M' \colon \mathbb{C} \to \mathbb{C}$. But since $T_j/T_k \notin \mathbb{R}$, there exists *only one* $\mathbb{R}$-linear map $M = M'$ that takes (T_j, T_k) into (T'_j, T'_k), which therefore automatically maps the complete collection $\boldsymbol{T}$ into $\boldsymbol{T}'$. □

27H. Further results: topological equivalence of linear Siegel-type foliations with Jordan blocks. If the matrix A of Siegel type is nondiagonalizable and "otherwise" hyperbolic (i.e., if the ratio of any two eigenvalues is nonreal unless they are equal and occur in the same Jordan block), then the topological classification of the corresponding holomorphic foliations is even more rigid, as was discovered by L. Ortiz Bobadilla [**OB96**].

As before, the key result is low-dimensional. Consider the class of linear systems in $\mathbb{C}^4$ whose matrices have one (2×2)-block with the eigenvalue λ_1, and two other eigenvalues λ_2, λ_3 are such that the triangle $\lambda_1, \lambda_2, \lambda_3$ contains the origin in the interior.

Two foliations $\mathcal{F}$ and $\mathcal{F}'$ generated by systems of this class, are topologically equivalent if the two corresponding tuples of eigenvalues are proportional over $\mathbb{C}$, i.e., if

$$\boldsymbol{\lambda} = c\boldsymbol{\lambda}', \qquad \boldsymbol{\lambda} = (\lambda_1, \lambda_2, \lambda_3),\ \boldsymbol{\lambda}' = (\lambda'_1, \lambda'_2, \lambda'_3), \qquad 0 \neq c \in \mathbb{C}. \tag{27.26}$$

The topological equivalence H in this case can be made *linear*, of the form $x \mapsto Cx$. Indeed, from the proportionality of the eigenvalues (27.26) and identical Jordan structure it follows that one can find a linear transformation such that the matrices A and $CA'C^{-1}$ would differ only by the scalar multiple c. But the leaves of two foliations $\mathcal{F}$ and $\mathcal{F}'$ defined by the *proportional* matrices, simply coincide.

It turns out that this is *the only* case where foliations of the considered class are topologically equivalent. In other words, the following result asserts the maximal topological rigidity of Siegel type foliations having Jordan blocks.

Theorem 27.27 (see [**OB96**]). *Two holomorphic foliations generated by Siegel-type linear vector fields in $\mathbb{C}^4$ having one Jordan block and hyperbolic otherwise, are topologically equivalent if and only if their eigenvalues are proportional over $\mathbb{C}$, in which case they are linear equivalent.*

Returning to the (truly) hyperbolic hence diagonalizable case, one may ask whether the study of holomorphic foliations generated by *nonlinear* vector fields, brings any new phenomena. In a surprising way, the answer is negative, as was established by M. Chaperon [**Cha86**] who proved the following complex analog of the Grobman–Hartman theorem.

Theorem 27.28 (M. Chaperon [**Cha86**]). *If the spectrum of a matrix A is hyperbolic and Siegel-type, then the singular holomorphic foliation by solutions of any nonlinear holomorphic vector field $\dot{x} = A(x) + \cdots$, is topologically linearizable (topologically equivalent to the foliation $\mathcal{F}'$ by solutions of the linearized field $\dot{x} = Ax$).*

The complete proofs of these results go beyond the scope of this book, though all the main tools required for the proof of, say, Theorem 27.27, were already described in this section.

28. Global properties of generic polynomial foliations of the complex projective plane $\mathbb{P}^2$

In this section, largely based on the article [**Ily78**], we consider polynomial singular holomorphic foliations on $\mathbb{P}^2$ having an invariant line, and their generic properties. These properties are in a stark contrast with the properties of *real polynomial foliations* on $\mathbb{R}P^2$. After describing the precise meaning of the word "generic", we will prove the following results, reduced for clarity into a table (Table V.1).

Complex holomorphic foliations from the class $\mathcal{A}_r$, $r \geqslant 2$	**Real polynomial foliations on the Poincaré sphere**
Leaves of a generic foliation are everywhere dense in $\mathbb{P}^2$ (except for the invariant line)	Leaves of a generic foliation can accumulate only to limit cycles and singular points
Foliations generically have infinite number of "complex limit cycles" (defined later)	Foliations generically have only finitely many limit cycles
Generic foliations are rigid (do not admit nontrivial homeomorphisms conjugating them with nearby foliations)	Generic foliations are structurally stable (all nearby foliations have the same topological type)

Table V.1. Comparison between generic properties of complex holomorphic foliations of $\mathbb{P}^2$ and real foliations of the 2-sphere.

The principal genericity assumption behind these results is existence of a separatrix with sufficiently rich holonomy group. This separatrix is the invariant line at infinity, which occurs generically in the class $\mathcal{A}_r$. Investigation of the holonomy group, a finitely generated subgroup of $\mathrm{Diff}(\mathbb{C}, 0)$ and dynamics of its orbits, is the main tool in establishing the properties summarized in Table V.1.

Recently a parallel theory was developed for generic singular analytic (nonpolynomial) foliations on $\mathbb{C}^2$ [**Fir06, GK06**].

28A. Foliations of the class $\mathcal{A}'_r$: holonomy at infinity. In Definition 25.49 we have introduced the class (denoted by $\mathcal{A}'_r$) of polynomial foliations on $\mathbb{P}^2$ tangent to the infinite line $\mathbb{I} \subset \mathbb{P}^2$ and having exactly $r+1$ singular points on $\mathbb{I}$. This class constitutes a Zariski open subset of the corresponding projective space $\mathcal{A}_r$ of foliations which in the fixed affine neighborhood $\mathbb{C}^2 \subset \mathbb{P}^2$ are defined by a polynomial vector field or a polynomial 1-form of degree $\leqslant r$ and isolated singularities.

Choose two nonsingular points on $\mathbb{I}$ and consider an affine chart $\mathbb{C} \subset \mathbb{I}$ in which these points correspond to the origin and infinity respectively. The fundamental group of the punctured infinite leaf $\mathbb{I} \smallsetminus \Sigma$, $\Sigma = \operatorname{Sing}(\mathcal{F})$, for foliations from the class $\mathcal{A}'_r$ is generated by a system of $r+1$ canonical loops γ_i around the singular points z_i (Definition 18.1). With each loop $\gamma_i \subseteq \mathbb{I} \smallsetminus \Sigma$ in the standard way the holomorphic holonomy germ $\Delta_i = \Delta_{\gamma_i} \in \operatorname{Diff}(\mathbb{C}, 0)$ is associated.

Definition 28.1. The *holonomy group at infinity*, or simply the *holonomy group* of a foliation $\mathcal{F} \in \mathcal{A}'_r$ is the subgroup $G \subseteq \operatorname{Diff}(\mathbb{C}, 0)$ generated by the germs $\Delta_0, \dots, \Delta_r$.

This group is obviously an invariant of the foliation in the following sense.

Proposition 28.2. *If two polynomial foliations $\mathcal{F}, \mathcal{F}' \in \mathcal{A}_r$ are conjugated by a homeomorphism, diffeomorphism or biholomorphism $H \colon \mathbb{P}^2 \to \mathbb{P}^2$ preserving the infinite line $\mathbb{I}$, than the corresponding holonomy groups G, G' are conjugated in the following sense.*

There exist two collections of generators $f_0, \dots, f_r \in G$, $f'_0, \dots, f'_r \in G'$ of these two groups and a homeomorphism (resp., diffeomorphism, biholomorphism) $h \colon (\mathbb{C}, 0) \to (\mathbb{C}, 0)$ such that

$$f_i \circ h = h \circ f'_i, \qquad i = 0, \dots, r.$$

Proof. Let $\gamma_1, \dots, \gamma_n$ be any collection of loops generating the fundamental group $\pi_1(\mathbb{I} \smallsetminus \operatorname{Sing} \mathcal{F}, a)$, and $f_j = \Delta_{\gamma_j}$ the corresponding holonomy germs. Then the images $\gamma'_j = H(\gamma_j)$ are the loops generating the fundamental group $\pi_1(\mathbb{I} \smallsetminus \operatorname{Sing} \mathcal{F}', a')$, $a' = H(a)$. Denote by h the restriction of H on a cross-section τ_a to $\mathbb{I}$ at a: h can be considered as a homeomorphism between two *analytic* cross-sections τ_a and $\tau_{a'}$, $a' = H(a)$ (the image $H(\tau_a)$ can be homeomorphically identified with $\tau_{a'}$ by projection along local leaves of $\mathcal{F}'$). Then the homeomorphism h conjugates f_j with the corresponding holonomy germs $f'_j = \Delta'_{\gamma'_j}$. □

Remark 28.3. One set of generators (say, for the group G) can be chosen as the holonomy operators of the canonical loops (see Definition 18.1), $f_i = \Delta_i$. However, the generators f'_i of the second group in this case will be holonomy germs corresponding to the loops $\gamma'_j = H(\gamma_i)$ which in general do not form a canonical system and even are not isotopic to a canonical system. This observation will play an important role in the discussion of rigidity of foliations.

Remark 28.4. By Theorem 25.56, a generic foliation from the class $\mathcal{A}_r$ has a single topologically uniquely defined algebraic separatrix, hence for most pairs of foliations the additional assumption that H preserves the infinite line, may be dropped.

Example 28.5 (Homogeneous vector fields on the plane)**.** Assume that the 1-form $\omega = p\,dx + q\,dy$ defining the foliation $\mathcal{F}$ in affine coordinates, is homogeneous: both $p(x,y)$ and $q(x,y)$ are homogeneous polynomials of degree r in (x,y).

Then from the computations (25.2) it follows that in the coordinates $u = 1/x$, $z = y/x$ the foliation is defined by a rational 1-form ω' with separated variables,

$$\omega' = \frac{du}{u} - R(z)\,dz, \qquad R(z) = \sum_0^r \frac{\lambda_i}{z - z_i} \in \mathbb{C}(z), \tag{28.1}$$

with a rational function $R(z)$ having *simple* poles exactly at the singularities $z_0, \dots, z_r$ of $\Sigma \cap \mathbb{I}$. The complex numbers $\lambda_0, \dots, \lambda_r$ are the characteristic exponents, cf. with (25.6).

The Pfaffian differential equation $\omega' = 0$ is linear with respect to the variable u and can be hence explicitly integrated: in particular, the monodromy (holonomy) operator corresponding to the path γ_i, is a linear map $\Delta_i \colon u \mapsto \mu_i u$, where $\mu_i = \exp 2\pi i \lambda_i$, $i = 0, \dots, r$.

Thus the whole holonomy group for a homogeneous foliation is a subgroup of the commutative (multiplicative) group $\mathbb{C}^*$ of linear invertible maps $\mathbb{C} \to \mathbb{C}$.

This description allows us to describe behavior of leaves of a *generic* homogeneous foliation: by Example 6.39 all noninfinite leaves of a generic homogeneous foliation are dense in $\mathbb{C}^2$.

Example 28.6. The holonomy group of a generic Hamiltonian foliation $\mathcal{F} \in \mathcal{A}_r$ is generated by commuting cyclical elements of order $r+1$, since this foliation has a meromorphic first integral $u^{-(r+1)}\big(h_r(v) + O(u)\big)$ near the infinite line $\mathbb{I} = \{u = 0\}$ (Exercise 11.11).

Thus all points are cycles for the holonomy (pseudo)group as they were defined in Definition 6.32, but none of them is a limit cycle.

The holonomy group of a homogeneous foliation from Example 28.5 is linear, hence commutative, and consequently all points are cycles. The same assertion applies to Darboux integrable foliations discussed in §25**G**.

In §6**D**–§6**I** we established the properties of density, countable number of limit cycles and rigidity for finitely generated groups of conformal germs. This section deals with translating these properties into the parallel properties of polynomial foliations as they appear in Table V.1, using the holonomy at infinity. Yet this translation is by no means trivial.

Theorems from this section are proved under different but repeating genericity assumptions. For the reader's convenience we label them by mnemonic labels explained in the footnotes.

28B. Density of leaves for polynomial foliations. The density of orbits of the holonomy (pseudo)group is the easiest to translate into the density of leaves of a foliation. Recall that the characteristic number (or characteristic exponent) of a singular point is the ratio of the eigenvalues of its linearization; it is well defined as an element of $\mathbb{C}^*$ modulo passing to the reciprocal.

Theorem 28.7. *Assume that a foliation $\mathcal{F}$ from the class $\mathcal{A}'_r$ satisfies the following two conditions*[8]:

(H) *$\mathcal{F}$ has only complex hyperbolic singularities on the infinite line* (*the characteristic numbers are nonreal*), *and*

(D) *the holonomy group meets the density condition* (6.17).

Then each leaf of $\mathcal{F}$ is either algebraic or dense in $\mathbb{P}^2$.

Since a generic foliation from the class $\mathcal{A}_r$ satisfies all assumptions of this theorem and in addition has no algebraic leaves by Theorem 25.56, we immediately have the following corollary.

Corollary 28.8. *A generic foliation from $\mathcal{A}_r$ has all leaves dense in $\mathbb{C}^2$.*

The proof of Theorem 28.7 is based on the fact that any leaf of a polynomial foliation on $\mathbb{P}^2$ must have points accumulating to infinity. Then one can derive the density of leaves near infinity from the density of orbits of the holonomy pseudogroup.

Lemma 28.9. *Any leaf L of an arbitrary polynomial foliation on $\mathbb{P}^2$ must accumulate to the infinite line*: $\overline{L} \cap \mathbb{I} \neq \varnothing$.

Proof of the lemma. The leaf L of a foliation defined by the Pfaffian equation $p(x,y)\,dx + q(x,y)\,dy = 0$ in the affine plane can be represented

[8](H) stands for hyperbolicity, (D) for density.

as the graph of a multivalued function $y = \varphi(x)$ which has ramification points only when intersecting an algebraic curve $\{q(x,y) = 0\}$. These intersections and their projections on the x-plane (ramification points) form at most countable sets, hence there exists a ray $R = \{x = x_0 + tv\colon t \in \mathbb{R}_+\}$, $v \in \mathbb{C}^*$ free from the ramification points and projections of singularities of $\mathcal{F}$.

The function $\varphi(x)$ can be continued analytically along the ray R. Consider the *maximal* interval $[x_0, x_0 + t_0 v)$ on which this continuation is possible. From the local rectification theorem it follows that either $\lim_{t\to t_0^-} \varphi(x_0 + tv) = \infty$, or $t_0 = +\infty$ itself. In both cases the leaf L accumulates to some point on the infinite line. $\square$

In the assumptions of Theorem 28.7 we can describe the intersection $\overline{L} \cap \mathbb{I}$.

Lemma 28.10. *If all singular points of the foliation $\mathcal{F}$ on the infinite line are complex hyperbolic, then each leaf of $\mathcal{F}$ is either algebraic, or contains the entire infinite line in its closure, $\mathbb{I} \subset \overline{L}$.*

Proof. If $\overline{L} \cap \mathbb{I}$ contains a *nonsingular* point a of the infinite line, then it contains the entire line as well. Indeed, let $b \in \mathbb{I} \smallsetminus \operatorname{Sing}(\mathcal{F})$ be any other nonsingular point. Choose any two cross-sections τ_a, τ_b to $\mathbb{I}$ respectively. By assumption, the intersections between L and τ_a contain an infinite sequence converging to a. The holonomy map $\Delta_{a,b}$ along an arbitrary path in $\mathbb{I} \smallsetminus \operatorname{Sing}(\mathcal{F})$ connecting a with b maps these points into an infinite sequence converging to b. By construction, all points of this sequence belong to L.

The only other remaining possibility is that the intersection $\overline{L} \cap \mathbb{I}$ consists of only singular points of $\mathcal{F}$. We show that in this case the leaf L is (a part of) an algebraic curve.

First, we note that in the assumptions of the theorem, all singular points $a_0, \dots, a_r \in \operatorname{Sing}(\mathcal{F}) \cap \mathbb{I}$ on the infinite line are linearizable complex saddles. Each saddle has exactly two local separatrices, one of them on the infinite line, another, denoted by S_i, is transversal to it. We claim that locally near each point a_i the leaf L must coincide with the separatrix S_i. Indeed, all other local leaves of the linear foliation intersect a close cross-section to the second local separatrix belonging to $\mathbb{I}$. Yet the orbit of the corresponding intersection point by the linear holonomy map $z \mapsto (\exp 2\pi i\lambda)z$ (associated with a small loop around a_i on $\mathbb{I}$) accumulates to the infinite line (recall that by our assumptions $\lambda \notin \mathbb{R}$, hence the holonomy map is hyperbolic). But this accumulation contradicts the choice of L. Thus the leaf L locally coincides with S_i.

Consider an affine chart (x, y) on $\mathbb{C}^2 \subset \mathbb{P}^2$ chosen so that the vertical direction is nonsingular for $\mathcal{F}$ (the point $[0:1:0] \in \mathbb{I}$ at infinity is nonsingular). In this chart the separatrices S_i are graphs of analytic functions $y = \varphi_i(x)$ which are well defined, holomorphic and growing no faster than linear, $|\varphi_i(x)| = O(|x|)$ for $|x|$ sufficiently large. Since L is the leaf of a foliation $\mathcal{F}$ hence a holomorphic nonvertical curve, the total number of intersections between L and any vertical line $\ell_c = \{x = c\}$, counted with their multiplicities, remains locally the same for all lines not passing through finite singularities of $\mathcal{F}$. This allows us to continue the symmetric functions $\sigma_1(x) = \varphi_0(x) + \cdots + \varphi_r(x)$, ..., $\sigma_{r+1}(x) = \varphi_0(x) \cdots \varphi_r(x)$ of the intersection set $L \cap \ell_x = \{\varphi_0(x), \dots, \varphi_r(x)\}$ as well-defined functions, holomorphic and bounded outside the finite set of x-coordinates of the singularities of $\mathcal{F}$. By construction, $\sigma_j(x)$ grows no faster than $O(|x|^j)$ as $|x| \to \infty$. Therefore the expression $\prod_{j=0}^{r}(y - \varphi_j(x))$ is in fact a polynomial of degree $\leqslant r+1$ in x, y, and L is an algebraic curve of degree $\leqslant r+1$. □

Proof of Theorem 28.7. We prove that in the assumptions of the theorem the closure of any nonalgebraic leaf L contains an arbitrary finite point $a \in \mathbb{C}^2$ (for points on the infinite line the inclusion $a \in \overline{L}$ follows from Lemma 28.10).

By Corollary 25.35, if a foliation $\mathcal{F}$ has a nonalgebraic leaf L, then almost all leaves of $\mathcal{F}$ are in fact nonalgebraic. Hence through a point a' arbitrarily close to a passes a nonalgebraic leaf L' that accumulates to one (hence to all) nonsingular points of the infinite line in the same way that the initial nonalgebraic leaf L does (Lemma 28.10).

By Corollary 6.40, under the density assumption (6.17) the closures of L and L' both contain some common neighborhood of the infinite line. Thus the leaf L' intersects the closure $\overline{L}$ and therefore $a' \in \overline{L}$. Since a' can be chosen arbitrarily close to a, we conclude that $a \in \overline{L}$ as well. □

28C. Infinite number of complex limit cycles. In this subsection we derive from Theorem 6.41 (on abundance of limit cycles for the holonomy pseudogroup) the assertion on abundance of complex limit cycles for generic polynomial foliations. Theorem 6.41 implies that leaves of a generic foliation from the class $\mathcal{A}_r$ carry an infinite number of closed loops ("real cycles"). The problem is to show that these cycles are *homologically independent* even if they accidentally happen to belong to the same leaf; cf. with Remark 6.42.

Definition 28.11. A *complex cycle* of a holomorphic foliation $\mathcal{F}$ is a free homotopy class $[\gamma]$ of real cycles (oriented closed loops) representing a loop γ on a leaf L.

A complex *limit cycle* is a free homotopy class of a complex cycle with a nonidentical associated holonomy map Δ_γ.

This definition obviously matches that of a complex limit cycle for a pseudogroup (Definition 6.32). On the other hand, it obviously extends the definition of a (real) limit cycle for real analytic vector fields (Definition 9.11). Note that the limit cycles cannot be contractible on the respective leaves, otherwise their holonomy must be trivial (see also Problem 28.1).

Since a leaf of a holomorphic foliation can carry many (even infinitely many) nonhomotopic closed loops, we will impose a stronger condition to distinguish between "truly different" complex limit cycles.

Definition 28.12. A collection of complex limit cycles $\gamma_1, \dots, \gamma_k, \dots$ of a foliation $\mathcal{F}$ is called *homologically independent*, if for any leaf L the cycles that belong to this leaf are homologically independent on L, i.e., no nontrivial integer combination $\sum_{\gamma_k \subseteq L} c_k \gamma_k$, $c_k \in \mathbb{Z}$, is homologous to zero on L.

The main result of this subsection asserts that generically a polynomial foliation from the class $\mathcal{A}_r$ on $\mathbb{P}^2$ has *infinitely many* homologically independent complex limit cycles.

Consider a foliation from the class $\mathcal{A}'_r$ and its linearization along the infinite line, which in the suitable coordinates (z, w) is defined by the linear equation; cf. with (28.1).

$$\frac{dw}{w} = \sum_{k=0}^{r} \frac{\lambda_k \, dz}{z - z_k}. \tag{28.2}$$

Choose a nonsingular point (the origin in the chart chosen as in §28**A**) and the loops $\gamma_0, \dots, \gamma_r$ on the z-plane, which begin and end at this point while encircling the respective singularity z_k. With each such loop we associate the complex number

$$I_k = \frac{1}{e^{-4\pi i \lambda_k} - 1} \int_{\gamma_k} \frac{dz}{w^2}, \qquad k = 1, \dots, r, \tag{28.3}$$

where the integral is computed using the branch of the multivalued function $w = w(z)$ obtained by continuation along $z \in \mu_k$ of the solution of (28.2) with the initial value $w(0) = 1$.

Theorem 28.13. *Assume that a foliation $\mathcal{F}$ from the class $\mathcal{A}'_r$ satisfies the following two conditions*[9],

(NC) *the holonomy group at infinity G is noncommutative, and*

(D) *G meets the density condition* (6.17).

If in these assumptions the integrals (28.3) *satisfy the inequalities $I_j \neq \pm I_k$ for all $j \neq k$, then the foliation $\mathcal{F}$ has infinitely many homologically independent complex limit cycles.*

[9](NC) stands for noncommutativity.

Proof. Theorem 6.41 immediately implies existence of infinitely many limit cycles, so it remains only to show that one can choose infinitely many of these cycles that would be homologically independent; cf. with Remark 6.42.

More precisely, consider the holonomy pseudogroup Γ of the foliation $\mathcal{F}$, associated with the cross-section at the point $\{z = 0\}$. This group is generated by the maps $f_k^{\pm} = (\Delta_k^{\pm 1}, U_k, \gamma_k^{\pm 1})$, $k = 0, \dots, r$, where Δ_k are the holonomy maps along the loops γ_k and U_k coincide with a fixed small disk $D_\varepsilon = \{|w| < \varepsilon\}$. The whole pseudogroup consists of the triples $(f_\alpha, U_\alpha, \gamma_\alpha)$, where $\gamma_\alpha \in \pi_1(\mathbb{I} \smallsetminus \Sigma, 0)$ is a closed loop on the infinite leaf of $\mathcal{F}$, f_α is the holonomy map associated with the loop γ_α and U_α the natural domain of f_α such that for any point $w \in U_\alpha$ the lift of the path γ on the leaf of $\mathcal{F}$ passing through $(0, w)$ remains in the specified tubular ε-neighborhood of infinity.

By Theorem 6.41 for an arbitrary small $\rho > 0$ there exists a point w_ρ with $|w_\rho| < \rho$ and an element $f_\rho \in \Gamma$ of the pseudogroup, such that $f_\rho(w_\rho) = w_\rho$. By construction, this means that the lift of the corresponding loop γ_ρ on the leaf L_ρ passing through the point $(0, w_\rho)$, is a closed loop ℓ_ρ on this leaf. Choosing a sequence of positive values ρ_m converging to zero sufficiently fast, we can guarantee that the cycles ℓ_{ρ_m} are pairwise disjoint.

We need to show that the freedom in constructing the maps $f_\rho \in \Gamma$ can be used to guarantee that the cycles ℓ_ρ are homologically independent. This will be achieved by choosing these maps so that the integrals of the rational 1-form $\omega = dz/w^2$ along the cycles ℓ_ρ tend to infinity.

Recall that the elements f_ρ in the proof of Theorem 6.41 were constructed as follows. First, a hyperbolic generator, say, $f_1 = f$, was chosen; without loss of generality it can be assumed to be contracting with the multiplicator ν, $|\nu| < 1$. Second, another expanding hyperbolic generator, say, $f_2 = g$, with the multiplicator μ, $|\mu| > 1$ is chosen so that the multiplicative subgroup $\langle \nu, \mu \rangle \subset \mathbb{C}^*$ is dense. Finally, a third element h, essentially nonlinear (in the linearizing chart for f) tangent to the identity is selected. Then for any $w'_\rho \neq 0$ arbitrarily close to the origin the coefficient $c_\rho = h(w'_\rho)/w'_\rho \approx 1$ of the linear map $w \mapsto c_\rho w$, is approximated by ratios $\mu^j \nu^k$, $k, j \to +\infty$, and the element f_ρ is constructed under the form of the composition

$$f_{\rho,j,k,n} = h^{-1} f^{-n} g^j f^{n+k}, \qquad \text{as } j,k,n \to +\infty, \quad \mu^j \nu^k \to c_\rho \approx 1. \tag{28.4}$$

This composition has an isolated fixed point w_ρ near w'_ρ: $|w_\rho| < \rho$, and w_ρ tends to w'_ρ in the limit as j, k, n tend to infinity. Clearly, since all contracting maps are collected first, this composition is well defined in the pseudogroup.

Let $\ell_{\rho,j,k,n}$ be the lift of the loop associated with the composition (28.4), on the leaf of the foliation $\mathcal{F}$ passing through the point w_ρ. We claim that in the assumptions of the theorem, integrals of the form ω over $\ell_{\rho,j,k,n}$ diverge[10].

Lemma 28.14.

$$\lim_{\rho\to 0^+,\ j,k,n\to\infty} \oint_{\ell_{\rho,j,k,l}} \frac{dz}{w^2} = \infty. \tag{28.5}$$

Assuming that the limit (28.5) is indeed equal to infinity, we can always choose from the family of cycles $\{\ell_{\rho,j,k,l}\}$ an infinite sequence of homologically independent cycles. Indeed, without loss of generality we may assume that the cycles are pairwise disjoint (considering a sufficiently fast decreasing sequence of the values $\rho_m \to 0^+$) and without self-intersections. Such cycles, if belonging to the same leaf, can be homologically dependent if and only if the coefficients of this dependence are ± 1, i.e., if they bound a domain on the leaf, eventually after changing their orientation. Yet under (28.5) we can construct an infinite sequence of cycles $\{\ell_{\rho_m}\}$ such that the integral of the form ω along each cycle is greater than the sum of absolute values of integrals over all preceding cycles,

$$\left|\oint_{\ell_{\rho_m}} \omega\right| > \sum_{s=1}^{m-1}\left|\oint_{\ell_{\rho_s}} \omega\right|, \qquad \text{for all } m = 2,3,\dots.$$

Clearly, this implies that the cycles ℓ_{ρ_m} cannot be homologically dependent on the same fiber with the coefficients ± 1. The proof of Theorem 28.13 is complete modulo Lemma 28.14. □

Sketch of the proof of Lemma 28.14. Together with the initial foliation $\mathcal{F}$ consider its linearization $\mathcal{F}'$ described by the Pfaffian equation (28.2). Because of the linearity, $\mathcal{F}'$ is invariant by the linear maps $(z,w)\mapsto(z,cw)$, $c\in\mathbb{C}^*$. Note also that the form ω is homogeneous. This implies that if α is an arc (closed or not) on a leaf of $\mathcal{F}'$, then its image by the above map, denoted by $c\alpha$, is again an arc on the other leaf, and $\int_{c\alpha}\omega = c^{-2}\int_\alpha \omega$.

Denote by $\ell'_{j,k,n}$ the lifts of the loops associated with the compositions (28.4) from the separatrix $\mathbb{I}$ to the leaves of the linear foliation $\mathcal{F}'$ passing through the same point $(0,1)$. Let α,β be two arcs through the point $(0,1)$ which are lifts of the loops $\gamma_1,\gamma_2\in\pi_1(\mathbb{I}\smallsetminus\Sigma,0)$; the holonomy maps of the foliation $\mathcal{F}'$ associated with these two loops, are both linear, $w\mapsto \nu w$ and $w\mapsto\mu w$ respectively.

Because of the linearity, the loop $\ell'_{j,k,n}$, except for its final part γ corresponding to the map h^{-1}, consists of arcs homothetic to α and β with different coefficients. Denote $A=\int_\alpha\omega$, $B=\int_\beta\omega$, $C=\int_\gamma\omega$. Then by homogeneity the overall integral modulo a

[10] Note that in the initial affine coordinates on the plane ω takes the form $\frac{1}{2}(x\,dy - y\,dx)$, so that if the foliation is real and the cycle $\ell_{\rho,j,k,n}$ were real, then the integral of the form ω would be the area bounded by this cycle. As the cycles are converging to the infinite line, clearly this area tends to infinity.

constant term can be expressed as several partial sums of geometric progressions,

$$\begin{aligned}\int_{\ell'_{j,k,n}} \omega = {} & A(1+\nu^{-2}+\nu^{-4}+\cdots+\nu^{-2(n+k)}) \\ & + \nu^{-2(n+k)}B(1+\mu^{-2}+\mu^{-4}+\cdots+\mu^{-2j}) \\ & - \nu^{-2(n+k)}\mu^{-2j}A(1+\nu^{2}+\cdots+\nu^{2n})+C.\end{aligned} \tag{28.6}$$

Note that $|\nu|<1<|\mu|$ and $\nu^k\mu^j \approx 1$, so that the geometric progression in the first line in (28.6) is diverging, whereas the progressions from the second and the third lines converge as $j,n,k\to\infty$. Computing the leading terms of the above sums, we conclude that the integral above grows to infinity asymptotically as

$$\frac{\nu^{-2(n+k+1)}}{\nu^{-2}-1}A+\nu^{-2(n+k)}\frac{1}{1-\mu^{-2}}B=\nu^{-2(n+k)}\left[\frac{A}{1-\nu^{-2}}+\frac{B}{1-\mu^{-2}}\right].$$

The expression in the square brackets is the difference of the respective integrals I_k from (28.3) and hence is nonzero by the assumptions of the theorem. Thus the limit of the integral in (28.6) is infinite. If instead of the loops $\ell'_{j,k,n}$ we lift on the leaves of the linear foliation $\mathcal{F}'$ the loop $w_\rho\ell'_{j,k,n}$, then the above result will be further multiplied by w_ρ^{-2} which is greater or equal to ρ^{-2} in the absolute value.

The integrals above were computed along arcs on the leaves of the *linearized* foliation $\mathcal{F}'$ which admits homothetic symmetries. Yet one can show that also for the initial foliation $\mathcal{F}$, whose linearization is $\mathcal{F}'$, the above computation yields the principal term of the integral $\oint_{\ell_{\rho,j,k,n}}\omega \cong \oint_{w_\rho\ell'_{j,k,n}}\omega$. Divergence of this principal term proves the lemma. We omit the detailed error estimates caused by the nonlinearity of $\mathcal{F}$; cf. [**SRO98**]. □

Remark 28.15. The paper [**SRO98**] contains the detailed proof of a stronger result: *the infinite number of homologically independent complex limit cycles exists for all polynomial foliations from the class* $\mathcal{A}_r$ *except for a nowhere dense real analytic subset of (real) codimension* $\geqslant 2$. The assumptions imposed on the foliations, are nonsolvability of the holonomy group and *absence* of certain identities between the integrals of the type (28.3).

28D. Deformational rigidity of polynomial foliations. The remaining part of this section deals with the rigidity phenomenon for polynomial foliations.

28D$_1$. *Deformations and triviality.* Speaking loosely, rigidity means that both the analytic and even the *affine type* of the foliation is completely determined by its topological type, at least in some small neighborhood of the corresponding point in the space $\mathcal{A}_r$. (Two foliations are *affine equivalent* if one of them can be transformed into the other by an affine map $A\colon \mathbb{C}^2\to\mathbb{C}^2$.) This happens when the topology (associated with qualitative dynamic properties of foliations) is rich enough to distinguish any two different foliations.

Definition 28.16. A polynomial foliation $\mathcal{F}$ from the class $\mathcal{A}_r$ is *ideally rigid*, if there exists a neighborhood U of $\mathcal{F}$ in $\mathcal{A}_r$ such that any other foliation $\mathcal{F}'\in U$ topologically equivalent to $\mathcal{F}$ is affine equivalent to $\mathcal{F}$.

Unfortunately, the ideal rigidity is practically unknown property, at least for polynomial foliations. In reality we have to operate with relaxed notions of rigidity, which require some additional properties of the homeomorphism which realizes the topological equivalence between $\mathcal{F}$ and $\mathcal{F}'$.

Probably, the weakest form of rigidity is the *deformational rigidity* which means that any *continuous deformation* preserving the topology of the given foliation $\mathcal{F}_0$, entirely consists of foliations affine equivalent to $\mathcal{F}_0$. The accurate definitions follow.

Definition 28.17. A *deformation* of a foliation $\mathcal{F} \in \mathcal{A}_r$ is the germ of a nonconstant analytic map $(\mathbb{C}^p, 0) \to (\mathcal{A}_r, \mathcal{F})$, $t \mapsto \mathcal{F}_t$.

A deformation $\{\mathcal{F}_t\}_{t\in(\mathbb{C}^p,0)}$ is *topologically trivial*, if there exists a *continuous* family of homeomorphisms $\{H_t\} \subset \operatorname{Homeo}(\mathbb{P}^2)$, deforming the identical homeomorphism $H_0 = \mathrm{id}$, such that for all $t \in (\mathbb{C}^p, 0)$ the homeomorphism H_t conjugates the foliation $\mathcal{F}_t$ with $\mathcal{F}_0$.

The analytic deformation $\{\mathcal{F}_t\}$ is *holomorphically trivial* if there exists an analytic family of biholomorphisms $\{H_t\}_{t\in(\mathbb{C}^p,0)}$, $H_0 = \mathrm{id}$, of $\mathbb{P}^2$ onto itself conjugating $\mathcal{F}_t$ with $\mathcal{F}_0$.

In fact, a generic foliation admits very few holomorphically trivial deformations.

Lemma 28.18. *A holomorphically trivial deformation of a foliation from the class $\mathcal{A}_r$ without algebraic leaves other than the infinite line consists of affine equivalent foliations: there exists an analytic family of affine maps $\{\mathbf{A}_t\colon \mathbb{C}^2 \to \mathbb{C}^2\}_{t\in(\mathbb{C}^p,0)}$ such that $\mathbf{A}_t$ conjugates $\mathcal{F}_t$ with $\mathcal{F}_0$.*

Proof. Assume first that the deformation is one-parametric, i.e., $p = 1$, and consider the velocity vector field $\frac{d}{dt}H_t$ of the holomorphic family of biholomorphisms H_t. This is a well-defined holomorphic vector field V_t on $\mathbb{P}^2$ for all values of $t \in (\mathbb{C}^p, 0)$. Since the infinite line $\mathbb{I}$ is the unique algebraic leaf of each $\mathcal{F}_t$, the field V_t is tangent to the infinite line.

But the only polynomial vector fields that extend holomorphically at the infinite line, are *affine vector fields* of degree $\leqslant 1$. Indeed, a vector field $P(x,y)\frac{\partial}{\partial x} + Q(x,y)\frac{\partial}{\partial y}$ in the coordinates $u = 1/x$, $v = y/x$, takes the form $-u^2P(u^{-1}, vu^{-1})\frac{\partial}{\partial u} + u\big(Q(u^{-1}, vu^{-1}) - v\,P(u^{-1}, vu^{-1})\big)\frac{\partial}{\partial v}$ which is holomorphic on the infinite line $\{u = 0\}$ and tangent to it if and only if $\deg P, \deg Q \leqslant 1$.

By a suitable translation one can make the vector field V_t linear; in these coordinates H_t is also linear.

The multiparametric case is reduced to the single parameter case by restricting the deformation on all possible lines through the origin in the parameter space $(\mathbb{C}^p, 0)$. □

Definition 28.19. A foliation $\mathcal{F} \in \mathcal{A}_r$ is called *deformationally rigid*, if any topologically trivial analytic deformation $\{\mathcal{F}_t\}$ of $\mathcal{F} = \mathcal{F}_0$ is holomorphically trivial and hence consists of foliations affine equivalent to $\mathcal{F}$.

As the first step in the investigation of rigidity, we prove that a generic foliation $\mathcal{F} \in \mathcal{A}_r$ is deformationally rigid.

Theorem 28.20. *Assume that a foliation $\mathcal{F} = \mathcal{F}_0$ from the class $\mathcal{A}_r$ satisfies the following assumptions.*

(T) *$\mathcal{F}$ has no algebraic leaves except for the infinite line,*

(H^+) *the foliation $\mathcal{F}$ has only hyperbolic singularities with nonreal characteristic numbers,*[11]

(D) *the holonomy group G of the infinite leaf satisfies the density condition* (6.17),

(NC) *the group G is noncommutative.*

Then $\mathcal{F}_0$ is deformationally rigid.

The proof of this theorem is given in §28$\mathbf{D}_4$: it requires appropriate tools which will be introduced in the intermediate subsections.

28$\mathbf{D}_2$. *Transversal holomorphy.* To prove Theorem 28.20, we introduce the notion of transversal holomorphy. Recall that any topological equivalence H conjugating two (singular) foliations $\mathcal{F}$ and $\mathcal{F}'$, maps plaques of $\mathcal{F}$ near any nonsingular point $a \notin \operatorname{Sing}(\mathcal{F})$ into plaques of $\mathcal{F}'$ near $a' = H(a) \notin \operatorname{Sing}(\mathcal{F}')$ (cf. with Definition 2.5). Therefore for any two cross-sections (holomorphic arcs transversal to the foliation) τ, τ' at the points a and $a' = H(a)$ respectively, there exists a unique germ, denoted by $H_a^{\pitchfork}\colon (\tau, a) \to (\tau', a')$, such that H sends a leaf of $\mathcal{F}$ passing through a point $z \in (\tau, a)$ to the leaf of $\mathcal{F}'$ passing through the point $z' = H_a^{\pitchfork}(z) \in (\tau', a')$. We will refer to the germ $H_a^{\pitchfork}$ as the *normal component* of the homeomorphism H at the point a.

Definition 28.21. The homeomorphism H conjugating two singular foliations $\mathcal{F}, \mathcal{F}'$ is *transversally holomorphic*, if for any nonsingular point a the normal component $H_a^{\pitchfork}\colon (\tau, a) \to (\tau', a')$ of H is holomorphic.

Clearly, this definition does not depend on the choice of the holomorphic cross-sections. Note that a transversally holomorphic map restricted on each leaf $L \in \mathcal{F}$, in general, is only a homeomorphism between L and $L' = H(L) \in \mathcal{F}'$.

Definition 28.22. A topologically trivial deformation $\{\mathcal{F}_t\}$, $t \in (\mathbb{C}^p, 0)$, is called *transversally holomorphic*, if all trivializing homeomorphisms H_t

[11] The condition (H^+) is a stronger version of the hyperbolicity condition (H) used earlier and requires hyperbolicity of all (not just infinite) singular points.

are transversally holomorphic and the respective normal components $H^{\pitchfork}_{t,a}$ depend analytically on $t \in (\mathbb{C}^p, 0)$ and $a \notin \operatorname{Sing}(\mathcal{F}_t)$.

For transversally holomorphic deformations, one can aggregate the individual foliations $\mathcal{F}_t$ into a *single* singular *holomorphic* foliation on the product space $\mathbb{P}^2 \times (\mathbb{C}^p, 0)$. This foliation, called the *trail foliation* and denoted by $T(\mathcal{F}_0)$, can always be easily constructed as a topological (singular) foliation in the sense of Definition 2.3. Proving that this foliation is holomorphic is possible if the deformation is transversally holomorphic.

For each deformation $\{\mathcal{F}_t\}$ of holomorphic foliations on $\mathbb{P}^2$, not necessarily topologically trivial, we can construct a singular one-dimensional (i.e., of complex codimension $p+1$) holomorphic foliation (denoted by $\mathscr{F}$) of the Cartesian product $\mathbb{P}^2 \times (\mathbb{C}^p, 0)$. The leaves of this foliation are tangent to the fibers of the Cartesian projection $\pi \colon \mathbb{P}^2 \times (\mathbb{C}^p, 0) \to (\mathbb{C}^p, 0)$ (i.e., each fiber is invariant by $\mathscr{F}$) and in each fiber $\pi^{-1}(t)$ coincide with the leaves of the respective foliation $\mathcal{F}_t$. If $F_t = a(x, y, t)\frac{\partial}{\partial x} + b(x, y, t)\frac{\partial}{\partial y} \in \mathcal{D}(\mathbb{C}^2)$ are the polynomial vector fields defining the foliations $\mathcal{F}_t$, then the field $F = F_t + 0 \cdot \frac{\partial}{\partial t} \in \mathcal{D}\big(\mathbb{C}^2 \times (\mathbb{C}^p, 0)\big)$ is the holomorphic vector field defining the foliation $\mathscr{F}$.

Denote by Σ the singular locus of $\mathscr{F}$: by definition, this is the union of individual singular loci $\Sigma_t = \operatorname{Sing} \mathcal{F}_t$,

$$\Sigma = \bigcup_{t \in (\mathbb{C}^p, 0)} \Sigma_t \times \{t\} \subseteq \mathbb{P}^2 \times (\mathbb{C}^p, 0), \quad \Sigma_t = \operatorname{Sing}(\mathcal{F}_t) \subset \mathbb{P}^2. \tag{28.7}$$

If the deformation $\{\mathcal{F}_t\}$ is topologically trivial and $H_t \colon \mathbb{P}^2 \to \mathbb{P}^2$ is the family of homeomorphisms conjugating $\mathcal{F}_t$ with $\mathcal{F}_0$ and preserving the infinite line $\mathbb{I}$, then one-dimensional leaves of the foliation $\mathscr{F}$ can be integrated into the leaves of a single *topological* foliation of complex *co*dimension 1 on $\mathbb{P}^2 \times (\mathbb{C}^p, 0)$ outside the singular locus of Σ. Consider the *trails* of the leaves of the foliation $\mathcal{F}_0$ by the deformation, that is, the sets of the form

$$T(L) = \bigcup_{t \in (\mathbb{C}^p, 0)} H_t^{-1}(L) \times \{t\}, \qquad L \in \mathcal{F}_0. \tag{28.8}$$

The union $\bigcup_{L \in \mathcal{F}_0} T(L)$ of all these trails forms a partition of the product space. We show in an instant that this partition is in fact a *topological foliation* of complex codimension 1. Indeed, the homeomorphism

$$\widetilde{H} \colon \mathbb{P}^2 \times (\mathbb{C}^p, 0) \to \mathbb{P}^2 \times (\mathbb{C}^p, 0), \qquad (a, t) \mapsto (H_t(a), t), \tag{28.9}$$

transforms the cylinders $C(L) = L \times (\mathbb{C}^p, 0)$, $L \in \mathcal{F}_0$, into the sets $T(L)$. Since the partition into the cylinders $C(L)$ is a holomorphic singular foliation of codimension 1 on the total space $\mathbb{P}^2 \times (\mathbb{C}^p, 0)$, and $\widetilde{H}$ is a homeomorphism, the sets (28.8) constitute a topological foliation on $\mathbb{P}^2 \times (\mathbb{C}^p, 0)$. The *trail foliation* $T(\mathcal{F})$ is the topological foliation by trails of the leaves as in (28.8),

$$T(\mathcal{F}) = \{T(L) \colon L \in \mathcal{F}_0\}. \tag{28.10}$$

In general $T(\mathcal{F})$ is not holomorphic (only topological), yet if the topological deformation $\{\mathcal{F}_t\}$ is transversally holomorphic, $T(\mathcal{F})$ turns out to be a singular *holomorphic* foliation of codimension 1.

Lemma 28.23. *If a topologically trivial analytic deformation $\{\mathcal{F}_t\}$ is transversally holomorphic, then the trail foliation $T(\mathcal{F})$ of the product space $\mathbb{P}^2 \times (\mathbb{C}^p, 0) \smallsetminus \Sigma$ (28.8) is a holomorphic singular foliation of complex codimension* 1. *The leaves of the trail foliation $T(\mathcal{F})$ intersect transversally each fiber $\{t = \text{const}\}$ by the leaves of the corresponding foliation $\mathcal{F}_t$.*

Proof. The assertion is local and obvious in suitably chosen local coordinates. Indeed, consider an arbitrary nonsingular point $a \notin \Sigma$. By the Rectification Theorem 1.18, the vector field $F = F_t + 0 \cdot \frac{\partial}{\partial t}$ can be rectified in some neighborhood $U \subset \mathbb{P}^2 \times (\mathbb{C}^p, 0)$, moreover, the rectifying biholomorphism can be chosen so that it preserves the t-coordinates.

In the corresponding local coordinates leaves of the foliation $\mathscr{F}$ are parallel lines $\{y = \text{const},\ t = \text{const}\}$. If we choose the y-axis as the local cross-section transversal to the lines, then from the transversal holomorphy of the family we conclude that the H_t-image of the leaf passing through the point $y = b$ at $t = 0$, is the line $\{x = \text{const}\}$ that passes through the point $y = H_t^{\pitchfork}(b)$, where $H_t^{\pitchfork}$ is the normal component occurring in the definition of the transversal holomorphy. If this component is a holomorphic function of $t \in (\mathbb{C}^p, 0)$ and $b \in (\mathbb{C}^1, 0)$ in some initial chart, then the same is true in the rectifying chart as well. Therefore the topological leaves $T(L)$ in the rectifying chart are cylinders over graphs of the corresponding holomorphic function $y = h_t(b)$, $x \in (\mathbb{C}^1, 0)$, for different values of the point $b \in (\mathbb{C}^1, 0)$. These leaves are given by the level curves of one holomorphic function $\psi(x, y, t) = h_t^{-1}(y)$ (in fact, independent of x). The transversality assertion of the lemma follows from the fact that the leaves are graphs of holomorphic functions of t. $\square$

28$\mathbf{D}_3$. *Deformational rigidity theorem: outline of the proof.* The proof of deformational rigidity of generic foliations can be split into two steps. First, we show that generic foliations can be topologically conjugated only by transversally holomorphic homeomorphisms.

Lemma 28.24. *Assume that a foliation $\mathcal{F} = \mathcal{F}_0$ from the class $\mathcal{A}'_\mathrm{T}$ satisfies the following conditions*[12]:

[12](T) stands for transcendence.

(T) *$\mathcal{F}$ has no algebraic leaves except for the infinite line,*

(H) *the infinite line $\mathbb{I}$ carries only hyperbolic singularities with nonreal characteristic numbers,*

(D) *the holonomy group G of the infinite leaf satisfies the density condition* (6.17),

(NC) *the group G is noncommutative.*

Then any homeomorphism conjugating $\mathcal{F}$ with another foliation $\mathcal{F}' \in \mathcal{A}_r$ is necessarily transversally holomorphic. Moreover, any topologically trivial deformation of $\mathcal{F}$ is transversally holomorphic in the sense of Definition 28.22.

This lemma is proved below by deriving the transversal holomorphy of H near the infinite line from Theorem 6.45, and then extending this transversal holomorphy to all nonsingular points by using the density of leaves (Theorem 28.7).

From the transversal holomorphy established in Lemma 28.24, by Lemma 28.23 it follows that the trail foliation $T(\mathcal{F})$ is holomorphic. Being a foliation of codimension 1, it is defined by a suitable 1-form Ω on $\mathbb{C}^2 \times (\mathbb{C}^p, 0)$, polynomial in (x, y)-coordinates and holomorphic in t.

The second step of the proof is deals with one-parametric transversally holomorphic deformations ($p = 1$). We will show that the constant vector field $\frac{\partial}{\partial t}$ on the base $(\mathbb{C}^1, 0)$ can be lifted to a holomorphic vector field on the product space $\mathbb{P}^2 \times (\mathbb{C}^1, 0)$ tangent to the trail foliation $T(\mathcal{F})$. Then the flow maps of this field will holomorphically trivialize the deformation $\mathcal{F}_t$. This will prove the Deformational Rigidity theorem.

Remark 28.25. The construction of the lift is nontrivial despite the fact that the foliation is transversal to the fibers $\{t = \text{const}\}$.

Indeed, any vector v tangent to the base at a point t, can be lifted to a vector tangent to each leaf $T(L)$ passing through a nonsingular point (a, t), thus producing a necessary vector field on the product space. Yet this construction is essentially nonunique, since one can add to the lift of v any multiple of the vector $F_t(a)$ tangent to L. Thus lifting a holomorphic vector field amounts to construction of a holomorphic section of some affine bundle, which in general is possible only locally. To construct a global section of this bundle is not always possible because of the global topological obstructions. These obstructions can sometimes be shown to vanish (see [**GM88, GM89**]), yet not always. In Remark 28.28 below we will give an example when there is an obstruction to such a lift.

We will construct the lift in a different way, using the hyperbolicity assumption concerning singularities of the foliations in the finite part $\mathbb{C}^2 \subset \mathbb{P}^2$.

The following assertion can be considered as a complex analog of the "path method" widely used in the smooth classification theory; see §5**F** and [**IY91**]. In order to stress independence of this lemma from the specific choice of the foliations, we change slightly the notation.

Lemma 28.26. *Consider the total space $\mathbb{P}^2 \times (\mathbb{C}^1, 0)$ with a holomorphic singular foliation $\mathcal{G}$ of codimension 1 on it. Assume that $\mathcal{G}$ is transversal to the fibers of the projection $\pi\colon (a,t) \mapsto t$, and tangent to the "infinite cylinder" $\mathbb{I} \times (\mathbb{C}^1, 0)$.*

If each singular foliation $\mathcal{G}_t = \mathcal{G}|_{\pi^{-1}(t)}$ obtained by restriction of $\mathcal{G}$ on the fiber $\{t = \text{const}\}$, has only hyperbolic singular points (with nonreal ratios of eigenvalues) in the finite part, then all these foliations are affine equivalent: there exists a holomorphic family of affine maps $\{\mathbf{A}_t\}_{t \in (\mathbb{C}^1, 0)}$ conjugating $\mathcal{G}_t$ with $\mathcal{G}_0$ and analytically depending on the parameter t.

28$\mathbf{D}_4$. *Demonstration of the Deformational Rigidity theorem.* In this subsection we prove all results formulated in §28$\mathbf{D}_1$–§28$\mathbf{D}_3$.

Proof of Lemma 28.24. Because of the assumption (T), the infinite line $\mathbb{I}$ is the unique algebraic leaf of both foliations; it is topologically distinguished by the fact that its closure does not coincide with the whole plane $\mathbb{P}^2$. Hence a homeomorphism H must map the infinite line into itself.

1°. By virtue of Theorem 6.45, in the assumptions (D), (H), and (NC) any topological conjugacy $H\colon \mathcal{F} \to \mathcal{F}'$ is transversally holomorphic "near infinity". Indeed, the corresponding normal component $H_a^{\pitchfork}(\tau, a) \to (\tau', a')$ conjugates the holonomy groups of the foliations $\mathcal{F}$ and $\mathcal{F}'$ associated with arbitrary holomorphic cross-sections τ and τ' at any nonsingular points $a \in \mathbb{I}$ and $a' = H(a)$ and is therefore holomorphic. Moreover, if $\mathcal{F}_t$ is an analytic deformation, then $H_{t,a}^{\pitchfork}$ depends analytically on a and t.

2°. It remains to show that H is transversally holomorphic at any other nonsingular point $b \notin \mathbb{I}$. By the Density Theorem 28.7, the leaf of $\mathcal{F}$ passing through b must cross (transversally) the cross-section τ constructed above. Choose an arbitrary holomorphic cross-section σ to $\mathcal{F}$ at b, denote $\sigma' = H(\sigma)$ and consider two holonomy maps, $\Delta_{b,a}\colon (\sigma, b) \to (\tau, a)$ and $\Delta'_{b',a'}\colon (\sigma', b') \to (\tau', a')$ for the foliations $\mathcal{F}, \mathcal{F}'$ respectively, along an arbitrary path γ and its image $\gamma' = H(\gamma)$.

From the definition of the normal component and the holonomy map it follows that

$$H_a^{\pitchfork} \circ \Delta_{b,a} = \Delta'_{b',a'} \circ H_b^{\pitchfork}\colon (\sigma, b) \to (\tau', b'). \tag{28.11}$$

The holonomy maps Δ, Δ' are holomorphic and depend holomorphically on the end points $a, b, a', b' \in \mathbb{P}^2$ (and any additional parameters $t \in (\mathbb{C}^p, 0)$ if they are present). Hence holomorphy of one of the normal components (in our case $H_a^{\pitchfork}$) in all the variables implies holomorphy of the other normal component and vice versa. □

Remark 28.27 (important). Lemma 28.24 can be relaxed as follows: *Any deformation $\mathcal{F}_t$ topologically trivial only in a neighborhood of the infinite line*

$\mathbb{I}$, *is transversally holomorphic there provided that* $\mathcal{F}_0$ *meets the assumptions* (H), (D) *and* (NC). The assumption (T) can be dropped in this case.

The first steps of the proof remain almost the same. The only change required concerns transversal holomorphy of H at a point $b \notin \mathbb{I}$ which belongs to the invariant manifold of a singular point $a_i \in \Sigma \subset \mathbb{I}$ at infinity: in this case the leaf of $\mathcal{F}$, passing through b, does not intersect τ, and hence on the second step of the proof one has to replace the holonomy along a leaf by the Dulac map between two cross-sections to two different leaves as follows.

2°_{alt}. Let σ be the cross-section to the separatrix leaf S of $\mathcal{F}$, and τ another cross-section to $\mathbb{I}$ at a point $a \in \mathbb{I}$ sufficiently close to the singularity $a_i \in \Sigma \cap \mathbb{I}$. Denote by S', σ', τ' their images by H, the homeomorphism conjugating $\mathcal{F} = \mathcal{F}_0$ near $\mathbb{I}$, with $\mathcal{F}' = \mathcal{F}_t$: S' is a uniquely defined separatrix of $\mathcal{F}'$ at $a'_i = H(a_i)$.

By the hyperbolicity assumption (H), both $\mathcal{F}, \mathcal{F}'$ are holomorphically linearizable near a_i and a'_i respectively. Consider the Dulac maps $\Delta\colon (\sigma, b) \to (\tau, a)$ and $\Delta'\colon (\sigma', b') \to (\tau', a')$, the multivalued holonomy maps (Δ sends a point $z \in \sigma$ to the intersection of the leaf $L_z \in \mathcal{F}$ with τ, Δ' does the same for $\mathcal{F}'$). For the linearized foliations, each map is (a branch of) the power function $w = z^\lambda$ (resp., $w' = z'^\lambda$) with the same λ for a suitable choice of the linearizing charts. Moreover, in these charts the normal component $H_a^\pitchfork$ of H on a cross-section τ_a to $\mathbb{I}$ at $a \in \mathbb{I}$, as before, is a linear map, $z' = cz$. Indeed, this is the only holomorphic map conjugating two linear rotations $z \mapsto (\exp 2\pi i\lambda) \cdot z$ and $z' \mapsto (\exp 2\pi i \lambda') \cdot z'$, with $\lambda, \lambda' \notin \mathbb{R}$, possible only if $\lambda = \lambda'$.

The normal component $H_b^\pitchfork$ (computed with respect to the linearizing chart) satisfies the analog of the equation (28.11), namely,

$$\Delta \circ H_a^\pitchfork = H_b^\pitchfork \circ \Delta'\colon (\sigma, b) \to (\tau', b'). \tag{28.12}$$

Altogether these arguments imply that $H_b^\pitchfork(z) = (cz^\lambda)^{1/\lambda} = c^{1/\lambda} z$ is again a linear map, which is obviously holomorphic. Thus the transversal holomorphy of H is established at all points of the neighborhood of $\mathbb{I}$. Since the normalizing charts depend analytically on $\mathcal{F}'$, so does the normal component.

Proof of Lemma 28.26. 1°. First we show that the foliation $\mathcal{G}$ can be globally defined by a Pfaffian 1-form $\Omega = \omega_t + R\,dt$, where $\omega_t \in \Lambda^1[\mathbb{C}^2]$ are the given polynomial forms, $\omega_t = P_t\,dx + Q_t\,dy$, defining the individual foliations $\mathcal{G}_t$, which depend analytically in t, and $R \in \mathbb{C}[x, y] \otimes \mathcal{O}(\mathbb{C}^1, 0)$ is a holomorphic function in x, y, t which is a polynomial of degree $\leqslant r + 1$ in (x, y) for every $t \in (\mathbb{C}^1, 0)$.

Indeed, in a small neighborhood U_a of each nonsingular point $a \notin \Sigma = \operatorname{Sing}(\mathcal{G})$ the foliation $\mathcal{G}$ is defined by a holomorphic Pfaffian form $\Omega_a =$

$p_a\,dx+q_a\,dy+r_a\,dt$ with holomorphic coefficients. Since the 2-plane $\{\Omega_a=0\}$ is transversal to the 2-plane $\{dt=0\}$ in the tangent bundle $\mathbf{T}_a\mathbb{P}^2\times(\mathbb{C}^1,0)$, the wedge product $p_a\,dx\wedge dt+q_a\,dx\wedge dt$ is nonsingular near a, hence the restriction of $p_a\,dx+q_a\,dy$ on $\pi^{-1}(t)$ is nonsingular in $U_a\cap\pi^{-1}(t)$.

The form Ω_a restricted on the fiber $\pi^{-1}(t)$ passing through a must vanish on the leaves of $\mathcal{G}_t$. Since the 1-forms ω_t and $\Omega_a|_{\pi^{-1}(t)}$ are both nonsingular near a, they must be proportional: $\varphi_a(p_a\,dx+q_a\,dy)=\omega_t$ for some invertible factor $\varphi_a\in\mathcal{O}(\mathbb{C}^3,a)$. The form $\varphi_a\Omega_a$ is holomorphic and its restriction on $\pi^{-1}(t)$ coincides with ω_t.

The local forms $\varphi_a\Omega_a$ define the same foliation $\mathcal{G}$ outside the singular locus Σ and their restrictions on the fibers $\pi^{-1}(t)$ coincide on pairwise intersections of the neighborhoods U_a. Therefore together they aggregate in a single holomorphic 1-form $\Omega=\omega_t+R(x,y,t)\,dt$ on the product space with the deleted locus Σ. Since Σ has complex codimension 2, the function R together with the form Ω extend onto it, thus defining a holomorphic 1-form on the product $\mathbb{C}^2\times(\mathbb{C}^1,0)$.

The foliation $\mathcal{G}$ is holomorphic also near the "infinite cylinder" $\mathbb{I}\times(\mathbb{C}^1,0)$ and tangent to it. In the coordinates $u=x^{-1}$ and $v=yx^{-1}$ the form Ω has the structure $u^{-(r+2)}\omega_t'+R(u^{-1},vu^{-1})\,dt$ with some polynomial 1-form ω_t'; cf. with (25.4). In order for the "infinite cylinder" to be invariant for $\mathcal{G}$, the function $R(u^{-1},vu^{-1})$ must have a pole of order $\leqslant r+2$ and still be divisible by u after multiplication by u^{r+2}. This is possible only if for every value of $t\in(\mathbb{C}^1,0)$ the function R is a polynomial of degree $\leqslant r+1$ in (x,y).

Note that the distribution $\Omega=0$ is integrable: its integral foliation is $\mathcal{G}$. By Frobenius Theorem 2.9, this means that $\Omega\wedge d\Omega\equiv 0$.

2°. In the second step we prove that for any t the polynomial $R_t=R(\cdot,\cdot,t)\in\mathbb{C}[x,y]$ belongs to the ideal spanned by the components of the form ω_t.

Indeed, consider an arbitrary singular point $b\in\Sigma_t$ of the form ω_t and assume that R_t does not vanish at this point. Then the form Ω is nonsingular at b; being integrable, Ω admits a nontrivial local analytic first integral $u\in\mathcal{O}\big(\mathbb{P}^2\times\mathbb{C}^1,b\big)$. The restriction of u on the fiber $\pi^{-1}(t)$ is therefore an analytic first integral $u_t\in\mathcal{O}(\mathbb{P}^2,b)$ of the component ω_t. Yet because of the hyperbolicity of ω_t at b, the foliation $\mathcal{F}_t$ does not admit nontrivial first integrals. Indeed, the foliation $\mathcal{F}_t$ is locally holomorphically equivalent to the foliation with the leaves $w=cz^\lambda$, $\lambda\notin\mathbb{Q}$, which are not relatively closed. This foliation is not simple in the sense of Definition 11.20 and hence cannot be integrable by Theorem 11.21.

The contradiction implies that $R_t(b)=0$. Again, because of the hyperbolicity, the ideal generated by the coordinates of the form ω_t in $\mathcal{O}(\mathbb{P}^2,b)$, is radical, so vanishing of R_t at b implies that the germ of R_t at each point b of the intersection $\Sigma \cap \pi^{-1}(t)$ belongs to the ideal generated by the germs of the components of ω_t in the local ring $\mathcal{O}(\mathbb{P}^2, b)$.

From the complex hyperbolicity of the singularities on the infinite line, the principal homogeneous components of the coefficients of the form ω_t have no common roots (the polynomials $q_r(1, v)$ and $h_{r+1}(1, v)$ in (25.4) cannot vanish simultaneously).

By the Max Noether theorem [**GH78**, Chapter 5, §3], this is sufficient to guarantee that the polynomial R_t globally belongs to the ideal $\langle P_t, Q_t\rangle \subset \mathbb{C}[x, y]$, generated by the components of $\omega_t = P_t\,dx + Q_t\,dy$: there exist polynomials $A_t, B_t \in \mathbb{C}[x, y]$, $\deg A_t, B_t \leqslant 1$, analytically depending on t, such that $R_t = P_tA_t + Q_tB_t$.

3°. The above computations show that the polynomial 1-form Ω defining the foliation $\mathcal{G}$, is representable under the form

$$\Omega = \omega_t + R\,dt = P_t(x, y)\big(dx + A_t(x, y)\,dt\big) + Q_t(x, y)\big(dy + B_t(x, y)\,dt\big).$$

This 1-form obviously vanishes on the vector field

$$Z = -A_t(x, y)\tfrac{\partial}{\partial x} - B_t(x, y)\tfrac{\partial}{\partial y} + 1 \cdot \tfrac{\partial}{\partial t}$$

which is transversal to the fibers $\{t = \text{const}\}$ and extends as a holomorphic vector field on the entire product space $\mathbb{P}^2 \times (\mathbb{C}^1, 0)$; cf. with Lemma 28.18. The flow maps of the field Z preserve the foliation $\mathcal{G}$ and the fibers of the bundle $\pi\colon \mathbb{P}^2 \times (\mathbb{C}^1, 0) \to (\mathbb{C}^1, 0)$, and are all affine. In other words, we constructed a collection of affine maps $\{\mathbf{A}_t : \pi^{-1}(t) \to \pi^{-1}(0)\}$ which conjugate the foliations $\mathcal{G}_t$ with $\mathcal{G}_0$. □

Proof of Theorem 28.20. Theorem 28.20 follows almost immediately from the three lemmas. Indeed, by Lemma 28.24, a deformation $\{\mathcal{F}_t\}$ satisfying the assumptions of the theorem, is transversally holomorphic. By Lemma 28.23 the topological foliation $T(\mathcal{F})$, aggregating the leaves of individual foliations $\mathcal{F}_t$ for different $t \in (\mathbb{C}^p, 0)$, is holomorphic. Restricting this foliation to any complex line $(\mathbb{C}^1, 0) \subseteq (\mathbb{C}^p, 0)$ in the base, we can apply Lemma 28.26 requiring only the assumption (H) of the theorem, and conclude that any foliation $\mathcal{F}_t$ is affine equivalent to $\mathcal{F}_0$. □

Remark 28.28. The assumption of complex hyperbolicity of singularities in Lemma 28.26 cannot be completely dropped. Indeed, consider an analytic deformation of the *Hamiltonian foliation* $\mathcal{G}_t = \{dS_t = 0\}$, where $S_t = S_t(x, y)$ is a generic deformation of the polynomial Hamiltonian transversal to infinity. In this case the foliation $\mathcal{G} = \{dS = 0\}$, $S = S(x, y, t)$, satisfies all assumptions of the lemma except that all its singular points are locally (even globally) integrable saddles. The form $\Omega = dS$ defining $\mathcal{G}$ is polynomial, but the coefficient $R_t = \partial S/\partial t$ does not necessarily vanish at the singular points, hence construction of the vector field Z fails globally (though is still possible locally). Note that in this case the deformation $\mathcal{G}_t$ is topologically trivial (cf. with §26**F**) but in general is not affine trivial.

Obviously, the hyperbolicity assumption in Lemma 28.26 can be relaxed to the local nonintegrability. *If the foliation $\mathcal{G}_0$ is not locally analytically integrable near any finite singularity*, then all nearby foliations $\mathcal{G}_t$ are also not analytically integrable, which is sufficient for the proof to work under this relaxed assumption.

28E. Rigidity of polynomial foliations. To pass from the deformation rigidity discussed in §28**D** to stronger forms of rigidity, an additional effort is required. We explain first why the ideal rigidity is so difficult to establish.

Assume for simplicity that the foliation $\mathcal{F} \in \mathcal{A}_r$ has only one algebraic separatrix at infinity. Then for any other foliation $\mathcal{F}'$ topologically equivalent to $\mathcal{F}$, the homeomorphism $H: \mathbb{P}^2 \to \mathbb{P}^2$ conjugating these foliations, necessarily preserves the infinite line $\mathbb{I} \subset \mathbb{P}^2$ and maps the singular loci $\Sigma_{\mathbb{I}} = \operatorname{Sing}(\mathcal{F}) \cap \mathbb{I}$ and $\Sigma'_{\mathbb{I}} = \operatorname{Sing}(\mathcal{F}') \cap \mathbb{I}$ at infinity into each other. The problem is that in absence of extra information, the restriction homeomorphism $H_{\mathbb{I}} : (\mathbb{I}, \Sigma_{\mathbb{I}}) \to (\mathbb{I}, \Sigma'_{\mathbb{I}})$ can be very wild.

For instance, consider a group of self-homeomorphisms of an abstract sphere $\mathbb{P}^1 \cong \mathbb{S}^2$ with a marked finite set of $r+1$ points. The infinite braid group on $r+1$ strains acts on all such homeomorphisms by isotopy. Each such homeomorphism h induces an automorphism $h_\diamond: \pi_1(\mathbb{P}^1 \smallsetminus S, a) \to \pi_1(\mathbb{C}P^1 \smallsetminus S, a)$ of the fundamental groups, which in a given basis of, say, canonical loops (as they were introduced in §28**A**) may have arbitrarily high *combinatorial complexity* (we define this complexity as the maximal length of the word representing the loop $h_\diamond(\gamma_i)$, $i = 1, \dots, r$ in the alphabet $\gamma_1, \dots, \gamma_r$ generating the free group $\pi_1(\mathbb{P}^1 \smallsetminus S, a)$; see §28**G**$_1$ below).

Yet without knowing the automorphism $h_\diamond$, it is impossible to find out how precisely the corresponding infinite holonomy groups are topologically conjugated (cf. with Proposition 28.2). This incertitude devaluates our principal tool, the holonomy of a foliation.

In order to set aside these difficulties, we introduce a relaxed notion of rigidity as follows.

Definition 28.29. Let $S = \{a_0, \dots, a_r\} \subset \mathbb{P}^1$ be a finite set of $r+1$ distinct points, $D_0, \dots, D_r$ a collection of $r+1$ disjoint open disks covering the set S and $b \notin D$, $D = \bigcup_i D_i$ a point outside these disks.

A homeomorphism $h: \mathbb{P} \to \mathbb{P}$ is called *homotopically trivial* over $\mathbb{P} \smallsetminus D$, if $h(b) = b$, for each point a_i its image $h(a_i)$ belongs to the same open disk, $h(a_i) \in D_i$, and the images $h(\lambda_j)$ of all the segments $\lambda_j = [b, a_j]$ connecting the base point b with each point $a_j \in S$, are homotopic to the corresponding segments λ_j in the class of homotopy with the fixed endpoint b and free endpoint $a_{i,t} \in D_i$ restricted to the respective disk.

A homeomorphism is said to be homotopically trivial without specifying the system of disks, if it is topologically trivial over *some* system of disks.

Intuitively a homeomorphism is homotopically trivial if it "preserves" the system of canonical loops to the extent possible for nonlinear homeomorphisms. The homotopically trivial homeomorphisms can be rather far from the identity, yet their action $h_\diamondsuit$ on the fundamental group is trivial.

Definition 28.30. A foliation $\mathcal{F} \in \mathcal{A}'_r$ will be called *reasonably rigid*, if there exists a neighborhood U of it in $\mathcal{A}_r$ such that any foliation $\mathcal{F}' \in U$ topologically equivalent to $\mathcal{F}$, is affine equivalent to $\mathcal{F}$ provided that the topological equivalence between $\mathcal{F}$ and $\mathcal{F}'$ induces a homotopically trivial homeomorphism of the infinite line $\mathbb{I}$ into itself.

In other words, reasonable rigidity of $\mathcal{F}$ does not exclude existence of topologically equivalent while not affine equivalent to $\mathcal{F}$ foliations near $\mathcal{F}$, but asserts that the conjugating homeomorphism in such cases must be rather "exotic".

The principal result of this section can be formulated as follows.

Theorem 28.31. *A generic foliation from the class $\mathcal{A}'_r$ is reasonably rigid.*

The exact genericity assumptions are listed in Theorem 28.20 formulated below. Later we will formulate some generalizations of this result.

28**E**$_1$. *Three varieties: an outline of the proof.* For a given foliation $\mathcal{F}_0 \in \mathcal{A}'_r$ we will introduce three subsets of $\mathcal{A}_r$, which consist of (a) foliations topologically equivalent to $\mathcal{F}_0$, (b) foliations whose holonomy group is topologically conjugate to that of $\mathcal{F}_0$ and (c) foliations affine equivalent to $\mathcal{F}_0$. The rigidity theorem will follow from the fact that the germs of all three sets at the point $\mathcal{F}_0 \in \mathcal{A}'_r$ coincide.

More precisely, for a foliation $\mathcal{F}_0 \in \mathcal{A}_r$ consider the singular locus at infinity $S_0 = \operatorname{Sing}\mathcal{F}_0 \cap \mathbb{I}$. Since all singularities are distinct, there exists a system of disjoint circular disks $D_0, \dots, D_r$ each covering only one of these points. There exists an open neighborhood $\widetilde{U} \subset \mathcal{A}_r$ such that for any other foliation $\mathcal{F}' \in \widetilde{U}$, the respective singular locus at infinity $S_{\mathcal{F}}$ belongs to the union of the disks $D = \bigcup D_j$ and each disk D_j contains exactly one singularity.

For all foliations from this "large" neighborhood, one can uniquely enumerate the singularities at infinity in a consistent way continuously depending on the foliation.

Denote by $\operatorname{Topo}(\mathcal{F}_0)$ the set of foliations from the neighborhood $\widetilde{U} \subset \mathcal{A}_r$, which are topologically conjugated to $\mathcal{F}_0$ by a homeomorphism H whose restriction on the infinite line is homotopically trivial over complement to the above system of open disks.

Denote by $\operatorname{Isohol}(\mathcal{F}_0) \subseteq \widetilde{U}$ the set of foliations $\mathcal{F}'$ from the same neighborhood, whose holonomy group at infinity is topologically conjugate to that of $\mathcal{F}_0$. More precisely, $\mathcal{F}' \in \operatorname{Isohol}(\mathcal{F}_0)$ if and only if there exists a homeomorphism conjugating the holonomy maps Δ_j for a system of canonical loops for $\mathcal{F}_0$, with the holonomy operators Δ'_j along "the same" loops for the foliation $\mathcal{F}'$ (the identification is possible since each disk D_j has only one singularity for both $\mathcal{F}_0$ and $\mathcal{F}'$).

Finally, denote by $\operatorname{Aff}(\mathcal{F}_0) \subset \widetilde{U}$ the set of all foliations from the "large" neighborhood, which are *affine equivalent* to $\mathcal{F}_0$, under the assumption that the affine conjugacy preserves the enumeration of singularities on the infinite line.

Clearly, for an arbitrary foliation $\mathcal{F}_0 \in \mathcal{A}'_r$,

$$\operatorname{Isohol}(\mathcal{F}_0) \supseteq \operatorname{Topo}(\mathcal{F}_0) \supseteq \operatorname{Aff}(\mathcal{F}_0). \tag{28.13}$$

This follows from Definition 28.29 and the enhanced form of Proposition 28.2. In principle, all inclusion can be strict. The foliation $\mathcal{F}_0$ is reasonably rigid in the sense of Definition 28.30 if and only if there exists a smaller neighborhood $U \subset \widetilde{U}$ of $\mathcal{F}_0$ such that the intersection of all three sets with U coincide. In other words, $\mathcal{F}_0$ is reasonably rigid if and only if the germs of these three sets at $\mathcal{F}_0$ coincide,

$$(\operatorname{Isohol}(\mathcal{F}_0), \mathcal{F}_0) = (\operatorname{Topo}(\mathcal{F}_0), \mathcal{F}_0) = (\operatorname{Aff}(\mathcal{F}_0), \mathcal{F}_0).$$

Theorem 28.32. *A foliation $\mathcal{F}_0 \in \mathcal{A}_r$ satisfying the assumptions of Theorem 28.20, is reasonably rigid.*

The proof of Theorem 28.32 is based on investigation of the largest set $\operatorname{Isohol}(\mathcal{F}_0)$. First, we show that for a generic $\mathcal{F}_0$ the conditions of topological conjugacy of the holonomy groups is *analytic*, i.e., that the germ $(\operatorname{Isohol}(\mathcal{F}_0), \mathcal{F}_0)$ is *analytic*.

On the second step we show that the analytic family $\operatorname{Isohol}(\mathcal{F}_0)$ parameterized by points of the space $\mathcal{A}_r$, is *topologically trivial near the infinite line*: for any point $\mathcal{F} \in \operatorname{Isohol}(\mathcal{F}_0)$ sufficiently close to $\mathcal{F}_0$, there exists a homeomorphism $H^\infty_{\mathcal{F}} \colon (\mathbb{P}^2, \mathbb{I}) \to (\mathbb{P}^2, \mathbb{I})$ conjugating $\mathcal{F}$ with $\mathcal{F}_0$ in this neighborhood. Moreover, we construct this homeomorphism so that it is automatically transversally holomorphic. The construction is not completely trivial and rests upon rigidity of the holonomy group and hyperbolicity of the singularities of $\mathcal{F}$ on $\mathbb{I}$.

By the standard extension theorem for functions of several complex variables, the map $H^\infty_{\mathcal{F}}$ extends from a neighborhood of infinity to the whole affine plane $\mathbb{C}^2$ as an analytic map. By analyticity, it remains a conjugacy between $\mathcal{F}$ and $\mathcal{F}_0$ for all $\mathcal{F} \in \operatorname{Isohol}(\mathcal{F}_0)$ sufficiently close to $\mathcal{F}_0$. Obviously, $\mathcal{F}$ can be connected with $\mathcal{F}_0$ by a one-parametric family $\{\mathcal{F}_t\}$, $\mathcal{F}_1 = \mathcal{F}$, entirely belonging to $\operatorname{Isohol}(\mathcal{F}_0)$. Lemma 28.18 guarantees that in this case $\mathcal{F}$ and $\mathcal{F}_0$ are affine equivalent.

28**E**$_2$. *Demonstration of the reasonable rigidity theorem.* We proceed with detailed arguments now. In order to avoid exotic notations, we consider the set of all foliations $\mathcal{A}_r$ as the parameter space and denote by $\mathcal{F}_t$ the foliation corresponding to the variable point $t \in \mathcal{A}_r$. The initial foliation is already labelled as $\mathcal{F}_0$.

Lemma 28.33. *If the holonomy group of the foliation $\mathcal{F}_0 \in \mathcal{A}'_r$ satisfies the conditions* (D) *and* (NC) *of Theorem 28.20, then the germ of the set* $\operatorname{Isohol}(\mathcal{F}_0) \subseteq \mathcal{A}_r$ *at $t = 0$ is the germ of an analytic subvariety of $\mathcal{A}_r$.*

Proof. By definition, the two holonomy groups G_0 and G_t of the foliations $\mathcal{F}_0$ and $\mathcal{F}_t$ are conjugated by a homeomorphism h_t which conjugates each generator $f_{0,j}$ of G_0 with the corresponding generator $f_{t,j}$ of G_t for all $j = 0, 1, \dots, r$, where the generators $f_{t,j}$ analytically depend on t.

By Theorem 6.45, in the assumptions of the lemma *any* homeomorphism conjugating G_0 with the group $G_t = \langle f_{t,1}, \dots, f_{t,n} \rangle \subseteq \operatorname{Diff}(\mathbb{C}^1, 0)$ with the prescribed action on the generators, must (after linear rescaling) be identical in the two charts linearizing two hyperbolic generators of G_0 and G_t respectively. In the assumptions of the lemma these charts depend analytically on t, so does the conjugating biholomorphism h_t.

In other words, for any $\mathcal{F}_t \in \operatorname{Isohol}(\mathcal{F}_0)$ there exists a holomorphic germ $h_t \colon (\mathbb{C}^1, 0) \to (\mathbb{C}^1, 0)$ analytically depending on t, such that $h_t \circ f_{0,t} = f_{0,1} \circ h_t$. The two groups G and G_t are topologically conjugate if and only if h_t also conjugates all other generators $f_{t,k}$ with $f_{0,k}$ for all $k = 1, \dots, n$. The conditions $h_t \circ f_{t,k} = f_{0,k} \circ h_t$ with known h_t impose infinitely many identities between the Taylor coefficients of the germs $f_{t,k}$ which are all analytic functions of finite-dimensional parameter $t \in \mathcal{A}_r$. Since the ring of germs of analytic functions is Noetherian, these infinitely many conditions together define the germ of an analytic subset which coincides with $\operatorname{Isohol}(\mathcal{F}_0)$. □

As was already noted, in general the subset $\operatorname{Topo}(\mathcal{F}_0)$ may be a proper subset of $\operatorname{Isohol}(\mathcal{F}_0)$. Yet in the assumptions of Theorem 28.20 we can prove that $\operatorname{Isohol}(\mathcal{F}_0)$ can be considered as a topologically trivial family (deformation) *near the infinite line.*

Lemma 28.34. *Assume that the foliation $\mathcal{F}_0 \in \mathcal{A}'_r$ satisfies the assumptions* (D), (NC) *and* (H). *Then for any $\mathcal{F}_t \in \operatorname{Isohol}(\mathcal{F}_0)$ sufficiently close to $\mathcal{F}_0$ there exists a homeomorphism $H_t^\infty \colon \mathbb{P}^2 \to \mathbb{P}^2$, $H_0^\infty = \mathrm{id}$, defined in a neighborhood of the infinite line $\mathbb{I}$ in the projective plane, fixing this line, continuously depending on t and conjugating the foliations $\mathcal{F}_t$ and $\mathcal{F}_0$ in the respective neighborhoods.*

Except for the relatively small effort necessary to extend the homeomorphisms to the neighborhood of singular points, the lemma claims that near the algebraic leaf $\mathbb{I} \smallsetminus \Sigma$, the only topological invariant of a holomorphic foliation is its holonomy group.

Proof. We first construct explicitly the homeomorphism H_t^∞ away from the singular points of $\mathcal{F}_t$ as follows.

1°. Consider the canonical projection $\sigma \colon \mathbb{C}^2 \smallsetminus \{0\} \to \mathbb{I}$, $(x,y) \mapsto [x:y:0]$. This projection after restriction on a small tubular neighborhood $V \subset \mathbb{P}^2$ of $\mathbb{I}$ defines topological bundle with a fiber homeomorphic to the disk.

Denote by $S_t = \Sigma_t \cap \mathbb{I}$ the collection of singular points of $\mathcal{F}_t$ on the infinite line and consider the open set

$$V' = V \cap \sigma^{-1}(\mathbb{I} \smallsetminus D), \qquad D = \bigcup_j D_j, \tag{28.14}$$

the tubular neighborhood of the infinite line with the deleted disks D_j which contain the singular points of all foliations $\mathcal{F}_t \in \widetilde{U}$. By construction, all foliations $\mathcal{F}_t$ are nonsingular in V' and transversal to the fibers $\sigma^{-1}(a)$ if the tubular neighborhood V was chosen sufficiently thin.

Denote by $\tau = \sigma^{-1}(s_0) \subset V'$ the cross-section to the infinite line at a point $s_0 \notin D$ which is nonsingular for the foliations $\mathcal{F}_t \in \widetilde{U}$. Assume that the homeomorphisms conjugating the holonomy groups of the foliations $\mathcal{F}_t$ and $\mathcal{F}_0$, are all associated with the same cross-section, $h_t \colon (\tau, s_0) \to (\tau, s_0)$.

The homeomorphisms H_t^∞ are completely determined in V' by the following conditions,

(1) all H_t^∞ map the cross-section τ into itself and their restriction on τ coincides with h_t, i.e., $H_t^\infty|_\tau = h_t$,

(2) each H_t^∞ preserves all fibers $\sigma^{-1}(s)$ which belong to V', and

(3) H_t^∞ conjugate $\mathcal{F}_t$ with $\mathcal{F}_0$ in V'.

Indeed, to construct the image of a point $a \in V'$, consider an arbitrary path γ connecting $s = \sigma(a)$ with $s_0 = a_0$ in $\mathbb{I} \smallsetminus D$. Consider the holonomy map $\Delta_t \colon \sigma^{-1}(s) \to \sigma^{-1}(s_0)$ along γ, and define $H_t^\infty = \Delta_0^{-1} \circ h_t \circ \Delta_t$ (corresponding to the travel along the leaf of $\mathcal{F}_t$ over γ, action by h_t and the backwards travel along the leaf of $\mathcal{F}_0$ back to the same fiber $\sigma^{-1}(s)$).

Obviously, this construction uniquely *defines* the homeomorphism H_t^∞ if h_t conjugates the holonomy groups of the foliations $\mathcal{F}_t$ and $\mathcal{F}_0$; indeed, in this case the result does not depend on the choice of the path γ.

This standard construction allows us to define H_t^∞ in a tubular neighborhood of the infinite line outside preimages $\sigma^{-1}(D)$ of the disks covering the singularities. Moreover, the resulting family is in fact biholomorphic and analytically depends on t.

2°. It remains to extend H_t^∞ onto the cylinders $\sigma^{-1}(D_j)$ around the singular points. The problem is purely local: we consider a small bidisk $B = \sigma^{-1}(D_j)$ with the foliations $\mathcal{F}_t$ having a unique hyperbolic singularity in this bidisk at the point a_t analytically depending on t. The problem is to extend the homeomorphism H_t^∞ constructed above, from the boundary $\sigma^{-1}(\partial D_j)$ on the interior of B.

Because of the hyperbolicity assumption, we may assume that the family $\mathcal{F}_t$ is analytically linearized: for all small t it is given by the linear forms $x\,dy - \lambda_t y\,dx = 0$ in the unit bidisk $\{|x| < 1,\ |y| < 1\}$.

We further claim that in fact, λ_t is independent of t. Indeed, the holonomy map corresponding to the small loop around the singular point a_t on $\mathbb{I}$, remains holomorphically conjugate to that computed at $t = 0$. This immediately implies that the multiplicator $\exp 2\pi i\lambda_t$ does not depend on t. Since λ_t varies analytically together with t this is possible if and only if λ_t (which is defined uniquely $\operatorname{mod}\mathbb{Z}$) also does not depend on t. Denote by $\mathcal{L}$ the linear foliation defined by the Pfaffian equation $x\,dy - \lambda y\,dx = 0$.

After all these simplifications without loss of generality we may assume that the problem of extending the foliation is as follows. We are given the standard linear foliation $\mathcal{L}$ in the bidisk $B = \{|x| < 1+\varepsilon,\ |y| < 1\}$ and a biholomorphism $H_t = H_t^\infty$ analytically depending on t, defined in the bicircular domain $C = \{\varepsilon < |x| < 1+\varepsilon,\ |y| < 1\}$, which is an automorphism of the restriction $\mathcal{L}|_C$ (this means that H_t permutes leaves of the linear foliation $\mathcal{L}$ in C). The problem is to extend H_t onto the entire bidisk B as an automorphism of $\mathcal{L}$ there so that it would remain a homeomorphism continuously depending on t. Then the extension H_t would automatically remain transversally holomorphic.

Consider the cross-section $\tau = \{x = 1\}$ at the point $b = (1, 0)$, the circular loop $\gamma \in \pi_1(\mathbb{C} \smallsetminus \{0\}, b)$ on the x-plane around the origin, the corresponding holonomy map $\Delta_\gamma \colon y \mapsto \mu y$, $\mu = \exp 2\pi i\lambda$ and the normal component h_t of H_t (cf. Definition 28.21), considered as a germ $h_t \colon (\tau, b) \to (\tau, b)$. The germ h_t is holomorphic and commutes with the hyperbolic germ Δ_γ, since H_t is an automorphism of $\mathcal{L}$ restricted on C. Thus h_t must necessarily be linear itself (while still analytically depending on t) and by suitably choosing

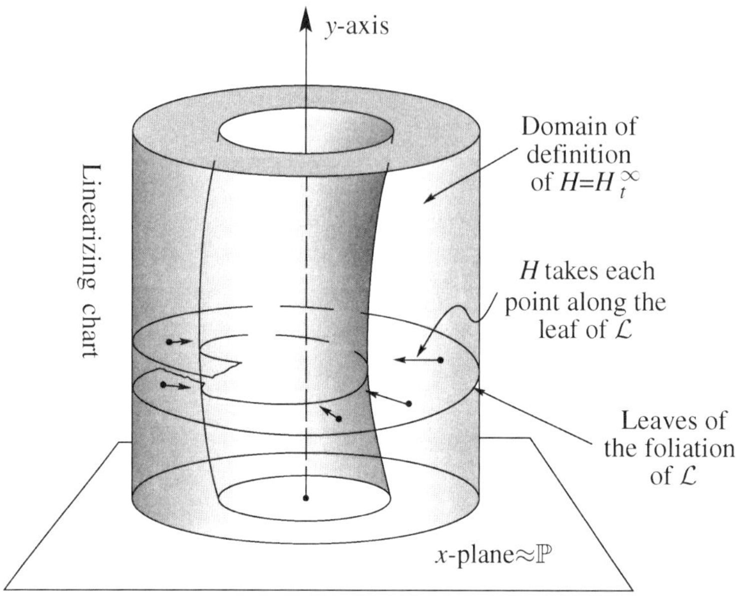

Figure V.8. Extension of the transversally holomorphic homeomorphism to a neighborhood of a singular point on the infinite line

the charts $y \mapsto y_t$ (analytically depending on t) one can assume that h_t is the identity, $h_t \colon y \mapsto y$, and independent of t.

The initial holomorphism H_t does not need to preserve the fibers $\{x = \text{const}\}$, but in the new chart its normal component can be assumed identical. Hence it can be described as the flow map along the linear vector field F generating the foliation $\mathcal{L}$,

$$H_t = \exp\bigl(\psi \cdot F\bigr), \qquad F = x\,\frac{\partial}{\partial x} + \lambda y\,\frac{\partial}{\partial y}, \tag{28.15}$$

for some variable complex time $\psi = \psi(t, x, y)$. The time function ψ is holomorphic outside the cylinder C.

Using a smooth partition of unity, we may extend the function ψ inside the bidisk so that it would be identically equal to zero in a sufficiently small neighborhood of the y-axis. The corresponding flow map (defined by the same formula (28.15) but now on the entire bidisk B) is a C^∞-diffeomorphism preserving the linear foliation $\mathcal{L}$ and extending H_t as required. Repeating this construction for all singularities, we extend H_t^∞ as a family of homeomorphisms, topologically trivializing the family $\operatorname{Isohol}(\mathcal{F}_0)$. □

Remark 28.35. The homeomorphism H_t constructed in the proof of Lemma 28.34, is essentially nonanalytic along the leaves. Indeed, H_t restricted on the infinite line $\mathbb{I} \cong \mathbb{P}$, is a homeomorphism of the Riemann sphere which maps the set S_t into S_0. If these sets contain more than three points, this would be in general impossible to achieve by a conformal map.

Proof of Theorem 28.32. Let $\mathcal{F} \in \operatorname{Isohol}(\mathcal{F}_0)$ be a foliation sufficiently close to $\mathcal{F}_0$ with the topologically conjugate holonomy group at infinity. Because of the analyticity of the set $\operatorname{Isohol}(\mathcal{F}_0)$ established in Lemma 28.33, either $\mathcal{F}_0$ is an isolated point of it and hence is automatically rigid, or (if dimension of $\operatorname{Isohol}(\mathcal{F}_0)$ is greater than zero) we can always find an analytic one-parameter deformation $\{\mathcal{F}_t\}$ connecting $\mathcal{F}_0$ with $\mathcal{F} = \mathcal{F}_1$. For simplicity we will assume that this deformation is parameterized by points of the unit disk $\{|t| < 1\}$.

By Lemma 28.34, the deformation $\{\mathcal{F}_t\}$ is topologically trivial in a neighborhood of the infinite line. By Lemma 28.24 and Remark 28.27 this deformation is also transversally holomorphic (in fact, the transversal holomorphy follows directly from the construction of the family of homeomorphisms H_t^∞ conjugating $\mathcal{F}_t$ with $\mathcal{F}_0$ in the proof of Lemma 28.34).

This topological triviality together with the transversal holomorphy allow us to conclude that there exists a singular foliation $T(\mathcal{F})$ defined on the cylinder $V \times \{|t| < 1\}$ by the formulas (28.8), (28.10). Here, as in the proof of Lemma 28.34, $V \subset \mathbb{P}^2$ is a tubular neighborhood of the infinite line and $\{|t| < 1\} \subset \mathbb{C}$ is the parameter space of the deformation $\{\mathcal{F}_t\}$.

We extend the foliation $T(\mathcal{F})$ from the cylinder $V \times \{|t| < 1\}$ over V to the cylinder $\mathbb{P}^2 \times \{|t| < 1\}$ over the entire projective plane $\mathbb{P}^2$ using the standard arguments as follows. As was shown in the first step of the proof of Lemma 28.26 (see p. 583), the foliation $T(\mathcal{F})$ is defined in $V \times \{|t| < 1\}$ by a holomorphic 1-form $\omega = \omega_t + R\,dt$, where ω_t is the polynomial form defining the foliation $\mathcal{F}_t$, and $R = R(x, y, t)$ is a function holomorphic in $V \times \{|t| < 1\}$ and having a pole of order not exceeding $r+1$ on the infinite line $\mathbb{I}$ for each t. By the Hartogs–Poincaré theorem on erasing compact singularities [**Sha92**, Theorem 3, §11], $R(\cdot, \cdot, t)$ can be extended to $\mathbb{C}^2$ as a polynomial in x, y of degree $\leqslant r + 1$.

Thus the foliation $T(\mathcal{F})$ gets extended on the cylinder $\mathbb{P}^2 \times \{|t| < 1\}$. By Lemma 28.26, this means that the deformation is affine trivial. $\square$

28F. Holonomy of a singular foliation is not a complete topological invariant. The task of extension of a conjugacy between holonomy operators to a topological conjugacy between the respective singular foliations, which played the key role in the proof of Rigidity Theorem 28.32 may be nontrivial even for the simplest cases of singularities.

For instance, consider two hyperbolic linear singular foliations $\mathcal{F} = \{x\,dy - \lambda y\,dx = 0\}$ and $\mathcal{F}' = \{x\,dy - \lambda' y\,dx = 0\}$ in $(\mathbb{C}^2, 0)$ with the characteristic ratios $\lambda, \lambda' \notin \mathbb{R}$. The hyperbolicity in two dimensions means that both singularities are in the Poincaré domain, and the main result of §27**B**, Theorem 27.12 asserts that they are *always* topologically equivalent regardless of the relationship between λ and λ'. In particular, the normal component of the respective homeomorphism conjugates the linear holonomy operators

$y \mapsto 2\pi i\lambda y$ and $y \mapsto 2\pi i\lambda' y$ associated with the standard loop $x = \exp 2\pi i t$, $t \in [0,1]$, and the cross-section $\{x = 1\}$ for these foliations. However, not every homeomorphism conjugating the linear holonomy operators can be extended to a homeomorphism between the foliations in the bidisk $\{|x| < 1,\ |y| < 1\}$.

Assume that $A\colon \mathbb{C} \to \mathbb{C}$ is an $\mathbb{R}$-linear map such that $A1 = 1$ and $A\lambda = A\lambda' + m$, $m \in \mathbb{Z}$ as in Remark 6.49 (we drop the hat from notation to simplify it). Then the homeomorphism $h(y) = y\,|y|^\beta$ covered by A, i.e., such that

$$h(\exp 2\pi i y) = \exp 2\pi i A y \quad \forall y, \qquad h(0) = 0, \tag{28.16}$$

conjugates the holonomy germs.

Proposition 28.36. *The homeomorphism h between the holonomy operators does not admit extension as a homeomorphism conjugating $\mathcal{F}$ with $\mathcal{F}'$ in the neighborhoods of the origin, unless $m = 0$.*

Proof. Assume that such an extension is possible and denote it by H. Replacing H by the composition $\exp(\psi(x,y)F') \circ H$, where ψ is a suitable smooth complex-valued function and F' the linear vector field generating $\mathcal{F}'$, without loss of generality we may assume that H takes each fiber $\{x = \text{const}\}$ over the unit circle $\{|x| = 1\}$ into itself.

This additional assumption uniquely defines H on the (real three-dimensional) cylinder $C = \{t \in \mathbb{R}/\mathbb{Z},\ |y| < 1\}$. Indeed, H must conjugate solutions of the linear systems

$$\dot y = 2\pi i \lambda y \quad \text{and} \quad \dot y = 2\pi i \lambda' y, \qquad t \in [0,1], \tag{28.17}$$

describing the traces of the foliations $\mathcal{F}, \mathcal{F}'$ on the cylinder C: $H(t,y) = \bigl(t, \mathbf{h}(t,y)\bigr)$, $\mathbf{h}(0,\cdot) = \mathbf{h}(2\pi,\cdot) = h$. The function $\mathbf{h}(t,y)$ can be explicitly computed by solving the equations (28.17):

$$\mathbf{h}(t,y) = e^{2\pi i t\lambda'} \cdot h\left(e^{-2\pi i t\lambda} y\right), \qquad t \in [0,1] \tag{28.18}$$

(correctness of this definition follows from the choice of h).

Consider the circle $\gamma_b = \{y = b,\ t \in \mathbb{R}/\mathbb{Z}\}$ for some value of b with $0 \ne |b| < 1$. This is a closed curve in C which is *not linked* with the curve $\gamma_0 = \{y = 0,\ t \in \mathbb{R}/\mathbb{Z}\}$. The image $H(\gamma_b)$ can be described as the t-parameterized curve $\widetilde\gamma_b$ on the y-plane, covered (in the sense of the exponential map $z \mapsto \exp(2\pi i z)$) by the t-parameterized line segment

$$t \mapsto \frac{1}{2\pi i} \ln \mathbf{h}(t,b) = t\lambda' + \frac{1}{2\pi i} \ln h\bigl(e^{-2\pi i t\lambda + 2\pi i\beta}\bigr) = t(\lambda' - A\lambda) + A\beta, \quad \beta = \frac{\ln b}{2\pi i},$$

by virtue of the identity (28.16) for any continuous choice of determination for the logarithm. If the integer number $m = \lambda' - A\lambda$ is nonzero, the curve $\widetilde\gamma_b$ is m times winding around the origin, hence the linking number between $H(\gamma_b)$ and $H(\gamma_0)$ in the cylinder C is $m \ne 0$ for $b \ne 0$.

Note that the traces of the foliations $\mathcal{F}, \mathcal{F}'$ on the 3-cylinder C and the unit 3-sphere $\mathbb{S}^3 = \{|x|^2 + |y|^2 = 1\}$ are diffeomorphic near $\gamma_0 \subseteq C \cap \mathbb{S}^3$ and any homeomorphism conjugating $\mathcal{F}$ with $\mathcal{F}'$ defines at the same time a conjugacy between the traces of $\mathcal{F}$ and $\mathcal{F}'$ on the sphere. Hence if the extension H of h onto the bidisk is possible, then there would exist two unlinked curves in $\mathbb{S}^3$ whose images by a homeomorphism of the sphere have nonzero linking number. This is impossible. □

Example 28.37 (Continuation of Example 28.5). As an application of this result we can establish a rigidity theorem for homogeneous foliations. If two such foliations defined by two homogeneous 1-forms of the same degree have dense (linear) holonomy groups and are topologically equivalent, then their holonomy groups are topologically equivalent. By Remark 6.49, there exists an $\mathbb{R}$-linear map of $\mathbb{C}$ into itself, which conjugates the respective characteristic exponents λ_j and λ'_j of the foliations modulo integers, $\widehat{A}\lambda_j = \lambda'_j \bmod \mathbb{Z}$, $j = 1, \dots, r+1$. Applying Proposition 28.36, we can get rid of the eventual integer terms and immediately derive the following more accurate result.

Theorem 28.38 (cf. with N. Ladis [**Lad79**]). *If two homogeneous foliations of the same degree with dense holonomy groups are topologically equivalent, then there exist an $\mathbb{R}$-linear map of $\mathbb{C}$ into itself, which maps 1 into 1 and the respective characteristic exponents λ_j into λ'_j for all $j = 0, 1, \dots, r$.* □

In fact, the density assumption can be omitted, as well as the homogeneity condition.

Theorem 28.39 ([**Ily78, Naĭ81**]). *If two polynomial foliations from the same class $\mathcal{A}'_r$ are topologically conjugated by a homeomorphism mapping the infinite line into itself, then the respective characteristic exponents are conjugated by an $\mathbb{R}$-linear map as in Theorem 28.38.*

The most difficult part of the proof in the nonhomogeneous case is covered by the Naĭshul Theorem 6.51.

28G. Further results on rigidity. Here we collect a few references to rigidity-type results for foliations.

28**G**$_1$. *Stronger rigidity.* One can further relax the condition of homotopic triviality on the conjugating homeomorphism between foliations in Definition 28.29, thus arriving at stronger and stronger forms of rigidity.

For instance, one can choose a finite bound N and consider only foliations $\mathcal{F}'$ conjugated to the given foliation $\mathcal{F}_0 \in \mathcal{A}'_r$ by a homeomorphism H of combinatorial complexity not exceeding N on the infinite line. By definition, this means that the corresponding automorphism $h_\diamond \colon \pi_1(\mathbb{I} \smallsetminus S, a) \to \pi_1(\mathbb{I} \smallsetminus S, a)$ of the fundamental group, maps all canonical loops $\gamma_0, \dots, \gamma_r$ into loops represented by words of length not exceeding N. (The combinatorial complexity of the homotopically trivial homeomorphism is 1.)

After the corresponding modification of the definitions the locus $\operatorname{Isohol}_N(\mathcal{F}_0)$ will consist of several (though finitely many) analytic components, and the corresponding locus $\operatorname{Topo}_N(\mathcal{F}_0)$ of foliations N-topologically conjugated to $\mathcal{F}_0$ (i.e., by a homeomorphism of complexity $\leqslant N$) will no longer be connected. Yet a minor modification of the demonstration given in §28**E** allows us to prove that a sufficiently small neighborhood of a generic foliation $\mathcal{F}$ from $\mathcal{A}'_r$ contains only finitely many different types of foliations, N-topologically equivalent but not topologically equivalent to $\mathcal{F}$ (Yu. Ilyashenko, 2006). The conjecture, however, is that for a generic foliation one can drop completely all assumptions on the homeomorphism at the price of having only finitely many types of nontrivial homeomorphisms.

Conjecture 28.40. *A generic foliation from the class $\mathcal{A}'_r$ is* almost ideally rigid, *i.e., in a sufficiently small neighborhood of $\mathcal{F}$ there is only finitely many different types of topologically equivalent but not affine equivalent to each other foliations.*

28**G**$_2$. *Weaker assumptions on the foliation.* Other rigidity results for pseudogroups from §6**I** can be translated into theorems on rigidity of foliations.

Theorem 28.41. *There exists a real algebraic subset $\Sigma_r \subset \mathcal{A}_r$ and a nowhere dense real analytic subset $\Sigma'_r \subset \mathcal{A}_r$ of real codimension at least 2, such that any foliation $\mathcal{F} \in \mathcal{A}_r \smallsetminus (\Sigma_r \cup \Sigma'_r)$ has the following properties:*

(1) *each leaf of the foliation $\mathcal{F}$ is dense in $\mathbb{C}^2$,*

(2) *$\mathcal{F}$ is reasonably rigid,*

(3) *for $r \geqslant 3$, the foliation $\mathcal{F}$ has a countable number of homologically independent complex limit cycles.*

The first statement follows from [**Shc82, Shc84**]. The third statement was announced in [**Shc86**] an proved in [**SRO98**]. The detailed proofs can be found in [**Shc06**].

The following conjecture was formulated in [**Ily79b**] for foliations by analytic curves in higher dimension spaces as well as for generic foliation from class $\mathcal{B}_r$.

Conjecture 28.42. *The majority (an open and dense subset of foliations from the class $\mathcal{B}_r$) possess the three properties listed in Theorem 28.41 above.*

The same is true for the majority of polynomial foliations on $\mathbb{P}^n$, $n \geqslant 3$, defined by polynomial vector fields in $\mathbb{C}^n$.

This conjecture is still not proved. A weaker result that claims existence of an open (though not dense) set with the density and rigidity properties was proved recently by F. Loray and J. Rebelo [**LR03**].

28$\mathbf{G}_3$. *Different classes of foliations.* Theorem 28.20 (deformational rigidity of generic foliations from the class $\mathcal{A}_r$) was generalized by X. Gomez–Mont [**GM88**]. He proved deformational rigidity for generic foliations having an algebraic separatrix (not necessarily a line, as is the case for the class $\mathcal{A}_r$).

Theorem 28.43. *Any topologically trivial deformation (parameterized by a reduced analytic space) of a generic homogeneous foliation in $\mathbb{C}^3$ having an algebraic separatrix, is holomorphically trivial.*

A topologically trivial transversally holomorphic deformation of a generic homogeneous foliation in $\mathbb{C}^{n+1}$, $n \geqslant 2$, is holomorphically trivial.

Here by homogeneous foliation we mean a singular complex one-dimensional foliation of $\mathbb{C}^n$ generated by a homogeneous polynomial vector field. The author's proof is by homological methods, yet it seems that Theorem 28.43 can be proved by the same type of arguments that were used to prove Theorem 28.20 above. Moreover, we believe that Theorem 28.43 may be further improved from deformational to (absolute) rigidity. Denote by $\mathcal{B}_r(C)$ the collection of foliations from the class $\mathcal{B}_r$ having a *fixed* algebraic separatrix $C \subset \mathbb{P}^2$. This class can be parameterized by a projective variety in the way similar to the "unrestricted" classes $\mathcal{A}_r, \mathcal{B}_r$.

Conjecture 28.44. *A generic equation from the class $\mathcal{B}_r(C)$ is rigid.*

Exercises and Problems for §28.

Problem 28.1. Prove that *any closed trajectory* γ of a polynomial vector field on the real plane $\mathbb{R}^2$ is noncontractible on its complexification L (the leaf of the holomorphic foliation on $\mathbb{P}^2$).

Hint. Show that γ cannot be homologous to zero, i.e., cannot be the boundary of a domain in L.

Problem 28.2. Find necessary and sufficient conditions for the closures of all leaves defined by (28.1) in $\mathbb{P} \times \mathbb{C}$ to be compact Riemann surfaces. Prove that under this condition an autonomous polynomial system corresponding to equation (28.1) has at least one singular point of the Poincaré type.

Problem 28.3 (see [**Lad79**]). Find a topological classification of equations (28.1) with dense leaves.

Problem 28.4. Find necessary and sufficient conditions for planar homogeneous foliations to be dense in $\mathbb{P}^2$ (Example 28.5).

Problem 28.5. Give a complete proof of Theorem 28.38.

Exercise 28.6. For any r construct a polynomial foliation of the projective degree r such that the closures of almost all leaves have real dimension 3.

Problem 28.7. Prove that any small deformation of an ultra-Morse polynomial results in a topologically trivial deformation of the corresponding foliation by the level curves of the polynomial.

Problem 28.8. Consider a holomorphic family of ultra-Morse polynomials H_t with the same (not depending on the parameter t) higher order terms. Prove that it is transversally holomorphic in some neighborhood of infinity.

Problem 28.9. Give an example of a topologically trivial and transversally holomorphic deformation which is nevertheless not analytically trivial.

Exercise 28.10. Give an example of a homeomorphism $h\colon \mathbb{P} \to \mathbb{P}$ which preserves a given set S of $r+1$ points, $r \geqslant 2$, and has arbitrarily high combinatorial complexity in the sense explained in §28$\mathbf{G}_1$.

First aid

A. Crash course on functions of several complex variables

In this appendix we collect several facts about holomorphic functions of several variables. They can be found in a number of sources, among which we recommend the books [**Hör00**], [**GH78**], [**GR65**], [**Sha92**], [**Chi89**], and more recently the textbooks [**FG02**] and especially [**Ebe07**].

A.1. Holomorphic functions of several variables. A complex function $f(z_1, \dots, z_n)$ defined on an open domain U of the complex n-space $\mathbb{C}^n$ is *holomorphic* or *analytic* (these words will be used as complete synonyms) in U, if the real and imaginary parts of the function are differentiable at every point $a \in U$, and the differential $df_a \colon \mathbf{T}_a U \to \mathbb{C}$, is $\mathbb{C}$-linear:

$$df_a(\lambda\xi) = \lambda \cdot df_a(\xi) \qquad \forall \xi \in \mathbf{T}_a U \cong \mathbb{C}^n. \tag{A.1}$$

This condition can be written in the form of a system of partial differential equations called the Cauchy–Riemann equations,

$$\frac{\partial}{\partial \bar{z}_j} f = 0, \quad j = 1, \dots, n, \qquad \frac{\partial}{\partial \bar{z}_j} = \frac{1}{2}\left(\frac{\partial}{\partial x_j} - \frac{1}{i}\frac{\partial}{\partial y_j}\right). \tag{A.2}$$

Functions holomorphic in the domain U form a linear space which will be denoted by $\mathcal{O}(U)$.

We will often use the space $\mathcal{A}(U)$ of functions holomorphic in U and continuous on the closure $\overline{U}$; cf. with §1**B**. This space is equipped with the norm

$$\mathcal{A}(U) = \mathcal{O}(U) \cap C(\overline{U}), \qquad \|f\|_U = \max_{z \in \overline{U}} |f(z)| \quad \forall f \in \mathcal{A}(U). \tag{A.3}$$

A function of several variables is holomorphic if and only if it is holomorphic in each variable separately (Hartogs theorem).

A.2. Holomorphic maps and their inversion. A map $f\colon U \to \mathbb{C}^m$ is holomorphic, if all its components are holomorphic. Differentials of holomorphic maps are $\mathbb{C}$-linear maps from $\mathbf{T}_a U \cong \mathbb{C}^n$ to $\mathbf{T}_{f(a)}\mathbb{C}^m \cong \mathbb{C}^m$ for all $a \in U$. Since the composition of $\mathbb{C}$-linear maps is again $\mathbb{C}$-linear, composition of holomorphic maps is again a holomorphic map.

If the differential df_a of a holomorphic map $f\colon U \to \mathbb{C}^n$, $U \subseteq \mathbb{C}^n$, is invertible (as a $\mathbb{C}$-linear map of $\mathbb{C}^n$ into itself), then the map is locally invertible: there exists a holomorphic map g, defined in some neighborhood of $f(a)$, such that $g \circ f = \mathrm{id}$.

For holomorphic maps the implicit function theorem holds. If $U \subset \mathbb{C}^{n+m}$ and $f\colon U \to \mathbb{C}^n$ is a holomorphic map, $f = f(z,w)$, such that the differential of f with respect to the first variable is invertible at some point $(a,b) \in U$, then the system of equations $f(z,w) = 0$ determines z as a holomorphic (vector) function of w in some neighborhood of $b \in \mathbb{C}^m$, such that $f(z(w),w) \equiv 0$. The condition of invertibility means that the matrix of partial derivatives $\bigl(\partial f_i/\partial z_j\bigr)_{i,j=1}^n$, has nonzero determinant at $(a,b) \in \mathbb{C}^{n+m}$.

A.3. Cauchy formula and its consequences. Let $D_r = D_r(a)$ be a polydisk of *polyradius* $r = (r_1, \dots, r_n)$, $r_j > 0$,

$$D_r(a) = \{(z_1, \dots, z_n) \in \mathbb{C}^n \colon |z_j - a_j| < r_j,\ j = 1, \dots, n\},$$

and D_r° its *skeleton*, the Cartesian product of the boundary circles,

$$D_r^\circ(a) = \{(z_1, \dots, z_n) \in \mathbb{C}^n \colon |z_j - a_j| = r_j,\ j = 1, \dots, n\}.$$

Note that the skeleton forms only a small fraction of the boundary ∂D_r.

Similarly to functions of one complex variable, a function holomorphic in a polydisk D_r as above and continuous on its closure, can be obtained from its values on the skeleton of the polydisk by the Cauchy integral formula,

$$f(a) = \frac{1}{(2\pi i)^n} \int\!\!\cdots\!\!\int_{D_r^\circ(a)} \frac{dz_1 \wedge \cdots \wedge dz_n}{(z_1 - a_1)\cdots(z_n - a_n)} \tag{A.4}$$

(the integral can be understood as an iterated integral).

The Cauchy integral formula implies numerous corollaries, the most important among them the possibility of expanding a holomorphic function in a *converging* Taylor series.

We use the standard multi-index notation: for an integer vector $\alpha = (\alpha_1, \dots, \alpha_n) \in \mathbb{Z}_+^n$ we denote

$$\alpha! = \alpha_1! \cdots \alpha_n!, \quad z^\alpha = z_1^{\alpha_1} \cdots z_n^{\alpha_n}, \quad \frac{\partial^\alpha}{\partial z^\alpha} = \frac{\partial^{\alpha_1}}{\partial z^{\alpha_1}} \cdots \frac{\partial^{\alpha_n}}{\partial z^{\alpha_n}}.$$

In these notations the integral representation (A.4) implies the *Cauchy inequalities*

$$\left|\frac{\partial^\alpha}{\partial z^\alpha} f(a)\right| \leqslant \frac{\|f\|_{D_r(a)}}{\alpha!\, r^\alpha}, \qquad \forall \alpha \in \mathbb{Z}_+^n,$$

which in turn guarantee that the Taylor series for f converges on $D_r(a)$, and this convergence is uniform on any smaller polydisk centered at a,

$$\forall z \in D_r(a), \quad f(z) = \sum_{|\alpha|=0}^{\infty} c_\alpha (z-a)^\alpha, \qquad c_\alpha = \frac{1}{\alpha!} \cdot \frac{\partial^\alpha f}{\partial z^\alpha}(a).$$

A.4. Weierstrass compactness principle. Another consequence of the Cauchy inequalities is the *Weierstrass compactness principle*. It asserts that a sequence of holomorphic functions $\{f_k\}_{k=1}^\infty \subseteq \mathcal{O}(U)$ uniformly convergent on a bounded domain $U \subset \mathbb{C}^n$ (with compact closure), has a holomorphic limit. This principle implies that the space $\mathcal{A}(U) = \mathcal{O}(U) \cap C(\overline{U})$ introduced in (A.3), is a Banach (*complete* normed) space. This completeness plays a central role throughout the book.

A.5. Germs of analytic functions. Germs of analytic functions at a given point, say, at the origin $0 \in \mathbb{C}^n$, form a commutative algebra over $\mathbb{C}$, denoted by $\mathcal{O}(\mathbb{C}^n, 0)$. This algebra is *local*: its unique maximal ideal $\mathfrak{m} \subset \mathcal{O}(\mathbb{C}^n, 0)$ consists of germs vanishing at the origin. Usually we ignore the difference between germs and their representatives (defined in sufficiently small domains) both in argumentation and in notations.

The ring of germs $\mathcal{O}(\mathbb{C}^n, 0)$ is *Noetherian*: any ascending chain of ideals in this ring eventually stabilizes. This implies that any ideal in this ring has finite basis (Hilbert's theorem).

Any germ can be factored as a product of finitely many irreducible germs; the irreducible factors are defined uniquely modulo multiplication by units (elements from $\mathcal{O}(\mathbb{C}^n, 0) \smallsetminus \mathfrak{m}$). A germ is *square-free*, if all its irreducible factors are pairwise distinct (modulo units).

A.6. Analytic sets. A subset $X \subset \mathbb{C}^n$ is analytic if in a neighborhood of each point $a \in \mathbb{C}^n$ it can be represented as common zero locus of several functions analytic at a. By Hilbert's theorem, the number of such functions can always be assumed finite. Analytic sets are sometimes referred to as analytic varieties; they are always closed.

A set is an analytic submanifold of codimension $k \leqslant n$, if near each point $a \in X$ it is a common zero locus of k functions holomorphic at a with linearly independent (over $\mathbb{C}$) differentials.

Analytic sets have rather regular structure even in the case where they are not submanifolds of $\mathbb{C}^n$. In particular, every analytic variety can be *stratified*, i.e., represented as (locally) finite union of strata X_k of different dimensions, such that

(1) each stratum X_k is an analytic submanifold in $\mathbb{C}^n$ of certain dimension d_k, and

(2) the closure of each stratum consists of itself and several strata of lower dimensions.

One may in fact guarantee that the tangent planes to strata near the boundary points have certain limit positions compatible with that of tangent planes to the adjoining strata (Conditions A and B of Whitney). For most purposes one can use the characteristic property formulated in terms of transversality: any smooth map transversal to a stratum X_k at a point $a \in X_k$, is transversal also to all strata of higher dimensions which have a at their closure, at all points sufficiently close to a.

The principal stratum of highest dimension is called the *regular part* or set of *regular points* of X and denoted $\operatorname{Reg} X$.

The germ (X, a) of an analytic set X at a point $a \in \mathbb{C}^n$ is *irreducible*, if it cannot be represented as the union of two germs of analytic sets $X = X_1 \cup X_2$, such that $\operatorname{Reg} X_i \subsetneq \operatorname{Reg} X$. The germ of a hypersurface $X = \{f = 0\}$ generated by an irreducible germ $f \in \mathcal{O}(\mathbb{C}^n, a)$, is irreducible. Regular parts of irreducible sets are locally connected.

Any germ of an analytic hypersurface admits an *irreducible decomposition* into the union of uniquely defined irreducible components of codimension 1. This follows from an irreducible factorization of holomorphic germs; see §**A.5**.

A.7. Uniformization. An analytic submanifold X of codimension k in $\mathbb{C}^n$ admits local uniformization near each point $a \in X$: there exists a holomorphic map $(\mathbb{C}^{n-k}, 0) \to (X, a)$ which is one-to-one.

Among singular analytic varieties, only *analytic curves*, varieties of complex dimension 1 admit *uniformization*. Any *irreducible* germ of an analytic curve $(X, a) \subset (\mathbb{C}^n, 0)$ can be parameterized by a holomorphic one-to-one map.

A.8. Forced analytic continuation: erasing of singularities. For some domains $U \subset \mathbb{C}^n$, any function holomorphic in U, can be extended as a function analytic in a larger domain. This phenomenon is peculiar for holomorphic functions in more than one variable.

If $U \subseteq \mathbb{C}^n$ is an open domain and $K \Subset U$ its compact subset, then any function analytic in $U \smallsetminus K$, extends on the whole of U. This means that *compact holes in the domain can be always erased* (Poincaré–Hartogs).

If (X, a) is the germ of an analytic variety of codimension 1 (hypersurface) and f is a *locally bounded* (i.e., bounded in some neighborhood of every point) function holomorphic in the complement to X, then f can be extended on X while remaining analytic. This can be proved by a straightforward application of the Cauchy integral formula exactly in the one-dimensional case, if X is a nonsingular hypersurface.

If X is the germ of analytic variety of codimension $\geqslant 2$, then the condition of local boundedness can be dropped: any function holomorphic in the complement to X, can be extended on X while remaining holomorphic. For instance, any isolated singular point of a holomorphic function on the plane $\mathbb{C}^2$ can be erased.

A.9. Meromorphic functions. The ring of holomorphic germs has no divisors of zero, hence admits extension to the field of fractions denoted by $\mathcal{M}(\mathbb{C}^n, a)$. A representative of a meromorphic germ $f = g/h$, $g, h \in \mathcal{O}(U)$, is a holomorphic function on the complement to the zero locus $\{h = 0\}$ of the denominator, which can be extended as a holomorphic map to the Riemann sphere $\mathbb{P} = \mathbb{C} \cup \{\infty\}$ on the complement to the *indeterminacy locus* $\{f = 0,\ g = 0\}$ (common zeros of the numerator and denominator).

A *meromorphic function* in a domain U is a collection of local representations $f_\alpha = g_\alpha/h_\alpha$ in charts of an open covering $\mathfrak{U} = \{U_\alpha\}$ of U, such that on the intersections $U_{\alpha\beta}$ the equalities $g_\alpha h_\beta - g_\beta h_\alpha = 0$ (this definition can be literally used for holomorphic manifolds). Under certain global assumptions on U, there exists a single global representation $f = g/h$ with holomorphic $g, h \in \mathcal{O}(U)$. Meromorphic functions form a field denoted by $\mathcal{M}(U)$.

A function meromorphic on $U \smallsetminus Y$, a complement to an analytic variety Y of codimension $\geqslant 2$, can be extended as a meromorphic function on U (Lévi theorem).

A.10. Analyticity vs. algebraicity. An analytic subvariety of a complex projective space $\mathbb{P}^n$ is an algebraic variety (Chow theorem). A meromorphic function on $\mathbb{P}^n$ is *rational*, ratio of two homogeneous polynomials of the same degree in the homogeneous coordinates on $\mathbb{P}^n$.

B. Elements of the theory of Riemann surfaces.

B.1. Riemann surfaces and algebraic curves. A Riemann surface is a complex manifold of dimension one. The principal examples are the complex line $\mathbb{C}$ itself, open domains in $\mathbb{C}$, the Riemann sphere $\mathbb{P} = \mathbb{C} \cup \{\infty\}$, smooth affine and projective algebraic curves.

A map $f\colon C \to C'$ between two Riemann surfaces is holomorphic if it is locally defined by a holomorphic function $z' = f(z)$ for any local holomorphic charts z, z' on C, C' respectively.

The zero locus C of a bivariate polynomial $\{P(x,y) = 0\} \subset \mathbb{C}^2$ is called an *affine algebraic curve*. It may be nonsmooth, yet there always exists a Riemann surface $\widetilde{C}$ and a map $\varphi\colon \widetilde{C} \to C$ such that any smooth point $b \in C$ has a unique preimage $a \in \widetilde{C}$ and the germ $\varphi_a\colon (\widetilde{C}, a) \to (C, b)$ is biholomorphic. The curve $\widetilde{C}$ is called *normalization* of C.

Existence of normalization for any algebraic curve (normalization theorem) may be easily proved using the local uniformization theorem from §**A.7** and the irreducible decomposition theorem for curves from §**A.6**. For curves with normal crossings see Problem 25.1.

The closure of an affine algebraic curve in the projective plane is called the *projective* algebraic curve. Projective curves also admit normalization which is a *compact* Riemann surface.

Conversely, *any compact Riemann surface is algebraic.* There are many ways to formalize this statement. One of them is the following. For any abstract compact Riemann surface S there exists a projective algebraic curve $C \subset \mathbb{P}^2$ for which S is a normalization: $S = \widetilde{C}$.

B.2. Genus and degree of an algebraic curve. For an affine algebraic curve C, there exists a unique (modulo constant factor) polynomial of minimal degree whose locus is C, which is called the minimal polynomial of C and of the projective closure of C. The *degree* of an affine (projective) curve is the degree of this minimal polynomial.

The degree of a projective algebraic curve $C \in \mathbb{P}^2$ is equal to the number of intersections between this curve and a generic line $\ell \subset \mathbb{P}^2$.

The *genus* of a projective algebraic curve is the (topological) genus of its normalization considered as a smooth 2-dimensional surface. The genus of the affine algebraic curve is the genus of its projective closure.

If $f\colon C \to C'$ is a holomorphic map between two compact Riemann surfaces, them it defines a ramified covering of C' over the set of *critical values* of f. Near each critical value $a \in C$ the map f has the form $z \mapsto z^k = z'$ for suitable choices of local charts $z, z' \in (\mathbb{C}, 0)$ on C, C' and some natural number $k = k_a \geqslant 1$ (we set for convenience $k_a = 1$ for a regular point $a \in C$). If m is the number of sheets of this covering and g, g' the genuses of C and C' respectively, then these numbers are related by the *Riemann–Hurwitz formula*

$$2(g-1) = 2m(g'-1) + \sum_{a \in C} (k_a - 1). \tag{B.1}$$

For any affine curve C the Cartesian projection $\pi : \mathbb{C}^2 \to \mathbb{C}$, $(z, w) \mapsto z$, restricted on C, extends to a holomorphic map between the projective closure of C and the Riemann sphere $\mathbb{P}$. If C is smooth, the number of sheets m of this covering is equal to the degree of C. The genus of the projective line is one. An easy computation of the total ramification index yields the formula $g = \frac{1}{2}(m-1)(m-2)$ for the genus of a smooth algebraic curve of degree m.

B.3. Meromorphic functions on Riemann surfaces. By the maximum modulus principle, there are no holomorphic maps from a compact Riemann surface to $\mathbb{C}$, hence there are no globally defined holomorphic functions. A map $f \colon C \to \mathbb{P}$ is called a *meromorphic function* on C. Locally any meromorphic function can be represented by a ratio of two holomorphic functions. For any point $a \in C$ the *order* $\operatorname{ord}_a f \in \mathbb{Z}$ is an integer number equal to the order of zero or the negative order of pole of f at a. This order is well defined independently of the choice of the local chart on (C, a) used for its computation (we assign the value $\operatorname{ord}_a f = 0$ if a is neither a root nor the pole of f).

Meromorphic functions on the projective line $\mathbb{P}$ are *rational* (i.e., polynomial in z and z^{-1} in the affine chart $\mathbb{C} \subset \mathbb{P}$). Holomorphic functions on an affine algebraic curve that are meromorphic on its closure, are restrictions of polynomials in two variables onto this curve. A meromorphic function on a projective algebraic curve $C \subset \mathbb{P}^2$ is always the restriction of some rational function $P(x, y)/Q(x, y)$ onto this curve.

For any meromorphic function on a compact Riemann surface,

$$\forall f \in \mathfrak{M}(C) \qquad \sum_a \operatorname{ord}_a f = 0. \tag{B.2}$$

B.4. Holomorphic and meromorphic forms on Riemann surfaces. A differential 1-form ω on a Riemann surface C is holomorphic (resp., meromorphic) if in any local chart z it has the form $\omega_z = f(z)\,dz$, where the coefficient f is holomorphic (resp., meromorphic). Poles of the coefficient f are called the poles of the form. The Cauchy-Riemann equation $\frac{\partial \bar z}{\partial =}0$ implies that any holomorphic 1-form on a Riemann surface is closed, $d\omega = 0$. Hence by the Stokes formula the integral of a holomorphic 1-form over a cycle on a Riemann surface depends on the homology class of the cycle only.

In particular, the integral

$$\operatorname{res}_a \omega = \frac{1}{2\pi i} \oint_\gamma \omega \tag{B.3}$$

of a meromorphic 1-form over any small loop around a point $a \in C$, does not depend on the loop and is called *the residue* of the form at a (the residue is zero if ω is holomorphic at a). By the Stokes theorem, for any meromorphic 1-form ω,

$$\sum_{a\in C} \operatorname{res}_a \omega = 0. \tag{B.4}$$

Applied to the logarithmic derivative $\omega = f^{-1}df$ of a meromorphic function $f \in \mathcal{M}(C)$, this identity implies (B.2).

B.5. Uniformization. There are three examples of simply connected Riemann surfaces that are not pairwise conformally equivalent: an open disc $\mathbb{D} = \{|z| < 1\}$, the complex line $\mathbb{C}$ and the Riemann sphere $\mathbb{P}$. The Poincaré–Kœbe uniformization theorem claims that these are the only possibilities: any simply connected Riemann surface is biholomorphically equivalent either to $\mathbb{D}$, $\mathbb{C}$ or $\mathbb{P}$. This implies, in particular, that any Riemann surface with a cyclic fundamental group is conformally equivalent either to $\mathbb{C}^* = \{0 < |z|\}$, to an annulus $\{\varepsilon < |z| < 1\}$, $\varepsilon > 0$, or to a punctures disc $\{0 < |z| < 1\}$. This trichotomy lies in the background of the study of parabolic germs in Chapter IV.

Bibliography

[AB60] L. Ahlfors and L. Bers, *Riemann's mapping theorem for variable metrics*, Ann. of Math. (2) **72** (1960), 385–404. MR0115006 (22 #5813)

[AB94] D. V. Anosov and A. A. Bolibruch, *The Riemann-Hilbert problem*, Vieweg Publ., Braunschweig, 1994. MR 95d:32024

[AGV85] V. I. Arnold, S. M. Guseĭn-Zade, and A. N. Varchenko, *Singularities of differentiable maps.*, vol. I. The classification of critical points, caustics and wave fronts, Birkhäuser Boston Inc., Boston, Mass., 1985. MR 86f:58018

[AGV88] ———, *Singularities of differentiable maps*, vol. II, Monodromy and asymptotics of integrals, Birkhäuser Boston Inc., Boston, MA, 1988. MR 89g:58024

[AI85] V. I. Arnold and Yu. S. Ilyashenko, *Ordinary differential equations*, Dynamical systems—I, Encyclopaedia of Mathematical Sciences, vol. 1, Springer, Berlin, 1985, translated from *Current problems in mathematics. Fundamental directions, Vol. 1*, 7–149, VINITI, Moscow, 1985, pp. 1–148. MR 87e:34049

[ALGM73] A. A. Andronov, E. A. Leontovich, I. I. Gordon, and A. G. Maĭer, *Qualitative theory of second-order dynamic systems*, Halsted Press (A division of John Wiley & Sons), New York-Toronto, Ont., 1973. MR 50 #2619

[And62] A. F. Andreev, *On Frommer's method of studying a singular point of a first-order differential equation*, Vestnik Leningrad. Univ. **17** (1962), no. 1, 5–21. MR 25 #5228

[And65a] ———, *On the number of operations used in Frommer's method for investigation of a singular point of a differential equation*, Differencial′nye Uravnenija **1** (1965), 1155–1176. MR 32 #5972

[And65b] ———, *Remarks on a paper of S. Lefschetz*, Differencial′nye Uravnenija **1** (1965), 199–203. MR 33 #2865

[Arn69] V. I. Arnold, *Remarks on singularities of finite codimension in complex dynamical systems*, Functional Anal. Appl. **3** (1969), no. 1, 1–5. MR 41 #4573

[Arn70a] ———, *Algebraic unsolvability of the problem of Ljapunov stability and the problem of the topological classification of the singular points of an analytic system of differential equations*, Funkcional. Anal. i Priložen. **4** (1970), no. 3, 1–9. MR 42 #7989

[Arn70b] ———, *Local problems of analysis*, Vestnik Moskov. Univ. Ser. I Mat. Meh. **25** (1970), no. 2, 52–56. MR 43 #633

[Arn78] ———, *Ordinary differential equations*, MIT Press, Cambridge, Mass., 1978. MR 58 #22707

[Arn83] ———, *Geometrical methods in the theory of ordinary differential equations*, Springer-Verlag, New York, 1983. MR 84d:58023

[Arn92] ———, *Ordinary differential equations*, Springer-Verlag, Berlin, 1992. MR 93b:34001

[Arn97] ———, *Mathematical methods of classical mechanics*, Graduate Texts in Mathematics, vol. 60, Springer-Verlag, New York, 1997, Corrected reprint of the second (1989) edition. MR1345386 (96c:70001)

[Arn04] ———, *Arnold's problems*, Springer-Verlag, Berlin, 2004, Translated and revised edition of the 2000 Russian original, With a preface by V. Philippov, A. Yakivchik and M. Peters. MR2078115 (2005c:58001)

[AVL91] D. V. Alekseevskiĭ, A. M. Vinogradov, and V. V. Lychagin, *Basic ideas and concepts of differential geometry*, Geometry, I, Encyclopaedia Math. Sci., vol. 28, Springer, Berlin, 1991, pp. 1–264. MR 95i:53001b

[Bau39] N. Bautin, *Du nombre de cycles limites naissant en cas de variation des coefficients d'un état d'équilibre du type foyer ou centre*, C. R. (Doklady) Acad. Sci. URSS (N. S.) **24** (1939), 669–672. MR 2,49a

[Bau54] N. N. Bautin, *On the number of limit cycles which appear with the variation of coefficients from an equilibrium position of focus or center type*, American Math. Soc. Translation **1954** (1954), no. 100, 19. MR 15,527h

[BBI01] D. Burago, Yu. Burago, and S. Ivanov, *A course in metric geometry*, Graduate Studies in Mathematics, vol. 33, American Mathematical Society, Providence, RI, 2001. MR1835418 (2002e:53053)

[BC93] M. Berthier and D. Cerveau, *Quelques calculs de cohomologie relative*, Ann. Sci. École Norm. Sup. (4) **26** (1993), no. 3, 403–424. MR1222279 (94k:32053)

[BCLN96] M. Berthier, D. Cerveau, and A. Lins Neto, *Sur les feuilletages analytiques réels et le problème du centre*, J. Differential Equations **131** (1996), no. 2, 244–266. MR 98a:58128

[BD00] P. Bonnet and A. Dimca, *Relative differential forms and complex polynomials*, Bull. Sci. Math. **124** (2000), no. 7, 557–571. MR 1 793 909

[Bel79] G. R. Belitskiĭ, Нормальные формы, инварианты и локальные отображения (*Normal forms, invariants and local maps*), "Naukova Dumka", Kiev, 1979, in Russian. MR 81h:58014

[Ben01] I. Bendixson, *Sur les courbes définies par des équations différentielles*, Acta Math. **24** (1901), 1–88.

[Bib79] Yu. N. Bibikov, *Local theory of nonlinear analytic ordinary differential equations*, Lecture Notes in Mathematics, vol. 702, Springer-Verlag, Berlin, 1979. MR 83a:34004

[BL88] J. Bernstein and V. Lunts, *On nonholonomic irreducible D-modules*, Invent. Math. **94** (1988), no. 2, 223–243. MR958832 (90b:58247)

[BLL97] M. Belliart, I. Liousse, and F. Loray, *Sur l'existence de points fixes attractifs pour les sous-groupes de* $\mathrm{Aut}(\mathbf{C}_{,0})$, C. R. Acad. Sci. Paris Sér. I Math. **324** (1997), no. 4, 443–446. MR1440964 (98c:58134)

[BM94] M. Berthier and R. Moussu, *Réversibilité et classification des centres nilpotents*, Ann. Inst. Fourier (Grenoble) **44** (1994), no. 2, 465–494. MR1296740 (95h:58103)

[Bol92] A. A. Bolibruch, *Sufficient conditions for the positive solvability of the Riemann-Hilbert problem*, Mat. Zametki **51** (1992), no. 2, 9–19, 156. MR 93g:34007

[Bol94] ———, *On an analytic transformation to the standard Birkhoff form*, Trudy Mat. Inst. Steklov. **203** (1994), 33–40. MR 97c:34066

[Bol00] ———, *Fuchsian differential equations and holomorphic vector bundles*, Modern lecture courses, Moscow Center for Continuous Mathematical Education, Moscow, 2000, (Russian).

[Bon99] P. Bonnet, *Description of the module of relatively exact 1-forms modulo a polynomial f on $\mathbb{C}^2$*, Preprint no. **184**, Laboratoire de Topologie, Université de Bourgogne, Dijon, 1999.

[Bru71] N. N. Brushlinskaya, *A finiteness theorem for families of vector fields in the neighborhood of a singular point of Poincaré type*, Funkcional. Anal. i Priložen. **5** (1971), no. 3, 10–15. MR 45 #8920

[BV89] D. G. Babbitt and V. S. Varadarajan, *Local moduli for meromorphic differential equations*, Astérisque (1989), no. 169-170, 217. MR1014083 (91e:32017)

[Can97] J. Cano, *Construction of invariant curves for singular holomorphic vector fields*, Proc. Amer. Math. Soc. **125** (1997), no. 9, 2649–2650. MR 97j:32027

[Car38] H. Cartan, *Sur le premier problème de Cousin*, C. R. Acad. Sci. Paris **207** (1938), 558–560.

[Car94] M. Carnicer, *The Poincaré problem in the nondicritical case*, Ann. of Math. (2) **140** (1994), no. 2, 289–294. MR1298714 (95k:32031)

[CC03] A. Candel and L. Conlon, *Foliations. I, II*, Graduate Studies in Mathematics, vol. 23, 60, American Mathematical Society, Providence, RI, 2000, 2003. MR1732868 (2002f:57058), MR1994394 (2004e:57034)

[CG93] L. Carleson and T. W. Gamelin, *Complex dynamics*, Universitext: Tracts in Mathematics, Springer-Verlag, New York, 1993. MR 94h:30033

[Cha86] M. Chaperon, *C^k-conjugacy of holomorphic flows near a singularity*, Inst. Hautes Études Sci. Publ. Math. (1986), no. 64, 143–183. MR 88m:58161

[Chi89] E. M. Chirka, *Complex analytic sets*, Mathematics and its Applications (Soviet Series), vol. 46, Kluwer, Dordrecht, 1989. MR 92b:32016

[CKP76] C. Camacho, N. H. Kuiper, and J. Palis, *La topologie du feuilletage d'un champ de vecteurs holomorphes près d'une singularité*, C. R. Acad. Sci. Paris Sér. A-B **282** (1976), no. 17, Ai, A959–A961. MR 54 #1301

[CKP78] ———, *The topology of holomorphic flows with singularity.*, Publ. Math., Inst. Hautes Etud. Sci. **48** (1978), 5–38 (English).

[CL00] C. Christopher and J. Llibre, *Integrability via invariant algebraic curves for planar polynomial differential systems*, Ann. Differential Equations **16** (2000), no. 1, 5–19. MR1768817 (2001g:34001)

[CLN91] D. Cerveau and A. Lins Neto, *Holomorphic foliations in* CP(2) *having an invariant algebraic curve*, Ann. Inst. Fourier (Grenoble) **41** (1991), no. 4, 883–903. MR1150571 (93b:32050)

[CLNS84] C. Camacho, A. Lins Neto, and P. Sad, *Topological invariants and equidesingularization for holomorphic vector fields*, J. Differential Geom. **20** (1984), no. 1, 143–174. MR772129 (86d:58080)

[CLO97] D. Cox, J. Little, and D. O'Shea, *Ideals, varieties, and algorithms*, second ed., Undergraduate Texts in Mathematics, Springer-Verlag, New York, 1997, An introduction to computational algebraic geometry and commutative algebra. MR 97h:13024

[CLP07] Colin Christopher, Jaume Llibre, and Jorge Vitório Pereira, *Multiplicity of invariant algebraic curves in polynomial vector fields*, Pacific J. Math. **229** (2007), no. 1, 63–117. MR2276503

[CNR00] A. Capani, G. Niesi, and L. Robbiano, `CoCoA`*, a system for doing computations in commutative algebra*, Available from `ftp://cocoa.dima.unige.it`, 2000, version 4.0.

[CS82] C. Camacho and P. Sad, *Invariant varieties through singularities of holomorphic vector fields*, Ann. of Math. (2) **115** (1982), no. 3, 579–595. MR 83m:58062

[CW79] Lan Sun Chen and Ming Shu Wang, *The relative position, and the number, of limit cycles of a quadratic differential system*, Acta Math. Sinica **22** (1979), no. 6, 751–758. MR559742 (81g:34031)

[DFN85] B. A. Dubrovin, A. T. Fomenko, and S. P. Novikov, *Modern geometry—methods and applications. Part II*, Graduate Texts in Mathematics, vol. 104, Springer-Verlag, New York, 1985, The geometry and topology of manifolds. MR807945 (86m:53001)

[Dul08] H. Dulac, *Détermination and intégration d'une certaine classe d'équations différentielles ayant pour point singulier un centre*, Bull. Sci. Math., Sér. 2 **32** (1908), no. 1, 230–252.

[Dul23] ———, *Sur les cycles limites*, Bull. Soc. Math. France **51** (1923), 45–188.

[Dum77] F. Dumortier, *Singularities of vector fields on the plane*, J. Differential Equations **23** (1977), no. 1, 53–106. MR 58 #31276

[Dum93] ———, *Techniques in the theory of local bifurcations: blow-up, normal forms, nilpotent bifurcations, singular perturbations*, Bifurcations and periodic orbits of vector fields (Montreal, PQ, 1992), NATO Adv. Sci. Inst. Ser. C Math. Phys. Sci., vol. 408, Kluwer Acad. Publ., Dordrecht, 1993, pp. 19–73. MR 94j:58123

[Ebe07] W. Ebeling, *Functions of several complex variables and their singularities*, Graduate Studies in Mathematics, vol. 83, American Mathematical Society, Providence, RI, 2007. MR2319634

[Eca85] J. Ecalle, *Les fonctions résurgentes. Tome III*, Publications Mathématiques d'Orsay, vol. 85, Université de Paris-Sud, Département de Mathématiques, Orsay, 1985. MR852210 (87k:32009)

[Eca92] J. Ecalle, *Introduction aux fonctions analysables et preuve constructive de la conjecture de Dulac*, Hermann, Paris, 1992. MR 97f:58104

[Edw79] R. E. Edwards, *Fourier series. A modern introduction. Vol. 1*, second ed., Graduate Texts in Mathematics, vol. 64, Springer-Verlag, New York, 1979. MR 80j:42001

[EISV93] P. M. Elizarov, Yu. S. Ilyashenko, A. A. Shcherbakov, and S. M. Voronin, *Finitely generated groups of germs of one-dimensional conformal mappings, and invariants for complex singular points of analytic foliations of the complex plane*, Nonlinear Stokes phenomena, Adv. Soviet Math., vol. 14, Amer. Math. Soc., Providence, RI, 1993, pp. 57–105. MR 94e:32055

[Eli88] P. M. Elizarov, *The orbital topological classification of analytic differential equations in a neighborhood of a degenerate elementary singular point in the two-dimensional complex plane*, Trudy Sem. Petrovsk. (1988), no. 13, 137–165, 257. MR961432 (90b:58198)

[FG02] K. Fritzsche and H. Grauert, *From holomorphic functions to complex manifolds*, Graduate Texts in Mathematics, vol. 213, Springer-Verlag, New York, 2002. MR1893803 (2003g:32001)

[Fir06] T. Firsova, *Topology of analytic foliations in $\mathbb{C}^2$. The Kupka-Smale property*, Proceedings of the Steklov Institute, vol. 254, 2006, pp. 152–168.

[FLLL89] W. W. Farr, Chengzhi Li, I. S. Labouriau, and W. F. Langford, *Degenerate Hopf bifurcation formulas and Hilbert's 16th problem*, SIAM J. Math. Anal. **20** (1989), no. 1, 13–30. MR 90g:58085

[FM98] I. Feldman and A. Markus, *On some properties of factorization indices*, Integral Equations Operator Theory **30** (1998), no. 3, 326–337. MR 99f:47023

[For91] O. Forster, *Lectures on Riemann surfaces*, Springer-Verlag, New York, 1991. MR 93h:30061

[Fra96] J. P. Françoise, *Successive derivatives of a first return map, application to the study of quadratic vector fields*, Ergodic Theory Dynam. Systems **16** (1996), no. 1, 87–96. MR1375128 (97a:58131)

[FY97] J.-P. Francoise and Y. Yomdin, *Bernstein inequalities and applications to analytic geometry and differential equations*, J. Funct. Anal. **146** (1997), no. 1, 185–205. MR 98h:34009c

[Gan59] F. R. Gantmacher, *The theory of matrices. Vols. 1, 2*, Chelsea Publishing Co., New York, 1959. MR 21 #6372c

[Gav98] L. Gavrilov, *Petrov modules and zeros of Abelian integrals*, Bull. Sci. Math. **122** (1998), no. 8, 571–584. MR 99m:32043

[Gav05] Lubomir Gavrilov, *Higher order Poincaré-Pontryagin functions and iterated path integrals*, Ann. Fac. Sci. Toulouse Math. (6) **14** (2005), no. 4, 663–682. MR2188587 (2006i:34074)

[GH78] P. Griffiths and J. Harris, *Principles of algebraic geometry*, Wiley-Interscience [John Wiley & Sons], New York, 1978, Pure and Applied Mathematics. MR 80b:14001

[GI06] A. Glutsyuk and Yu. Ilyashenko, *Restricted infinitesimal Hilberts Sixteenth Problem*, Doklady Mathematics **73** (2006), no. 2, 185–189.

[GI07] ———, *Restricted Version of the Infinitesimal Hilbert 16th Problem*, Moscow Math. J. **7** (2007), no. 2, 281–325.

[GK60] I. C. Gohberg and M. G. Kreĭn, *Systems of integral equations on a half line with kernels depending on the difference of arguments*, Amer. Math. Soc. Transl. (2) **14** (1960), 217–287. MR 22 #3954

[GK06] T. Golenishcheva-Kutuzova, *A generic analytic foliation in $\mathbb{C}^2$ has infinitely many cylindrical leaves*, Proceedings of the Steklov Institute, vol. 254, 2006, pp. 180–183.

[GLCP96] J.-M. Gambaudo, P. Le Calvez, and É. Pécou, *Une généralisation d'un théorème de Naishul*, C. R. Acad. Sci. Paris Sér. I Math. **323** (1996), no. 4, 397–402. MR1408775 (97h:58096)

[Glu06] A. A. Glutsyuk, *An explicit formula for period determinant*, Annales de l'institut Fourier **56** (2006), no. 4, 887–917.

[GM88] X. Gómez-Mont, *The transverse dynamics of a holomorphic flow*, Ann. of Math. (2) **127** (1988), no. 1, 49–92. MR924673 (89d:32049)

[GM89] ______, *Unfoldings of holomorphic foliations*, Publ. Mat. **33** (1989), no. 3, 501–515. MR1038486 (91d:32026)

[GR65] R. Gunning and H. Rossi, *Analytic functions of several complex variables*, Prentice-Hall Inc., Englewood Cliffs, N.J., 1965. MR 31 #4927

[Gra62] H. Grauert, *Über Modifikationen und exzeptionelle analytische Mengen*, Math. Ann. **146** (1962), 331–368. MR0137127 (25 #583)

[Gro62] D. M. Grobman, *Topological classification of neighborhoods of a singularity in n-space*, Mat. Sb. (N.S.) **56 (98)** (1962), 77–94. MR 25 #2270

[Guc72] J. Guckenheimer, *Hartman's theorem for complex flows in the Poincaré domain*, Compositio Math. **24** (1972), 75–82. MR 46 #920

[Har82] P. Hartman, *Ordinary differential equations*, second (reprinted) ed., Birkhäuser, Boston, Mass., 1982.

[Her63] M. Hervé, *Several complex variables. Local theory*, Published for the Tata Institute of Fundamental Research, Bombay by Oxford University Press, London, 1963. MR 27 #1616

[Hil00] D. Hilbert, *Mathematical problems*, Bull. Amer. Math. Soc. (N.S.) **37** (2000), no. 4, 407–436, Reprinted from Bull. Amer. Math. Soc. **8** (1902), 437–479.

[HKM61] M. Hukuhara, T. Kimura, and T. Matuda, *Equations différentielles ordinaires du premier ordre dans le champ complexe*, Publications of the Mathematical Society of Japan, 7. The Mathematical Society of Japan, Tokyo, 1961. MR0124549 (23 #A1861)

[Hör00] L. Hörmander, *An introduction to complex analysis in several variables*, 2nd ed., North Holland, 2000.

[HRT99] H. Hauser, J.-J. Risler, and B. Teissier, *The reduced Bautin index of planar vector fields*, Duke Math. J. **100** (1999), no. 3, 425–445. MR 2001f:34054

[Ily69] Yu. S. Ilyashenko, Возникновение предельных циклов про возмущении уравнения $dw/dz = -R_z/R_w$, где $R(z,w)$—многочлен (*Appearance of limit cycles by perturbation of the equation $dw/dz = -R_z/R_w$, where $R(z,w)$ is a polynomial*), Mat. Sbornik (New Series) **78 (120)** (1969), no. 3, 360–373.

[Ily72a] ______, *Algebraic unsolvability and almost algebraic solvability of the problem for the center-focus*, Funkcional. Anal. i Priložen. **6** (1972), no. 3, 30–37. MR 47 #3749

[Ily72b] ______, *Foliations by analytic curves*, Mat. Sb. (N.S.) **88 (130)** (1972), 558–577. MR 47 #503

[Ily77] ______, *Remarks on the topology of the singular points of analytic differential equations in a complex domain, and Ladis' theorem*, Funkcional. Anal. i Priložen. **11** (1977), no. 2, 28–38, 95. MR 56 #755

[Ily78] ______, *Topology of phase portraits of analytic differential equations on a complex projective plane*, Trudy Sem. Petrovsk. (1978), no. 4, 83–136. MR524528 (84k:58164)

[Ily79a] ______, *Divergence of series that reduce an analytic differential equation to linear normal form at a singular point*, Funktsional. Anal. i Prilozhen. **13** (1979), no. 3, 87–88. MR 82d:34007

[Ily79b] ______, *Global and local aspects of the theory of complex differential equations*, Proceedings of the International Congress of Mathematicians, Helsinki, 1978 (Berlin), vol. 2, Springer–Verlag, 1979, pp. 821–826.

[Ily84] ———, *Limit cycles of polynomial vector fields with nondegenerate singular points on the real plane*, Functional Anal. Appl. **18** (1984), no. 3, 199–209. MR757247 (86a:34054)

[Ily85] ———, *Dulac's memoir "On limit cycles" and related questions of the local theory of differential equations*, Uspekhi Mat. Nauk **40** (1985), no. 6(246), 41–78, 199. MR 87j:34052

[Ily91] ———, *Finiteness theorems for limit cycles*, American Mathematical Society, Providence, RI, 1991. MR 92k:58221

[Ily02] Yu. Ilyashenko, *Centennial history of Hilbert's 16th problem*, Bull. Amer. Math. Soc. (N.S.) **39** (2002), no. 3, 301–354 (electronic). MR1898209 (2003c:34001)

[Ily04] Yu. S. Ilyashenko, *Three gems in the theory of linear differential equations (in the work of A. A. Bolibrukh)*, Uspekhi Mat. Nauk **59** (2004), no. 6(360), 73–84. MR2138468 (2006j:34015)

[Inc44] E. L. Ince, *Ordinary Differential Equations*, Dover Publications, New York, 1944. MR 6,65f

[IY91] Yu. Ilyashenko and S. Yakovenko, *Finitely smooth normal forms of local families of diffeomorphisms and vector fields*, Uspekhi Mat. Nauk **46** (1991), no. 1(277), 3–39, 240. MR 92i:58165

[IY95] ———, *Finite cyclicity of elementary polycycles in generic families*, Concerning the Hilbert 16th problem, Amer. Math. Soc., Providence, RI, 1995, pp. 21–95. MR 96f:34042

[IY96] ———, *Counting real zeros of analytic functions satisfying linear ordinary differential equations*, J. Differential Equations **126** (1996), no. 1, 87–105. MR 97a:34010

[Kal03] V. Kaloshin, *The existential Hilbert 16-th problem and an estimate for cyclicity of elementary polycycles*, Invent. Math. **151** (2003), no. 3, 451–512. MR1961336 (2004e:34054)

[Kel67] A. Kelley, *The stable, center-stable, center, center-unstable, unstable manifolds*, J. Differential Equations **3** (1967), 546–570. MR 36 #4096

[Kho91] A. Khovanskii, *Fewnomials*, American Mathematical Society, Providence, RI, 1991. MR 92h:14039

[Kle95] O. Kleban, *Order of the topologically sufficient jet of a smooth vector field on the real plane at a singular point of finite multiplicity*, Concerning the Hilbert 16th problem, Amer. Math. Soc. Transl. Ser. 2, vol. 165, Amer. Math. Soc., Providence, RI, 1995, pp. 131–153. MR 96d:58095

[Kos92] V. P. Kostov, *Fuchsian linear systems on $\mathbb{C}P^1$ and the Riemann-Hilbert problem*, C. R. Acad. Sci. Paris Sér. I Math. **315** (1992), no. 2, 143–148. MR 94a:34007

[KP06] I. A. Khovanskaya (Pushkar′), *The weakened infinitesimal Hilbert 16th problem*, Tr. Mat. Inst. Steklova **254** (2006), no. Nelinein. Anal. Differ. Uravn., 215–246. MR2301007

[Kra97] I. S. Krasil′shchik, *Calculus over commutative algebras: a concise user guide*, Acta Appl. Math. **49** (1997), no. 3, 235–248, Algebraic aspects of differential calculus. MR 1 603 164

[KY96] A. Khovanskii and S. Yakovenko, *Generalized Rolle theorem in $\mathbb{R}^n$ and $\mathbb{C}$*, J. Dynam. Control Systems **2** (1996), no. 1, 103–123. MR 97f:26016

[Lad77] N. N. Ladis, *Topological equivalence of hyperbolic linear systems*, Differencial′nye Uravnenija **13** (1977), no. 2, 255–264, 379–380. MR 58 #1396

[Lad79] ———, *The integral curve of a complex homogeneous equation*, Differentsial′nye Uravneniya **15** (1979), no. 2, 246–251, 380. MR527325 (80d:34009)

[Lef56] S. Lefschetz, *On a theorem of Bendixson*, Bol. Soc. Mat. Mexicana (2) **1** (1956), 13–27. MR 18,481g

[Lef68] ———, *On a theorem of Bendixson*, J. Differential Equations **4** (1968), 66–101. MR 36 #2879

[Lev80] B. Ja. Levin, *Distribution of zeros of entire functions*, revised ed., Translations of Mathematical Monographs, vol. 5, American Mathematical Society, Providence, R.I., 1980. MR 81k:30011

[Lor99] F. Loray, *Cinq leçons sur la structure transverse d'une singularité de feuilletage holomorphe en dimension 2 complexe*, vol. 1, Cursos de integración y monografías de la Red Europea "Singularidades de ecuaciones diferenciales y foliaciones", Tordesillas, 1999.

[Lor06] ———, *A preparation theorem for codimension-one foliations*, Ann. of Math. (2) **163** (2006), no. 2, 709–722. MR2199230

[LR03] F. Loray and J. C. Rebelo, *Minimal, rigid foliations by curves on* $\mathbb{CP}^n$, J. Eur. Math. Soc. (JEMS) **5** (2003), no. 2, 147–201. MR1985614

[Mar91] P. Mardešić, *An explicit bound for the multiplicity of zeros of generic Abelian integrals*, Nonlinearity **4** (1991), no. 3, 845–852. MR1124336 (92h:58163)

[Med06] N. B. Medvedeva, *On the analytic solvability of the problem of distinguishing between center and focus*, Trudy Mat. Inst. Steklov. **256** (2006), 7–93.

[Mil99] J. Milnor, *Dynamics in one complex variable*, Friedr. Vieweg & Sohn, Braunschweig, 1999, Introductory lectures. MR1721240 (2002i:37057)

[Mir95] R. Miranda, *Algebraic curves and Riemann surfaces*, Graduate Studies in Mathematics, vol. 5, American Mathematical Society, Providence, RI, 1995. MR1326604 (96f:14029)

[MM80] J.-F. Mattei and R. Moussu, *Holonomie et intégrales premières*, Ann. Sci. École Norm. Sup. (4) **13** (1980), no. 4, 469–523. MR 83b:58005

[MMJR97] P. Mardešić, L. Moser-Jauslin, and C. Rousseau, *Darboux linearization and isochronous centers with a rational first integral*, J. Differential Equations **134** (1997), no. 2, 216–268. MR1432095 (98h:34061)

[Mou82] R. Moussu, *Une démonstration géométrique d'un théorème de Lyapunov-Poincaré*, Bifurcation, ergodic theory and applications (Dijon, 1981), Astérisque, vol. 98, Soc. Math. France, Paris, 1982, pp. 216–223. MR 85g:58012

[Mou98] ———, *Sur l'existence d'intégrales premières holomorphes*, Ann. Scuola Norm. Sup. Pisa Cl. Sci. (4) **26** (1998), no. 4, 709–717. MR1648570 (99i:32040)

[MR83] J. Martinet and J.-P. Ramis, *Classification analytique des équations différentielles non linéaires résonnantes du premier ordre*, Ann. Sci. École Norm. Sup. (4) **16** (1983), no. 4, 571–621 (1984). MR740592 (86k:34034)

[Mum76] D. Mumford, *Algebraic geometry. I*, Springer-Verlag, Berlin, 1976, Complex projective varieties, Grundlehren der Mathematischen Wissenschaften, No. 221. MR0453732 (56 #11992)

[Naĭ81] V. A. Naĭshul′, *Topological equivalence of differential equations in* $\mathbf{C}^2$ *and* $\mathbf{C}P^2$, Vestnik Moskov. Univ. Ser. I Mat. Mekh. (1981), no. 4, 8–11, 84. MR631498 (83g:34042)

[Naĭ82] ———, *Topological invariants of analytic and area-preserving mappings and their application to analytic differential equations in* $\mathbf{C}^2$ *and* $\mathbf{C}P^2$, Trudy Moskov. Mat. Obshch. **44** (1982), 235–245. MR656288 (84f:58092)

[Nak94] I. Nakai, *Separatrices for nonsolvable dynamics on* $\mathbf{C}, 0$, Ann. Inst. Fourier (Grenoble) **44** (1994), no. 2, 569–599. MR1296744 (95j:58124)

[NN57] A. Newlander and L. Nirenberg, *Complex analytic coordinates in almost complex manifolds*, Ann. of Math. (2) **65** (1957), 391–404. MR0088770 (19,577a)

[Nov02] D. Novikov, *Modules of Abelian integrals and Picard-Fuchs systems*, Nonlinearity **15** (2002), no. 5, 1435–1444.

[NS60] V. V. Nemytskii and V. V. Stepanov, *Qualitative theory of differential equations*, Princeton Mathematical Series, No. 22, Princeton University Press, Princeton, N.J., 1960, Reprinted by Dover Publications in 1989. MR0121520 (22 #12258)

[NY99] D. Novikov and S. Yakovenko, *Trajectories of polynomial vector fields and ascending chains of polynomial ideals*, Ann. Inst. Fourier (Grenoble) **49** (1999), no. 2, 563–609. MR 2001h:32054

[NY01] ———, *Redundant Picard-Fuchs system for Abelian integrals*, J. Differential Equations **177** (2001), no. 2, 267–306. MR 1 876 646

[NY03] ———, *Quasialgebraicity of Picard–Vessiot fields*, Mosc. Math. J. **3** (2003), no. 2, 551–591.

[NY04] ———, *Lectures on meromorphic flat connections*, Normal forms, bifurcations and finiteness problems in differential equations, NATO Sci. Ser. II Math. Phys. Chem., vol. 137, Kluwer Acad. Publ., Dordrecht, 2004, pp. 387–430. MR2085816 (2005f:34255)

[OB96] L. Ortiz Bobadilla, *Topological equivalence of linear autonomous equations in* C^m *with Jordan blocks*, Trans. Moscow Math. Soc. **57** (1996), 67–91. MR 99a:58127

[Otr54] N. F. Otrokov, *On the number of limit cycles of a differential equation in the neighborhood of a singular point*, Mat. Sbornik N.S. **34(76)** (1954), 127–144. MR0063506 (16,130f)

[Per01] L. Perko, *Differential equations and dynamical systems*, third ed., Texts in Applied Mathematics, vol. 7, Springer-Verlag, New York, 2001. MR1801796 (2001k:34001)

[Pet96] I. G. Petrowsky, *Selected works. Part II; differential equations and probability theory*, Classics of Soviet Mathematics, vol. 5, Gordon and Breach Publishers, Amsterdam, 1996. MR1677648 (99m:01106b)

[Pha67] F. Pham, *Introduction à l'étude topologique des singularités de Landau*, Mémorial des Sciences Mathématiques, Fasc. 164, Gauthier-Villars Éditeur, Paris, 1967. MR0229263 (37 #4837)

[PL55] I. G. Petrovskiĭ and E. M. Landis, *On the number of limit cycles of the equation* $dy/dx = P(x,y)/Q(x,y)$, *where* P *and* Q *are polynomials of 2nd degree*, Mat. Sb. N.S. **37(79)** (1955), 209–250. MR 17,364d

[PL57] ———, *On the number of limit cycles of the equation* $dy/dx = P(x,y)/Q(x,y)$, *where* P *and* Q *are polynomials*, Mat. Sb. N.S. **43(85)** (1957), 149–168. MR 19,746c

[Ple64] J. Plemelj, *Problems in the sense of Riemann and Klein*, Interscience Publishers John Wiley & Sons Inc. New York-London-Sydney, 1964, Interscience Tracts in Pure and Applied Mathematics, No. 16. MR 30 #5008

[PM01] R. Pérez-Marco, *Total convergence or general divergence in small divisors*, Comm. Math. Phys. **223** (2001), no. 3, 451–464. MR 2003d:37063

[PMY94] R. Pérez Marco and J.-C. Yoccoz, *Germes de feuilletages holomorphes à holonomie prescrite*, Astérisque (1994), no. 222, 7, 345–371, Complex analytic methods in dynamical systems (Rio de Janeiro, 1992). MR 96b:58090

[Poi90] H. Poincaré, *Sur le problème des trois corps et les équations de la dynamique*, Acta Math. **XIII** (1890), 1–270.

[Pon34] L. Pontryagin, *On dynamical systems close to hamiltonian ones*, Zh. Exp. & Theor. Phys. **4** (1934), no. 8, 234–238.

[PS70a] J. Palis and S. Smale, *Structural stability theorems*, Global Analysis (Proc. Sympos. Pure Math., Vol. XIV, Berkeley, Calif., 1968), Amer. Math. Soc., Providence, R.I., 1970, pp. 223–231. MR 42 #2505

[PS70b] C. Pugh and M. Shub, *Linearization of normally hyperbolic diffeomorphisms and flows*, Invent. Math. **10** (1970), 187–198. MR 44 #1055

[Pus97] I. A. Pushkar′, *A multidimensional generalization of Il′yashenko's theorem on abelian integrals*, Funktsional. Anal. i Prilozhen. **31** (1997), no. 2, 34–44, 95. MR 98k:58183

[Rou89] R. Roussarie, *Cyclicité finie des lacets et des points cuspidaux*, Nonlinearity **2** (1989), no. 1, 73–117. MR980858 (90m:58169)

[Rou98] ———, *Bifurcation of planar vector fields and Hilbert's sixteenth problem*, Birkhäuser Verlag, Basel, 1998. MR 99k:58129

[RY97] N. Roytwarf and Y. Yomdin, *Bernstein classes*, Ann. Inst. Fourier (Grenoble) **47** (1997), no. 3, 825–858. MR 98h:34009a

[Sad87] P. Sad, *Central eigenvalues of complex saddle-nodes*, Dynamical systems and bifurcation theory (Rio de Janeiro, 1985), Pitman Res. Notes Math. Ser., vol. 160, Longman Sci. Tech., Harlow, 1987, pp. 387–397. MR907900 (88k:58133)

[Sav82] V. I. Savel′ev, *Zero-type imbedding of a sphere into complex surfaces*, Vestnik Moskov. Univ. Ser. I Mat. Mekh. (1982), no. 4, 28–32, 85. MR671883 (84d:32007)

[Sch93] D. Schlomiuk, *Algebraic and geometric aspects of the theory of polynomial vector fields*, Bifurcations and periodic orbits of vector fields (Montreal, PQ, 1992), NATO Adv. Sci. Inst. Ser. C Math. Phys. Sci., vol. 408, Kluwer Acad. Publ., Dordrecht, 1993, pp. 429–467. MR 95a:34042

[Sch04] ———, *Aspects of planar polynomial vector fields: global versus local, real versus complex, analytic versus algebraic and geometric*, Normal forms, bifurcations and finiteness problems in differential equations, NATO Sci. Ser. II Math. Phys. Chem., vol. 137, Kluwer Acad. Publ., Dordrecht, 2004, pp. 471–509. MR2085818 (2005f:34081)

[Sei68] A. Seidenberg, *Reduction of singularities of the differential equation* $A\,dy = B\,dx$, Amer. J. Math. **90** (1968), 248–269. MR 36 #3762

[Sha92] B. V. Shabat, *Introduction to complex analysis. Part II*, Translations of Mathematical Monographs, vol. 110, American Mathematical Society, Providence, RI, 1992, Functions of several variables, Translated from the third (1985) Russian edition by J. S. Joel. MR1192135 (93g:32001)

[Sha94] I. R. Shafarevich, *Basic algebraic geometry. 1*, second ed., Springer-Verlag, Berlin, 1994, Varieties in projective space, Translated from the 1988 Russian edition and with notes by Miles Reid. MR 95m:14001

[Shc82] A. A. Shcherbakov, *Density of the orbit of a pseudogroup of conformal mappings and generalization of the Khudaĭ-Verenov theorem*, Vestnik Moskov. Univ. Ser. I Mat. Mekh. (1982), no. 4, 10–15, 84. MR671879 (84m:30015)

[Shc84] ———, *Topological and analytic conjugation of noncommutative groups of germs of conformal mappings*, Trudy Sem. Petrovsk. (1984), no. 10, 170–196, 238–239. MR778885 (86g:58083)

[Shc86] ———, *Complex limit cycles of the equation $dw/dz = P_n/Q_n$*, Uspekhi Mat. Nauk **41** (1986), no. 1(247), 211–212. MR832430 (87j:58077)

[Shc06] ———, *Dynamics of local groups of conformal mappings and generic properties of differential equations on $\mathbb{C}^2$*, Trudy Mat. Inst. Steklov. **256** (2006), 103–120.

[Shi80a] Song Ling Shi, *A concrete example of the existence of four limit cycles for plane quadratic systems*, Sci. Sinica **23** (1980), no. 2, 153–158. MR 81f:34037

[Shi80b] ———, *A concrete example of the existence of four limit cycles for plane quadratic systems*, Sci. Sinica **23** (1980), no. 2, 153–158. MR574405 (81f:34037)

[Sib62] Y. Sibuya, *Simplification of a system of linear ordinary differential equations about a singular point*, Funkcial. Ekvac. **4** (1962), 29–56. MR 27 #2694

[Sib90] ———, *Linear differential equations in the complex domain: problems of analytic continuation*, American Mathematical Society, Providence, RI, 1990. MR 92a:34010

[Siu77] Y.-T. Siu, *Every Stein subvariety admits a Stein neighborhood*, Invent. Math. **38** (1976/77), no. 1, 89–100. MR0435447 (55 #8407)

[Šoš72] A. N. Šošitaĭšvili, *Bifurcations of topological type of singular points of vector fields that depend on parameters*, Funkcional. Anal. i Priložen. **6** (1972), no. 2, 97–98. MR 45 #6036

[Šoš75] ———, *The bifurcation of the topological type of the singular points of vector fields that depend on parameters*, Trudy Sem. Petrovsk. (1975), no. Vyp. 1, 279–309. MR 57 #17724

[SRO98] A. A. Shcherbakov, E. Rosales-González, and L. Ortiz-Bobadilla, *Countable set of limit cycles for the equation $dw/dz = P_n(z,w)/Q_n(z,w)$*, J. Dynam. Control Systems **4** (1998), no. 4, 539–581. MR1662926 (99m:58150)

[SZ02] E. Stróżyna and H. Żołądek, *The analytic and formal normal form for the nilpotent singularity*, J. Differential Equations **179** (2002), no. 2, 479–537. MR1885678 (2003g:37091)

[Tak71] F. Takens, *Partially hyperbolic fixed points*, Topology **10** (1971), 133–147. MR 46 #6399

[Tak01] ———, *Forced oscillations and bifurcations*, Global analysis of dynamical systems, Inst. Phys., Bristol, 2001, Reprint from Comm. Math. Inst. Rijksuniv. Utrecht, No. 3-1974, **1974**, pp. 1–61. MR 2002i:37081

[Tam92] I. Tamura, *Topology of foliations: an introduction*, Translations of Mathematical Monographs, vol. 97, American Mathematical Society, Providence, RI, 1992, Translated from the 1976 Japanese edition and with an afterword by Kiki Hudson, With a foreword by Takashi Tsuboi. MR1151624 (93c:57021)

[Tit39] E. Titchmarsh, *The theory of functions*, Oxford University Press, 1939.

[Tre83] A. Treibich, *Un résultat de Plemelj*, Mathematics and physics (Paris, 1979/1982), Birkhäuser Boston, Boston, MA, 1983, pp. 307–312. MR 87c:32016

[Tri90] S. I. Trifonov, *Divergence of Dulac's series*, Mat. Sb. **181** (1990), no. 1, 37–56. MR1048829 (91k:58109)

[Tsu59] M. Tsuji, *Potential theory in modern function theory*, Maruzen Co. Ltd., Tokyo, 1959. MR 22 #5712

[Van89] A. Vanderbauwhede, *Centre manifolds, normal forms and elementary bifurcations*, Dynamics reported, Vol. 2, Dynam. Report. Ser. Dynam. Systems Appl., vol. 2, Wiley, Chichester, 1989, pp. 89–169. MR 90g:58092

[Var84] V. S. Varadarajan, *Lie groups, Lie algebras, and their representations*, Graduate Texts in Mathematics, vol. 102, Springer-Verlag, New York, 1984, Reprint of the 1974 edition. MR 85e:22001

[Var89] A. N. Varchenko, *Critical values and the determinant of periods*, Russian Math. Surveys **44** (1989), no. 4, 209–210 (English. Russian original).

[Var96] V. S. Varadarajan, *Linear meromorphic differential equations: a modern point of view*, Bull. Amer. Math. Soc. (N.S.) **33** (1996), no. 1, 1–42. MR 96h:34011

[vdD88] L. van den Dries, *Alfred Tarski's elimination theory for real closed fields*, J. Symbolic Logic **53** (1988), no. 1, 7–19. MR 89h:01040

[vdE79] A. van den Essen, *Reduction of singularities of the differential equation* $Ady = Bdx$, Équations différentielles et systèmes de Pfaff dans le champ complexe (Sem., Inst. Rech. Math. Avancée, Strasbourg, 1975), Lecture Notes in Math., vol. 712, Springer, Berlin, 1979, pp. 44–59. MR 82m:34007

[vdPS03] M. van der Put and M. F. Singer, *Galois theory of linear differential equations*, Grundlehren der Mathematischen Wissenschaften, vol. 328, Springer-Verlag, Berlin, 2003. MR1960772 (2004c:12010)

[VK75] A. M. Vinogradov and I. S. Krasil′ščik, *What is Hamiltonian formalism?*, Uspehi Mat. Nauk **30** (1975), no. 1(181), 173–198. MR 58 #31235

[VR04] Kleptsyn V. and B. Rabinovich, *Analytic classification of Fuchsian singular points*, Matem. Zametki **76** (2004), no. 3, 372–383.

[War83] F. Warner, *Foundations of differentiable manifolds and Lie groups*, Springer-Verlag, New York, 1983, Corrected reprint of the 1971 edition. MR 84k:58001

[Was87] W. Wasow, *Asymptotic expansions for ordinary differential equations*, Dover Publications Inc., New York, 1987, Reprint of the 1976 edition. MR 88i:34003

[Wel80] R. O. Wells, Jr., *Differential analysis on complex manifolds*, second ed., Graduate Texts in Mathematics, vol. 65, Springer-Verlag, New York, 1980. MR608414 (83f:58001)

[Wol96] Wolfram Research, Inc., `Mathematica`, Champaign, Illinois, 1996, Version 3.0.

[Yak95] S. Yakovenko, *A geometric proof of the Bautin theorem*, Concerning the Hilbert 16th problem, Amer. Math. Soc., Providence, RI, 1995, pp. 203–219. MR 96j:34056

[Yak00] ______, *On zeros of functions from Bernstein classes*, Nonlinearity **13** (2000), no. 4, 1087–1094. MR 2001e:30008

[Yak02] ______, *Bounded decomposition in the Brieskorn lattice and Pfaffian Picard–Fuchs systems for Abelian integrals*, Bull. Sci. Math **126** (2002), no. 7, 535–554.

[Yak05] ______, *Quantitative theory of ordinary differential equations and the tangential Hilbert 16th problem*, On finiteness in differential equations and Diophantine geometry, CRM Monogr. Ser., vol. 24, Amer. Math. Soc., Providence, RI, 2005, pp. 41–109. MR2180125 (2006g:34062)

[Yak06] ———, *Oscillation of linear ordinary differential equations: on a theorem of A. Grigoriev*, J. Dyn. Control Syst. **12** (2006), no. 3, 433–449. MR2233029

[Yoc88] J.-C. Yoccoz, *Linéarisation des germes de difféomorphismes holomorphes de* $(\mathbf{C}, 0)$, C. R. Acad. Sci. Paris Sér. I Math. **306** (1988), no. 1, 55–58. MR 89i:58123

[Yoc95] ———, *Théorème de Siegel, nombres de Bruno et polynômes quadratiques*, Astérisque (1995), no. 231, 3–88, Petits diviseurs en dimension 1. MR 96m:58214

[Yom99] Y. Yomdin, *Global finiteness properties of analytic families and algebra of their Taylor coefficients*, The Arnoldfest (Toronto, ON, 1997), Amer. Math. Soc., Providence, RI, 1999, pp. 527–555. MR 1 733 591

[Žiž61] A. B. Žižčenko, *Homology groups of algebraic varieties*, Izv. Akad. Nauk SSSR Ser. Mat. **25** (1961), 765–788. MR0136615 (25 #83)

[Żoł94] H. Żołądek, *Quadratic systems with center and their perturbations*, J. Differential Equations **109** (1994), no. 2, 223–273. MR 95b:34047

[Żoł06] ———, *The monodromy group*, Instytut Matematyczny Polskiej Akademii Nauk. Monografie Matematyczne (New Series) [Mathematics Institute of the Polish Academy of Sciences. Mathematical Monographs (New Series)], vol. 67, Birkhäuser Verlag, Basel, 2006. MR2216496

Index